THIN FILM MATERIALS

Stress, defect formation and surface evolution

Thin films play an important role in many technological applications including microelectronic devices, magnetic storage media and surface coatings. This book provides a comprehensive coverage of the major issues and topics dealing with stress, defect formation and surface evolution in thin films. Physical phenomena are examined from the continuum down to the sub-microscopic length scales, with the connections between the structure of the material and its behavior described wherever appropriate. While the book develops a comprehensive scientific basis with which stress, deformation and failure in thin film materials can be characterized, an attempt is also made to link the scientific concepts to a broad range of practical applications through example problems, historical notes, case studies and exercises. Of particular interest to engineers, materials scientists and physicists, this book will be essential reading for senior undergraduate and graduate courses on thin films.

L. BEN FREUND is the Henry Ledyard Goddard University Professor in the Division of Engineering at Brown University.

SUBRA SURESH is the Ford Professor of Engineering and Head of the Department of Materials Science and Engineering, and Professor of Mechanical Engineering at Massachusetts Institute of Technology.

THIN FILM MATERIALS

Stress, Defect Formation and Surface Evolution

L. B. FREUND

Division of Engineering
Brown University

S. SURESH

Department of Materials Science and Engineering
Massachusetts Institute of Technology

CAMBRIDGE
UNIVERSITY PRESS

PUBLISHED BY THE PRESS SYNDICATE OF THE UNIVERSITY OF CAMBRIDGE
The Pitt Building, Trumpington Street, Cambridge, United Kingdom

CAMBRIDGE UNIVERSITY PRESS
The Edinburgh Building, Cambridge CB2 2RU, UK
40 West 20th Street, New York, NY 10011–4211, USA
477 Williamstown Road, Port Melbourne, VIC 3207, Australia
Ruiz de Alarcón 13, 28014 Madrid, Spain
Dock House, The Waterfront, Cape Town 8001, South Africa

http://www.cambridge.org

First published 2003

Printed in the United Kingdom at the University Press, Cambridge

Typeface Times 11/14 pt.   *System* LATEX 2_ε [TB]

A cataloge record for this book is available from the British Library

Library of Congress Cataloging in Publication data

Freund, L. B.
Thin film materials: stress, defect formation, and surface evolution/L.B. Freund, S. Suresh.
p.   cm.
Includes bibliographical references and index.
ISBN 0 521 82281 5
1. Thin films.   I. Suresh, S. (Subra)   II. Title
TA418.9.T45F74 2003
621.3815′2 – dc21   2003043937

ISBN 0 521 82281 5 hardback

To our families

Contents

Preface

Within a period of a few decades, the field of materials science and engineering has emerged as a focal point for developments in virtually all areas of engineering and applied science. The study of *thin film materials* has been one of the unifying themes in the development of the field during this period. As understood here, the area encompasses films bonded to relatively thick substrates, multilayer materials, patterned films on substrates and free-standing films. Significant advances in methods for synthesizing and processing these materials for ever more specific purposes, as well as in instrumentation for characterizing materials at ever diminishing size scales, have been key to modern engineering progress.

At the dawn of the 21st century, the United States National Academy of Engineering reported the outcome of a project intended to identify the twenty most significant engineering achievements of the preceding century. It is evident from the list compiled that achievements of the second half of the twentieth century – electronics, computers, health technologies, laser and fiber optics, for example – were all based on the creative and efficient exploitation of materials; thin film materials represent a major component of this advance in materials technology. In fact, the impact of advances in the specialized uses of materials was so pervasive in the achievements being recognized by the Academy that the development of high-performance materials itself was included as one of the most significant achievements.

The goal of this book is to summarize developments in the area of thin film materials that have occurred over the past few decades, with emphasis on the generation of internal stress and its consequences. Internal stress can induce a variety of undesirable consequences including excessive deformation, fracture, delamination, permanent deformation and microstructural alterations. In spite of these possibilities, thin films have been inserted into engineering systems in order to accomplish a wide range of practical service functions. Among these are microelectronic devices and packages; micro-electro-mechanical systems or MEMS; and surface coatings intended to impart a thermal, mechanical, tribological, environmental, optical, electrical, magnetic or biological function. To a large extent, the success of this endeavor has been enabled by research leading to reliable means for estimating stress in small material systems and by establishing frameworks in which to assess the integrity or functionality of the systems. The prospect for material failure due to stress continues to be a technology-limiting barrier, even in situations in which load-carrying capacity of the material is not among its primary functional characteristics. In some circumstances, stress has desirable consequences, as in bandgap engineering for electronic applications and in the self-assembly of small structures driven by stored elastic energy. It is our hope that the information included in this book will be useful as an indicator of achievements in the field and as a guide for further advances in a number of new and emerging directions.

The first chapter is devoted largely to a discussion of the origins of residual stress in thin film materials and to identification of relationships between processing methods and generation of stress. The consequences of stress are discussed in subsequent chapters, with the presentation generally organized according to the size scale of the dominant physical phenomena involved. Overall deformation of film–substrate systems or multilayer structures is considered in Chapters 2 and 3. This is followed by examination of the general failure modes of fracture, delamination and buckling of films in Chapters 4 and 5. The focus then shifts to a smaller scale to discuss conditions for dislocation formation in Chapter 6 and inelastic deformation of films in Chapter 7. Finally, the issues of stability of material surfaces and evolution of surface morphology or alloy composition are considered in Chapters 8 and 9. The consequences of stress in thin films are linked to the structure of the film materials wherever possible.

It is recognized that each of the principal topics covered in the book could itself be developed into a substantial monograph, but the goal here is not the exhaustive treatment of a topic of limited scope. The area is inherently interdisciplinary, and the intention is to provide a comprehensive coverage of issues relevant to stress and its consequences in thin film materials. Adoption of this approach meant that many choices had to be made along the way about depth of coverage of specific topics and balance among different topics; we hope that readers will judge the choices made to be reasonable. The main purpose of the book is the coherent presentation of a sound scientific basis for describing the origins of stress in films and for anticipating the consequences of stress in defect formation, surface evolution and allied effects. Many references to original work are included as a guide to the archival literature in the area. In addition, the fundamental concepts developed are made more concrete through implementation in sample calculations and through discussion of case studies of practical significance. The description of experimental methods, results and observations is included as an integral part of developing the conceptual structure of the topics examined. Each chapter concludes with a set of exercises that further extend the material discussed and which can challenge newcomers to the area at applying concepts. It is our hope that, with this structure, the book will serve as a research reference for those pursuing the area at its frontiers, as a useful compilation of readily applicable results for practicing engineers, and as a textbook for graduate students or advanced undergraduate students wishing to develop background in this area.

The idea for the book grew out of a course on thin films that has been offered for students in solid mechanics and materials science at Brown University since 1992, as a natural outgrowth of emerging research activity in the area. We are grateful to the many students, postdoctoral research associates and colleagues who attended these lectures and whose enthusiasm gave this project its initial impetus.

We are also grateful to many colleagues who have contributed in various ways to the preparation of this book. We particularly thank John Hutchinson who used a draft of parts of the book for a course for graduate students at Harvard and MIT, and who provided valuable feedback on this material. Both John Hutchinson and Bill Nix kindly shared with us their own course materials on thin films. Several colleagues read drafts of various sections of the book and offered helpful recommendations; they include Ilan Blech, Eric Chason, Ares Rosakis, Vivek Shenoy and Carl Thompson. Several graduate students who took courses based in part on draft chapters, particularly Yoonjoon Choi and Nuwong Chollacoop, suggested a number of clarifications and improvements in the presentation. Finally, we are grateful to the many colleagues who provided figures and micrographs from their own work; in these cases, acknowledgments are noted along with the included material. Tim Fishlock at Cambridge University Press offered us considerable flexibility in the formulation of the scope of this book and in the preparation of the document. We also thank Desiree Soucy who secured the agreements necessary to reproduce proprietary material and who proofread the entire manuscript.

LBF is grateful to the Materials Research Science and Engineering Center, funded by the National Science Foundation at Brown University, for long-term support of research in the general area of thin films and for the collaborations fostered through the Center. He is also thankful to the Division of Engineering and Applied Sciences at Caltech for hosting a sabbatical leave; the kind hospitality and congenial environment afforded an opportunity for pursuing the book writing project at its early critical stage. SS is grateful to the Defense University Research Initiative in Nano-Technology, funded by the Office of Naval Research at MIT, and the Programme on Advanced Materials for Micro and Nano Systems, funded by the Singapore-MIT Alliance, for their support for research in the areas covered by the book.

A project of this magnitude would not have been possible without the support and encouragement of the members of our families. We are extremely grateful for their enduring patience and understanding during our long hours of immersion in this project over the past several years.

1

Introduction and Overview

Thin solid films have been used in many types of engineering systems and have been adapted to fulfill a wide variety of functions. A few examples follow.

— Great strides in thin film technology have been made in order to advance the rapid development of miniature, highly integrated electronic circuits. In such devices, confinement of electric charge relies largely on interfaces between materials with differing electronic properties. Furthermore, the need for thin materials of exceptionally high quality, reproducible characteristics and reliability has driven film growth technology through a rapid succession of significant achievements. More recently, progress in the physics of material structures that rely on quantum confinement of charge carriers continues to revolutionize the area. These systems present new challenges for materials synthesis, characterization and modeling.

— The use of surface coatings to protect structural materials in high temperature environments is another thin film technology of enormous commercial significance. In gas turbine engines, for example, thin surface films of materials chosen for their chemical inertness, stability at elevated temperatures and low thermal conductivity are used to increase engine efficiency and to extend significantly the useful lifetimes of the structural materials that they protect. Multilayer or continuously graded coatings offer the potential for further progress in this effort.

— The useful lifetimes of components subjected to friction and wear due to contact can be extended substantially through the use of surface coatings or surface treatments. Among the technologies that rely on the use of thin films in this way are internal combustion engines, artificial hip and knee implants, and computer hard disks for magnetic data storage.

— Thin films are integral parts of many micro-electro-mechanical systems designed to serve as sensors or actuators. For example, a piezoelectric or piezoresistive thin film deposited on a silicon membrane can be used to detect electronically a deflection of the membrane in response to a pressure applied on its surface or by an acceleration of its supports. Devices based on thin film technology are used as microphones in hearing aids, monitors of blood pressure during exercise, electronically positioned thin film mirrors on flexible supports in optical display systems, and probes for detecting the degree of ripeness of fruits.

Numerous other technologies rely on thin film behavior. An immediate observation that follows from the foregoing list is that the principal function of the thin film components in these applications is often not structural. Consequently, load-carrying capacity may not be a principal consideration for design or material selection. However, fabrication of thin film configurations typically results in internal stress in the film of a magnitude sufficient to induce mechanical deformation, damage or failure. A tendency for stress-driven failure of a thin film structure can be a disabling barrier to incorporation of that film into a system, even when load-carrying capacity is of secondary importance as a functional characteristic. The presence of an internal stress in a thin film structure may also influence the electrical or magnetic properties in functional devices.

This chapter provides an overview of commonly used deposition and processing methods for the synthesis and fabrication of thin film structures. This is followed by a discussion of the fabrication of small volume structures, where examples of the basic steps involved in lithography, surface micromachining, bulk micromachining and molding processes are considered in the context of the manufacture of microelectronic devices, as well as small structures encountered in the development of micro-electro-mechanical systems (MEMS) and nano-electro-mechanical systems (NEMS). Attention is devoted to the effects of processing on the nucleation and growth of monocrystalline and polycrystalline thin films on substrates, the evolution of film microstructure in polycrystalline films, and the generation of internal stresses through processing. In the chapters that follow, consequences of stress in thin film and multilayer materials are studied, organized according to the deformation or failure phenomena that can be induced by internal stress.

1.1 A classification of thin film configurations

As a guide for the development and application of concepts that are useful for describing mechanical behavior of solid thin film and multilayer materials, it is convenient to classify structures in terms of their geometrical configurations and the nature of constraint on their deformation imposed by their surroundings. For this purpose, configurations are classified in terms of the relative extent of the

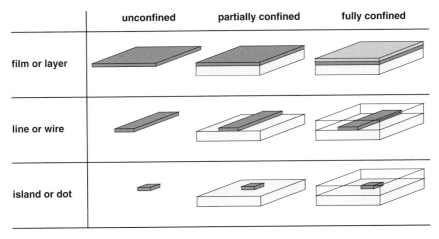

Fig. 1.1. A categorization of small volume structures in terms of their general shapes and levels of constraints. A film is often called a two-dimensional structure, a line is called a one-dimensional structure and a dot a zero-dimensional structure, but this terminology is not standard and it varies significantly among different technical specialities.

solid bodies in three orthogonal directions, with the orientation of the reference coordinate system being dictated by the configuration. The degree of constraint is determined by the interaction of the thin film structure with other deformable solids to which it may be attached or with which it may otherwise be in contact. The former situation requires compatibility of deformation, while the latter requires some restriction on its motion. The categories of configuration are termed film (or layer), line (or wire) and island (or dot); the categories of constraint are termed unconfined, partially confined and fully confined. These classes are illustrated in Figure 1.1. There is nothing fundamental about such a classification scheme, but its adoption can facilitate understanding of the ranges of applicability of the various ideas in the field.

   With reference to Figure 1.1, a structure of an extent that is small in one direction compared to its extent in the other two directions is termed a thin film; in structural mechanics, such configurations are identified as plates or shells. The qualifier 'small' as used here means that the largest dimensions are at least twenty times greater than the small dimension, and more commonly are hundreds of times greater than the small dimension. A structure that has small extent in two directions compared to its extent in the third direction is termed a line or wire; such configurations are usually identified as rods or bars in structural mechanics. Lastly, a structure that has small extent in all three directions, compared to the dimensions of its surroundings in this case, is termed an island or a dot.

   Concerning the degree of constraint on deformation, a small structure is said to be unconfined if the boundaries associated with its thin dimensions are free to displace without restriction. On the other hand, it is said to be fully confined if all boundaries associated with its thin dimensions are constrained against deformation. In virtually all cases, the constraint at a boundary is due to another material which shares that boundary as a common interface. The structure is said to be partially confined if displacement of its boundaries associated with some, but not all, directions of thinness are unconstrained.

The classification matrix in Figure 1.1 includes some illustrations of varying degrees of confinement and thinness. As a specific example, consider a layer of a SiGe alloy $1\,\mu$m in thickness, which is deposited on a 1 cm by 1 cm area of a Si substrate that is 0.5 mm thick. This configuration results in a partially confined thin film structure. A stripe of copper with a square cross-section that is $0.5\,\mu$m on a side and a length of 5 mm deposited on a relatively thick Si substrate is a partially confined line. If the surfaces of the Si substrate and the wire are then completely covered over with a $1\,\mu$m thick coating of SiO_2 to electrically isolate the wire, the structure becomes a fully confined line. An InAs quantum dot being formed by stress-driven surface diffusion is a partially confined island. If this configuration is then covered over by a blanket deposit of AlAs, the final structure is a fully confined island or quantum dot configuration.

The idea of the classification scheme is intended only as a conceptual guide. There are many situations in which the structure exhibits behavior represented by more than one entry in the categorization matrix in Figure 1.1. For example, suppose that a thin film bonded to a relatively thick substrate supports a compressive stress. This is a partially confined thin film configuration. If the stress becomes large enough in magnitude, the film will tend to buckle by debonding from the substrate over some portion of the interface and then deflecting away from the substrate over this portion. The region of the buckle becomes an unconfined film but the remainder of the film is partially confined.

The categorization of thin film systems as summarized in Figure 1.1 is based on relative physical dimensions and makes no reference to any length scale reflecting the underlying structure of the material. There is usually a heirarchy of such length scales associated with a material of given chemical composition, and the scales can depend on the processing methods used to form the material structure. For example, for a polycrystalline film, intrinsic length scales include the size of the atomic unit cell, the spacing of crystalline defects and the size of the crystal grains, at the very least. Thus, a further subcategorization of film configurations based on a comparison of the small dimension of the thin structure to the *absolute* length scale characteristic of the constituent material must be considered.

When the thickness of the film is small compared to that of the substrate (typically by a factor of 50 or more), it represents a *mechanically thin film*. In this case, the film material either has no intrinsic structural length scales, as in the case of an amorphous film, or the film thickness is much larger than all the characteristic microstructural length scales such as the grain size, dislocation cell size, precipitate or particle spacing, diameter of the dislocation loops, mean free path for dislocation motion, or the magnetic domain wall size. Such structures, typically tens or hundreds of micrometers in thickness, are deposited onto substrates by plasma spray or physical vapor deposition, or layers bonded to substrates through welding, diffusion bonding, explosion cladding, sintering or self-propagating high temperature combustion synthesis. This definition, of course, holds only when the size scale of the microstructure is small compared to the film thickness. The continuum mechanics approach to be presented for the analysis of stress, substrate curvature and fracture in such mechanically thin films applies to a broad range of practical situations.

When the small dimension of the material structure is comparable to the characteristic microstructural size scale, the film is considered to be a *microstructurally thin film*. Most metallic thin films used in microelectronic devices and magnetic storage media are examples of microstructurally thin films, where the film thickness is substantially greater than atomic or molecular dimensions. Although the film thickness normally includes only a few microstructural units in these cases, the plane of the film has dimensions significantly larger than the characteristic microstructural size scale. The mechanical properties of these films are much more strongly influenced by such factors as average grain size, grain shape, grain size distribution and crystallographic texture than in the case of mechanically thin films. Grain to grain variations in crystallographic orientation as well as crystalline anisotropy of thermal, electrical, magnetic and mechanical properties also have a more pronounced effect on the overall mechanical response of microstructurally thin films. The mechanisms and mechanics of microstructurally thin films are considered extensively throughout this book. A microstructurally thin film can be patterned into lines or stripes on the substrate surface, in which case the cross-sectional dimensions of each line are comparable to the microstructural unit dimension. For single crystal films epitaxially bonded to relatively thick substrates, the only microstructurally significant dimension is the lattice spacing. Consequently, such films can usually be treated as microstructurally thin films even though the film thickness may be as small as several times the atomic unit cell dimension. Such structures are studied in Chapter 6.

Atomically thin films constitute layers whose thicknesses are comparable to one or a few atomic layers. An adsorbed monolayer of gas or impurity atoms on a surface is an example of an atomically thin layer. Here the mechanical response of the thin layer is likely to be more influenced by interatomic potentials and surface energy than by macroscopic mechanical properties or by micromechanisms of deformation.

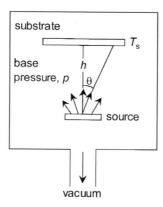

Fig. 1.2. Schematic showing the basic features of an evaporative deposition system.

1.2 Film deposition methods

Physical vapor deposition (PVD) and chemical vapor deposition (CVD) are the most common methods for transferring material atom by atom from one or more sources to the growth surface of a film being deposited onto a substrate. Vapor deposition describes any process in which a solid immersed in a vapor becomes larger in mass due to transference of material from the vapor onto the solid surface. The deposition is normally carried out in a vacuum chamber to enable control of the vapor composition. If the vapor is created by physical means without a chemical reaction, the process is classified as PVD; if the material deposited is the product of a chemical reaction, the process is classified as CVD. Many variations of these basic vapor deposition methods have been developed in efforts to balance advantages and disadvantages of various strategies based on the requirements of film purity, structural quality, the rate of growth, temperature constraints and other factors. In this section, the salient features of these processing methods are briefly described. This is of general interest because the state of stress in a film can be strongly influenced by its deposition history, as described in the later sections of this chapter.

1.2.1 Physical vapor deposition

Physical vapor deposition is a technique whereby physical processes, such as evaporation, sublimation or ionic impingement on a target, facilitate the transfer of atoms from a solid or molten source onto a substrate. Evaporation and sputtering are the two most widely used PVD methods for depositing films.

Figure 1.2 schematically illustrates the basic features of evaporative deposition. In this process, thermal energy is supplied to a source from which atoms are evaporated for deposition onto a substrate. The vapor source configuration is intended to concentrate heat near the source material and to avoid heating the surroundings.

Heating of the source material can be accomplished by any of several methods. The simplest is resistance heating of a wire or stripe of refractory metal to which the material to be evaporated is attached. Larger volumes of source material can be heated in crucibles of refractory metals, oxides or carbon by resistance heating, high frequency induction heating or electron beam evaporation. The evaporated atoms travel through reduced background pressure p in the evaporation chamber and condense on the growth surface. The deposition rate $\dot{R}$ of the film is commonly denoted by the number of atoms arriving at the substrate per unit area of the substrate per unit time, by the time required to deposit a full atomic layer of film material, or by the average normal speed of the growth surface of the film. The deposition rate or flux is a function of the travel distance from the source to the substrate, the angle of impingement onto the substrate surface, the substrate temperature T_s and the base pressure p. If the source material (such as Cr, Fe, Mo, Si and Ti) undergoes sublimation, sufficiently large vapor pressures may be obtained below its melting temperature so that a solid source could be employed for evaporative deposition. On the other hand, for most metals in which a sufficiently large vapor pressure ($\sim 10^{-3}$ torr, or 0.13 Pa) cannot be achieved at or below the melting temperature, the source is heated to a liquid state so as to achieve proper deposition conditions.

Metal alloys, such as Al–Cu, Co–Cr or Ni–Cr, can generally be evaporated directly from a single heated source. If two constituents of the alloy evaporate at different rates causing the composition to change in the melt, two different sources held at different temperatures may be employed to ensure uniform deposition. Unlike metals and alloys, inorganic compounds evaporate in such a way that the vapor composition is usually different from that of the source. The resulting molecular structure causes the film stoichiometry to be different from that of the source. High purity films of virtually all materials can be deposited in vacuum by means of electron beam evaporation.

Molecular beam epitaxy (MBE) is an example of an evaporative method. This growth technique can provide film materials of extraordinarily good quality which are ideal for research purposes. However, the rate of growth is very low compared to other methods, which makes it of limited use for production of devices. In MBE, the deposition of a thin film can be accurately controlled at the atomic level in an ultra-high vacuum (10^{-10} torr, or 1.33×10^{-8} Pa). A substrate wafer is placed in the ultra-high vacuum chamber. It is sputtered briefly with a low energy ion beam to remove surface contamination. This step is followed by a high temperature anneal to relax any damage done to the growth surface during preparation. The substrate is then cooled to the growth temperature, typically between 400 and 700 °C, and growth commences by directing atomic beams of the film material, as well as a beam of dopant material if necessary, toward the growth surface of the substrate. The

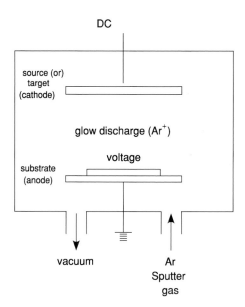

Fig. 1.3. Schematic showing the basic features of a dc sputter deposition system.

beams are emitted from crucibles of the growth materials which have been heated to temperatures well above the substrate temperature to induce evaporation and condensation. The films may be examined by transmission electron microscopy or x-ray diffraction after cooling. The complete history of evolution of internal stress in the film during deposition can be obtained *in situ* by monitoring the changes in curvature of the substrate on which the film is deposited as described in detail in Chapter 2.

In sputter deposition, ions of a sputtering gas, typically Ar, are accelerated toward the target at high speed by an imposed electric field. The initial concentration of charge carriers in the system is significantly increased with an increase in the dc voltage, as the ions collide with the cathode, thereby releasing secondary electrons, and with the neutral gas atoms. As critical numbers of electrons and ions are created through such avalanches, the gas begins to glow and the discharge becomes self-sustaining. Gaseous ions striking the target or the source material from which the film is made dislodge surface atoms which form the vapor in the chamber. The target is referred to as the cathode since it is connected to the negative side of the direct current power supply. Figure 1.3 schematically shows the basic elements of a sputter deposition system. The chamber is evacuated and then Ar gas, at a pressure of approximately 13.3 Pa (10^{-1} torr), is introduced for the purpose of maintaining a visible glow discharge. The Ar^+ ions bombard the target or cathode, and the ensuing momentum transfer causes the neutral atoms of the target source to be dislodged. These atoms transit through the discharge and condense onto the substrate, thus providing film growth.

Several different sputtering methods are widely used for the deposition of thin films in different practical applications: (i) dc sputtering (also commonly referred to as cathodic or diode sputtering), (ii) radio frequency (rf) sputtering with frequencies typically in the 5–30 MHz range, (iii) magnetron sputtering, where a magnetic field is applied in superposition with a parallel or perpendicularly oriented electric field between the substrate and the target source, and (iv) bias sputtering, where either a negative dc or rf bias voltage is applied to the substrate so as to vary the energy and flux of the incident charged species.

There are many distinctions between the sputtering process and the evaporative process for film deposition, as described by Ohring (1992) for example. Evaporation is a thermal process where the atoms of the material to be deposited arrive at the growth surface with a low kinetic energy. In sputtering, on the other hand, the bombardment of the target source by Ar^+ ions imparts a high kinetic energy to the expelled source atoms. Although sputter deposition promotes high surface diffusivity of arriving atoms, it also leads to greater defect nucleation and damage at the deposition surface because of the high energy of the atoms. While evaporation occurs in a high vacuum (10^{-6} to 10^{-10} torr, or 1.33×10^{-4} to 1.33×10^{-8} Pa), sputtered atoms transit through a high pressure discharge zone with a pressure of approximately 0.1 torr (13.33 Pa). Sputter-deposited films generally contain a higher concentration of impurity atoms than do films deposited by evaporation and are prone to contamination by the sputtering gas. As a result, sputter deposition is not well suited for epitaxial growth of films.

For polycrystalline films, the film grain structure resulting from sputter deposition typically has many crystallographic orientations without preferred texture. However, evaporative deposition leads to highly textured films for which the grain size is typically greater than that of the sputtered films. Sputter deposition offers better control in maintaining stoichiometry and film thickness uniformity than evaporative deposition and has the flexibility to deposit essentially any crystalline and amorphous materials. These issues are discussed in more detail in Section 1.8.

1.2.2 Chemical vapor deposition

Chemical vapor deposition is a versatile deposition technique that provides a means of growing thin films of elemental and compound semiconductors, metal alloys and amorphous or crystalline compounds of different stoichiometry. The basic principle underlying this method is a chemical reaction between a volatile compound of the material from which the film is to be made with other suitable gases so as to facilitate the atomic deposition of a nonvolatile solid film on a substrate, as indicated schematically in Figure 1.4. The chemical reaction in a CVD process may involve pyrolysis or reduction.

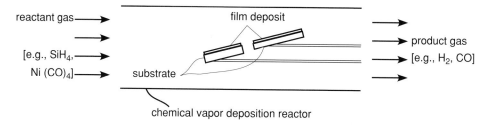

Fig. 1.4. Schematic showing the basic features of an open reactor system for chemical vapor deposition.

Consider the production of amorphous or polycrystalline Si films on Si substrates, where pyrolysis at 650 °C leads to the decomposition of silane gas according to the reaction

$$SiH_4(g) \rightarrow Si(s) + 2H_2(g).$$

High-temperature reduction reactions where hydrogen gas is used as a reducing agent are also employed to produce epitaxial growth of Si films on monocrystalline Si substrates at 1200 °C according to the reaction

$$SiCl_4(g) + 2H_2(g) \rightarrow Si(s) + 4HCl(g).$$

The nature of epitaxy is described in detail later in this chapter.

In CVD, as in PVD, vapor supersaturation affects the nucleation rate of the film whereas substrate temperature influences the rate of film growth. These two factors together influence the extent of epitaxy, grain size, grain shape and texture. Low gas supersaturation and high substrate temperatures promote the growth of single crystal films on substrates. High gas supersaturation and low substrate temperatures result in the growth of less coherent, and possibly amorphous, films. Low-pressure CVD (LPCVD), plasma-enhanced CVD (PECVD), laser-enhanced CVD (LECVD) and metal-organic CVD (MOCVD) are variants of the CVD process used in many situations to achieve particular objectives.

1.2.3 Thermal spray deposition

The thermal spray process of thin film fabrication refers broadly to a range of deposition conditions wherein a stream of molten particles impinges onto a growth surface. In this process, which is illustrated schematically in Figure 1.5, a thermal plasma arc or a combustion flame is used to melt and accelerate particles of metals, ceramics, polymers or their composites to high velocities in a directed stream

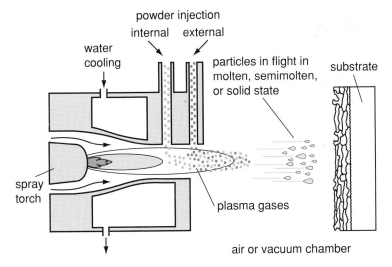

Fig. 1.5. Schematic illustration of the thermal spray process.

toward the substrate. The sudden deceleration of a particle upon impact at the growth surface leads to lateral spreading and rapid solidification of the particle forming a 'splat' in a very short time. The characteristics of the splat are determined by the size, chemistry, velocity, degree of melting and angle of impact of the impinging droplets, and by the temperature, composition and roughness of the substrate surface. Successive impingement of the droplets leads to the formation of a lamellar structure in the deposit. The oxidation of particles during thermal spray of metals also results in pores and contaminants along the splat boundaries. Quench stresses and thermal mismatch stresses in the deposit are partially relieved by the formation of microcracks or pores along the inter-splat boundaries and by plastic yielding or creep of the deposited material. There are several different types of thermal spray processes, a review of which can be found in Herman et al. (2000).

Continuous or step-wise gradients in composition through the thickness of the layer can be achieved by use of multiple nozzles whereby the flow rate of the constituent phases of the deposited composite can be modulated during spray, as demonstrated by Kesler et al. (1997). Alternatively, the feed rate of the powders of the different constituent phases can also be controlled appropriately during thermal spray so as to deposit a graded layer onto a substrate. Gradients in porosity can be introduced through the thickness of the deposited layer by manipulating the processing parameters and deposition conditions.

The *plasma spray* technique offers a straightforward and cost-effective means to spray deposits of metals and ceramics that are tens to hundreds of micrometers in thickness onto a variety of substrates in applications involving thermal-barrier or

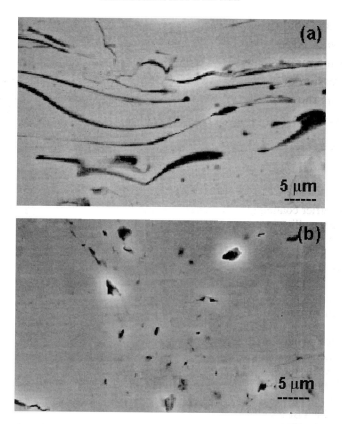

Fig. 1.6. Scanning electron micrographs showing the microstructures of plasma-spray coating of NiCrAlY on a 1020 steel substrate. (a) Air-plasma-spray coated layer with inter-splat cracks whose origin can be traced to the oxidation of Al in the coated material during deposition. (b) Vacuum plasma sprayed coating of the same material without inter-splat microcracks. Reproduced with permission from Alcala et al. (2001).

insulator coatings. Typical plasma-spray deposits are porous, with only 85–90% of theoretical density.

For applications requiring higher density coatings with a strong adhesion to the substrate, *low-pressure plasma spray* is employed where spraying is done in an inert gas container operating at a reduced pressure. *Vacuum plasma spray* is another thermal spray process which is used to improve purity of the deposited material and to reduce porosity and defect content, albeit at a higher cost than air plasma spray.

Figure 1.6(a) is a representative micrograph of the cross-section of an air-plasma-sprayed NiCrAlY coating, commonly used as a bondcoat between a ceramic thermal barrier coating and a nickel-base superalloy substrate in gas turbine engines, as described in the example in the next subsection. This coating was deposited onto a 1020 steel substrate. The dark streaks are the inter-splat boundaries along which microcracks and voids have formed. The origins of these defects could be traced to the formation of Al_2O_3 during deposition (Alcala et al. 2001). On the other hand, vacuum plasma spray deposition of the same material onto the steel substrate results in the suppression of such oxidation and the attendant cracking of the inter-splat boundaries, as shown in Figure 1.6(b). The resulting coating has a more uniform microstructure with a significantly reduced pore density.

1.2.4 Example: Thermal barrier coatings

The thermal barrier coating (TBC) system is a multilayer arrangement introduced to thermally insulate metallic structural components from the combustion gases in gas turbine engines. The design of TBCs along with appropriate internal cooling of the high-temperature metallic components has facilitated the operation of gas turbine engines at gas temperatures well in excess of the melting temperature of the turbine blade alloy. This thermal protection system, which reduces the surface temperature of the alloy by as much as 300 °C, leads to better engine efficiency, performance, durability and environmental characteristics.

The performance requirements for TBCs are stringent. The selection of materials for TBCs and the design of the layered coating structure inevitably require consideration of highly complicated interactions among such phenomena and processes as phase transformation, microstructural stability, thermal conduction, diffusion, oxidation, thermal expansion mismatch between adjoining materials, radiation, as well as damage and failure arising from interface delamination, film buckling, subcritical fracture, foreign object impact, erosion, thermal and mechanical fatigue, inelastic deformation and creep. Summaries and reviews of such issues for TBCs have been reported by Evans et al. (2001) and Padture et al. (2002).

A representative TBC system for a gas turbine engine comprises four layers: (a) a metallic substrate, that is, the turbine blade itself, (b) a metallic interlayer or bondcoat, (c) a thermally grown oxide (TGO), and (d) a ceramic outerlayer or topcoat. A schematic of the turbine blade along with a scanning electron micrograph of the cross-section of a layered TBC coating is shown in Figure 1.7.

The turbine blade is commonly made of a nickel-base or cobalt-base superalloy which is investment-cast as a single crystal or polycrystal. The bondcoat, typically 75 to 150 μm in thickness and made of an oxidation-resistant alloy of NiCrAlY or NiCoCrAlY, is deposited onto the substrate using plasma spray or electron-beam PVD. In some cases, the bondcoats are deposited by electroplating along with either diffusion-aluminizing or CVD with layers of Ni and Pt aluminides. Occasionally, the bondcoats consist of sublayers of different phases or compositions. The bondcoat is a critical component of the TBC system that determines the spallation resistance of the TBC.

The oxidation of the bondcoat at an operating temperature as high as 1050 °C results in a 1 to 10 μm thick TGO layer between the bondcoat and the topcoat. This oxidation process is aided by the transport of oxygen from the surrounding hot gases in the engine environment through the porous TBC top layer. The TGO layer is also commonly engineered to serve as a diffusion barrier so as to suppress further oxidation of the bondcoat; for this purpose, the *in situ* formation of a uniform, defect-free α-Al_2O_3 TGO interlayer is facilitated by controlling the composition of the bondcoat.

Y_2O_3-stabilized Zr_2O_3 (YSZ), typically with a Y_2O_3 concentration of 7 to 8 wt%, is the common material of choice for the ceramic topcoat in light of its many desirable properties as a TBC (Padture et al. 2002):

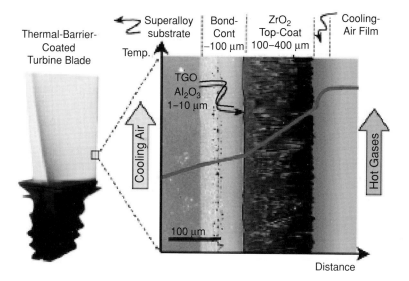

Fig. 1.7. Illustration of the thermal-barrier coated turbine blade which is air-cooled internally along hollow channels. The outer surface of the blade is coated for thermal protection
from the hot gases so that there exists a temperature gradient through the cross-section of
the TBC. Also shown is a scanning electron micrograph of the cross-sectional view of the
coating layers which comprise a ceramic topcoat deposited by electron-beam PVD, an alumina TGO layer, and a NiCrAlY bondcoat on a nickel-base superalloy substrate. Reprinted
with permission from Padture et al. (2002).

— the thermal expansion coefficient of YSZ has a high value of approximately $11 \times 10^{-6}\,°C^{-1}$, which is closer to that of the metallic layer beneath it ($\sim 14 \times 10^{-6}\,°C^{-1}$) than to that of most ceramic materials. Consequently, the stresses generated by thermal expansion mismatch with the underlying layers during the thermal cycles generated by the operation of the turbine engine would be minimized;

— YSZ has a very low thermal conductivity of approximately $2.3\,W(m\ K)^{-1}$ at $1000\,°C$ when fully dense;

— the low density of YSZ, typically about $6.4 \times 10^3\,kg\ m^{-3}$, improves the performance of the rotating engine component in which it is used as a coating;

— the high melting temperature of approximately $2700\,°C$ makes YSZ a desirable thermal barrier material;

— with a myriad of available processing methods, the YSZ topcoat can be deposited with controlled pore content and distribution in such a way that it can be made compliant with an elastic modulus of approximately 50 GPa; and

— fully dense YSZ has a hardness of roughly 14 GPa at room temperature, which renders it resistant to damage from foreign object impact and erosion; however, the hardness value can be low as 1 GPa due to porosity or elevated temperature.

The two most common methods of producing the TBC topcoat entail air plasma spray (see
Section 1.2.3) and electron-beam physical vapor deposition, EB-PVD (see Section 1.2.1).
The former deposition method produces 15 to 25% porosity whereby low values of thermal
conductivity and elastic modulus result. The spray coating, typically 300 to 600 μm thick,
contains pancake-shaped splats with a diameter of 200–400 μm and thickness of 1–5 μm.

The pores and cracks along the inter-splat boundaries are generally oriented parallel to the interface with the bondcoat and normal to the direction of heat flow, a consequence of which is the low thermal conductivity of 0.8–1.7 $W(m\ K)^{-1}$. The air plasma spray method provides an economical means for the large-scale production of TBCs. However, the structural defects inherent in this process and the rough interface between the sprayed coating and the material beneath it make this deposition technique more suitable for less critical parts. The EB-PVD deposited TBCs, on the other hand, are typically 125 μm thick and are more durable and costly compared to the plasma-sprayed coatings. They are engineered to comprise the following microstructural features: (a) an equiaxed structure of YSZ grains with a diameter of 0.5–1.0 μm near the interface with the bondcoat, (b) columnar grains of YSZ, 2–10 μm in diameter which extend from the equiaxed grains to the top surface of the coating, with the boundaries between the columnar grains amenable to easy separation to accommodate thermal stresses that develop during service, and (c) nanometer-size pores inside the columnar grains.

1.3 Modes of film growth by vapor deposition

There is enormous variation in the microstructures of films formed by deposition of atoms on the surfaces of substrates from vapors. Final structures can range from single crystal films, through polycrystalline films with columnar or equiaxed grains, to largely amorphous films. Some materials can be deposited in ways that yield any of these structures, with the final microstructure depending on the materials involved, the deposition method used and the environmental constraints imposed. The purpose in this section is to discuss some general ideas in film growth by vapor deposition that transcend issues of final microstructure. The discussion is largely descriptive, and it is couched in the terminology of thermodynamics. The principles of thermodynamics provide powerful tools for deciding whether or not some particular change in a material system can occur. On the other hand, in those cases in which the change considered can indeed take place, thermodynamics is silent on whether or not it does take place and, if it does occur, on how it proceeds. Nonetheless, the thermodynamic framework provides a basis for establishing connections between deposition circumstances and film formation. Progress toward understanding physical processes of film growth is discussed in depth by Tsao (1993), Pimpinelli and Villain (1998) and Venables (2000).

1.3.1 From vapor to adatoms

In this section, the factors that control the very early stages of growth of a thin film on a substrate are described in atomistic terms. The process begins with a clean surface of the substrate material, which is at temperature T_s, exposed to a vapor of a chemically compatible film material, which is at the temperature T_v. To form a single crystal film, atoms of the film material in the vapor must arrive at the substrate

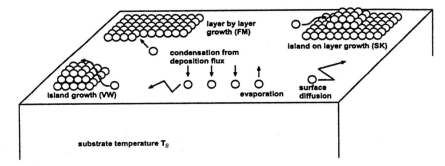

Fig. 1.8. Schematic showing the atomistics of film formation on substrates.

surface, adhere to it, and settle into possible equilibrium positions before structural defects are left behind the growth front. To form an amorphous film, on the other hand, atoms must be prevented from seeking stable equilibrium positions once they arrive at the growth surface. In either case, this must happen in more or less the same way over a very large area of the substrate surface for the structure to develop. At first sight, this outcome might seem unlikely, but such films are produced routinely.

Atoms in the vapor come into contact with the substrate surface where they form chemical bonds with atoms in the substrate. The temperature of the substrate must be low enough so that the vapor phase is supersaturated in some sense with respect to the substrate, an idea that will be made more concrete below. There is a reduction in energy due to formation of the bonds during attachment. Some fraction of the attached atoms, which are called *adatoms*, may return to the vapor by evaporation if their energies due to thermal fluctuations are sufficient to overcome the energy of attachment, as suggested in the schematic diagram in Figure 1.8. To make the discussion a bit more specific, a simple hexagonal close-packed crystal structure is assumed for convenience in counting bonds.

It has already been recognized that, for film growth to be possible, it is necessary that the vapor in contact with the growth surface is supersaturated with respect to the substrate at its temperature T_s. For a homogeneous crystal at some temperature that is in contact with its own vapor at the same temperature, the *equilibrium vapor pressure* p_e of the system is defined as the pressure at which condensation of vapor atoms onto the solid surface and evaporation of atoms from the surface proceed at the same rate. At equilibrium, the entropic free energy per atom in the vapor equals the free energy per atom in the interior of the crystal. The lower internal energy of atoms in the *interior* of the crystal compared to those in the vapor, due to chemical bonding, is offset by the lower entropic energy within the crystal. For net deposition on the substrate surface, it is essential that the pressure p in the vapor exceeds the equilibrium vapor pressure p_e at the substrate temperature, that is, the vapor must be *supersaturated*. For the pressure p, the entropic free energy per atom of the vapor, over and above the free energy at pressure p_e, is estimated as the work needed per atom to increase the vapor pressure from p_e to p at constant temperature. According to the ideal gas law, this work equals $k T_v \ln(p/p_e)$ where $k = 1.38 \times 10^{-23} \text{ J K}^{-1} = 8.617 \times 10^{-5} \text{ eV K}^{-1}$ is the Boltzmann constant and T_v is the absolute temperature of the vapor. If the vapor becomes supersaturated, a free energy difference between the vapor and the interior of the crystal exists, providing a *chemical potential* for driving the advance of the interface toward the vapor. As the interface advances in a self-similar way, a remote layer of vapor of some mass is converted into an interior layer of crystal of the same mass. The interface does not advance or retreat when these energies are identical.

In film deposition, the situation is often complicated by the fact that the vapor and the substrate are not phases of the same material, and by the fact that the temperature of the substrate is usually lower than that of the vapor. The definition of equilibrium vapor pressure is not so clear in this case. However, in most cases there will be some level of vapor pressure below which deposition of film material onto the growth surface will not occur; this serves as an operational definition of p_e for film growth.

Once adatoms become attached to the substrate, their entropic free energy is reduced from that of the vapor. The adatoms form a distribution on the substrate surface having the character of a two-dimensional vapor. The deposited material is believed to thermalize quickly and to take on the temperature T_s. On a crystal surface, there is some density ρ_{ad} of adatoms that is in equilibrium with a straight surface step or ledge bordering a partial monolayer of film atoms, that is, for which the step neither advances nor recedes. The step is a boundary between two phases in a homogeneous material system. Consequently, the notion of equilibrium vapor pressure or *equilibrium vapor density* ρ_{ad}^e can be invoked in an analogous way. For film growth to occur by condensation, the free energy per atom of the two-dimensional gas must exceed the free energy of a fully entrained surface atom by the amount $\mathcal{E}_{ph} = k T_s \ln(\rho_{ad}/\rho_{ad}^e)$.

1.3.2 From adatoms to film growth

Each adatom presumably resides within an equilibrium energy well on the surface most of the time, and this well is separated from adjacent energy wells by a barrier of height $\mathcal{E}_d > 0$ with respect to the equilibrium position. The atoms oscillate in their wells due to thermal activation and, occasionally, they acquire sufficient energy to hop into adjacent equilibrium wells. Surface diffusion results from such jumps. If the surface is spatially uniform, diffusion occurs randomly with no net mass transport on a macroscopic scale. The hopping rate of any given atom increases with increasing substrate temperature. Occasionally, attached atoms may also change positions with substrate atoms, but this possibility is not pursued further in this discussion. If the surface is not uniform, perhaps as a result of a strain gradient or a structural gradient, then surface diffusion can be directionally biased, resulting in a net mass transport along the surface on a macroscopic scale. Consequences of such transport are considered in Chapters 8 and 9. If the temperature is very low or if the diffusion barrier is very high, adatoms stick on the growth surface where they arrive, and the film tends to grow with an amorphous or very fine grained polycrystalline structure.

The growth surface invariably has some distribution of surface defects – crystallographic steps, grain boundary traces and dislocation line terminations, for example – which provide sites of relatively easy attachment for adatoms. The spacing between defects represents a length scale for comparison to the extent of random walk diffusion paths. If the diffusion distance is large compared to the defect spacing, then adatoms tend to encounter these defects and become attached to them, giving up some free energy in the process. This is the case of heterogeneous nucleation and growth of films. General statements about such processes are also difficult to make. In some cases, surface steps are relatively transparent to migrating adatoms, that is, adatoms often pass over the step in either the up direction or the down direction without attachment. For other materials, virtually every encounter of an adatom with the step results in attachment. In yet other cases, adatoms easily bypass steps in the up direction but not in the down direction, a manifestation of the so-called Schwoebel barrier (Schwoebel 1969). If the spacing of defects is large compared to the diffusion distance, on the other hand, then migrating adatoms have the potential for lowering the energy of the system by binding together upon mutual encounters to form clusters. The case of formation of such stable clusters is called homogeneous nucleation of film growth. That there is some minimum cluster size necessary for formation of a stable nucleus can be demonstrated by appeal to the following argument based on classical nucleation theory.

It was noted in Section 1.3.1 that the free energy of an adatom in the supersaturated distribution on the surface is $\mathcal{E}_{ph}$ with respect to its state once entrained within the surface layer. In other words, this is the amount of energy reduction per atom associated with the phase change from a two-dimensional gas of adatoms on the surface to a completely condensed surface layer. Suppose that a planar cluster of n atoms is formed on the surface. Will it tend to grow into a larger cluster, and eventually into a film, or will the cluster tend to disperse? If all n atoms are fully entrained within the cluster then the free energy reduction due to cluster formation would be $-n\mathcal{E}_{ph}$. However, those atoms on the periphery of the cluster are not fully incorporated. They possess an excess free energy compared to those that are fully entrained; this energy can be estimated to be roughly $4\sqrt{\pi n}\mathcal{E}_{f}$ for an equiaxed cluster, if n is fairly large compared to unity. This quantity has the character of a surface step energy. The free energy change associated with cluster formation is then

$$\mathcal{F} = -n\mathcal{E}_{ph} + 4\sqrt{\pi n}\mathcal{E}_f. \tag{1.1}$$

A graph of $\mathcal{F}$ versus n for fixed $\mathcal{E}_{ph}$ and $\mathcal{E}_f$ shows a maximum value for a cluster size of $n_* = 4\pi(\mathcal{E}_f/\mathcal{E}_{ph})^2$. The value of $\mathcal{F}$ at n_*, which is the activation energy for cluster formation, is $\mathcal{F}_* = 4\pi\mathcal{E}_f^2/\mathcal{E}_{ph}$. The implication is that clusters smaller than the size n_* are unstable and that they tend to disperse, whereas clusters larger than this size tend to grow, driven by a corresponding reduction in free energy. It is likely that many clusters form and disperse for each one that evolves into an island. In fact, the classical nucleation theory says nothing about such processes, other than that they are possible. In any case, for a film to form on the substrate surface, it is necessary that either nuclei formed by such homogeneous processes are able to grow or that a sufficient number of surface defects are available to serve as sites of heterogeneous nucleation.

The mode of film formation is determined by the relative values of the various energies involved in the process, and this mode largely determines the eventual structure of the film. There are two main comparisons to be considered. One of these contrasts the height of the diffusion barrier $\mathcal{E}_d$ to the background thermal energy. If $\mathcal{E}_d$ is large compared to the background thermal energy then surface mobility of adatoms is very low. Under such conditions, adatoms more or less stick where they arrive on the substrate surface.

For growth of crystalline films, it is important that $\mathcal{E}_d$ be less than the background thermal energy so that adatoms are able to seek out and occupy virtually all available equilibrium sites in the film crystal lattice as it grows. This requires the substrate temperature and/or the degree of supersaturation of the vapor to be high enough to insure such mobility. Suppose that this is indeed so and that the adatoms are able to migrate over the surface.

The other important energy comparison concerns the propensity for atoms of film material to bond to the substrate. This is represented by the magnitude of $\mathcal{E}_{fs}$, relative to their tendency to bond to other, less well-bound, atoms of film material, as represented by $\mathcal{E}_f$. Two kinds of growth processes can be distinguished, one with $\mathcal{E}_{fs}$ larger in magnitude than $\mathcal{E}_f$ and a second with the relative magnitudes reversed. If $\mathcal{E}_{fs}$ is the larger of the two energy changes in magnitude, then film growth tends to proceed in a layer by layer mode, as indicated in the schematic diagram in Figure 1.8. Adatoms are more likely to attach to the substrate surface than to other film material surfaces. Once small stable clusters of adatoms form on the surface, other adatoms tend to attach to the cluster at its periphery where they can bond with both substrate and film atoms, thereby continuing the planar growth, as indicated in Figure 1.8. This layer-by-layer film growth mode is often called the *Frank–van der Merwe growth mode* or FM mode, according to a categorization of growth modes proposed by Bauer (1958) on the basis of more macroscopic considerations of surface energy. This alternate point of view will be considered in Section 1.3.5.

On the other hand, if $\mathcal{E}_f$ is larger in magnitude than $\mathcal{E}_{fs}$, then it is energetically favorable for adatoms to form three-dimensional clusters or islands on the surface of the substrate. Film growth proceeds by the growth of islands until they coalesce; this type of growth is commonly called the *Volmer–Weber growth mode* or the VW mode; see Figure 1.8.

A third type of growth, which combines features of both the Frank–van der Merwe and the Volmer–Weber modes, is called the *Stranski–Krastanov growth mode* or SK mode. In this mode, the film material tends to prefer attachment to the growth surface rather than the formation of clusters on the growth surface; that is, $\mathcal{E}_{fs}$ is greater in magnitude than $\mathcal{E}_f$. However, after a few monolayers of film material are formed and after the structure of the film becomes better defined as a crystal in conformity with the substrate, the tendency is reversed. In other words, once the planar growth surface becomes established as film material, subsequent adatoms tend more to gather into clusters than to continue planar growth. The magnitude of $\mathcal{E}_{fs}$ appears to depend on the thickness of the film in the early stages of growth, decreasing from values larger than the magnitude of $\mathcal{E}_f$ to values that are smaller. The occurrence of this mode is most likely when the first few layers of film material are heavily strained due to the constraint of the substrate. This issue is revisited in Section 1.3.5.

1.3.3 Energy density of a free surface or an interface

In the preceding discussion, the early stages in the growth of a film from a vapor were considered in a qualitative way in terms of the behavior of atoms. Many of the inferences drawn can also be made in terms of surface energy and interface energy of solid materials. These quantities are macroscopic measures attributable to the discreteness of the material. However, they represent only ensemble averages of behavior at the atomistic level and do not incorporate discreteness of the material in any direct way. In many cases of microstructure evolution, knowledge of these quantities provides an adequate basis for understanding the film growth process. In this section, these quantities are discussed in general terms, and implications for film growth are considered.

Surface energy is an important concept in considering the evolution of microstructure in small-scale material systems. The free energy of the bounding surface of a crystal is a macroscopic quantity representing, in some sense, the net work that had to be done to create that surface. This energy is distributed over a specific mathematical surface which has no thickness and which approximates the physical boundary between the crystal and its surroundings. From this definition of surface energy it is clear that the reference level for energy of a free surface is the state of the material on that same crystallographic surface when it is embedded deep within a perfect crystal. Presumably, creation of the surface was accomplished by dividing a larger crystal into two parts by separating these parts along a common bounding surface. The total work done per unit area is assumed to be evenly divided between the two bounding surfaces created. The concept of specific surface energy, or energy of cohesion, appears to have its origins in a theory of fluid surfaces due to Young (1805) and Rayleigh (1890). The concept was subsequently extended to crystal surfaces, where it was given a central role in the work of Gibbs (1878). A useful geometrical interpretation was provided subsequently by Wulff (1901). Later important contributions to the equilibrium theory of crystal surfaces and to the description of the energetics of surface evolution were provided by Herring (1953).

In general, there is a free energy reduction (increase) associated with forming (breaking) a chemical bond. It follows that there is an increase in free energy associated with creating a free surface that is more or less proportional to the area of surface created or the number of chemical bonds per unit area that are directly involved. Once the surface is created, the corresponding free energy is, first and foremost, in the form of chemical bonding potential, but other physical factors can contribute to surface energy as well. At the surface of a coherent crystal, the atoms retain the general arrangement that defines the crystal structure. However, the spacings in this arrangement are altered once bonds are broken; this is necessary to restore equilibrium following the 'disappearance' of interactions across the surface. Through this change in spacing, atoms at or very near the surface are able to relax to energy states somewhat lower than those in the natural lattice sites in the interior of the crystal; this change in positional energy may reduce the surface energy from the level predicted by bond counting alone. The effect is due primarily to the longer range interactions between atoms in the crystal. If only nearest neighbor interactions are taken into account, then interaction forces are typically zero at equilibrium. However, if next nearest neighbor interactions are also taken into account, then nearest neighbor interaction forces are compressive and next nearest neighbor forces are tensile at equilibrium. Consequently, if some of the interactions are eliminated to form a surface, unbalanced forces remain in the configuration unless rearrangements occur.

There is yet another factor that comes into play in considering the energies of surfaces, namely, surface *reconstruction* or *rebonding*. Chemical bonds that are broken in creating a free surface are free or 'dangling' in the process described up to this point. Atoms in surfaces with bonding potential may seek out other atoms within the surface with the potential to form bonds. Bond formation is likely to occur in such cases if it results in a net lowering of the energy of the system. Usually, more than one such reconstruction is possible for a given crystallographic surface. Each reconstructed configuration represents a local minimum of free energy with respect to that configuration. Presumably, the system seeks out the absolute minimum from among these as the actual reconstructed configuration; the choice might depend on temperature, lattice strain or other factors. Reconstructed surfaces are typically periodic, but with lattice vectors that are very different from those of the underlying crystal. Reconstruction may also result in surface anisotropy that is not representative of the lattice itself.

In summary, surface free energy represents the net energy increase of the system due to work that must be done to overcome chemical bonds in creating the surface, reduced by any relaxation of atomic positions of lattice sites adjacent to the surface, and further reduced by surface reconstructions which take place to match up available bonds in the surface. It is a macroscopic quantity represented mathematically by a scalar function of position over a material surface. The value of surface energy per unit area of a given crystallographic surface orientation is determined by the fine scale structure of that surface. The fine scale structure of the 'perfect' crystal may include steps only one atom spacing in height separating terraces, combinations of facets, various rebounding configurations and other features unique to surface structure.

Consider an isolated small crystal of some particular initial shape. Knowledge of the dependence of surface energy on surface orientation is sufficient to predict a minimum energy shape for the small crystal when surface energy is the only free energy contribution that changes when shape changes. This shape is most readily determined by means of the Wulff construction, as illustrated schematically in two dimensions in Figure 1.9. In this diagram, the solid lobed curve represents the variation of surface free energy with orientation; the radial distance in some direction from the origin of the polar plot to the solid curve is the free energy per unit area of a planar surface with normal vector in the direction of the radial line. A representative radial line is shown in the figure. To form the surface shape, a line perpendicular to the radial line is drawn through the point where the radial line intersects the curve of surface energy. Several examples of such lines are shown in the figure, with the particular line perpendicular to the radial line shown in heavier weight. The low temperature equilibrium shape of the small crystal is then geometrically similar to the region completely enclosed by the totality of these perpendicular lines as the angle of the radial line varies continuously over its full range. This equilibrium shape is shown as the heavy dashed line in the figure. A proof that this construction leads to the shape with minimum surface energy is outlined by Herring (1953). The construction is readily generalized to three space dimensions.

Many additional observations can be made on the basis of the same construction (Herring 1951). Among the most useful of these concerns the possibility of a macroscopically flat surface assuming a corrugated or peak-and-valley shape in order to reduce its free energy. For example, if a macroscopically flat surface has the

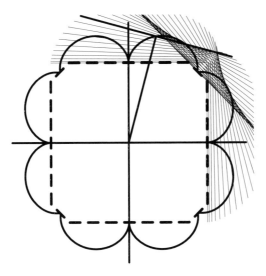

Fig. 1.9. Representative polar plot of surface free energy in a symmetry plane of a crystal; the free energy of a surface at a certain orientation is the radial distance from the origin to the lobed curve in the direction of the surface normal. The Wulff construction is illustrated by the family of lighter weight lines, with one member of the family and its corresponding perpendicular radial line shown in a heavier weight. The constructed equilibrium shape of the crystal is shown by the dashed line.

orientation of one of the flat surfaces in the equilibrium shape, then no corrugated surface composed microscopically of other orientations can be more stable. Conversely, if a given macroscopically flat surface of a crystal does not coincide in orientation with some portion of the boundary of the equilibrium shape, then there will always be a corrugated structure, with the same average orientation as the initial flat surface, which has a lower free energy than the flat surface; this corrugated structure will be made up of segments of surface coinciding in orientation with the flat faces of the equilibrium shape. The scale of the surface corrugations is indeterminate unless an energy is associated with the edges where the different facets of the configuration intersect.

Like free surfaces, interface surfaces at which materials are joined can also be represented macroscopically as discrete surfaces and can be characterized by an interface surface energy per unit area. The physical origin of this energy is essentially the same as that of a free surface. The energy density of an interface is presumably less than the sum of the free surface energies of the two materials joined at that interface, and greater than the energy densities within either of the materials at interior points remote from the interface.

Suppose that a small cluster or island of film material of volume V is deposited onto the surface of the substrate. With this deposition, there is a decrease in the area

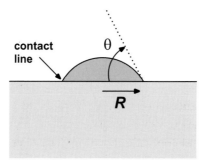

Fig. 1.10. Schematic diagram of a spherical cap island on a substrate. The radius of the
circular contact line is R and the angle between the substrate surface and the tangent plane
to the island surface at the contact line is θ.

of substrate free surface, a change in the area of free surface of the film material, and
an increase in the area of a shared interface. There is a surface free energy density
associated with both the substrate and film materials, say γ_s and γ_f, respectively.
Likewise, there is also a characteristic interface free energy per unit area, say γ_{fs},
associated with the interface. What effect does the change in free energy associated
with these surfaces have on the equilibrium shape of the island, assuming that the
substrate surface remains flat? The equilibrium shape is that particular shape that
minimizes the free energy for a given island volume.

To illustrate the idea, assume that γ_f is independent of orientation of the island
surface. Furthermore, assume that the shape of the island is a spherical cap of
constant volume. Two geometrical parameters are needed to specify the shape of
the island. Denote the radius of the circular contact line where the island surface
meets the substrate surface by R, and denote the angle between the substrate surface
and the tangent plane to the island surface at the contact line by θ. These parameters
are indicated in the sketch in Figure 1.10. In terms of R and θ, the island volume V
and the island surface area A are

$$V = \frac{\pi R^3}{3}\left(\frac{2+\cos\theta}{1+\cos\theta}\right)\left(\frac{1-\cos\theta}{\sin\theta}\right), \quad A = \frac{2\pi R^2}{1+\cos\theta}. \tag{1.2}$$

Up to an additive constant, the free energy change associated with island formation
is

$$\mathcal{F}(V,\theta) = A\gamma_f + \pi R^2(\gamma_{fs} - \gamma_s) \tag{1.3}$$

where R is understood to depend on V and θ according to (1.2)$_1$. The requirement
that θ must take on a value that minimizes the free energy $\mathcal{F}$, subject to the constraint

that V is constant, leads to the condition that†

$$\gamma_{\mathrm{f}} \cos\theta + \gamma_{\mathrm{fs}} - \gamma_{\mathrm{s}} = 0 \quad \Rightarrow \quad \cos\theta = \frac{\gamma_{\mathrm{s}} - \gamma_{\mathrm{fs}}}{\gamma_{\mathrm{f}}}. \tag{1.4}$$

This result is frequently said to follow from a 'force balance' among surface energies at the contact line, but the correctness of the result is only fortuitous; the basis for the result is constrained energy minimization. The expression in (1.4) is known commonly as Young's equation for the *wetting angle* of a fluid droplet on a rigid solid surface (Young 1805); the dependence of the wetting angle on the surface energies does not hinge on the assumption of a spherical shape. If additional forms of energy depend on island shape, the argument must be altered accordingly and the form of the edge condition may be affected. An example arises in the next section through the influence of elastic energy.

1.3.4 Surface stress

In the foregoing discussion, the values of free surface energy of a solid are assumed to be independent of the deformation. In fact, the value of free surface energy density at a particular material point as introduced in Section 1.3.3 depends on the elastic strain of the material at that point, in general. As a matter of convention throughout this discussion, surface energy density is interpreted as energy per unit area in the *unstrained* reference configuration, unless explicitly stated otherwise. The unconstrained configuration is determined by the equilibrium lattice dimensions in the interior of a large unconstrained crystal. The elastic strain implies relatively small changes in positions of atoms of the material making up the surface, but there is no rearrangement of the surface atoms as a result of elastic deformation. In other words, the same material particles make up the surface at all levels of elastic strain imagined here; this is the feature that distinguishes a solid from a fluid. If the surface energy density depends on surface strain, then the change in surface energy per unit amount of strain represents a configurational force, commonly called the *surface stress*. This tensor-valued configurational force is work-conjugate to surface strain with respect to surface energy. It is a macroscopic quantity representing aspects of the discrete nature of the material in the vicinity of a surface. From the definition of surface stress, the work done by it during an infinitesimal change in surface strain is the change in surface energy density associated with that change in strain. At a point on the surface, surface strain can be represented by a second rank symmetric tensor, say $\epsilon_{ij}^{\mathrm{S}}$, the components of which are the extensional and shear strains with respect to

---

† The result (1.4) is enclosed in a box to highlight it as a result of broad significance and utility. The practice of identifying principal results in this way is followed throughout the book.

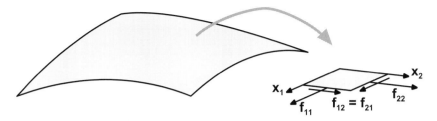

Fig. 1.11. A schematic diagram showing the components of surface stress f_{ij}^{S} acting on an infinitesimal element of a material surface. Surface stress is interpreted as a second rank tensor, and the components are referred to a local rectangular coordinate system in a plane tangent to the surface at the point of interest.

a locally rectangular coordinate system defined in the tangent plane of the surface at the material point of interest. For a given level of strain with corresponding surface stress f_{ij}^{S}, an additional increment $d\epsilon_{ij}^{\mathrm{S}}$ in strain results in a change dU_{S} in surface energy per unit area, so that

$$dU_{\mathrm{S}} = f_{ij}^{\mathrm{S}} d\epsilon_{ij}^{\mathrm{S}}. \tag{1.5}$$

In the context of the discussion in Section 1.3.3, U_{S} might represent γ_{s} or γ_{f}. It follows that surface stress is derivable from the strain dependent surface energy density according to

$$\boxed{f_{ij}^{\mathrm{S}} = \frac{\partial U_{\mathrm{S}}}{\partial \epsilon_{ij}^{\mathrm{S}}}.} \tag{1.6}$$

Development of the ideas underlying the concept of a surface stress in solids was pioneered by Shuttleworth (1950), Herring (1951) and Cahn (1980).

An interpretation of surface stress is suggested in Figure 1.11. Consider the plane that is tangent to the smooth surface at the point of interest. A rectangular $x_1 x_2$-coordinate system is introduced in the tangent plane. Then, the state of surface stress in the surface at the point of tangency can be expressed in terms of the components of membrane force acting on the boundary of an infinitesimal surface element with edges aligned with the coordinate directions. The physical dimensions of the components of surface stress are force/length. In the case of a cut through the surface at the point to tangency with orientation defined by the unit vector τ_j^{S} in the tangent plane, the components of surface stress in the coordinate directions are $f_{ij}^{\mathrm{S}} \tau_j^{\mathrm{S}}$. This physical quantity has the same relationship to surface stress that traction has to the bulk stress tensor.

The local change in area from the undeformed configuration to the strained configuration represented by $\epsilon_{ij}^{\mathrm{S}}$ is the trace of the surface strain tensor $\epsilon_{kk}^{\mathrm{S}}$. It follows that the surface energy per unit area in the deformed configuration is $U_{\mathrm{S}}(1 + \epsilon_{kk}^{\mathrm{S}}) = \hat{U}_{\mathrm{S}}$.

In terms of this measure of surface energy density, (1.6) becomes

$$f_{ij}^S = (1 + \epsilon_{kk}^S)\frac{\partial \hat{U}_S}{\partial \epsilon_{ij}^S} + \delta_{ij}\hat{U}_S \approx \frac{\partial \hat{U}_S}{\partial \epsilon_{ij}^S} + \delta_{ij}\hat{U}_S. \quad (1.7)$$

If the surface energy density per unit *deformed* surface area remains constant, then the surface stress is an isotropic second rank tensor with components numerically equal to $\hat{U}_S$. On the other hand, if the surface energy density per unit *undeformed* surface area remains constant, the surface stress vanishes.

If the surface of interest is a bonded interface between two materials, rather than a free surface, surface stress analogous to f_{ij}^S arises. This interface surface stress f_{ij}^I is related to the interface surface energy density according to

$$f_{ij}^I = \frac{\partial U_I}{\partial \epsilon_{ij}^I} \quad (1.8)$$

where U_I is the interface energy density in a reference configuration and ϵ_{ij}^I is the surface strain with respect to that reference configuration. In the context of Section 1.3.3, U_I might represent γ_{fs}. One aspect of interface surface stress that requires special attention is the identification of the reference configuration. If two materials are joined without mismatch strain then the configuration with both bulk materials being unstrained is a natural choice. On the other hand, if there is a mismatch strain of one material forming the interface with respect to the other material, then the surface stress will depend on this mismatch strain. In describing mismatch strain, it is customary to interpret it as the elastic strain in one of the materials when the elastic strain in the other material is zero. The configuration in which this is the case provides a convenient and reasonable choice of a reference configuration for identification of U_I. Aside from this particular feature, the character of interface surface stress is no different from that of free surface stress.

1.3.5 Growth modes based on surface energies

With the notions of free surface energy and interface surface energy in place, a more macroscopic description of the difference among the modes of film growth that were identified in Section 1.3.2 can be pursued. Once again, consider a small island or cluster of film material on the surface of the substrate as depicted in Figure 1.10. The volume of the cluster is again denoted by V but, in this instance, the volume is not held constant; instead, the possibility of *growth* of the cluster is considered. This growth is achieved by incorporating adatoms drawn from the surrounding substrate surface, as described in atomistic terms in Section 1.3.1. In that discussion, the free energy per adatom relative to the state of an atom entrained in the surface layer was denoted by $\mathcal{E}_{ph}$, and this notation is retained here. As a free energy change, the quantity $\mathcal{E}_{\mathrm{ph}}$ has the character of a chemical potential that acts as a driving force for entrainment of adatoms into clusters. In this discussion, which is based on continuum concepts, it is more convenient to represent this tendency in terms of a material volume density rather than on a per atom basis. If the atomic volume of the film material is denoted by Ω, then the reduction in free energy per unit volume of the island or cluster is $\mathcal{E}_{\mathrm{ph}}/\Omega$.

As before, it will be assumed that the shape of the cluster is a spherical cap. Following the development in Section 1.3.3, the free energy change associated with its formation is

$$\mathcal{F}(V,\theta) = A\gamma_{\mathrm{f}} - \pi R^2(\gamma_{\mathrm{s}} - \gamma_{\mathrm{fs}}) - V\mathcal{E}_{\mathrm{ph}}/\Omega, \qquad (1.9)$$

where the geometrical parameters A, V and R are again related as in (1.2). The appearance of both volume and surface energies in the free energy implies a natural length scale for the system, which is taken to be $\zeta = \gamma_{\mathrm{f}}\Omega/\mathcal{E}_{\mathrm{ph}}$. In terms of this length parameter, the normalized free energy change (1.9) can be expressed as

$$\frac{\mathcal{F}(v,\theta)}{\zeta^2\gamma_{\mathrm{f}}} = \left[\frac{2}{1+\cos\theta} - \frac{\gamma_{\mathrm{s}} - \gamma_{\mathrm{fs}}}{\gamma_{\mathrm{f}}}\right]\pi r(\theta)^2 v^{2/3} - v, \qquad (1.10)$$

where $v = V/\zeta^3$ and $r(\theta) = R/V^{1/3}$. When written in this way, it is clear that the free energy change can be viewed as a function of the two variables v and θ, and the nature of the behavior is determined by the nondimensional system parameter $c_\gamma = (\gamma_{\mathrm{s}} - \gamma_{\mathrm{fs}})/\gamma_{\mathrm{f}}$.

Qualitative implications for the behavior of the system are evident from the form of (1.10). If the quantity enclosed within brackets is positive, then there is an activation barrier to formation of the cluster. The height of the energy activation barrier is $\mathcal{F}_* = \mathcal{F}(v_*,\theta_*)$ where $v_* = 8\pi(2 - 3c_\gamma + c_\gamma^3)/3$ and $\theta_* = \cos^{-1}(c_\gamma)$ are the roots of $\partial\mathcal{F}/\partial v = 0$, $\partial\mathcal{F}/\partial\theta = 0$. The point $v = v_*$, $\theta = \theta_*$ is a saddle point of the function $\mathcal{F}(v,\theta)$, with paths of descent away from the saddle point in directions of constant θ and paths of ascent in directions of constant v; this behavior is illustrated on the left in Figure 1.12.

The ratio $2/(1 + \cos\theta)$ varies between one and two for $0 \le \theta \le \pi/2$. Consequently, an activation barrier exists for $c_\gamma < 1$ or $\gamma_{\mathrm{s}} < \gamma_{\mathrm{f}} + \gamma_{\mathrm{fs}}$. This implies that clusters nucleate with a nonzero contact angle θ_*. Consequently, the film nuclei form on the substrate as islands and the ensuing film growth mode is the Volmer–Weber island growth mode, as was noted in Section 1.3. This mode is preferred when the atoms or molecules deposited onto the substrate are more strongly bound to one another than to the underlying substrate, as in the case of many metals on insulator substrates.

If the quantity in brackets in (1.10) is negative, on the other hand, there is no activation barrier to cluster formation. This is the case for $c_\gamma > 1$ or $\gamma_{\mathrm{s}} > \gamma_{\mathrm{f}} + \gamma_{\mathrm{fs}}$. Examination of the free energy function in this case reveals that there is no stationary

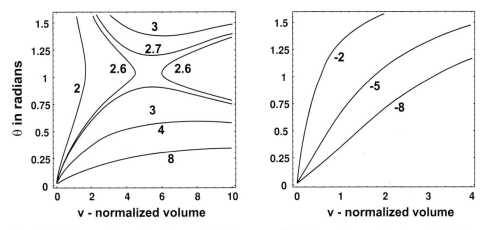

Fig. 1.12. Illustration of level curves of the normalized free energy function defined in (1.10) for $c_\gamma = 0.5$ (left), showing a saddle point activation barrier, and for $c_\gamma = 2$ (right), showing a path of steepest descent for small values of θ.

point and that the cluster size increases most readily with θ essentially equal to zero or, in other words, that cluster growth is planar; this behavior is illustrated on the right in Figure 1.12. The layer simply 'wets' the surface of the substrate forming a planar sheet layer. This mode of film growth on the substrate is the Frank–van der Merwe growth mode, as was noted in Section 1.3. This growth mode is favored when the atoms arriving at the substrate surface are more tightly bound to the substrate surface than to one another. Examples of this mode include epitaxial growth of monocrystalline semiconductor films.

If film growth begins in the layer-by-layer mode, the situation may change once the film material becomes established as the growth surface material. With reference to (1.10), parameter values reduce to $\gamma_s = \gamma_f$ and $\gamma_{fs} = 0$, or $c_\gamma = 1$, once the film material comprises the growth surface and if growth continues with the crystal structure of the substrate. In this case, there is an energy activation barrier to growth of a cluster with $\theta > 0$, but no barrier at $\theta = 0$. On the other hand, neither is there a strong driving force for planar growth with $\theta = 0$, as there is for $c_\gamma > 1$. In a sense, the situation is neutral or indeterminate, and other physical factors that have not yet been incorporated dictate the particular growth mode adopted by the material. For example, suppose that the film material must grow with a residual elastic strain in order to retain atomic registry with the substrate, a phenomenon discussed in detail in Section 1.4.1. In this case, the free energy expression (1.10) would include an additional term proportional to v. The coefficient of this term would be positive (because elastic strain energy is positive) but would be decreasing in value with increasing θ (because the amount of the elastic energy released increases with increasing θ; see Section 8.9). Then, once planar growth of the film material was established, nonplanar growth with cluster formation would become more likely.

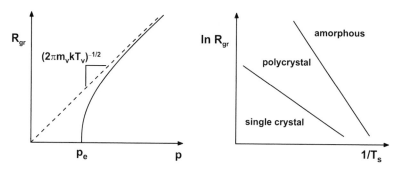

Fig. 1.13. The dependence of growth flux R_{gr} on vapor pressure p is illustrated schematically on the left, where p_e is the equilibrium pressure of the vapor, m_v is the atomic mass of the vapor, T_v is the temperature of the vapor, and k is the Boltzmann constant. The diagram on the right is a schematic microstructure map suggesting the influence of substrate temperature T_s and growth flux on film structure.

A growth mode with these characteristics is identified as the Stranski–Krastanov mode, as was noted in Section 1.3. It occurs through a combination of the Frank–van der Merwe and Volmer–Weber modes, as shown schematically in Figure 1.8, when one or more of the surface energy terms in the balance given in (1.4) varies with film thickness, in effect. This mode is exhibited in the growth of many metal–metal and semiconductor–semiconductor systems, especially in the presence of lattice mismatch and strained layer growth.

1.4 Film microstructures

The preceding section was concerned primarily with the onset of film growth on the substrate surface by condensation from a vapor of film material. From the perspective of thermodynamics, it was seen that several different growth modes are possible; under ideal growth conditions, the growth mode that dominates in a given situation depends primarily on the materials involved, the temperature of the substrate and the degree of supersaturation of the vapor. In general, the evolution of film structure depends on kinetic processes which are still not completely understood. In this section, some observations are made on the final structure of films grown by vapor deposition.

Perhaps the two most important parameters that influence the final grain structure of a vapor-deposited film for a given material system are substrate temperature T_s and growth flux R_{gr}. The growth flux is a quantity that represents the mass or volume flow rate onto the growth surface, and it is closely related to the degree of supersaturation of the vapor. The flux of film material onto the growth surface can be estimated by appeal to statistical mechanics (Venables 2000), and its qualitative behavior is illustrated on the left in Figure 1.13. The diagram on the right in the figure is a schematic illustration of the way in which final film microstructure depends on the growth flux R_{gr} and the substrate temperature T_s. In general, increasing the rate of deposition tends to promote formation of a finer scale microstructure, while increasing the substrate temperature tends to promote a coarser microstructure or single crystal growth. The different structures represented on the right in Figure 1.13 are discussed in the remainder of this section, again in a qualitative way.

1.4.1 Epitaxial films

The word *epitaxy* derives from the Greek words *epi*, meaning 'located on', and *taxis*, meaning 'arrangement'. Epitaxial growth refers to continuation of the registry or alignment of crystallographic atom positions in the single crystal substrate into the single crystal film. More precisely, an interface between film and substrate crystals is *epitaxial* if atoms of the substrate material at the interface occupy natural lattice positions of the film material and vice versa. The two materials need not be of the same crystal class for this to be the case, but they commonly are so. When the film material is the same as the substrate material, crystallographic registry between the film and the substrate is commonly referred to as *homoepitaxy*. The epitaxial deposition of a film material that is different from the substrate material is known as *heteroepitaxy*.

Epitaxial growth technology offers major advantages in material fabrication for microelectronic and optoelectronic applications. It enables the preparation of thin films of extremely good crystalline quality. It also makes possible the fabrication of thin films of compound materials that have desirable electronic or optical properties but that do not exist naturally. There are many factors that drive selection of materials and choice of processing methods in epitaxial growth. These include: chemical compatibility of the film material and substrate material; the magnitude of the energy bandgap of the film material and its relationship to the bandgap and the band edge energies of the substrate material; whether or not the conduction band energy minimum is aligned with respect to carrier wave number with the valence band energy maximum in the active material, an important factor in optical applications; and chemical compatibility of the dopants required to produce desired functional behavior.

In heteroepitaxial film growth, the substrate crystal structure provides a template for positioning the first arriving atoms of film material, and each atomic layer of film material serves the same function for the next layer to be formed by FM growth, as described in the previous section. This is a common growth mode if the substrate is a single crystal of good quality, if the degree of vapor supersaturation is moderate and if adatoms have fairly high mobility on the growth surface. If the mismatch in lattice parameter is not large, say below 0.5% or so, then growth tends to be planar. If the mismatch is large, the material tends to gather into islands on the surface but to remain epitaxial; island growth is described in detail in Chapters 8 and 9.

Planar growth proceeds by attachment of adatoms at the edges of steps, causing the steps to migrate over the growth surface. In general, the stress-free lattice dimension of the film material in a direction parallel to the interface, say a_f, will be different from that of the substrate, say a_s; this difference may be as large as several percent. Nonetheless, the atoms of the film material position themselves in alignment with those of the substrate, continuing its atomic structure, with the film material taking on whatever strain is necessary along the interface to make this possible. In terms of lattice parameters, this mismatch strain is

$$\epsilon_m = \frac{a_s - a_f}{a_f}. \tag{1.11}$$

The definition of mismatch strain in (1.11) is consistent with the standard definition of extensional elastic strain of materials with respect to the stress-free configuration. Sometimes a measure of lattice mismatch is adopted with a denominator of a_s rather than a_f, in which case the value of mismatch strain is slightly different. The definition (1.11), which is consistent with the notion of extensional strain as the change in length of a material element with respect to initial length, will be used in all cases here. The process of epitaxial film growth is illustrated schematically in Figure 1.14. While the process leads to potentially large elastic strain in the film, the energy cost of adding material with strain is easily offset for very thin films by the energy gain associated with the underlying chemical bonding effect in the process. That this is so can be made plausible by means of a simple argument which relies only on the values of familiar bulk parameters.

Consider an elementary cubic lattice with elastic modulus E and free surface energy γ. The elastic strain energy per atom due to biaxial strain of magnitude ϵ_m is on the order of $E\epsilon_m^2 a_f^3$. On the other hand, the in-plane chemical bonding energy per atom is on the order of $2\gamma a_f^2$. For a choice of parameters of $E = 10^{11}\,\mathrm{N\,m^{-2}}$, $\gamma = 1\,\mathrm{J\,m^{-2}}$, $\epsilon_m = 0.02$ and $a_f = 0.3\,\mathrm{nm}$, the ratio of the elastic energy per atom to bonding energy per atom is roughly

$$\frac{E\epsilon_m^2 a_f^3}{2\gamma a_f^2} \approx 0.5 \times 10^{-2}. \tag{1.12}$$

Thus, the elastic effect is very small compared to the chemical effect on an atom-by-atom basis. The elastic effect can come into play only when the chemical effect is sufficiently diluted by symmetry or scale so that it is no longer the dominant effect.

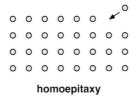

homoepitaxy

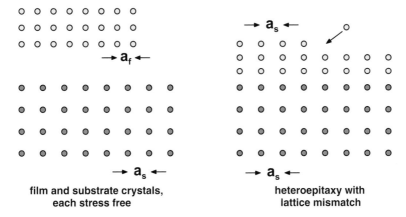

film and substrate crystals,   **heteroepitaxy with**
each stress free   **lattice mismatch**

Fig. 1.14. Schematic illustration of heteroepitaxial film growth with lattice mismatch. The substrate thickness is presumed to be large compared to film thickness, and the structure extends laterally very far compared to any thicknesses. Under these circumstances, the lattice mismatch is accommodated by elastic strain in the deposited film.

Carrying this simple comparison one step further, both elastic energy and bonding energy might be compared to kT_s, the Boltzmann constant times the absolute temperature of the substrate. This quantity is an indicator of the mean energy per adatom on the substrate. At the temperature of 800 K, a typical deposition temperature, the value of kT_s is on the order of 10^{-20} J or 0.1 eV. Interestingly, this value is an order of magnitude smaller than the bonding energy estimate but an order of magnitude larger than the elastic energy estimate. Although these are crude estimates, the calculation is instructive.

As long as the crystallographic registry between the film and substrate crystals is maintained, the interface is said to be fully *coherent* and the macroscopic interface energy density is relatively low compared to the free surface energy density of either the film or the substrate materials. An example of a fully coherent interface shared by two cubic crystals with aligned axes is shown in Figure 1.15. The figure shows a high-resolution transmission electron microscope (HRTEM) image of a coherent (001) interface between Al and $MgAl_2O_4$, a crystal known commonly as spinel. The lattice constant of face-centered cubic (fcc) Al is $a_{Al} = 0.405$ nm. The

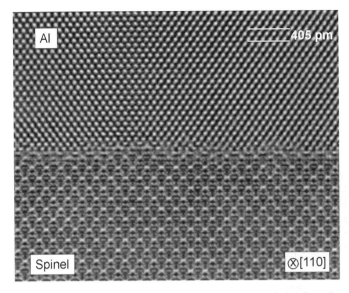

Fig. 1.15. A HRTEM image showing the [1̄10] cross-sectional view of a coherent interface that develops with no interface defects during the epitaxial growth of an Al film on a $MgAl_2O_4(001)$ substrate. (Courtesy of M. Rühle, Max-Planck Institut für Metallforschung, Stuttgart, Germany. Reproduced with permission.)

arrangement of the O ions in $MgAl_2O_4$ is nearly a face-centered cubic structure with a lattice constant of 0.404 nm. The Mg and Al ions occupy tetrahedral and octahedral sites, respectively, within this structure. In light of the structure produced by the occupation of interstices in the O substructure by the cations, the lattice constant of spinel is $a_{Sp} = 0.808$ nm, or nearly twice the lattice parameter of the O substructure. The epitaxial misfit strain at the coherent interface between Al and the O substructure of spinel is

$$\epsilon_m = \frac{\frac{1}{2}a_{Sp} - a_{Al}}{a_{Al}} = -0.0025 \tag{1.13}$$

where Al is interpreted as the film material according to (1.11). The small magnitude of this misfit strain facilitates the development of a coherent interface between the Al film and the underlying spinel substrate. Schweinfest et al. (1999) report such coherent interfaces without defects for Al films with thicknesses of up to 100 nm.

As the coherent film becomes thicker, it grows with essentially uniform elastic strain over most of the growth area. The strain energy per unit *volume* is constant; consequently, over most of the growth area, the stored strain energy per unit *area* of interface increases linearly with film thickness beyond thicknesses of a few atomic dimensions. This increasing energy reservoir is available to drive various physical mechanisms for relaxation of elastic strain.

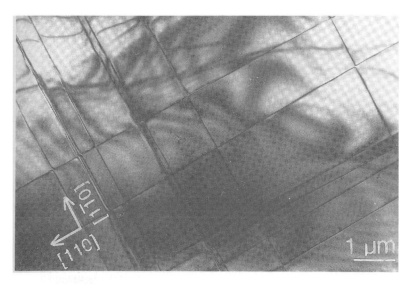

Fig. 1.16. A plan-view, bright-field transmission electron microscopy (TEM) image of misfit dislocations within the interface between a 200-nm thick $Si_{0.9}Ge_{0.1}$ film and a much thicker (001) Si substrate. The dislocations form a crossed-grid pattern which is oriented along two $\langle 110 \rangle$ directions on the (001) interface plane. Reproduced with permission from Fukuda et al. (1988).

Among the most common mechanisms of relaxation of the elastic mismatch strain in epitaxial films is the formation of glide dislocations at the film–substrate interface, the so-called *misfit dislocations*. An illustration of a semi-coherent interface populated with misfit dislocations is shown in the micrograph in Figure 1.16. Once such dislocations are formed, the interface is no longer completely coherent; it is said to be *semi-coherent*. Conditions necessary for the formation and growth of strain-relieving dislocations are discussed in detail in Chapter 6. Some circumstances favor the formation of partial dislocations, in which case the stacking fault energy of the material is brought into consideration. Another common mechanism of strain relaxation is surface morphology change, that is, the initially flat surface of a highly strained crystal will tend to become wavy in order to reduce the stored elastic energy in the system. Conditions for this phenomenon to occur are discussed in Chapters 8 and 9. Other strain-relieving mechanisms can be driven by the elastic energy stored in an epitaxial film, depending on material, temperature and fabrication conditions.

Perhaps the most studied of the strained heteroepitaxial systems is the SiGe system. The two elements Si and Ge are adjacent to each other in column IV of the periodic table. They share the diamond cubic crystal structure, have the same chemical bonding characteristics, and have similar indirect bandgap electronic characteristics. Furthermore, they are chemically compatible, which means that the two elements can be intermixed freely in order to form a substitutional solid solution of any composition. They are completely miscible so there is no significant chemical driving force tending to cause segregation of the elements in the alloy. The influence of a nonzero energy of mixing is considered in Chapter 9. A single crystal SiGe alloy is usually denoted by $Si_{1-x}Ge_x$ where x represents the fraction of sites in the lattice occupied by Ge atoms over a sample sufficiently large to be representative of the crystal; the value of x is in the range $0 \leq x \leq 1$. If a_{Si} and a_{Ge} are the lattice parameters of the stress-free Si and Ge crystals, then the stress-free average lattice parameter of the $Si_{1-x}Ge_x$ alloy is approximated commonly by a linear rule of mixtures as

$$a_{SiGe} = (1 - x)\, a_{Si} + x\, a_{Ge}. \tag{1.14}$$

If a film of $Si_{1-x}Ge_x$ is grown epitaxially onto a (100) surface of a relatively thick single crystal Si substrate, then the mismatch in lattice parameter is accommodated entirely within the film, as discussed in Chapter 3. The film grows with the elastic strain

$$\epsilon_m = \frac{a_{Si} - a_{SiGe}}{a_{SiGe}} = x\,\frac{a_{Si} - a_{Ge}}{a_{SiGe}}. \tag{1.15}$$

Values of the elemental lattice parameters are $a_{Si} = 0.54306$ nm and $a_{Ge} = 0.56574$ nm at room temperature, so that the elastic strain due to lattice mismatch is $\epsilon_m = -0.0418\,x/(1+0.0418\,x)$ for any Ge fractional concentration. The mismatch of Ge with respect to Si is then -0.040 according to the consistent strain definition of (1.11). Finally, it is noted that the same SiGe alloy discussed above could be represented as Si_yGe_{1-y} where y represents the fraction of lattice sites occupied by Si atoms in a representative sample of the material. The alternative representation is identical to the former with the substitution $x \rightarrow 1 - y$.

Semiconductor materials are commonly grouped into families according to the column or columns of the periodic table in which the constituent elements are located. Those in column IV, such as Si, Ge and C, are called elemental semiconductors. Those compound semiconductors formed from one material from column III and one material from column V, such as GaAs, AlAs, InP and GaN, are called III–V semiconductors, while those formed from elements from column II and column VI, such as CdTe and ZnSe, are called II–VI semiconductors. For other purposes, grouping according to crystal structure is advantageous. The most common structures are diamond cubic, such as Si and C; zinc blende cubic, such as GaAs and ZnSe; and wurtzite hexagonal, such as CdSe and GaN. Some compounds can exist in more than one crystal structure, depending on temperature and processing conditions.

A graphical representation of lattice dimensions and bandgap energies for a range of semiconductor crystals is shown in Figure 1.17, where the energy gap is represented in electron volts and the lattice dimensions in nanometers. The points

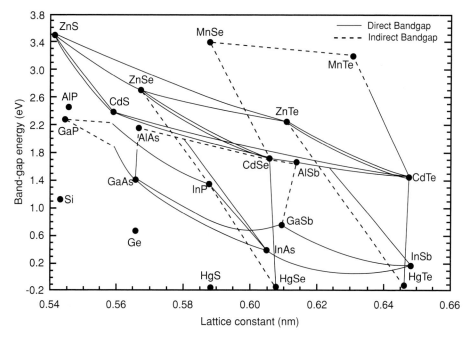

Fig. 1.17. Variation of bandgap with lattice constant for III–V, II–VI and IV semiconductors. Adapted from V. Keramidas and R. Nahory. Lucent Technologies, Murray Hill, New Jersey, 2001.

represent elemental semiconductors and binary compound semiconductors that can be formed and grown as bulk substrates. Non-stoichiometric binary and ternary alloys are represented as lines connecting elementary and binary compounds that are solid solutions with varying degrees of miscibility. For example, InP_xAs_{1-x} ternary solid solution alloys are represented by the line that connects InP and InAs. The solid lines denote direct bandgap ternary compounds whereas the dashed lines indicate indirect bandgap materials. The immiscible alloys cannot be easily grown as bulk crystals due to divergence of the compositions of the liquid and the solid during cooling, but they can be grown as films in some cases.

The III–V compound semiconductors AlAs and GaAs both have the zinc blende structure and are chemically compatible. Furthermore, as can be seen from Figure 1.17, the stress-free lattice dimensions of the two materials are nearly identical. Another pair of III–V materials having nearly identical lattice dimensions is AlP and GaP. In either of these cases, a thin film of one material can be deposited onto a substrate of the other material without inducing significant elastic strain due to mismatch.

Epitaxial growth of thin films becomes increasingly difficult as the lattice mismatch between the film material and the substrate material increases in magnitude.

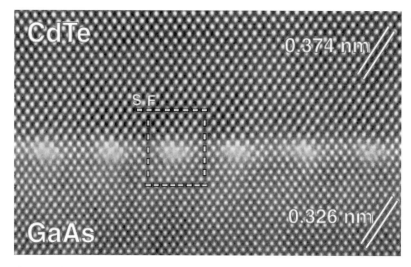

Fig. 1.18. Atomic resolution transmission electron micrograph of an interface between CdTe and GaAs where the misfit strain is relaxed by the introduction of edge dislocations. The letters 'S' and 'F' refer to 'start' and 'finish' for the Burgers circuit around the dislocation. The bright white spots correspond to atomic positions and the fuzzy white clusters at the interface correspond to the cores of the dislocations. Reproduced with permission from Schwartzman and Sinclair (1991).

For example, the difference between the lattice constants of GaAs and InAs is large enough so that it is difficult to create high quality epitaxial layers of any consequential thickness. However, by using intervening layers with gradual transitions in mismatched constituents as a means of stress management, a wide range of semiconductor material compositions between the pure compositions becomes accessible. Graded layer systems are discussed in Chapters 2 and 6.

There are many practical microelectronic device configurations that are based on epitaxial semiconductor thin films. Heteroepitaxial thin films are also used in magnetic recording media, such as computer hard disks; an example is described in Section 1.4.4. Epitaxially grown multilayer structures are also used in planar array laser diodes, which serve as optical interconnnects for short- and medium-range information transfer, and which offer potential for the development of new devices for such applications as solid state lighting, optical storage, optical scanning, lithography and projection-based display.

A semi-coherent interface with relatively large lattice mismatch between the film and substrate is shown in cross-sectional view in Figure 1.18. This atomic resolution transmission electron microscopy image shows a CdTe film on a GaAs substrate, both with zinc blende structure. The mismatch strain in this case is approximately -0.13. However, nearly all of the elastic strain due to lattice mismatch in the film has been relaxed through the formation of a remarkably regular array of interface misfit dislocations, each having its Burgers vector in the plane of the interface. For this to be so, the dislocations would have to be spaced at intervals of approximately eight lattice spacings along the interface, which is indeed the case.

For epitaxial growth to occur, a thin film need not necessarily have the same crystal structure as the underlying substrate. Crystallographic orientation relationships in which the two-dimensional interface most closely approaches the three-dimensional film–substrate lattices are commonly preferred in epilayer growth involving different crystal structures. For example, a $\{111\}$ face-centered cubic (fcc) film on a $\{110\}$ body-centered cubic (bcc) substrate can exhibit epitaxial growth at the Nishiyama–Wasserman orientations with $[\bar{2}11]_{\mathrm{fcc}} \parallel [\bar{1}10]_{\mathrm{bcc}}$ or $[0\bar{1}1]_{\mathrm{fcc}} \parallel [001]_{\mathrm{bcc}}$ (Nishiyama 1934, Wasserman 1933), or the Kurdjumov–Sachs orientations with $[1\bar{1}0]_{\mathrm{fcc}} \parallel [1\bar{1}1]_{\mathrm{bcc}}$ for which there exist minima in the interface free energies (Kurdjumov and Sachs 1930).

1.4.2 *Example: Vertical-cavity surface-emitting lasers*

A periodic arrangement of many epitaxially grown thin layers with lattice mismatch constitutes a strained-layer superlattice. An example of such a superlattice structure can be found in the vertical-cavity surface-emitting laser (VCSEL). As discussed by Choquette (2002) and Nurmikko and Han (2002), the control of layer thickness, elastic strain due to lattice mismatch, stress-driven crack formation and processing induced defects in the superlattice presents major scientific and technological challenges in the development of these devices.

The VCSEL strained superlattice structure comprises as many as one hundred epitaxially grown layers of compound semiconductors surrounding an optical cavity. The VCSEL device, schematically shown in Figure 1.19(a), is composed of two so-called distributed Bragg reflector (DBR) mirrors which are made up of alternating thin film layers of high- and low-refractive index materials; the optical cavity separates the DBRs. These components collectively constitute an optical resonator that transmits a spectrally narrow emission of a circular beam of light. For example, pairs of lattice-matched $Al_{0.2}Ga_{0.8}As$ and $AlAs$ films with a relatively high mismatch in refractive index are used in the DBR layers of VCSELs emitting light at 850 nm wavelength. Similarly, VCSELs intended to operate in the blue, violet and near-ultraviolet wavelength ranges employ layered AlGaN and GaN films.

A VCSEL emits light in the direction normal to the surface of the layers. Proper VCSEL operation requires precise alignment of (a) the composition and thickness of the DBR layers for achieving necessary reflectance wavelengths, (b) the composition and thickness of the optical cavity and the DBR layers for achieving necessary cavity wavelength, and (c) the composition and thickness of the quantum wells for achieving necessary laser gain spectrum. These stringent requirements dictate the reliance on precise film growth methods such as molecular beam epitaxy and metal-organic vapor-phase epitaxy, along with *in-situ* monitoring of film geometry and stress during the growth of the film. The DBR mirror is designed in such a way that it exhibits maximum reflectance at a particular wavelength λ_{d}. The thickness of each pair of films with high- and low-refractive index n_{H} and n_{L}, respectively, in the DBR structure is designed to equal $\lambda_{\mathrm{d}}/4n_{\mathrm{H}}$ and $\lambda_{\mathrm{d}}/4n_{\mathrm{L}}$, respectively, so that the film thickness constitutes a quarter-wavelength distance of the optical path inside that film. With the interfaces of high- and low-index film pairs thus separated by one-half

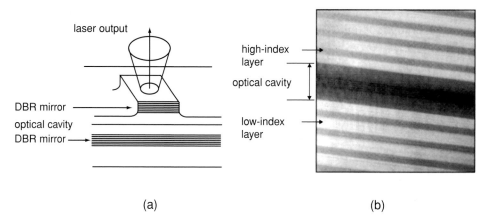

(a)   (b)

Fig. 1.19. (a) Schematic of the vertical cavity surface emitting laser. (After Choquette (2002).) (b) TEM image of the optical cavity and the surrounding DBR mirror region which comprises alternating layers of high- and low-refractive index material. The thickness of the optical cavity is 250 nm. (Photograph courtesy of D. Mathes and R. Hull. Reproduced with permission.)

of the wavelength of light in the device, mirrors with a reflectance in excess of 99% are produced, as the reflectivities of these interfaces add constructively. Figure 1.19(b) is a TEM image showing a section of the DBR mirror that surrounds the active cavity. An AlAs film constitutes the low-index layer, and an $Al_{0.2}Ga_{0.8}As$ film is used for the high index layer.

Longitudinal confinement of light is facilitated in the VCSEL structure shown in Figure 1.19. In addition, transverse confinement of photons and carriers is required for optimal device performance. This is commonly achieved through ion implantation, whereby protons are implanted into the top DBR mirror rendering the material surrounding the laser cavity nonconductive and channeling the injection current into the active region.

The VCSEL semiconductor laser diode is distinctly different from the earlier edge-emitting laser designs, particularly the so-called Fabry–Pérot lasers in which an elliptical beam of light is transmitted in the plane of the epitaxial films parallel to the wafer surface and reflected from mirrors which are produced through the cleavage of a crystal facet across the active layers. While the edge-emitting laser diode has the advantage of small output aperture and high power density, the VCSEL device offers high beam quality, single-mode coupling to optical fibers, and batch-processing fabrication; these advantages, along with low drive currents and less demanding temperature control requirements, facilitate high-volume, low-cost manufacture of VCSEL components for optoelectronic applications at low power.

1.4.3 Polycrystalline films

Polycrystalline films are commonly described in terms of their grain structure, which is usually understood to include some measure of grain size, grain boundary morphology (typical grain shape), and film texture (distribution of crystallographic orientations). The connection between these characteristics – size, morphology and texture – and processing conditions is understood qualitatively for common low-energy vapor deposition processes involving relatively pure film materials and substrates for which epitaxial interface formation is not likely. The onset of epitaxial growth is not possible if the substrate is amorphous, of course, and it is unlikely if the substrate surface is covered with an adsorbed contaminant layer or if the mismatch strain is very large. In such cases, adatoms have a greater propensity for bonding together than for bonding to the substrate.

Important factors in controlling the structure of a growing film are the growth flux or deposition rate and the substrate temperature. Suppose that the growth flux is represented by the time required to deposit a single monolayer of film material. A second material-specific characteristic time that is important in determining the outcome of a growth process is the representative time interval between encounters of an adatom with other adatoms diffusing randomly on the growth surface. It is the ratio of these times that is important in determining film structure. The ratio of the substrate temperature to the melting temperature of the film material is a second important factor in determining the structure of a polycrystalline film. These factors determine the degree to which adatoms are able to seek out minimum energy positions and grain boundaries are able to adopt minimum energy morphologies.

When the growth flux R_{gr} is very large and the substrate temperature T_s is moderate, the common mode of film growth initiation is the Volmer–Weber mode. Deposited material agglomerates into clusters on the substrate surface; the clusters increase in size until they impinge on each other to form a continuous film. If the crystallite clusters do not have sufficient mobility to align crystallographically at this point, or if the thermodynamic driving force for elimination of the grain boundary is too low, then the clusters join with an intervening grain boundary. This is invariably the case for growth of metal films on insulating substrates when lattice mismatch is far too large for epitaxial growth to occur. Even in cases for which epitaxial growth might be achieved, say for Al films grown on Si, the substrate temperature may be too low to permit the atomic migration necessary to establish an epitaxial film–substrate interface. For semiconductor films, the epitaxial structure is essential to the electronic performance of the material. For metal films, on the other hand, the electrical conductivity of a polycrystalline film is nearly as large as that of a single crystal film, so that the extra effort required to grow an epitaxial structure is not worthwhile from this point of view. A phenomenon for which grain boundary structure is potentially important, however, is mass transport by electromigration, as described in Section 9.7. In magnetic materials, on the other hand, a uniform and well-controlled grain structure is essential for proper functional performance; an illustration is given in Section 1.4.4.

Growth of polycrystalline films generally begins with the thermally activated nucleation of islands or clusters of film material on the growth surface; this was described in atomistic terms in Section 1.3.2 and in terms of more macroscopic quantities in Section 1.3.5. The activation energy $\mathcal{F}_*$ and critical cluster size, represented in terms of number n_* of film atoms or volume v_* of film material, were estimated in both cases. The number N_c of clusters nucleated per unit area of the growth surface per unit time is customarily assumed to follow an Arrhenius rule. While the Arrhenius relationship provides a useful basis for correlation of data, fundamental understanding of the processes involved remains elusive.

Some additional qualitative observations on cluster formation and growth can be made. Initial crystallographic orientations of the nascent film that minimize surface and interface energies are favored over others, and the nucleation rates for clusters with lower energy orientations are greater than those for others. This can lead to a strong texture in the microstructure of the film; this point is pursued in greater detail in Section 1.8.6.

Another potentially significant feature of distributions of surface clusters follows from the observation in Section 1.3 that the free energy per atom or per unit volume decreases as the island size increases following nucleation. This implies that larger islands provide a stronger driving force for incorporation of nearby surface adatoms than do smaller islands. Furthermore, this free energy difference implies the existence of a chemical potential for transfer of material from the smaller islands to larger islands. The result of such transfer is that larger islands grow at the expense of smaller islands, a type of process commonly identified as *Ostwald ripening* after the discoverer of such processes in colloidal dispersions (Ostwald 1887, Zinke-Allmang et al. 1992).The process of ripening is discussed in Section 8.9. When coalescence occurs between two islands of different sizes and very different crystallographic orientations, the texture of the final cluster is often determined by the initial orientation of the larger island.

Numerous experimental studies such as those reported by Klokholm and Berry (1968), Wilcock and Campbell (1969), Wilcock et al. (1969), Abermann et al. (1978), Abermann and Koch (1986), Abermann (1990) and Koch (1994), for example, have contributed to understanding some general trends in the evolution of structure during the growth of polycrystalline thin films. This information is represented schematically in Figure 1.20, which illustrates the various stages of growth of a thick polycrystalline film on a substrate. The processes discussed in the preceding section in connection with the formation of adatom clusters, island nucleation and growth, and initial film structure also apply to the case of polycrystalline films. However, as is implied by Figure 1.20, the formation of a grain structure and continued structure evolution during the thickening of the polycrystalline film are also influenced by the surface diffusivity of the film material which is strongly dependent on the ratio of T_s to its melting temperature, as was noted above.

For many materials of practical interest, the density of nucleation sites on the growth surface can be quite high. As a result, the average spacing between nuclei

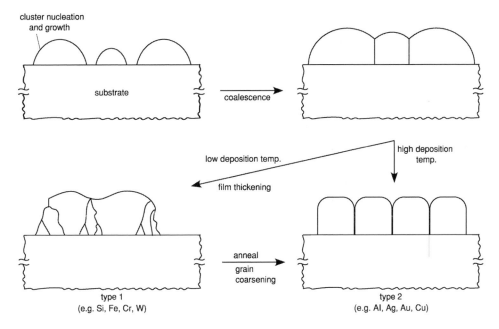

Fig. 1.20. A schematic representation of several stages in the formation of a thick polycrystalline film on a substrate. Cluster formation and growth leads to impingement of clusters to form grain boundaries. The development of grain structure and subsequent film growth is determined largely by surface mobility of adatoms relative to the deposition rate and by the substrate temperature relative to the melting temperature of the film material. Adapted from Thompson (2000).

and the average grain size upon the impingement of nuclei can be as small as 10 nm. Since the thicknesses of the polycrystalline films required in most practical applications are greater than 10 nm, substantial further growth of the film is needed.

Experimental studies of evaporative deposition of a variety of polycrystalline thin films on substrates, such as those reported by Abermann (1990), indicate that there are two distinctly different types of grain structure evolution. A class of materials with low diffusivity, such as refractory metals and elements with a diamond cubic crystal structure, exhibit a polycrystalline grain structure which resembles the type 1 structure depicted in Figure 1.20. For these materials, the surface diffusivity is relatively low even at high substrate temperatures T_s. Because diffusivity generally scales with the melting temperature T_m, the homologous temperature T_s/T_m is a better indicator of diffusivity than the absolute temperature. As a result, in materials such as Si, W, Cr, Fe and Ta, the grain boundaries that form due to island coalescence are essentially immobile even at $T_s/T_m \approx 0.5$. The structure that originally develops during island nucleation, growth and impingement is essentially 'frozen in'; any further growth of the polycrystalline film occurs predominantly as epitaxial growth on this initial structure leading to the formation of columnar grains.

The ensuing average grain size parallel to the plane of the interface with the substrate is significantly smaller than the average grain size measured through the film thickness. Post-deposition annealing or increases in relative substrate temperatures T_s/T_m approaching values of 0.8–0.9, however, could lead to a transition from a columnar grain structure to an equiaxed grain structure (Thompson 2000).

In face-centered-cubic (fcc) metals, such as Al, Ag, Au and Cu, the higher surface mobility of the adatoms facilitates continued structure evolution during film thickening. The resulting equiaxed grain structure is termed type 2 and is represented schematically in Figure 1.20. These polycrystalline fcc metals tend to grow on substrates in such a way that most of the grains in the film develop a $\{111\}$ texture, that is, most grains are oriented so that the crystallographic direction that is normal to the film–substrate interface is a $\langle 111 \rangle$ direction. Otherwise, the in-plane orientations of the grains are random with respect to the $\langle 111 \rangle$ axis normal to the film–substrate interface. The final in-plane grain size is typically on the order of the film thickness.

As a rule, higher substrate temperatures and lower deposition rates lead to a larger initial grain size as the islands impinge to form a polycrystalline film. For a fixed substrate temperature, materials with high melting temperatures and low surface diffusivities develop smaller initial grain sizes during island impingement. The driving force for grain growth is the reduction in the total grain boundary surface area and the attendant reduction in the total energy associated with grain boundary surfaces; this issue is discussed further in Section 1.8.6. Another mechanism for the development of structure in polycrystalline films involves recrystallization, the driving force for which is the minimization of energy associated with defects, such as pre-existing dislocations, in addition to grain boundary energy. Minimization of stored elastic energy arising from intrinsic and mismatch strains in the film can also serve as an additional factor contributing to grain growth and recrystallization.

Although the foregoing discussion of structure evolution in polycrystalline films has focused on diffusive processes at the surface, it is generally recognized that the final structure is sensitive to the specific deposition method employed and to the various parameters that influence deposition conditions. As noted earlier, the diffusive processes, which include surface mobility, bulk diffusion and desorption of adatoms, scale with the ratio of the substrate temperature to the melting temperature of the film material. In addition to these factors, it is known that a low homologous substrate temperature T_s/T_m promotes structures that are strongly influenced by the geometric constraint of the roughness of the film surface on which new deposits are made and by the 'line of flight' of the incoming atoms at the growth surface. This so-called 'shadowing effect' influences the structure development in both amorphous and crystalline films at low substrate temperatures where the diffusivity of atoms at the surface is limited.

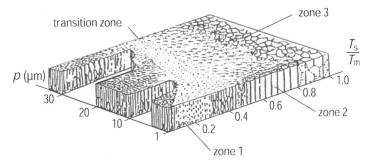

Fig. 1.21. Structure evolution in polycrystalline films as a function of substrate temperature and argon gas pressure, following the four-zone model of Thornton (1977).

Categorization of grain morphology evolution was first proposed by Movchan and Demchishin (1969). Using experimental observations of structural evolution in magnetron-sputtered metal films 20 to 250 μm in thickness, Thornton (1977) proposed a four-zone model which is illustrated in Figure 1.21. The Thornton model is similar to that of Movchan and Demchishin, but it includes an additional transition zone to account for the influence of the inert gas pressure and surface diffusion. In zone 1, tapered or fibrous grains and voided boundaries form where the structure is influenced by the shadowing effect and by the roughness of the substrate surface on which deposition is made. The transition zone is marked by dense grain boundary arrays and a film with a high dislocation density. Columnar grains evolve in zone 2, where the evolution of the structure is aided by surface diffusion processes. The transition from surface diffusion to bulk diffusion, as well as recrystallization and grain growth, lead to the formation of an equiaxed grain structure in zone 3. As a general trend, similar structures evolve at higher temperatures in sputtered films as compared to evaporated films (Thornton 1977, Thompson 2000).

1.4.4 Example: Films for magnetic storage media

The concept of epitaxial film growth on a substrate was introduced in Section 1.4.1, and some consequences of epitaxial mismatch on the evolution of film stress and substrate curvature will be considered in the next chapter. In this section, some materials issues involving grain structure and epitaxial film growth for magnetic thin film multilayers are presented. Magnetic materials have long been used in data recording and storage media. Early magnetic tape media made use of γ-Fe_2O_3 particles. Thin magnetic films have gradually replaced magnetic particles for data storage on disks, and this trend applies to the case of magnetic tapes as well.

Thin film magnetic storage media can be broadly classified into two groups: those in which the magnetization lies primarily within the plane of the film (so-called longitudinal recording) and those in which it lies out of the plane of the film (so-called perpendicular recording). The vast majority of thin film media of current practical interest involve longitudinal recording. Major developments in the evolution of longitudinal thin film magnetic

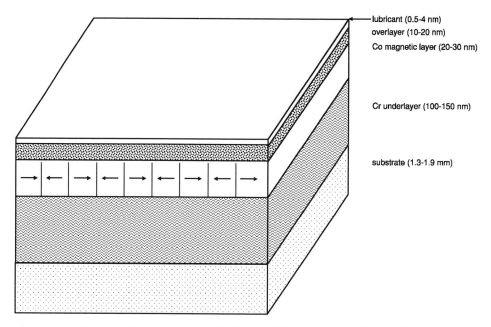

lubricant (0.5-4 nm)
overlayer (10-20 nm)
Co magnetic layer (20-30 nm)

Cr underlayer (100-150 nm)

substrate (1.3-1.9 mm)

Fig. 1.22. Schematic diagram showing the layered film structure of longitudinal recording magnetic media.

materials respond to a primary need, namely, to increase the storage density significantly while maintaining or improving upon the signal to noise ratio.

An increase in the density of longitudinal recording in thin film media requires an increase in the in-plane coercivity† for smaller transition lengths and for smaller bit size, and a decrease in the grain size for enhancing the signal to noise ratio. A widely used magnetic thin film is a CoCrTaPt alloy with a hexagonal close-packed (hcp) crystal structure. When Cr is added to the alloy, it segregates to the grain boundaries. This segregation reduces the exchange coupling between the grains which, in turn, increases the signal to noise ratio. The addition of Ta aids in the segregation of Cr to the grain boundary, and Pt raises the magneto-crystalline anisotropy of the alloy (McHenry and Laughlin 2000).

Structure of a longitudinal thin film recording medium consists of a substrate, typically a borosilicate glass or an Al–Mg alloy about 1–2 mm in thickness, coated with an amorphous non-magnetic layer of NiP that is 5–10 μm in thickness, a Cr underlayer approximately 100–150 nm in thickness, a CoCrTaPt magnetic layer approximately 20–30 nm in thickness and a carbon overcoat approximately 10–20 nm thick. Such a structure is depicted schematically in Figure 1.22 which represents a typical cross-sectional view of the layered structure of a hard disk medium.

An important consideration for the function of the magnetic layer is its crystallographic texture. When a Co alloy with an hcp structure is sputter-deposited on an amorphous layer such as NiP or borosilicate glass, its crystallographic texture develops such that the c-axis of the film is perpendicular to the plane of the film, as the closest-packed basal plane of the hcp alloy aligns with the substrate surface so as to minimize the total surface free energy. This is undesirable if the c-axis is required to lie in the plane of the film, although such texture

† Coercivity of a magnetic material is a measure of the degree of stability of the magnetized state, that is, it provides a measure of the magnetic field required to 'demagnetize' the material.

evolution could be optimal for perpendicular recording media. In order to force the *c*-axis of the magnetic layer to be along the plane of the film, the Cr layer is first deposited on a NiP or glass substrate. If the Cr deposition takes place at room temperature, the {110} planes of the bcc Cr underlayer, which are the closest-packed planes, align with the substrate surface and minimize the surface energy. As a result, the Cr layer forms on the substrate with a [110] texture. If the CoCrTaPt is then deposited on the Cr underlayer, it grows epitaxially such that its {10$\bar{1}$1} plane is parallel to a (110) plane of Cr. However, this epitaxial relationship places the *c*-axis of the magnetic film at an angle of 28° out of the plane of the film, that is, with respect to the {10$\bar{1}$1} plane, which is not desirable for longitudinal recording media.

On the other hand, if the Cr underlayer is deposited on a heated amorphous substrate, it grows with a [200] crystallographic texture. When the CoCrTaPt alloy is next deposited on the Cr underlayer, its texture evolves epitaxially in such a way that a {200} plane of the Cr is parallel to a {11$\bar{2}$0} plane of CoCrTaPt. This forces the *c*-axis of the Co alloy to be in-plane, which is the magnetic easy axis. This example serves to illustrate how epitaxial growth and the attendant crystallographic texture can be engineered through the appropriate choice of processing conditions in order to achieve the desired magnetic property in a layered structure.

Another important microstructural feature that determines functionality is the grain size of the magnetic thin film (McHenry and Laughlin 2000). In order to ensure that a sufficiently good signal to noise ratio is obtained, written bits of information should encompass at least about 50 grains. For a given amount of noise, more information could be stored in the magnetic thin film if the grain size is reduced and the same number of grains are involved per written bit. It should be noted that there is a limit to the use of grain refinement since smaller thermal excursions become more significant with smaller grain sizes. As discussed in Section 1.4.3, evolution of grain size in a sputter-deposited film is strongly influenced by factors such as substrate temperature, atomic mobility and thermal diffusivity. When a thin film is sputtered onto a crystalline underlayer, the grain size of the underlayer influences the grain size evolution of the sputter-deposited film. As noted earlier, Cr underlayers are commonly used for CoCrTaPt alloys because of the favorable epitaxial conditions promoted by them. It has been found that thin films with NiAl underlayers inherently develop grain sizes smaller by a factor of two than Cr. The NiAl underlayer develops a [112] texture when sputter-deposited on a glass substrate. When the Co alloy is deposited on NiAl, epitaxial growth occurs with the orientation relationship {101$\bar{1}$0}$_{Co}$ ∥ {112}$_{NiAl}$. This epitaxial relationship is also considered more favorable than {11$\bar{2}$0}$_{Co}$ ∥ {002}$_{Cr}$ because only one orientation variant per NiAl underlayer grain is obtained. In addition to the smaller grain size, the narrower distribution of grain sizes facilitated by NiAl can be desirable for magnetic performance because grains of the same size would switch at the same time causing square-shaped hysteresis loops to evolve. These results indicate a way in which the deposition conditions, grain size, crystallographic texture and choice of materials can be exploited to tailor the functionality of magnetic thin films.

1.5 Processing of microelectronic structures

The discussion up to this point has focused on the deposition of continuous thin films onto substrates and on the relationships between deposition conditions and the evolution of structure within the film. Attention is now turned to processes by means of which fine features and patterns can be introduced in the thin film.

The most widespread examples of such patterning arise in the use of lithography and the damascene process, both of which are practiced in the manufacture of microelectronic devices.

The creation of planar patterns in thin films with highly integrated mechanical and electrical functions is a well developed technology which is employed in the production of silicon-based microelectronic devices using the complementary metal–oxide–semiconductor (CMOS) processes. These devices comprise doped single crystal silicon substrates on which multiple levels of metal interconnects, insulators and passivating layers are deposited and patterned. Several examples of the configurations of integrated circuits used in computer microchips are presented in subsequent chapters; in these circuits, very large scale integration (VLSI) involving sub-micron feature dimensions is achieved through deposition, patterning and etching processes.

1.5.1 Lithography

Microelectronics device fabrication employs lithography in order to transfer a layout pattern onto the surface of a substrate, such as a silicon wafer. In this technique, the surface of a Si wafer is oxidized, typically at 800–1200 °C in steam or dry oxygen, to form a thin layer of silicon oxide as shown in the upper left portion of Figure 1.23. Then, a thin film of a polymer known as a *photoresist* is deposited on the oxide layer. This material is sensitive to ultraviolet radiation but resistant to chemical attack by etchants and, in some cases, to electrons and x-rays. A photomask, which comprises a glass plate with a metal pattern, is then placed above the photoresist, and the top surface is subjected to ultraviolet radiation, as indicated in the upper right portion of Figure 1.23. The glass in the mask is transparent to ultraviolet light, whereas the metal blocks the radiation from being transmitted to the photoresist. In the case of a positive photoresist, the polymer is weakened by the chemical reaction induced by ultraviolet light in regions of the surface not protected by the metal pattern, and it is washed away in a developing solution, as depicted in the left central portion of Figure 1.23. The exposed silicon oxide layer, that is, the region no longer covered by the photoresist, is etched away by an acid such as hydrofluoric acid which attacks the oxide but not the photoresist. Once the oxide is removed, the remnant photoresist is stripped off, leaving the bare Si substrate with patterns of oxide on it which reproduce the metal pattern in the photomask, as illustrated in the left lower portion of Figure 1.23. Similarly, a negative photoresist deposited on the oxide layer strengthens when exposed to ultraviolet radiation, and the developing solution in this case removes those areas of the photoresist unexposed to ultraviolet light. In this case, the photoresist develops a negative image of the metal pattern on the mask, as shown in the right lower portion of Figure 1.23.

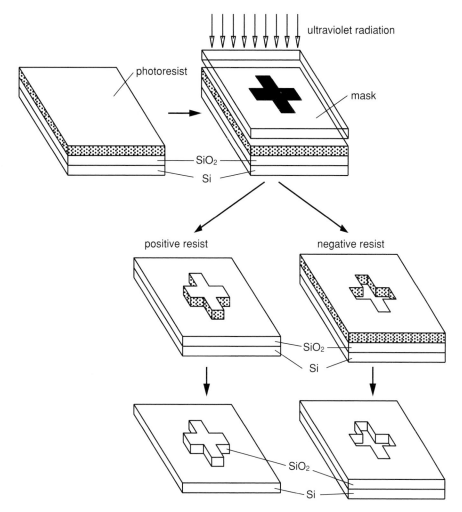

Fig. 1.23. Illustration of the steps involved in developing patterns of oxide on a substrate using lithography by recourse to either a positive or negative photoresist.

Such CMOS-compatible processes and materials facilitate the fabrication of devices with highly integrated mechanical and electronic functions. In the case of microelectronic devices, the oxide pattern serves as a mask that modulates the spatial distribution of dopants such as boron (for p-doping) and phosphorus (for n-doping) which diffuse into the substrate through the openings in the oxide layer to create regions of conductance or resistance. Electrical contacts, gates and metal interconnect lines are then added within the level of the oxide layer, after which the entire surface is polished and again covered by a blanket deposit of an oxide layer, repeating the process at this level. Circuits may comprise numerous levels of patterns aligned one over the other to dimensional tolerances as small as 0.1 μm in large scale fabrication involving millions of microelectronic devices on large Si wafers. In such a case, the second insulating layer would be patterned in preparation for laying down the second level of metallization, and so on.

1.5.2 The damascene process for copper interconnects

Metal wiring is necessary to electrically connect the many individual elements of an integrated circuit. Metallization technology has evolved along with other advances in integrated circuit fabrication. Aluminum and its alloys were the standard choice for metallization in integrated circuits for many years for a variety of reasons. The thin layer of native oxide, typically 5–10 nm in thickness, which forms on Al is capable of protecting the metal during thermal processes used in the manufacture of the chip. Aluminum is also a low cost material with a relatively low electrical resistivity. Furthermore, the volatile halide compounds which it forms are amenable to interconnect patterning through the process of reactive ion etching (RIE). There are, however, limitations in the use of Al as the interconnect metal. The relatively low melting temperature of aluminum, in conjunction with its susceptibility to failure processes such as stress-voiding and electromigration, pose uncertainties about the mechanical integrity of aluminum lines, especially as the ultra large scale integration (ULSI) processing of sophisticated integrated circuits has forced the minimum interconnect dimension well into the sub-micron range. With the attendant increase in current density in the metal lines, degradation by electromigration, which is aggravated by stress effects, also assumes a more prominent role in circuit failure.

Copper has a lower electrical resistivity and higher melting point than does aluminum and, for the same electrical design, it is expected to function as effectively as an aluminum interconnect with roughly half the cross-sectional dimensions. Such reduced dimensions favor further miniaturization of integrated circuits and facilitate total planarization of interconnects in the multilayers of the integrated circuit. Research into copper interconnect wiring indicates superior performance and reliability of copper, particularly in the context of failure against stress-voiding and electromigration. Despite these advantages, copper interconnect technology has drawbacks: copper does not adhere well to silicon oxide and diffuses rapidly into silicon; the native oxide on copper is not as protective as that on aluminum; and reactive ion etching technology for copper lines less than 0.5 μm in size, involving high-temperature etching and inorganic masking materials, remains elusive (Mizawa et al. 1994).

In view of these considerations, the damascene process has been developed as an alternative to reactive ion etching for the introduction of copper interconnects in integrated circuit multilayers. This process is shown schematically in Figure 1.24. Here, trenches with dimensions that conform to the geometry of the copper interconnect

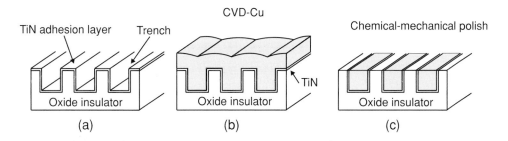

Fig. 1.24. A schematic of the damascene process involving (a) dry-etching of trenches in the silicon oxide insulator and the deposition of an adhesion layer of TiN which also serves as an oxidation and diffusion barrier for Cu, (b) plugging the trenches with CVD-Cu, and (c) chemical–mechanical polishing of Cu.

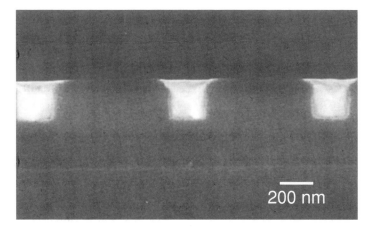

Fig. 1.25. A scanning electron micrograph of the as-polished damascene copper interconnections, approximately 0.25 μm in cross-section. The copper lines appear as the white regions in the figure. Reproduced with permission from Mizawa et al. (1994).

wires in the circuit are dry-etched into an insulator such as silicon oxide grown on the silicon substrate, as sketched in Figure 1.24(a). The side and bottom surfaces of the trenches are then coated with a TiN adhesion layer that serves to prevent both copper oxidation and diffusion into the adjacent insulator. These trenches are then filled with copper, by means of chemical vapor deposition or electroplating, to form wires. The material microstructure and the state of stress in the metal lines following fabrication depend on the entire processing history and are influenced by the elastic stiffnesses of the surrounding materials. The excess copper is removed by chemical–mechanical planarization, as described by Singh and Bajaj (2002), leaving the surface ready for beginning deposition of the next layer of features with the necessary interconnect metallization.

In ULSI devices, multilevel metallization is required. As a result, the first-formed interconnect lines may undergo several thermal cycles between room temperature and about 400 °C as subsequent layers of metallization are formed. This temperature range is dictated by the fact that the deposition of the passivation layer is commonly done at about 400 °C.

1.6 Processing of MEMS structures

Micro-electro-mechanical systems, which are miniature devices with integrated mechanical and electrical functions, are designed to fulfill an array of applications as optical switches, sensors, actuators, biomedical components, chemical reactors, powering devices and consumer products. Specific examples of MEMS devices include: array of tilting metal mirrors used to modulate light beams in portable digital projectors (illustrated in Figure 1.27), nozzles used in inkjet printers, accelerometers used for the rapid deployment of airbags in automobiles, fuel atomizers, micromotors and micro-turbine engines. These devices, typically a few micrometers to hundreds of micrometers in dimensions, are fabricated by means of planar processing technologies similar to those used in microelectronic devices. In the case of MEMS, a greater emphasis is placed on the simultaneous machining of large numbers of integrated devices with mechanical and non-mechanical functions with low production costs per unit rather than on the precise control of micro-scale dimensions themselves, as in microelectronics. The processing methods for MEMS structures can be broadly classified into one or a combination of three basic groups: (a) bulk micromachining, (b) surface micromachining, and (c) molding processes. A brief description of each of these processes is given the following subsections.

1.6.1 Bulk micromachining

The process of fabricating very small mechanical parts by etching a wafer of Si, or perhaps some other material, is called bulk micromachining. Generally, etching refers to any process by which material is removed from a workpiece literally by being dissolved or eroded from its surface. The most common etching processes are those in which a wet chemical etchant is used. An isotropic etchant removes material from a surface at the same rate in all directions, whereas an anisotropic etchant dissolves material much faster in particular crystallographic directions than in others.

In the bulk micromachining of a Si wafer, a thin film of SiO_2 is blanket deposited onto the wafer surface, after which the oxide is partially removed by lithographic techniques as discussed in Section 1.5.1, leaving a prescribed pattern of exposed Si surface. The wafer is then exposed to the etchant which acts through the openings in the SiO_2 film that serves as a mask. The action of the etchant creates a rounded

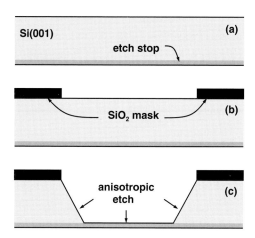

Fig. 1.26. A schematic illustration of the use of a surface mask, an anisotropic etchant and an etch stop in the fabrication of a supportive membrane by means of bulk micromachining.

depression in the Si if it is isotropic with respect to that material and an angular pattern with nearly flat surfaces if it is anisotropic.

A common device for limiting the range of an etchant is a so-called *etch stop*. For example, suppose that a wafer is to be etched until the remaining thickness over a certain area is a predetermined value. This can be accomplished by diffusing a suitable dopant into the wafer to the predetermined depth from the face opposite to the face to be etched and then using an etchant having the corresponding *selectivity*, that is, the capacity to dissolve the undoped material but not the doped material. Thus, the etchant becomes ineffective once the surface being etched reaches the doped region. An etch stop can also be induced by means of an applied electric potential in some cases. A schematic of anisotropic etching of Si(001) through an SiO$_2$ mask to an etch stop as a means of creating a very thin membrane of large area is shown in Figure 1.26. Such membrane structures are important for mechanical testing of membrane films, and for application in pressure transducers, miniature condenser microphones and other sensing devices.

The structure of a MEMS device typically involves a part that has been micromachined from a wafer integrated with other components to result in a three-dimensional configuration. A technique for integration of components with basic planar geometry is wafer bonding, usually achieved by pretreatment of the wafer to assure planarity and cleanliness, followed by a high temperature anneal with the surfaces to be bonded in firm contact.

1.6.2 Surface micromachining

Whereas bulk micromachining of MEMS structures involves the removal of material from a workpiece until the desired shape is achieved, surface micromachining

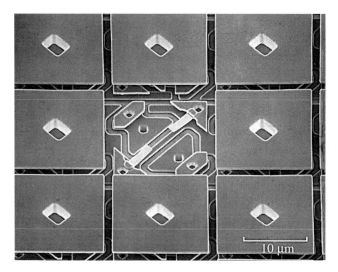

Fig. 1.27. An array of surface-micromachined mirrors used in a portable digital projector. The center mirror has been removed to expose the remaining segments of the rectangular torsion bar and the electrodes used for angular positioning. (Image provided by Texas Instruments Corp., Attleboro, Massachusetts and reproduced with permission.)

involves the *addition* of planar thin film features onto a substrate surface to build up the structure. The processing methods are essentially those of microelectronics as described in Section 1.5, and devices that are an order of magnitude smaller than bulk micromachined devices are achievable. A significant advantage of surface micromachined structures is that they are readily integrated with microelectronic circuit elements. The small size, on the other hand, implies very small mass which diminishes their potential for motion sensing. Small size also limits the range of accessible force or displacement in capacitively driven actuators to extremely small magnitudes.

An aspect distinguishing processing of surface machined MEMS devices from microelectronic structures is the use of sacrificial layers in the structure, that is, layers that are removed (usually by etching) after subsequent layers have been added. Removal is necessary so that some portion of the structure is sufficiently unrestrained or compliant to fulfill the mechanical function of the device. A persistent problem in surface machined MEMS structures is *stiction*, or the tendency for nearby surfaces to stick together due to capillary forces, electrostatic forces or other causes.

A particular surface micromachined structure for which neither low inertia nor small actuator force is a drawback is the digital micromirror array shown in Figure 1.27. Each $4\,\mu$m $\times$ $4\,\mu$m aluminum mirror in the array pivots on a diagonal torsion bar and can be positioned at an angle up to ten degrees in either direction from the position parallel to the base, independently of the orientations of other mirrors. An array might include up to 10^6 mirrors, and the microelectronic control elements are embedded in the CMOS substrate. The structure is basically a flat plate attached to a torsion bar fixed at both ends. A deflecting torque is applied capacitively by means of electrodes on the base which are positioned on either side of the torsion bar, and a restoring torque arises from elastic resistance to deformation of the torsion bar. In order to achieve angular deflections of significant magnitude, the plate is actually separated from the torsion bar by means of a hollow post; the square indent near the center of each mirror seen in Figure 1.27 is the inside of that hollow post. Sacrificial layers are required in the fabrication process to separate the torsion bar from the base and to separate the mirror from the torsion bar. Both the mirror and torsion bar are aluminum in this case, and the sacrificial material is customarily a hardened photoresist that can be removed by a dry-etching process.

1.6.3 Molding processes

Another approach to forming small mechanical components of a device is to deposit the material into a suitable microfabricated mold. The procedure involves the creation of a polymer mold by recourse to lithography, following which a metal is electroplated to fill the mold cavities. High aspect ratio features are created through x-ray lithography. The most commonly practiced mold process is referred to by the acronym LIGA which stems from the German phrase lithography, galvanoformung (electroforming) and abformung (molding). Figure 1.28 shows an example of the basic steps involved in a LIGA process.

The versatility of the molding process rests in its flexibility to deposit materials other than through electrodeposition as, for example, by employing chemical vapor deposition (CVD) and slurry processing, thereby including a wider array of materials. Polycrystalline silicon and silicon carbide (of interest as a MEMS material for harsh environments involving high temperature, shock, vibration, erosion and corrosion) can be deposited using CVD, whereas refractory ceramics can be molded through slurry processing methods.

Hybrid methods that combine features of the above basic processes have also been employed to fabricate MEMS devices. For example, SiC MEMS structures can be fabricated through a combination of micromolding and surface micromachining methods. Wafer bonding provides another hybrid method for fabricating MEMS structures where multilevel integration can be achieved.

1.6.4 NEMS structures

Mechanical devices, with dimensions on the order of tens of nanometers and with integrated electronic, chemical, fluidic, medical, biological, data storage and optical

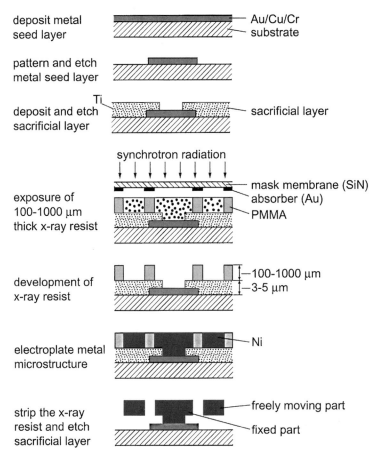

Fig. 1.28. An example of a flowchart illustrating the fabrication steps involved in the LIGA process. Adapted from Spearing (2000).

functions, can be fabricated with capabilities for sensing or imposing motion on the nanometer scale. Figure 1.29(a) shows an array of cantilever beams, each approximately 10 nm by 100 nm in cross-sectional dimensions and 500 nm in length. The diagram in part (b) of the figure suggests that a coating of DNA on one side of such a beam will result in a detectable change in curvature, a response that could serve as a signal for drug release control or other effects. For example, the top surface of these cantilevers can be coated with DNA chains of oligonucleotides. These coated cantilevers in solution can then be exposed to oligonucleotides with a complementary sequence of base pairs. The intermolecular forces exerted by the matched pairs tend to make the coating expand, much like a thermal expansion, which results in a curvature in the cantilever beam; see Section 2.1. The amount of bending, which can be measured with a scanning laser, serves as a highly sensitive probe for particular DNA sequences.

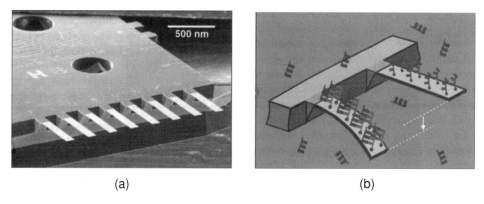

(a)   (b)

Fig. 1.29. (a) Image of a nanosized structure comprising a series of cantilever beams, and (b) a schematic showing the bending of a DNA-coated cantilever in response to the binding of strands with their complements. Reprinted with permission from Hellemans (2000).

Nanoscale devices are also being fabricated with the objective of electronically observing biological systems in their natural environments, with dimensional precision of a nanometer size range. Since the diameter of a DNA molecule is approximately 2 nm, nanoscale devices offer the prospect for manipulating or monitoring biological molecules.

With some modifications, surface micromachining, as discussed in Section 1.6.2 in the context of MEMS fabrication, can also be used to produce NEMS devices. Motion at the nanometer scale is commonly induced through the application of an electric field. Lorentz forces generated by passing an alternating current in a wire placed transversely to a strong magnetic field or resonance induced by piezoelectric elements which oscillate the supports of NEMS devices are other examples of methods used to accomplish motion at the nanoscale. The detection of nanoscale motion is generally achieved by recourse to optical interference methods, by deflecting a laser beam, by measuring the motion-induced electromotive force or voltage, or by electron tunneling. Nanoscale actuation can also be induced using the scanning tunneling microscope (STM) where motion is induced in the mechanical system by the application of an alternating current voltage to the axial piezoelectric drive of the STM probe tip. The surface can also be imaged using the STM so that the surface features are correlated with the mechanical response.

The very small feature sizes of NEMS structures also give rise to mechanical devices with reduced mass, lowered force constants and significantly enhanced resonant frequencies. Figure 1.30(a) shows components of a compound torsional oscillator made of single crystal silicon, where the mirror width is approximately 2 μm and the width of the supporting wires is as small as 50 nm. Figure 1.30(b) shows nanowires of varying lengths, made of silicon single crystal using e-beam

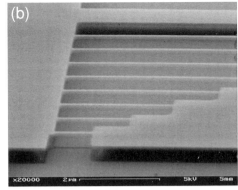

Fig. 1.30. Electron micrographs of silicon NEMS structures made by surface micromachining and e-beam lithography. (a) A compound torsional oscillator tilting mirror. (b) Silicon nanowire resonant structures. Reprinted with permission from Craighead (2000).

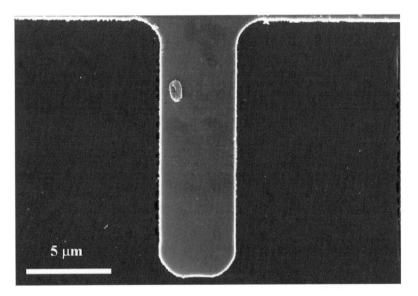

Fig. 1.31. An electron micrograph showing the presence of a single *E. coli* bacterium on the surface of a silicon nitride cantilever oscillator which is coated with antibodies for this bacterium. Reproduced with permission from Ilic et al. (2000).

lithography and surface micromachining. For a wire approximately 2 μm in length, the resonant frequency is on the order of 400 MHz.

Changes in absorbed mass as small as 1 pg on a nanomechanical oscillator, such as a nanocantilever beam, can be used for biological studies. Figure 1.31 is an electron micrograph showing a single *E. coli* bacterium on an antibody-coated silicon nitride cantilever oscillator. When the cantilever vibrates in air, the change in the resonance frequency can be detected with sufficient accuracy to identify the presence of this single bacterium.

1.6.5 Example: Vibrating beam bacterium detector

Suppose that a microfabricated cantilever beam is to be used as a sensor for detecting the presence of a certain bacterium in a liquid solution; see Figure 1.31. Following its fabrication, the beam is first coated with an antibody that is specific to the particular bacterium of interest and its free vibration characteristics are determined. The beam is then dipped into the solution to be tested. If the bacteria of interest are present, they will attach themselves to the surface of the beam, attracted by the antibody. The beam is then removed from the solution and its free vibration characteristics are re-examined. Estimate the influence of an added mass due to the attached bacterium cells, say n_b cells each of mass m_b, on the fundamental natural frequency of the beam. (The effectiveness of this approach for detection of bacteria has been demonstrated by Ilic et al. (2000).)

Solution:

For a uniform, slender cantilever beam of length L, total mass m and elastic bending stiffness EI, the small amplitude transverse deflection $w_b(x, t)$ of the midplane is governed by the partial differential equation

$$EI\frac{\partial^4 w_b}{\partial x^4} + \frac{m}{L}\frac{\partial^2 w_b}{\partial t^2} = 0 \tag{1.16}$$

according to Bernoulli–Euler beam theory. The elastic bending stiffness is the elastic modulus E times I, the cross-sectional area moment of inertia; the latter geometrical property of the cross-section is height$^3 \times$width/12 for a rectangular cross-section. This equation, along with the boundary conditions that the midplane slope and deflection vanish at the cantilevered end of the beam and that the internal shear force and bending moment vanish at the free end, lead to an eigenvalue problem for free transverse vibration in which the eigenvalues are the squares of the natural frequencies of vibration. The fundamental natural frequency is

$$\omega = \frac{1.875^2}{L^2}\sqrt{\frac{EI\,L}{m}}. \tag{1.17}$$

Assume that the attached bacterium cells are distributed uniformly over the surface of the beam so that the total mass increases by $\Delta m = n_b m_b$. Furthermore, assume that the presence of the cells does not affect the bending stiffness of the beam. To lowest order in the change in mass due to the bacteria, the perturbed natural frequency of the beam is

$$\omega + \Delta\omega = \omega\left(1 - \frac{1}{2}\frac{\Delta m}{m}\right) \tag{1.18}$$

where ω is the value given in (1.17). This implies a reduction in natural frequency that is proportional to the mass added. Suppose that the properties of the beam are such that $L = 200\,\mu\mathrm{m}$, $m = 1980\,\mathrm{pg}$ and $EI = 8192\,\mu\mathrm{N}\,\mu\mathrm{m}^2$. The fundamental natural frequency of vibration is then 7.99×10^4 radians per second. If 100 bacterium cells, each of mass $m_b = 0.5\,\mathrm{pg}$, become attached to the beam as a result of dipping it into the solution, then the resulting shift in frequency is approximately -10^4 radians per second.

The sensitivity of the instrument is defined to be $\Delta\omega/m_b$. Mass added to the beam near the cantilevered end has very little influence on the frequency shift, which is dominated by the mass added near to the free end. If the antibody coating were applied to the beam only near the free end, the total mass change would be lower than for uniform coverage but the sensitivity of the sensor would be higher.

1.7 Origins of film stress

In general, residual stress refers to the internal stress distribution present in a material system when all external boundaries of the system are free of applied traction. Virtually any thin film bonded to a substrate or any individual lamina within a multilayer material supports some state of residual stress over a size scale on the order of its thickness. The presence of residual stress implies that, if the film would be relieved of the constraint of the substrate or an individual lamina would be relieved of the constraint of its neighboring layers, it would change its in-plane dimensions and/or would become curved. If the internal distribution of mismatch strain is incompatible with a stress-free state, then some residual stress distribution will remain even under these conditions.

1.7.1 Classification of film stress

Film stresses are usually divided into two broad categories. One category is *growth stresses*, which are those stress distributions present in films following growth on substrates or on adjacent layers. Growth stresses are strongly dependent on the materials involved, as well as on the substrate temperature during deposition, the growth flux and growth chamber conditions. Advances in nonintrusive observational methods for *in-situ* stress measurement and growth surface monitoring have made it possible to follow the evolution of growth surface features and the corresponding evolution of average stress levels in the course of film formation. These capabilities have provided new insights into the origins of film stress and have led to a subdivision of the category of growth stresses into those stresses which arise during various phases of the growth process and those which are present at the end of the growth process. Usually, growth stresses are reproducible for a given process and the values at the end of growth persist at room temperature for a long time following growth. Growth stresses are also commonly called *intrinsic stresses*, a term of limited use as a descriptor of what is represented.

A second category of film stress represents those stress conditions arising from changes in the physical environment of the film material following its growth. Such externally *induced stresses* are commonly called *extrinsic stresses*. In many cases, these stresses arise only when the film is bonded to a substrate, and the distinction between growth stresses and induced stresses becomes hazy at times.

The classification scheme has no fundamental significance, and the lack of a clear distinction between categories is largely immaterial. The purpose in this section is to identify examples of both types of stress.

The development of growth or intrinsic stresses for a particular material system, substrate temperature and growth flux depends on many factors. Perhaps the most important among these are the bonding of the deposit to the substrate (epitaxial or not, for example), the mobility of adatoms on the film material itself, and the mobility of grain boundaries formed during growth. Except for the case of ideal epitaxy, the final growth structure is inevitably metastable. Because of the huge number of degrees of freedom involved in establishing this metastable structure, the extent of departure from a completely stable equilibrium structure can be significant. Many mechanisms for stress generation during film material deposition have been proposed, and the more common among these have been reviewed by Doerner and Nix (1988). These mechanisms include:

— surface and/or interface stress,
— cluster coalescence to reduce surface area,
— grain growth, or grain boundary area reduction,
— vacancy annihilation,
— grain boundary relaxation,
— shrinkage of grain boundary voids,
— incorporation of impurities,
— phase transformations and precipitation,
— moisture adsorption or desorption,
— epitaxy,
— structural damage as a result of sputtering or other energetic deposition process.

Stresses induced in films arising from external influences following growth, or extrinsic stresses, can arise from a multitude of physical effects. Among these are:

— temperature change with a difference in coefficients of thermal expansion between bonded elements,
— piezoelectric or electrostrictive response to an electric field,
— electrostatic or magnetic forces,
— gravitational or inertial forces,
— compositional segregation by bulk diffusion,
— electromigration,
— chemical reactions,
— stress induced phase transformations,
— plastic or creep deformation.

Reasonably complete models leading to quantitative estimates of film stress arising from external influences are available, and many examples of phenomena included in the list above are discussed in the chapters that follow. The understanding
of growth stresses and their origins is not nearly as well developed. All of the effects listed above surely give rise to growth stresses of some magnitude, but robust
models for comparative estimates of magnitudes have remained elusive. The obvious exception is the case of epitaxial film growth, which is discussed briefly in
the subsection to follow. The more complex issue of sources of growth stress in
polycrystalline films is addressed in Section 1.8.

1.7.2 Stress in epitaxial films

In general, stress in epitaxial films arises as a result of the constraint of the substrate
on the film material dictated by the requirement of coherency. Estimates of stress
magnitude as discussed in Section 1.4.1 and in subsequent chapters are generally
reliable in this case, particularly for films of a thickness that is both spatially uniform
and much larger than atomic spacing in the lattice. For very thin epitaxial films or
for coherent films that are not uniform in thickness, some interesting and significant
differences emerge. For example, Jesson et al. (1991) observed that when a Si(001)
surface was terminated by a single atomic layer of Ge, the surface energy was
reduced significantly from the value for the Si surface itself. This effect implies the
existence of a driving force for segregation of Ge to the growth surface during the
early stages of growth of a $Si_{1-x}Ge_x$ film on a Si(001) substrate surface, at least until
the surface layer is composed entirely of Ge. For the case of $x = 0.25$, this could
occur once four atomic layers had been deposited, at which point the thickness is a
full unit cell of the lattice. If the surface layer of Ge were two atomic layers thick,
then it could not become fully established until eight atomic layers of the alloy
had been deposited. In the measurement of film stress during the early stages of
SiGe/Si(001) film growth, Floro and Chason (1996) observed anomalous deviations
in film stress from the behavior anticipated in Section 1.4.1 on the basis of a spatially
uniform film composition, which they attributed to Ge segregation at the surface
and its influence on surface stress. The stress measurements involved substrate
curvature observations as described in Chapter 2; the capability of measuring film
stress changes as small as surface stress or as small as the stress induced by a partial
monolayer of growth has been demonstrated by Martinez et al. (1990) and Schell-
Sorokin and Tromp (1990). The issue of spontaneous segregation of the constituents
of an alloy, particularly the interaction of deformation and segregation, is discussed
in Chapter 9.

A strained epitaxial film has a natural tendency to become nonuniform in thickness, even under well-controlled growth conditions, as discussed in Chapter 8. The

origin of the driving force for this tendency may be the prospect of partially relaxing elastic energy due to mismatch in the course of creating additional free surface, or it may be the strain induced destabilization of the crystal surface in the original orientation. The issue of growth surface stability is pursued in detail in Chapters 3 and 8. Once such a change in surface morphology occurs, the average film stress is reduced and the stress field becomes spatially nonuniform. Local stress measurement becomes extraordinarily difficult due to the small size of such epitaxial structures. Also, because of the small size of the film features, surface stress can come into play in establishing the mechanical state of the system.

1.8 Growth stress in polycrystalline films

The circumstances for epitaxial growth, as described in the preceding section, occur only under very special combinations of material selection, growth temperature and growth chamber vacuum conditions. For most film–substrate material combinations, films grow in the Volmer–Weber (VW) mode which leads to a polycrystalline microstructure. As described in Sections 1.3.2 and 1.3.5, the distinguishing feature of this growth mode is that deposited material gathers into discrete clusters or islands on the substrate surface from the outset, with no tendency toward planar growth. Thereafter, the microstructure evolves through a sequence of stages that is characteristic of a wide range of material combinations. Following the initial nucleation of islands of film material, successive stages typically include: island growth, island-to-island contact and coalescence into larger islands, establishment of large area contiguity, and filling in of the remaining gaps in the structure to form a continuous film. Once islands begin to interact to form grain boundaries, the process of grain coarsening can also contribute to structural evolution.

The role of stress in these stages of microstructure evolution is not yet fully understood. However, the application of real-time stress measurement techniques has made it possible to identify some general trends in behavior. The stress measurement techniques are usually based on *in-situ* monitoring of the curvature of the substrate (Flinn et al. 1987, Koch et al. 1990, Floro et al. 1997); the connection between substrate curvature and the film force or average stress inducing this curvature is discussed in detail in Chapter 2. The substrate curvature is interpreted in terms of an equivalent isotropic membrane force on the substrate surface that is necessary to induce that curvature. If it is assumed that the known amount of deposited material is uniformly distributed over the substrate growth surface, then there is a unique equi-biaxial volume-average stress corresponding to the observed curvature. The calculation of this volume-average stress for a particular film microstructure is discussed in Section 8.6. Real-time measurements of the history of this stress parameter

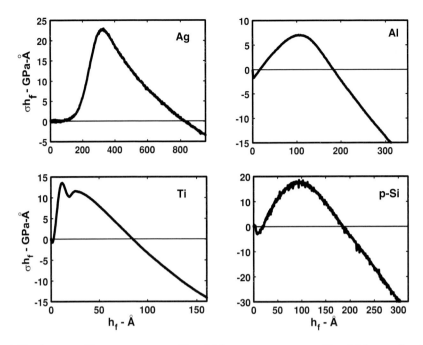

Fig. 1.32. Average film stress×mean film thickness versus mean film thickness for deposition of Ag(0.24), Al(0.32), Ti(0.15) and polycrystalline Si(0.46) on a SiO_2 surface. The number included in parentheses with each deposited material represents the corresponding homologous deposition temperature. From Floro et al. (2001); reprinted with permission.

have provided insights into the nature of the dynamic interplay between different physical mechanisms of structural evolution that induce or relax internal stress.

Experimental results based on deposition of four different materials, each grown on a Si substrate with a thick SiO_2 surface layer under ultrahigh vacuum conditions, are shown in Figure 1.32. The film materials with the corresponding homologous deposition temperatures shown in parentheses are: Ag(0.24), Al(0.32), Ti(0.15) and p-Si(0.46). Each curve shows the volume-average stress in the film times the mean film thickness, as inferred from *in-situ* substrate curvature measurements, versus mean film thickness. In each case, the ordinate value of the graph at any thickness provides the average stress, while the slope of the curve represents the incremental stress associated with an increment in mean thickness. Because of the evolutionary nature of the film microstructure, the incremental stress is not necessarily the stress in the increment in film material deposited. While the graphs in Figure 1.32 show obvious differences in detail, they share a common general character in several respects. This common character is represented in a schematic diagram of the time evolution of volume-average film stress versus film thickness during steady deposition shown in Figure 1.33. In the course of forming a polycrystalline film by vapor deposition, the volume-average stress first becomes compressive, then tensile, and

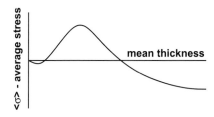

Fig. 1.33. Schematic diagram of volume-average film stress versus mean film thickness during growth with a steady deposition flux. The features of initial compressive stress, then tensile stress, and then again compressive stress with a plateau value are common to a wide range of film materials.

then again compressive, as indicated in the diagram. In the discussion that follows, mechanisms that can account for the observed behavior are described.

1.8.1 Compressive stress prior to island coalescence

The compressive stress that arises early in the deposition process is usually attributed to the action of surface and/or interface stress. The origin of the idea lies in the observation that the lattice spacing in a very small isolated crystallite is smaller than the spacing in a bulk crystal of the same material at the same temperature. An elementary estimate of the magnitude of the effect can be obtained by considering a homogeneous and isotropic spherical crystallite of radius R. If f is the isotropic surface stress acting within the spherical surface of the crystallite, then equilibrium requires that there must be an internal equi-triaxial stress with magnitude

$$\sigma = -\frac{2f}{R} \tag{1.19}$$

acting throughout the interior of the crystal.

Mays et al. (1968) estimated the surface stress generated during the evaporative deposition of gold onto an amorphous carbon substrate in ultra-high vacuum on the basis of (1.19). Using electron diffraction experiments, they documented lattice contraction of gold nuclei with diameters as small as 3.5 to 12.5 nm as a function of the nucleus diameter R. If the lattice spacing in the interior of a bulk gold crystal is a, and the lattice spacing within the gold nano-crystal is determined to be $a + \Delta a$, then the equi-triaxial elastic strain is $-\Delta a/a$. The corresponding value of surface stress determined according to (1.19) is

$$f = -\frac{1}{2}R\sigma = -\frac{RE_f}{2(1 - 2\nu_f)}\frac{\Delta a}{a} \tag{1.20}$$

where ν_f and E_f are the Poisson ratio and elastic modulus of the isotropic film material.

Mays et al. (1968) observed that the average lattice constant varied from 0.4075 nm to 0.4063 nm as the radius of curvature of the nano-crystal varied from very large values down to about 2 nm. Using (1.20), they estimated the surface stress f to be 1.175 N m^{-1}. The surface energy of gold has been estimated from independent measurements to be about 1.4 J m^{-2}; see the review by Cammarata (1994), for example. Presumably, the difference between these values arises from the strain dependence of surface stress.

The hypothesis for generation of compressive stress in the early stage of film growth is that crystallites forming on the growth surface become firmly 'attached' to the substrate at an early stage of their growth. Upon further growth in the volumes of the clusters, the internal elastic strain in the crystallites tends to relax because the surface radius becomes larger and the corresponding internal stress becomes smaller, as suggested by (1.19). However, it is prevented from doing so due to the constraint of the substrate.

An elementary model leading to a stress estimate due to this effect can be constructed in the following way. Suppose that a representative crystallite becomes attached to the substrate, or becomes 'locked-down' against further relaxation, at some radius R_{ld}. The internal equi-triaxial strain in the grain at this point is

$$\epsilon_{\mathrm{ld}} = -\frac{1 - 2\nu_{\mathrm{f}}}{E_{\mathrm{f}}} \frac{2f}{R_{\mathrm{ld}}}. \tag{1.21}$$

As the crystallite continues to grow to larger radius $R > R_{\mathrm{ld}}$, the strain is prevented from relaxing further. In other words, as the grain becomes larger, it is subject to a mismatch strain

$$\epsilon_{\mathrm{m}} = \frac{1 - 2\nu_{\mathrm{f}}}{E_{\mathrm{f}}} \left(\frac{2f}{R} - \frac{2f}{R_{\mathrm{ld}}} \right) \tag{1.22}$$

for $R > R_{\mathrm{ld}}$. The remaining elastic strain is balanced by surface stress, but the portion ϵ_{m} is unbalanced within the grain itself. Its effect is transferred to the substrate through tractions acting across the film–substrate interface. The unbalanced stress that tends to deform the substrate is

$$\sigma = \frac{2f}{R_{\mathrm{ld}}} \left(\frac{R_{\mathrm{ld}}}{R} - 1 \right). \tag{1.23}$$

This quantity is representative of the stress magnitude that can be expected to arise in the film according to the proposed mechanism. Note that it has the feature that it continues to increase as R becomes very large, approaching an asymptotic limit of $-2f/R_{\mathrm{ld}}$ as $R/R_{\mathrm{ld}} \to \infty$.

This mechanism for generation of the initial compressive stress range indicated in Figure 1.33 was apparently first proposed by Laugier (1981). It was subsequently adopted by Abermann and Koch (1986) and Floro et al. (2001) as a basis for interpretation of observations on growth of metal films on substrates. A detailed estimate of the stress in (1.23) based on the same general idea has been proposed by Cammarata et al. (2000); the latter estimate provides additional numerical factors of order unity for the elementary result shown in (1.23) which itself follows from dimensional considerations.

Although this model is appealing for several reasons, it is neither complete nor particularly well-founded. Perhaps the most significant unresolved questions concern the nature of load transfer between the strained crystallite and its substrate that is necessary for the mechanism to operate, an issue discussed in the context of stress measurement by Floro et al. (2001). How does this load transfer occur? Is it possible that there is slippage between the crystallite and the substrate in the course of island growth, perhaps in the form of glide, so that the load is only partially transferred and more relaxation of surface stress induced mismatch occurs than is usually presumed? Are there observations that could yield *direct* evidence in support of the lock-down model? Questions of load transfer based on a balance of surface and interface stresses across the interface have been addressed for a liquid surface island (Spaepen 1996) and for a solid surface island (Spaepen 2000). If the adatoms or surface islands would simply reduce the surface stress of the growth surface, but not interact significantly otherwise, then this effect itself could account for the apparent compressive stress in the film material.

In the interpretation of observational data in the context of the type of model represented by (1.23), an issue that is commonly overlooked is the importance of surface area coverage or density of islands on the growth surface. The stress field associated with an *isolated* island on the substrate surface is self-equilibrating; overall, the stress field has no resultant force or moment and its magnitude decays to very small values within a short distance from the island. The prospect that a dilute distribution of tiny islands on a relatively thick substrate of comparable stiffness could result in a significant curvature is remote, no matter how large the mismatch stress might be. For a distribution of islands to lead to a perceptible curvature of the substrate, the distribution must be sufficiently dense so that the fields of the islands interact. This is essential if the rapid spatial decay of the self-equilibrating stress fields of individual islands is to be overridden to a degree that results in measurable curvature. This issue is studied in some detail in Chapter 3 for patterned films on substrates, but an elementary estimate of the role of surface distribution of islands on the connection between mismatch stress in an island array and corresponding substrate curvature can be obtained on the basis of elementary elastic beam theory. This is illustrated by means of an example calculation in Section 1.8.2 below.

The apparent compressive stress present in the early stage of the growth process is active from the onset of growth until the islands establish large area contiguity across the growth surface. It has been observed that, before this occurs, some growing

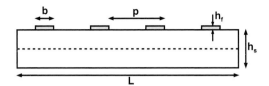

Fig. 1.34. A model configuration of a distribution of islands on the substrate. The islands are subject to an elastic mismatch strain with respect to the substrate, which gives rise to substrate curvature.

islands are in the form of bicrystals or tricrystals. This is presumed to occur because the islands nucleated in close proximity and subsequently grew together at an early stage in the process. Alternatively, it might suggest that tiny islands are mobile on the growth surface in the early stages of deposition. *In-situ* experimental measurements of stress during the Volmer–Weber growth of polycrystalline Cu thin films indicate that the compressive stress arising prior to the coalescence of islands is largely relaxed during growth interruptions (Friesen and Thompson 2002). In any case, the incremental stress tends to become positive or tensile once mutual impingement of the growing islands occurs on a scale such that a connected structure is established over relatively large portions of the substrate surface. The origin of this tensile stress is considered in Section 1.8.3.

1.8.2 Example: Influence of areal coverage

Consider an elastic beam of length L, height h_s and unit depth as illustrated in the sketch in Figure 1.34. For simplicity, the deformation is assumed to be consistent with a one-dimensional state of stress, typical of elementary beam theory (although plane strain deformation might be a bit more appealing physically). The elastic modulus of the isotropic beam material is E_s. Attached to the surface of the beam are N identical and equally spaced 'islands', each of length b and height h_f. The elastic modulus of the isotropic island material is E_f. If the islands are deposited onto the substrate surface with mismatch strain ϵ_m, determine how the curvature κ of the substrate midline depends on the distribution of the islands on the surface, as represented by b and p.

Solution:

Note that the center-to-center spacing of the islands is $p = L/N$. The relationship between substrate curvature and mismatch strain is readily obtained by appeal to the principle of minimum potential energy. The elastic energy stored in the beam at uniform curvature κ is $\frac{1}{24}E_s h_s^3 \kappa^2 L$. Suppose that positive curvature corresponds to a reduction in extensional strain on the beam surface to which the islands are firmly attached. Thus, the uniform extensional strain in each island in the direction of the interface is $\epsilon_m - \frac{1}{2}\kappa h_s$, assuming that $h_f \ll h_s$. It follows that the total elastic energy of the system as a function of beam curvature κ is

$$\mathcal{E}(\kappa) = \tfrac{1}{24}E_s h_s^3 \kappa^2 L + \tfrac{1}{2}E_f(\epsilon_m - \tfrac{1}{2}\kappa h_s)^2 Nbh_f. \qquad (1.24)$$

It follows that the minimum value in energy under variations in curvature κ occurs for the curvature value

$$\kappa h_s \approx 6 \frac{b}{p} \frac{\epsilon_m E_f}{E_s} \frac{h_f}{h_s}. \tag{1.25}$$

A term on the right side that is higher order in the small ratio h_f/h_s has been omitted. The areal coverage is represented by the factor b/p, a ratio that can be very small compared to unity for a dilute distribution of islands but which approaches one in value at full coverage. A more detailed examination of these issues is pursued in Section 3.10.

1.8.3 Tensile stress due to island contiguity

The most widely adopted mechanistic model for the origin of the tensile stress is based on the work of Hoffman (1966, 1976) and Doljack and Hoffman (1972). They hypothesized that small gaps between adjacent grains could be closed by forming grain boundaries, and that the energy released through reduction in surface area could be converted to elastic deformation of the participating grains as a result of the deformation needed to effect the gap closure. The maximum distance across which the gap between surfaces of crystallites could be closed was estimated by drawing an analogy between the energy of interaction between atoms and that between crystallite surfaces. It was concluded that the maximum gap that could be closed in this way is on the order of $\delta_{gap} = 0.17$ nm.

A simple expression for the stress generated when islands impinge on each other can be obtained on the basis of this concept. Consider a flat substrate that is covered with a doubly periodic array of square islands, each of thickness h and of lateral dimensions $d_{gr} \times d_{gr}$, the grain size. The substrate thickness is large compared to d_{gr}. The side faces of each island are separated from the parallel faces of neighboring islands by the distance δ_{gap} on all four sides. The free surface energy of all surfaces has the value γ_f and the islands are free of stress. Suppose that the islands deform spontaneously to form a continuous film by closing all the gaps in the array. In doing so, each island must be strained equi-biaxially by an amount δ_{gap}/d_{gr} in the plane of the interface. The interfacial energy of the newly formed interfaces has the value γ_i. If the material is an isotropic elastic material with elastic modulus E_f and Poisson ratio ν_f, the resulting increase in elastic strain energy of each island is $h d_{gr}^2 \times (\delta_{gap}/d_{gr})^2 E_f/(1 - \nu_f)$ where any influence of the substrate on deformation has been ignored. The corresponding change in surface/interface energy is $-4 d_{gr} h \times (\gamma_f - \frac{1}{2}\gamma_i) < 0$, where it is assumed that one-half of the interface energy γ_i is associated with each of the islands forming an interface and that $2\gamma_f > \gamma_i$ is a necessary condition for the coalescence to occur. The uniform tensile stress based on this simple model is

$$\sigma = \frac{E_\mathrm{f}}{1 - \nu_\mathrm{f}} \frac{\delta_\mathrm{gap}}{d_\mathrm{gr}}. \tag{1.26}$$

For representative values of the parameters of $E_\mathrm{f}/(1 - \nu_\mathrm{f}) = 100\,\mathrm{GPa}$, $\gamma_\mathrm{f} - \frac{1}{2}\gamma_i = 1\,\mathrm{J\,m^{-2}}$ and $d_\mathrm{gr} = 17\,\mathrm{nm}$, (1.26) leads to a stress estimate of $\sigma = 1\,\mathrm{GPa}$, a large stress magnitude.

An approach aimed at overcoming some shortcomings in this approach was introduced by Nix and Clemens (1999). They imagined the island boundaries to be rounded surfaces. The islands would then grow until each made contact at a single point with each of its adjacent islands. Interfaces then take on the physical characteristics of an elastic crack, with the contact zone boundary being the crack edge. The system can lower its net free energy by closing up this crack, thereby replacing free surface with its relatively high surface energy γ_f by an interface with a relatively low interfacial energy γ_i. In the course of 'zipping up' the interface, the participating islands become strained elastically. The island size at which impingement occurs is generally influenced by process parameters such as growth flux, substrate temperature and surface diffusivity.

The idea of zipping up a grain boundary between crystallites, once they make contact, provides a viable and physically appealing framework for estimating the magnitude of this tensile stress generated when islands impinge. To pursue this issue, the same general concept is examined in Section 8.6.2 within the framework of a particular model which can capture the zipping up process in a realistic way. The model is based on the theory of elastic contact of solids with rounded surfaces known as Hertz contact theory, with the added feature that the contacting surfaces cohere.

As noted, an estimate of the tensile stress magnitude based on the zipping model often leads to a stress magnitude that is larger than any observed. However, as soon as the stress level reaches the yield stress of a polycrystalline material, glide dislocations can form to modulate the amplitude. This line of reasoning has been discussed in some detail by Machlin (1995) who pointed out that the mechanism as described is athermal except for the possible dependence of yield stress on temperature. At elevated temperatures, mechanisms associated with increased surface mobility of atoms can come into play to diminish the effectiveness of the Hoffman–Nix mechanism in generating stress, and evidence for the influence of such mechanisms is apparent in experimental data.

1.8.4 Compressive stress during continued growth

Shortly after it first arises during growth of a polycrystalline film, the tensile stress is observed to decrease in magnitude, implying a negative or compressive incremental stress. For the high mobility materials that are the subject of the present discussion, the average stress continues to decrease until the average stress in the film takes on compressive values. This fact alone indicates that the decrease involves more than merely a mechanism of relaxation of the tensile stress. Ultimately, the compressive average stress approaches a steady value for a fixed growth flux, as depicted in Figure 1.33, with this value depending on temperature and growth flux magnitude for a given material system.

Two additional observations may provide clues as to the origin of the compressive stress. First of all, if the growth is interrupted while the stress magnitude is at its plateau value, the stress magnitude falls off rapidly. Then, upon resumption of the deposition flux, the falloff in stress magnitude is fully reversed and the same compressive stress plateau is eventually re-established (Shull and Spaepen 1996, Floro et al. 2001). The second observation concerns the role of grain boundaries. It has been demonstrated that the stress in Pd films deposited on polycrystalline Pt substrates becomes compressive while the stress in otherwise identical films deposited onto single crystal Pt surfaces remains tensile (Ramaswamy et al. 2001).

Presumably, the compressive stress that arises in polycrystalline films is due to an excess number of atoms comprising the film. The available observations suggest that this excess probably depends on the presence of a deposition flux and on the presence of grain boundaries in the film microstructure. Spaepen (2000) has pointed out that only a small number of excess atoms is necessary to result in the stress levels observed experimentally. From the point of view of thermodynamics, there is always a driving force tending to expel excess atoms from the crystal under equilibrium conditions. However, this may not be the case under the circumstances of deposition that maintain nonequilibrium conditions on the growth surface. It was observed in Section 1.3.3 that the growth flux provides a supersaturated distribution of adatoms on the growth surface. These atoms have free energies in excess of the energies of surface atoms already incorporated into the crystal which, in turn, have free energies higher than those of interior atoms. The energies of the nonequilibrium surface atoms may not be sufficient to drive them into interstitial sites in the crystal. However, the surface atoms could reduce their energies by migrating into the grain boundaries; a mechanistic model based on this idea has been proposed by Chason et al. (2002). The incorporation of excess atoms into the grain boundaries creates a compressive stress in the film. This compressive stress increases the energies of atoms within the grain boundaries; the associated grain boundary chemical potential is discussed in Section 8.2.2. The driving force for introducing additional excess grain boundary atoms decreases with increasing magnitude of compressive stress, leading eventually to a steady-state balance between these effects. This model, which is based on the observed central role of nonequilibrium processes at the growth surface in generating compressive stress, is not yet fully developed. However, its

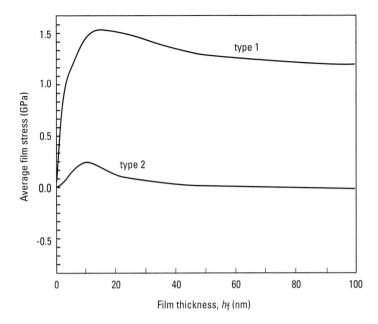

Fig. 1.35. The evolution of intrinsic stresses during the evaporative deposition of a range of type 1 and type 2 materials as a function of film thickness. After Abermann (1990) and Schneeweiss and Abermann (1992).

behavior does account for the relaxation observed in films when the growth flux is interrupted and the effect of growth rate on the steady-state deposition compressive stress level is captured.

1.8.5 Correlations between final stress and grain structure

Correlations have been observed between the growth stress in polycrystalline films and the final grain structure of the films as represented by the two types of microstructure depicted in Figure 1.20. By measuring *in situ* the bending of cantilever beams onto which a wide range of thin films were vapor deposited, Abermann (1990), Thurner and Abermann (1990), Schneeweiss and Abermann (1992) estimated the volume-average internal stress as a function of film thickness in both type 1 and type 2 microstructures. Representative results are plotted in Figure 1.35. This figure shows that type 1 materials, in which film thickening essentially occurs by homoepitaxial growth on top of grains nucleated at the film–substrate interface, large tensile residual stresses are generated. By contrast, the higher surface diffusivity of type 2 materials promotes greater stress relaxation and, in these cases, much smaller tensile stresses are estimated. As the film thickens, the tensile stresses essentially vanish; for thick films, the film stress becomes compressive, for reasons discussed above.

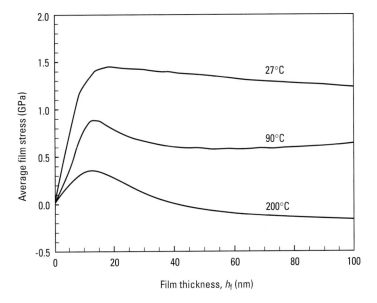

Fig. 1.36. Experimental observations, derived from the work of Abermann (1990), of average film stress as a function of film thickness in Fe films vapor-deposited on MgF_2 substrates at $T_s = 27$, 90 and 200 °C.

It is also noted that, for a given material, the transition from type 1 to type 2 response can occur as the substrate temperature T_s is raised. An example of this effect and the associated evolution of intrinsic stress in Fe films vapor-deposited on MgF_2 substrates is shown in Figure 1.36. At room temperature, the Fe film thickens without a precipitous drop in the initial tensile stress because of the low mobility of adatoms at the film surface. However, as the substrate temperature is raised to 90 °C or 200 °C, a marked reduction in the tensile stress occurs in the film with increasing adatom mobility. For $T_s = 200$ °C, the tensile stresses vanish at a film thickness of about 40 nm and, thereafter, growth occurs with the film stressed in compression.

1.8.6 Other mechanisms of stress evolution

The compression–tension–compression sequence of stress evolution during deposition of thin polycrystalline films, as described in Sections 1.8.1, 1.8.3 and 1.8.4, appears to be characteristic of many growth processes. Furthermore, once the films become continuous across the growth surface, careful measurements reveal very little alteration in average film stress during interruptions in the deposition process; in most cases, stress evolution upon resumption of deposition resumed continuously from the behavior observed prior to the interruption (Floro et al. 2001, Shull and Spaepen 1996). This observation suggests a relatively minor role for time-dependent stress relaxation processes that are not associated with the deposition process itself.

However, time-dependent processes of stress relaxation such as grain boundary area reduction, texture evolution and vacancy annihilation may contribute to stress development in some cases. Consequently, the energetics of these mechanisms are briefly described in this section. The prospect for tensile stress development by reduction of grain boundary area, a mechanism that is commonly called *grain growth*, is considered first. This is followed by observations on the relationship between film stress and texture evolution and on stress generation as a result of elimination of excess vacancies. Each of these processes is time-dependent but no effort is made here to incorporate specialized kinetic relations as a basis for describing evolution. Instead, a rough estimate of the magnitude of stress that could be achieved through operation of each mechanism separately is obtained by assuming that thermodynamic equilibrium is reached from some initial nonequilibrium state.

Stress arising from reduction in grain boundary area

Consider a two-dimensional polycrystalline material, perhaps a portion of a film with columnar grain structure, with its outer bounding surface constrained against normal displacement. The boundary constraint presumably arises through interaction of the sample with statistically identical adjacent portions of the film, in which case the bounding surface is a symmetry surface. A model introduced by Chaudhari (1971) for development of tensile stress in this representative portion of the film is based on a competition between two physical effects. First, associated with the grain boundary surface is an energy per unit area, say γ_{gb}, which represents the excess energy of the grain boundary structure over and above that of the regular lattice structure within the grains; this is an interface surface energy as described in Section 1.3.3. It follows that the free energy of the sample can be *reduced* by decreasing the total area of grain boundary surface within the representative material sample. For a two-dimensional polycrystalline material, as assumed for this discussion, this is equivalent to reduction in the net length of the trace of grain boundary surfaces in any plane of the film parallel to the film–substrate interface.

The second feature is that the mass density of the material in the immediate vicinity of a grain boundary is less than the density within the grains themselves, a feature also arising from the relative disorder of the atomic structure along a grain boundary. This volume deficiency is represented as a volume per unit area of grain boundary, say Δv, that can be viewed in the following way. If a grain boundary lying on a cross-section of a prismatic bicrystal of unit cross-sectional area migrates out of the material, the resulting reduction in length of the solid upon becoming a single crystal is Δv. It follows that the volume of the representative sample of the material is reduced by Δv per unit area of grain boundary eliminated. Returning attention to the representative material sample with many interior grain boundaries, the sample undergoes a stress-free isotropic contractive strain upon reduction in grain boundary

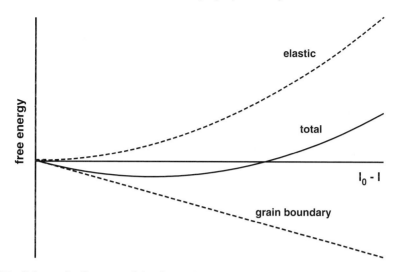

Fig. 1.37. Schematic diagram of the dependence of free energy of the film with columnar grains on grain boundary area, as represented by a net length of the surface trace of grain boundaries. Contributions due to elastic deformation and to grain boundary area reduction are shown as dashed lines.

area. The boundary of the sample is constrained against normal displacement due to symmetry, however. Consequently, an elastic extensional strain must be induced to satisfy this constraint. Associated with the elastic strain is an elastic energy density throughout the material. The elastic energy of the sample then *increases* parabolically with grain boundary area reduction.

The nature of the competition between these two effects for a portion of the film with surface area $L \times L$ is depicted in Figure 1.37. If the total length of the surface trace of a grain boundary at the outset is l_0 and the subsequent length is l, then the change in free energy per unit thickness of the film due to reduction in grain boundary area is $-\gamma_{gb}(l_0 - l)$. The stress-free change in total volume per unit thickness of the sample due to grain boundary area reduction is $-\Delta v(l_0 - l)$. This strain must be negated by a uniform equi-biaxial elastic strain $\frac{1}{2}\Delta v(l_0 - l)/L^2$ which gives rise to the elastic energy of the system. The change in free energy per unit thickness of the sample due to change $l_0 - l$ in the grain boundary trace length is then

$$\mathcal{E} = -\gamma_{gb}(l_0 - l) + \frac{E_f}{2(1 - \nu_f)} \frac{\Delta v^2}{L^2}(l_0 - l)^2 \qquad (1.27)$$

where E_f is the modulus of the elastically isotropic film material and ν_f is the Poisson ratio. The two terms in this equation are sketched separately in Figure 1.37 as dashed curves, and the sum is shown as the solid curve. The illustration of elastic energy presumes that the stress is zero when the grain boundary trace length is l_0; the argument requires minor modification if this is not the case. The stress at which total energy change is minimum is found to be

$$\sigma = \frac{\gamma_{gb}}{2\Delta v}.$$ (1.28)

The expressions are conveniently expressed in terms of grain size, rather than grain boundary trace length, by noting that

$$\frac{l_0 - l}{2L^2} \sim \frac{1}{d_{0gr}} - \frac{1}{d_{gr}}$$ (1.29)

where d_{0gr} is the initial grain size d_{gr} is the subsequent grain size and the factor two accounts for the fact that each grain boundary is shared by two grains.

Insight into the magnitudes of grain size at which the energy reaches a minimum value and the corresponding internal stress are gained by considering an example with $\gamma_{gb} \sim 0.4\,\mathrm{J\,m^{-2}}$, $E_f/(1 - v_f) = 100\,\mathrm{GPa}$ and $\Delta v \sim 0.2\,\mathrm{nm}$. For these values, grain growth leads to an equilibrium stress of 1 GPa, as seen from (1.28). The grain size at which the energy minimum is reached is $d_{gr} = (d_{0gr}^{-1} - 0.5)^{-1}$ where the initial and final grain sizes are given in units of nm. For an initial grain size $d_{0gr} = 10\,\mathrm{nm}$, the minimum in energy occurs at a grain size of $d_{gr} = 20\,\mathrm{nm}$. If the initial grain boundary dimension is already larger than this value at zero stress, then grain growth cannot occur by this mechanism.

For an initial tensile stress in the film, there is a linear dependence of elastic energy on grain boundary area reduction, in addition to the quadratic increase illustrated in Figure 1.37. As a result, the minimum energy value of grain boundary area reduction decreases, and grain growth becomes impossible if the value reaches zero. On the other hand, a compressive initial stress is expected to move the minimum energy position toward larger values of grain boundary area reduction, thus promoting a greater degree of grain growth than in the absence of initial stress.

The presence of impurities and cavities on the grain boundaries can also alter grain growth in thin films and the associated stress evolution. The presence of interstitials and cavities changes the density of the grain boundary region, while the interaction of these defects with the surrounding atoms modifies the total energy change during grain growth. Thus, as reviewed by Doerner and Nix (1988) such phenomena as oxidation, diffusion, inert gas entrapment, or moisture ingress during the deposition of the film would be expected to alter the evolution of internal stress.

Texture evolution in polycrystalline thin films

Departures from normal grain growth in thin films are introduced as a result of the formation of grooves where the grain boundaries meet the free surface of the film (Mullins 1958, Frost et al. 1990). Such abnormal grain growth can occur with additional contributions to the driving force for grain growth that is specific to the crystallographic orientation of the grains. Consider the case of fcc metal films deposited on amorphous substrates such as oxidized Si. The free surface energy density of the film γ_f, as well as the energy density of the interface between the film and the substrate γ_{fs}, are minimized when the grains of the film have a $\{111\}$ texture, that is, when the grains are oriented with the $\langle 111 \rangle$ crystallographic directions being normal to the interface. This suggests an energetic preference for growth of grains with the $\{111\}$ texture over growth at other crystallographic orientations.

For fcc metals, consider the evolution of texture during grain growth, as represented by a particular reorientation of film grains, due to thermal excursions such as those arising from passivation of the material or from post-patterning anneal. There can be a resulting energy change due to orientation dependence of elastic moduli, as well as an energy change due to orientation dependence of the surface/interface energy densities. Elastic strain in the film in the case of temperature excursion ΔT arises from the difference in coefficients of thermal expansion between the film and substrate materials. It is established in Chapter 2 that the thermal mismatch strain must be accommodated entirely within the film if the substrate is relatively thick. If the coefficients of thermal expansion of the film and substrate materials are denoted by α_f and α_s, respectively, then the stress-free thermal strain of the film with respect to the substrate is $(\alpha_f - \alpha_s)\Delta T$. Because the film is constrained against such strain by the substrate, an equal but opposite elastic strain is generated to satisfy the constraint. If the thermal expansion of the film material is isotropic, and if the biaxial elastic modulus of the film material changes from M_0 in the 'initial' orientation to M in the 'final' orientation, then the change in elastic energy per unit volume of film material due to reorientation is

$$\mathcal{E}_e = (M - M_0)(\alpha_f - \alpha_s)^2 \Delta T^2 . \tag{1.30}$$

The change in surface energy per unit volume due to both the free surface of the film and the film–substrate interface is

$$\mathcal{E}_s = \frac{(\gamma - \gamma_0)}{h_f} \tag{1.31}$$

where $\gamma = \gamma_f + \gamma_{fs}$ is the surface energy, free surface and interface together, per unit volume in the 'final' orientation and γ_0 is the corresponding value in the 'initial' orientation.

For fcc metals on amorphous substrates, the $\{100\}$ texture, which minimizes the strain energy density, dominates during grain growth when $\mathcal{E}_e > \mathcal{E}_s$. In this case, grain growth serves as a strain relaxation mechanism. For $\mathcal{E}_e \leq \mathcal{E}_s$, on the other hand, the $\{111\}$ texture, which minimizes the surface energy, would be expected to develop preferentially. The condition for the transition from the $\{100\}$ to the $\{111\}$

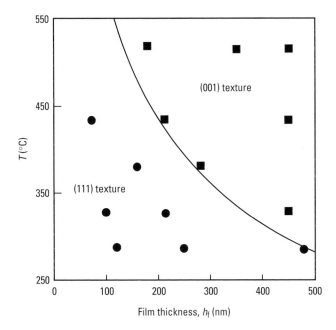

Fig. 1.38. Experimentally observed texture following grain growth induced by a temperature change ΔT from the deposition temperature in Ag thin films, up to 500 nm in thickness. The films were deposited on MgO(100) substrates coated with an amorphous silicon dioxide layer. The solid line corresponds to the texture transition prediction of (1.32). Adapted from Thompson and Carel (1995).

texture formation during grain growth can then be written as

$$\left(M_{\{111\}} - M_{\{100\}}\right)(\alpha_s - \alpha_f)^2 \, \Delta T^2 \approx \frac{\gamma_{\{111\}} - \gamma_{\{100\}}}{h_f} \qquad (1.32)$$

in terms of material parameters specific to these orientations.

Thompson and Carel (1995) have experimentally studied texture evolution in Ag thin films and have shown that the competition between strain energy minimization and surface energy minimization leads to film textures that are consistent with the transition predicted by (1.32). Figure 1.38 shows their experimental results on texture evolution in Ag polycrystalline thin films of different thicknesses as a function of temperature excursion ΔT from the deposition temperature. Here, the Ag films were deposited on MgO(100) substrates coated with amorphous silicon oxide. The solid line in this figure denotes the texture transition condition predicted by (1.32).

It should be noted that this discussion has presumed from the outset that the deformation of the polycrystalline films involved only elastic strains and thermal expansion strains. With the progression of grain growth, the yield strength of the film is expected to decrease. For the case of face-centered cubic materials, grains with {110} or {210} texture promote plastic flow at lower stresses than do {100} or {111} textured grains. The onset of plastic yielding may favor the growth of the former textures over the latter two possibilities due to the lower yield strength in the former instance.

Excess vacancy reduction

The deposition of a crystalline film onto a substrate at relatively low temperature can result in a distribution of vacancies throughout the film that is in excess of the equilibrium concentration at the deposition temperature. This structure is essentially a substitutional solid solution with the vacancies dissolved in the film grains. There is an energetic driving force acting on a vacancy near the growth surface of the film during deposition that tends to draw the vacancy out of the material. An excess vacancy concentration can be incorporated into the material if the rate of advance of the growth front is greater than the diffusive flux of vacancies through the instantaneous position of the growth surface; the vacancy diffusivity is strongly dependent on growth temperature. If a crystallite containing an excess concentration of vacancies would be free of boundary constraint, then the concentration could decay in time to the equilibrium concentration by vacancy diffusion, with excess vacancies leaving the crystal through its boundary. In effect, the vacancies are annihilated at the crystal boundary. No stress is generated through vacancy annihilation within an unconstrained grain as long as vacancy concentrations are spatially uniform, but the volume of the grain decreases by an amount proportional to the number of vacancies removed. For purposes of this discussion, it is assumed that this volume contraction is isotropic.

If the boundary surface of the grain in a film is constrained against displacement due to its attachment to the substrate or to adjacent grains, it follows immediately that stress must be generated in the grain upon diffusion of vacancies out of the grain. It is also possible for a grain boundary atom to combine with a vacancy, resulting in annihilation of the vacancy and generation of tensile stress in the grains forming the grain boundary as the interface reconstructs to account for the atom removed. In any of these cases, the number of atoms in a grain remains unchanged as the number of vacancies is reduced.

The vacancy volume Ω_{vac} can be defined in the following way. If an initially perfect crystal is altered by inserting a number density ρ of vacancies per unit volume throughout its interior without changing the number of atoms in the crystal, then the crystal undergoes a stress-free isotropic expansion if its boundaries are unconstrained. If the change in volume per unit volume is expressed as $\rho\Omega_{vac}$, then Ω_{vac} is the volume of a vacancy; its value is commonly assumed to be between one-half and one times the atomic volume of the material. If the distribution is not dilute, then the value of vacancy volume could depend on the density itself.

The vacancy creation energy of the film material is essentially the cohesion energy. In terms of the equilibrium energy $\mathcal{E}_f$ for pair potential chemical bonding introduced in Section 1.3.2, the energy of cohesion under the circumstances assumed is on the order of $6\mathcal{E}_f$ for a close-packed structure, and the energy change associated with movement of an interior atom to the surface may be half as large. If ρ_0 is an initial number density of vacancies, then the change in free energy associated with a reduction in vacancy number density to ρ is estimated to be roughly $\mathcal{E}_{vac} = -3(\rho_0 - \rho)\mathcal{E}_f$.

The stress-free isotropic extensional strain associated with reduction in vacancy concentration from ρ_0 to ρ is $-\frac{1}{3}(\rho_0 - \rho)\Omega_{vac}$. If the film is free to contract in the thickness direction but is constrained from changing its in-plane dimensions by the substrate, then an equal but opposite equi-biaxial elastic extensional strain is generated by the reduction in vacancy concentration. An estimate of the associated increase in elastic energy is

$$\mathcal{E}_e = \frac{1 - \nu_f}{9E_f}(\rho_0 - \rho)^2\Omega_{vac}^2 \tag{1.33}$$

where E_f is the elastic modulus and ν_f is the Poisson ratio of the elastically isotropic film. If the free energy change $\mathcal{E}_e + \mathcal{E}_{vac}$ is minimized with respect to variations in vacancy concentration, then the tensile stress corresponding to the minimum value in free energy is approximately

$$\sigma \approx 4.5\frac{\mathcal{E}_f}{\Omega_{vac}}. \tag{1.34}$$

For example, if $\mathcal{E}_f = 0.5\,\text{eV}$ and $\Omega_{vac} = 0.1\,\text{nm}^3$ then $\sigma \approx 3.6\,\text{GPa}$. The estimate of stress is indeed very large, indicating that the vacancy reduction mechanism is potentially significant. However, there are several other mechanisms by which the effect of this mechanism could be mitigated. These include formation of interior or grain boundary cavities with increasing volume which could absorb vacancies without generating stress, motion of stress relieving dislocations through the grains, atomic diffusion along grain boundaries and so on.

Stress generation in sputtering deposition

The deposition of thin films by sputtering processes leads predominantly to compressive stresses. Two features distinguishing sputtering deposition, as described in Section 1.2.1, from deposition by vapor condensation are the relatively high kinetic energies of atoms arriving at the growth surface and the argon atmosphere of the growth chamber. The bombardment of the growth surface by the energetic atoms

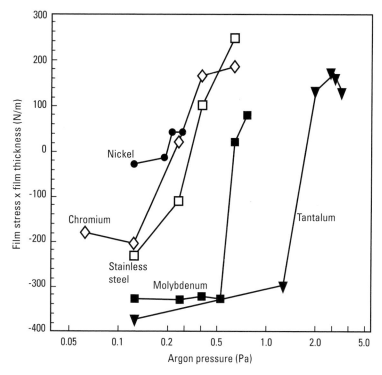

Fig. 1.39. Average internal stress versus argon pressure in Cr, Mo, Ni, Ta and stainless steel sputter-deposited films. Adapted from Thornton (1977), Hoffman and Thornton (1979) and Thornton et al. (1979).

arriving there can lead to the generation of excess interstitials in the near surface region. It has been observed that the amount of damage induced at the surface depends on both the energy of the arriving atoms and the background pressure of the inert gas.

Figure 1.39 shows results of average film stress multiplied by film thickness, which were derived from the deposition of thin films of a wide range of materials onto glass wafer substrates surrounding dc cylindrical magnetron sputtering sources. The film stresses were inferred from interferometric observations of the deflection of the glass wafer substrate with respect to an optical flat. Large compressive stresses, up to several GPa in magnitude, evolve in these materials at argon pressures typical of common sputter deposition processes. It was found that the occurrence of compressive internal stresses in the sputtered films correlates with the presence of entrapped argon in the films. However, large changes in argon content (where the atomic concentration of argon varied from 0.02% to 2%) do not lead to appreciable changes in the magnitude of the compressive stresses (Thornton et al. 1979). As seen in Figure 1.39, low sputtering pressures and low deposition rates

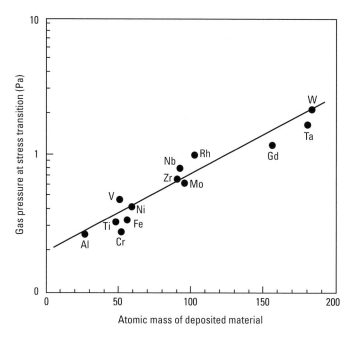

Fig. 1.40. The variation of critical gas pressure at which the intrinsic stresses change from compression to tension, as a function of the atomic mass of the deposited material. The results are taken from the works of Thornton and Hoffman (1977) and Thornton et al. (1979).

lead to the evolution of compressive internal stresses. An examination of the zone model illustrated in Figure 1.21 reveals that the formation of columnar grains is restricted at low gas pressures. Lower gas pressures during flight of the target atoms lead to less loss of energy through collisions and, as a result, the arriving source atoms would be expected to have a higher energy.

On the basis of these observations, Thornton and coworkers postulated that the 'atomic peening' process which induces damage at the surface, either during forward bombardment of high energy atoms or during atomic recoil, is the primary cause of compressive intrinsic stresses during sputter deposition. It is also interesting to note that the argon gas pressure at which the transition from compressive intrinsic stresses to tensile stresses occurs directly correlates with the atomic mass of the deposited material; an increase in atomic mass leads to an increase in the gas pressure at which the stresses change from compression to tension, as shown in Figure 1.40. This trend also provides evidence for the mechanism of atom peening, with materials with larger masses producing more compressive stresses through increased damage.

Argon entrapment is believed to contribute to the intrinsic stress build-up; however, it is not regarded as the principal mechanism. Although mechanisms underlying the evolution of compressive stresses in sputter-deposited thin films remain largely unexplored, atom peening of surfaces is commonly regarded as the main source of intrinsic stresses and as the chief reason for a stress generation process that is distinctly different from that seen in vapor-deposited films.

1.9 Consequences of stress in films

Several inferences can be drawn from the foregoing discussion. For one thing, it is very difficult to avoid the introduction of residual stress in the course of fabrication of thin film and multilayer material structures. Furthermore, the understanding of the relationship between states of residual stress and the processing conditions through which they were introduced is only qualitative in most cases. This residual stress can induce a variety of consequences which are undesirable in most circumstances – excessive deformation, fracture, delamination and microstructural changes in the materials. Nonetheless, such structures have been inserted into engineering systems in order to accomplish a wide range of practical service functions. The success of this endeavor has been enabled, to a large extent, through research leading to the development of ways to measure residual stress in small structures and by the identification of frameworks for characterizing the resistance of material systems to failure. At this juncture, the need to ensure mechanical integrity continues to be a technology limiting factor in adoption of small material structures, even in situations where load carrying capacity is not an eventual functional factor.

The purpose of the chapters that follow is to summarize the results of the research effort underlying the use of these material systems. The presentation is loosely organized according to the scale of the dominant physical phenomena involved. Accordingly, the discussion begins with the study of overall deformation of film–substrate systems or multilayer structures in Chapters 2 and 3. This is followed by examination of the general failure modes of fracture, delamination and buckling of films in Chapters 4 and 5. Finally, on a smaller scale, the focus is on the ways in which stress can affect microstructural features through dislocation motion in Chapter 6, plastic deformation in Chapter 7, and evolution of surface morphology and alloy composition in Chapters 8 and 9.

1.10 Exercises

1. A practical example of *homoepitaxy* is the deposition of a single crystal Si film on a Si substrate. This is a common step in the manufacture of bipolar transistors and some metal–oxide semiconductor (MOS) transistors. Explain the benefit of epitaxially depositing a Si film on the Si substrate when the Si substrate alone could well suffice as the material for the required electronic and mechanical response?

2. Body-centered cubic Fe can be epitaxially grown on GaAs which has the zinc-blende structure. An Fe film is grown on a GaAs(110) substrate. The lattice constant of Fe is 0.2866 nm while that of GaAs is 0.5653 nm. Consequently, two Fe unit cells could be accommodated into one unit cell of GaAs. Noting that the epitaxial geometry

of Fe(110) ∥ GaAs(110) is equivalent to Fe[200] ∥ GaAs[100], calculate the misfit strain for Fe(001) grown on a GaAs(001) substrate.

3. There is a similarity between the hexagonal close-packed (hcp) and body-centered cubic (bcc) crystal structures.

   (a) Identify the plane in the bcc structure that is similar in structure to the basal plane in the hcp structure.
   (b) Draw the arrangement of atoms in this plane and determine the stacking arrangement normal to this plane in the bcc structure. Is the bcc stacking arrangement the same as the hcp arrangement?

4. Ge and Si form a completely miscible binary solid solution, commonly denoted by Ge_xSi_{1-x} where the atomic fraction of Ge is in the range $0 \le x \le 1$. Using the binary phase diagram of Ge–Si and a specific composition, say $x = 0.5$, explain why it is practically difficult to grow high quality Ge–Si alloy single crystals of uniform composition using bulk crystal growth methods.

5. Refer to Figure 1.17 and estimate the following quantities.

   (a) The misfit strain between a layer of GaAs and AlAs.
   (b) The bandgap energy of a cladding layer of $Al_{0.3}Ga_{0.7}As$.
   (c) The misfit strain between a layer of $Al_{0.3}Ga_{0.7}As$ and a GaAs substrate.

6. The volume change associated with the annihilation of vacancies located at grain boundaries oriented normally to the plane of a thin film was discussed in Section 1.8.6. Although a volume change occurs at any site in a thin film where a vacancy is annihilated, this process need not always result in the generation of an internal stress. Provide a simple mechanistic justification for each of the following statements.

   (a) The annihilation of a vacancy located at the free surface of the film, or at the internal surface of a large void (whose dimensions are much larger than the atomic dimensions), or at the film–substrate interface and grain boundaries oriented parallel to the plane of the film, does not lead to a significant intrinsic stress.
   (b) If an atom located on the so-called extra half plane of an edge dislocation replaces a nearby vacancy site, a tensile stress is generated if the Burgers vector **b** of the dislocation has a component within the plane of the film; no stress is produced if **b** is normal to the film–substrate interface.

7. The epitaxial growth of magnetic thin films on crystalline underlayers was discussed in Section 1.4.4.

   (a) For the CoCrTaPt alloy grown epitaxially on the [110]-textured Cr under-layer, show that the resulting epitaxial geometry is: $\{10\bar{1}1\}_{Co} \parallel \{110\}_{Cr}$.
   (b) Illustrate how the above epitaxial geometry forces the c-axis of the Co alloy to be out of the plane of the film.
   (c) For the CoCrTaPt alloy grown epitaxially on the [112]-textured NiAl under-layer, show that the resulting epitaxial geometry is: $\{10\bar{1}0\}_{Co} \parallel \{112\}_{NiAl}$.

8. The derivation of Young's equation (1.4) for the wetting angle was based on the assumption of a spherical cap island on a rigid substrate. For some general ax-isymmetric shape with surface height $w(r)$ versus radial distance r from the axis of

symmetry over the portion of the substrate surface within $0 \leq r \leq R$, the island surface area A and volume V are given by

$$A = 2\pi \int_0^R \sqrt{1 + w'(r)^2}\, r\, dr\,, \quad V = 2\pi \int_0^R w(r) r\, dr \qquad (1.35)$$

The free energy change is still given by (1.3).

    (a) For alterations in shape of a constant volume island, show that the edge condition has the form $\gamma_f \sqrt{1 + w'(R)^2} = \gamma_s - \gamma_{fs}$ and verify that this is once again Young's equation.

    (b) Determine the differential equation for the shape $w(r)$ that minimizes the free energy under constant volume variations in shape.

9. For cubic metals at relatively high temperature, the free surface energy density γ does not vary significantly with surface orientation and the surface stress magnitude is relatively small. Under such circumstances, the surface energy density can be measured by the so-called method of zero creep (Blakely 1973). Circular wires of sub-millimeter radius R with bamboo grain structure are each subjected to a fixed deadweight tensile loading while being heated. For relatively large weights, the wires elongate by creep deformation whereas, for relatively small weights, the wires contract. For some weight force W, the bulk stress and creep rate are both zero. Assuming that the work of this force is required to balance the increase free surface energy and decrease grain boundary energy at zero stress and constant grain volume, show that W is related to the surface energy density, the radius R and length L of a representative cylindrical grain, and grain boundary surface energy γ_{gb} according to $W = \pi R(\gamma - \gamma_{gb} R/L)$. In some cases, it is possible to relate γ and γ_{gb} according to the Young relationship (1.4) through observation of grooving along the surface traces of the grain boundaries.

10. Suppose that the separation u between adjacent planes of atoms of a crystal, measured from the stress free spacing a, is resisted by a force per unit area of

$$\sigma(u) = \frac{Eb}{\pi a} \sin \frac{\pi u}{b} \qquad (1.36)$$

for $0 \leq u \leq b$ where E is the bulk elastic modulus. Note that this force law matches the linear stress–strain behavior of the material for $u/a \ll 1$. The resistance to separation is assumed to be zero for $u > b$.

    (a) Obtain an estimate of surface energy density for the material in terms of E, a and b.

    (b) Choose values of the parameters appropriate for Si(100) and compare the value of the corresponding surface energy estimate to measured values of surface energy density for this material.

2

Film stress and substrate curvature

The existence of residual stress in films deposited on substrates and the effects of such stress on delamination and cracking were recognized as early as the nineteenth century. For example, an antimony film deposited on a substrate was found to be prone to cracking as a result of vibration or local application of heat (Gore 1858, Gore 1862). Gore ascribed the bending of the deposited layers to the development of 'unequal states of cohesive tension' through the thickness of the deposit. The Earl of Rosse (circa 1865, cited in *Nature*, Aug. 20, 1908, p. 366) attempted to make flat mirrors by chemically coating glass with silver and then electroplating with copper. It was noted that the contraction of copper film caused it to be detached from the glass. Stoney (1909) found that copper electrodeposited as a protective layer on silver films in searchlight reflectors easily 'peeled off' when the thickness of the copper layer was in excess of 10 μm.

Stoney (1909) made the observations that a metal film deposited on a thick substrate was in a state of tension or compression when no external loads were applied to the system and that it would consequently strain the substrate so as to bend it. He suggested a simple analysis to relate the stress in the film to the amount of bending in the substrate. In an attempt to assess the implications of his analysis, Stoney electrolytically deposited thin layers of nickel, 5.6 μm to 46.2 μm in thickness, on 0.31 mm thick steel rulers which were 102 mm long and 12 mm wide. Using measurements of the film thickness, the amount of bending of the steel ruler and the elastic modulus of steel, Stoney estimated from his analysis the tensile stress in the nickel film to be in the range from 152 to 296 MPa for film thickness ranging from 46.2 to 5.6 μm, respectively. He also observed that when the rulers were heated to 'a red heat so as to anneal them', they 'straightened out to a considerable extent'.

Although highly sophisticated experimental tools for measuring the curvature of substrates with thin film deposits have become available since Stoney's work, the concepts underlying his analysis have remained the basis for the estimation of film stresses in a wide variety of modern applications including microelectronics, opto-electronics and surface coatings in structural components. Stoney's results for curvature and stress in a film–substrate system can be derived using a number of different methods. In the next section an energy method, which facilitates direct extensions to more complex multilayer geometries without modifications, is described in detail.

2.1 The Stoney formula

Suppose that a thin film is bonded to one surface of a substrate of uniform thickness h_s. It will be assumed that the substrate has the shape of a circular disk of radius R, although the principal results of this section are independent of the actual shape of the outer boundary of the substrate. A cylindrical r, θ, z-coordinate system is introduced with its origin at the center of the substrate midplane and with its z-axis perpendicular to the faces of the substrate; the midplane is then at $z = 0$ and the film is bonded to the face at $z = h_s/2$. The substrate is thin so that $h_s \ll R$, and the film is very thin in comparison to the substrate. The film has an incompatible elastic mismatch strain with respect to the substrate; this strain might be due to thermal expansion effects, epitaxial mismatch, phase transformation, chemical reaction, moisture absorption or other physical effect. Whatever the origin of the strain, the goal here is to estimate the curvature of the substrate, within the range of elastic response, induced by the stress associated with this incompatible strain. For the time being, the mismatch strain is assumed to be an isotropic extension or compression in the plane of the interface, and the substrate is taken to be an isotropic elastic solid with elastic modulus E_s and Poisson ratio ν_s: the subscript 's' is used to denote properties of the substrate material. The elastic shear modulus μ_s is related to the elastic modulus and Poisson ratio by $\mu_s = \frac{1}{2}E_s/(1 + \nu_s)$.

Imagine that the deformation is produced in the following way. The substrate is initially separate from the film, stress-free and undeformed, as shown in Figure 2.1. The isotropic membrane force f, with physical dimensions of force/length, is maintained in the film by some external means. The magnitude of f is such that it induces the prescribed elastic mismatch strain in the free film; f may be positive or negative. If the incompatibility arises from a stress-free transformation strain of the film material with respect to the substrate material, then the elastic mismatch strain and the stress-free transformation strain are equal in magnitude but opposite in sign. The strained film is then brought into contact with the substrate surface and bonded to it, after which the external means of maintaining the film tension is relaxed. It is in the course of this relaxation that the substrate becomes strained.

Substrate deformation is analyzed by invoking a number of assumptions, the range of validity of which will be probed in subsequent sections in this chapter. It is assumed that:

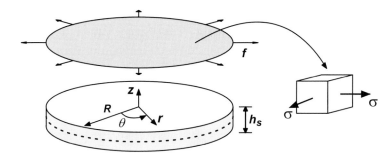

Fig. 2.1. Film and substrate separated, but with distributed force f acting on the film edge so that its strain is exactly the mismatch strain. This loading gives rise to an equi-biaxial state of stress at each material point in the film; such a state of stress with magnitude σ is illustrated on the right side.

- the substrate deforms according to the Kirchhoff hypothesis of thin plate theory, with the general expectation that the normal stress component $\sigma_{zz} = 0$ everywhere, and material lines that are straight and perpendicular to the midplane of the substrate prior to deformation remain so after deformation, so that $\epsilon_{rz} = \epsilon_{\theta z} = 0$;
- all components of displacement gradient are very small compared to unity in the substrate so that the linear theory of elasticity can be applied;
- the properties of the film–substrate system are such that the film material contributes negligibly to the overall elastic stiffness;
- the membrane force f in the film is a system parameter determined by the mismatch strain alone; and
- the change in the magnitude of the membrane force due to the deformation of the substrate is small compared to the magnitude of f.

Certain assumptions are also invoked concerning the deformation itself. It is assumed that:

- the deformation is axially symmetric so that $\epsilon_{r\theta} = 0$ throughout and all fields are independent of θ;
- the curvature of the substrate midplane surface κ is spatially uniform;
- the in-plane strain of the midplane is a uniform, isotropic extension $\epsilon_{\rm o}$ so that $\epsilon_{rr}(r, 0) = \epsilon_{\theta\theta}(r, 0) = \epsilon_{\rm o}$; and
- localized edge effects around the periphery of the film, the principal site of load transfer between the film and substrate, are ignored for the time being. The load transfer mechanism associated with film edge effects will be discussed in Chapter 4.

As a result of symmetry and translational invariance, the deformed shape of the midplane is spherical. This is a good approximation provided that the material is nearly isotropic and the deformation is small.

For the system features outlined above, the strain energy per unit volume at any point in the substrate material is expressible in terms of the nonzero strain components as

$$U(r, z) = \frac{\mu_s}{1 - \nu_s} \left[\epsilon_{rr}^2 + \epsilon_{\theta\theta}^2 + 2\nu_s \epsilon_{rr} \epsilon_{\theta\theta} \right]. \tag{2.1}$$

For small deformation, the elastic strains appearing in (2.1) are conveniently expressed in terms of $u(r)$ and $w(r)$, the radial and out-of-plane displacement components, respectively, of points on the substrate midplane, as

$$\epsilon_{rr}(r, z) = u'(r) - zw''(r), \quad \epsilon_{\theta\theta}(r, z) = \frac{1}{r}u(r) - \frac{z}{r}w'(r), \tag{2.2}$$

where the prime denotes differentiation with respect to the argument.

The strategy adopted is to select plausible parametric forms for $u(r)$ and $w(r)$, and then to invoke the principle of stationary potential energy to determine optimal values of the parameters involved to minimize potential energy. For small deflections, the radial deformation and the out-of-plane or transverse deformation are uncoupled and a choice for the midplane displacement is

$$u(r) = \epsilon_0 r, \quad w(r) = \tfrac{1}{2}\kappa r^2, \tag{2.3}$$

where κ represents the curvature, or inverse of the radius of curvature, of this plane. This is a reasonable choice because, in the absence of the edge effects, points on the substrate midplane are indistinguishable and therefore the curvature should be spatially uniform over the midplane. The in-plane normal strain is $\epsilon_{\theta\theta} = \epsilon_{rr} = \epsilon_0 - \kappa z$ where ϵ_0 represents the extensional strain of the substrate midplane and κz denotes the extensional strain due to the bending at any position z with respect to the midplane of the substrate. Because the state of deformation is equi-biaxial strain at each point of the substrate, the resistance to the deformation can be represented in terms of the biaxial elastic modulus of the substrate material, which is denoted by M_s. In an isotropic material, the biaxial elastic modulus is the ratio of equi-biaxial stress to equi-biaxial strain. In the present instance, its dependence on the elastic modulus E_s and Poisson ratio ν_s becomes evident by noting that the extensional strain ϵ_{rr} in the substrate midplane is given by

$$\epsilon_{rr} = \frac{\sigma_{rr}}{E_s} - \nu_s \frac{\sigma_{rr}}{E_s} = \frac{\sigma_{rr}}{M_s}, \tag{2.4}$$

and similarly for the circumferential direction or for any other direction in the substrate midplane. Thus, the biaxial modulus of the substrate material is $M_s = E_s/(1 - \nu_s)$.

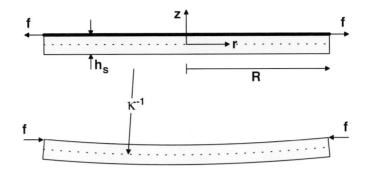

Fig. 2.2. Development of curvature in a film–substrate system with mismatch is represented as the sum of the two deformation states depicted here.

The strain energy density throughout the substrate is $U(r, z) = M_s (\epsilon_0 - \kappa z)^2$. The total potential energy of the substrate in the state represented by the lower portion of Figure 2.2, following relaxation of the external agent, is

$$V(\epsilon_0, \kappa) = 2\pi \int_0^R \int_{-h_s/2}^{h_s/2} U(r, z)\, r \, dz \, dr + 2\pi f \, u_r(R, h_s/2)\, R$$

$$(2.5)$$

$$= \pi R^2 M_s h_s \left(\epsilon_0^2 + \tfrac{1}{12}\kappa^2 h_s^2 \right) + 2\pi R^2 f \left(\epsilon_0 - \tfrac{1}{2}\kappa h_s \right),$$

where the sign of the external potential energy term is determined by the fact that the substrate is strained by *negating* the membrane force f as indicated in Figure 2.2. Within the class of permissible deformations, the equilibrium midplane deformation, represented by ϵ_0 and κ, is that which renders the total potential energy V stationary with respect to variations in its arguments, that is, $\partial V / \partial \epsilon_0 = 0$ and $\partial V / \partial \kappa = 0$. It follows from (2.5) that

$$\epsilon_0 = -\frac{f}{M_s h_s},$$

$$(2.6)$$

and

$$\boxed{\kappa = \frac{6f}{M_s h_s^2}.}$$

$$(2.7)$$

The sign of curvature is the same as the sign of f. For the case of a tensile mismatch stress, the face of the substrate bonded to the film becomes concave whereas, for a compressive mismatch stress, the face of the substrate bonded to the film becomes convex. If h_f is the thickness of the film, then the mean stress σ_m in the film is

$$\sigma_m = \frac{f}{h_f}. \tag{2.8}$$

The expression for curvature in (2.7) is the famous *Stoney formula* relating curvature to stress in the film (Stoney 1909). Stoney's original analysis of the stress in a thin film deposited on a rectangular substrate was based on a uniaxial state of stress. Consequently, his expression for curvature did not involve use of the substrate biaxial modulus M_s. However, the relationship (2.7) is based on Stoney's concept as outlined in this section, and this equation has become known as the Stoney formula. It has the important property that the relationship between curvature κ and membrane force f does not involve the properties of the film material. Consequently, (2.7) can be applied in situations in which mismatch derives from inelastic effects. The elastic mismatch strain ϵ_m corresponding to the stress σ_m given in (2.8) is

$$\epsilon_m = \frac{f}{M_f h_f} \tag{2.9}$$

where M_f is the biaxial elastic modulus of the film material and h_f is the film thickness. The Stoney formula (2.7) serves as a cornerstone of experimental work in which stress values are inferred from curvature measurement in thin films bonded to substrates; several experimental techniques are discussed in later sections of this chapter.

As a secondary result, the location of the unstrained *neutral plane* in the substrate, at $z = z_{np}$ say, can be determined from (2.6), (2.7) and (2.2). The condition that $\epsilon_{rr}(r, z_{np}) = 0$ implies that

$$z_{np} = \frac{\epsilon_o}{\kappa} = -\tfrac{1}{6}h_s. \tag{2.10}$$

Thus, the neutral plane always lies at a distance $\tfrac{1}{6}h_s$ from the midplane of the substrate and in the direction *away* from the film, no matter what the sign or magnitude of f might be. It is also noted that $\kappa h_s \gg 1$, to the level of approximation assumed in deriving the Stoney formula (2.7). This implies that the differences among the curvatures of the midplane, neutral plane, top face and bottom face of the substrate are negligibly small.

The results in (2.7) can be obtained by a number of different approaches. For example, the lower portion of Figure 2.2 implies that the deformation of the substrate is due to a compressive force f per unit length around its periphery and a bending moment $fh_s/2$ per unit length around its periphery. The Stoney formula follows from these observations and a force balance approach to derivation of the result is developed more fully in later sections. However, use of the energy method outlined above is justified by the fact that it can be applied without modifications in order to handle more complicated cases. Much of the discussion in the remainder of this chapter is devoted to examining various extensions of the simple development in this section. This discussion should lead to a better understanding of the limitations on the use of the Stoney formula (2.7) and to more detailed results which relax some of the restrictions on the range of validity of the Stoney formula.

2.1.1 Example: Curvature due to epitaxial strain

Suppose that a single crystal film of thickness $h_f = 100$ nm of a SiGe alloy is grown epitaxially on a Si substrate of thickness $h_s = 1$ mm. The alloy is composed of 80 atomic percent Si and 20 atomic percent Ge, and is commonly designated by $Si_{0.8}Ge_{0.2}$. The mean lattice parameter at room temperature of $Si_{0.8}Ge_{0.2}$ alloy is $a_{SiGe} \approx 0.8a_{Si} + 0.2a_{Ge} = 0.5476$ nm, while the lattice parameter of Si at room temperature is $a_{Si} = 0.5431$ nm. The interface between the alloy and substrate is the (100) plane of each material. The biaxial moduli M_f and M_s for the film and substrate, respectively, are estimated by using the elastic stiffnesses, shear moduli and fractional concentrations for the principal crystallographic directions, and by using linear interpolation between the values for Si and Ge separately to determine the properties of the SiGe alloy. The properties of the constituent materials are $M_{Ge(100)} = 142$ GPa and $M_{Si(100)} = 180.5$ GPa, from which it is found by linear interpolation that $M_f = 173$ GPa and $M_s = 180.5$ GPa. Determine (a) the mismatch strain of the film with respect to the substrate at room temperature, (b) the stress in the film due to the mismatch strain, and (c) the curvature of the substrate supporting the film.

Solution:

(a) The mismatch strain at room temperature is

$$\epsilon_m = \frac{a_{Si} - a_{SiGe}}{a_{SiGe}} = -0.0082. \tag{2.11}$$

The stress-free lattice parameter of Ge is larger than that of Si. Therefore, the alloy is compressed in directions parallel to the interface, and consequently, the mismatch strain is found to be negative.

(b) From (2.9), the corresponding uniform mismatch stress is

$$\sigma_m = \epsilon_m M_f = -1.43 \text{ GPa}. \tag{2.12}$$

(c) From (2.7), the change in curvature of the substrate is

$$\kappa = \frac{6f}{M_s h_s^2} = \frac{6\sigma_m h_f}{M_s h_s^2} = -4.74 \times 10^{-3} \text{ m}^{-1}. \tag{2.13}$$

Therefore, the change in radius of curvature is $\rho = \kappa^{-1} = -211$ m. The result that the curvature is negative implies that the substrate is concave on the face away from the bonded film. This is consistent with compressive mismatch strain in the film with respect to the substrate.

2.1.2 Example: Curvature due to thermal strain

Consider a thin film of aluminum, 1 μm in thickness, which is deposited uniformly on the (100) surface of a Si substrate, which is 500 μm thick and 200 mm in diameter, at a temperature of 50 °C. The thermoelastic properties of the film are: elastic modulus, $E_f = E_{Al} = 70$ GPa, Poisson's ratio, $\nu_f = \nu_{Al} = 0.33$, and coefficient of thermal expansion, $\alpha_f = \alpha_{Al} = 23 \times 10^{-6}$ °C^{-1}. The corresponding properties of the Si(100) substrate are: $M_s = M_{Si} = 181$ GPa and $\alpha_s = \alpha_{Si} = 3 \times 10^{-6}$ °C^{-1}. The film–substrate system is stress-free at the deposition temperature. Determine (a) the mismatch strain of the film with respect to the substrate at room temperature, that is, at 20 °C, (b) the stress in the film due to the mismatch strain, and (c) the radius of curvature of the substrate.

Solution:

(a) The mismatch strain in the Al film with respect to the substrate, at room temperature, is

$$\epsilon_m = (\alpha_s - \alpha_f)\Delta T = 6.0 \times 10^{-4}. \tag{2.14}$$

(b) The equi-biaxial stress in the film is determined from (2.9) to be

$$\sigma_m = \epsilon_m M_f = 62.7 \text{ MPa}. \tag{2.15}$$

Note that this is substantially smaller than a typical plastic yield stress for aluminum thin films, indicating that the stress in the film is elastic at room temperature.

(c) From (2.7), the curvature of the substrate is

$$\kappa = \frac{6f}{M_s h_s^2} = \frac{6\sigma_m h_f}{M_s h_s^2} = 8.31 \times 10^{-3} \text{ m}^{-1}. \tag{2.16}$$

The radius of curvature is $\rho = 120$ m. The positive curvature implies that the substrate is concave on the surface on which the film is deposited, which is consistent with a state of residual tension in the film.

Values of linear thermal expansion coefficients of commonly used nonmetallic thin film, interlayer or substrate materials are given in Table 2.1 over a broad range of temperatures of practical interest. Table 2.2 provides corresponding values of linear thermal expansion coefficients for polycrystalline metals. Table 2.3 lists the room temperature values of elastic modulus and Poisson's ratio for a wide variety of polycrystalline and amorphous materials with isotropic elastic properties, which are commonly used as thin films, interlayers or substrates. The anisotropic elastic properties of cubic and hexagonal single crystals are given in Tables 3.1 and 3.2, respectively, in the next chapter.

2.2 Influence of film thickness on bilayer curvature

In the preceding section, an estimate was made of the curvature caused by the mismatch strain when a very thin film is bonded to the surface of a substrate. It was

Table 2.1. *Representative values of thermal expansion coefficient α of nonmetallic materials used as substrates, films and layers.*

Material[a]	Linear thermal expansion coefficient α in units of $(10^{-6} \,^\circ C^{-1})$; T in units of $^\circ C$	Temp. range ($^\circ C$)
Si (100)	$3.084 + 0.00196\,T$	20 to 700
Ge (100)	$6.05 + 0.0036\,T - 3.5 \times 10^{-7} T^2$	20 to 810
GaAs (100)	$5.35 + 0.008\,T$	10 to 300
C (diamond)	$0.87 + 0.00923\,T + 7 \times 10^{-6} T^2$	0 to 555
polycrystalline diamond film	2.5	30 to 350
DLC film (diamond-like C)	2.3	30 to 250
TiN	$4.9 + 0.0077\,(T+273) - 2.6 \times 10^{-6}(T+273)^2$	20 to 1327
polysilicon	$-0.15 - 3.1 \times 10^{-3}(T+273) + 4.6 \times 10^{-5}(T+273)^2$	−253 to 20
	$1.6 + 4.4 \times 10^{-3}(T+273) - 1.6 \times 10^{-6}(T+273)^2$	20 to 1327
fused SiO$_2$	$-1.479 + 0.0111\,(T+273) - 1.432 \times 10^{-5}(T+273)^2$	−193 to 20
	$0.3968 + 9.332 \times 10^{-4}(T+273) - 1.034 \times 10^{-6}(T+273)^2$	20 to 727
Ta$_2$O$_5$	$0.027 + 0.0094\,(T+273) - 5.4 \times 10^{-6}(T+273)^2$	25 to 937
α-SiC	$3.0 + 0.0028\,(T+273) - 4.5 \times 10^{-7}(T+273)^2$	20 to 2500
β-SiC	$1.6 + 0.0042\,(T+273) - 5.9 \times 10^{-7}(T+273)^2$	20 to 900
Si$_3$N$_4$	$-3.4 + 0.018\,(T+273) - 1.2 \times 10^{-5}(T+273)^2$	20 to 227
	$1.7 + 0.0024\,(T+273) - 6.3 \times 10^{-7}(T+273)^2$	227 to 1727
Al$_2$O$_3$	$-2.5 + 0.033\,(T+273) - 2.0 \times 10^{-5}(T+273)^2$	−173 to 20
	$4.5 + 0.0062\,(T+273) - 1.5 \times 10^{-6}(T+273)^2$	20 to 1627
ZrO$_2$	$13 - 0.018\,(T+273) + 1.2 \times 10^{-5}(T+273)^2$	20 to 1127
ThO$_2$	$5.1 + 0.0075\,(T+273) - 2.3 \times 10^{-6}(T+273)^2$	−123 to 1727
Cr$_2$O$_3$	$10.4 - 0.0062\,(T+273) + 3.2 \times 10^{-6}(T+273)^2$	20 to 1127
BeO	$4.4 + 0.0066\,(T+273) - 8.3 \times 10^{-7}(T+273)^2$	20 to 2000

[a] The entries in this table are taken from the following sources: (1) R. S. Krishnan, R. Srinivasan and S. Devanarayanan, *Thermal Expansion of Crystals*, Pergamon Press, New York (1979), (2) Y. S. Touloukian, R. K. Kirby, R. E. Taylor and T. Y. R. Lee, *Thermophysical Properties of Matter: Thermal Expansion; Nonmetallic Solids*, vol. 13, IFI/Plenum Press, New York (1977), and (3) Tencor Instruments, Film Stress Applications Note 3, Mountain View, CA (1993).

assumed that the change in film stress due to substrate deformation was negligible, and that the stiffness of the system depended only on the properties of the substrate. These assumptions led to the Stoney formula (2.7) relating the membrane force in the film to the curvature of the midplane of the substrate. The value of film thickness h_f entered the derivation only peripherally. The issue is re-examined in this section for cases where the film thickness h_f is not necessarily small compared to the substrate thickness h_s.

Detailed analyses of the effects of film thickness on substrate curvature in bimaterials date back to the early twentieth century, when interest in the use of thermostatic bimetals began to expand rapidly, as described in the historical note on thermostatic bimetals in Section 2.2.3. Timoshenko (1925) and Rich (1934) derived thermoelastic solutions for curvature and stress evolution in a bimetallic strip as a function of temperature change, for arbitrary variations in the relative thickness and elastic properties of the two layers in the strip. Their analyses involved a local equilibrium

Table 2.2. *Representative values of thermal expansion coefficient α of polycrystalline metals used as substrates, films and layers.*

Material[a]	Linear thermal expansion coefficient, α in units of (10^{-6} °C^{-1}); T in units of °C	Temp. range (°C)
W	$4.266 + 1.696 \times 10^{-3}(T - 20) - 5.922 \times 10^{-7}(T - 20)^2$	20 to 1122
	$5.416 + 3.904 \times 10^{-4}(T - 1122) + 1.327 \times 10^{-6}(T - 1122)^2$	1122 to 2222
	$7.451 + 3.308 \times 10^{-3}(T - 2222) + 2.27 \times 10^{-7}(T - 2222)^2$	2222 to 3327
Mo	$4.697 + 1.951 \times 10^{-3}(T - 20) + 2.821 \times 10^{-7}(T - 20)^2$	20 to 1272
	$7.583 + 2.658 \times 10^{-3}(T - 1272) + 3.447 \times 10^{-6}(T - 1272)^2$	1272 to 2527
Al	$1.415 + 0.1615(T + 268) - 2.60 \times 10^{-4}(T + 268)^2$	−268 to 27
	$23.64 + 8.328 \times 10^{-3}(T - 27) + 2.481 \times 10^{-5}(T - 27)^2$	27 to 627
Cu	$10.73 + 0.0581(T + 173) - 1.364 \times 10^{-4}(T + 173)^2$	−173 to 20
	$16.85 + 5.404 \times 10^{-3}(T - 20) + 3.447 \times 10^{-6}(T - 20)^2$	20 to 1027
Au	$11.67 + 0.02694(T + 173) - 6.351 \times 10^{-5}(T + 173)^2$	−173 to 20
	$14.51 + 2.426 \times 10^{-3}(T - 20) + 5.043 \times 10^{-6}(T - 20)^2$	20 to 1027
Ag	$16.47 + 7.478 \times 10^{-3}(T + 273) + 1.885 \times 10^{-6}(T + 273)^2$	−73 to 927
steel	$4.337 + 25.46 \times 10^{-3}(T + 273) - 1.334 \times 10^{-5}(T + 273)^2$	−223 to 727

[a] The entries in this table are taken from Y. S. Touloukian, R. K. Kirby, R. E. Taylor and T. Y. R. Lee, *Thermophysical Properties of Matter: Thermal Expansion; Nonmetallic Solids*, vol. 13, IFI/Plenum Press, New York (1977), where complete references to original sources of the data can be found.

approach that invoked translational invariance and symmetry. Similar approaches have since been exploited to extract curvature and elastic stress fields as well as certain conditions governing the onset of plastic deformation in multilayers with arbitrary combinations of layer thicknesses and thermoelastic properties (Freund 1993, Suresh et al. 1994, Giannakopoulos et al. 1995). Further discussion of the local equilibrium approach is taken up in Section 2.4.1 in the context of multilayered and compositionally graded materials, and in Chapter 7 in the context of inelastic deformation in layered materials.

2.2.1 Substrate curvature for arbitrary film thickness

The issue of film thickness effects on substrate curvature evolution is pursued by recourse to the energy minimization method which was introduced in Section 2.1 for the derivation of the Stoney formula. All other features of the system introduced in that section are retained in this discussion, which follows the work of Freund et al. (1999). It is assumed that the film material carries an elastic mismatch strain in the form of an isotropic extension ϵ_m (or contraction if ϵ_m is negative) in the plane of the interface; the physical origin of the mismatch strain is immaterial. The mismatch strain is spatially uniform throughout the film material. In this case, ϵ_m is a system parameter but it does not necessarily represent the actual strain anywhere in the material.

The system is depicted in Figure 2.3. The top portion of the figure shows the film–substrate system with an artificial, externally applied traction acting to maintain the

Table 2.3. *Elastic properties of polycrystalline and amorphous materials.*

Material[a]	Elastic modulus E (GPa)	Poisson ratio ν
polycrystalline diamond	974	0.10
diamond-like C (DLC)	110	0.13
TiC	370 to 380	0.19
polysilicon	170 ± 10	0.22
amorphous silicon	80 ± 20	0.22
SiO_2	71	0.16
SiC	400 to 445	0.22
Si_3N_4	280 to 310	0.22
Al_2O_3	372	0.25
ZrO_2	160 to 240	0.26
Cr_2O_3	105	0.20
WC	450 to 650	0.22
W	385	0.30
Mo	324	0.31
Al	70	0.35
Cu	130	0.34
Cr	279	0.21
Au	78	0.44
Ag	83	0.37
Ta and alloys	150 to 186	0.34
Mg and alloys	41 to 45	0.29
steels	190 to 214	0.30
Ni-base alloys	130 to 234	0.31
Co-base alloy	200 to 248	0.30
Zr and alloys	96	–
polycarbonate	2.6	–
polypropylene	0.9	–
polymethyl methacrylate	3.4	–

[a] The data are room temperature values taken from: (1) G. W. C. Kaye and T. H. Laby, *Tables of Physical and Chemical Constants*, 14th edition, Longman, London (1973), p. 31. (2) M. F. Ashby and D. R. H. Jones, *Engineering Materials I*, Second edition, Butterworth–Heinemann, Cornwall, UK (1997), p. 34. (3) Tencor Instruments, Film Stress Applications Note 3, Mountain View, CA (1993), where details on original sources of the data can be found. The elastic properties are essentially insensitive to film geometry, but variations in properties could result from variations in processing conditions, impurity content or property measurement techniques.

mismatch strain ϵ_m; the substrate is unstrained in this state. The magnitude of the traction necessary to maintain this configuration is $\sigma_m = M_f \epsilon_m$, and its effect is to render the film compatible with respect to the undeformed substrate. There is no interaction between the film and the substrate across the shared interface in this state. Then, the artificial externally applied traction is relaxed. The result is the deformed configuration shown in the lower part of Figure 2.3. As before, the goal is to relate the curvature κ induced in the substrate midplane to the mismatch strain ϵ_m and other system parameters.

The description of deformation on which the development in Section 2.1 is based is retained. However, if the thickness of the film is to be taken into account, then the strain energy of the film material must be included in the calculation of total potential energy. This is accomplished by adopting the strain expression (2.2) for

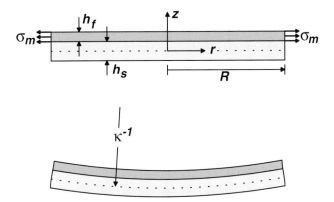

Fig. 2.3. In the upper diagram, the elastic mismatch is maintained by externally applied traction of magnitude σ_m; there is no interaction between the film and substrate in this condition and the substrate is unrestrained. If the externally applied traction is relaxed, the mismatch strain in the film induces a curvature in the substrate as shown in the lower diagram.

the film as well as the substrate, but augmenting it by the elastic mismatch strain ϵ_m in the former case. The strain energy density throughout the system is then

$$
U(r, z) = \begin{cases} M_s \, (\epsilon_o - \kappa z)^2 & \text{for } -\tfrac{1}{2}h_s < z < \tfrac{1}{2}h_s, \\[2mm] M_f \, (\epsilon_o - \kappa z + \epsilon_m)^2 & \text{for } \tfrac{1}{2}h_s < z < \tfrac{1}{2}h_s + h_f. \end{cases} \tag{2.17}
$$

The total potential energy of the film–substrate system is

$$
V(\epsilon_o, \kappa) = 2\pi \int_0^R \int_{-\frac{1}{2}h_s}^{\frac{1}{2}h_s + h_f} U(r, z) \, r \, dz \, dr. \tag{2.18}
$$

There is no external potential energy for the deformed state represented in the lower part of Figure 2.3 as there was in (2.5). In effect, the loading is now represented by ϵ_m rather than by an external force comparable to f.

As in Section 2.1, the actual midplane deformation within the class of admissible deformations is that which renders the total potential energy V stationary with respect to variations in ϵ_o and κ. It follows from the requirements that $\partial V / \partial \epsilon_o = 0$ and $\partial V / \partial \kappa = 0$ that the curvature is given by

$$
\frac{\kappa}{\kappa_{St}} = \left(1 + \frac{h_f}{h_s}\right) \left[1 + 4\frac{h_f}{h_s}\frac{M_f}{M_s} + 6\frac{h_f^2}{h_s^2}\frac{M_f}{M_s} + 4\frac{h_f^3}{h_s^3}\frac{M_f}{M_s} + \frac{h_f^4}{h_s^4}\frac{M_f^2}{M_s^2}\right]^{-1}, \tag{2.19}
$$

and that the midplane extensional strain is given by

$$\frac{\epsilon_0}{\epsilon_{0,\mathrm{St}}} = \left(1 + \frac{h_\mathrm{f}^3}{h_\mathrm{s}^3}\frac{M_\mathrm{f}}{M_\mathrm{s}}\right)\left[1 + 4\frac{h_\mathrm{f}}{h_\mathrm{s}}\frac{M_\mathrm{f}}{M_\mathrm{s}} + 6\frac{h_\mathrm{f}^2}{h_\mathrm{s}^2}\frac{M_\mathrm{f}}{M_\mathrm{s}} + 4\frac{h_\mathrm{f}^3}{h_\mathrm{s}^3}\frac{M_\mathrm{f}}{M_\mathrm{s}} + \frac{h_\mathrm{f}^4}{h_\mathrm{s}^4}\frac{M_\mathrm{f}^2}{M_\mathrm{s}^2}\right]^{-1}. \quad (2.20)$$

The factors

$$\kappa_{\mathrm{St}} = \frac{6\epsilon_\mathrm{m}}{h_\mathrm{s}}\frac{h_\mathrm{f}}{h_\mathrm{s}}\frac{M_\mathrm{f}}{M_\mathrm{s}} \quad \text{and} \quad \epsilon_{0,\mathrm{St}} = -\epsilon_\mathrm{m}\frac{h_\mathrm{f}}{h_\mathrm{s}}\frac{M_\mathrm{f}}{M_\mathrm{s}} \quad (2.21)$$

appearing on the left sides of (2.19) and (2.20) are the curvature and midplane extensional strain for this configuration according to (2.6) and (2.7). Consequently, any departure from a value of unity on the right side of (2.19) or (2.20) represents the influence of film thickness on the ratio $\kappa/\kappa_{\mathrm{St}}$ or $\epsilon_0/\epsilon_{0,\mathrm{St}}$. The curvature implied by the Stoney formula is a limiting case of (2.19), that is, $\kappa/\kappa_{\mathrm{St}} \to 1$ in the limit as $h_\mathrm{f}/h_\mathrm{s} \to 0$, and similarly for the extensional strain. Once the midplane curvature κ and the midplane extensional strain ϵ_0 are known, the components of elastic strain in the substrate can be determined according to $\epsilon_{rr} = \epsilon_{\theta\theta} = \epsilon_0 - z\kappa$; the strain components in the film are given by the same expression with the value of ϵ_m added to account for elastic mismatch.

The influence of film thickness or, more precisely, the influence of the ratio $h_\mathrm{f}/h_\mathrm{s}$ on substrate curvature, is considered here in two ways. First, the leading two terms in a series expansion of curvature κ in powers of $h_\mathrm{f}/h_\mathrm{s}$ for any value of $M_\mathrm{f}/M_\mathrm{s}$ are identified. The result is

$$\kappa \approx \frac{6\epsilon_\mathrm{m}}{h_\mathrm{s}}\frac{M_\mathrm{f}}{M_\mathrm{s}}\frac{h_\mathrm{f}}{h_\mathrm{s}}\left[1 + \left(\frac{M_\mathrm{s} - 4M_\mathrm{f}}{M_\mathrm{s}}\right)\frac{h_\mathrm{f}}{h_\mathrm{s}}\right], \quad (2.22)$$

which provides some indication of the range of validity of the Stoney formula. For example, if $M_\mathrm{f} = M_\mathrm{s}$, the second 'correction' term within the square brackets is $-3h_\mathrm{f}/h_\mathrm{s}$. Thus, if $h_\mathrm{f}/h_\mathrm{s} = 0.05$, the error inferred in using the Stoney formula for curvature in this case is 15%. However, this way of examining the expression for curvature does not reflect the influence of the stiffness ratio.

A second way to represent the influence of film thickness on substrate curvature is to establish the range of both parameters $M_\mathrm{f}/M_\mathrm{s}$ and $h_\mathrm{f}/h_\mathrm{s}$ for which the error incurred in using the Stoney formula is less than some prescribed value, say 10%. For example, the range of these parameters for which

$$\left|\frac{\kappa h_\mathrm{s}}{6\epsilon_\mathrm{m}}\frac{h_\mathrm{s}}{h_\mathrm{f}}\frac{M_\mathrm{s}}{M_\mathrm{f}} - 1\right| \le \frac{1}{10} \quad (2.23)$$

can be calculated from (2.19); the result is illustrated in Figure 2.4. From this graph, it is clear that the limitation on the use of the Stoney formula depends on film stiffness as well as thickness. If the film material is more (less) stiff than

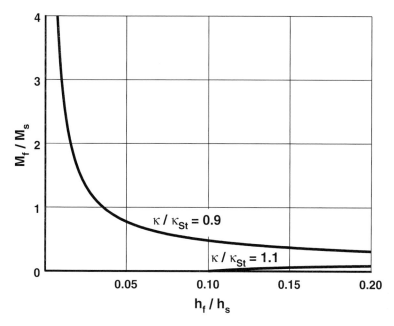

Fig. 2.4. With reference to the expression (2.19) for substrate curvature for any values of h_f/h_s and M_f/M_s, the error in the Stoney formula for a system corresponding to parameter values in the region bounded by the two curves and the coordinate axes is less than 10%.

the substrate material, then the restriction on film thickness becomes more (less) severe.

The equilibrium elastic strain in both the film and the substrate is independent of r; the edge effects being ignored will be taken up in Chapter 4. The strain components vary linearly with z, with a discontinuity of magnitude ϵ_m in both ϵ_{rr} and $\epsilon_{\theta\theta}$ across the interface between the film and substrate. The variation of ϵ_{rr}/ϵ_m with z/h_s is shown in Figure 2.5 for three ratios of film thickness to substrate thickness in h_f/h_s for the case when $M_f/M_s = 1$; recall that $\epsilon_{\theta\theta} = \epsilon_{rr}$ in the range of small deformation. Perhaps the most significant feature of this figure is that the strain in the film is already very different from the mismatch strain for the case when $h_f/h_s = 0.1$. This observation highlights the restricted range of validity of one of the assumptions that underlies the Stoney formula. The results presented in Figure 2.5 make it clear that the elastic strain in the film can differ significantly from the mismatch parameter ϵ_m unless the film is indeed very thin compared to the substrate.

To probe the influence of modulus ratio, consider the sensitivity of κ to M_f/M_s for the case when $h_f = h_s$. Then, from (2.19), the curvature is given by

$$\kappa = \frac{12\epsilon_m}{h_s}\frac{M_f}{M_s}\left[1 + 14\frac{M_f}{M_s} + \frac{M_f^2}{M_s^2}\right]^{-1}. \qquad (2.24)$$

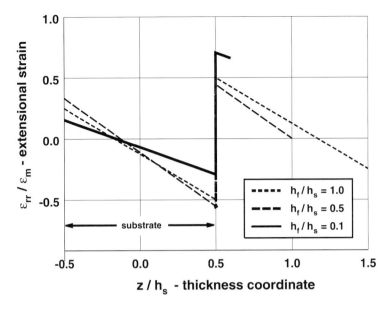

Fig. 2.5. The distribution of normalized strain ϵ_{rr}/ϵ_m versus normalized distance z/h_s across the thickness of a substrate–film system for three values of the ratio h_f/h_s. The neutral plane of the substrate is located by the value of z at which $\epsilon_{rr}/\epsilon_m = 0$. The material properties are such that $M_f/M_s = 1$.

For $M_f/M_s = 1$, this reduces to

$$\kappa = \frac{3}{4}\frac{\epsilon_m}{h_s},\tag{2.25}$$

whereas, for either $M_f/M_s = \frac{1}{2}$ or 2, it reduces to

$$\kappa = \frac{24}{33}\frac{\epsilon_m}{h_s}.\tag{2.26}$$

Thus, the difference in curvature between (2.25) and (2.26) is only 3 percent for a twofold increase or decrease in the modulus ratio.

The foregoing result suggests that κ passes through a local maximum as the modulus ratio varies from $\frac{1}{2}$ to 2 for the thickness ratio $h_f/h_s = 1$ and ϵ_m held fixed. To pursue this matter further, consider curvature κ for fixed ϵ_m and h_s as a function of thickness ratio h_f/h_s and modulus ratio M_f/M_s. To find the maximum curvature for fixed h_f/h_s under variations in M_f/M_s, set the partial derivative $\partial\kappa/\partial(M_f/M_s) = 0$ for the general result in (2.19), which is satisfied for

$$\frac{M_s^2}{M_f^2} = \frac{h_f}{h_s}.\tag{2.27}$$

Substituting this result into (2.19), the maximum value of curvature is found to be

$$\kappa_{max} = \epsilon_m \frac{3}{(h_f + h_s)} \quad \text{for fixed} \quad \frac{h_f}{h_s}. \tag{2.28}$$

On the other hand, to find the maximum curvature for fixed M_f/M_s under variations in h_f/h_s, with all other parameters being held fixed as before, the requirement that $\partial \kappa / \partial(h_f/h_s) = 0$ implies that

$$\frac{h_f^2}{h_s^2} \left(3 + 2\frac{h_f}{h_s} \right) = \frac{M_s}{M_f}. \tag{2.29}$$

The corresponding curvature is

$$\kappa_{max} = \epsilon_m \frac{3h_s + 2h_f}{(h_f + h_s)^2} \quad \text{for fixed} \quad \frac{M_f}{M_s}. \tag{2.30}$$

The expressions (2.27) and (2.29) have no intersections in parameter space, which implies that there is no *absolute* maximum in curvature over the full range of M_f/M_s and h_f/h_s.

2.2.2 *Example: Maximum thermal stress in a bilayer*

Consider a film–substrate bilayer system of circular geometry, where the film and the substrate have the same thicknesses and biaxial moduli: $h_f/h_s = 1$ and $M_f/M_s = 1$, and the bilayer diameter, $d \gg (h_s + h_f)$. Let the mismatch strain ϵ_m in this case be a consequence of a temperature change from an initial, stress-free temperature T_o to another temperature T, and let the thermal expansion coefficients of the film and the substrate be denoted by α_f and α_s, respectively. (a) Determine the variation of the radial stress σ_{rr} and circumferential stress $\sigma_{\theta\theta}$ across the thickness of the film and the substrate. (b) Find the magnitude and sign of the thermal mismatch stress as the interface is approached from the film and from the substrate. Show that the magnitude of the stress at the interface is independent of the thickness of the film or the substrate for a fixed thickness ratio.

Solution:

(a) Following the discussion immediately preceding (2.17), the variation of the radial stress σ_{rr} through the thickness of the film–substrate system is written as

$$\sigma_{rr}(z) = \begin{cases} M_s (\epsilon_o - \kappa z) & \text{for } -\frac{1}{2}h_s < z < \frac{1}{2}h_s, \\ M_f (\epsilon_o - \kappa z + \epsilon_m) & \text{for } \frac{1}{2}h_s < z < \frac{1}{2}h_s + h_f. \end{cases} \tag{2.31}$$

Since the in-plane dimension of the film–substrate system is much larger than its total thickness, an equi-biaxial state of stress can be assumed in the system (ignoring edge effects). Therefore, $\sigma_{\theta\theta}(z) = \sigma_{rr}(z)$.

(b) From (2.14), note that $\epsilon_m = (\alpha_s - \alpha_f) \cdot (T - T_0)$. From (2.19)–(2.21), it is seen that $\epsilon_0 = -\epsilon_m/8$ and $\kappa = 3\epsilon_m/(4h_s)$. At the interface between the film and the substrate at $z = h_s/2$, the stress values for $M_f = M_s = M$ are

$$\sigma_{rr} = \begin{cases} -\frac{1}{2}M\,(\alpha_s - \alpha_f) \cdot (T - T_0) & \text{for } z = \frac{1}{2}h_s \text{ in the substrate,} \\[2mm] \frac{1}{2}M\,(\alpha_s - \alpha_f) \cdot (T - T_0) & \text{for } z = \frac{1}{2}h_s \text{ in the film.} \end{cases} \quad (2.32)$$

If $(\alpha_s - \alpha_f) < 0$ and $(T - T_0) > 0$, the normal stress components in the lateral direction, σ_{rr} and $\sigma_{\theta\theta}$, are both compressive as the interface is approached from the film side, and they are both tensile as the interface is approached from the substrate. Thus, similar to the trends shown in Figure 2.5 for the strain components, the magnitude of the stress discontinuity at the interface can be estimated for the film–substrate system; the elastic stresses vary linearly with z in each layer. Equation (2.32) reveals that when the thickness ratio h_f/h_s is fixed, the peak stresses occur at the interface, and that these peak stresses are independent of the individual thickness of the film or the substrate; they depend only on the thickness ratio. When the biaxial moduli of the two materials are the same, the jump in stress across the interface is $M(\alpha_s - \alpha_f)(T - T_0) = M\epsilon_m$.

2.2.3 Historical note on thermostatic bimetals

A thermostatic bimetal is a layered component made by bonding plates of two metal alloys with differing coefficients of thermal expansion (CTE). The two alloy layers, usually of comparable thicknesses, are typically joined together by brazing or welding. In addition to their differing values of thermal expansion coefficient, the alloys are chosen so as to minimize deformation hysteresis during thermal excursions and to have adequate strength. When the bimetal thermostat is subjected to a temperature change, it undergoes a change of shape which, in turn, is exploited for the accurate control of temperature or some other function.

The first documented application of a thermostatic bimetal was in balance wheel compensators dating back to 1775 (Eskin and Fritze 1940). A United States patent (No. 24896) was granted to Wilson (1858) for his invention of a thermostatic bimetal comprising brass and steel layers. Since these alloys do not have a high thermal expansion mismatch, the shape change or bending produced by the bimetal during thermal excursion was not sufficiently large for many practical applications. With the development of a special class of Ni–Fe alloys, known as 'Invar', by Guillaume (1897), thermostatic bimetals became the focus of an increasing number of industrial applications for automatic use of regulating and indicating devices intended to facilitate enhanced control accuracy, energy savings and operational efficiency. Herrman (1920) published a detailed report of the thermal properties of bimetal strips based on his work conducted for the General Electric Company. Timoshenko (1925) presented a comprehensive analysis of the thermoelastic deformation and curvature evolution in bimetallic strips as functions of the bilayer geometry and mismatch in properties. By the 1930s, bimetals in the shapes of spirals, helices, disks as well as flat and U-shaped strips had found a wide range of industrial, automotive, aviation and marine applications.

Thermostatic bimetals were primarily designed to produce a deflection or rotation when subjected to a temperature change. In many applications, such as thermometers, the chief function of the bimetal was that of moving an indicator dial or pointer. In this case, the

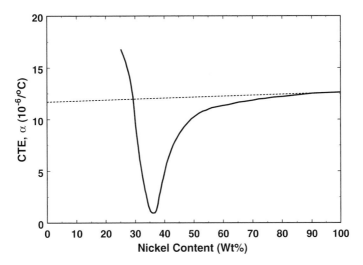

Fig. 2.6. The coefficient of linear thermal expansion α at room temperature of Ni–Fe alloys containing 0.4% Mn and 0.1% C. The dotted line denotes the value of α predicted by a linear rule of mixtures. (Adapted from Eskin and Fritze (1940).)

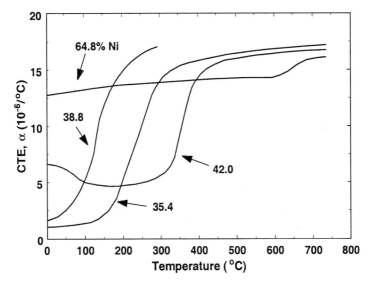

Fig. 2.7. The variation of thermal expansion coefficient with temperature in several Ni-containing steels. (Adapted from Eskin and Fritze (1940).)

only work to be performed was that needed to overcome the friction of the bearings of the pointer, and the deflection constant was the principal mechanical property which entered into design. There were also many other applications where the bilayer strip was used to perform considerable work such as lifting components or obtaining snap actions by overcoming stored elastic or magnetic energy in a system. In such situations, the force constant of the bilayer strip was an important design input.

The thermostatic bimetal consists of a low expansivity alloy bonded to a high expansivity alloy. In his investigation of Ni–Fe alloys, Guillaume (1897) discovered that the addition of 35.4 wt% Ni to Fe (typically containing 0.4% Mn and 0.1% C) led to a very low value of thermal expansion coefficient, as shown in Figure 2.6. Since the dimensions of this alloy were nearly invariant during ordinary fluctuations in atmospheric temperature, it was given the name 'Invar'. It is seen from Figure 2.6 that the thermal expansion coefficient of pure Fe is $11.9 \times 10^{-6} \, °\text{C}^{-1}$, while that of pure Ni is $13 \times 10^{-6} \, °\text{C}^{-1}$. The dashed line connecting these two limiting values denotes the theoretical thermal expansion coefficient values obtained from a simple law of mixture. As seen by the trend exhibited by the solid line, however, the true thermal expansion coefficient values deviate significantly from those predicted by the rule of mixture.

Figure 2.7 shows the variation of thermal expansion coefficient in Ni–Fe alloys as a function of temperature for different Ni concentrations. Invar, with a concentration of 35.4% Ni in Fe, has the lowest thermal expansion coefficient in the temperature range 0–100 °C. Beyond 100 °C, however, its thermal expansion coefficient increases precipitously, making it unsuitable for high temperature applications. The alloy with 42% Ni in Fe maintains its thermal expansion coefficient at a nearly fixed value up to a temperature of about 300 °C, although its thermal expansion coefficient is much higher at room temperature than that of Invar. Thus Invar found widespread use as a low expansivity alloy for a thermostatic bimetal for low temperature applications, whereas the 42% Ni alloy became the choice as a low expansivity metal for a variety of applications involving medium to high temperatures. For the high expansivity layer of the bimetal, brass became the common choice for use in conjunction with Invar. Subsequent developments led to the adoption of Monel (for example, commercial alloy Monel 400 with composition: Ni (63.0% min.), Cu (28.0–34.0 %), Fe (2.5% max), Mn (2.0% max), Si (0.5% max), C (0.3% max), and S (0.024% max)), and Ni–Cr–Fe alloys (typically with 18–27% Ni, 3–11% Cr and the balance, Fe) as high expansivity alloys in bimetal strips for medium to high temperature regimes.

2.3 Methods for curvature measurement

The change in substrate curvature induced during film deposition or temperature excursion provides valuable insight into the evolution of mismatch stress in the thin film. As noted earlier, a particularly appealing feature of curvature measurement is that extraction of the membrane force f from substrate curvature by recourse to the Stoney formula (2.7) does not involve the material properties of the film, provided that the film is sufficiently thin compared to the substrate. Substrate curvature measurements also provide a means to assess the functional properties of thin films in photonic and microelectronic applications. For example, strain in films can modify the electronic transport characteristics of layered semiconductor systems through modification of the band structure of the material (Singh 1993).

Methods to measure changes in substrate curvature during stress evolution in a layered material can be broadly classified into the following groups: mechanical methods, capacitance methods, x-ray diffraction methods, and optical methods. All these techniques, with the exception of x-ray diffraction, have the common feature that they provide a measure of the out-of-plane deflection of the curved film–substrate system.

The most common mechanical method of estimating curvature involves a stylus which scans the surface along the radial direction by making physical contact with it. The resulting record of the out-of-plane displacement $w(r)$ versus distance r from some reference point is converted to the radius of curvature ρ, or curvature κ, using the relation

$$\kappa = \frac{1}{\rho} = \frac{d^2 w(r)}{dr^2}. \tag{2.33}$$

Since the original substrate is not generally flat, it is essential to measure the substrate radii of curvature ρ_1 and ρ_2 before and after film deposition, respectively. Use of the Stoney formula (2.7) provides the average film stress in terms of the measured values of the radii of spherical curvature of the substrate as

$$\sigma_m = \frac{M_s h_s^2}{6 h_f} \left\{ \frac{1}{\rho_2} - \frac{1}{\rho_1} \right\} \tag{2.34}$$

provided that the film is sufficiently thin compared to the substrate.

The capacitance method involves a non-contact probe which records the changes in capacitance between a reference point in the probe and the surface it scans as discussed by Martinez et al. 1990, for example. The radial variation of capacitance change is then converted to $w(r)$, from which the curvature is estimated using (2.33). Although the capacitance and mechanical stylus techniques can provide accurate measures of curvature, they are not particularly well suited to monitor curvature changes at elevated temperature or during fluctuations in temperature. They are also generally unsuitable for curvature measurements in constrained space, such as inside a deposition chamber or a high temperature furnace. Since the geometry and mounting requirements for the specimen can be restrictive, these methods are not widely used for *in-situ* probing of curvature evolution during film deposition or passivation.

The x-ray diffraction method, which will be discussed in more detail in Section 3.6, can also be used to estimate the curvature of a crystalline substrate by successively determining the Bragg angle for the maximum diffraction intensity at different radial locations on the substrate surface. For example, consider the geometrical center on the surface of a circular substrate of (100) Si. Using x-ray diffraction, the exact Bragg angle at which a peak occurs in diffraction intensity for the particular crystallographic texture of the substrate is determined by oscillating the beam through the Bragg angle. The beam is then translated radially to a new location on the substrate, and the new angular position of maximum intensity is determined. The radius of curvature of the substrate is then the radial distance between the two locations on the surface at which the diffraction measurements are made

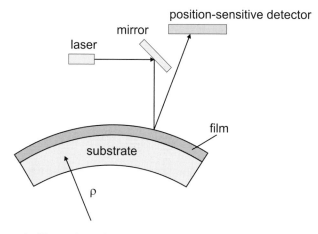

Fig. 2.8. Schematic illustration of the scanning laser method for measuring substrate curvature.

divided by the difference in the angular position of the intensity maxima (Blech and Meieran 1967, Van Mellaert and Schwuttke 1972, Blech and Cohen 1982). While this method can be fully computerized to provide accurate measures of substrate curvature in crystalline materials, it is not commonly used because of the cost associated with the x-ray diffraction setup and the safety precautions needed for the use of x-rays.

Many of the limitations of the above techniques are overcome by optical methods which offer the convenience, accuracy and flexibility to measure curvature through remote sensing capabilities. In this section, several different optical techniques for the measurement of substrate curvature are described, and their advantages and limitations are examined.

2.3.1 Scanning laser method

The optical method for measuring substrate curvature is generally convenient for *in-situ* measurement of film deposition stress in CVD and MBE systems, provided that optical access is provided to the substrate and that the specimen is mounted in a manner that facilitates unconstrained curvature evolution in one direction. One of the most common optical methods of estimating thin film stress, particularly for films deposited on Si wafers used in microelectronics, involves the reflection of a laser beam from the substrate; this technique is often simply referred to as the wafer curvature measurement method, and it has been discussed by Retajczyk and Sinha (1980), Pan and Blech (1984), Flinn et al. (1987), Volkert (1991), and Shull and Spaepen (1996), among others. The substrate is fixed at one point such that its position and orientation there are known. Then, a laser beam incident on the substrate surface is scanned along a straight line, and the angular deflection 2Θ of the reflected beam from the incident is measured as a function of distance r from a reference point. The scanning mirror arrangement, schematically sketched in Figure 2.8, utilizes a mirror and laser to scan a single laser beam along the specimen surface. The deflection of the scanned beam is monitored through a position-sensitive detector. The curvature is related to the angle of reflection of the beam by the relation

$$\kappa = \frac{1}{\rho} = \frac{d^2 w(r)}{dr^2} = \frac{d\Theta(r)}{dr}. \tag{2.35}$$

A linear regression analysis of the Θ versus r data then provides an estimate of κ from which the mean stress can be determined using the Stoney formula for a sufficiently thin film. The effect of initial curvature of the substrate is factored into the stress calculation by comparing the curvatures before and after film deposition or thermal cycling, as indicated in (2.34). Commercial scanning laser systems typically are capable of measuring radii of curvature as large as several km.

The time involved in scanning a large area of a substrate could also limit the use of this method as a real-time probe for curvature evolution in some applications. In addition, a major limitation of this technique is the sensitivity of this serial scanning method to mechanical vibrations, such as those commonly encountered in ultra-high vacuum chambers. This limitation could be overcome by the introduction of a beam splitter in the path of the laser to introduce multiple incident beams on the specimen, as described by Schell-Sorokin and Tromp (1990) and Martinez et al. (1990). The reflections of these beams could then be independently monitored by a CCD camera.

2.3.2 Multi-beam optical stress sensor

This method, schematically shown in Figure 2.9, is an optical technique which exploits the advantages of a multi-beam output and which obviates the need for a multitude of position-sensitive detectors for vibration isolation. In this approach (Floro and Chason 1996), an incident beam from a HeNe laser is directed through a spatial filter with a focusing objective lens and an *etalon* which generates multiple parallel beams that are targeted at the specimen whose curvature is to be determined. The beams reflected from the curved specimen are directly captured by a CCD camera, thus circumventing the need for position-sensitive detectors. Since the CCD camera could completely image the reflected spots, these spots are easily monitored on a computer screen for focusing the objective lens. The curvature of a line in the specimen can be determined from the relative deflection of the adjacent beams reflected from points along that line. This constitutes a distinct advantage over optical measurements that involve a beam splitter where only a single beam

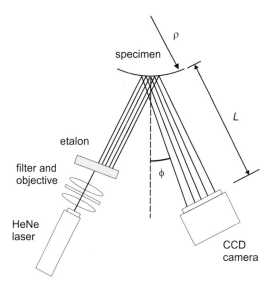

Fig. 2.9. Schematic of the setup of a multi-beam optical stress sensor.

spacing is captured. Information from a frame grabber can then be numerically analyzed to define and capture the pixels around each beam. A two-dimensional grid of reflected spots, which provides the mapping of the surface of the curved substrate, can also be extracted using this method by introducing a second etalon which is oriented orthogonally to the first.

If ϕ is the angle of reflection of the laser beams, L is the distance between the specimen and the CCD camera, D is the average spacing between adjacent reflected beams, and D_0 is the average initial spacing, the substrate curvature is calculated from the expression

$$\kappa = \frac{1}{\rho} = \frac{\cos\phi}{2L}\left\{1 - \left(\frac{D}{D_0}\right)\right\}. \tag{2.36}$$

A typical resolution of 15 km is achieved for the radius of curvature ρ for an array consisting of four or five parallel laser beams. From the substrate curvature, the film stress is calculated using the Stoney formula as a function of film thickness, thereby providing a complete history of intrinsic stress evolution during the deposition of the film on a substrate. For a typical setup, the deposition of a 1 nm thick $Si_{1-x}Ge_x$ film with a stress of 50 MPa (i.e., a membrane force of 50 MPa nm) on a 100 μm thick Si substrate is routinely detected (Floro et al. 1997).

If the reflected spots are merely translated because of vibration or rigid-body motion of the specimen (as, for example, during film deposition or heating), re-alignment of the camera is not necessary because the CCD array has a large active area. Since all the laser spots move in unison during vibration, such external noise

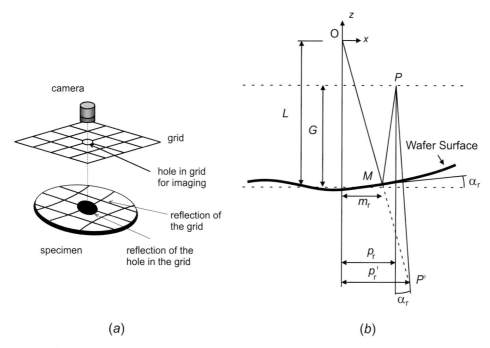

(a) *(b)*

Fig. 2.10. The grid reflection method. (a) Overall setup. (b) Schematic and definition of key dimensions.

does not introduce any change in the measured curvature. The setup shown in Figure 2.9 provides a compact apparatus which is given optical access to CVD and MBE systems through standard viewports.

A limitation of the serial and parallel laser scanning methods and the multi-beam optical stress sensor is that they employ monochromatic beams, which would lead to difficulties in measuring the curvature of transparent films with thicknesses of one-fourth the wavelength. These optical methods also require reflective surfaces for curvature determination.

2.3.3 *Grid reflection method*

The grid reflection method is a simple optical method which provides a full-field reflection of a grid or grating from the surface of a curved substrate (Finot et al. 1997). The working principle of this method is schematically sketched in Figure 2.10. A flat surface containing a periodic pattern or grating is placed at a distance G from the reflective surface of a wafer or a substrate with deposited film, with the grid facing the surface of the wafer. The grid is appropriately illuminated so as to achieve the best image quality on the wafer. A CCD camera images the reflection of the grid on the substrate through a hole in the grid plane, as shown in Figure 2.10(a).

As the substrate curves under the influence of the film stress, the reflected grid pattern captured by the camera becomes distorted. Analysis of continuous changes in the grid pattern during film deposition or thermal cycling provides *in-situ* and real-time records of radial variations in curvature over the entire surface of the substrate.

Let the camera be located at a distance L from the reflective surface of the substrate, Figure 2.10(b). Consider a point P on the plane of the grid whose image is located at P'. A straight line linking the point O, which is the point at which the reference coordinate axes and the camera are located, with P' intersects the curved substrate surface at M. Provided that the out-of-plane deflection of the curved substrate is much smaller than both G and L, the orientation of the substrate at point M is given by the angle

$$\alpha_r = \frac{p_r' - p_r}{2G}, \tag{2.37}$$

where p_r' and p_r are the radial distances of the points P' and P, respectively, from the center of the substrate. It is readily shown from simple geometrical considerations that

$$\alpha_r = \frac{\psi m_r - p_r}{2G}, \qquad \psi = \frac{L + G}{L}, \tag{2.38}$$

where m_r is the radial distance of point M from the center. For a flat substrate, for which $\alpha_x = 0$, let $m_r = m_{rF}$. With this definition, (2.38) is rewritten as

$$\alpha_r = \frac{\psi}{2G} (m_r - m_{rF}). \tag{2.39}$$

In other words, the orientation angle of the surface at point M for the curved substrate is proportional to the deviation of the grid point image from that reflected from a flat mirror. The substrate curvature is determined to be

$$\kappa = \frac{\Delta \alpha_r}{\Delta m_r} = \frac{\psi}{2G} \left\{ 1 - \frac{\Delta m_{rF}}{\Delta m_r} \right\}, \tag{2.40}$$

by considering small variations Δm_r and Δm_{rF} in m_r and m_{rF}, respectively. The accuracy of the curvature values determined by the grid reflection method is strongly influenced by the distances L and G, and the sensitivity with which optical images of grid distortion are digitally converted to small variations in grid point locations, which is essentially determined by the resolution of the digital camera.

The scanning laser method or the multi-beam optical stress sensor are convenient for curvature measurement during film deposition and thermal cycling when the out-of-plane deflection of the specimen is much smaller than the substrate thickness, that is, when deformation is in the geometrically linear range. On the other hand,

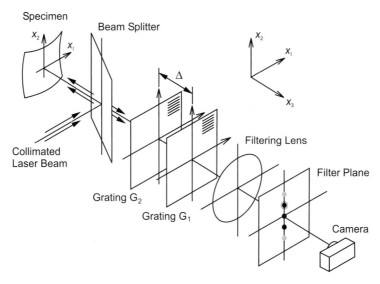

Fig. 2.11. Schematic illustration of the experimental setup of CGS in the reflection mode. (After Rosakis et al. (1998).)

the grid reflection method does not require the use of a monochromatic optical beam; it is found to be a particularly useful and simple method for measuring full-field, nonuniform curvature evolution over large wafers and flat panels where the out-of-plane deflection is greater than the substrate thickness. To compensate for errors introduced by large deflections, the solution is iterated until a self-consistent deflection value is obtained. Variations in the design of this method also include the use of an array of light-emitting diodes (LEDs) to provide grid points instead of the grating (Giannakopoulos et al. 2001). An example of the use of the grid reflection method for curvature determination is included in Section 2.6.

2.3.4 Coherent gradient sensor method

The coherent gradient sensor (CGS) method is a full-field interferometric technique that produces fringe patterns by laterally shearing an incident wavefront. This method, developed by Rosakis et al. (1998) for curvature measurement in film–substrate systems, is amenable for use in a variety of experimental configurations in either a reflection or a transmission arrangement.

In a typical reflection experimental setup, shown in Figure 2.11, a collimated laser beam illuminates the region of interest on the specimen which must be specularly reflective. In a transmission configuration, the beam would be transmitted through the specimen instead. After penetrating through or reflecting off the specimen, the

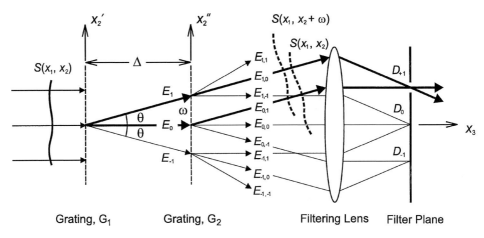

Fig. 2.12. Schematic illustration of the optical path in the CGS method. After Rosakis et al. (1998).

collimated laser light traverses through two parallel sets of identical, high-density gratings, marked G_1 and G_2, that are separated by a distance Δ.

Figure 2.12 is a two-dimensional schematic illustrating the principles of the CGS method. The figure shows an optical wavefront, $S(x_1, x_2)$, incident on the two gratings in which the lines are taken to be oriented along x_1. At the first grating G_1, the incident wavefront is diffracted into several wavefronts, E_0, E_1, E_{-1}, E_2, E_{-2}, etc., of which only the first three are drawn in Figure 2.12. Each of these wavefronts, in turn, is diffracted by the second grating to generate additional wavefronts, such as $E_{0,1}$, $E_{1,0}$, $E_{0,-1}$, etc. The diffracted beams are combined by a filtering lens to produce diffraction spots, such as D_{+1}, D_0, D_{-1}, etc., in the focal plane of the lens. One of the diffraction spots, typically the first diffraction order, that is, the D_{+1} spot, is chosen for imaging in a camera with an aperture onto the film plane.

The presence of the two gratings in the path of the optical wavefront generates a lateral shift or shearing or 'optical differentiation' of the front. For example, the diffracted beam $E_{1,0}$, denoted as $S(x_1, x_2 + \omega)$, is shifted from the beam $E_{0,1}$, denoted as $S(x_1, x_2)$, by a distance ω in the x_2 direction. If $\theta = \arcsin(\lambda/p)$ is the diffraction angle, where λ is the wavelength of light and p is the pitch of the gratings, the shift distance is $\omega = \Delta \tan\theta$. For small angles of diffraction, $\omega \approx \Delta \cdot \theta \approx (\Delta \cdot \lambda)/p$.

The condition for constructive interference of the original and shifted wavefronts is given by

$$S(x_1, x_2 + \omega) - S(x_1, x_2) = n^{(2)}\lambda, \qquad n^{(2)} = 0, \pm1, \pm2, \dots K, \qquad (2.41)$$

where $n^{(2)}$ is an integer that represents fringes associated with shearing along the x_2 direction, and K is a positive integer. Dividing (2.41) by ω, and taking ω to be sufficiently small, it is seen that

$$\frac{\partial S\,(x_1,\,x_2)}{\partial x_2} = \frac{n^{(2)}\lambda}{\omega} = \frac{n^{(2)}p}{\Delta}, \qquad n^{(2)} = 0,\ \pm 1,\ \pm 2, \ldots K. \qquad (2.42)$$

This equation can be generalized to represent the shearing of the wavefront in either the x_1 or the x_2 direction:

$$\frac{\partial S\,(x_1,\,x_2)}{\partial x_\alpha} = \frac{n^{(\alpha)}p}{\Delta}, \qquad n^{(\alpha)} = 0,\ \pm 1,\ \pm 2, \ldots K, \qquad (2.43)$$

where $n^{(\alpha)}$ denotes the fringes generated as a result of shearing along the x_α direction, with $\alpha \in \{1,\ 2\}$. The interferograms formed in the CGS method are governed by this set of equations.

The relation between the optical wavefront and the reflector surface topography can be found by expressing the specularly reflective, curved surface as

$$x_3 = f\,(x_1,\,x_2). \qquad (2.44)$$

Using this relation, it can be shown that the wavefront $S\,(x_1,\,x_2)$ can be reasonably well approximated by $2f\,(x_1,\,x_2)$. Consequently, the alternating dark and bright interference fringes correspond to constant values of components of the in-plane gradient of $f(x_1,\,x_2)$ as follows:

$$\frac{\partial f\,(x_1,\,x_2)}{\partial x_\alpha} \approx \frac{n^{(\alpha)}p}{2\Delta}, \qquad n^{(\alpha)} = 0,\ \pm 1,\ \pm 2, \ldots K. \qquad (2.45)$$

This set of equations thus relates the CGS fringe patterns to the in-plane gradients of the specimen surface described by the functional form of (2.44).

In the present context of thin films on substrates, the quantity of interest is the local curvature. The curvature tensor field, $\kappa_{\alpha\beta}$, can be determined directly from the CGS patterns recorded in reflection by differentiating the fringes of the in-plane gradient:

$$\kappa_{\alpha\beta}\,(x_1,\,x_2) \approx \frac{\partial^2 f\,(x_1,\,x_2)}{\partial x_\alpha\,\partial x_\beta} \approx \frac{p}{2\Delta}\left\{\frac{\partial n^{(\alpha)}\,(x_1,\,x_2)}{\partial x_\beta}\right\}, \qquad (2.46)$$

where $\alpha,\ \beta = 1, 2$. In order to determine the full curvature tensor, it is clear that the gradient field in two orthogonal directions must be recorded. Equations (2.46) represent the governing equations that determine the instantaneous curvature tensor field at any in-plane location $(x_1,\,x_2)$ and they enable the full-field measurement of curvature for the film–substrate system. The stress tensor can be calculated from the experimentally determined curvature field with appropriate assumptions regarding, for example, the deformation state. For the simple case of a continuous thin film on a substrate, the Stoney formula (2.7) provides a direct link between the substrate curvature and the film stress. Similar connections are derived in Section 3.9 for the more complex situation involving patterned thin stripes on substrates.

A system based on the CGS method offers several advantages for curvature measurements for thin films and layered solids. The measurement provides all the normal and shear components of the curvature tensor. It also provides full-field information from the entire area of the substrate–film system. The measurement area could also be scaled as necessary from a few millimeters to hundreds of millimeters so that large wafers and flat panels with thin film deposits are tested. The method involves non-contact measurements which are carried out with an adjustable working distance, and performed *in-situ* and in real time as, for example, during thermal cycling, or film deposition, or passivation. The measurement sensitivity is adjustable by varying the grating separation. Since the CGS technique measures gradients of displacement rather than displacements, no special provisions are necessary to isolate the system from vibration. The method, however, requires a detailed computer software package, such as that involving Fourier optics, to provide an automatic and digital record of fringe geometry and spacing from the optical images so that the curvature tensor could be readily extracted. Examples of the use of the CGS method for determining substrate curvature will be presented in Section 2.6.

2.4 Layered and compositionally graded films

In some engineering applications involving thin films and multilayers, diffusion of atomic species occurring during processing and/or service often leads to local spatial variations in composition and microstructure. Such variations could have a marked influence on the evolution of stress, damage and cracking as well as on the overall mechanical integrity of the layered structure, as described by Suresh and Mortensen (1998), Kaysser (1999), and Suresh (2001), for example.

Compositional gradients are also purposely introduced in a variety of engineering situations with the specific objective of enhancing the mechanical performance by controlling stress, inelastic deformation or damage at specific sites such as surfaces, corners and interfaces. Examples include tribological protection of gear teeth surfaces through carburizing and nitriding processes which lead to controlled gradations in carbon and nitrogen content, respectively, beneath the surface, and ion implantation used in microelectronic devices and components. Gradations in composition and microstructure in one or more dimensions can be induced in a controlled manner by recourse to a variety of deposition and synthesis methods such as PVD, CVD, MBE, thermal spray, sintering, three-dimensional printing and combustion synthesis.

The motivation for the introduction of compositional gradients in thin films and multilayers arises through efforts to achieve various technical objectives.

— Eliminating abrupt transitions in thermal, elastic and plastic mismatch across interfaces between dissimilar solids through the introduction of continuous or step-wise gradients in composition can, in some cases, result in marked reductions in thermal and residual stress at critical crack nucleation sites such as interfaces and sharp corners where interfaces intersect free surfaces. The onset of consequent inelastic deformation and damage can also be suppressed or delayed by such graded interfaces. It is, however, important to note that beneficial effects of grading which promote enhanced mechanical performance at some locations can also lead to a more detrimental response at some other location of the same layered structure. Discussion of methods to compute stress in graded multilayers is taken up in the next section.

— The corners at which sharp interfaces between dissimilar materials intersect free surfaces are prime sites of crack nucleation where singular stress fields, similar to those at crack tips, develop. The mechanics of stress evolution and cracking at such corners will be described in Chapter 4. An insightful discussion on the use of fine scale layering to influence stress levels near a geometrical stress concentration is given by Budiansky et al. (1993). Blurring of the interface by grading the composition may, in some situations, mitigate stress concentrations at such corners (Erdogan 1995).

— Applications requiring surface protection against thermal degradation or wear necessarily require the deposition of a 'thick' coating, with a surface layer thickness of 1 mm or higher, on a substrate. The deposition of 'thick' ceramic coatings on metallic substrates for thermal protection in the piston head of diesel engines used in ground vehicles or marine applications is a case in point. In such situations, strong adherence between the coating and the substrate is achieved through the introduction of step-wise or continuous gradients in composition across the interface (Beardsley 1997).

— In optoelectronic devices, control of misfit dislocation nucleation and of the kinetics of threading dislocations is essential for the optimal design of light-emitting diodes and quantum wells. Strategies for manipulating dislocation densities and mobility in such applications will be discussed in Section 7.3.2. Purposely introducing compositionally graded buffer layers between semiconductor substrates and quantum wells, by recourse to CVD or MBE processing methods, is a practical method for enhancing the performance of optoelectronic devices.

— Spatial gradients in elastic and/or plastic properties at contact surfaces are known to have a marked effect on the resistance of the surface to normal indentation and sliding contact damage (Suresh 2001). Analysis and experiment show that appropriate choices of gradients in composition beneath the surface could redistribute contact stress in such a way that the onset of Hertzian cone cracks during normal indentation or herringbone cracks during frictional sliding could be suppressed (Giannakopoulos and Suresh 1997, Jitcharoen et al. 1998, Suresh et al. 1999).

In this section, the effect on substrate curvature of the variation of mismatch strain and material properties through the thickness of layered films is analyzed. The derivation of the Stoney formula (2.7) in Section 2.1 refers only to the resultant membrane force in the film; any through-the-thickness variation of mismatch strain in the film is considered only peripherally. Film thickness was taken into account explicitly in Section 2.2, but it was assumed there that mismatch strain and elastic properties of the material were uniform throughout the film. However, there are situations of practical significance for which this is not the case. Two of the most common cases are compositionally graded films in which the mismatch strain and the elastic properties vary continuously through the thickness of the film, and multilayered films for which the mismatch strain and the elastic properties are discontinuous, but piecewise constant, from layer to layer throughout the thickness of the film. In both cases, the mismatch strain and the material properties are assumed to be uniform in the plane of the interface. With reference to the cylindrical r, θ, z-coordinate system introduced in Section 2.1, the mismatch strain and film properties are now assumed to vary with z for fixed r and θ, but both are invariant with respect to r and θ, for fixed z.

2.4.1 Nonuniform mismatch strain and elastic properties

All features of the film–substrate system introduced in Section 2.2 are retained. In addition, it is assumed that the film material carries a mismatch strain in the form of an isotropic equi-biaxial extensional strain $\epsilon_m(z)$ parallel to the interface. This strain may depend on distance from the interface in an arbitrary way, as suggested by writing the mismatch as a function of z; this function need not be continuous in z. Similarly, the biaxial elastic modulus of the film $M_f(z)$ may vary through the thickness of the film.

The system is depicted in Figure 2.13 which shows the undeformed substrate. An externally applied traction acts to maintain the mismatch $\epsilon_m(z)$ in the state shown. The magnitude of this traction at a given z is $\sigma_m(z) = M_f(z)\epsilon_m(z)$, and its effect is to render the film compatible with respect to the stress-free substrate. Upon release of this artificial externally applied traction, the substrate takes on a curvature which is to be estimated. This is first pursued on the basis of the principle of minimum

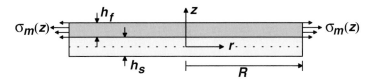

Fig. 2.13. An elastic mismatch strain, which varies in an arbitrary way through the thickness of the film, is maintained by an externally applied traction. Release of this traction induces curvature in the substrate.

potential energy, followed by a discussion of an equilibrium approach leading to the formula relating curvature and mismatch strain.

The strain energy density of the system can be written just as in (2.17), if it is understood that ϵ_m and M_f are now functions of z. To make the results useful for a broader class of systems, it is convenient to adopt a slightly modified notation. First, suppose that the definition of mismatch strain is modified so that $\epsilon_m(z) \equiv 0$ for $-\frac{1}{2}h_s \leq z < \frac{1}{2}h_s$. In addition, suppose that the biaxial modulus throughout the system is denoted by

$$
M(z) =
\begin{cases}
M_s & \text{for } -\frac{1}{2}h_s \leq z < \frac{1}{2}h_s \\[2ex]
M_f(z) & \text{for } \frac{1}{2}h_s \leq z \leq \frac{1}{2}h_s + h_f.
\end{cases}
\tag{2.47}
$$

Then, the strain energy density can be written as

$$
U(r, z) = M(z)\{\epsilon_o - z\kappa + \epsilon_m(z)\}^2
\tag{2.48}
$$

throughout the system. The total potential energy is then

$$
V(\epsilon_o, \kappa) = 2\pi \int_0^R \int_{-\frac{1}{2}h_s}^{\frac{1}{2}h_s + h_f} U(r, z)\, r\, dz\, dr
\tag{2.49}
$$

$$
= \pi R^2 \left[\epsilon_o C_{0,0} - 2\epsilon_o\kappa C_{1,0} + \kappa^2 C_{2,0} - 2\kappa C_{1,1} + C_{0,2} + 2\epsilon_o C_{0,1} \right],
$$

where

$$
C_{m,n} = \int_{-\frac{1}{2}h_s}^{\frac{1}{2}h_s + h_f} z^m M(z)\epsilon_m(z)^n\, dz.
\tag{2.50}
$$

For any particular system, the quantity $C_{m,n}$ is constant for any choice of integer values for m and n.

If the total potential energy is minimized with respect to variations in ϵ_o and κ, it is found that, at the minimum value,

$$
\kappa = \frac{C_{1,1}C_{0,0} - C_{0,1}C_{1,0}}{C_{2,0}C_{0,0} - C_{1,0}^2}, \qquad \epsilon_o = \frac{C_{1,1}C_{1,0} - C_{0,1}C_{2,0}}{C_{2,0}C_{0,0} - C_{1,0}^2}.
\tag{2.51}
$$

While these results appear to be simple in form, they are not sufficiently transparent to permit easy interpretation for the general case. Thus, several particular cases are pursued in detail. First, the result for curvature obtained in Section 2.2 is recovered. To do so, it is noted that if $M_f(z)$ and $\epsilon_m(z)$ have the constant values M_f and ϵ_m throughout the film, the constants in (2.51) are

$$C_{0,0} = h_f M_f + h_s M_s, \qquad C_{1,0} = \tfrac{1}{2} a_f M_f (h_f + h_s)$$
$$C_{0,1} = \epsilon_m h_f M_f, \qquad C_{1,1} = \tfrac{1}{2} \epsilon_m h_f M_f (h_f + h_s) \qquad (2.52)$$
$$C_{2,0} = \tfrac{1}{12} \left(4 h_f^3 M_f + 6 h_f^2 h_s M_f + 3 h_f h_s^2 M_f + h_s^3 M_s \right),$$

and the result (2.19) is recovered.

Energy minimization is a powerful tool for establishing global deformation characteristics of a strained film–substrate system on the basis of a particular set of assumptions and approximations, as demonstrated above. Essentially, the same assumptions and approximations can be used as a basis for a local equilibrium approach to arrive at the same results. Indeed, by exploiting the features of translational invariance and symmetry, it is revealed that some of the 'assumptions' on which the energy analysis is based are known *a priori* to be true. These include the vanishing of the normal stress component σ_{zz}, the shear stress component $\sigma_{r\theta}$ and the shear strain components ϵ_{rz} and $\epsilon_{\theta z}$. This approach was adopted by Freund (1993) and the main features are sketched here for future reference.

Consider an observation point in the midplane of the substrate. For the conditions adopted, the system is invariant under translation of the observation point in any direction along the midplane so long as edge effects are ignored. It follows immediately that all components of stress are independent of position in the midplane. Furthermore, both the top surface at $z = h_f + h_s/2$ and the bottom surface at $z = -h_s/2$ are free of traction, so that all shear components of stress vanish identically, and the normal stress in the z-direction is zero everywhere. Furthermore, the normal components of stress in the lateral direction are identical for isotropic material response, that is,

$$\sigma_{rr}(z) = \sigma_{\theta\theta}(z) \qquad (2.53)$$

in cylindrical coordinates. The distribution $\sigma_{rr}(z)$ is subject to restrictions imposed by overall equilibrium. The absence of any externally applied loads on the system implies that the net force and net moment on any cross-section due to internal stress must vanish. In view of the translational invariance noted above, this condition requires that

$$\int_{-h_s/2}^{h_f+h_s/2} \sigma_{rr}(z)\,dz = 0, \qquad \int_{-h_s/2}^{h_f+h_s/2} z\sigma_{rr}(z)\,dz = 0. \qquad (2.54)$$

These conditions do not depend on the particular choice of reference plane $z = 0$.

The materials involved are isotropic and linearly elastic; they are also laterally homogeneous but the properties can vary with z in the thickness direction. Thus, the extensional strain in any lateral direction is related to the corresponding stress through the local value of bilateral elastic modulus $M(z)$. In particular, for the radial direction,

$$\sigma_{rr}(z) = M(z)\epsilon_{rr}(z), \qquad (2.55)$$

where $M(z)$ is a piecewise continuous function of z.

The total strain in the system is not compatible, which is the feature that gives rise to residual stress. The incompatibility is readily handled explicitly, however, so that a compatible difference strain can be defined. Without loss of generality, the midplane of the substrate is chosen as the reference plane for definition of the mismatch distribution $\epsilon_m(z)$. Any strain of the material in either the film or the substrate with respect to the reference plane at the reference temperature is included in $\epsilon_m(z)$. The equi-biaxial difference strain $\epsilon_{dif}(z)$ is simply

$$\epsilon_{dif}(z) = \frac{\sigma(z)}{M(z)} - \epsilon_m(z). \qquad (2.56)$$

When $\sigma(z) = \sigma_m(z)$, the difference strain is zero (but the system is not in equilibrium, in general).

The strain components for any deformation are subject to the conditions of strain compatibility which assure that the strain field can be derived from a physically realizable displacement field. In the present case, the only compatibility condition which is not satisfied identically is $\epsilon_{dif}''(z) = 0$. This condition follows directly from the symmetry and translational invariance characteristics of the system and is independent of material response. If the notation introduced above is preserved here, whereby κ is the spherical curvature of the substrate midplane and z_{np} is the location of the plane for which the difference strain is zero with respect to the reference plane, then

$$\epsilon_{dif}(z) = -\kappa(z - z_{np}). \qquad (2.57)$$

It then follows from (2.56) that

$$\sigma_{rr}(z) = M(z)\left[\epsilon_m(z) - \kappa(z - z_{np})\right]. \qquad (2.58)$$

The values of the two parameters κ and z_{np} then follow from imposition of the equilibrium conditions (2.54). The results for curvature κ and strain of the substrate midplane ϵ_o are again given by (2.51). The parameters ϵ_o and z_{np} are related through $\kappa z_{np} = \epsilon_o$. The equilibrium approach and the energy approach provide precisely the same results as long as the approaches are based on the same systems with the same physical features.

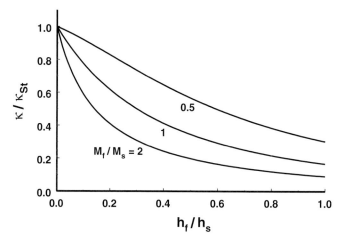

Fig. 2.14. Substrate curvature due to a mismatch strain which increases linearly through the film thickness for three values of stiffness ratio. The curvature is normalized by the value κ_{St} based on the Stoney formula for this case.

2.4.2 Constant gradient in mismatch strain

Next, the case of a film with a constant gradient in mismatch strain is considered, that is, the case of a film for which

$$\epsilon_m(z) = \frac{\epsilon_{top}}{h_f} \left(z - \tfrac{1}{2}h_s\right), \quad \tfrac{1}{2}h_s \le z \le \tfrac{1}{2}h_s + h_f \tag{2.59}$$

where the mismatch strain increases linearly from zero at the film–substrate interface to ϵ_{top} at the film surface. The curvature κ of the midplane of the substrate is compared to the curvature κ_{St} based on the Stoney formula (2.7). The Stoney formula is based on the *total stress resultant f* in the film due to the mismatch, discounting any relaxation of strain in the film due to deformation of the substrate. Thus, in this case, this force is

$$f = \int_{\frac{1}{2}h_s}^{\frac{1}{2}h_s + h_f} \frac{\epsilon_{top}}{h_f} M_f(z) \left(z - \tfrac{1}{2}h_s\right) dz. \tag{2.60}$$

For the present example, it will be assumed that $M_f(z) = M_f$, a constant, throughout the film. In this case, $f = \tfrac{1}{2}\epsilon_{top}M_f h_f$.

If the curvature given by (2.51) is evaluated for the linearly varying mismatch strain (2.59) and constant film stiffness M_f, then

$$\frac{\kappa}{\kappa_{St}} = \frac{3 + 4\eta + m\eta^2}{3(1 + 4m\eta + 6m\eta^2 + 4m\eta^3 + m^2\eta^4)} \tag{2.61}$$

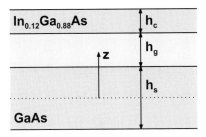

Fig. 2.15. A compositionally graded InGaAs layer sandwiched between homogeneous layers of GaAs and $In_{0.12}Ga_{0.88}As$.

where $m = M_f/M_s$ and $\eta = h_f/h_s$. This result is illustrated in Figure 2.14 for three values of the parameter m. It is evident from the figure that the normalized curvature is strongly influenced by both the modulus ratio and the thickness ratio.

2.4.3 Example: Stress in compositionally graded films

Consider a ternary epitaxial structure in which a linearly graded layer provides a transition between a uniform substrate and a uniform capping layer of composition different from that of the substrate. In this example, the substrate is a gallium arsenide layer of thickness h_s and biaxial modulus M_s, and the capping layer is an indium–gallium arsenide film of thickness h_c and biaxial modulus M_c. Twelve percent of the gallium sites in GaAs have been replaced by indium to form the capping layer compound, which is indicated by writing $In_{0.12}Ga_{0.88}As$. In the intermediate layer of thickness h_g, the indium content of the material varies linearly from zero at the interface with the substrate to 12 percent replacement of gallium sites in the compound at the interface with the capping layer. All materials are assumed to be isotropic elastic materials for simplicity, and both the mismatch strain and elastic modulus are assumed to vary linearly within the graded layer. The stress-free lattice parameters of GaAs and InAs at room temperature are 5.653 Å and 6.058 Å, respectively. The extensional elastic moduli along any cubic axis are 85.3 MPa and 51.4 MPa, and the contraction ratios (essentially, Poisson ratios) along orthogonal cubic axes are 0.31 and 0.35, respectively. Find the distribution of biaxial stress characterized by $\sigma_{rr}(z) = \sigma_{\theta\theta}(z)$ as a function of position z through the thickness of the tri-layer for $h_c = 0.1h_s$ and for the three values of thickness of the graded layer $h_g = 0, 0.5h_s, h_s$.

Solution:

The biaxial modulus of the substrate is $M_s = E_s/(1 - \nu_s) = 85.3/0.69 = 123$ MPa. The biaxial modulus of InAs is $51.4/0.65 = 79.1$ MPa. By linear interpolation, the modulus of the capping layer is estimated to be $M_c \approx (0.88 \times 123) + (0.12 \times 79.1) = 118$ MPa. The variation of modulus through the thickness of the structure is then

$$M(z) = \begin{cases} M_s, & -\tfrac{1}{2}h_s \leq z < \tfrac{1}{2}h_s \\ M_s\left(\tfrac{1}{2}h_s + h_g - z\right)/h_g \\ \quad + M_c\left(z - \tfrac{1}{2}h_s\right)/h_g, & \tfrac{1}{2}h_s \leq z < \tfrac{1}{2}h_s + h_g \\ M_c, & \tfrac{1}{2}h_s + h_g \leq z < \tfrac{1}{2}h_s + h_g + h_c. \end{cases} \qquad (2.62)$$

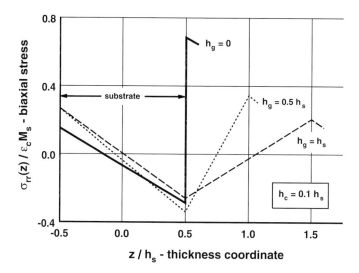

Fig. 2.16. Stress distribution across the tri-layer depicted in Figure 2.15 for three values of graded layer thickness $h_g = 0$, $0.5h_s$, h_s. The capping layer thickness is $h_c = 0.1h_s$ for all cases. Adapted from Freund (1993).

The stress-free mean lattice parameter in the capping layer is estimated to be $(0.88 \times 5.653\,\text{Å}) + (0.12 \times 6.058\,\text{Å}) = 5.702\,\text{Å}$. The uniform mismatch strain in the capping layer, with respect to the substrate, is denoted by ϵ_c and is estimated to be

$$\epsilon_c \approx \frac{5.653 - 5.702}{5.702} = -0.00859. \tag{2.63}$$

Thus, by linear interpolation, the variation of mismatch strain through the thickness of the structure is

$$\epsilon_m(z) = \begin{cases} 0, & -\tfrac{1}{2}h_s \leq z < \tfrac{1}{2}h_s \\ \epsilon_c \left(z - \tfrac{1}{2}h_s\right)/h_g, & \tfrac{1}{2}h_s \leq z < \tfrac{1}{2}h_s + h_g \\ \epsilon_c, & \tfrac{1}{2}h_s + h_g \leq z < \tfrac{1}{2}h_s + h_g + h_c. \end{cases} \tag{2.64}$$

With the variation of modulus and mismatch strain known explicitly in (2.62) and (2.64), respectively, the values of $C_{m,n}$ which are required for determining deformation according to (2.51) can be determined from (2.50); these are

$$C_{0,0} = \frac{h_s M_s}{2}\left[2 + \frac{M_c}{M_s}\frac{h_g(2h_c + h_g)}{h_s^2}\right] \tag{2.65}$$

$$C_{0,1} = \frac{\epsilon_c h_s M_s}{6}\left[\frac{h_g}{h_s} + 2\frac{M_c}{M_s}\frac{3h_c + h_g}{h_s}\right]$$

$$C_{1,0} = \frac{h_s^2 M_s}{12}\left[\frac{h_g(3h_s + 2h_g)}{h_s^2} + \frac{M_c}{M_s}\frac{6h_c(h_c + h_s) + h_g(12h_c + 3h_s + 4h_g)}{h_s^2}\right]$$

$$C_{1,1} = \frac{\epsilon_c h_s^2 M_s}{12}\left[\frac{h_g(h_s + h_g)}{h_s^2} + \frac{M_c}{M_s}\frac{6h_c(h_c + h_s) + h_g(12h_c + 2h_s + 3h_g)}{h_s^2}\right]$$

$$C_{2,2} = \frac{h_s^3 M_s}{24} \left[2 + 3\frac{h_g}{h_s} + 4\frac{h_g^2}{h_s^2} + 2\frac{h_g^3}{h_s^3} + \frac{M_c}{M_s} \times \right.$$
$$\left. \frac{8h_c^3 + 12h_c^2(h_s + 2h_g) + 6h_c(h_s + 2h_g)^2 + h_g(3h_s^2 + 8h_gh_s + 6h_g^2)}{h_s^3} \right].$$

The magnitude of the equi-biaxial strain for any value of z in the structure is

$$\epsilon_{rr}(z) = \epsilon_{\theta\theta}(z) = \epsilon_o - \kappa z + \epsilon_m(z) \tag{2.66}$$

and the biaxial stress follows from (2.55). Results are shown graphically in Figure 2.16 for $M_c/M_s = 1$ in the form of $\sigma_{rr}(z)/\epsilon_c M_s$ versus z/h_s.

2.4.4 Periodic multilayer film

Consider a periodic multilayered film consisting of N_λ periods, each of thickness λ, bonded to a substrate of thickness h_s; the overall film thickness is then $h_f = \lambda N_\lambda$. A period consists of a single layer of a material with modulus M_a, thickness Δh_a and mismatch strain ϵ_{ma} with respect to the substrate plus another single layer of material with modulus M_b, thickness $\Delta h_b = \lambda - \Delta h_a$ and mismatch strain ϵ_{mb} with respect to the substrate. The quantity $C_{m,n}$, defined in (2.50), can then be evaluated in terms of the parameters $N_\lambda, h_s, M_s, \Delta h_a, M_a, \Delta h_b, M_b, \epsilon_{ma}$ and ϵ_{mb}. The resulting curvature will again be normalized with respect to the approximate form based on the Stoney formula (2.7). For this case, the unrelaxed stress resultant f appearing in the Stoney formula is

$$f = N_\lambda (f_a + f_b) = N_\lambda (\Delta h_a \epsilon_{ma} M_a + \Delta h_b \epsilon_{mb} M_b). \tag{2.67}$$

For purposes of illustration, suppose that the substrate is also composed of material 'b' and that all layers have the same thickness, so that $\Delta h_a = \Delta h_b = \frac{1}{2}\lambda$, $M_b = M_s$ and $\epsilon_{mb} = 0$. In terms of the remaining parameters $\lambda, h_f, h_s, M_s, M_a$, and ϵ_{ma}, evaluation of $C_{m,n}$ in (2.50) leads to the values

$$C_{0,0} = h_s M_s + \frac{1}{2}h_f(M_s + M_a), \quad C_{0,1} = \frac{1}{2}h_f \epsilon_{ma} M_a$$

$$C_{1,0} = \frac{1}{8}h_f [2(h_s + 2h_f)(M_s + M_a) + \lambda(M_s - M_a)]$$

$$C_{1,1} = \frac{1}{8}\epsilon_{ma} h_f M_f [2h_s + 2h_f - \lambda] \tag{2.68}$$

$$C_{2,0} = \frac{1}{12}h_s^3 M_s + \frac{1}{8}h_f h_s [h_s(M_s + M_a) + \lambda(M_s - M_a)].$$

The curvature is then given by (2.51). Also, for this case,

$$\kappa_{St} = \frac{3\epsilon_{ma}}{h_s}\frac{h_f}{h_s}\frac{M_a}{M_s}. \tag{2.69}$$

It can be confirmed that, if the number of periods in the multilayer is very large, the expression for curvature is consistent with the result obtained for a homogeneous layer in (2.19). This can be demonstrated by letting $\lambda/h_s \to 0$, identifying the film modulus M_f in (2.19) with $\frac{1}{2}(M_s + M_a)$, and identifying the mismatch strain ϵ_m in (2.19) with $\frac{1}{2}\epsilon_{ma}$.

2.4.5 Example: Overall thermoelastic response of a multilayer

Copper is an important material for electronic applications because of its electrical properties. To compensate for its relatively low yield strength σ_y and moderate coefficient of thermal expansion α it is used in some applications in multilayer form along with molybdenum, which has a relatively high yield strength and low coefficient of thermal expansion. The geometrical arrangement is usually in the form of a Cu/Mo/Cu tri-layer with thickness ratios depending on the application. The Cu/Mo/Cu is sometimes bonded to a ceramic, such as Al_2O_3, to provide a ceramic chip carrier assembly for use in avionics and telecommunication systems. The Al_2O_3 ceramic is brittle and it fails under a tensile stress σ_{TS} as low as 140 MPa. The requisite material properties for Cu, Mo and Al_2O_3 are given in the accompanying table.

The basic multilayer geometry under consideration is shown schematically in Figure 2.17. The Poisson ratio of the metallic layers is 0.33 and that of Al_2O_3 is 0.27. Suppose that the thickness of each Cu layer is $h_{Cu} = 95.3\ \mu$m, and that of the Al_2O_3 layer is $h_{Al_2O_3} = 0.762$ mm. (a) What should be the thickness h_{Mo} of each Mo layer so that any extensional strain which tends to develop along the interface in the Al_2O_3 layer during any temperature change ΔT is compensated by an equal strain in the multilayer film? (b) Assuming that the entire multilayer is stress-free at room temperature, what is the stress state in each layer when the laminate is subjected to a temperature increase of $40\,^{\circ}$C?

Material	E (GPa)	$\alpha\ (10^{-6}\,^{\circ}C^{-1})$	σ_y (MPa)	σ_{TS} (MPa)
Cu	117	16.7	140	–
Mo	324	4.8	690	–
Al_2O_3	300	6.9	–	140

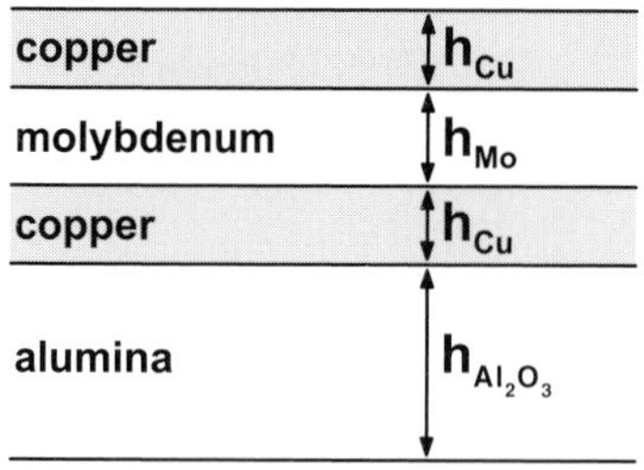

Fig. 2.17. A schematic of the multilayer film comprising Cu and Mo on an alumina substrate.

Solution:

(a) Note that the layered Cu–Mo film is symmetric under reflection in its midplane. If it is not constrained by the substrate during temperature change, there would be no gradients in stress across the thickness in any layer and no tendency for the film to change its curvature. To render the Al_2O_3 layer stress-free for any temperature change ΔT, the resultant biaxial force in the film must vanish. This will be the case if $2h_{Cu}\sigma_{Cu} + h_{Mo}\sigma_{Mo} = 0$ for any ΔT, or

$$2h_{Cu}M_{Cu}\left(\alpha_{Al_2O_3} - \alpha_{Cu}\right) + h_{Mo}M_{Mo}\left(\alpha_{Al_2O_3} - \alpha_{Mo}\right) = 0, \qquad (2.70)$$

where a common factor ΔT has been canceled from each term. Solving (2.70) for h_{Mo} yields

$$h_{Mo} = \frac{2\,h_{Cu}M_{Cu}\left(\alpha_{Cu} - \alpha_{Al_2O_3}\right)}{M_{Mo}\left(\alpha_{Al_2O_3} - \alpha_{Mo}\right)} = 321.2\,\mu m, \qquad (2.71)$$

which is the thickness necessary to keep the Al_2O_3 substrate free of stress. In effect, h_{Mo} has been chosen to make the effective coefficient of thermal expansion of the multilayer equal to that of the substrate. It is clear from (2.71) that this can be achieved only if the coefficient of thermal expansion of the Al_2O_3 substrate is intermediate between the coefficients of thermal expansion of the two materials comprising the film.

(b) For $\Delta T = 40\,°C$, the strain common to all layers is $\alpha_{Al_2O_3}\,\Delta T = 2.76 \times 10^{-4}$. The biaxial stress in the Cu and Mo layers, respectively, are then

$$\sigma_{Cu} = M_{Cu}\left(2.76 \times 10^{-4} - \alpha_{Cu}\,\Delta T\right) = -68.4\,\text{MPa}$$

$$\sigma_{Mo} = M_{Mo}\left(2.76 \times 10^{-4} - \alpha_{Mo}\,\Delta T\right) = 40.6\,\text{MPa}. \qquad (2.72)$$

The layer geometry is such that there is no stress in the Al_2O_3 layer.

2.4.6 Multilayer film with small total thickness

As another example of the result obtained in Section 2.4, consider the case of a thin film of total thickness $h_f \ll h_s$ which consists of a number N_f of separate layers deposited on a substrate of thickness h_s and modulus M_s. Suppose that the i-th layer has thickness Δh_i, modulus $M_{f,i}$ and mismatch strain $\epsilon_{m,i}$ with respect to the substrate, where i ranges from 1 to N_f. No assumptions on periodicity of the layering or on other relationships among the properties of the individual layers are presumed.

If the expression for curvature κ given in (2.51) is evaluated for this case, and if only those terms of first order in $\Delta h_i/h_s$ are retained, then it is found that

$$\kappa = \frac{6}{M_s h_s^2}\sum_{i=1}^{N_f}\Delta f_i = \sum_{i=1}^{N_f}\Delta\kappa_{St,i} \qquad (2.73)$$

where $\Delta f_i = M_{f,i}\,\epsilon_{m,i}\,\Delta h_i$ is the membrane force within the i-th layer. Thus, to first order in the small parameters $\Delta h_i/h_s$, the total curvature is equal to the simple sum of the curvatures that would be induced if each individual layer would be deposited *by itself* on the substrate. Each individual curvature $\Delta\kappa_{St,i}$ is given by the Stoney formula (2.7). The average stress in the multilayered film is

$$\sigma_{ave} = \frac{1}{h_f}\sum_{i=1}^{N_f}\Delta f_i, \qquad h_f = \sum_{i=1}^{N_f}\Delta h_i. \qquad (2.74)$$

In the foregoing development, it is tacitly assumed that the individual layers in the film are added sequentially and that the mismatch strain in each layer depends only on the substrate but not on the order in which the layers are formed. If this is the case, then (2.73) yields the cumulative curvature after N_f layers have been added. However, it is noted that the same reasoning can be applied if the layers are *removed* sequentially. In the latter situation, the change in total substrate curvature due to removal of the last layer added is $-\Delta\kappa_{\text{St},N_f}$. This can be a useful viewpoint, for example, in a situation in which curvature change cannot be measured during deposition of the layers, but it can be measured as material is removed from the film surface by chemical or other means (Freund 1996).

2.4.7 Example: Stress in a thin multilayer film

A $0.614\,\mu m$ thick tetraethylorthosilane (TEOS) film is deposited on a relatively thick Si wafer. The biaxial stress in the TEOS film, say σ_{TEOS}, is estimated from wafer curvature measurements to be -114 MPa. A $0.6\,\mu m$ thick silicon nitride film is then deposited on the TEOS film. The average stress in the bilayer film (composite TEOS and silicon nitride layers) is estimated from curvature measurements to be -190 MPa. (a) If a $0.6\,\mu m$ thick silicon nitride film is then deposited on a Si wafer, estimate the mismatch stress σ_{SiN} in this film. (b) Suggest an experimental strategy for determining the stress in each film for a two-film stack deposited on a substrate, without relying on the superposition formula given in (2.74).

Solution:

(a) Using (2.74), σ_{SiN} is determined to be

$$\sigma_{\text{SiN}} = \frac{1}{h_{\text{SiN}}}\left[\sigma_{\text{ave}}\left(h_{\text{TEOS}} + h_{\text{SiN}}\right) - \sigma_{\text{TEOS}}h_{\text{TEOS}}\right] = -268 \text{ MPa}. \qquad (2.75)$$

(b) Consider a substrate on which a single layer thin film has been deposited. A second thin layer is then deposited. Assuming that the first deposited layer deforms elastically during deposition of the second layer, it is possible to measure the stress in the second film by measuring the substrate radius change before and after the second film deposition. The stress in the second film is then obtained simply by measuring the membrane force in the two-film stack and then subtracting the membrane force of the first film, as was done in (a). There are, however, many material systems and deposition conditions where the stress in the first film could change *irreversibly* due to the temperature cycle employed to deposit the second film. For this situation, the stress in the second film could not be extracted using the simple superposition method given by (2.74). Instead, the second film could be preferentially etched, using a wet chemical, without damaging the first film. The curvature change of the substrate is then measured before and after the removal of the second film by the wet chemical etch; the stress in the second film is then estimated from this curvature change. Once the stress in the second film is known, the stress in the first film could be calculated from knowledge of the average stress in the two-film stack using (2.74). One may also compare the radius of curvature of the substrate after the second film is etched away with the original radius of curvature of the bare substrate to calculate

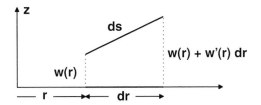

Fig. 2.18. Illustration of extensional strain in an infinitesimal radial line element in the substrate midplane due only to transverse displacement of the midplane.

the stress in the first film. If the two-film stack comprises a crystalline film and an amorphous film, the stress in the crystalline film could be measured using x-rays whereas that in the amorphous film could be estimated afterwards using (2.74).

2.5 Geometrically nonlinear deformation range

Because of their relatively low bending resistance (compared to extensional resistance at comparable strain) and relatively large lateral extent, plate-like solids are readily deformed into a regime where rotations of material line elements are no longer small compared to unity, even though strains remain small. In such cases, it becomes necessary to adopt nonlinear kinematics for a proper description of deformation. In the present context, this can be seen by considering the radial extensional strain induced at the substrate midplane due to transverse deflection alone. An elementary geometrical construction illustrated in Figure 2.18 shows this strain to be essentially $ds/dr - 1 \approx \frac{1}{2}\{w'(r)\}^2$. An instructive exercise is to compare the magnitude of this quantity to ϵ_o at $r = R$ for the circumstances in which the Stoney formula (2.7) is 'valid', that is, for $w'(R) = \kappa R$ and $\epsilon_o = \frac{1}{6}\kappa h_s$. In this case,

$$\frac{\{w'(R)\}^2}{2\epsilon_o} = 3\kappa R \frac{R}{h_s}, \tag{2.76}$$

which is not always small in magnitude. For example, for $h_s = 100\,\mu$m, $R = 1$ cm and radius of curvature of $\kappa^{-1} = 10$ m, this number is 0.3, a significant value compared to 1.

This effect arises because the initially flat substrate does not deform into a developable surface, that is, a surface for which at least one principal curvature is zero. As a result, the substrate cannot deform into a spherical cap shape without stretching or compressing portions of its midplane. The substrate is very stiff in extension compared to bending, as noted above, and this coupling between curvature and stretching tends to stiffen the system response in comparison to behavior in the linear range.

2.5.1 Limit to the linear range

The effect of midplane stretching is incorporated into the model by including the nonlinear term $\frac{1}{2}\{w'(r)\}^2$ in the expression for radial strain in (2.2), which then takes the form

$$\epsilon_{rr}(r, z) = u'(r) - zw''(r) + \frac{1}{2}\{w'(r)\}^2. \tag{2.77}$$

The origin of the additional term as the extensional strain due to transverse deflection is evident from the geometrical construction illustrated in Figure 2.18. It should be noted that the strains are still small and that Hooke's law is still appropriate as a description of material behavior. The main point is that the second-order contribution of rotation to strain can be as significant as the first-order linear effect. (In terms of the nonlinear strain tensor commonly known as Lagrange strain in material coordinates, the second-order contributions due to stretching are ignored but the second-order contributions due to rotation are retained; see Fung (1965), for example.)

In this case, the midplane displacement is assumed to be

$$u(r) = \epsilon_{\mathrm{o}}r + \epsilon_1 r^3 + \epsilon_{\mathrm{m}}r, \quad w(r) = \frac{1}{2}\kappa r^2, \tag{2.78}$$

where it is understood that $\epsilon_{\mathrm{m}} \equiv 0$ in the substrate. The term in $u(r)$ which is cubic in r contributes to ϵ_{rr} in the same way as $\{w'(r)\}^2$ and, in a sense, these two terms compete in minimizing the total potential energy. Also, once the nonlinear description is adopted, the lateral dimension R must also enter into an estimate of curvature. Whereas the kinematic 'assumption' in (2.2) turned out to be strictly true within the linear deformation range, as noted in Section 2.4, there is no basis for expecting (2.78) to be so. The point in adopting this form of deformation is that it offers an additional degree of freedom in the model to probe the issue of departure from linear behavior. There is, however, no *a priori* reason to expect (2.78) to provide an accurate picture of deformation into the range of nonlinear elastic response.

The requirement that the potential energy must be stationary under variations in κ, ϵ_{o} and ϵ_1 at equilibrium leads to a nonlinear relationship between curvature and mismatch strain which is too involved to warrant presentation here in its most general form. However, there are two useful special cases which can both be cast into the form

$$\bar{\epsilon}_{\mathrm{m}} = \bar{\kappa}\left[1 + (1 - \nu_{\mathrm{s}})\bar{\kappa}^2\right], \tag{2.79}$$

where $\bar{\epsilon}_{\mathrm{m}}$ is a normalized mismatch strain and $\bar{\kappa}$ is a normalized curvature (Freund 2000). For the case when the materials have the same elastic properties ($M_{\mathrm{f}} = M_{\mathrm{s}}$)

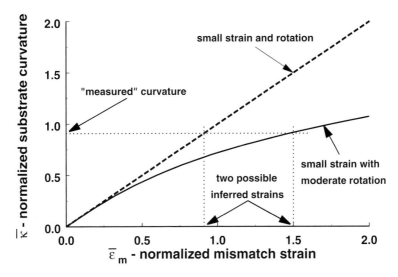

Fig. 2.19. Normalized curvature versus normalized mismatch strain according to the linear and nonlinear theories. The definitions of $\bar{\epsilon}_m$ and $\bar{\kappa}$ for two particularly useful cases are given in (2.80) and (2.81).

but h_f/h_s is arbitrary,

$$\bar{\epsilon}_m = \frac{3}{2}\epsilon_m \frac{R^2 h_s h_f}{(h_s + h_f)^4}, \quad \bar{\kappa} = \frac{R^2 \kappa}{4(h_s + h_f)}. \tag{2.80}$$

For the case when $h_f/h_s \ll 1$ but the material properties are arbitrary,

$$\bar{\epsilon}_m = \frac{3}{2}\epsilon_m \frac{R^2}{h_s^2} \frac{M_f}{M_s} \frac{h_f}{h_s}, \quad \bar{\kappa} = \frac{R^2 \kappa}{4 h_s}. \tag{2.81}$$

A graph of $\bar{\kappa}$ versus $\bar{\epsilon}_m$ applicable to either case is shown in Figure 2.19, where it is compared to the linear relationship between curvature and strain based on the same assumptions but ignoring nonlinear kinematics. The most important observation to be made on the basis of this calculation is that the relationship between curvature and mismatch strain begins to depart from linearity at about $\bar{\kappa} \approx 0.3$ or $\bar{\epsilon}_m \approx 0.3$. As an illustration of the significance of this departure, consider the value of mismatch strain to be inferred from a particular curvature observation. The two values of normalized strain inferred for a 'measured' curvature of $\bar{\kappa} \approx 0.9$, approximately $\bar{\epsilon}_m \approx 0.9$ for linear deformation and $\bar{\epsilon}_m \approx 1.5$ for nonlinear deformation, are indicated in the figure. The value $\bar{\epsilon}_m = 1.5$ of normalized strain corresponds, for example, to $\epsilon_m = 0.01$, $R/h_s = 100$, $h_f/h_s = 0.01$ and $M_f/M_s = 1$. The actual behavior of the system in the range of geometrically nonlinear deformation cannot be established on the basis of this simple model due to the unfounded assumption of spatially uniform curvature. The case of general nonlinear deformation without this restriction is considered next.

2.5.2 Axially symmetric deformation in the nonlinear range

The assumption that underlies the derivation of (2.79), that the radial curvature is uniform everywhere in the film–substrate system, is an essential feature of the deformation in the linear range. In the nonlinear deformation range, on the other hand, there is no basis for expecting the curvature to be uniform. The finite element method of numerical analysis can be used to determine the deformed shape of the substrate midplane in the nonlinear range without *a priori* assumptions on the distribution of curvature. The deformation is constrained to be consistent with the Kirchhoff hypothesis but it is otherwise general. In particular, transverse deflections which are large compared to the substrate thickness h_s are accommodated.

As an illustration, the following situation is considered. A thin film is bonded to a circular substrate for which $R/h_s = 100$. The film thickness is such that $h_s/h_f = 100$ and the film carries a spatially uniform mismatch strain ϵ_m with respect to the substrate. The elastic properties of the two materials are assumed to be identical.

The problem was analyzed in the following way. The approximate deflection was computed for values of mismatch strain ϵ_m that spanned a range of practical interest. An eighth-order polynomial in r was fit to the computed transverse deflection $w(r)$ for each value of ϵ_m, and the polynomial was differentiated to determine the radial curvature $\kappa(r) = w''(r)/(1 + \{w'(r)\}^2)^{3/2}$. The overall dependence of local normalized radial curvature $\bar{\kappa}(r) = R^2\kappa(r)/4h_s$ on mismatch strain and radial position was then represented by means of a contour plot of $\bar{\kappa}$ on the plane of r/R versus $\bar{\epsilon}_m$. The result is shown in Figure 2.20 for the case when $h_s/h_f = 100$ and $R/h_s = 100$ (Freund 2000). The approximation $\kappa(r) = w''(r)$ leads to results that are nearly indistinguishable from those shown.

With reference to Figure 2.20, the curvature is essentially constant over the entire substrate at a given level of mismatch strain as long as the level curves of curvature are parallel to the radial distance axis for $0 \le r/R \le 1$. This is indeed the case for normalized curvature varying between zero and a value of about 0.3. Because of the normalization convention for curvature and mismatch strain adopted here from (2.80), this implies that the corresponding normalized mismatch strain is also in the range $0 \le \bar{\epsilon}_m \le 0.3$. Thus, the implied limit on the range of linear response found here is consistent with the result represented in Figure 2.19.

As the normalized mismatch strain $\bar{\epsilon}_m$ is increased above the value 0.3, the curvature distribution becomes increasingly nonuniform. The general trend is that the curvature assumes values substantially *below* the average curvature for portions of the substrate near its center, and it takes on values substantially *above* the average

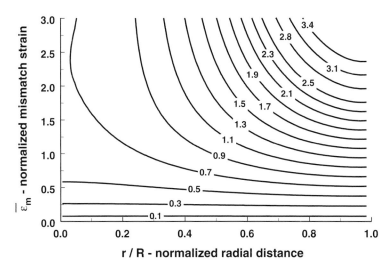

Fig. 2.20. Contour plot showing level curves of normalized curvature $\bar{\kappa}(r)$ for axially symmetric deformation in the plane with normalized distance r/R as coordinate on the horizontal axis and normalized mismatch strain $\bar{\epsilon}_m$ on the vertical axis for $\nu_f = \nu_s = 1/4$. Uniform curvature requires the level curves to be parallel to the horizontal axis.

value near the periphery of the substrate. For example, for a normalized mismatch strain of $\bar{\epsilon}_m = 2$, the normalized curvature varies from about 0.7 at the substrate center to a value of about 2.5 at the substrate edge. The result shown in Figure 2.20 is typical as long as $h_f/h_s \ll 1$, $h_s/R \ll 1$ and the elastic properties of the film and substrate are similar.

Substrate curvature in the geometrically nonlinear range was studied experimentally by Finot et al. (1997) and Lee et al. (2001). In both instances, the systems studied comprised metal films on Si substrates. In the former study, the substrates had diameters of 15 cm or 20 cm, and thicknesses of 675 μm or 730 μm, respectively. A thin SiO_2 layer was first deposited on a {100} face of the substrate, followed by a 0.5 μm thick layer of an Al alloy with 0.5 wt% Cu. Tungsten films of thickness 0.6 μm, 0.9 μm or 2.4 μm were deposited on the Al–Cu layer. To broaden the parameter range over which nonlinear deformation occurred, several substrates were then thinned by grinding the back side to final thicknesses between 325 μm and 415 μm. Substrate curvatures were observed by a standard laser scanning method (see Section 2.3.1) or, for cases in which the deformations were too large for use of such a method, by the grid reflection method (see Section 2.3.3). Lee et al. (2001) studied Si wafers with nominal diameters of 25 mm and nominal thicknesses of 105 μm. An aluminum film of thickness 5 μm was deposited onto a {100} surface of the substrate at a temperature of about 75 °C. A wide range of elastic behavior was probed by varying the temperature between room temperature and 75 °C, with the curvature measured experimentally by recourse to the coherent gradient sensor

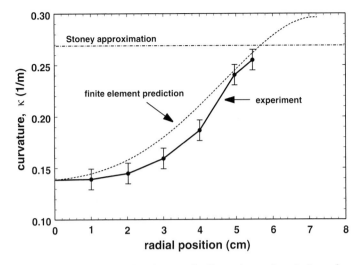

Fig. 2.21. Experimentally observed and numerically estimated variation of curvature as a function of radial position, measured from the center of a Si substrate with a W film deposit. After Finot et al. (1997).

method (see Section 2.3.4). In all cases observed in these two sets of experiments, behavior consistent with that represented in Figure 2.20 was reported.

Figure 2.21 shows an example from Finot et al. (1997) of the variation of radial curvature along a radial direction in a 15 cm diameter, $350\,\mu$m thick circular Si substrate on which the presence of a $0.9\,\mu$m thick continuous W film led to a membrane force $f \approx 10^3\,\mathrm{N\,m^{-1}}$. Also plotted in this figure is the result of a finite element simulation of the evolution of nonuniform curvature for comparison with the experimental observation. It is evident that the curvature $\kappa_x = \kappa_y = 0.14\,\mathrm{m^{-1}}$ at the center of the substrate is approximately one half of that predicted by the Stoney approximation; nonlinear deformation is not taken into account in the latter result.

2.6 Bifurcation in equilibrium shape

Consider a thin film in which a uniform mismatch strain is imposed, with the magnitude of the strain being increased from an initial value of zero. For relatively small values of the mismatch strain, the deformed shape of the substrate is essentially spherical as long as the response remains within the range of linear deformation. This is the range of behavior that was discussed in Section 2.2. Once ϵ_m becomes large enough in magnitude to bring the system into the range of geometrically nonlinear response, the deformed shape may continue to be axially symmetric, as described in Section 2.5. However, this deformation mode requires that the substrate must deform in extension as well as in bending, and the stiffness against such deformation is very large compared to bending stiffness at comparable levels of surface strain. In contrast, *cylindrical* bending, or generalized plane strain bending, can occur with only very limited midplane extension. This suggests that, as the magnitude of ϵ_m increases, the system may begin a transition, at some value of ϵ_m, from axially symmetric deformation, as the only possible equilibrium shape, toward cylindrical bending deformation as an alternate shape which is energetically favorable. Conditions at which departure from spherical curvature of the substrate gives way to other more complicated modes of deformation mark the onset of the geometrically nonlinear range of behavior. Such conditions were discussed quantitatively in the preceding section. In this range of behavior, the local deformation response of the system at any material point depends on the overall size and shape of the substrate. If the plan-view shape of the substrate is a regular geometric shape, such as a square, regular octagon or circle, the curvature distribution retains the symmetry of the substrate shape well into the nonlinear range. Eventually, a bifurcation occurs at which the highly symmetric deformation gives way to a more complex deformed shape. If the substrate does not have a shape that is highly regular, say a rectangle or a semi-circle, the curvature deformation departs from symmetry as soon as nonlinear effects become evident but no sharply defined bifurcation occurs (Finot and Suresh 1996). In this section, attention is focused mainly on the highly symmetric shapes for which a well-defined bifurcation state exists. Quantitative estimates are obtained for specifying this state in terms of system parameters. The same approaches can also be used to study curvature in film–substrate systems with less symmetric substrate shapes.

The influence of nonlinear response on equilibrium shape has important implications for the design and use of layered materials in a variety of structural and functional applications. Figure 2.22(a)–(c) show three examples of shapes in three distinctly different engineering situations. The evolution of a cylindrical shape at room temperature, upon cooling from the curing temperature, is shown in Figure 2.22(a) for an unsymmetric, laminated graphite–polyimide composite. The four-layer 0.46 m × 0.67 m polyimide matrix laminate has a graphite fiber reinforcement stacking sequence of [0/–45/90/45]. The driving force for this large deformation is the mismatch strain induced by differential thermal contraction between the layers upon slowly cooling from the curing temperature of approximately 170 °C. Figure 2.22(b) is an optical micrograph showing the development of cylindrical shape in an Al_2O_3–ZrO_2 tapecast ceramic bilayer which was cooled from a bonding temperature of approximately 1500 °C. The thermal mismatch as well as shrinkage mismatch between the two layers led to curling and cracking. Figure 2.22(c) is a micrograph revealing the post-bifurcation, non-spherical shape of a polysilicon disk which is connected at its center to a monocrystalline Si substrate in a MEMS device.

The issue of bifurcation in equilibrium shape is pursued in two steps. First, a simple energy approach is taken. The deformed shape of the substrate midplane is assumed to be ellipsoidal so that the shape is characterized completely by two principal curvatures which need not be the same. Consistent in-plane displacements,

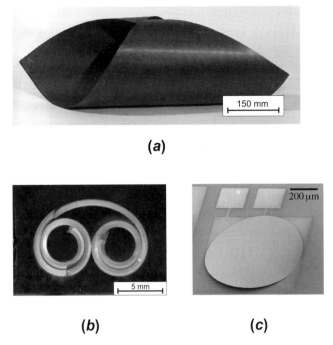

(a)

(b)   **(c)**

Fig. 2.22. (a) Cylindrical shape of an unsymmetric graphite–polyimide laminate at room temperature. (NASA photograph L-79-1771. Courtesy of National Aeronautics and Space Administration, Langley Research Center, Langley, VA. Reprinted with permission.) (b) Photograph showing curling and cracking in an alumina–zirconia bilayer ceramic. (Courtesy of D. Rowcliffe, The Royal Institute of Technology, Stockholm, Sweden. Reprinted with permission.) (c) Large deformation of a polysilicon disk attached at its center to a Si wafer. (Courtesy of H. Kahn, Case Western Reserve University, Cleveland, Ohio. Reprinted with permission.)

which account for midplane stretching, are also assumed. The principle of stationary potential energy is then invoked to determine the relationship of the principal curvatures to system parameters to ensure that the system is in equilibrium. A more detailed examination of bifurcation on the basis of a finite element simulation, without *a priori* restrictions on deformation beyond the Kirchhoff hypothesis, is described subsequently. The goal is to extract fairly simple analytical results which are merely representative of the phenomena of interest. As before, attention is restricted to systems for which the elastic response of both the film and substrate materials is isotropic and the moduli of the film and substrate are the same.

2.6.1 Bifurcation analysis with uniform curvature

As described in Section 2.1, the basic idea is to adopt a parametric family of deformed shapes for the substrate midplane, along with a consistent strain distribution which incorporates the mismatch strain, and then to determine values of the parameters which represent stationary points of the total potential energy. Even though the undeformed substrate is a circular disk in the present case, the deformed shape is not axially symmetric, in general, and rectangular coordinates are thus more convenient than polar coordinates as independent variables. The origin of coordinates is still in the midplane of the substrate and on the axis of rotational symmetry of the undeformed substrate. The x-axis and y-axis are taken to coincide with the orthogonal directions of principal curvature; circumstances for which this is not so are considered in Section 2.6.2. The transverse deflection of the substrate midplane is assumed to be

$$w(x, y) = \frac{1}{2}\left(\kappa_x x^2 + \kappa_y y^2\right) \tag{2.82}$$

where κ_x and κ_y are the (spatially uniform) principal curvatures of the deformed shape. The in-plane displacement components are assumed to be

$$u_x(x, y) = a_1 x + a_2 \frac{x^3}{3R^2} + a_3 \frac{xy^2}{R^2} + \epsilon_m x$$

$$u_y(x, y) = b_1 y + b_2 \frac{y^3}{3R^2} + b_3 \frac{x^2 y}{R^2} + \epsilon_m y, \tag{2.83}$$

where $a_1, a_2, \ldots, b_3$ are six additional dimensionless parameters characterizing the deformation. Values of the principal curvatures and of these six parameters are to be determined. A compatible strain distribution consistent with the deflection distribution (2.83) is determined according to

$$\epsilon_{xx} = \frac{\partial u_x}{\partial x} + \frac{1}{2}\left(\frac{\partial w}{\partial x}\right)^2 - z\frac{\partial^2 w}{\partial x^2}$$

$$\epsilon_{yy} = \frac{\partial u_y}{\partial y} + \frac{1}{2}\left(\frac{\partial w}{\partial y}\right)^2 - z\frac{\partial^2 w}{\partial y^2} \tag{2.84}$$

$$\epsilon_{xy} = \frac{1}{2}\left(\frac{\partial u_x}{\partial y} + \frac{\partial u_y}{\partial x} + \frac{\partial w}{\partial x}\frac{\partial w}{\partial y}\right) - z\frac{\partial^2 w}{\partial x \partial y},$$

which generalizes (2.77) in such a way that strain of the substrate midplane surface due to transverse deflection w is taken into account. The strain components are

$$\epsilon_{xx} = a_1 + a_2 \frac{x^2}{R^2} + a_3 \frac{y^2}{R^2} + \kappa_x^2 \frac{x^2}{2} - z\kappa_x + \epsilon_m$$

$$\epsilon_{yy} = b_1 + b_2 \frac{y^2}{R^2} + b_3 \frac{x^2}{R^2} + \kappa_y^2 \frac{y^2}{2} - z\kappa_y + \epsilon_m \tag{2.85}$$

$$\epsilon_{xy} = (a_3 + b_3)\frac{xy}{R^2} + \kappa_x\kappa_y \frac{xy}{2}.$$

As before, the mismatch strain is uniform throughout the film and it is identically zero in the substrate. This deformation is identical to that represented by (2.78) when $\kappa_x = \kappa_y = \kappa$, $a_1 = b_1 = \epsilon_0$, $a_2 = b_2 = \epsilon_1$ and $a_3 = b_3 = 0$.

With the explicit spatial dependence of strains in (2.85), the strain energy density can be integrated over the volume of the film–substrate system to determine the total potential energy $V(\kappa_x, \kappa_y, a_1, b_1, a_2, b_2, a_3, b_3)$ as a function of the eight parameters characterizing the deformation. Then, equilibrium configurations are determined by finding the values of the characterizing parameters for which V is stationary with respect to its arguments, that is, any set of values of κ_x, κ_y, a_1, b_1, a_2, b_2, a_3, b_3 that satisfies the equations $\partial V/\partial \kappa_x = 0, \ldots, \partial V/\partial b_3 = 0$ defines an equilibrium configuration.

For the present case, the equilibrium conditions can be obtained in relatively simple form for the case when the elastic properties of the film and substrate materials are the same (Masters and Salamon 1993, Freund 2000). In particular, for a given mismatch strain and geometric parameters, it is possible to express equilibrium values of κ_x and κ_y in terms of ϵ_{m}. If ϵ_{m} is then eliminated, a relationship between κ_x and κ_y is obtained which represents the locus of equilibrium states for the system in the plane of κ_x versus κ_y, namely,

$$\left(\kappa_x - \kappa_y\right) \left[\kappa_x \kappa_y R^4 (1 + \nu) - 16(h_{\mathrm{s}} + h_{\mathrm{f}})^2\right] = 0, \qquad (2.86)$$

where ν is the value of the Poisson ratio common to both materials. This result is exact within the class of deformations (2.85). An important consequence of admitting the possibility of finite deflections is evident in this expression. It is clear that a spherical deformed shape of the system with $\kappa_x = \kappa_y$ is an equilibrium shape. The new feature is the possibility of an alternate *asymmetric* equilibrium shape represented by the vanishing of the term in square brackets in (2.86).

The locus of possible equilibrium curvatures is plotted in Figure 2.23 for $\nu = 1/4$ in terms of curvature normalized according to $(2.80)_2$. The straight line bisecting the quadrant represents the spherical shape with $\kappa_x = \kappa_y$. The curved branch of the locus is obtained by setting the second factor in (2.86) equal to zero, and it represents asymmetric deformation, that is, $\kappa_x \neq \kappa_y$. The intersection point of these two branches is a bifurcation point. For values of spherical curvature on the branch with $\kappa_x = \kappa_y$ that are less than the curvature at the bifurcation point, it is found that the equilibrium value of potential energy is a local minimum under variations in curvature. Thus, that part of the symmetric branch represents *stable* equilibrium configurations. On the other hand, for values of spherical curvature that are larger than the curvature at bifurcation, the stationary value of potential energy is found to be a saddle point, so that part of the symmetric branch represents *unstable* equilibrium configurations. All equilibrium configurations on the asymmetric branch in

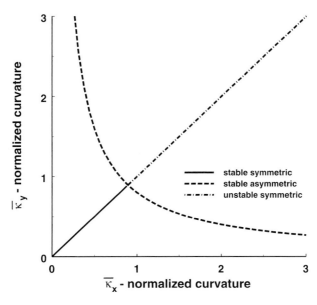

Fig. 2.23. Relationship between normalized principal curvatures $\bar{\kappa}_x$ and $\bar{\kappa}_y$ implied by the equilibrium condition (2.86) for $\nu = 1/4$, based on the assumed deformed shape (2.82) for the substrate midplane. The intersection between the branch $\bar{\kappa}_x = \bar{\kappa}_y$ corresponding to spherical curvature and the hyperbolic branch is the bifurcation point.

Figure 2.23 (except that corresponding to the bifurcation point itself) are found to be stable configurations.

If the strain magnitude is increased beyond the value corresponding to the bifurcation point (a value that will be given below) then the deformation becomes asymmetric with the curvature increasing in some direction and decreasing in an orthogonal direction. The principal directions of the asymmetric deformation are completely arbitrary. The bifurcation is stable, in the sense that an increasing strain is required to move the equilibrium configuration away from the bifurcation point along the asymmetric branch in Figure 2.23. As the strain magnitude increases further, the equilibrium shapes become more and more asymmetric, approaching a cylindrical limiting shape, that is, $\kappa_x/\kappa_y \to 0$ or $\to \infty$ as $|\epsilon_m| \to \infty$.

The locus of equilibrium shapes depicted in Figure 2.23 provides no information on the actual variation of mismatch strain ϵ_m along the equilibrium paths, so this is illustrated in Figure 2.24. The mismatch strain is normalized according to (2.80)$_1$. The nonlinear curvature and mismatch strain prior to bifurcation is identical to that shown in Figure 2.19.

The results in Figure 2.23 are valid for the full range of geometrical parameters for systems which meet the general characteristics of compliant free-standing layers. To give an impression of the magnitudes of parameters involved, consider

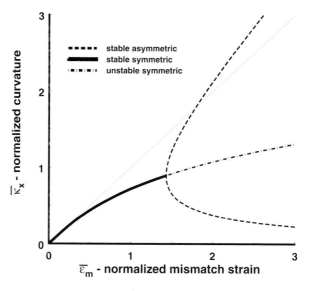

Fig. 2.24. Relationship between one normalized principal curvature $\bar{\kappa}_x$ and the normalized mismatch strain $\bar{\epsilon}_m$ corresponding to the behavior illustrated in Figure 2.23 for $\nu = 1/4$. The bifurcation occurs at approximately $\bar{\epsilon}_{m(\mathrm{bif})} = 1.43$. The dotted line is the linear relationship between curvature and mismatch strain which is valid for small deformation.

the state represented by the bifurcation point in Figure 2.23. If the geometry of the system is characterized by h_s/R, the ratio of substrate thickness to radius, and by h_f/h_s, the ratio of film thickness to substrate thickness, then the spherical curvature $\kappa = \kappa_x = \kappa_y$ at bifurcation is given by

$$\kappa_{(\mathrm{bif})} = \frac{4}{\sqrt{1+\nu}} \frac{h_s}{R^2} \left(1 + \frac{h_f}{h_s} \right) \quad \Rightarrow \quad \bar{\kappa}_{(\mathrm{bif})} = \frac{1}{\sqrt{1+\nu}} \tag{2.87}$$

and the critical magnitude of mismatch strain in the film which leads to bifurcation is given by

$$\epsilon_{m(\mathrm{bif})} = \frac{4}{3(1+\nu)^{3/2}} \frac{h_s^2}{R^2} \frac{h_s}{h_f} \left(1 + \frac{h_f}{h_s} \right)^4 \quad \Rightarrow \quad \bar{\epsilon}_{m(\mathrm{bif})} = \frac{2}{(1+\nu)^{3/2}}. \tag{2.88}$$

For $\nu = 1/4$, the value of $\bar{\epsilon}_{m(\mathrm{bif})}$ is about 1.43.

Figure 2.25 shows the results of observations of pre-bifurcation and post-bifurcation curvatures in the central portion of Si wafers with different in-plane

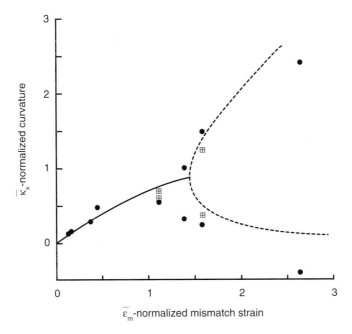

Fig. 2.25. Measured normalized curvature versus normalized mismatch strain for Si wafers with W films. The data points represent experiments conducted for different combinations of wafer diameter, wafer thickness and film thickness. The filled circles correspond to curvature measurements made by Finot et al. (1997) using the scanning laser method for the pre-bifurcation data points or the grid reflection method for the post-bifurcation data points. The other symbols denote experiments using the coherent gradient sensor method. Superimposed on the experimental data are the predicted trends for the stable symmetric and stable asymmetric cases replotted from Figure 2.24 with $\nu = 0.26$.

dimensions and thicknesses and with different thicknesses of W thin film deposits. The critical values of normalized curvature and normalized mismatch strain at which bifurcation in equilibrium shape is triggered in the experiments are close to those predicted by (2.87) and (2.88). The pre-bifurcation nonlinear response matches the analytical estimates more closely than the near- or post-bifurcation response. The observed discrepancy between the simple theory and the experimental measurements could possibly arise from the nonuniform curvature evolution, which will be discussed in Section 2.6.3, or perhaps from the anisotropic elastic properties of the substrate material; these factors have not been taken into account in the analysis up to this point.

The data shown in Figure 2.25 were obtained by means of the wafer curvature scan, the grid reflection method, or the coherent gradient sensor method. The last of these will be discussed further in Section 2.6.2 in the context of a visualization

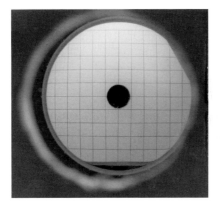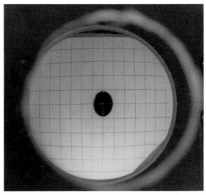

Fig. 2.26. Micrographs obtained using the grid reflection method which illustrate large deformation of Si wafers containing W films; the wafer diameter is 150 mm. Nonlinear deformation with axially symmetric curvature, prior to bifurcation, is shown on the left. A post-bifurcation asymmetric shape is shown on the right. Note that the axially symmetric curvature of the wafer on the left causes the hole in the grid plane to be reflected as a dark circle in the center of the wafer, whereas the post-bifurcation shape of the wafer on the right causes the hole to be reflected with an elliptical shape. Reproduced with permission from Finot et al. (1997).

of deformation via Mohr's circle for curvature. The grid reflection method was described in Section 2.3.3, and a typical set of images obtained with this method is shown in Figure 2.26.

Once the deformation of the substrate progresses into the geometrically nonlinear range, the state of curvature is no longer uniform over most of the area of the substrate. Within the pre-bifurcation range of nonlinear behavior, roughly $0.3 < \bar{\epsilon}_m < 1.5$, the state of curvature remains spherical only near the center of the substrate and it tends to diverge further from being spherical with increasing distance from the center. Within the post-bifurcation range of nonlinear behavior, roughly $1.5 < \bar{\epsilon}_m$, the state of curvature is locally ellipsoidal everywhere but with principal curvatures varying from point to point. In spite of the complexity of this spatially nonuniform curvature in the range of nonlinear response, the deformation of the substrate is often discussed in terms of pre-bifurcation spherical curvature and post-bifurcation ellipsoidal curvature. In such cases, the reported values of curvature must be understood to be only representative or average values over a fairly large portion of the substrate near the center. It is in this spirit that the Mohr's circle method of visualizing substrate curvature is discussed briefly in the next section.

A number of additional observations on film–substrate deformation in the geometrically nonlinear range can be made on the basis of modeling of the kind introduced here. Among these are:

– A sharp bifurcation in equilibrium shape, such as that shown in Figure 2.24, is possible only for symmetric substrate configurations, such as a circle, a square and an equilateral triangle.

– Estimates for the critical curvature $\kappa_{(bif)}$ and the mismatch strain $\epsilon_{m(bif)}$ at which bifurcation occurs were given in (2.87) and (2.88) for a circular substrate of radius R. The corresponding results for a film–substrate system of square shape with lateral dimensions $L \times L$ are obtained by replacing R^2 in (2.87) and (2.88) with L^2/π (Giannakopoulos et al. 2001).

– Without a high degree of symmetry, the principal curvatures of the substrate midplane are unequal from the onset of deformation. This trend is best described by considering a rectangular shape of lateral dimension $L_x \times L_y$, where $L_x > L_y$. Instead of a sharp bifurcation seen when $L_x = L_y$, the rectangular shape exhibits a more gradual transition from a uniform spherical curvature in the linear deformation regime to an asymmetric curvature in the nonlinear regime as the mismatch strain is increased. This transition is accompanied by a larger principal curvature along the longer dimension L_x. An increase in the shape aspect ratio L_x/L_y causes a reduction in the mismatch strain at which a transition occurs from nominally symmetric deformation to asymmetric deformation, similar to the behavior shown in Figure 2.30 for the asymmetry in mismatch strain (Finot and Suresh 1996).

– While the emphasis in the discussion of nonlinear elastic deformation of film–substrate systems has been on situations in which mismatch strain is the source of internal stress, large diameter wafers and thin plates in flat panel displays show a propensity for deformation in the geometrically nonlinear range due to weight alone. Giannakopoulos et al. (2001) have extended the energy minimization method introduced in Section 2.6 to obtain solutions for the effects of gravity on large deformation and bifurcation in equilibrium shape. Some general trends predicted by their analysis are: (a) the extent of large deformation is strongly influenced by the manner in which the film–substrate system is supported; (b) the magnitude of curvature at which a sharp bifurcation occurs for circular and square shapes of the system is unaffected by gravitational forces; (c) the critical mismatch strain at which bifurcation is triggered is strongly influenced by whether the film is facing up or down as the system is placed on a fixed set of support points; for example, the critical mismatch strain at which bifurcation occurs under the influence of gravity is higher when the back side of the substrate is supported than when the free surface of the film is held on support points; and

(d) the radial variation of curvature during large deformation is significantly influenced by gravity.

2.6.2 *Visualization of states of uniform curvature*

In the discussion in Section 2.6.1 of substrate curvature in the range of behavior where the deformation is not axially symmetric, it was assumed *a priori* that the coordinate axes coincided with the axes of principal curvature. This assumption was incorporated in writing the transverse deflection $w(x, y)$ in the form given in (2.82). On the other hand, in determining curvature from measurements in this range of behavior, the directions of principal curvature are not known in advance. How is data to be interpreted in order to extract complete curvature information under the circumstances? An approximation based on measurements obtained with the CGS method, as discussed in Section 2.3.4, is briefly considered here on the basis of a graphical construction known commonly as Mohr's circle.

As a starting point, consider a set of synthetic fringes obtained by means of the CGS method, as illustrated on the left in Figure 2.27. The two fringe patterns were obtained by aligning the diffraction gratings with two orthogonal but otherwise arbitrary directions in the plane of the sample. These directions will be denoted as the x-direction and the y-direction. With the gratings normal to the x- or y-direction, a gradient in slope in the x- or y-direction is sensed optically. The region of the sample represented by Figure 2.27 will typically be only a part of the entire substrate area in which the curvatures can be viewed as being uniform.

The components κ_x, κ_y and κ_{xy} referred to the coordinate directions are determined in the region by measuring the fringe spacing distances in Figure 2.27 as indicated. The quantity κ_{xy} is called the *twist* of the surface in this discussion, although there is no standard terminology for this geometrical quantity. In analogy with the definition of curvature, the term derives from the fact that the quantity $-\kappa_{xy}$ is the local gradient of the x-component (y-component) of a unit vector normal to the surface in the y-direction (x-direction). As such, the term twist is nicely descriptive. The gradient of the x-component of the normal vector in the x-direction is $-\kappa_x$, and similarly for $-\kappa_y$. The synthetic fringes then yield the values

$$\kappa_x = \frac{p}{2\Delta} \frac{1}{\Delta x^{(x)}}, \quad \kappa_y = \frac{p}{2\Delta} \frac{1}{\Delta y^{(y)}},$$

$$\kappa_{xy} = \kappa_{yx} = \frac{p}{2\Delta} \frac{1}{\Delta x^{(y)}}$$

(2.89)

on the basis of (2.46). The symmetry indicated in extracting the value of twist is a check that the deformed surface is locally smooth on the scale of the observation.

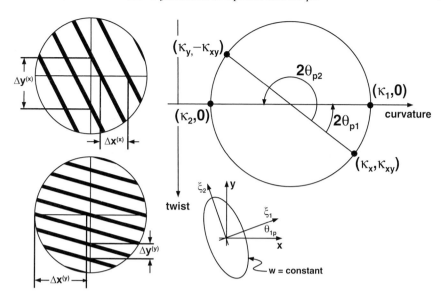

Fig. 2.27. The two circles on the left are idealized schematics of CGS fringe patterns. The upper (lower) circle represents a changing substrate slope in the x-direction (y-direction); the uniformity of fringes implies a constant gradient. The curvature components inferred from the fringe spacing in coordinate directions according to (2.89) are used to construct Mohr's circle, as shown on the upper right; the construction provides values of principal curvatures κ_1 and κ_2, as well as directions of principal curvature. A constant deflection contour with respect to the tangent plane to the substrate at the point of measurement is shown in the inset on the lower right, with ξ_1, ξ_2-axes aligned with the directions of principal curvature. The figures are drawn for the case of a substrate deformed into an ellipsoidal shape, but other shapes are possible.

The parameters p and Δ are the pitch and spacing of the diffraction grating plates used in the experiment.

The four quantities identified in (2.89) are the components of a second rank tensor with the same mathematical characteristics as the tensor representing a state of plane stress. It follows immediately that the geometrical construction known as Mohr's circle can be used to determine the principal curvatures and the directions of principal curvature for the deformation represented by Figure 2.27, as was noted by Hyer (1981). The construction is illustrated in Figure 2.27 which is drawn for a particular case when all components of the curvature tensor are positive. The points (κ_x, κ_{xy}) and $(\kappa_y, -\kappa_{xy})$ are plotted in the plane with horizontal axis representing curvature and vertical axis representing twist. The convention whereby the twist is positive downward in the plane is adopted so that rotations in both the physical plane and in the Mohr's circle plane corresponding to a given rotation of coordinates are in the same sense. The line connecting these two points is a diameter of Mohr's circle. In the plane of Mohr's circle, the principal curvatures are readily identified at $(\kappa_1, 0)$ and $(\kappa_2, 0)$. The directions of the larger and smaller principal curvatures are at a counterclockwise angle of θ_{p1} and θ_{p2}, respectively, from the x-axis in the experiment where both angles must satisfy

$$\tan 2\theta_p = \frac{2\kappa_{xy}}{\kappa_x - \kappa_y}. \tag{2.90}$$

The values of principal curvature are given in terms of the measured components by

$$\kappa_{1,2} = \tfrac{1}{2}(\kappa_x + \kappa_y) \pm \sqrt{\tfrac{1}{4}(\kappa_x - \kappa_y)^2 + \kappa_{xy}^2}. \tag{2.91}$$

In a rectangular coordinate system aligned with the principal directions of curvature, with the coordinate axes labeled ξ_1 and ξ_2, say, the transverse deflection measured from a tangent plane at $\xi_1 = \xi_2 = 0$ is

$$w(x, y) = \tfrac{1}{2}\left(\kappa_1 \xi_1^2 + \kappa_2 \xi_2^2\right) \tag{2.92}$$

as in (2.82). A level curve $w(x, y) = $ constant of this deflection is illustrated in Figure 2.27.

In general, the values of principal curvature and the principal directions of curvature indicate the most essential features of the local substrate deformation. For example,

- if both κ_1 and κ_2 are nonzero and have the same algebraic sign, the shape is ellipsoidal, and Mohr's circle is completely confined to one side or the other of the twist axis;
- if both κ_1 and κ_2 are nonzero and have opposite signs, the shape is hyperboloidal or saddle-like, and Mohr's circle spans the twist axis;
- if either κ_1 or κ_2 is zero then the surface is cylindrical and Mohr's circle is tangent to the twist axes;
- if $\kappa_1 = \kappa_2$, the shape is spherical and Mohr's circle is a point on the curvature axis;
- if $\kappa_1 = \kappa_2 = 0$, Mohr's circle reduces to a point at the origin of the curvature–twist plane.

A limitation on the use of Mohr's circle to interpret curvature observations is that it applies only over regions of the surface small enough so that the components of the curvature tensor are essentially constant within the region. For conditions in which the deformation is in the geometrically nonlinear range, the curvature is no longer uniform over the entire area of the substrate and interpretation of observations becomes less straightforward than it is in the linear range.

Figure 2.28 shows fringe patterns obtained using the coherent gradient sensor method described in Section 2.3.4 to observe the large deformation of a 150-mm diameter Si wafer with a W film, a system similar to that used to obtain the results plotted in Figure 2.21. The fringe patterns associated with deformation prior to bifurcation in equilibrium shape reveal an axisymmetric shape for a substrate thickness of 350 μm, a film thickness of 0.9 μm and a film stress of 1.1 GPa. Fringe patterns in the geometrically nonlinear range were obtained by keeping the directions in the two CGS gratings parallel to each other and horizontal (see Figure 2.11) and by rotating the wafer in increments of 45 degrees. Prior to bifurcation, the fringe patterns at all orientations are essentially the same, which confirms that the shape of the wafer is axially symmetric. On the other hand, for a Si wafer of 415 μm thickness with a 2.4 μm thick W film with a mismatch stress of 0.98 GPa, an asymmetric shape evolves; the principal curvatures κ_x and κ_y associated with this post-bifurcation deformation mode show different numbers of fringes parallel to the gratings. When this Si wafer is rotated 45 degrees counterclockwise from either principal curvature orientation, the fringes are no longer parallel to the gratings, which indicates that twist curvature can be detected in this direction. Rotation in opposite directions from either principal curvature direction results in distortion of the fringe patterns in the same general manner but in the opposite sense, indicating the onset of substrate twist of opposite signs. The use of the CGS method, in conjunction with Mohr's circle construction, provides global information on the shapes of surfaces with different combinations of curvature and twist.

2.6.3 Bifurcation for general curvature variation

The foregoing discussion of bifurcation is based on the assumption of spherical curvature of the substrate midplane prior to bifurcation and on the assumed transverse deflection (2.82) with spatially uniform principal curvatures following bifurcation. It was noted in the discussion of axially symmetric deformation in the previous section that the deformed shape of the substrate midplane can depart significantly from a shape with uniform curvature. Therefore, the bifurcation analysis is repeated in this section, but without *a priori* assumptions on the deformed shape of the substrate midplane, by means of the numerical finite element method.

The calculations were carried out under the assumption that the deformation has at least one plane of reflective symmetry. The midplane of the 180 degree sector of the substrate was covered with a regular radial–circumferential mesh. An eight-noded plate element was prescribed in each mesh segment. The element adopted admits through-the-thickness variation of material properties, and the film–substrate system was defined by prescribing the appropriate variation. Mismatch strain was imposed by specifying a coefficient of thermal expansion for the film material relative to the substrate material and by making temperature the imposed loading parameter. The

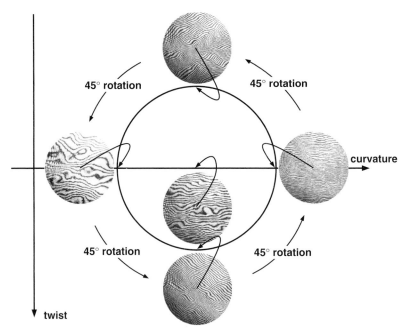

Fig. 2.28. A series of images showing the correlation of CGS fringe patterns with points on Mohr's circle for curvatures. For axially symmetric surface deformation in the geometrically linear range, the twist is zero and the principal curvatures are identical, so the state corresponds to a point on the horizontal axis. The four images around the outside of the circle show the same deformation, but in a way that senses the larger principal curvature, mean curvature with twist, smaller principal curvature, and mean curvature with opposite twist, going around the circle in a counterclockwise sense. These images were taken simply by rotating the entire sample at a fixed condition. CGS fringe patterns provided by T.-S. Park, Massachusetts Institute of Technology (2002).

temperature was gradually increased from zero initial value, and the equilibrium shape for large deflections was computed.

To precipitate stable deformation beyond the point of bifurcation, a slight imperfection in the system was introduced in the form of an anisotropic mismatch strain. Typically, the mismatch strain in the x-direction (y-direction) was taken to be 0.01% larger (smaller) than the nominal value ϵ_m. With this level of imperfection, the deformation prior to bifurcation was essentially indistinguishable from results based on an *a priori* assumption of axial symmetry, the bifurcation point was sharply defined in each case and reproducible from case to case, and the post-bifurcation behavior was stable and reproducible.

The general response observed as the mismatch strain was increased was a range of axially symmetric deformation, with the substrate midplane curvature becoming ever more nonuniform. Then, over a very narrow range of values of nominal mismatch strain, the midplane showed first a slight waviness in the circumferential

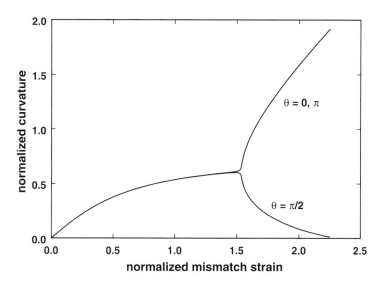

Fig. 2.29. Plots of normalized radial curvature $\bar{\kappa}$ in the directions $\theta = 0$, $\pi/2$ and π versus normalized mismatch strain $\bar{\epsilon}_m$ in a circular film–substrate system for $\nu = 1/4$. These results were obtained by finite element calculation, assuming identical elastic constants for the film and substrate materials. They involve no *a priori* assumptions on the deformed shape. Bifurcation occurs when $\bar{\epsilon}_{m(bif)} \approx 1.54$. The normalized curvature $\bar{\kappa}$ and mismatch strain $\bar{\epsilon}_m$ are defined in (2.81). The results obtained for $R/h_s = 50$, 100 and 200 are identical when expressed in terms of these parameters.

direction (compared to substrate thickness) followed by large amplitude waviness in the circumferential direction. This behavior is illustrated in Figure 2.29 which shows plots of the normalized radial curvature $\bar{\kappa}$ near the center of the substrate in the directions $\theta = 0$, $\pi/2$ and π versus normalized mismatch strain $\bar{\epsilon}_m$. As $\bar{\epsilon}_m$ increases from zero, the deformation remains axially symmetric within the resolution of the graph. As $\bar{\epsilon}_m$ increases through the value of about 1.54, a fairly sharp transition in response occurs. The radial curvatures in the directions $\theta = 0$ and $\theta = \pi$ increase dramatically for a very small increase in $\bar{\epsilon}_m$, while the curvature in the direction $\theta = \pi/2$ decreases dramatically. In effect, this is the same phenomenon as illustrated in Figure 2.24. In the present case, however, the curvature is not restricted to be radially uniform.

There are other significant features of the behavior illustrated in Figure 2.29. Among these are: the axially symmetric response is nonlinear for values of $\bar{\epsilon}_m$ beyond about 0.3, consistent with the behavior observed in Figure 2.20; the maximum deflection at the substrate periphery reaches a value of about two times the substrate thickness before bifurcation occurs; the post-bifurcation deformation becomes more like cylindrical bending as $\bar{\epsilon}_m$ increases to values substantially beyond 1.54. Also, as noted in the preceding section, the substrate midplane surface is less curved at interior points and more curved near its outer edge.

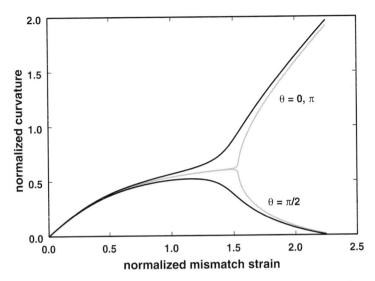

Fig. 2.30. Plots of normalized radial curvature $\bar{\kappa}$ in the directions $\theta = 0$, $\pi/2$ and π versus normalized mismatch strain $\bar{\epsilon}_m$ for $\nu = 1/4$. The difference in mismatch strains in orthogonal directions divided by the sum of these strains was assumed to have a value of 0.01. The calculation was identical to that leading to Figure 2.29 (results shown here as curves in lighter shade) except that the normalized difference in mismatch strains for that case had a value of 0.0001.

Another noteworthy aspect of the result illustrated in Figure 2.29 is the apparent insensitivity of the behavior to the aspect ratio of the substrate. Calculations were carried out for $R/h_s = 50$, 100 and 200. The plots for the three cases, when expressed in terms of the normalized parameters used in Figure 2.29, are indistinguishable. Note that the ratios $h_s/h_f = 100$ and $M_s/M_f = 1$ were maintained in all calculations so that the film is always relatively thin and the effects of modulus difference are not considered.

For a circular substrate, the elementary model yielded $\bar{\epsilon}_{m(bif)} = 1.43$ as the critical value for bifurcation, while the finite element analysis yielded the estimate $\bar{\epsilon}_{m(bif)} = 1.54$ for a range of values of R/h_s. Most calculations were carried out for $h_s/h_f = 100$, that is, the dependence of the value of bifurcation strain on variations of h_s/h_f was not examined systematically. On the basis of a few calculations, it appears that the bifurcation value of $\bar{\epsilon}_m$ is lowered slightly as the value of h_s/h_f is increased. In any case, the magnitude of mismatch strain needed for bifurcation to occur which is implied by $\bar{\epsilon}_{m(bif)} = 1.54$ with $R/h_s = 100$, $h_s/h_f = 100$ and $M_s/M_f = 1$, is roughly 0.01.

A systematic experimental study of deformation of film–substrate systems in the range of nonlinear deformation that covers a wide range of system parameters is not yet available. However, some observations have been reported in the literature which are generally consistent with the models introduced here. Finot et al. (1997) described experiments in which a bilayer film consisting of an Al–Cu alloy layer and a W layer was deposited on a circular Si substrate wafer. Principal curvatures of the substrate were measured for various relative thicknesses of the layers, and the results reported were consistent with the behavior indicated in Figure 2.24. The value of normalized effective mismatch strain $\bar{\epsilon}_m$ at bifurcation, as determined from the net membrane force in the bilayer film, was found to be approximately 1.55. In a similar study, Lee et al. (2001) conducted experiments with an Al film on a Si substrate. In this case, the film was observed to be nonuniform in thickness, a feature which was incorporated into the simulation of the experiment. They also observed a mode of deformation that is consistent with that depicted in Figure 2.24 and a value of normalized mismatch strain of about 1.55 upon onset of asymmetric deformation.

Finally, it is noted that the behavior of thin structures of the kind being discussed here is very sensitive to imperfections in the system. To illustrate the point in the present context, the calculation which led to Figure 2.29 (shown by curves in a lighter shade here) was redone with a mismatch strain that is 1% larger (smaller) than the nominal value ϵ_m in the x-direction (y-direction). The result is shown in Figure 2.30, where it can be seen that a 1% imperfection in mismatch strain obliterates the sharp bifurcation transition. Instead, the system undergoes a long, gradual transition from axially symmetric deformation to asymmetric deformation as $\bar{\epsilon}_m$ increases.

2.6.4 A substrate curvature deformation map

The results obtained for film–substrate systems in Section 2.5 and in the present section provide connections between substrate curvature and film mismatch strain that define boundaries between regimes of behavior. For the case of a very thin film on a relatively thick substrate, it was shown in Section 2.5.1 that the response is linear with spherical curvature for normalized mismatch strain in the range $0 \leq |\bar{\epsilon}_m| \leq 0.3$. Furthermore, for a circular substrate, the response is geometrically nonlinear but axially symmetric for $0.3 \leq |\bar{\epsilon}_m| \leq 1.5$, as shown in Section 2.6. For magnitudes of $|\bar{\epsilon}_m|$ greater than roughly 1.5, the deformation is asymmetric.

Following Finot et al. (1997), these ranges of behavior can be represented graphically in parameter space as a general guide to understanding behavior. The parameters involved in the definition of $\bar{\epsilon}_m$ include ϵ_m, R/h_s, h_f/h_s and M_f/M_s. Often, the values of some of these parameters are fixed by material or geometric constraints. For example, in a case where the origin of mismatch is the constraint of epitaxy in a particular material system, both ϵ_m and M_f/M_s are specified at the outset. For purposes of illustration, suppose that the values are $M_f/M_s = 1$ and $\epsilon_m = 0.01$. Then

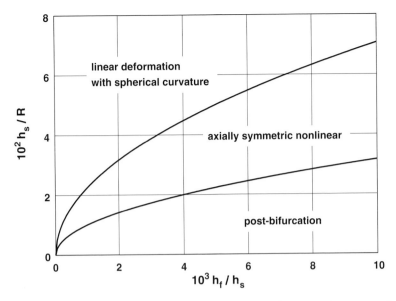

Fig. 2.31. An illustration of a substrate curvature map for the case when $\epsilon_m = 0.01$ and $M_f/M_s = 1$. The regimes of behavior are separated by curves which represent the locus of conditions for which geometrically nonlinear effects come into play and for which asymmetric bifurcation occurs in a circular substrate.

the behavior can be represented in a plane of h_s/R versus h_f/h_s that is divided into regions by the curves $|\bar{\epsilon}_m| = 0.3$ and $|\bar{\epsilon}_m| = 1.5$. The result is shown in Figure 2.31. Other types of curvature maps can be constructed in the same spirit.

2.6.5 Example: A curvature map for a Cu/Si system

Consider a circular Si wafer of radius R and thickness h_s on which a thin Cu film is deposited. The film thickness is h_f. Consider wafer diameters of 200 mm and 300 mm, each with a film thickness of 1 μm. Determine the minimum thickness that each of these wafers can have if the deformation is to remain in the linear range of behavior for a temperature change of 100 °C from the reference temperature. Assume that the system is stress-free at the reference temperature and that the response is thermoelastic.

Solution:

For a temperature change of ΔT from the reference temperature, the mismatch strain is $\epsilon_m = (\alpha_s - \alpha_f)\Delta T$ and the corresponding film stress is $\sigma_m = M_f \epsilon_m$. The boundary of the geometrically linear range of behavior is defined by $|\bar{\epsilon}_m| = 0.3$. In the present instance, this equation takes the form

$$1 = 5|(\alpha_s - \alpha_f)\Delta T| \frac{R^2}{h_s^2} \frac{M_f}{M_s} \frac{h_f}{h_s} \qquad (2.93)$$

which is a relationship among material and geometrical parameters that defines the limit of the linear range. The lower curve in Figure 2.31 illustrates such a relationship. For given values of the material parameters, wafer radius and film thickness, this relationship gives the minimum thickness $(h_s)_{min}$ in terms of $|\Delta T|$.

To obtain a numerical estimate for the minimum admissible value $(h_s)_{min}$ for each wafer diameter, assume that $\alpha_f = 17 \times 10^{-6}\,{}^\circ C^{-1}$, $\alpha_s = 3 \times 10^{-6}\,{}^\circ C^{-1}$, $M_f = 197 \times 10^9\,N\,m^{-2}$ and $M_s = 227 \times 10^9\,N\,m^{-2}$. For $|\Delta T| = 100\,{}^\circ C$ and $h_f = 1\,\mu m$, (2.93) implies that $(h_s)_{min} = 390\,\mu m$ for $R = 100\,mm$ and $(h_s)_{min} = 515\,\mu m$ for $R = 150\,mm$.

The features of deformation of the film–substrate system observed here are indicative of the deformation modes which can be observed. Although the range of parameter space which has been considered in detail is quite limited, any departure from a uniformly strained layer can be taken into account through the net membrane force in the film in most cases, provided that the film is thin compared to the substrate. Other modes of deformation may be found by considering noncircular substrates, materials with different moduli, films with gradation in mismatch strain or properties, and other possibilities.

2.7 Exercises

1. A tungsten silicide (WSi) film, 0.193 μm in thickness, was deposited on a (100) Si wafer, 200 mm in diameter and 730 μm in thickness. The radius of curvature of the wafer before and after film deposition was measured to be -350.4 m and 162.9 m, respectively. If the biaxial modulus of the (100) Si film is 180.5 GPa, estimate the average film stress.

2. A simple procedure for determining the biaxial modulus M_f and the thermal expansion coefficient α_f of a thin film involves the so-called *two-substrate method*. In this method, a thin film of the material for which the properties are to be determined is first deposited on a substrate with thermal expansion coefficient α_{s1}. This film–substrate system is then subjected to a temperature change ΔT and the change in the film mismatch stress $\Delta\sigma_{m1}$ which occurs as a result of the temperature change is inferred by recourse to substrate curvature or x-ray diffraction techniques. Then, the same film material is deposited on a substrate of a different material with thermal expansion coefficient α_{s2}. The change in film mismatch stress $\Delta\sigma_{m2}$ for this system is again inferred for the same temperature change ΔT. Assume that all materials remain elastic during thermal excursions and that all material properties are essentially independent of temperature over the range of the observations. Derive expressions for α_f and M_f in terms of α_{s1}, α_{s2}, $\Delta\sigma_{m1}$, $\Delta\sigma_{m2}$ and ΔT.

3. A thin film of W is deposited on a Si substrate with an unknown mismatch stress. When this film–substrate system is heated from 100 °C to 300 °C, it is estimated by curvature measurement that the film stress decreases by 120 MPa. Another thin film of W is now deposited on a GaAs substrate, again at unknown mismatch stress. When this system is heated from 100 °C to 300 °C, the film stress is observed to increase by 205 MPa. The coefficients of thermal expansion of the substrate materials are $\alpha_{Si} = 3.5 \times 10^{-6}\,{}^\circ C^{-1}$ and $\alpha_{GaAs} = 6.4 \times 10^{-6}\,{}^\circ C^{-1}$. Infer values of

biaxial modulus and thermal expansion coefficient of W from these experimental observations, assuming the incremental deformations are elastic and the materials are isotropic.

4. A thin film made of a new polymeric material is bonded to a metallic substrate. Prolonged exposure to a moist atmosphere causes the polymer to swell. The moisture intake increases the stress-free volume of the polymer by 3%. Derive an expression for the biaxial mismatch stress in the polymer due to this swelling under the assumption that the biaxial modulus M_f is not affected by the swelling.

5. The rationale for the choice of materials used in thermostatic bimetals was discussed in Section 2.2.3.

    (a) Explain why lead cannot be used as one of the two metals in the bimetallic strip in thermostats.
    (b) Derive an expression for the variation of curvature of a brass (90% Cu–10% Zn)–Invar thermostatic bimetal as a function of temperature change, ΔT. Assume that the brass and Invar layers have the same thickness.
    (c) Discuss why very rapid fluctuations in temperature could affect the ability of the thermostatic bimetals to provide accurate measures of instantaneous temperature.

6. The variation of curvature as a function of mismatch strain, geometry and material properties can be analyzed for a general bilayer using the energy minimization method, Section 2.2, or the local equilibrium method, Section 2.4.

    (a) Starting with the force and moment balance equations, (2.54), derive (2.19) and (2.20), for the bilayer shown in Figure 2.3 for arbitrary values of h_f, h_s, M_f, M_s and ϵ_m.
    (b) Derive (2.6) and (2.7) by taking the limit $h_f/h_s \to 0$ in (2.19)–(2.21).

7. In Section 2.6.1, a model bifurcation problem was considered for the departure from axially symmetric deformation of a circular film–substrate system due to increasing mismatch strain in the film. Consider the same basic problem except that the shape of the substrate is a square of extent L on a side, rather than a circle of radius R. The undeformed substrate midplane coincides with the xy-plane, and the coordinate axes are perpendicular to the sides of the square.

    (a) Assume the displacement field of the form given by (2.82) and (2.83) with R^2 replaced by L^2, and derive a result equivalent to (2.86) to study bifurcation for the case when the elastic properties of the film and the substrate are the same.
    (b) Estimate the curvature and mismatch strain at which bifurcation occurs and compare the estimates to the corresponding results (2.87) and (2.88) for the case of a circular substrate.

8. Consider the special case of the issue studied in Section 2.4.1 when the film and substrate have spatially uniform and identical elastic properties. Suppose the film is subjected to a mismatch strain with through-the-thickness variation given by $\epsilon_m(z)$.

    (a) Show that the general result for curvature given in (2.51) reduces to in this case.

$$\kappa(h_f) = -\frac{6}{(h_f + h_s)^2} \int_0^{h_f} \epsilon_m(\zeta + \tfrac{1}{2}h_s)\,d\zeta$$

$$+ \frac{12}{(h_f + h_s)^3} \int_0^{h_f} (\zeta + \tfrac{1}{2}h_s)\epsilon_m(\zeta + \tfrac{1}{2}h_s)\,d\zeta \qquad (2.94)$$

(b) Observation of response as material is added or removed can be used to
   investigate a state of residual stress in a solid (Freund 1996). Suppose that
   curvature $\kappa(\eta)$ has been observed for all values of thickness η in the range
   $0 \leq \eta \leq h_f$. Show that the curvature expression given in part (a) can then
   be inverted to yield the variation of mismatch strain in terms of curvature
   history as

$$\epsilon_m(\eta + \tfrac{1}{2}h_s) = \tfrac{2}{3}(\eta + h_s)\kappa(\eta) + \tfrac{1}{6}(\eta + h_s)^2\kappa'(\eta) + \frac{1}{3}\int_0^\eta \kappa(\zeta)\,d\zeta. \quad (2.95)$$

3

Stress in anisotropic and patterned films

In the previous chapter, fundamental issues that arise when considering stress and deformation in film–substrate systems were examined, with attention restricted to systems for which the materials are isotropic and the film geometry is continuous. Some film materials exhibit anisotropy in their elastic properties. In other cases, an isotropic film may be patterned on a small scale in such a way that it gives rise to an apparent anisotropy on a larger scale. These effects can have a significant influence on mechanical response. Geometric inhomogeneity can arise as a result of film patterning, composite films, island formation, distributed film cracks or other circumstances. This chapter deals with the role of material anisotropy and geometric nonuniformity in influencing stress and deformation in layered solids.

The chapter begins with an overview of elastic anisotropy in crystalline materials. Anisotropy of elastic properties in materials with cubic symmetry, as well as other classes of material symmetry, is described first. Also included here are tabulated values of typical elastic properties for a variety of useful crystals. Examples of stress measurements in anisotropic thin films of different crystallographic orientation and texture by recourse to x-ray diffraction measurements are then considered. Next, the evolution of internal stress as a consequence of epitaxial mismatch in thin films and periodic multilayers is discussed. Attention is then directed to deformation of anisotropic film–substrate systems where connections among film stress, mismatch strain and substrate curvature are presented. A Stoney-type formula is derived for an anisotropic thin film on an isotropic substrate. Anisotropic curvature due to mismatch strain induced by a piezoelectric film on a substrate is also analyzed.

Focus is then shifted to the evolution of curvature in layered systems in response to the introduction of geometric nonuniformity. For this purpose, a detailed derivation is provided for the Stoney-type formula for films with a regular array of cracks or films patterned into thin lines on substrates where explicit connections are made between curvatures parallel to or normal to the lines or cracks. These results also facilitate the determination of volume averaged stress in patterned film structures directly from curvature experiments. The implications of the analyses presented in the chapter for curvature and stress evolution in a periodic array of patterned lines or stripes are evaluated through comparisons with available experimental results and finite element simulations. Commonly used measurement methods, such as the substrate curvature method, the x-ray diffraction method and the micro-Raman spectroscopic technique, for the determination of average stress in periodically patterned thin films are described; the advantages and shortcomings of each method are briefly summarized.

3.1 Elastic anisotropy

The states of stress and strain in a deformed crystal being idealized as a continuum are characterized by symmetric second rank tensors σ_{ij} and ϵ_{ij}, respectively, each comprising six independent components. Hooke's law of linear elasticity for the most general anisotropic solid expresses each component of the stress tensor linearly in terms of all components of the strain tensor in the form

$$\sigma_{ij} = c_{ijkl}\,\epsilon_{kl}, \tag{3.1}$$

where c_{ijkl} is the array of elastic stiffness constants. Alternately, the inverse form of Hooke's law is written to express each component of the strain tensor linearly in terms of all components of the stress tensor as

$$\epsilon_{ij} = s_{ijkl}\,\sigma_{kl}, \tag{3.2}$$

where s_{ijkl} is the array of elastic compliance constants. In (3.1) or (3.2), each of the nine equations for a stress or strain component involves nine material parameters. Each of the fourth-order tensors c_{ijkl} and s_{ijkl} comprise 81 components. The symmetry of the stress and strain tensors, that is, $\sigma_{ij} = \sigma_{ji}$ and $\epsilon_{ij} = \epsilon_{ji}$, further implies that the components of the stiffness tensor must satisfy $c_{ijkl} = c_{ijlk} = c_{jikl}$; likewise, $s_{ijkl} = s_{ijlk} = s_{jikl}$. As a consequence, the number of independent elastic constants is reduced from 81 to 36 in either case.

The tensor form of the constitutive equation in (3.1) provides a concise and effective statement of Hooke's law for use in theoretical developments. For purposes of measurement and calculation, however, it is often more convenient to adopt a matrix form of the constitutive equation. Such a form is suggested naturally by the fact that there are six independent components of stress, six independent components of strain, and 36 material parameters representing the relationship between stress and strain. For this purpose, a contracted notation is commonly introduced (see Kelly and Groves (1970), Hosford (1993) for example) whereby a six-component array σ_i is constructed by means of the replacements $\sigma_{11} \rightarrow \sigma_1$, $\sigma_{22} \rightarrow \sigma_2$, $\sigma_{33} \rightarrow \sigma_3$, $\sigma_{23} = \sigma_{32} \rightarrow \sigma_4$, $\sigma_{13} = \sigma_{31} \rightarrow \sigma_5$ and $\sigma_{12} = \sigma_{21} \rightarrow \sigma_6$. A similar contracted notation for the strain components makes it possible to recast (3.1) into the form

$$\sigma_i = c_{ij}\epsilon_j. \tag{3.3}$$

The one-dimensional strain array is constructed by means of the replacements of ϵ_{ij} as $\epsilon_{11} \rightarrow \epsilon_1$, $\epsilon_{22} \rightarrow \epsilon_2$, $\epsilon_{33} \rightarrow \epsilon_3$, $\epsilon_{23} = \epsilon_{32} \rightarrow \epsilon_4/2$, $\epsilon_{13} = \epsilon_{31} \rightarrow \epsilon_5/2$ and $\epsilon_{12} = \epsilon_{21} \rightarrow \epsilon_6/2$. Thus, the quantities ϵ_4, ϵ_5 and ϵ_6 represent components of engineering shear strain; by definition, these are equal to two times the corresponding components of tensorial shear strain. The form (3.3) is appealing because of its simplicity, but the tensor character of Hooke's law has been sacrificed to achieve it. The connection between components of c_{ijkl} and c_{ij} is evident through comparison of (3.1) with (3.3).

In order to write the inverse relationship of (3.3) as

$$\epsilon_i = s_{ij}\sigma_j, \tag{3.4}$$

the definitions of the components s_{ij} in terms of the components of s_{ijkl} must be chosen with a numerical factor of 1 whenever both indices of s_{ij} are among 1,2,3 (for example, $s_{22} = s_{2222}$), with a numerical factor of 2 whenever one and only one index of s_{ij} is among 1,2,3 (for example, $s_{24} = 2s_{2223}$), and with a numerical factor of 4 whenever neither index of s_{ij} is among 1,2,3 (for example, $s_{55} = 4s_{1313}$).

The number of independent constants comprising either c_{ij} or s_{ij} can be further reduced. This is enforced by requiring that the work done by the stress to achieve any homogeneous state of deformation from the undeformed reference configuration is independent of the deformation path followed in reaching the final state or any of several equivalent conditions implying path-independence. This condition implies the existence of a strain energy density function that serves as a potential function of strain from which stress can be derived as a conjugate variable for any possible state of deformation; the existence of a strain energy function assures the required path-independence of stress work. For a material with a strain energy density function, both the stiffness matrix and the compliance matrix are symmetric 6×6 matrices, that is, $c_{ij} = c_{ji}$ and $s_{ij} = s_{ji}$. In other words, each of these matrices involves only 21 independent elastic constants for the most general anisotropic material behavior consistent with the existence of a strain energy function.

For crystals with a triclinic structure, which possess only a center of symmetry and exhibit the most general elastic anisotropy, a complete characterization requires all 21 independent constants. The existence of a higher degree of crystal symmetry can further reduce the number of independent elastic constants needed for proper description. In such cases, some elastic constants may vanish and members of some subsets of constants may be related to each other in some definite way, depending on the crystal symmetry. Finally, it should be noted that, while the total number of material constants required to characterize a material of a certain class is independent of the coordinate axes used to represent components of stress or strain, the values of particular components of c_{ij} do depend on the material reference axes chosen for their representation.

3.2 Elastic constants of cubic crystals

The simplest material symmetry beyond isotropy is cubic symmetry, a property of crystals that possess three fourfold axes of rotational symmetry, the cube axes, and four threefold axes of rotational symmetry, the cube diagonals. Alternatively, cubic symmetry may be described as invariance of material structure under a translation of a certain distance in any of three mutually orthogonal directions; these directions are usually identified as the cube axes. Consider a cubic material for which the [100], [010] and [001] cube axes are parallel to the axes of an underlying rectangular x_1, x_2, x_3-coordinate system. For this case, it is evident that

$$c_{11} = c_{22} = c_{33}, \quad c_{12} = c_{23} = c_{31}, \quad c_{44} = c_{55} = c_{66}. \tag{3.5}$$

All the other elastic constants vanish because of the fourfold rotational symmetry of the reference axes. This point can be demonstrated by reasoning that the remaining constants must vanish so as to avoid introducing a change in the sign of the shear strain for a given shear stress merely to account for rotation. It follows from (3.5) that the elastic response of any cubic crystal is characterized by three independent elastic constants and that the complete stiffness matrix is

$$[c_{ij}] = \begin{bmatrix} c_{11} & c_{12} & c_{12} & 0 & 0 & 0 \\ c_{12} & c_{11} & c_{12} & 0 & 0 & 0 \\ c_{12} & c_{12} & c_{11} & 0 & 0 & 0 \\ 0 & 0 & 0 & c_{44} & 0 & 0 \\ 0 & 0 & 0 & 0 & c_{44} & 0 \\ 0 & 0 & 0 & 0 & 0 & c_{44} \end{bmatrix}, \tag{3.6}$$

where components are referred to the *natural* coordinate axes, that is, the rectangular axes aligned with the cube edges.

Similarly, the components of the compliance matrix are related by

$$s_{11} = s_{22} = s_{33}, \quad s_{12} = s_{23} = s_{31}, \quad s_{44} = s_{55} = s_{66} \tag{3.7}$$

for a material with cubic symmetry; all other components vanish. Values of elastic stiffness and compliance constants for a variety of cubic crystals are listed in Table 3.1.

Unique relationships among the compliance constants and stiffness constants for cubic crystals can be derived by considering simple states of stress and deformation of the crystal. For example, consider a uniaxial tensile stress σ_1 applied in the direction of the x_1-axis, which is aligned with one of the cube axes. From (3.3), (3.4) and (3.6), it follows that

$$\epsilon_1 = s_{11}\sigma_1, \qquad \epsilon_2 = s_{12}\sigma_1, \qquad \epsilon_3 = s_{12}\sigma_1,$$

$$\sigma_1 = c_{11}\epsilon_1 + c_{12}\epsilon_2 + c_{12}\epsilon_3. \tag{3.8}$$

If these relations are to be applied for arbitrary values of σ_1, then elimination of the strain components implies the condition that

$$c_{11}s_{11} + 2c_{12}s_{12} = 1. \tag{3.9}$$

On the other hand, imposition of a hydrostatic tension of magnitude σ on the material results in an extensional strain ϵ in each of the coordinate directions, so that

$$\epsilon = (s_{11} + 2s_{12})\sigma \quad \text{and} \quad \sigma = (c_{11} + 2c_{12})\epsilon. \tag{3.10}$$

If strain is eliminated from these equations, and the resulting expression is required to be valid for arbitrary σ, it follows that

$$(s_{11} + 2s_{12})(c_{11} + 2c_{12}) = 1. \tag{3.11}$$

The two equations (3.9) and (3.11) can be solved for s_{11} and s_{12} in terms of c_{11} and c_{12} to yield

$$s_{11} = \frac{c_{11} + c_{12}}{(c_{11} - c_{12}) \cdot (c_{11} + 2c_{12})}, \qquad s_{12} = \frac{-c_{12}}{(c_{11} - c_{12}) \cdot (c_{11} + 2c_{12})}. \tag{3.12}$$

Similarly, if the cubic crystal is subjected only to a pure shear stress σ_4, it is apparent from (3.3) and (3.6) that

$$\sigma_4 = c_{44}\epsilon_4 = c_{44}s_{44}\sigma_4 \qquad \Rightarrow \qquad s_{44} = \frac{1}{c_{44}}. \tag{3.13}$$

Together, (3.12) and (3.13) provide relationships between the three independent constants of the stiffness matrix s_{ij} and the three independent constants of the compliance matrix c_{ij}; these provide relationships between alternate representations of the response of a cubic material.

3.2.1 Directional variation of effective modulus

In thin films and multilayers that have cubic symmetry due to crystal structure, geometrical texturing or some other origin, it is of interest to find an effective elastic modulus that is identified with a specific crystallographic or material direction within the film. The *effective elastic modulus* associated with a particular direction in the material is defined as the ratio of the magnitude of a uniaxial stress to the magnitude of the resulting extensional strain in that direction. In this section, a general expression is derived for the effective elastic modulus for uniaxial loading

Table 3.1. *Elastic constants for cubic crystals at room temperature.*

Crystal structure	Material[a]	c_{11}	c_{44}	c_{12}	s_{11}	s_{44}	s_{12}	AR
		(in units of GPa)			(in units of GPa^{-1} × 10^3)			
Face-centered cubic	Ag	124.0	46.1	93.4	22.86	21.69	−9.82	3.01
	Al	107.3	28.3	60.9	15.82	35.36	−5.73	1.22
	Au	192.9	41.5	163.8	23.55	24.10	−10.81	2.85
	Cu	168.4	75.4	121.4	15.00	13.26	−6.28	3.21
	Ir	580.0	256.0	242.0	2.29	3.91	−0.07	1.51
	Ni	246.5	124.7	147.3	7.34	8.02	−2.74	2.51
	Pb	49.5	14.9	42.3	94.57	67.11	−43.56	4.14
	Pd	227.1	71.7	176.0	13.63	13.94	−5.95	2.81
	Pt	346.7	76.5	250.7	7.34	13.07	−3.08	1.59
Body-centered cubic	Cr	339.8	99.0	58.6	3.10	10.10	−0.46	0.70
	Fe	231.4	116.4	134.7	7.56	8.59	−2.78	2.41
	K	4.14	2.63	3.31	833.0	380.0	−370.0	6.34
	Li	13.5	8.78	11.44	332.8	113.9	−152.7	8.52
	Mo	440.8	121.7	172.4	2.91	8.22	−0.818	0.91
	Na	6.15	5.92	4.96	581.0	168.9	−259.4	9.95
	Nb (Cb)	240.2	28.2	125.6	6.5	35.44	−2.23	0.49
	Ta	260.2	82.6	154.5	6.89	12.11	−2.57	1.56
	V	228.0	42.6	118.7	6.82	23.5	−2.34	0.78
	W	522.4	160.8	204.4	2.45	6.22	−0.69	1.01
Diamond cubic	C	949.0	521.0	151.0	1.1	1.92	−1.51	1.31
	Ge	128.4	66.7	48.2	9.80	15.0	−2.68	1.66
	Si	166.2	79.8	64.4	7.67	12.54	−2.14	1.57
Compounds	GaAs	118.8	59.4	53.7	11.72	16.82	−3.65	1.82
	GaP	141.2	70.5	62.5	9.73	14.19	−2.99	1.79
	InP	102.2	46.0	57.6	16.48	21.74	−5.94	2.06
	KCl	39.5	6.3	4.9	26.0	158.6	−2.85	0.36
	LiF	114.0	63.6	47.7	11.65	15.71	−3.43	1.92
	MgO	287.6	151.4	87.4	4.05	6.6	−0.94	1.51
	NaCl	49.6	12.9	12.4	22.4	77.5	−4.48	0.69
	TiC	500.0	175.0	113.0	2.18	5.72	−0.40	0.90

[a] Data from G. Simmons and H. Wang, *Single Crystal Elastic Constants and Calculated Aggregate Properties: A Handbook*, Second edition, Massachusetts Institute of Technology Press, Cambridge, MA (1970), where complete references to original sources of these data can be found. The axes of reference are taken parallel to the cube axes of the crystals. The reported values can vary considerably as a result of differences in crystal growth methods, chemical purity, defect content, and testing methods.

along an arbitrary crystallographic direction defined by its Miller indices [*hkl*]. The uniaxial stress is denoted by σ_{hkl} and the corresponding effective modulus is denoted by E_{hkl}.

Suppose that a cubic crystal is subjected to a uniaxial tensile stress σ_{hkl} along the crystallographic direction [*hkl*]. The direction cosines of the oriented line represented by [*hkl*], with respect to the underlying rectangular $x_1 x_2 x_3$-coordinate axes, are denoted by $\mathcal{A}$, $\mathcal{B}$ and $\mathcal{C}$, respectively; in terms of the Miller indices,

$$\mathcal{A} = \frac{h}{\sqrt{h^2 + k^2 + l^2}}, \qquad \mathcal{B} = \frac{k}{\sqrt{h^2 + k^2 + l^2}}, \qquad \mathcal{C} = \frac{l}{\sqrt{h^2 + k^2 + l^2}}.$$

(3.14)

From the stress transformation rules consistent with the tensorial character of stress, it follows that the normal and shear stress acting in the coordinate directions are related to σ_{hkl} by

$$\sigma_1 = \mathcal{A}^2\sigma_{hkl}, \quad \sigma_2 = \mathcal{B}^2\sigma_{hkl}, \quad \sigma_3 = \mathcal{C}^2\sigma_{hkl},$$

$$\sigma_4 = \sigma_{23} = \mathcal{BC}\sigma_{hkl}, \quad \sigma_5 = \sigma_{31} = \mathcal{CA}\sigma_{hkl}, \quad \sigma_6 = \sigma_{12} = \mathcal{AB}\sigma_{hkl}. \quad (3.15)$$

From (3.4) it follows that the corresponding strains are

$$
\begin{aligned}
\epsilon_1 &= \left[\mathcal{A}^2 s_{11} + \mathcal{B}^2 s_{12} + \mathcal{C}^2 s_{12}\right]\sigma_{hkl}, & \epsilon_4 &= \left[\mathcal{BC} s_{44}\right]\sigma_{hkl}, \\
\epsilon_2 &= \left[\mathcal{A}^2 s_{12} + \mathcal{B}^2 s_{11} + \mathcal{C}^2 s_{12}\right]\sigma_{hkl}, & \epsilon_5 &= \left[\mathcal{CA} s_{44}\right]\sigma_{hkl}, \qquad (3.16)\\
\epsilon_3 &= \left[\mathcal{A}^2 s_{12} + \mathcal{B}^2 s_{12} + \mathcal{C}^2 s_{11}\right]\sigma_{hkl}, & \epsilon_6 &= \left[\mathcal{AB} s_{44}\right]\sigma_{hkl}.
\end{aligned}
$$

The component transformation rules consistent with the tensorial character of strain can then be used to represent the component of strain in the $[hkl]$ direction as

$$\epsilon_{hkl} = \left[\mathcal{A}^2\epsilon_1 + \mathcal{B}^2\epsilon_2 + \mathcal{C}^2\epsilon_3 + \mathcal{BC}\epsilon_4 + \mathcal{CA}\epsilon_5 + \mathcal{AB}\epsilon_6\right]\sigma_{hkl} \qquad (3.17)$$

$$= \left(\mathcal{A}^4 + \mathcal{B}^4 + \mathcal{C}^4\right)s_{11} + \left[\left(\mathcal{B}^2\mathcal{C}^2 + \mathcal{C}^2\mathcal{A}^2 + \mathcal{A}^2\mathcal{B}^2\right)\left(2s_{12} + s_{44}\right)\right]\sigma_{hkl}.$$

This representation can be simplified through the use of trigonometric identities to yield

$$\frac{1}{E_{hkl}} = \frac{\epsilon_{hkl}}{\sigma_{hkl}} = s_{11} + \frac{(2s_{12} - 2s_{11} + s_{44}) \cdot \left(k^2 l^2 + l^2 h^2 + h^2 k^2\right)}{\left(h^2 + k^2 + l^2\right)^2}. \qquad (3.18)$$

The maximum and minimum values of the effective elastic modulus, from among all possible line orientations in a cubic crystal, are found to exist for orientations along the $\langle 100 \rangle$ and $\langle 111 \rangle$ directions, respectively. The extreme values and their ratio are

$$E_{100} = \frac{1}{s_{11}}, \quad E_{111} = \frac{3}{s_{11} + 2s_{12} + s_{44}}, \quad \frac{E_{111}}{E_{100}} = \frac{3s_{11}}{s_{11} + 2s_{12} + s_{44}}. \qquad (3.19)$$

The ratio (3.19)$_3$ provides a measure of the degree of departure from isotropy in any particular cubic material. The value of the ratio may be either greater than or less than unity; $E_{111} > E_{100}$ for some cubic materials, such as Al, Cu, Si, Fe and Ni, while the reverse is true for other cubic materials, such as Cr, Mo and Nb.

The modulus E_{hkl} corresponding to any particular orientation $[hkl]$ can also be expressed in terms of the extreme values E_{100} and E_{111} for cubic materials by rewriting (3.18) as

$$\frac{1}{E_{hkl}} = \left[\frac{1}{E_{100}}\right] + \left\{\frac{3\left(h^2 k^2 + k^2 l^2 + l^2 h^2\right)}{\left(h^2 + k^2 + l^2\right)^2} \cdot \left[\frac{1}{E_{111}} - \frac{1}{E_{100}}\right]\right\}. \qquad (3.20)$$

3.2.2 *Isotropy as a special case*

For an isotropic crystal, E_{hkl} in (3.18) is independent of orientation. This is the case if the second term on the right side of (3.18) is independent of line orientation or, equivalently, if the ratio $(3.19)_3$ is equal to 1. Either condition is assured if $s_{44} = 2(s_{11} - s_{12})$. Equivalently, the special case of isotropy of a cubic material can be expressed in terms of components of the stiffness matrix c_{ij} as the condition that $2c_{44} = (c_{11} - c_{12})$. The degree of departure from isotropy in the response of a cubic material can be characterized by the anisotropy ratio AR in either of the equivalent forms

$$AR = \frac{2c_{44}}{c_{11} - c_{12}}, \qquad \text{or} \qquad AR = \frac{2(s_{11} - s_{12})}{s_{44}}. \qquad (3.21)$$

Note that $AR = 1$ for an isotropic crystal. The physical significance of this ratio is that it represents a measure of the relative response of cubic crystals to two different types of shear strain: c_{44} is an indicator of shear resistance in the $\langle 100 \rangle$ directions on $\{100\}$ planes, whereas the quantity $(c_{11} - c_{12})/2$ is an indicator of shear resistance in the $\langle 110 \rangle$ directions on $\{110\}$ planes. Values of anisotropy ratios for various cubic crystals are listed in Table 3.1. Among cubic metals, W is essentially isotropic and Al is very close to being so. Many materials listed in Table 3.1 have anisotropy ratios which depart significantly from unity.

Elastic deformation in isotropic materials is fully characterized in terms of two elastic constants, such as elastic modulus E and Poisson's ratio ν, or in terms of the Lamé constants, μ (the shear modulus) and λ. For an isotropic material, the various elastic constants are related by

$$E = \frac{1}{s_{11}}, \qquad \nu = -\frac{s_{12}}{s_{11}}, \qquad s_{12} = -\frac{\nu}{E},$$

$$\lambda = c_{12}, \qquad \mu = c_{44} = \frac{1}{s_{44}} = \frac{1}{2}(c_{11} - c_{12}) = \frac{1}{2(s_{11} - s_{12})},$$

$$\mu = \frac{E}{2(1 + \nu)}, \qquad \lambda = \frac{\nu E}{(1 + \nu)(1 - 2\nu)}. \qquad (3.22)$$

3.3 Elastic constants of non-cubic crystals

Five elastic constants are required to characterize the elastic response of a material with hexagonal symmetry. A convenient material coordinate frame in this case is a rectangular coordinate system with the x_3-axis aligned with the c-axis of the material, which is the axis with respect to which the material has sixfold rotational invariance. The x_1-direction and x_2-direction can then be chosen to coincide with any pair of orthogonal axes in the basal plane so as to form a right-handed coordinate system. The five elements of the matrix defining elastic response can be chosen to be c_{11}, c_{33}, c_{44}, c_{12} and c_{13}. Several other elements can be expressed in terms of these according to

$$c_{22} = c_{11}, \quad c_{55} = c_{44}, \quad c_{23} = c_{13}, \quad c_{66} = \tfrac{1}{2}(c_{11} - c_{12}) \tag{3.23}$$

and all other elements of c_{ij} vanish for hexagonal symmetry. The compliance constants s_{ij} are related to the stiffness constants c_{ij} of the hexagonal crystal by

$$
\begin{aligned}
s_{11} &= \frac{c_{33}c_{11} - c_{13}^2}{\left[c_{33}\left(c_{11}+c_{12}\right) - 2c_{13}^2\right]\left(c_{11}-c_{12}\right)}, \quad s_{66} = 2\left(s_{11} - s_{12}\right), \\
s_{33} &= \frac{c_{11} + c_{12}}{\left[c_{33}\left(c_{11}+c_{12}\right)\right] - 2c_{13}^2}, \quad s_{13} = -\frac{c_{13}}{\left[c_{33}\left(c_{11}+c_{12}\right) - 2c_{13}^2\right]}, \\
s_{12} &= \frac{-c_{33}c_{12} + c_{13}^2}{\left[c_{33}\left(c_{11}+c_{12}\right) - 2c_{13}^2\right]\left(c_{11}-c_{12}\right)}, \quad s_{44} = \frac{1}{c_{44}}.
\end{aligned}
\tag{3.24}
$$

Table 3.2 lists the room temperature elastic stiffness constants of some commonly encountered hexagonal crystals. The corresponding compliance constants are easily determined from (3.24).

For materials with orthorhombic symmetry (or orthotropic symmetry), there are three mutually orthogonal axes, each of twofold rotational symmetry. Translation of the material in the direction of any of these axes leaves the material behavior unaltered, but the translation distances required to recover the same lattice in crystalline materials are distinct, in general. Nine independent constants $(c_{11}, c_{22}, c_{33}, c_{44}, c_{55}, c_{66}, c_{13}, c_{23}, c_{12})$ are required to specify the elastic response of a general orthorhombic material. For a material in which one of the three symmetry axes has fourfold rotational symmetry (or two of the translational invariance distances are equal), the number of independent constants is reduced to six; this is achieved by requiring that $c_{22} = c_{11}, c_{55} = c_{44}$ and $c_{23} = c_{12}$, for example.

Elastic constants of bulk crystalline materials are commonly measured by means of the ultrasound method where sound velocities in specific crystallographic orientations are monitored (Kittel 1971). For example, the case of a cubic crystal for which the longitudinal wave speed along the [100] direction is $\sqrt{c_{11}/\rho}$, where ρ is the density of the crystal. Likewise, the shear wave speed along the [100] direction is $\sqrt{c_{44}/\rho}$. The speed of the longitudinal wave along the [111] direction is $\sqrt{c_{11} + 4c_{44} + 2c_{12}/3\rho}$, from which the elastic constant c_{12} can be determined. Other techniques have been developed in recent years to measure the elastic constants of thin films on substrates. For example, line-focus acoustic microscopy

Table 3.2. *Elastic constants for hexagonal crystals.*

Material[a]	c_{11}	c_{33}	c_{44} (in units of GPa)	c_{12}	c_{13}	Ref.
Be	292.3	336.4	162.5	26.7	14.0	a
C (graphite)	1160.0	46.6	2.3	290.0	109.0	b
Cd	115.8	51.4	20.4	39.8	40.6	c
Co	307.0	358.1	78.3	165.0	103.0	d
Hf	181.1	196.9	55.7	77.2	66.1	e
Mg	59.7	61.7	16.4	26.2	21.7	d
Ti	162.4	180.7	46.7	92.0	69.0	e
Zn	161.0	61.0	38.3	34.2	50.1	d
Zr	143.4	164.8	32.0	72.8	65.3	e
ZnO	209.7	210.9	42.5	121.1	105.1	f

[a] For the results shown in this table, the x_3 axis of reference is taken parallel to the c-axis and with the x_1 and x_2 reference axes taken on the basal (0001) plane of the crystal. All data, which represent room temperature elastic properties, are taken from the list provided in Kelly and Grove (1970). Individual references to the data are: [a] J. F. Smith and C. L. Arbogast, *J. Appl. Phys.*, **31**, p. 99 (1960). [b] G. B. Spence, Proc. Fifth Conference on Carbon, **2**, p. 531, Pergamon Press (1961). [c] C. W. Garland and J. Silverman, *Phys. Rev.*, **119**, p. 1218 (1960). [d] H. B. Huntington, *Solid State Phys.*, **7**, p. 213 (1958). [e] E. S. Fisher and C. J. Renken, *Phys. Rev.*, **135A**, p. 482 (1964). [f] T. B. Bateman, *J. Appl. Phys.*, **33**, p. 3309 (1962).

(LFAM) senses the velocity of a surface acoustic wave (SAW) on an elastic material (Kim et al. 1993) from which the elastic constants can be extracted.

3.4 Elastic strain in layered epitaxial materials

If two single crystals share a plane interface and if every boundary atom on one side of the interface occupies a natural lattice site of the crystal on the other side, then the interface is said to exhibit *epitaxial* bonding, as described in Section 1.4.1. In the most common situation, the two crystals belong to the same crystal class, for example, diamond cubic or zinc blende, and have a common crystallographic orientation with respect to the interface plane. In epitaxial growth of a thin film on a substrate, the substrate crystal is used as a growth template and the film material is deposited atom by atom on the surface at a temperature sufficiently high to permit the deposited atoms to seek out the positions that are most favorable energetically, presumably lattice sites. If this occurs in a layer by layer fashion, then the film becomes a perfect crystal that continues the structure of the substrate. If the deposited material is the same as the substrate material, then the process is called *homoepitaxy*; if the materials are different the process is called *heteroepitaxy*. An interface without extended crystal defects is often described as being *fully coherent*; if the interface incorporates some extended defects it is described as being partially coherent.

In the case of heteroepitaxy, the natural or stress-free lattice spacing of the film material a_f in some direction along the interface is different from the natural lattice parameter of the substrate a_s in that direction, in general. As a result, the film is

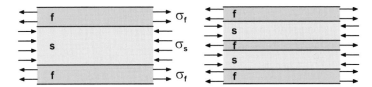

Fig. 3.1. Illustrations of two epitaxial multilayers comprising alternating layers of materials identified as film 'f' and substrate 's' materials for convenience. The multilayers are chosen to be symmetric to avoid overall bending of the structures and to each contain film material of total thickness h_f and substrate material of total thickness h_s. Otherwise, the arrangement of layers is arbitrary. For lattice mismatch ϵ_m, the elastic mismatch strains are given in (3.28).

said to have a lattice mismatch with respect to the substrate in that direction with the magnitude of the mismatch represented by the parameter

$$\epsilon_m = \frac{a_s - a_f}{a_f}. \tag{3.25}$$

If the film thickness h_f, which is assumed to be uniform for the time being, is very small compared to the substrate thickness h_s, then the elastic mismatch is accommodated entirely within the film. The result is that the film is subjected to a uniform extensional elastic strain in the direction being considered and that the magnitude of that elastic strain is ϵ_m. It is important to notice the distinction between lattice mismatch and elastic strain that results from mismatch; the former is a system parameter independent of any particular configuration, whereas the latter is the actual elastic strain field due to the presence of lattice mismatch. In more complex configurations, the mismatch elastic strain can be a spatially nonuniform field while the lattice mismatch is simply a number. If the crystal structure is such that ϵ_m is the lattice mismatch in orthogonal directions in the interface, then the mismatch elastic strain of the thin film is equi-biaxial and of magnitude ϵ_m. The generation of mismatch strain due to lattice mismatch is illustrated schematically in Figure 1.14.

To expand on the general idea of lattice mismatch and the elastic mismatch strain induced by it, consider a multilayer structure comprising alternating layers of cubic crystals. One material will be designated as the film material and the other as the substrate material, although these terms are now simply identifiers rather than literal descriptors. Two such structures are shown in Figure 3.1 where the layering differs between the two cases but the total amount of each material is the same. The layering is chosen with reflective symmetry with respect to the midplane in order to avoid the complications of bending of the structure in the presence of lattice mismatch. No externally applied forces or moments act on the structure, and edge effects are considered to be remote and negligible. The lattice mismatch of the film material with respect to the substrate material is ϵ_m. However, neither material supports an elastic mismatch strain equal to ϵ_m, in general.

Denote the equi-biaxial mismatch elastic strain in the film material by ϵ_f and in the substrate by ϵ_s. Compatibility of deformation requires that

$$\epsilon_f - \epsilon_s = \epsilon_m. \tag{3.26}$$

The equi-biaxial stress associated with ϵ_f and ϵ_s is $\sigma_f = M_f \epsilon_f$ and $\sigma_s = M_s \epsilon_s$, respectively, where M_f and M_s are the appropriate biaxial moduli for the materials at hand. Equilibrium requires that the net force on any cross-section must be zero, so that

$$M_f \epsilon_f h_f + M_s \epsilon_s h_s = 0. \tag{3.27}$$

The net bending moment on any section vanishes due to symmetry. Together, the two conditions (3.26) and (3.27) imply that

$$\epsilon_f = \epsilon_m \frac{h_s M_s}{h_f M_f + h_s M_s} , \quad \epsilon_s = -\epsilon_m \frac{h_f M_f}{h_f M_f + h_s M_s} . \tag{3.28}$$

Note that $\epsilon_f \to \epsilon_m$ as $h_f / h_s \to 0$, so that the thin film limit is recovered, and that $\epsilon_s \to -\epsilon_m$ as $h_f / h_s \to \infty$, which provides the proper thin substrate limit. The common lattice parameter a_{int} shared by the two materials in this configuration is intermediate between a_f and a_s, and it has the value

$$a_{int} = a_f (1 + \epsilon_f) = a_s (1 + \epsilon_s) . \tag{3.29}$$

Evidently, $a_{int} \to a_s$ as $\epsilon_s \to 0$ and $a_{int} \to a_f$ as $\epsilon_f \to 0$, as expected.

The in-plane extensional strain in the thin film layers is given by (3.28). The extensional strain in the direction perpendicular to the interfaces, however, depends on their crystallographic orientation and anisotropy, as will be illustrated in the next section. If the material behavior is assumed to be isotropic, then these extensional strains in the z-direction, commonly denoted ϵ_{zz}, in the two materials are

$$\epsilon_f^{\perp} = -\frac{2\nu_f}{1 - \nu_f} \epsilon_f , \quad \epsilon_s^{\perp} = -\frac{2\nu_s}{1 - \nu_s} \epsilon_s . \tag{3.30}$$

It is this strain component that is commonly detected by x-ray diffraction observations of thin film deformation. Note that a_{int} refers only to the lattice spacing in a direction parallel to the interface and that the lattice spacing perpendicular to the interface has a different value. In general, the spacing of lattice planes that are parallel to the interface, which is measured in a direction *perpendicular* to the interface, is

$$a_f^{\perp} = a_f \left(1 + \epsilon_f^{\perp}\right), \quad a_s^{\perp} = a_s \left(1 + \epsilon_s^{\perp}\right). \tag{3.31}$$

3.5 Film stress for a general mismatch strain

In this section, the question of film stress in an anisotropic film is addressed for cases in which the mismatch strain may be more general than equi-biaxial strain and in which the film anisotropy may be quite general. The anisotropy can arise from single crystal behavior, film material inhomogeneity or patterning, or other effects. For example, consider the case of a single crystal thin film deposited epitaxially on the plane surface of a relatively thick crystalline substrate. A rectangular coordinate system is introduced in this configuration with its origin at a point on the film–substrate interface. The axes of this coordinate system, which will be referred to as the global coordinate system, are oriented so that the interface is the x_1x_2-plane and the x_3-axis extends into the film.

The mismatch strain is that strain which must be imposed on the film material in order to bring it into perfect atomic registry with the unrestrained substrate. Because this involves straining only in directions parallel to the interface, the specification of mismatch strain involves only the components of $\epsilon_{11}^{(m)}$, $\epsilon_{22}^{(m)}$ and $\epsilon_{12}^{(m)}$ with respect to the global coordinate frame. The magnitudes of these three components are customarily given in terms of relative lattice dimensions, in the case of epitaxy, or in terms of coefficients of thermal expansion and temperature change, in the case of thermal strain. In any case, three strain components are presumed to be known at the outset. The film is very thin compared to the substrate, so that the strain in the film is $\epsilon_{ij}^{(m)}$ after it is bonded to the substrate, with three of the six independent components being known. The mismatch strain may depend on position through the thickness of the film, but this possibility will not be taken into account explicitly in the course of the following discussion.

The surface of the film at $x_3 = h_f$ is traction-free, so that

$$\sigma_{33}^{(m)} = \sigma_{13}^{(m)} = \sigma_{23}^{(m)} = 0 \tag{3.32}$$

on that surface for any mismatch strain. If the film is invariant under translation in any direction parallel to the interface (edge effects are being neglected), it follows immediately that these three stress components are zero everywhere throughout the film, as discussed in Section 2.4. Therefore, in addition to three components of mismatch strain being known *a priori*, three components of mismatch stress $\sigma_{ij}^{(m)}$ are also known. These observations are based only on geometrical invariance under lateral translation, and the implications are valid for any orientation of the (possibly anisotropic) material.

3.5.1 Arbitrary orientation of the film material

Suppose that the properties of the film material are given with respect to some 'natural' rectangular $x_1^*x_2^*x_3^*$-coordinate system referred to the material itself. For example, the natural coordinate axes might be the cube edge axes for a material with cubic symmetry, or they might be normal to the planes of reflective symmetry for an orthotropic material. In any case, the elastic stiffness properties of the material are given by the components c_{ij}^*, where the star '*' is used here and elsewhere to indicate that the material constants are referred to the material coordinate frame. The coordinates at a point in a natural material when referred to the material coordinate system x_i^* are linearly related to the coordinates of the same physical point when referred to the underlying global coordinate system x_i by

$$x_k^* = Q_{ki} x_i, \tag{3.33}$$

where the ki-component of Q_{ki} is the direction cosine of the x_k^*-axis with respect to the x_i-axis. If $\mathbf{e}_i$ and $\mathbf{e}_i^*$ are unit vectors aligned with the coordinate axes in the global frame and the material frame, respectively, then $Q_{ki} = \mathbf{e}_k^* \cdot \mathbf{e}_i$. The elements of Q_{ki} are the components of a rotation tensor with the property that $Q_{ij} Q_{kj} = \delta_{ik}$, where δ_{ik} is the identity matrix. Once the two coordinate systems are specified, the components of Q_{ij} can be determined.

The components of the strain tensor, which are $\epsilon_{ij}^{(m)}$ when referred to the global coordinate system, can then be expressed with respect to the material coordinate system as

$$\epsilon_{ij}^{*(m)} = Q_{ik} \epsilon_{kl}^{(m)} Q_{jl}. \tag{3.34}$$

In the natural material coordinate system, the contracted notation whereby $\epsilon_1^{*(m)}, \ldots, \epsilon_6^{*(m)} = \epsilon_{11}^{*(m)}, \ldots, \epsilon_{12}^{*(m)}$ is merely an alternative way to represent the mismatch strain. It follows that the corresponding mismatch stress, also in contracted notation, is

$$\sigma_i^{*(m)} = c_{ij}^* \epsilon_j^{*(m)}. \tag{3.35}$$

In light of (3.33) through (3.35), the components of the full stress tensor $\sigma_{ij}^{*(m)}$ referred to material coordinates can then be reconstructed from the six components in contracted notation as

$$\sigma_{ij}^{*(m)} = \begin{bmatrix} \sigma_1^{*(m)} & \sigma_6^{*(m)} & \sigma_5^{*(m)} \\ \sigma_6^{*(m)} & \sigma_2^{*(m)} & \sigma_4^{*(m)} \\ \sigma_5^{*(m)} & \sigma_4^{*(m)} & \sigma_3^{*(m)} \end{bmatrix}. \tag{3.36}$$

Finally, the components of the stress tensor $\sigma_{ij}^{(m)}$ in the global coordinate system are given by

$$\sigma_{ij}^{(m)} = Q_{ki} \sigma_{kl}^{*(m)} Q_{lj}. \tag{3.37}$$

This provides an expression for the six independent components of mismatch stress in terms of the six independent components of mismatch strain in the global coordinate system. At this point, only three components of $\epsilon_{ij}^{(m)}$ are known. However, the conditions (3.32), when applied to (3.37), provide three linear equations for $\epsilon_{33}^{(m)}$, $\epsilon_{13}^{(m)}$ and $\epsilon_{23}^{(m)}$ in terms of the known components $\epsilon_{11}^{(m)}$, $\epsilon_{22}^{(m)}$ and $\epsilon_{12}^{(m)}$. Once these linear equations are solved and the results are substituted back into (3.37), the mismatch stress is known completely.

It is possible to express the mismatch stress in terms of mismatch strain for general anisotropy without reference to the substrate because the issue has been pursued under the same set of assumptions that underlie the derivation of the Stoney formula as outlined in Section 2.1. Recall the earlier assumptions that the film is very thin compared to the substrate and that the change in film strain associated with curvature of the substrate is small compared to the mismatch strain itself. Under these conditions, the film strain and film properties determine the stress which results in substrate curvature. The details of the resulting curvature do indeed depend on the properties of the substrate, and this issue will be taken up in Section 3.7.

Alternatively, the description of material properties which is known *a priori* in the material coordinate system could be transformed into the global coordinate system. The sequence of steps that accomplish this transformation are as follows: c_{ijkl}^{*} is constructed from the components of c_{ij}^{*}, then c_{ijkl} is determined from it by the rules for transformation of components of the tensor under coordinate transformations according to

$$c_{ijkl} = Q_{mi} Q_{nj} Q_{pk} Q_{ql} c_{mnpq}^{*} \tag{3.38}$$

and c_{ij} is constructed from this result. With this result in hand, the expression of Hooke's law in contracted notation referred to the global coordinate system is

$$\sigma_i^{(m)} = c_{ij} \epsilon_j^{(m)}. \tag{3.39}$$

Once again, it is possible to determine mismatch stress from the three known components of $\epsilon_i^{(m)}$ and from the conditions that $\sigma_3^{(m)} = \sigma_4^{(m)} = \sigma_5^{(m)} = 0$. The reason that these approaches appear to be cumbersome is that $\sigma_i^{(m)}$, $\epsilon_i^{(m)}$ and c_{ij}, representations chosen for their convenience for calculation and ease of interpretation rather than their underlying mathematical structure, are not components of tensors. In particular, their components do not transform under transformation of coordinate systems as do the components of tensors. Consequently, before components of any physical quantity represented by a tensor of any order can be transformed from one coordinate system to another, the array must be put into its *bona fide* tensor form.

As a simple illustration, consider a thin film of cubic material deposited on a planar substrate surface. Furthermore, suppose that the cube axes coincide with the axes of the global coordinate system. In other words, the material directions [100],

[010] and [001] coincide with the x_1-axis, the x_2-axis and the x_3-axis in the global coordinate system. Then $Q_{ij} = \delta_{ij}$, the identity matrix. The mismatch strain is assumed to be equi-biaxial extensional strain so that $\epsilon_{11}^{(m)} = \epsilon_m$, $\epsilon_{22}^{(m)} = \epsilon_m$ and $\epsilon_{12}^{(m)} = 0$. The mismatch strain is then

$$\epsilon_i^{(m)} = \left\{ \epsilon_m, \ \epsilon_m, \ \epsilon_{33}^{(m)}, \ \epsilon_{23}^{(m)}, \ \epsilon_{13}^{(m)}, \ 0 \right\} \tag{3.40}$$

when referred to either the global or material coordinate axes. It follows that the mismatch stress is given by

$$\sigma_i^{(m)} = c_{ij} \epsilon_j^{(m)}. \tag{3.41}$$

The vanishing of the three stress components according to (3.32) implies that

$$c_{12}\epsilon_m + c_{12}\epsilon_m + c_{11}\epsilon_{33}^{(m)} = 0,$$
$$c_{44}\epsilon_{23}^{(m)} = 0, \quad c_{44}\epsilon_{13}^{(m)} = 0, \tag{3.42}$$

from which it follows that

$$\epsilon_{33}^{(m)} = -\frac{2c_{12}}{c_{11}}\epsilon_m, \quad \epsilon_{13}^{(m)} = \epsilon_{23}^{(m)} = 0. \tag{3.43}$$

These results, which generalize the corresponding results for isotropy given in (3.30), are then substituted back into (3.41) to find that

$$\sigma_{12}^{(m)} = 0, \quad \sigma_{11}^{(m)} = \sigma_{22}^{(m)} = \left(c_{11} + c_{12} - \frac{2c_{12}^2}{c_{11}} \right)\epsilon_m. \tag{3.44}$$

Thus, a cubic material that has its cube axes aligned with the global coordinate axes and that is subjected to an equi-biaxial mismatch strain carries an equi-biaxial mismatch stress. The response is isotropic in the plane of the interface with an *effective biaxial modulus* in the plane with normal in the (001) material direction of

$$M_{(001)} = \frac{\sigma_m}{\epsilon_m} = \left(c_{11} + c_{12} - \frac{2c_{12}^2}{c_{11}} \right). \tag{3.45}$$

3.5.2 Example: Cubic thin film with a (111) orientation

Suppose that a thin film of a cubic material is deposited on a substrate as a single crystal, oriented so that its (111) crystallographic plane is parallel to the substrate surface. The elastic constants of the material are c_{ij}^* in a coordinate system aligned with the cube axes of the material. A global coordinate system is introduced with the x_3-axis normal to the film–substrate interface, as in Section 3.5.1. In this frame, the known components of mismatch strain are $\epsilon_{11}^{(m)} = \epsilon_{22}^{(m)} = \epsilon_m$, $\epsilon_{12}^{(m)} = 0$. Determine

(a) the remaining components of $\epsilon_{ij}^{(m)}$ which are not dictated by the constraint of epitaxy,

(b) all components of mismatch stress $\sigma_{ij}^{(m)}$, and

(c) the effective biaxial modulus, say $M_{(111)}$, for this orientation and the imposed mismatch strain.

Solution:

The approach outlined in Section 3.5 will be followed to solve this problem. For definiteness, suppose that the material $x_1^* x_2^* x_3^*$-coordinate system is oriented so that $\mathbf{e}_1^* \cdot \mathbf{e}_2 = 0$ and that $\mathbf{e}_1^* \cdot \mathbf{e}_3 = \mathbf{e}_2^* \cdot \mathbf{e}_3 = \mathbf{e}_3^* \cdot \mathbf{e}_3 = 1/\sqrt{3}$; it follows immediately that $\mathbf{e}_1^* = \mathbf{e}_1 \sqrt{2/3} + \mathbf{e}_3/\sqrt{3}$. From the geometry of the configuration, it is evident that the other unit vectors must be of the form $\mathbf{e}_2^* = -p\mathbf{e}_1 + q\mathbf{e}_2 + \mathbf{e}_3/\sqrt{3}$ and $\mathbf{e}_3^* = -p\mathbf{e}_1 - q\mathbf{e}_2 + \mathbf{e}_3/\sqrt{3}$, where p and q are positive numbers to be determined. The orthogonality conditions $\mathbf{e}_2^* \cdot \mathbf{e}_1^* = 0$ and $\mathbf{e}_2^* \cdot \mathbf{e}_3^* = 0$ then yield the values $p = 1/\sqrt{6}$ and $q = 1/\sqrt{2}$. With the unit vectors in material coordinate directions known, the transformation matrix Q_{ki} is calculated to be

$$Q_{ki} = \begin{bmatrix} \sqrt{2/3} & 0 & 1/\sqrt{3} \\ -1/\sqrt{6} & 1/\sqrt{2} & 1/\sqrt{3} \\ -1/\sqrt{6} & -1/\sqrt{2} & 1/\sqrt{3} \end{bmatrix}. \tag{3.46}$$

(a) With the coordinate transformation known between global coordinates and material coordinates, the solution proceeds according to the steps outlined in Section 3.5.1. The expressions produced can be somewhat lengthy, but the procedure is straightforward and ideally suited for symbolic computation. The strain tensor in material coordinates can be determined from (3.34), from which the form $\epsilon_i^{*(m)}$ in contracted notation can be determined; for example, the first component is $\epsilon_1^{*(m)} = \frac{1}{3}\left(2\sqrt{2}\,\epsilon_{13}^{(m)} + \epsilon_{33}^{(m)} + 2\epsilon_m\right)$. The mismatch stress components $\sigma_i^{*(m)}$ in contracted notation follow from (3.35), whereupon the corresponding stress tensor referred to material coordinates can be calculated according to (3.36); the 11-component of this tensor is found to be

$$\sigma_{11}^{*(m)} = \frac{1}{3}\left[2c_{12}^*\left(-\sqrt{2}\,\epsilon_{13}^{(m)} + \epsilon_{33}^{(m)} + 2\epsilon_m\right) + c_{11}^*\left(2\sqrt{2}\,\epsilon_{13}^{(m)} + \epsilon_{33}^{(m)} + 2\epsilon_m\right)\right] \tag{3.47}$$

for example, where the elastic constants c_{ij}^* are written with a superposed $*$ to make clear that the components are referred to the material coordinate frame rather than the global frame. The components of mismatch stress in global coordinates are then given by (3.37). Recall that three particular components of stress are required to vanish in all cases according to (3.32); in the present case, this requires that

$$\sigma_{33}^{(m)} = \frac{1}{3}\left[4c_{44}^*(\epsilon_{33}^{(m)} - \epsilon_m) + c_{11}^*(\epsilon_{33}^{(m)} + 2\epsilon_m) + 2c_{12}^*(\epsilon_{33}^{(m)} + 2\epsilon_m)\right] = 0,$$

$$\sigma_{13}^{(m)} = \frac{2}{3}(c_{11}^* - c_{12}^* + c_{44}^*)\epsilon_{13}^{(m)} = 0, \tag{3.48}$$

$$\sigma_{23}^{(m)} = \frac{2}{3}(c_{11}^* - c_{12}^* + c_{44}^*)\epsilon_{23}^{(m)} = 0.$$

These equations can be solved for the components of $\epsilon_{ij}^{(m)}$ that are not specified by the constraint of epitaxy, with the result that

$$\epsilon_{33}^{(m)} = -\frac{2(c_{11}^* + 2c_{12}^* - 2c_{44}^*)\epsilon_m}{c_{11}^* + 2c_{12}^* + 4c_{44}^*},$$

$$\epsilon_{13}^{(m)} = 0, \quad \epsilon_{23}^{(m)} = 0. \tag{3.49}$$

(b) If the strain components (3.49) are substituted into the expression for mismatch stress $\sigma_{ij}^{(m)}$, then the mismatch stress is known in terms of ϵ_m. The full array of mismatch stress components is

$$\sigma_{ij}^{(m)} = \frac{6(c_{11}^* + 2c_{12}^*)c_{44}^*}{c_{11}^* + 2c_{12}^* + 4c_{44}^*} \begin{bmatrix} \epsilon_m & 0 & 0 \\ 0 & \epsilon_m & 0 \\ 0 & 0 & 0 \end{bmatrix}. \tag{3.50}$$

Note that the fact that all shear stress components vanish is a result of the calculation; it was not presumed to be true at the outset.

(c) It is evident from (3.50) that the mismatch stress in the film is invariant under rotation of the material about the particular {111} direction that is normal to the interface. In other words, the material appears to be transversely isotropic with effective biaxial modulus

$$M_{(111)} = \frac{6(c_{11}^* + 2c_{12}^*)c_{44}^*}{c_{11}^* + 2c_{12}^* + 4c_{44}^*} \tag{3.51}$$

in the plane of the interface. The equi-biaxial mismatch stress is $\sigma_m = M_{(111)}\epsilon_m$. It is readily confirmed that, for the special case of isotropy of the film material with $c_{44}^* = \frac{1}{2}(c_{11}^* - c_{12}^*)$, the biaxial modulus reduces to the form $E/(1-\nu)$.

3.6 Film stress from x-ray diffraction measurement

X-ray diffraction is a readily available experimental method that can be used to infer values of stress in thin crystalline films on substrates. With this method, the normal spacing d_{hkl} between adjacent crystallographic planes within a family of planes with indices (hkl) is determined by means of Bragg's law in the form

$$d_{hkl} = \frac{n\lambda_X}{2\sin\theta}, \tag{3.52}$$

where λ_X is the wavelength of the x-ray beam incident on the film and θ is the angle of incidence and reflection, as illustrated in Figure 3.2(a). Measurement is based on the first-order reflected beam, so that $n = 1$ in (3.52). Typically, the intensity $I(\theta)$ of the scattered x-ray beam is monitored as a function of 2θ, as indicated in Figure 3.2(b). For a crystal with low defect density, periodicity of the material results in fairly narrow intensity peaks in the scattered x-ray field at angles at which constructive interference occurs. The position 2Θ of the first-order peak in the $I(\theta)$ versus 2θ plot is used to determine d_{hkl} from values of λ_X and Θ according to

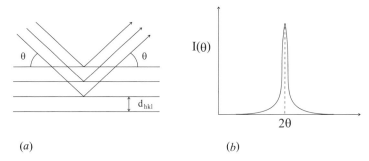

Fig. 3.2. Schematic showing (a) the reflection of x-rays by crystallographic planes, and (b) a plot of the intensity of x-ray beams versus 2θ.

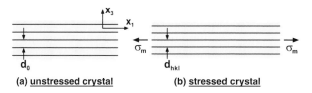

Fig. 3.3. Schematic showing d-spacings between crystallographic planes in (a) an unstressed crystal and (b) the same crystal subjected to an equi-biaxial mismatch stress σ_m.

(3.52) (Cullity 1978). This measured value is commonly identified as the 'd-spacing' for the crystallographic direction represented by (hkl).

3.6.1 Relationship between stress and d-spacing

Consider an isotropic film on a substrate for a case in which the equi-biaxial mismatch stress σ_m and the corresponding mismatch strain ϵ_m are to be estimated by recourse to d-spacing measurements involving x-ray diffraction. The state of strain in the thin crystalline film on its substrate alters the interplanar spacing of a particular family of planes from the value in an unstressed crystal, say d_0; see Figures 3.3(a) and (b). Thus, the difference in d-spacings of crystallographic planes within a family between the strained and unstrained states of a thin film, as determined by x-ray diffraction, provides a measure of the elastic strain components in the film.

Based on the discussion in Sections 2.1 and 3.5, the nonzero components of stress in the film, with reference to the $x_1x_2x_3$-coordinate system shown in Figure 3.3, are $\sigma_{11}^{(m)} = \sigma_{22}^{(m)} = \sigma_m$. The corresponding nonzero strain components are

$$\epsilon_{11}^{(m)} = \epsilon_{22}^{(m)} = \epsilon_m = \frac{\sigma_m}{M_f}, \quad \epsilon_{33}^{(m)} = -\frac{2\nu_f}{1 + \nu_f}\epsilon_m = -\frac{2\nu_f}{E_f}\sigma_m. \quad (3.53)$$

This provides a relationship between the mismatch stress σ_m and the out-of-plane strain $\epsilon_{33}^{(m)}$ in the film. Referring to Figure 3.3, the extensional elastic strain normal to the plane of the film, expressed in terms of the d-spacings in the deformed and undeformed states, is

$$\epsilon_{33}^{(m)} = \frac{d_{hkl} - d_0}{d_0}. \tag{3.54}$$

Note that calculation of $\epsilon_{33}^{(m)}$ using this equation presumes a knowledge of the reference d-spacing value d_0 of the film in the unstressed state. Equations (3.53) and (3.54) together yield the mismatch stress in the film in terms of the measured d-spacing as

$$\sigma_m = -\frac{E_f}{2\nu_f} \left(\frac{d_{hkl} - d_0}{d_0} \right). \tag{3.55}$$

A tensile mismatch stress in the film implies that $\sigma_m > 0$ and, therefore, $\epsilon_{33}^{(m)} < 0$.

The same line of reasoning applies for the case of an anisotropic material supporting an equi-biaxial extensional mismatch strain and zero shear strain. Consider a cubic single crystal thin film with a (001) film–substrate interface plane. The mismatch strain and mismatch stress in this film can be estimated using x-ray diffraction measurements along with (3.43) and (3.45) by noting that

$$\epsilon_m = -\frac{c_{11}}{2c_{12}} \left(\frac{d_{001} - d_0}{d_0} \right),$$

$$\sigma_m = -\frac{c_{11}}{2c_{12}} \left(c_{11} + c_{12} - \frac{2c_{12}^2}{c_{11}} \right) \left(\frac{d_{001} - d_0}{d_0} \right). \tag{3.56}$$

Similar expressions can be obtained for a cubic thin film with a (111) film–substrate interface orientation using (3.49) and (3.51).

A homogeneous strain in a film gives rise to a sharp peak in the intensity of the x-ray beam at the Bragg angle; see Figure 3.2. However, if the mismatch strain distribution in the film is spatially inhomogeneous, local gradients in strain within the region sensed by the incident x-ray beam would cause x-ray peak broadening, resulting in a more diffuse intensity plot in Figure 3.2(b) (Murakami 1978, Murakami 1982).

Grazing incidence x-ray scattering is a particular application of the x-ray diffraction method for stress measurement whereby the incident x-ray beams are made to impinge on the film surface at a very low angle of incidence, typically 0.2 degrees. This low angle of incidence results in reflection of the x-rays out of the film rather than penetration into the film. By varying the grazing angle, strain over different depths can be sampled by the x-rays. This method affords the possibility of measuring the d-spacings of planes that are oriented normally to the film–substrate interface, whereby the mismatch strain can be measured directly (Doerner and Brennan 1988).

3.6.2 Example: Stress implied by measured d-spacing

The residual strains in an aluminum film, 1 μm in thickness, on a Si substrate, 550 μm thick and 200 mm in diameter, were studied by x-ray diffraction. It was found that the d-spacing of crystallographic planes in the film, which were oriented parallel to the plane of the film, decreased by 0.054%, compared to the situation when the film was stress-free.

(a) If the film is assumed to be elastically isotropic, find the equi-biaxial mismatch stress and mismatch strain in the film. The elastic modulus and Poisson ratio of the film are 70 GPa and 0.33, respectively.
(b) Repeat the calculation assuming that the aluminum film has (001) texture.

Solution:

(a) From the observation that the d-spacing of crystallographic planes in the film, which were oriented parallel to the plane of the film, decreased by 0.054%, it follows that $\epsilon_{33}^{(m)} = -0.00054$. From (3.54), the mismatch stress in the film is determined to be

$$\sigma_m = -\frac{E_f}{2\nu_f} \epsilon_{33}^{(m)} = 57.3 \text{ MPa}. \tag{3.57}$$

The equi-biaxial mismatch strain in the film is

$$\epsilon_{11}^{(m)} = \epsilon_{22}^{(m)} = \sigma_m \frac{1 - \nu_f}{E_f} = 0.00055. \tag{3.58}$$

(b) From $(3.56)_1$, the mismatch strain in the film with the (100) texture is written as

$$\epsilon_m = -\frac{c_{11}}{2c_{12}} \epsilon_{33}^{(m)}. \tag{3.59}$$

Noting from Table 3.1 that $c_{11} = 107.3$ GPa and $c_{12} = 60.9$ GPa for aluminum, it follows that $\epsilon_m = 0.00047$. Substituting the values of c_{11} and c_{12} into (3.45), the biaxial modulus of the aluminum film with the (001) texture is found to be 99.1 GPa. This result, in conjunction with $(3.56)_2$, provides the mismatch stress in the film as $\sigma_m = 47.2$ MPa.

3.6.3 Stress-free d-spacing from asymmetric diffraction

The foregoing discussion deals with situations in which the family of diffracting planes is parallel to the film–substrate interface plane so that the incident and reflecting beams of x-rays are symmetrically oriented with respect to the normal to the film surface, as shown in Figure 3.2(a). The measurement of change in d-spacing from this symmetric diffraction leads to the determination of a single component of extensional strain in the film, namely $\epsilon_{33}^{(m)}$. The other components of both strain and

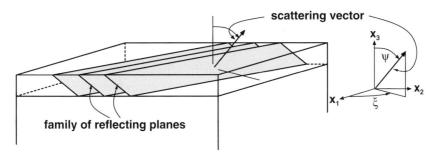

Fig. 3.4. Schematic diagram showing a family of diffracting crystallographic planes in a film. The planes are oriented so that their normal vector, or scattering vector, is at an angle ψ with respect to the normal to the film–substrate interface surface. The projection of the reflecting plane normal vector onto the plane of the interface is oriented at an angle ψ with respect to the x_1-axis. The incident and reflected x-ray beams are both inclined at an angle θ with respect to the reflecting planes, as in Figure 3.3.

stress are then determined according to (3.53) through (3.56). Such an approach presumes *a priori* knowledge of the unstressed or reference d-spacing value d_0 of the diffracting plane. Although values of d_0 are available in the literature for commonly used bulk crystals, it may be necessary in some situations to determine values of d_0 for thin films experimentally (Clemens and Bain 1992, Cornella et al. 1997). For example, variations in composition and impurity content arising from variations in film deposition methods, diffusion, interfacial reactions or environmental interactions can cause the unstrained lattice spacings for thin films to differ markedly from those determined for bulk materials. In addition, some materials are available *only* in thin film form, in which case d_0 values may not be available in the literature. It is necessary to determine the evolution of film stress as a function of temperature in some instances, and values of d_0 as functions of temperature are not commonly available for many materials. In addition, lattice spacing values for unstressed crystals are usually determined from known elastic constants. For some thin film materials, such estimates of elastic properties may not be readily available at the temperature of interest. Finally, the ability to determine d_0 at different temperatures through x-ray diffraction would also provide a means to estimate the coefficient of thermal expansion of the film without the need to isolate the film from its substrate and without information about the elastic properties of the film.

Nonsymmetric x-ray diffraction is an experimental technique that can be used to infer values of mismatch strain in a thin film by monitoring the intensity of x-rays diffracted from planes that are not parallel to the film–substrate interface. Figure 3.4 schematically shows a thin film on a substrate where the normal to the film–substrate interface plane is oriented at an angle ψ with respect to the normal to the diffracting planes; the unit vector normal to the diffracting planes is commonly known as the scattering vector. The extensional strain in the direction of the scattering vector is denoted by ϵ_ψ. This strain can be determined for any value of ψ in terms of the strain components referred to the coordinate axes, using the coordinate transformation procedure described in Section 3.5.1.

Consider once again a thin film with equi-biaxial mismatch strain $\epsilon_{11}^{(m)} = \epsilon_{22}^{(m)} = \epsilon_m$ in the plane of the film–substrate interface, out-of-plane extensional strain $\epsilon_{33}^{(m)}$ and zero shear strain everywhere. In this case, the extensional strain in the direction of the scattering vector is related to the strain components in the coordinate directions according to

$$\epsilon_\psi = \epsilon_m \sin^2 \psi + \epsilon_{33}^{(m)} \cos^2 \psi = \epsilon_{33}^{(m)} + \left(\epsilon_m - \epsilon_{33}^{(m)} \right) \sin^2 \psi . \tag{3.60}$$

Alternatively, this same strain can be expressed in terms of the d-spacing values of the diffracting planes as

$$\epsilon_\psi = \frac{d_\psi - d_0}{d_0}, \tag{3.61}$$

where, for convenience, the deformed spacing of the particular family of planes being interrogated is here denoted by d_ψ rather than by d_{hkl}. The two expressions (3.60) and (3.61) can be combined to yield an expression for the measured spacing of the strained film in terms of other system parameters as

$$d_\psi = d_0 \left(1 - \epsilon_{33}^{(m)} \right) + d_0 \left(\epsilon_m - \epsilon_{33}^{(m)} \right) \sin^2 \psi . \tag{3.62}$$

Suppose that the film material is an isotropic elastic material, so that the relationships between mismatch strain and mismatch stress noted in (3.53) apply. In this case, (3.62) takes the form

$$d_\psi = d_0 - \frac{2\nu_f}{E_f} \sigma_m d_0 + \frac{1 + \nu_f}{E_f} \sigma_m d_0 \sin^2 \psi . \tag{3.63}$$

If d_0 and the elastic constants of the film material would be known, it follows that the mismatch stress σ_m could be inferred from a measurement of d_ψ by means of this expression. Note that there is some direction in the material, say ψ_0 defined by

$$\sin^2 \psi_0 = \frac{2\nu_f}{1 + \nu_f}, \tag{3.64}$$

for which the extensional strain ϵ_ψ is zero or for which $d_\psi = d_0$. This is not the direction of the scattering vector in general, nor is it the orientation of any member of the family of diffracting planes.

The expression (3.63) can be used for the determination of thin film stress *without a priori* knowledge of either stress-free d-spacing of the set of crystallographic

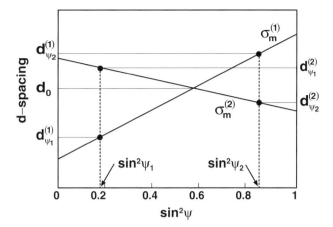

Fig. 3.5. Schematic illustration of the determination of stress-free d-spacing d_0 for a given family of crystallographic planes in thin films subjected to equi-biaxial mismatch stress. The procedure is based on the linear dependence of d_ψ on $\sin^2 \psi$ established in (3.63), (3.65) or (3.66) and the measurement of d-spacings for two members of a given family of crystallographic planes, each at two unknown values of stress. The discrete points represent the data from which the construction proceeds.

planes being interrogated or the elastic constants of the film material. The procedure exploits the linear relationship between d_ψ and $\sin^2 \psi$ that was identified in (3.63), and it requires that measurements be made for two different values of mismatch stress; the magnitudes of mismatch stress are unknown in both cases. Implementation of the procedure requires the measurement of d-spacings for two different members of the *same* family of crystallographic planes, say members with scattering vector orientations ψ_1 and ψ_2 which yield d-spacing values of d_{ψ_1} and d_{ψ_2}, respectively. This same measurement must be done for two different values of unknown mismatch stress, say $\sigma_m^{(1)}$ and $\sigma_m^{(2)}$. These four measurements of d-spacing at different stress levels are sufficient to determine the two coefficients in the linear relationship between d_ψ and $\sin^2 \psi$ for the two stress levels. The key observation is that, due to the linearity of material response, the direction ψ_0 of zero extensional strain will be the *same* for the two stress magnitudes. Consequently, if the two linear relationships obtained from the measurements are shown in a graph, the intersection of the two lines defines the value of d_0. This is illustrated in Figure 3.5. Cornella et al. (1997) have demonstrated the use of this method for the determination of stress-free d-spacing in 500 nm thick Al and Au films on relatively thick Si substrates. The experimental scatter in the measured d_0 values was as small as 0.121% for Al and 0.189% for Au, and the measured values compared well with known literature values. It is emphasized that the two lines used to determine the value of d_0 must result from measurements based on members of the same family of crystallographic planes.

If the thin film with an equi-biaxial extensional mismatch strain and zero shear mismatch strain is an anisotropic crystal with (001) orientation, the strain ϵ_ψ given in (3.60) can be rewritten in terms of the single crystal elastic constants, using (3.22) and (3.41)–(3.44), as

$$\epsilon_\psi = \frac{d_\psi - d_0}{d_0} = (s_{11} - s_{12})\,\sigma_{\mathrm{m}}\sin^2\psi + 2s_{12}\sigma_{\mathrm{m}}. \tag{3.65}$$

Similarly, the results presented in Section 3.5.2 can be used to show that, for a cubic film with a (111) orientation,

$$\epsilon_\psi = \frac{d_\psi - d_0}{d_0} = \sigma_{\mathrm{m}}\left\{\frac{2s_{11} - s_{44} + 4s_{12}}{3} + \frac{s_{44}\sin^2\psi}{2}\right\}. \tag{3.66}$$

In all cases, there is a particular orientation in the film with respect to the plane of the film–substrate interface for which the extensional strain is zero, as is evident from (3.62) which does not involve material constants.

Since the nonsymmetric diffraction method also facilitates determination of interplanar spacing as a function of $\sin^2\psi$ at different temperatures, the coefficient of thermal expansion of the film can also be determined without removing the film from the substrate or without knowledge of the elastic properties of the film material. This technique has been used by Kraft and Nix (1998) to determine the thermal expansion coefficient of a polycrystalline Al film with (111) fiber texture on a relatively thick Si substrate. They measured the d-spacing values of two scattering vectors for the {422} family of planes. The two data points so obtained for both tensile and compressive mismatch stress in the film provide the two intersecting d_ψ versus $\sin^2\psi$ lines at each temperature at which x-ray diffraction measurements are made. Determination of d_0 from the point of intersection of the two lines obtained for each temperature then provides the coefficient of thermal expansion as a function of temperature. The choice of this particular family of planes by Kraft and Nix (1998) was dictated by the consideration that different {422} planes can be identified at different angles to the normal to the plane of the film.

The results for characterizing stress in elastically anisotropic cubic single crystal films presented in Section 3.5 and in this section can also be adapted, under appropriate conditions, for textured polycrystalline films. An equi-biaxial mismatch strain in the (001) or (111) textured thin film of a cubic crystal results in an equi-biaxial mismatch stress. Because the response is transversely isotropic, the biaxial moduli derived for cubic crystals of these orientations can also be used to characterize polycrystalline thin films with perfect (001) or (111) fiber textures. However, a thin film of a cubic crystal oriented with its [011] axis perpendicular to the film–substrate interface plane does not exhibit transverse isotropy. Consequently, single crystal elastic response cannot be directly adapted to characterize polycrystalline films with (011) fiber texture.

3.6.4 Example: Determination of reference lattice spacing

A thin film of a polycrystalline fcc metal with {111} fiber texture was deposited onto a relatively thick Si substrate. Nonsymmetric x-ray diffraction measurements showed that the (242) and (24$\bar{2}$) planes had d-spacing values of 0.8268 and 0.8274 Å, respectively. The thin film material is known to be nearly isotropic. The elastic modulus and Poisson ratio of the film material are 70 GPa and 0.33, respectively.

(a) Determine the orientation ψ_0 in which extensional strain is zero.
(b) Determine the equi-biaxial mismatch stress σ_m and the strain $\epsilon_{33}^{(m)}$ normal to the plane of the film.

Solution:

(a) Let ψ_1 denote the angle between the normal to the (242) plane, or equivalently, the normal to the (121) plane and the normal to the plane of the film, which is along the [111] direction. From the geometry of the fcc crystal,

$$\cos\psi_1 = \frac{1}{\sqrt{6}}[121] \cdot \frac{1}{\sqrt{3}}[111] = 0.943 \quad \Rightarrow \quad \psi_1 = 19.44°. \qquad (3.67)$$

Similarly the angle ψ_2 between the normal to the (24$\bar{2}$) plane, or equivalently, the normal to the (12$\bar{1}$) plane and the [111] direction is found from geometry as

$$\cos\psi_2 = \frac{1}{\sqrt{6}}[12\bar{1}] \cdot \frac{1}{\sqrt{3}}[111] = 0.471 \quad \Rightarrow \quad \psi_2 = 61.9°. \qquad (3.68)$$

Because the film material can be considered elastically isotropic with Poisson ratio $v_f = 0.33$, the orientation of zero extensional strain can be determined from (3.64) to be

$$\sin^2\psi_0 = \frac{2v_f}{1+v_f} = 0.496 \quad \Rightarrow \quad \psi_0 = 44.7°. \qquad (3.69)$$

(b) The equation of the straight line representing the linear variation of d_ψ with $\sin^2\psi$ is given in (3.62). Inserting the d_ψ values of 0.8268 Å and 0.8274 Å for $\sin^2\psi_1 = 0.11$ and $\sin^2\psi_2 = 0.778$, respectively, it is readily seen that $d_0(1 - \epsilon_{33}^{(m)}) = 0.8267$ Å and that $d_0(\epsilon_m - \epsilon_{33}^{(m)}) = 0.8276$ Å. For the particular case of $\sin^2\psi = \sin^2\psi_0 = 0.496$, the reference spacing of planes in the {422} family is $d_0 = 0.8271$ Å. The equi-biaxial mismatch strain and mismatch stress in the film, respectively, are then

$$\epsilon_m = 6 \times 10^{-4}, \qquad \sigma_m = \frac{E_f}{1 - v_f}\epsilon_m = 62.7 \text{ MPa}. \qquad (3.70)$$

The strain component normal to the film–substrate interface plane is

$$\epsilon_{33}^{(m)} = -5 \times 10^{-4}. \qquad (3.71)$$

3.7 Substrate curvature due to anisotropic films

For all of the cases of substrate curvature induced by film stress that were considered in Section 2.1, it was assumed that both the film and substrate materials were isotropic. This provided a basis for a relatively transparent discussion of curvature phenomena and it led to results which have proven to be broadly useful. However, there are situations for which some understanding of the influence of material anisotropy of the film material or the substrate material, or perhaps of both materials, is important to know. Therefore, in this section, representative results on the influence of material anisotropy on substrate curvature are included. Results are established for two particular curvature formulae. In the first case, the film is presumed from the outset to be very thin compared to the substrate. Furthermore, the substrate is assumed to be isotropic and the film is considered to be generally anisotropic. In the second case discussed, no restriction is placed on the thickness of the film relative to the substrate, but both materials are assumed to be anisotropic. However, to obtain expressions for curvature which are not too complex to be interpretable, attention is limited to cases for which both the film and substrate materials are orthotropic, their axes of orthotropy are aligned with each other and one axis of orthotropy of each material is normal to the film–substrate interface. There is no connection between the values of the orthotropic elastic constants of the two materials. Consideration of these two cases illustrates the most useful approaches for anisotropic materials; generalizations for cases of greater complexity are evident.

3.7.1 Anisotropic thin film on an isotropic substrate

Certain key features of the analysis of curvature in Section 2.1 leading to the classic Stoney formula (2.7) when $h_f \ll h_s$ can be used to great advantage in the case of an anisotropic stressed film bonded to a substrate. Among these features is that it is only the resultant membrane force in the film prior to substrate deformation which contributes to substrate curvature. Any through-the-thickness variation of the film stress giving rise to this force is of secondary importance. A second feature is that the deformation due to curvature is sufficiently small so that the magnitude of the resultant force is essentially unaltered by the deformation of the substrate associated with curvature. This feature implies that the curvature of the substrate midplane can be calculated directly from the moment of this force (or, more precisely, from the moment of the force which negates the membrane force in the film) acting through the moment arm of length $h_s/2$, knowing only the elastic properties of the substrate. The overall bending resistance is not influenced to a significant degree by the presence of the film and, as a consequence, the elastic properties of the film material do not enter the calculation directly. This is the advantage of the Stoney formula; the same features were exploited in Section 2.1.

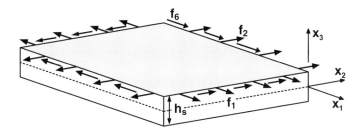

Fig. 3.6. Schematic diagram of a thin film bonded to the surface of a substrate. The film thickness h_f is much less than the thickness h_s of the substrate. Due to an elastic mismatch strain in the film, a membrane force per unit length must be imposed around the periphery of the film to balance the internal stress due to mismatch without deforming the substrate. Relaxation of this artificial membrane force resultant to render the edge free of applied moment induces curvature in the substrate.

To pursue this idea for the case of an anisotropic film, suppose for purposes of simplicity that the substrate is isotropic. Furthermore, suppose that the film is subjected to a mismatch strain with respect to the undeformed substrate. Following the discussion of Section 3.5, the mismatch strain is conveniently expressed with respect to the global coordinate axes shown in Figure 3.6 by the strain components $\epsilon_1^{(m)}$, $\epsilon_2^{(m)}$ and $\epsilon_6^{(m)}$ in the contracted notation introduced there. These are the two in-plane extensional strain components ($\epsilon_{11}^{(m)}$ and $\epsilon_{22}^{(m)}$ in the full strain matrix) and the in-plane shear strain component ($2\epsilon_{12}^{(m)}$ in the full strain matrix). Both the elastic constants of the film and the mismatch strain may vary with coordinate x_3 through the thickness of the film. An expression for the corresponding film mismatch stress components $\sigma_1^{(m)}$, $\sigma_2^{(m)}$ and $\sigma_6^{(m)}$, also in contracted notation, in terms of film elastic properties and mismatch strain is given in (3.39). The resultant force due to this stress, per unit length along the edge of the film prior to deformation of the substrate, is then given by the components

$$f_i = \int_{h_s/2}^{(h_s/2)+h_f} \sigma_i^{(m)}(x_3)\,dx_3, \quad i = 1, 2, 6. \tag{3.72}$$

If the substrate is to be maintained in a stress-free state, the components f_1 and f_2 must act on film edges defined by planes normal to the x_1 and x_2 directions, respectively, while the component f_6 represents the shear force inducing shear deformation between these two directions. If the film material is *homogeneous* and if the mismatch strain is *uniform* through the thickness then the integrands in (3.72) are independent of x_3 and the force expressions reduce to $f_i = \sigma_i^{(m)}h_f$. Furthermore, if the film material is isotropic and the mismatch is equi-biaxial with $\epsilon_1^{(m)} = \epsilon_2^{(m)} = \epsilon_m$, then both f_1 and f_2 reduce to the same value $f = M_f h_f \epsilon_m$ as defined in (2.9). The component f_6 is zero in this case.

The resultant forces f_1, f_2 and f_6 distributed along the film edges with magnitudes given in (3.72) are shown in the schematic diagram in Figure 3.6. Curvature of the substrate is induced upon *relaxation* of the artificially imposed forces. For example, relaxation of the force f_1 is equivalent to imposition of a compressive force resultant of magnitude f_1 along the edge of the substrate midplane plus a bending moment per unit length along the midplane edge of magnitude $\frac{1}{2}h_s f_1$. Examination of the substrate response leads to dependence of the components of curvature in the coordinate directions in Figure 3.6 on the mismatch force components in the form

$$\kappa_1 = \frac{6(f_1 - \nu_s f_2)}{E_s h_s^2}, \quad \kappa_2 = \frac{6(f_2 - \nu_s f_1)}{E_s h_s^2},$$

$$\kappa_6 = (1 - \nu_s)\frac{6 f_6}{E_s h_s^2}. \tag{3.73}$$

Here, κ_6 represents the *twist* of the substrate midplane, which was introduced in Section 2.6.2. These three parameters characterize the transverse midplane deflection of the substrate, which has the form $u_3(x_1, x_2, 0) = \frac{1}{2}\kappa_1 x_1^2 + \kappa_6 x_1 x_2 + \frac{1}{2}\kappa_2 x_2^2$ with respect to a plane tangent to the curved substrate at the point $x_1 = x_2 = 0$. Although the configuration is illustrated for a rectangular substrate in Figure 3.6, the curvature results are independent of the shape of the substrate.

It is interesting to note that, under any circumstances for which $f_1 = f_2$, the two curvature expressions both reduce to the Stoney formula (2.7). It should also be noted that the expressions in (3.73) are exact up to terms of order h_f/h_s. No approximations are involved in obtaining them other than those that follow directly from the restriction to situations for which $h_f/h_s \ll 1$. The expressions in (3.73) turn out to be useful for extracting information about patterned film configurations, as opposed to laterally homogeneous films, from curvature observations. The expression (3.72) makes clear that f_i can be viewed as a through-the-thickness average of the stress component $\sigma_i(z)$ in the film, say $\langle \sigma_i \rangle = f_i/h_f$, where the angle brackets denote a value averaged over some material volume. With this point of view, the expressions (3.73) provide a relationship between substrate curvature and average film stress. This idea is developed more completely in the Section 3.10 for films which are not uniform in thickness or which are cracked.

3.7.2 *Aligned orthotropic materials*

The physical system considered in this section is again a plate-like configuration depicted in Figure 3.7 with total thickness much less than the lateral dimensions. The

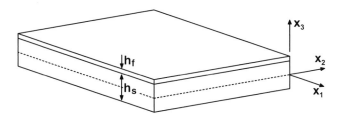

Fig. 3.7. Schematic diagram of a thin film bonded to the surface of a substrate. Both the film and substrate are assumed to be elastically orthotropic with axes of orthotropy aligned with the coordinate directions.

configuration consists of a substrate of uniform thickness h_s on which there is bonded a film of thickness h_f. Both the substrate and the film are assumed to be elastic and to have orthotropic symmetry, with the axes of orthotropy of the film and substrate being aligned. The origin of coordinates lies in the midplane of the substrate, and the x_3-direction is normal to the interface; the x_1- and x_2-axes lie in the midplane and are parallel to the axes of orthotropy. The substrate is homogeneous and its midplane would be flat in the absence of any applied loading. On the other hand, an elastic mismatch in extensional strain in the directions of orthotropy is assumed to be present in the film. This mismatch strain has components $\epsilon_k^{(m)}$ in contracted notation which vary in an arbitrary way through the thickness but which are laterally uniform. The elastic constants in the film, say the stiffness components $c_{ij}^f(x_3)$, can also vary in an arbitrary way through the film thickness. The mismatch strain and elastic constants may vary discontinuously with x_3 to represent a multilayer structure, or they may vary continuously to represent a graded layer.

All mechanical fields are referred to the underlying rectangular coordinate system; see Figure 3.7. The remote lateral edges of the configuration are free of applied loading, in the sense that the resultant force, resultant bending moment and resultant twisting moment, all measured per unit length along the edge of the substrate midplane, are zero on the lateral faces. Other details of the mechanical fields very close to the lateral edges are ignored. The conditions imposed on the mechanical fields everywhere except near the remote edges by the requirements of equilibrium, material behavior and geometric compatibility are now determined. The main result to be extracted is the relationship between the mismatch strain in the film and the curvature of the substrate.

Symmetry and invariance arguments were invoked in Section 3.5 to conclude that

$$\sigma_3 = \sigma_4 = \sigma_5 = 0 \tag{3.74}$$

throughout the system, in contracted notation for stress components. The invariance property also implies that the remaining stress components are independent of x_1 and x_2. In view of the absence of any remotely applied resultant force or bending moment on the edge of the configuration, it follows that the resultant force and bending moment must vanish on any interior cross-section as well, so that

$$\int_{-h_s/2}^{(h_s/2)+h_f} \sigma_k(x_3)\,dx_3 = 0\,, \quad \int_{-h_s/2}^{(h_s/2)+h_f} x_3\sigma_k(x_3)\,dx_3 = 0\,, \quad k = 1, 2. \quad (3.75)$$

The shear stress component σ_6 also vanishes in this case due to the assumed symmetry of the configuration and the absence of a mismatch in shear strain.

The total strain field in the system is not compatible, in general, due to the mismatch. However, the incompatibility is represented explicitly by the mismatch strain $\epsilon_k^{(m)}$, which is the elastic strain that must be imposed on the film material to fully counteract the incompatibility. The total strain minus the mismatch strain is compatible, from which it follows that the total strain components necessarily have the form

$$\epsilon_k(x_3) = \epsilon_k^{(m)}(x_3) - x_3\kappa_k + \epsilon_k^{(o)}\,, \quad k = 1, 2 \quad (3.76)$$

where $\epsilon_k^{(o)}$ are the in-plane extensional strain components of the substrate midplane $x_3 = 0$, κ_k are the curvature components of the midplane surface in the x and y directions for $k = 1$ and 2, respectively, and the mismatch strain vanishes in the substrate.

The response of the material is linearly elastic with orthotropic symmetry, so that

$$\sigma_i(x_3) = c'_{ij}(x_3)\epsilon_j(x_3)\,, \quad i, j = 1, 2 \quad (3.77)$$

where

$$c'_{11} = c_{11} - c_{13}^2/c_{33}\,, \quad c'_{22} = c_{22} - c_{23}^2/c_{33}\,, \quad c'_{12} = c_{12} - c_{13}c_{23}/c_{33} \quad (3.78)$$

are the plane stress compliances which account for the fact that $\epsilon_3 = -(c_{13}\epsilon_1 + c_{23}\epsilon_2)/c_{33}$ as a result of $\sigma_3 = 0$. In writing (3.77), it is assumed that the stiffness is $c_{ij} = c_{ij}^s$ when $-\frac{1}{2}h_s < x_3 < \frac{1}{2}h_s$ or $c_{ij} = c_{ij}^f$ when $\frac{1}{2}h_s < x_3 < \frac{1}{2}h_s + h_f$.

The stress in (3.77) is next expressed in terms of κ_k and $\epsilon_k^{(o)}$ by means of (3.76). A consequence of imposing the equilibrium conditions (3.75) on the result is a set of linear algebraic equations for the curvature and the strain of the substrate midplane,

$$J_{11}^0(\epsilon_1^{(m)}) + \epsilon_1^{(o)}J_{11}^0(1) - \kappa_1 J_{11}^1(1) + J_{12}^0(\epsilon_2^{(m)}) + \epsilon_2^{(o)}J_{12}^0(1) - \kappa_2 J_{12}^1(1) = 0$$

$$J_{11}^1(\epsilon_1^{(m)}) + \epsilon_1^{(o)}J_{11}^1(1) - \kappa_1 J_{11}^2(1) + J_{12}^1(\epsilon_2^{(m)}) + \epsilon_2^{(o)}J_{12}^1(1) - \kappa_2 J_{12}^2(1) = 0$$

$$\quad (3.79)$$

$$J_{12}^0(\epsilon_1^{(m)}) + \epsilon_1^{(o)}J_{12}^0(1) - \kappa_1 J_{12}^1(1) + J_{22}^0(\epsilon_2^{(m)}) + \epsilon_2^{(o)}J_{22}^0(1) - \kappa_2 J_{22}^1(1) = 0$$

$$J_{12}^1(\epsilon_1^{(m)}) + \epsilon_1^{(o)}J_{12}^1(1) - \kappa_1 J_{12}^2(1) + J_{22}^1(\epsilon_2^{(m)}) + \epsilon_2^{(o)}J_{22}^1(1) - \kappa_2 J_{22}^2(1) = 0\,,$$

where

$$J_{ij}^n(g) = \int_{-h_s/2}^{(h_s/2)+h_f} x_3^n c'_{ij}(x_3) g(x_3) \, dx_3 \,, \quad i, j = 1, 2 \qquad (3.80)$$

for any function $g(x_3)$. Thus, given the mismatch strain distribution $\epsilon_k^{(m)}(x_3)$, which is identically zero in the substrate, and the material parameters $c_{ij}(x_3)$, the system of equations (3.79) can be solved for substrate curvature components κ_1 and κ_2. While the expressions (3.79) appear to be complicated, they are quite simple in most cases of interest, in the sense that all details can be extracted by evaluation of a few integrals in terms of elementary functions.

This approach can be generalized to account for more complex geometrical arrangements of the layers or for other material symmetries. The essential features of the approach are the limitation to small deflections and the translational invariance of the configuration in any direction parallel to the substrate midplane. If the substrate is absent, that is, if $h_s/h_f \to 1$, the role played by the substrate midplane in the above discussion is naturally assumed by the plane face $x_3 = 0$ of the remaining film.

If the film and substrate materials are homogeneous and isotropic, then

$$c_{11}^f = \frac{(1 - \nu_f) E_f}{(1 + \nu_f)(1 - 2\nu_f)}, \quad c_{22}^f = c_{33}^f = c_{11}^f$$

$$c_{12}^f = \frac{\nu_f E_f}{(1 + \nu_f)(1 - 2\nu_f)}, \quad c_{13}^f = c_{23}^f = c_{12}^f \qquad (3.81)$$

for the film material, and similarly for the substrate material. If the mismatch strain is also equi-biaxial so that $\epsilon_1^{(m)} = \epsilon_2^{(m)} = \epsilon_m$, a constant, then solution of the system of equations (3.79) again yields the curvature expression in (2.19) for both κ_1 and κ_2. In addition, if the film is also very thin with respect to the substrate, then the Stoney expression in (2.7) is recovered for this case.

3.8 Piezoelectric thin film

Piezoelectricity is a physical effect exhibited by crystals that are electrically neutral but that do not have a center of reflective symmetry in crystal structure. The essential property of a point of reflective symmetry is that for each atomic position in the crystal lattice there is a matching atomic position at a point that is directly opposite the first with respect to the point of symmetry; this second position is at the same distance from the point of symmetry as the first. For example, the face-centered cubic structure has a point of reflective symmetry at the geometrical center of the unit cell whereas the cubic zinc blende structure does not have a center of symmetry.

When a stress is applied to such a crystal, it may become electrically polarized as a result of a charge separation on the atomic level, leaving unbalanced net charges of opposite sign on opposite faces of the crystal. A closely related phenomenon, which is known as the *piezoelectric effect*, is the production of mechanical strain when such a crystal is subjected to an electric field. Reversal of the field direction results in a reversal of the induced strain, and the relationship between field strength and strain is linear for strains as large as 0.01 for some materials. This effect, as well as the closely related effect of *electrostriction*, are central to the development of modern electromechanical sensors, actuators and other devices. The issue addressed briefly here is the generation of mismatch strain in a thin film system through the piezoelectric effect.

3.8.1 Mismatch strain due to an electric field

Phenomenologically, the relationship between imposed electric field $\mathcal{E}_i$ on a crystal and corresponding strain ϵ_{ij} induced in the crystal is of the form (Auld 1973)

$$\epsilon_{ij} = b_{ijk}\mathcal{E}_k \qquad (3.82)$$

where b_{ijk} is the array of piezoelectric constants. The physical dimensions of the components of electric field are energy/(length×electric charge) so that the physical dimensions of the components of b_{ijk} are (length×electric charge)/energy. For example, if electric field strength is expressed in units of J/coulomb·m then the components of b_{ijk} are expressed in units of coulomb/N. Symmetry of the strain array ϵ_{ij} under interchange of its indices implies that, of the 27 components of b_{ijk}, only 18 can be independent. Material symmetry reduces the number of independent components further for most real crystals. The array of constants b_{ijk} represents the components of a tensor, so that these components transform as tensor components between different sets of mutually orthogonal rectangular coordinate axes, as illustrated for the case of the stress tensor in (3.37).

For purposes of calculation for a given orientation of the material with respect to some underlying rectangular coordinate system, the reduced matrix representation of the relationship (3.82) is again found to be convenient, following the convention adopted in Section 3.1. As before, the cost of this convenience is that the tensor character of the arrays representing physical quantities is lost. A reduced form of (3.82) is

$$\epsilon_i = b_{ik}\mathcal{E}_k, \quad i = 1, \ldots, 6, \qquad (3.83)$$

where the same symbols are used for the reduced matrix quantities as were used for the tensor representations. The interpretation of the symbols should be evident from the context, and no confusion should result from this practice.

The reduced piezoelectric matrix b_{ik} is a 6×3 matrix. For a triclinic crystal, all 18 elements are required for a complete description of the material, in general.

For a tetragonal crystal, such as barium titanate or lead zirconate–titanate (PZT), only seven components are nonzero and these are specified by four independent constants according to

$$b_{31} = b_{32}, \quad b_{33}, \quad b_{15} = b_{24}, \quad b_{14} = -b_{25}. \tag{3.84}$$

This representation presumes that the c-axis is aligned with the x_3-direction. For a cubic crystal, such as gallium arsenide, only three components of the piezoelectric matrix are nonzero, and these are specified in terms of a single constant according to

$$b_{14} = b_{25} = b_{36}. \tag{3.85}$$

It is presumed that the cube axes are aligned with the rectangular coordinate axes in this case.

Suppose that the piezoelectric thin film is bonded to an isotropic elastic substrate that does not exhibit piezoelectric behavior. If this film–substrate system is exposed to an electric field $\mathcal{E}_k$, then the film material will tend to undergo the stress-free strain specified by (3.83). However, the film is constrained from deforming freely by the substrate. As a result, a stress is generated in the film, and this stress may result in a detectable curvature of the substrate. If the x_3-axis is normal to the film–substrate interface then the components of mismatch strain in the film are, in reduced notation,

$$\epsilon_1^{(m)} = -b_{1k}\mathcal{E}_k, \quad \epsilon_2^{(m)} = -b_{2k}\mathcal{E}_k, \quad \epsilon_6^{(m)} = -b_{6k}\mathcal{E}_k. \tag{3.86}$$

Once the components of mismatch strain are known, the mismatch stress can be established as in Section 3.5. For a thin film, the curvature and twist of the substrate follow from (3.73).

3.8.2 Example: Substrate curvature due to an electric field

In a transducer for use in a hearing aid, a PZT film of thickness $h_f = 5 \,\mu$m is deposited on a polysilicon substrate of thickness $h_s = 150 \,\mu$m. The tetragonal film is prepared with its c-axis coincident with the x_3-axis of the system coordinate frame, which is perpendicular to the film–substrate interface. The array of piezoelectric parameters for the PZT film material is

$$b_{ik} = \begin{bmatrix} 0 & 0 & -60 \\ 0 & 0 & -60 \\ 0 & 0 & 152 \\ 0 & 440 & 0 \\ 440 & 0 & 0 \\ 0 & 0 & 0 \end{bmatrix} \times 10^{-12} \text{ coulomb/N.} \tag{3.87}$$

(a) An electric field of magnitude 10^6 volts/m·coulomb is applied along the x_1-axis. Determine the mismatch strain induced in the film.

(b) Suppose that an electric field of the same magnitude as in part (a) is applied, but this time along the x_3-axis. Determine the mismatch strain induced in the film for this case.

(c) For the case described in part (b), estimate the radius of curvature in the substrate due to the mismatch strain in the film.

Solution:

(a) In this case, $\mathcal{E}_1 = 10^6$ N/coulomb and $\mathcal{E}_2 = \mathcal{E}_3 = 0$. Then, from (3.86), it follows that $\epsilon_1^{(m)} = \epsilon_2^{(m)} = \epsilon_6^{(m)} = 0$. Consequently, no mismatch strain with respect to the substrate is developed in the film. The strain component $\epsilon_5 = b_{51}\mathcal{E}_1$ takes on the value 0.00044, but this strain can develop freely without constraint from the substrate.

(b) In this case, $\mathcal{E}_1 = \mathcal{E}_2 = 0$ and $\mathcal{E}_3 = 10^6$ volts/m·coulomb. It follows from (3.86) that

$$\epsilon_1^{(m)} = \epsilon_2^{(m)} = b_{13}\mathcal{E}_3 = -60 \times 10^{-6}, \qquad \epsilon_6^{(m)} = b_{63}\mathcal{E}_3 = 0. \qquad (3.88)$$

The film tends to contract along the interface but is not able to do so because of the substrate. Consequently, a state of equi-biaxial tensile mismatch strain develops in the film with respect to the substrate.

(c) The in-plane biaxial modulus for PZT is approximately $M_f = 6.1 \times 10^{11}$ N m^{-2}. Thus, the membrane force in the fully constrained film is

$$f = \epsilon_1^{(m)} h_f M_f = 183 \, \text{N m}^{-1}. \qquad (3.89)$$

According to the Stoney formula (2.7), the substrate curvature is given in terms of this force, the biaxial modulus of the substrate and the thickness of the substrate. The biaxial modulus is approximately $M_s = 76 \times 10^9$ N m^{-2} for polysilicon. The resulting estimate of radius of curvature is $\kappa^{-1} = 1.56$ m.

3.9 Periodic array of parallel film cracks

The discussion up to this point has focused on the relationship between the curvature of an elastic substrate and the stress in a single layer or multilayer film in which the mismatch is invariant under any translation parallel to the interface. The films considered have also been continuous and of uniform thickness over the entire film–substrate interface. Within the range of small deflections, such an equi-biaxial film stress induces a spherical curvature in the substrate midplane, except very near the edge of the substrate. What is the deformation induced in the substrate if such a film does not have uniform thickness or if the mismatch stress varies with position along the interface? This question is addressed in this section for the cases when the nonuniformity in mismatch stress or thickness varies periodically along the interface, either in one direction or in orthogonal directions, and when the characterizing dimensions of the periodic variation are small compared to the substrate thickness h_s in some sense.

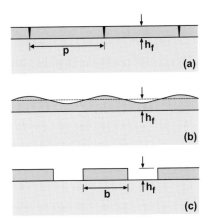

Fig. 3.8. Examples of film/substrate configurations for which the film is not uniform but instead has structure which is periodic along the interface, with the spatial period being p in each case.

Examples of such nonuniformly strained films are illustrated in Figure 3.8. Cracks may form in a film in tension, and Figure 3.8(a) shows the case where a single periodic array of parallel cracks has formed, with the crack planes being perpendicular to the interface and with crack length being equal to the film thickness. Any curvature in the substrate due to mismatch stress in the film prior to cracking will be altered by the formation of the cracks. The mechanics of formation of such cracks will be discussed in Chapter 5. Figure 3.8(b) suggests a situation where the flat surface of a strained semiconductor film of uniform thickness is thermodynamically unstable, and the free surface evolves into a periodic profile over time as a result of surface diffusion of material or evaporation/condensation, for example. Eventually, the valleys in the profile may reach the substrate, forming isolated islands of film material as will be discussed in Chapter 8. The third example is motivated by the idea that a means of assessing the reliability of metal interconnect lines is to form many parallel lines on the surface of a semiconductor wafer as depicted in Figure 3.8(c). Stress in the patterned lines during temperature excursions can be measured by means of substrate curvature methods.

The three configurations in Figure 3.8 are representative of systems being considered in this section. In all cases, it is assumed that the mean film thickness and the spatial period of the variation along the interface are small compared to h_s. The basic reasoning which leads to a connection between patterned film properties and substrate curvature is first outlined for two-dimensional plane strain bending, and then for three-dimensional biaxial bending. The goal is to understand the coupling between deformation at the level of film microstructure, often in the submicron range, and the substrate deformation at a size scale that is several orders of magnitude greater.

3.9.1 Plane strain curvature change due to film cracks

Consider a uniform elastic film of thickness h_f bonded to the surface of an elastic substrate of thickness h_s. The film supports a biaxial tensile mismatch stress σ_m but is constrained to deform in plane strain bending for the time being, so as to render discussion of the general approach as transparent as possible. The film is thin which, from Section 2.2, is understood to mean that $E_f h_f \ll E_s h_s$. If the deflections are small in this case, the uniaxial or cylindrical curvature of the substrate is

$$\kappa_{nc} = \frac{6\sigma_m h_f}{\bar{E}_s h_s^2}, \tag{3.90}$$

where $\bar{E} = E/(1 - \nu^2)$ is the plane strain modulus, with the subscript 's' denoting the substrate. The expression in (3.90) provides the plane strain bending analog of the Stoney formula (2.23). The subscript 'nc' denotes the substrate curvature of the continuous thin film with no cracks.

Suppose that a periodic array of parallel cracks with crack spacing p then forms in the film, as depicted in Figure 3.8(a), with the crack edges being parallel to the bending axis, which is the y-axis in this case. As a result of crack formation, the curvature of the substrate midplane will be reduced from the value in (3.90). The objective is to estimate the curvature following crack formation or, equivalently, the change in curvature due to crack formation. This will be done at the same level of approximation as the approach that underlies the Stoney formula (2.7), but the implementation of the assumptions requires somewhat more care in the present case. The main assumptions exploited here are: (i) the state of stress in the film is determined solely by elastic mismatch of the film with respect to the undeformed substrate and is not significantly altered by the deformation of the substrate as it becomes curved and (ii) the flexural stiffness of the substrate is not significantly influenced by the presence of the film material on its surface. Under these circumstances, the curvature is essentially the moment of this force with respect to the substrate midplane divided by the flexural stiffness of the substrate, as in (3.90).

The deformation associated with crack formation can be analyzed by considering a reduced problem which is defined to take advantage of the thinness of the film. It is important to recognize that the configuration of the film–substrate system under consideration is symmetric under reflection with respect to any of the planes at $x = 0, \pm p, \ldots$ or any of the planes $x = \pm\frac{1}{2}p, \pm\frac{3}{2}p, \ldots$ midway between the crack planes, far from the substrate edge. As a result, the shear traction transmitted across each of these surfaces is identically zero and these surfaces remain planar during deformation. The reduced problem is defined to be a single period of the

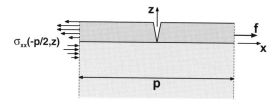

Fig. 3.9. A reduced boundary value problem for a periodic film/substrate configuration which includes a single period of extent p, showing the total force f which acts on the planes of reflective symmetry of the configuration; this force is the resultant of the normal stress distribution $\sigma_{xx}(-p/2, z)$ which acts on each symmetry plane. Determination of f makes it possible to estimate the curvature of the substrate according to (3.93).

overall configuration, which is bounded by symmetry planes as shown in Figure 3.9. The normal displacements of the faces $x = \pm\frac{1}{2}p$ on which the opposed forces of magnitude f act are *set equal to zero* in the reduced problem. This is justified by the feature noted above that the value of f is not significantly affected by the deformation of the substrate. It is precisely this deformation which is being ignored by imposing the boundary condition of vanishing displacement. The loading in the reduced problem still derives from the mismatch stress present in the film prior to formation of the cracks. Suppose that the magnitude of the resultant normal force (per unit length in the y-direction) acting on the opposite symmetry planes in Figure 3.9 is denoted by f, as shown on the right side of the segment; for equilibrium, it is necessary that

$$f = \int_{-\infty}^{h_f} \sigma_{xx}(-p/2, z)\, dz. \tag{3.91}$$

The symmetry of the single period of the configuration shown in Figure 3.9 has another important consequence. From overall equilibrium of the segment, it follows that the resultant force on any plane $z = $ constant is also zero. In particular, the shear force vanishes identically on any such plane and, by St. Venant's principle, it follows that the local shear stress becomes negligibly small for depths beyond about $2p$ into the substrate. The absence of shear traction on a section of the substrate is one of the hallmarks of pure flexural deformation. This reasoning supports the use of cylindrical or spherical substrate curvature concepts to study the behavior of patterned surface films.

The main calculation which must be done to estimate curvature in the presence of cracking is the determination of f, the force which plays the role here that $\sigma_m h_f$ served in the derivation of the Stoney formula. The value of f, or perhaps $\Delta f = f - \sigma_m h_f$, the change in force due to crack formation from the value for a uniform film, can usually be determined directly by means of the finite element method applied to the configuration illustrated in Figure 3.9. By choosing planes as boundaries of the reduced problem which do not include sharp geometric

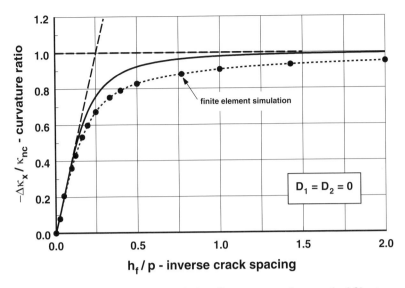

Fig. 3.10. Ratio of the change in plane strain bending curvature in a cracked film to curvature for an uncracked film at the same level of mismatch stress versus crack spacing p. The elastic properties of the film and substrate are identical. The discrete points connected by dotted lines follow from a finite element simulation while the approximation represented by the solid curve is obtained by means of elastic reciprocity (3.101). The asymptotes represented by dashed lines are evident on physical grounds.

features, the calculation of f by means of the finite element method is a routine task.

Alternatively, reliable results of broad applicability can often be found by other means. For example, if a solution is known for the configuration of interest but with different boundary conditions, then elastic reciprocity can be used to extract useful results with minimal effort. In any case, once f or Δf is known, the plane strain curvature κ, or the change in curvature $\Delta \kappa$, is given by

$$\kappa = \frac{6f}{\bar{E}_s h_s^2} \quad \text{or} \quad \Delta \kappa = \frac{6\Delta f}{\bar{E}_s h_s^2}. \tag{3.92}$$

Representative results for the ratio of substrate curvature change due to film cracking at a given level of equi-biaxial mismatch strain ϵ_m to the corresponding curvature for an uncracked film at the same level of mismatch strain are shown in Figure 3.10. Certain asymptotic results, namely,

$$\frac{\Delta \kappa}{\kappa_{nc}} \to -1 \text{ as } \frac{h_f}{p} \to \infty, \quad \frac{\Delta \kappa}{\kappa_{nc}} \to 0 \text{ as } \frac{h_f}{p} \to 0 \tag{3.93}$$

are evident in the plot. The discrete points shown in the plot were determined by means of the finite element method for the case when the elastic moduli of the film and the substrate are identical.

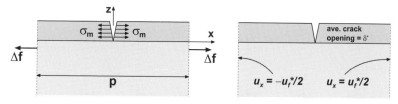

Fig. 3.11. The two boundary value problems for which the theorem of elastic reciprocity is applied. In the first case, normal compressive traction of magnitude σ_m acts on the crack faces and total force f acts on the planar symmetry surfaces. In the second case, the symmetry surfaces are subject to a uniform normal outward displacement of $u_f^*/2$ and the average opening between the crack faces is δ^*.

Next, application of elastic reciprocity to estimate f is illustrated. This requires use of two instances of an elastic solid which are identical in properties and shape, but which are subjected to boundary conditions chosen in a way to reveal certain information. When the only mechanical loading involved is applied to the elastic solid through its bounding surface, say C, then the Betti–Rayleigh reciprocal theorem relates the stress field σ_{ij} and displacement field u_i of one boundary value problem to the stress field σ_{ij}^* and displacement field u_i^* of the other problem through the global relationship

$$\int_C \sigma_{ij} n_j u_j^* \, dC = \int_C \sigma_{ij}^* n_j u_j \, dC, \qquad (3.94)$$

where n_i is the outward unit normal vector to C. For two compatible equilibrium elastic fields, the work done by the boundary traction of the first field through the boundary displacement of the second field equals the work done by the boundary tractions of the second field through the displacement of the first field.

The two boundary value problems to be considered here are illustrated in Figure 3.11. The instance on the left side of the figure is similar to that in Figure 3.9 except that no mismatch stress is present and a normal pressure of magnitude σ_m acts on the crack faces. Analysis of this problem provides the change in f, termed Δf above, due to crack formation in a uniform film. As before, the normal displacement on the bounding symmetry planes at $x = \pm \frac{1}{2}p$ is required to vanish. In the second instance of the same configuration shown in the right side of Figure 3.11, the faces of the crack are free of traction and the uniform normal displacements of the faces $x = \pm \frac{1}{2}p$ are specified to be $u_x = \pm \frac{1}{2}u_f^*$ where u_f^* is an arbitrary displacement amplitude. As a consequence of this normal displacement, the crack faces deform; the average opening of the crack over $0 < z < h_f$ is denoted by δ^* and is defined by

$$\delta^* = h_f^{-1} \int_0^{h_f} \left[u_x(0^+, z) - u_x(0^-, z) \right] dz. \qquad (3.95)$$

The superscript (*) is intended to serve as a reminder that the deflections u_f^* and δ^* are independent of the force f.

When applied to the two configurations in Figure 3.11, elastic reciprocity (3.94) then requires that

$$\Delta f\, u_f^* + \sigma_m h_f \delta^* = 0. \qquad (3.96)$$

There is no contribution from the right side of (3.94) because the boundary value problems have been defined so that $u_i = 0$ wherever on the boundary C that $\sigma_{ij}^* n_j \neq 0$. Due to linearity of the boundary value problem on the right on Figure 3.11, δ^* is proportional to u_f^*. If the proportionality factor is known in terms of material parameters and geometrical dimensions then f can be determined from (3.96). If the same finite element calculation procedure which led to the discrete points in Figure 3.10 would be used to estimate δ^*, then (3.96) would lead to exactly the same result for curvature change. For this case, however, estimates of δ^* are available (Wikström et al. (1999a); Xia and Hutchinson (2000)). These will be introduced once the asymptotic behavior of δ^*/u_f^* for both very large and very small crack spacing is described.

The asymptotic behavior of δ^*/u_f^* for very small and very large values of crack spacing compared to film thickness are evident on physical grounds. When the crack spacing is small compared to film thickness, the mismatch stress in the film is relaxed with virtually no load transfer to the substrate. In terms of the auxiliary problem in Figure 3.11, this implies that $\delta^*/u_f^* \to 1$ as $h_f/p \to \infty$. Therefore, it follows from (3.96) that $\Delta f \to -\sigma_m h_f$ or $f/\sigma_m h_f \to 0$ as $h_f/p \to \infty$. At the other extreme, when the crack spacing is very large compared to film thickness, the ratio δ^*/u_f^* must be equal to that for an isolated crack in a thin film. This result has been determined for materials with different elastic properties by Beuth (1992). When the elastic properties of the film and substrate are identical, this ratio is approximately $\delta^*/u_f^* \approx 1.258\pi h_f/p$ as $h_f/p \to 0$. It follows that $\Delta f \to 1.258\pi \sigma_m h_f^2/p$ or that $\Delta\kappa/\kappa_{nc} \to 1.258\pi h_f/p$ as $h_f/p \to 0$. Both asymptotes are indicated by dashed lines in Figure 3.10.

An approximation for all values of h_f/p for this proportionality factor has been given by Xia and Hutchinson (2000) for bimaterial systems as

$$\frac{\delta^*}{u_f^*} \approx \frac{\ell}{p}\tanh\left(\frac{p}{\ell}\right), \qquad (3.97)$$

where ℓ is a parameter with physical dimension of length that is proportional to h_f. This approximation was obtained for the case of a periodic array of cracks under plane strain conditions with a normal stress σ^* applied to the crack faces and tending to open the cracks. Through identification of σ^* with $\bar{E}_f u_f^*/p$, the approximation can be exploited in the present development. The proportionality factor has been tabulated by Beuth (1992) for various combinations of material parameters. The dependence of ℓ on elastic constants can be expressed in terms of the two dimensionless parameters defined by

$$D_1 = \frac{\bar{E}_f - \bar{E}_s}{\bar{E}_f + \bar{E}_s}, \quad D_2 = \frac{1}{4}\frac{\bar{E}_f(1-\nu_f)(1-2\nu_s) - \bar{E}_s(1-\nu_s)(1-2\nu_f)}{\bar{E}_f(1-\nu_f)(1-2\nu_s) + \bar{E}_s(1-\nu_s)(1-2\nu_f)} \quad (3.98)$$

and known as the Dundurs parameters (Dundurs 1969). Values of the nondimensional ratio ℓ/h_f are plotted versus D_1 in Figure 3.12 for the particular choices of $D_2 = 0$ and $D_2 = D_1/4$. The latter condition is equivalent to the condition $\nu_f = \nu_s$, whereas the former choice is of special interest in elastic fracture mechanics, a topic to be considered in Chapter 4. Level curves of D_1 in the plane of ν_f and ν_s are shown in Figure 3.13 for the case when $D_2 = 0$. When the film and substrate materials have essentially the same elastic properties, $D_1 = D_2 = 0$ and $\ell = 1.258\pi h_f$. In the case when the elastic properties of the film and the substrate are identical, an approximate expression for δ^*/u_f^* has also been given by Wikström et al. (1999a) in the form of the series

$$\frac{\delta^*}{u_f^*} \approx 1.258\,\pi\frac{h_f}{p}\Upsilon(h_f/p), \quad \Upsilon(s) \equiv \sum_{k=1}^{10}\frac{d_k}{(1+s)^k}, \quad (3.99)$$

where the coefficients in the series are given by

k	d_k
1	0.25256
2	0.27079
3	−0.49814
4	8.6296
5	−51.246
6	180.96
7	−374.29
8	449.59
9	−286.51
10	73.842

It follows from (3.96) that

$$\Delta f \approx -\frac{\ell h_f \sigma_m}{p}\tanh\left(\frac{p}{\ell}\right). \quad (3.100)$$

The bending moment that produces the change in substrate curvature is $\Delta f(h_s/2)$. The change in curvature from κ_{nc} due to crack formation is

$$\Delta\kappa \approx \frac{6\Delta f}{\bar{E}_s h_s^2} = -\kappa_{nc}\frac{\ell}{p}\tanh\left(\frac{p}{\ell}\right). \quad (3.101)$$

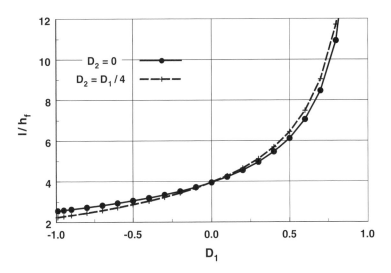

Fig. 3.12. Plots of ℓ/h_f versus D_1 for $D_2 = 0$ and $D_2 = D_1/4$.

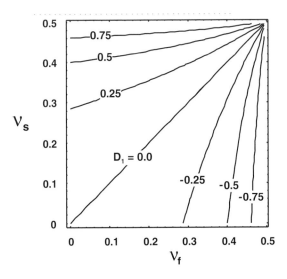

Fig. 3.13. Level curves of D_1 in the plane of ν_f and ν_s for the case when $D_2 = 0$.

The result (3.101) is also illustrated in Figure 3.10 in the form of $\Delta\kappa/\kappa_{nc}$ versus h_f/p for the case when $D_1 = D_2 = 0$. Note that the approximate result for $\Delta\kappa$ is asymptotically consistent as $h_f/p \to 0$ and $h_f/p \to \infty$ but that it shows some difference from the finite element results when h_f/p is of the order of unity. It is evident that the curvature falls off rapidly as widely spaced cracks appear in the film, becoming relatively small compared to the initial uncracked curvature once spacing has decreased to about $p \approx h_f$.

3.9.2 Biaxial curvature due to film cracks

Next, turn to the three-dimensional situation concerning the change in curvature of the substrate due to formation of a periodic array of parallel cracks in a strained film in tension. As before, the tensile mismatch stress is σ_m and the film thickness is h_f, and both the film and substrate materials are assumed to be elastic and isotropic. However, the restriction of the deformation to be plane strain bending which has been in force up to this point in the subsection is now relaxed. Formation of the cracks will still result in a relaxation of the curvature κ_x of the substrate in the x-direction due to a change in bending moment acting on the substrate midplane about the y-direction. Now the possibility of a change in curvature κ_y of the substrate due to a change in bending moment about the x-direction in the substrate midplane is also taken into account. The general relationship between the components of curvature of the substrate midplane and the bending moment resultants due to mismatch stress in the film is given by

$$\kappa_x = \frac{12}{E_s h_s^3}\left(m_x - v_s m_y\right), \quad \kappa_y = \frac{12}{E_s h_s^3}\left(m_y - v_s m_x\right), \tag{3.102}$$

where m_x is the resultant bending moment acting on a line $x = $ constant in the substrate midplane per unit length in the y-direction, and similarly for m_y. For the case of a uniform uncracked film, the unrelaxed resultant force in the film is $\sigma_m h_f$. The moment of this force acting through the distance $h_s/2$ must be balanced at a free edge of the substrate by resultant moments $m_x = m_y = \sigma_m h_f h_s/2$. The substrate curvature implied by (3.102) is then $\kappa_x^{nc} = \kappa_y^{nc} = \kappa_{St}$, as predicted by the Stoney formula (2.23).

The same approach is followed for this case as for plane strain bending. However, in the three-dimensional case, two resultant forces must be calculated to determine the changes in both m_x and m_y in order to compute curvature change by means of (3.102). The steps rely on the same assumptions that eventually led to the results for plane strain bending, so only the main intermediate steps are included. A reduced boundary value problem as depicted in the left portion of Figure 3.14 is introduced. It consists of a single period of the cracked film–substrate system of extent p in the x-direction and of arbitrary extent q in the y-direction. The shaded surfaces $x = \pm\frac{1}{2}p$ and $y = \pm\frac{1}{2}q$ both lie on planes of reflective symmetry, so shear traction is zero; these surfaces remain planar during deformation. In the reduced problem, the normal displacement on the shaded surfaces is constrained to be zero. A uniform normal pressure of magnitude σ_m acts on the crack faces. This loading produces normal forces $f_x q$ and $f_y p$ on the constrained surfaces, where f_x and f_y are the force resultants per unit length along the film edge which are sought; see Figure 3.14. As before, the changes in resultant force due to crack formation are $\Delta f_x = f_x - \sigma_m h_f$ and $\Delta f_y = f_y - \sigma_m h_f$.

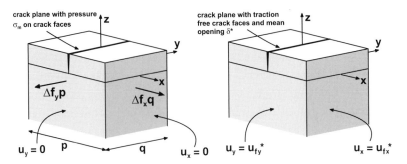

Fig. 3.14. The left part of the figure shows the reduced boundary value problem which serves as the basis for calculating the force resultants Δf_x and Δf_y, and thereby the change in substrate curvature due to crack array formation. The right portion shows the auxiliary boundary value problem for use with elastic reciprocity.

Before seeking simple approximate expressions for the change in curvature, results from direct determination of f_x and f_y obtained by means of the finite element method are illustrated in Figure 3.15. The estimates of curvature change $\Delta\kappa_x$ and $\Delta\kappa_y$ are determined from the force calculation according to

$$\Delta\kappa_x = \frac{6(\Delta f_x - \nu_s \Delta f_y)}{E_s h_s^2}, \qquad \Delta\kappa_y = \frac{6(\Delta f_y - \nu_s \Delta f_x)}{E_s h_s^2}. \tag{3.103}$$

The graph is drawn for the case when the two materials have identical elastic properties, that is, $D_1 = D_2 = 0$, and the curvature changes are normalized by κ_{St}. For this case, it is found that $\Delta\kappa_y/\kappa_{St} = 0$. The results here are somewhat different from the corresponding plane strain bending result in Figure 3.10, highlighting the influence of three-dimensional effects.

It is seen in Figure 3.15 that, for the case of closely spaced cracks, the change in curvature $\Delta\kappa_x$ in the x-direction, which is perpendicular to the array of cracks, is larger in magnitude than the curvature κ_x before formation of the cracks. This *curvature reversal* phenomenon is due to the Poisson contraction effect. Suppose that the mismatch strain in the film with respect to the substrate is tensile, so that curvature is positive with $\kappa_x = \kappa_y$ before crack formation. The direct effect of formation of a closely spaced array of cracks in the y-direction is to relieve the tensile stress due to mismatch in the x-direction, thereby reducing κ_x to zero. However, the tensile film stress in the y-direction is only partially relieved by crack formation. The remaining tensile stress in the film acting in the y-direction maintains the curvature κ_y. A consequence of this curvature is a compressive strain in the y-direction in the substrate material near the interface associated with the substrate curvature κ_y. This imposed compressive strain introduces an additional extensional strain in the x-direction in the substrate material near the interface; the magnitude of this

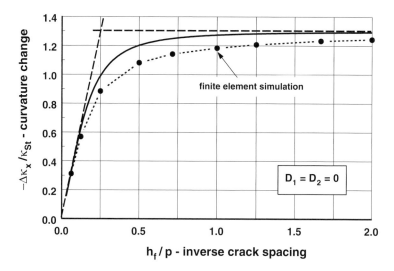

Fig. 3.15. Curvature change in the x-direction due to crack formation in a thin film with mismatch strain versus crack spacing for three-dimensional bending. The elastic properties of the film and substrate are identical in this illustration. The computed change in curvature in the y-direction for this case is essentially zero for all crack spacings. The solid curve is an estimate of the curvature change due to cracking based on application of elastic reciprocity (3.108).

additional strain is ν_s times the imposed strain in the y-direction. The net effect of this extensional strain is to induce an additional negative or reversed curvature in the x-direction. Indeed, the asymptotic behavior of curvature for vanishingly small crack spacing is seen in Figure 3.15.

Turning next to application of the reciprocal theorem to calculate Δf_x and Δf_y, the changes in force resultants due to crack formation, the auxiliary boundary value problem shown in the right side of Figure 3.14 is introduced. In this problem, the plane surfaces $x = \pm\frac{1}{2}p$ are given a normal displacement $u_x = \pm\frac{1}{2}u^*_{f_x}$ while the plane surfaces $y = \pm\frac{1}{2}q$ are given a normal displacement $u_y = \pm\frac{1}{2}u^*_{f_y}$, where $u^*_{f_x}$ and $u^*_{f_y}$ are arbitrary displacement amplitudes. As a result of these imposed boundary displacements, the crack experiences an opening which is independent of the y coordinate. As before, the average crack opening is denoted by δ^* as defined in (3.95).

Elastic reciprocity (3.94) then requires that

$$q\Delta f_x u^*_{f_x} + p\Delta f_y u^*_{f_y} + q\sigma_m h_f \delta^* = 0 \qquad (3.104)$$

from which Δf_x and Δf_y are to be determined. Note that δ^* is necessarily linear in $u^*_{f_x}$ and $u^*_{f_y}$. It is possible to determine both $u^*_{f_x}$ and $u^*_{f_y}$ from (3.104) by invoking the arbitrariness of $u^*_{f_x}$ and $u^*_{f_y}$. The linear dependence of δ^* on $u^*_{f_x}$ and $u^*_{f_y}$ can be established in the following way. It is noted that the interface traction in the complementary problem would vanish completely if a tensile normal stress σ^* of magnitude

$$\frac{\sigma^*}{\bar{E}_f} = \left(\frac{u^*_{f_x}}{p} + v_f \frac{u^*_{f_y}}{q} \right) \tag{3.105}$$

would act on the faces of the cracks. This observation provides a basis for generalizing the estimate of δ^* given in (3.97) or (3.99). If the factor u^*_f / p on the right side of (3.97) is replaced by the right side of (3.105) then (3.97) can be used in the present case in the form

$$\delta^* \approx \ell \left(\frac{u^*_{f_x}}{p} + v_f \frac{u^*_{f_y}}{q} \right) \tanh \left(\frac{p}{\ell} \right). \tag{3.106}$$

Substituting (3.106) into (3.104) and alternately setting either $u^*_{f_x} = 1$ and $u^*_{f_y} = 0$ or $u^*_{f_x} = 0$ and $u^*_{f_y} = 1$ yields

$$\Delta f_x \approx -\frac{\ell h_f \sigma_m}{p} \tanh \left(\frac{p}{\ell} \right), \qquad \Delta f_y \approx -\frac{\ell h_f \sigma_m v_f}{p} \tanh \left(\frac{p}{\ell} \right). \tag{3.107}$$

The resulting curvature changes of the substrate corresponding to these film force resultants are determined from (3.103) to be

$$\boxed{\frac{\Delta \kappa_x}{\kappa_{St}} \approx -\frac{(1 - v_f v_s)\ell}{(1 - v_s)p} \tanh \left(\frac{p}{\ell} \right), \quad \frac{\Delta \kappa_y}{\kappa_{St}} \approx -\frac{(v_f - v_s)\ell}{(1 - v_s)p} \tanh \left(\frac{p}{\ell} \right).} \tag{3.108}$$

Note that the curvature change in the x-direction always decreases due to crack formation whereas the curvature change in the y-direction can either increase or decrease, depending on whether v_f is smaller or larger than v_s. For widely spaced cracks with $p/h_f \gg 1$,

$$\frac{\Delta \kappa_x}{\kappa_{St}} \approx -\frac{(1 - v_f v_s)}{(1 - v_s)} \frac{\ell}{p}, \quad \frac{\Delta \kappa_y}{\kappa_{St}} \approx -\frac{(v_f - v_s)}{(1 - v_s)} \frac{\ell}{p}. \tag{3.109}$$

For closely spaced cracks, the asymptotic behavior as $p/h_f \to 0$ is

$$\frac{\Delta \kappa_x}{\kappa_{St}} \to -\frac{(1 - v_f v_s)}{(1 - v_s)}, \quad \frac{\Delta \kappa_y}{\kappa_{St}} \to -\frac{(v_f - v_s)}{(1 - v_s)}. \tag{3.110}$$

Notice that differences in elastic constants between the film and substrate are included in (3.108) and (3.109) through ℓ, as well as through the explicit dependence on the Poisson ratios.

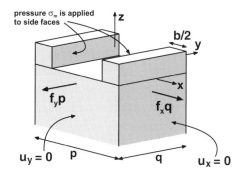

Fig. 3.16. Reduced boundary value problem to be solved to determine the force resultant increments necessary for estimating substrate curvature according to (3.103).

3.10 Periodic array of parallel lines or stripes

A second particular case of substrate curvature due to a surface film with periodic nonuniformity is considered in this section. Suppose that a film of thickness h_f is deposited onto a substrate in the configuration of a periodic array of identical parallel stripes of rectangular cross-section. Each stripe is of width b and the periodicity of the configuration is p, as depicted in Figure 3.8(c). If the film material in this case carries an equi-biaxial mismatch strain ϵ_m with respect to the substrate, what is the resulting curvature of the substrate midplane? The film behaves as an orthotropic material over distances large compared to p, and the curvatures are different, in general, in the directions parallel to and perpendicular to the stripes.

3.10.1 Biaxial curvature due to lines

Because the film material is not deposited with uniform thickness in this case, the parameter κ_{St} is perhaps less suitable as a normalizing factor for curvature than it was in the discussion of a periodically cracked film. The mean film thickness in this case is $h_f b/p$, so a more suitable normalizing factor for curvature might be $b\kappa_{St}/p$. Nevertheless, the interpretation of the quantity represented by κ_{St} is left unaltered for this development, that is, it depends on h_f but not on b or p.

The geometry of the configuration and the coordinate system adopted are illustrated in Figure 3.16. The origin of a rectangular xyz-coordinate system is located on the substrate surface at a point midway between two neighboring stripes. The x-axis lies in the interface in a direction perpendicular to the stripes, the y-axis lies in the interface parallel to the stripes, and the z-axis is in the direction of the outward normal vector to the substrate surface. The figure shows a segment of the configuration which includes a single period of length p in the x-direction, extending from the midplane of one stripe to the midplane of a neighboring stripe. The extent of the segment in the y-direction is some arbitrary but fixed q.

If both p and h_f are very small compared to substrate thickness h_s, then the configuration can be analyzed within the framework established in considering the Stoney formula (2.7). The main task here is to determine the resultant forces per unit length in coordinate directions that act on symmetry planes of the configuration with the substrate in its undeformed state; these resultant forces are again denoted by f_x and f_y. The calculation can be pursued on the basis of elastic reciprocity (3.94), as in the case of a periodic array of cracks, and only a few steps in the procedure will be included. In the present case, the force resultants can be written as

$$f_x = \sigma_m h_f + \Delta f_x, \quad f_y = \sigma_m h_f b / p + \Delta f_y, \tag{3.111}$$

where the force increments Δf_x and Δf_y are defined by the reduced boundary value problem illustrated in Figure 3.16.

The specification of a boundary value problem that is complementary to that illustrated in Figure 3.16 is based on the same reasoning that led to the complementary problem in Figure 3.14. Application of the reciprocal theorem (3.94) again leads to the relationship (3.104), where δ^* now represents the average relative displacement of the side faces of the thin film stripes at $x = \pm \frac{1}{2}(p - b)$, that is, in the present case

$$\delta^* = \frac{1}{h_f} \int_0^{h_f} \left[u_x^*((p-b)/2, z) - u_x^*(-(p-b)/2, z) \right] dz \tag{3.112}$$

where $u_x^*(x, z)$ is the x-component of material displacement in the complementary elastic field. In other words, δ^* represents the relative normal displacement of strip side faces on opposite sides of the gaps between the stripes in the film. The dependence of δ^* on p, b, h_f and material parameters is expected to be very complicated in detail. However, in order to obtain an estimate of the curvature it seems reasonable to assume that the deformed shapes of the side faces of the stripes are essentially the same as the deformed shapes of the cracked faces in Figure 3.14, except that now these faces are separated by film material of width b, rather than of width p in the case of cracks. If this guess for δ^* is adopted, then (3.106) is replaced by

$$\delta^* \approx \ell \left(\frac{u_{f_x}^*}{p} + \nu_f \frac{u_{f_y}^*}{q} \right) \tanh \left(\frac{b}{\ell} \right) + \frac{u_{f_x}^*}{p} (p - b) . \tag{3.113}$$

The second term on the right side of (3.113) accounts for the fact that the side faces of the stripes are separated by the gap of width $p - b$, so this contribution represents an additional relative displacement of these faces due to the roughly uniform strain $u_{f_x}^*/p$ in this gap region. When the elastic properties of the film and substrate are the same, the factor $\tanh(b/\ell)$ in (3.113) can be replaced by the alternative form $\Upsilon(h_f/b)$ as defined in (3.99).

Aside from the task of devising an estimate for δ^* in the case of film stripes, the analysis is otherwise identical to that followed in deriving (3.108). The results for principal curvatures of the midplane of a substrate of thickness h_s and elastic properties E_s and ν_s are

$$
\frac{p}{b}\frac{\kappa_x}{\kappa_{St}} \approx 1 - \frac{1 - \nu_f \nu_s}{1 - \nu_s}\frac{\ell}{b}\tanh\left(\frac{b}{\ell}\right),
$$

$$
\frac{p}{b}\frac{\kappa_y}{\kappa_{St}} \approx 1 - \frac{\nu_f - \nu_s}{1 - \nu_s}\frac{\ell}{b}\tanh\left(\frac{b}{\ell}\right).
$$

(3.114)

This estimate is remarkably simple, given the complexity of the underlying boundary value problem. The left sides of the expressions in (3.114) are the curvatures normalized by the effective Stoney curvature based on the average film thickness $h_f b/p$. Furthermore, the right sides of (3.114) do not involve the parameter p. In other words, p enters the estimate for curvature only through the effective Stoney curvature based on the amount of film material involved. When $b = p$, the response is identical to that for a periodic array of parallel cracks.

Implications of the foregoing elastic analyses of curvature of substrates with periodic thin stripes or lines are next considered in the context of experimental results reported by Shen et al. (1996) for a model system that comprises Si substrates with unidirectional SiO_2 lines patterned on them. This particular system is chosen for discussion here primarily because, during oxidation of Si to form a continuous SiO_2 film, during patterning of the continuous SiO_2 film into parallel stripes on the Si substrate, and during subsequent thermal excursions, the entire film–substrate system remains elastic. The validity of the predictions of the earlier analyses could thus be directly assessed by systematically examining curvature evolution in the SiO_2–Si system for controlled variations in the thickness, width and spacing of the SiO_2 lines.

A number of single crystal Si wafers, 100 mm in diameter and 525 μm or 515 μm in thickness, with a (111) surface orientation, were prepared as substrate materials. The curvature of each blank wafer was measured prior to deposition using the scanning laser technique described in Section 2.3.1. The wafers were then thermally oxidized in steam to form a blanket SiO_2 film on both sides of each wafer. Because both the film and substrate are elastic and the oxide adheres well to the substrate, curvature evolution during patterning and thermal cycling could be interpreted without complications arising from inelastic deformation and damage in the film or delamination at the film–substrate interface. The temperature and duration of oxidation was varied for different wafers so that the SiO_2 films of different thicknesses were grown on different Si wafers. Nominal oxide thicknesses of 100 nm, 300 nm and 700 nm were obtained for oxidation treatments of 920 °C for 60 min, 1030 °C for 23 min, and 1030 °C for 100 min, respectively. The oxide films on one side of the wafers were removed subsequently.

Net compressive stress develops in the remaining oxide films upon cooling to room temperature from the oxidation temperature because the oxidation of Si results in an increase in volume and because the coefficient of thermal expansion of the oxide film is less than that of Si. The curvature of the wafers was measured at room temperature both along and transverse to the prospective line direction. These measurements revealed that, as anticipated, the curvatures along the x and y directions as oriented in Figure 3.16 were essentially the same, that is, $\kappa_x = \kappa_y = \kappa_{St}$, and that the corresponding mismatch stress σ_m in the continuous oxide films, predicted using the Stoney formula (2.7), ranged from about -281 MPa for the thickest film to -366 MPa for the thinnest film.

Following this step, the oxide films were dry-etched to form unidirectional, patterned lines spanning essentially the entire surface of the wafers. Six specimens with lines of widths roughly in the range $1\,\mu m \le b \le 1.43\,\mu m$ and thicknesses in the approximate range $100\,nm \le h_f \le 700\,nm$ were prepared; the spacing between the lines p was varied only in the narrow range from $2.81\,\mu m$ to $3.02\,\mu m$.

Figure 3.17 is a plot of the ratio of κ_x/κ_y as a function of the line aspect ratio h_f/b upon etching the continuous oxide film to form a periodic array of lines. As was anticipated by (3.114) the normalized line spacing p/b has essentially no effect on the curvature ratio plotted in Figure 3.17. When the continuous oxide film is etched into parallel lines, the wafers show reduced curvatures both along and across the lines, compared to the corresponding case of the continuous oxide films. This effect is a consequence of the partial relief of mismatch stress in the oxide film during material removal by etching of the film into lines. The extent of such decrease in curvature was observed to be stronger in the direction perpendicular to the lines than parallel to the lines. When $h_f/b \to 0$ with $p/b = 1$, the response of the line–substrate system approaches that of the film–substrate system, that is,

$$\frac{\kappa_x}{\kappa_{St}} \to 1 \quad \text{and} \quad \frac{\kappa_y}{\kappa_{St}} \to 1 \quad \text{as} \quad \frac{h_f}{b} \to 0 \tag{3.115}$$

in this case. These results follow directly from (3.114). For $h_f/b \to 0$ and $p/b > 1$, curvature depends on the fraction of the surface of the substrate covered by the thin oxide lines, as can be seen from (3.114), which implies that

$$\frac{\kappa_x}{\kappa_{St}} \to \frac{b}{p} \quad \text{and} \quad \frac{\kappa_y}{\kappa_{St}} \to \frac{b}{p} \quad \text{as} \quad \frac{h_f}{b} \to 0 \tag{3.116}$$

in this case. Experimental results indicate that κ_y/κ_{St} increases as p/b decreases, and that it is relatively insensitive to the line aspect ratio h_f/b when $h_f/b > 0.5$ (Shen et al. 1996). When the lines are very closely distributed, κ_y can be greater than κ_{St}.

As h_f/b increases from zero, the effects of the traction-free side surfaces of the unpassivated lines become progressively more dominant, and the curvature ratio κ_y/κ_{St} decreases precipitously. Figure 3.17 also shows that at $h_f/b \approx 0.35$, the

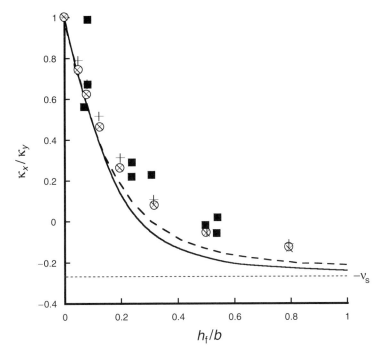

Fig. 3.17. The curvature ratio κ_x/κ_y versus h_f/b for several normalized values of line spacing p/b. The solid line represents the prediction based on (3.114), whereas the dashed line represents the prediction of the same behavior with $\tanh(b/\ell)$ in (3.114) replaced by $\Upsilon(b/h_f)$ from (3.99). The finite element predictions are denoted by the symbols $+$, $\times$ and $\circ$ for $p/b = 1.05, 2.10$ and 3.00, respectively. The filled square symbols represent the experimental data obtained by substrate curvature measurements by Shen et al. (1996). The analytical limit $\kappa_x/\kappa_y \to -\nu_s$ as $h_f/b \to \infty$ indicated by the dotted line is given in (3.117).

curvature ratio κ_x/κ_y changes from a positive value to a negative value. In other words, there occurs a transition from an ellipsoidal shape to a hyperboloidal or saddle shape in the line–substrate system as the line aspect ratio becomes greater than approximately 0.35 for the material combinations and geometries considered. This change in shape can be attributed to the Poisson effect which accounts for the coupling of strains in the in-plane orthogonal directions. Because of the traction-free side walls of the lines, it is reasonable to suppose that the normal stress acting in a direction across the lines $\sigma_{xx} \to 0$ throughout as $h_f/b \to \infty$. This is the limiting case considered in (3.110) from which it follows that

$$\frac{\kappa_x}{\kappa_y} \to -\nu_s, \quad \text{as} \quad \frac{h_f}{b} \to \infty. \tag{3.117}$$

The dashed line in Figure 3.17 shows this limit, which is approached by both the theoretical and experimental results at high line aspect ratios. The reversal of sign in curvature leading to a saddle shape for the Si substrate, when the geometry changes from 'short' to 'tall' oxide lines at $h_f/b \approx 0.35$ can, therefore, be regarded as a consequence of the substrate Poisson effect which causes anisotropic straining. The negative curvature transverse to the lines is only a small fraction of the curvature along the lines, and hence this Poisson effect, albeit experimentally detectable, is not very pronounced in the SiO_2–Si system.

3.10.2 *Volume averaged stress in terms of curvature*

The foregoing results relate the curvature of a substrate with a periodic array of lines or stripes on its surface to a mismatch strain in those lines. Some applications depend on the ability to make the inverse relationship, that is, to estimate stress levels in a patterned line on the basis of observation of substrate curvature. The stress levels are needed to understand evolution of damage or cracking in the lines, or to estimate service life of the patterned structure. This is the case in assessing damage due to electromigration in current carrying metal interconnects in integrated circuits as discussed in Section 9.7. The task of estimating stress in a periodically patterned film on a substrate is significantly more complex than for a uniform film. The reentrant corners, interfaces and free surfaces invariably result in large local stress gradients. Nonetheless, it is possible to draw inferences about the dependence of *volume averaged* stress components in periodic lines from substrate curvature observations. The relationship between the volume averaged stress and the corresponding curvatures may be useful in studying the evolution of stress in films during processing, patterning and temperature excursion.

As was noted in Section 3.7, the value of any stress component field σ_{ij} averaged over the region contained within a material volume V is denoted by $\langle \sigma_{ij} \rangle$ and is defined as

$$\langle \sigma_{ij} \rangle = \frac{1}{V} \int_V \sigma_{ij}(x_1, x_2, x_3)\, dV. \tag{3.118}$$

In the present context of periodically patterned films, the volume V is understood to be the volume of film material included within a single period of length p in a direction normal to the lines and an arbitrary distance q along the lines. The volume of film material included in Figure 3.16 is an example of such a volume V for the case of a periodic array of film stripes. With this definition, it becomes clear that the force resultants f_x, f_y introduced in Section 3.7 and Section 3.10.1 already represent the volume averaged stress components $\langle \sigma_{xx} \rangle$, $\langle \sigma_{yy} \rangle$ as long as the film is a single phase patterned material and the constraints of a thin film on a relatively thick substrate are satisfied. The case of a multiphase or composite film will be addressed subsequently. With reference to Figure 3.16, elementary force balance in the x-direction and y-direction leads to the relationships

$$f_x q = \langle \sigma_{xx} \rangle \frac{b h_f}{p} q , \quad f_y p = \langle \sigma_{yy} \rangle \frac{b h_f}{p} p, \tag{3.119}$$

respectively. These relationships presume that *all* of the force that induces substrate curvature arises from stress in the film material, but this is not strictly true; some load transfer occurs through the substrate near the line edges. Nonetheless, the relationships (3.119) will be adopted as representative of film stress levels.

Once the identification (3.119) is made, the means of deducing values of $\langle \sigma_{xx} \rangle$ and $\langle \sigma_{yy} \rangle$ from curvature measurements is straightforward. Consider the relationships (3.73), and associate the coordinate direction x transverse to the line direction with x_1 and associate the coordinate direction y parallel to the lines with x_2. Then, solving (3.73) for f_x and f_y and invoking (3.119), it follows that the volume averaged stress components in the lines transverse to and along the line directions are given in terms of substrate curvature by

$$\langle \sigma_{xx} \rangle \approx \frac{1}{6} \frac{h_s^2}{h_f} \bar{E}_s \frac{p}{b} \left(\kappa_x + \nu_s \kappa_y \right),$$

$$\tag{3.120}$$

$$\langle \sigma_{yy} \rangle \approx \frac{1}{6} \frac{h_s^2}{h_f} \bar{E}_s \frac{p}{b} \left(\kappa_y + \nu_s \kappa_x \right).$$

These expressions represent a reinterpretation of (3.73) based on the identification of the stress resultants f_x, f_y with average stress values for a thin film containing periodic parallel stripes as in (3.119).

In the limiting case of very tall stripes ($h_f/b \rightarrow \infty$), it is readily seen from the first of equations (3.120) that $\kappa_x/\kappa_y \rightarrow -\nu_s$ as $h_f/b \rightarrow \infty$. The theoretical and experimental results plotted in Figure 3.17 indicate that κ_x/κ_y asymptotically approaches the value $-\nu_s$, when the aspect ratio of the stripes h_f/b is approximately unity.

The effect of patterning the continuous film into periodic arrays of parallel stripes on stress evolution can be determined by combining (3.114) and (3.120) to obtain the volume averaged stress components in the lines in terms of the mismatch stress in the original continuous film from which the unpassivated lines are patterned, with the result that

$$\langle \sigma_{xx} \rangle \approx \sigma_m \left\{ 1 - \frac{\ell}{b} \tanh \left(\frac{b}{\ell} \right) \right\}, \quad \langle \sigma_{yy} \rangle \approx \sigma_m \left\{ 1 - \nu_f \frac{\ell}{b} \tanh \left(\frac{b}{\ell} \right) \right\}. \tag{3.121}$$

Here σ_m is the mismatch stress that would be present in a continuous film of uniform thickness h_f under the same conditions. As noted before, when the elastic properties of the patterned film and the substrate are comparable, the factor $\tanh (b/\ell)$ in (3.121)

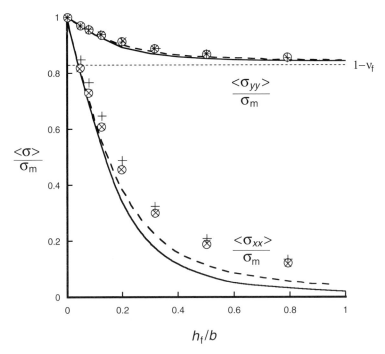

Fig. 3.18. Average values of the normalized principal stress transverse to the oxide line orientation $\langle\sigma_{xx}\rangle/\sigma_m$ and those along the line $\langle\sigma_{yy}\rangle/\sigma_m$, as a function of inverse strip width $\frac{h_f}{b}$, for several different normalized values of line spacing p/b. The solid lines represent the analytical predictions of average stress from (3.121), whereas the dashed lines are based on (3.122) for all values of p/b. Finite element predictions by Wikström et al. (1999a) are denoted by the symbols $+$, $\times$ and $\circ$ for $p/b = 1.05$, 2.10 and 3.00, respectively. The analytical limit, $\langle\sigma_{yy}\rangle/\sigma_m \rightarrow 1 - \nu_f$, is from (3.117) and is denoted by the dotted line for $h_f/b \rightarrow \infty$.

can be replaced by the alternative form $\Upsilon(b/h_f)$ defined in (3.99). The resulting values of average stress components in the latter case are

$$\langle\sigma_{xx}\rangle \approx \sigma_m \left\{ 1 - \Upsilon\left(\frac{h_f}{b}\right) \right\}, \quad \langle\sigma_{yy}\rangle \approx \sigma_m \left\{ 1 - \Upsilon\left(\frac{h_f}{b}\right) \nu_f \right\}. \qquad (3.122)$$

A comparison of the predictions of (3.121) with finite element simulations of average principal stress along and transverse to the SiO_2 line orientations is shown in Figure 3.18. Note that the approximate results given by (3.114) compare more favorably with the finite element results for $\langle\sigma_{yy}\rangle$ than for $\langle\sigma_{xx}\rangle$; the former is invariably the component that is larger in magnitude. Also, note that $\langle\sigma_{xx}\rangle \rightarrow \sigma_m$ and $\langle\sigma_{yy}\rangle \rightarrow \sigma_m$ as $h_f/b \rightarrow 0$, and that $\langle\sigma_{xx}\rangle \rightarrow 0$ and $\langle\sigma_{yy}\rangle \rightarrow (1 - \nu_f)\sigma_m$ in the limit $h_f/b \rightarrow \infty$.

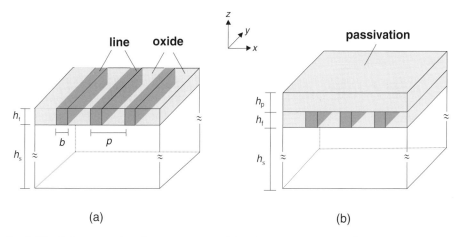

Fig. 3.19. Schematic showing a segment of the damascene copper line structure on a silicon substrate and the associated nomenclature. (a) Unpassivated structure. (b) Passivated structure.

3.10.3 Volume averaged stress in a damascene structure

The foregoing discussion of substrate curvature induced by stress in a periodic array of lines involved only unpassivated lines or stripes. In such a case, the patterned metal film is a single phase material formed by the subtractive process whereby the material originally between the stripes in a uniform film is removed. A more complex system is encountered in the damascene process described in Section 1.5.2 in which metal is deposited in an array of trenches formed in an originally uniform film of an insulating material. This results in a unidirectionally patterned composite thin film on a much thicker substrate. A common system of this type is formed by electrodepositing copper into trenches patterned into a silicon oxide layer, leading to the configuration depicted in Figure 3.19(a). In this section, the volume averaged mismatch stress in such a composite structure due to temperature change from an initially stress-free state is estimated. The stress distribution within the lines can be computed accurately by means of a finite element method for the structure. Alternately, the procedure developed in Section 3.10.1 could also be followed; however, the key approximation (3.113) which played a central role in that procedure is not yet available. Therefore, in this section, an estimate is obtained by following the approach of Wikström et al. (1999b); an alternate approach was described by Park and Suresh (2000).

Each copper line in Figure 3.19(a) has thickness h_f and width b, and the pitch of the periodic structure is p. The lines are laterally confined by stripes of silicon oxide of the same layer thickness h_f and width $(p - b)$. The Poisson ratio, elastic modulus and linear thermal expansion coefficient of the copper lines are denoted by v_f, E_f and α_f; the corresponding values for the surrounding oxide are v_o, E_o and α_o, and for the Si substrate of thickness $h_s \gg h_f$ are v_s, E_s and α_s. The stress is induced in response to a temperature change ΔT.

The volume averaged shear stress components in both the film material and the oxide material are identically zero due to the symmetry of the periodic structure. Therefore, the task is to estimate the volume averaged normal stress components in the coordinate directions. This task is facilitated by adopting the assumptions implicit in derivation of the Stoney formula (2.7) for a thin film on a relatively thick substrate. In particular, the mean strain in both the direction parallel to the lines and the direction perpendicular to the lines is required to be equal to the mean strain in the substrate surface, which is an isotropic extensional strain of magnitude $\alpha_s \Delta T$. These conditions imply that

$$\langle \sigma_{yy}^f \rangle - v_f \langle \sigma_{xx}^f \rangle - v_f \langle \sigma_{zz}^f \rangle + (\alpha_f - \alpha_s) \Delta T E_f = 0$$

$$\langle \sigma_{yy}^o \rangle - v_o \langle \sigma_{xx}^o \rangle - v_o \langle \sigma_{zz}^o \rangle + (\alpha_o - \alpha_s) \Delta T E_o = 0$$

(3.123)

$$\frac{(p-b)}{E_o} \left[\langle \sigma_{xx}^o \rangle - v_o \langle \sigma_{yy}^o \rangle - v_o \langle \sigma_{zz}^o \rangle \right] + (p-b)(\alpha_o - \alpha_s) \Delta T +$$

$$\frac{b}{E_f} \left[\langle \sigma_{xx}^f \rangle - v_f \langle \sigma_{yy}^f \rangle - v_f \langle \sigma_{zz}^f \rangle \right] + b(\alpha_f - \alpha_s) \Delta T = 0,$$

which provides three equations involving the six normal stress components to be determined.

Three additional equations are needed to establish the stress estimate. The condition

$$\langle \sigma_{xx}^f \rangle = \langle \sigma_{xx}^o \rangle$$

(3.124)

ensures that the two phases of the film exert equal and opposite forces on each other, but it does not require continuity of displacement across the interface. Finally, the free surface condition is satisfied by requiring the net force per period in the z-direction to be zero and the extensional strain in the z-direction to be the same in the line and oxide. These conditions are enforced by

$$\frac{b}{p} \langle \sigma_{zz}^f \rangle + \frac{p-b}{p} \langle \sigma_{zz}^o \rangle = 0,$$

$$\frac{1}{E_f} \left[\langle \sigma_{zz}^f \rangle - v_f \langle \sigma_{xx}^f \rangle - v_f \langle \sigma_{zz}^f \rangle \right] + \alpha_f \Delta T = \qquad (3.125)$$

$$\frac{1}{E_o} \left[\langle \sigma_{zz}^o \rangle - v_o \langle \sigma_{xx}^o \rangle - v_o \langle \sigma_{zz}^o \rangle \right] + \alpha_o \Delta T.$$

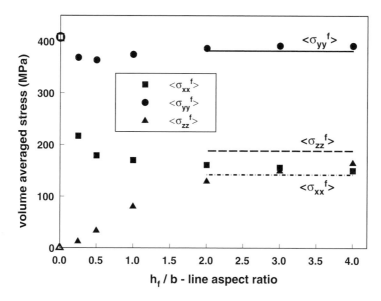

Fig. 3.20. Volume averaged stress in Cu lines embedded in SiO_2 on a Si substrate due to a temperature reduction of 180 °C computed by the finite element method. The horizontal lines represent the estimates for high aspect ratios. The open symbols on the $h_f/b = 0$ axis represent the low aspect ratio limits of stress components corresponding to the matching closed symbols. Finite element results were provided by T.-S. Park, Massachusetts Institute of Technology (2001).

This approximation is expected to be most effective for values of h_f larger than b and $p - b$.

The components of averaged stress in the film material can be determined by solving the set of linear equations defined by (3.123), (3.124) and (3.125). It is readily confirmed from these equations that, when the volume fraction of either of the phases approaches zero, the estimate of average stress in the remaining phase approaches the appropriate uniform stress limit. For example, when $E_o \rightarrow E_f$, $\nu_o \rightarrow \nu_f$ and $\alpha_o \rightarrow \alpha_f$, the stress estimate has the properties that $\langle \sigma_{xx}^f \rangle = \langle \sigma_{yy}^f \rangle \rightarrow -M_f(\alpha_f - \alpha_s)\Delta T$ and $\langle \sigma_{zz}^f \rangle \rightarrow 0$.

These expressions are easily evaluated for specified material properties and temperature change. For example, consider cooling of a system in which the film material is Cu for which $E_f = 110\,\text{GPa}$, $\nu_f = 0.30$ and $\alpha_f = 17.0 \times 10^{-6}\,°\text{C}^{-1}$, the oxide material is SiO_2 for which $E_o = 71.4\,\text{GPa}$, $\nu_o = 0.16$ and $\alpha_o = 0.524 \times 10^{-6}\,°\text{C}^{-1}$ and the substrate is Si for which $E_s = 130\,\text{GPa}$, $\nu_s = 0.28$ and $\alpha_s = 2.60 \times 10^{-6}\,°\text{C}^{-1}$. Consider the case when $b/p = 0.5$. For a temperature change of $\Delta T = -180\,°\text{C}$, the estimates of volume averaged stress components in this case are $\langle \sigma_{xx}^f \rangle = 141\,\text{MPa}$, $\langle \sigma_{yy}^f \rangle = 382\,\text{MPa}$ and $\langle \sigma_{zz}^f \rangle = 180\,\text{MPa}$. To develop an understanding of the quality of this approximation, the system depicted in Figure 3.19(a) was analyzed by means of the finite element method for a range of values of h_f/b with $b/p = 0.5$ (Wikström et al. 1999b). The computed results for average stress are shown in Figure 3.20 by means of discrete points for six values of aspect ratio h_f/b. The estimates obtained in this section, which are independent of h_f/b, are represented in the figure by horizontal lines. As was anticipated, the estimate is quite accurate for h_f/b greater than about two, and it is probably a useful approximation for h_f/b greater than about one.

The substrate curvature implied by the average stress estimates from (3.123)–(3.125) is also readily obtained. The effect of membrane force resultants in the composite film are obtained by means of (3.119) as

$$f_x = \langle \sigma_{xx}^f \rangle h_f = \langle \sigma_{xx}^o \rangle h_f, \qquad f_y = \frac{b}{p} \langle \sigma_{yy}^f \rangle h_f + \frac{p-b}{p} \langle \sigma_{yy}^o \rangle h_f. \qquad (3.126)$$

The induced curvature changes in the coordinate directions due to these stress resultants are obtained by means of (3.73) as

$$\kappa_x = \frac{6}{E_s h_s^2} \left(f_x - \nu_s f_y \right), \qquad \kappa_y = \frac{6}{E_s h_s^2} \left(f_y - \nu_s f_x \right). \qquad (3.127)$$

Finally, suppose that the patterned composite film structure on the left in Figure 3.19 is capped with a uniform thin passivation layer, as shown on the right in the figure. If both surface layers are thin compared to the substrate, then their contributions to substrate curvature are additive, as was demonstrated in Section 2.4.7. In such a case, the total curvature is the sum of the contribution from the patterned film as discussed in this section and that due to the uniform film as discussed in Chapter 2.

3.11 Measurement of stress in patterned thin films

As noted in previous sections, the development of life prediction models for the reliability of patterned features such as periodic lines on substrates inevitably requires knowledge of intrinsic stress and mismatch stress generated during film growth, patterning, passivation and service. In this section, three experimental methods for determining stress in thin films with patterned lines are considered: the substrate curvature method, the x-ray diffraction method, and the micro-Raman spectroscopic method. The advantages and limitations of each of these techniques are also briefly addressed.

3.11.1 The substrate curvature method

The substrate curvature method, described in Section 2.3, provides a simple experimental tool for *in-situ* measurement of changes in principal curvatures in a patterned thin film. In the scanning laser method for substrate curvature measurement discussed in Section 2.3.1, this is achieved by scanning the laser beam sequentially along the anticipated principal directions, that is, along and transverse to the line direction. With the grid reflection technique discussed in Section 2.3.3 or the coherent gradient sensor technique discussed in Section 2.3.4, variations in both the curvature κ_y along the lines and the curvature κ_x transverse to the lines can be monitored simultaneously and in real time as the mismatch strains evolve. Once the principal curvature values are known, the average values of in-plane normal stress components are estimated by recourse to analytical approximations, such as (3.121) or (3.122), or the finite element method of numerical simulation.

A drawback of this method is that measurements of changes in principal curvature components represent changes in average values of principal stress which are sampled over a large number of periodically spaced, nominally identical lines. The curvature method is thus incapable of assessing stress evolution in individual lines as they develop slit-like cracks or voids; line to line variations in stress also cannot be quantified using this method. Nevertheless, this technique offers a convenient means of monitoring small variations in substrate curvature from which average stress values in both crystalline and amorphous films comprising periodic patterned lines can be obtained using relatively simple-to-use, commercially available experimental systems. Examples of experimental results obtained through the substrate curvature method were illustrated in Figure 3.17.

3.11.2 The x-ray diffraction method

As described in Section 3.6, the x-ray diffraction method relies on measurements of elastic strains from changes in the lattice spacing caused by stress. Such elastic strains are converted to average stress values from known elastic constants of the material. Real time measurements of stress evolution (as, for example, during electromigration) in unpassivated and passivated patterned lines can be conducted with x-ray microdiffraction using synchrotron-based white beam x-rays where the beamline is confined by a pinhole for a narrow x-ray beam, typically several μms in diameter (Wang et al. 1998); diffraction from a few patterned lines or stripes are sampled and averaged in this manner. The changes in d-spacing of a particular set of crystal planes during stress evolution is then measured at different positions along the lines by translating the specimen. A particularly appealing feature of this technique is that the stress gradient which develops along the length of the lines can be monitored in real time.

As in the case of the substrate curvature method, the x-ray diffraction technique provides average values of stress that are sampled over many lines; it does not provide a means to estimate stress in a single line. Another shortcoming of the method is that the measured changes in strains can be very small for the patterned line structure. For example, periodically patterned Al interconnect lines on Si substrates which are subjected to electric currents that are typical of service conditions in microelectronic circuitry are known to develop strains smaller than 10^{-3}; see Wang et al. (1998), Flinn and Chiang (1990), DeWolf et al. (1999). Consequently, the accuracy of the standard x-ray diffraction units may not be adequate for the level of precision required in many applications involving patterned thin films; cost and safety issues are also additional limiting factors. The use of the x-ray diffraction method is restricted to crystalline thin films.

3.11.3 Micro-Raman spectroscopy

Micro-Raman spectroscopy has evolved as a possible experimental method for estimating local stress in nonplanar, submicron scale structures of interest in very large scale integration (VLSI) devices and MEMS; see, for example, Kobayashi et al. (1990) and DeWolf et al. (1992). Examples of the geometries studied by this method include: local isolation structures involving techniques such as local oxidation of polysilicon over silicon, isolation trenches, metal conductors in microelectronic devices, and V-groove structures.

For the case of interconnect line features patterned on a single crystal Si substrate, a common procedure for the use of micro-Raman spectroscopy entails determination of the stress in the interconnect from the distortion produced in the substrate and the dielectric which immediately surrounds the interconnect (DeWolf et al. 1992, Ma et al. 1995, DeWolf et al. 1999); the Raman spectra can be measured in the backscattering mode using an argon laser. The piezo-spectroscopic shift in the Raman spectrum, which is induced by such local distortions, can be measured to a spatial resolution on the order of 1 μm so that stress in individual lines can be estimated.

A single crystal Si substrate has three 'Raman-active' optical vibration modes: for incident optical direction along the [001] crystallographic axis, the one longitudinal mode coincides with [001] directions, and the two transverse modes are oriented along [100] and [010] directions. If the Si substrate is devoid of any stress, these three vibration modes have the same frequency, $\omega_0 \approx 520$ R cm^{-1}. When backscattering occurs from the (001) surface, it is known that the Raman signal is produced only by the longitudinal mode. Local straining of the Si substrate as a result of the mismatch stress in the interconnect line lowers the symmetry of the Si crystal. This causes the vibration frequencies of the three optical modes to shift from ω_0, with the amount of shift in each of the three modes dependent upon the nature of the strain tensor. These frequency shifts can easily be related to the strain tensor. (In addition, the

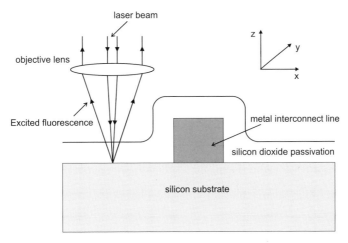

laser beam

objective lens

Excited fluorescence

metal interconnect line

silicon dioxide passivation

silicon substrate

Fig. 3.21. Schematic showing an experimental setup for determining the piezo-spectroscopic shifts in the Raman spectrum in a Si substrate immediately surrounding an interconnect line.

associated changes in the Raman polarizability tensors can be used to assess the scattering efficiency which signifies the ratio of the scattered power to the incident power of the laser beam.) These steps enable the determination of elastic strains from the experimentally monitored shifts in the Raman spectrum. Once the strain tensor is determined, the stress in the interconnect line can be approximated using the finite element method whereby the measured frequency shifts are matched with the stress distributions for a particular interconnect line geometry as a function of distance from the line.

Figure 3.21 schematically shows a typical experimental setup used for the micro-Raman spectroscopy. A laser beam, typically an argon-ion laser with an output power of no more than a few tens of mW (so as to avoid local heating of the substrate), is directed in a region immediately surrounding an interconnect line of interest. Note that the laser beam easily traverses through a passivation or capping layer of silicon dioxide or silicon nitride so that passivated line structures can also be studied. The beam is focused at the interface between the passivation layer and the substrate. The signal of the excited Raman fluorescence is gathered by the objective lens and then passed through a monochromator for frequency analysis. A series of Raman spectra is obtained at different points on either side of the interconnect so that the spatial variation of frequency shift is adequately captured.

A drawback of this method is that micro-Raman spectroscopy cannot generally be applied to metallic materials. Furthermore, stress estimation can only be made on the basis of Raman spectra gathered in the substrate and the dielectric on the sides of the interconnect lines. Such measurements cannot be made in the region immediately below the line, where the stress is more uniform and more sensitive to the stress in the lines themselves, if the substrate is nontransparent. Despite these limitations, the micro-Raman spectroscopy method has been successfully used to infer the average values of internal stress components in interconnect lines tested in the as-fabricated condition as well as after electromigration testing (Ma et al. 1995, DeWolf et al. 1999).

3.12 Exercises

1. A cubic crystal is subjected to uniaxial tension along the [112] direction. Show that the effective elastic modulus for that orientation is

$$E_{112} = \left\{ \frac{s_{11}}{2} + \frac{s_{12}}{2} + \frac{s_{44}}{4} \right\}^{-1}. \tag{3.128}$$

2. A thin film of a cubic crystal is deposited on a substrate. The orientation of the film material is such that the direction perpendicular to the interface coincides with the [101] crystallographic direction of each material. The elastic constants of the film referred to the natural coordinate axes are c_{11}^*, c_{12}^* and c_{44}^*.
   (a) Show that the film appears to be orthorhombic (or orthotropic), with five independent elastic stiffness constants, in a global coordinate system with axes coinciding with the crystallographic directions $[10\bar{1}]$, $[010]$ and $[101]$.
   (b) In the global coordinate system, show that the nonzero components of c_{ij} are

$$c_{11} = c_{33} = \tfrac{1}{2}c_{11}^* + \tfrac{1}{2}c_{12}^* + c_{44}^*, \qquad c_{13} = \tfrac{1}{2}c_{11}^* + \tfrac{1}{2}c_{12}^* - c_{44}^*,$$
$$c_{22} = c_{11}^*, \qquad c_{12} = c_{23} = c_{12}^*, \qquad c_{44} = c_{66} = c_{44}^*. \tag{3.129}$$

3. A thin single crystal film of a cubic material is grown epitaxially to thickness h_f on a substrate of another cubic material of thickness h_s. Both materials are oriented with the shared interface being the (101) plane of each crystal. As noted in the preceding exercise, both materials can be viewed as orthotropic materials in the global coordinate system adopted in Section 3.5, that is, they are aligned orthotropic materials. Assuming that the extensional mismatch strain in the film with respect to the substrate is ϵ_m in all directions in the interface and that the two materials have the same elastic constants, determine the ratio of curvatures κ_2/κ_1 in the global coordinate directions in the interface coinciding with the crystallographic directions $[10\bar{1}]$ and $[010]$. Express the result in terms of the three elastic constants adopted in (3.129).
4. The residual strains in an aluminum film, 1 μm in thickness, on a Si substrate, 550 μm thick and 200 mm in diameter, were studied by x-ray diffraction. It was found that the d-spacing of crystallographic planes in the film, which were oriented parallel to the plane of the film decreased by 0.054%, compared to the situation when the film had no residual strains.
   (a) If the film is assumed to be elastically isotropic, find the equi-biaxial mismatch strain and mismatch stress in the film.
   (b) Repeat your calculations assuming that the aluminum film has (100) texture.

5. A polycrystalline thin film of Cu is deposited on a thick Si substrate. The grains of the Cu film have a 'bamboo' structure, i.e., the grain boundaries extend through the entire thickness of the film with only one grain through the thickness. If the texture of the film is such that the [111] direction of the grains is oriented normal to the film–substrate interface for all the grains, describe how you would approximate the in-plane elastic modulus of the polycrystalline Cu film.

6. A metal thin film is deposited on a Si substrate at an elevated temperature. After cooling the film–substrate system to room temperature, it was found from x-ray diffraction measurement that the d-spacing in the direction normal to the film–substrate interface had decreased by 1 percent compared to that in the same crystallographic direction in the unstressed film material. Assume that the film is isotropic, and that $E_f = 70$ GPa, $h_f = 1\,\mu$m, $\nu_f = 0.33$, $\alpha_f = 23\times10^{-6}\,°\text{C}^{-1}$, $E_{Si} = 130$ GPa, $h_s = 500\,\mu$m, $\nu_{Si} = 0.28$, and $\alpha_{Si} = 3\times10^{-6}\,°\text{C}^{-1}$.
   (a) Calculate the mismatch strain and stress in the film.
   (b) Determine the curvature of the substrate.
   (c) After the x-ray diffraction test, it is determined that the curvature is too large for the Si wafer to be acceptable for the intended application. As a first suggestion to decrease the curvature, you recommend that the film thickness be reduced if possible. The Si wafer manufacturer notes that although the film thickness could be reduced, the ratio h_f/h_s should be kept constant for the particular application. Does this requirement change your recommendation?

7. A multilayer is made by growing alternate layers of single crystalline Si and Ge. The repeating layers of Si and Ge in the multilayer are (100) films with thicknesses of 8 nm and 2 nm, respectively. The lattice parameters of Si and Ge are 0.542 nm and 0.564 nm, respectively.
   (a) If the epitaxial mismatch strains between adjacent layers are not relieved by the nucleation of misfit dislocations, compute the equi-biaxial stress and strains in each Si and Ge layer.
   (b) What is the out-of-plane strain in each Si and Ge layer?

8. Consider the isotropic aluminum film on silicon substrate which is discussed in problem 4. This film is etched into unidirectional lines, of the same thickness as the film and with 1 μm width and 2 μm spacing.
   (a) Estimate the values of the volume averaged stress along and across the lines.
   (b) Determine the volume averaged shear stress in the lines.

9. Ten different thin film specimens are fabricated, with each specimen comprising a continuous thin film of Cu deposited uniformly on a Si substrate which is 200 mm in diameter and 600 μm in thickness. The film thickness in the ten specimens ranges from 100 nm to 1000 nm. X-ray diffraction experiments indicate that the mismatch stress in the films varies from about 200 MPa for the thinnest film to 40 MPa for the thickest, and that the stress for the in-between thicknesses can be approximated by linear interpolation between these two limits. The films in all ten specimens are then patterned into unidirectionally oriented lines which are 750 nm wide and 3,000 nm apart; the thicknesses of the lines in the different specimens are the same as those of the continuous films from which they are etched. Assuming that curvature evolution during the patterning of the film into lines arises solely from geometry changes in each specimen, determine the average stress in the lines and the curvatures along and across the lines.

10. The methods outlined in Section 3.10 for patterned thin films on substrates can also be used, with appropriate modifications, to estimate substrate curvature and volume-averaged elastic stress in multilayered patterned films. Figure 3.22 shows a

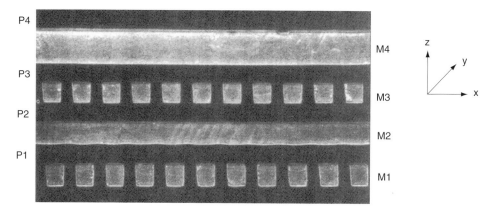

Fig. 3.22. A cross-sectional micrograph showing a four-level damascene copper line structure, where the lines appear as white regions. The silicon oxide passivation layers, $P_1, \ldots,$
P_4, are deposited on each level of metallization. The dark regions in the figure are the
silicon oxide trenches and passivation layers. The thickness, width and spacing of the Cu
lines and the surrounding oxide trenches in each metallization level and the thickness of
the passivation layers separating different levels of metallization are all 1 μm. Note that the
Cu lines in levels M_1 and M_3 are oriented along the y-direction whereas those in levels M_2
and M_4 are parallel to the x-direction. (Image provided by IBM Corp. and reproduced with
permission.)

four-level damascene copper line structure where the Cu lines fill the silicon oxide
trenches; different levels of metallization are separated by silicon oxide passivation
films.

(a) Consider a Si wafer substrate, 200 mm in diameter and 525 μm thick, which
has a uniform initial curvature of -0.001 m^{-1}, with the sign convention
from Section 2.1 that the face of the substrate bonded to the film becomes
concave for the case of a tensile mismatch stress. This substrate is then
oxidized at 400 °C in such a way that a uniform silicon oxide film, 1 μm
in thickness, is formed on one surface. After the film–substrate system is
cooled down to room temperature, the substrate curvature was measured
to be -0.008 m^{-1}. The oxide is then patterned at room temperature into
unidirectional trenches, 1 μm wide with a spacing of 1 μm. The elastic
modulus, Poisson ratio, and thermal expansion coefficient of the Si substrate are $E_s = 130$ GPa, $\nu_s = 0.28$ and $\alpha_s = 2.6 \times 10^{-6}$ °C^{-1}, respectively;
the corresponding values for silicon oxide are: $E_o = 71.4$ GPa, $\nu_o = 0.16$ and
$\alpha_o = 0.524 \times 10^{-6}$ °C^{-1}, respectively. Find the magnitude and sign of curvature parallel to and perpendicular to the orientation of the oxide trenches
on the Si substrate. Assume, in all parts of this problem, that the entire
film–substrate system remains elastic.

(b) The gap between the oxide trenches is now electrodeposited with pure Cu
at room temperature, and the excess Cu above the oxide trench is removed
by chemical–mechanical polishing. Any intrinsic stresses introduced during
metal deposition and polishing may be assumed to be negligible. The elastic
modulus, Poisson ratio, and thermal expansion coefficient of the Cu lines
are $E_f = 110$ GPa, $\nu_f = 0.3$ and $\alpha_f = 17 \times 10^{-6}$ °C^{-1}, respectively. A 1 μm

thick silicon oxide passivation layer, denoted as layer P_1 in Figure 3.22, is now deposited on the composite layer M_1 at 400 °C. It may be assumed that the formation of the continuous oxide film at 400 °C does not lead to any intrinsic stress buildup. Find the curvatures along and across the Cu line direction when the passivated film–substrate system is cooled to room temperature.

(c) Three more levels of metallization and passivation comprising layers M_2, P_2, M_3, P_3, M_4 and P_4 are now made by repeating the above procedure, with the lines in layer M_3 being parallel to those in layer M_1, and the lines in layer M_4 being parallel to those in layer M_2. Find the principal curvatures of this four-level metallization system at room temperature.

4

Delamination and fracture

Any material that transmits mechanical load from one place to another is susceptible to fracture. This susceptibility is enhanced if the geometry of the object includes reentrant corners, internal defects or other geometrical variations which serve as sites of *stress concentration*. At these sites, the local stress can be much larger than the nominal stress, which is loosely defined as the average stress transmitted at a cross-sectional area. As a consequence of stress concentration, the local stress can exceed the strength of the material and fracture ensues, even though the nominal stress is well below the fracture strength.

Mechanical interactions between a thin film and a substrate to which it is bonded were the focus of discussion in the preceding chapters. These interactions, when occurring in the absence of any failure or delamination processes, are manifested in a number of ways: the constraint of the substrate prevents the film from relaxing its internal stress, and the film–substrate system accommodates the internal stress by in-plane stretch or contraction, substrate curvature and/or plastic yielding of the film. However, when edge effects are neglected, the *traction* exerted by the film and substrate material on each other across the film–substrate interface is zero everywhere.† If this is the case, how do the film and substrate interact?

The paradox of interaction without traction on the shared interface is resolved by recalling that the thickness of the film–substrate system is much smaller than its in-plane dimensions. If the mechanical state of the material is examined at various points on the film–substrate interface more distant from the film edge than several times the film thickness h_f, these points appear to be indistinguishable. As a result, the mechanical stress and deformation fields are translationally invariant in any direction parallel to the interface over nearly the entire volume of the film and the

---

† In general, traction is a vector valued quantity representing the local force per unit area acting on a material surface, whereas stress is a tensor valued property of the local mechanical state in a material which is independent of any particular surface orientation. While the terms are commonly used interchangeably, the distinction is sometimes important for understanding.

substrate material, the exception being only a small region near the free edges. Thus, when global measures of deformation such as potential energy or curvature are considered, the fact that some relatively small volume of material near the film edge has been neglected has only very minor consequences.

The property of translational invariance cannot be invoked near the film edge, and this is precisely the region where mechanical load transfer between the film and substrate is effected. Because the free edge of a film is a site of an abrupt change in geometrical shape, its neighborhood is expected to be a region of stress concentration. Furthermore, this local stress field must be self-equilibrating because the translationally invariant internal stress field itself is an equilibrium field in the absence of any externally imposed loads. Finally, the abrupt change in geometry arising at the intersection of a film–substrate interface with a free edge is conceptually similar to the region of stress concentration introduced by a crack or the edge of a delamination zone. Indeed, a sharp crack in a solid is commonly regarded as the ultimate stress concentrating feature.

In this chapter, the goal is to discuss the stress concentration at the edge of a film bonded to a substrate in terms of material and geometrical parameters. Since stress concentration can lead to film fracture and interfacial delamination, attention is also directed to quantitative descriptions of the growth of a crack or delamination. The study of stress concentration is based on continuum constitutive behavior and equilibrium stress analysis, and its extension to the study of interface delamination and crack growth requires introduction of a fracture criterion as an additional physical postulate of material behavior. Such fracture criteria are inevitably linked to the mechanisms of material separation which, in turn, can be influenced by microstructure and environment.

The discussion of stress concentration near a film edge in the next section is followed by a brief review of linear elastic fracture mechanics concepts, a prelude to a discussion of delamination and cracking due to film residual stress. A survey of these topics set in the context of fracture mechanics has been presented by Hutchinson and Suo (1992). The chapter also includes descriptions of various experimental techniques for evaluating the fracture resistance of interfaces between films and substrates. In addition, representative experimental results on the interface fracture resistance, as a function of interface chemistry and environment, are presented for a variety of thin film and multilayer systems of scientific and technological interest.

4.1 Stress concentration near a film edge

The fundamental edge stress configuration for strained thin films is shown in Figure 4.1. This configuration can be imagined to arise in any number of different ways. For example, start with a relatively thick, stress-free substrate which has a

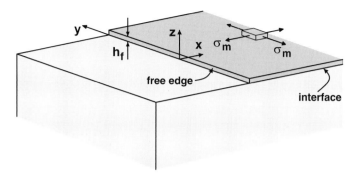

Fig. 4.1. Schematic diagram of a thin film with a free edge bonded to a thick substrate. The equi-biaxial stress in the film is σ_m at points far from the film edge compared to h_f. The planar edge of the film $x = 0$ is traction-free.

large planar surface that coincides with the xy-plane in the figure. Then, a uniform all-around normal traction of magnitude σ_m, either tension or compression, is applied to the edges of a film of thickness h_f, resulting in a homogeneous state of equi-biaxial stress throughout the film. The stressed film is then bonded to the substrate, forming a planar interface, after which the traction acting on the edge of the film coinciding with the plane $x = 0$ in Figure 4.1 is relaxed. This results in the edge of the film becoming a free surface. If attention is then focused on material points near the edge of the film where the film boundary in plan view is fairly smooth, the system has the features shown. The film edge lying along the y-axis appears to be straight at this scale, and the only pertinent physical dimension is the film thickness h_f. The edge of the film at $x = 0$ is traction-free. At points in the film that are remote from the edge compared to the distance h_f, the stress is unrelaxed; asymptotically, the stress state approaches the equi-biaxial stress of magnitude σ_m as x/h_f becomes large. Stress and deformation fields vary relatively slowly with distance along the film edge on this scale. Consequently, this state of deformation is essentially two-dimensional generalized plane strain in the xz-plane, assuming at least orthotropic material behavior with the axes of orthotropy aligned with the coordinate axes.

Near the free edge of the film, some traction is exerted on the film by the substrate across the common interface. Far from the edge, the film stress is uniform and, as a consequence of the requirement of local equilibrium, no traction acts across the interface. The goal here is to understand the spatial scale of the transition from the edge behavior to the remote interior behavior and the transfer of mechanical load from the substrate to the film in the region near the free edge of the film.

This edge problem can be studied at several scales of observation. It is instructive to consider states of stress for two extreme sites, namely, for points that are very close to the edge compared to film thickness and for points that are very far from the edge compared to film thickness. For points close to the interface edge for which $(x^2 + z^2)^{1/2} \ll h_f$, the problem of stress concentration along the film edge has no characteristic length scale. It can be formulated as a plane strain analysis of wedges that are joined along a common boundary (Bogy 1971). At this scale of observation, the film is represented by a 90° elastic wedge subjected to a mismatch strain. One face of the wedge is free of traction and the other is bonded to the surface of an elastic half space. In this case, the stress magnitude is found to be singular at the corner and to exhibit rapid oscillations in amplitude with distance from the corner, in general. For the special case when both materials behave as isotropic elastic solids with the same elastic constants, the oscillations vanish and the stress components vary with distance from the edge as $r^{-\alpha}$ where $\alpha \approx 0.4$.

At the other extreme, for points far from the edge compared to h_f, the film thickness is imperceptible and the influence of this length is again suppressed. In the remote field ($x \gg h_f$), the force per unit distance in the y-direction on any film section is $\sigma_m h_f$, and this force acts in the x-direction. At the film edge $x = 0$, the force on the film is zero. Consequently, by the condition of overall equilibrium, the resultant force due to the traction exerted by the substrate on the film has magnitude $-\sigma_m h_f$ and it acts in the x-direction. Thus, the stress in the substrate at points far from the origin compared to h_f is that due to a concentrated tangential line load of magnitude $\sigma_m h_f$ per unit distance in the y-direction, acting in the x-direction on the substrate surface. A main feature of this stress field, which is the classical Flamant solution for isotropic elastic solids (Johnson 1985), is that the magnitude of any stress component varies as r^{-1} with distance from the film edge. The decay in magnitude along the interface is expected to be more rapid than x^{-1} because the stress acting on this interface is identically zero in the Flamant solution. In any case, the configuration has no characteristic length at this scale of observation, and the stress components necessarily decay in magnitude inversely with distance from the load point at the origin.

While both of these asymptotic extremes provide useful information about film edge effects, they do not address the important question of load transfer posed above. Specifically, what is the character of the transition in response between these two limiting cases, and over what size scale does this transition occur? This issue is taken up next, beginning with the simplest system that reveals the nature of this transition; more realistic systems are discussed subsequently.

4.1.1 A membrane film

The issue of interface stress near a film edge is addressed first by idealizing the thin film as an elastic membrane. Such an idealization rests on the notion that thin

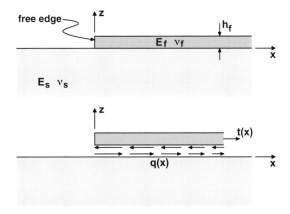

Fig. 4.2. A schematic diagram of a film with a free edge bonded to a substrate is shown in the upper portion. The lower portion depicts the same system but with the film and substrate separated to reveal the shear traction distribution $q(x)$ through which they interact across their interface and the internal membrane tension $t(x)$ in the film.

membrane structures are relatively more resistant to extensional loading than to bending; they are not capable of sustaining transverse loads during small deformation. A major outcome of this idealization is the realization that the interface traction between the film and the substrate near the free edge is a shear traction.

The problem is illustrated schematically in Figure 4.2, where the film and substrate have been separated to reveal the spatially nonuniform interfacial shear traction $q(x)$ exerted by each material on the other. The goal here is to determine $q(x)$ in terms of the film mismatch stress σ_m and film thickness h_f, for arbitrary combinations of elastic properties for the film and the substrate, both of which are assumed to be isotropic. Also shown in Figure 4.2 is the tensile force $t(x)$ acting per unit width of the section in the film located at distance x from the edge. On physical grounds, it is expected that $q(x) \rightarrow 0$ and $t(x) \rightarrow h_f \sigma_m$ as $x/h_f \rightarrow +\infty$ along the interface, as noted in the introduction to the chapter.

The interfacial shear stress is determined by requiring that the material is everywhere in equilibrium without the application of the external loading and that the deformation resulting from relaxation of the edge stress from the value σ_m at $x = 0$ is continuous across the interface. The steps involved in deriving the governing equation for $q(x)$ are outlined in Section 4.1.2. It is anticipated that the stress distribution $q(x)$ will be singular in the limit as $x/h_f \rightarrow 0^+$. By means of the J-integral of elastic fracture mechanics (Rice 1968) or other equivalent arguments, it can be established that

$$q(x) \sim \sigma_m \sqrt{\frac{k h_f}{2\pi x}} \qquad (4.1)$$

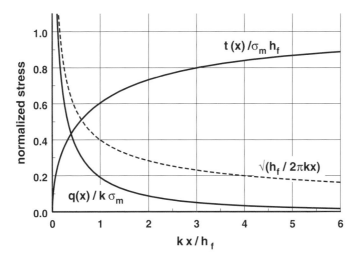

Fig. 4.3. The solid curve labeled $q(x)/k\sigma_m$ shows the shear traction versus distance kx/h_f, as determined from the numerical solution of the elastic membrane problem. The dashed curve shows the square root singular behavior of the shear traction anticipated in (4.1) and the curve labeled $t(x)/\sigma_m h_f$ is the normalized film tension that is in equilibrium with the shear traction $q(x)$.

asymptotically as $x/h_f \rightarrow 0^+$, where

$$k = \frac{\bar{E}_s}{\bar{E}_f} = \frac{E_s}{1-\nu_s}\frac{1-\nu_f}{E_f} \tag{4.2}$$

is the ratio of the plane strain elastic modulus of the substrate to that of the film.

The numerical solution of the equation governing $q(x)$ for $0 < x < \infty$, which is established in Section 4.1.2, is readily obtained by means of the technique developed by Erdogan and Gupta (1972). The result is graphed in nondimensional form $q(x)/k\sigma_m$ versus kx/h_f in Figure 4.3. Included in the same figure are graphs of the normalized extensional stress in the film $t(x)/\sigma_m h_f$ and of the square root singular asymptotic form (4.1). Several noteworthy features are evident in Figure 4.3. For one thing, the numerical solution matches the square root singular behavior very well as $kx/h_f \rightarrow 0^+$. For values of kx/h_f large compared to unity, the shear stress was anticipated above to decay more rapidly with distance from the edge than x^{-1} and this is indeed so. The action of load transfer between the film and the substrate is confined to a region of length only a few times h_f/k near the film edge. Thus, if the materials have approximately the same elastic properties, so that $k \approx 1$, the film membrane force $t(x)$ increases from zero to about 80% of its asymptotic value $\sigma_m h_f$ within a distance of about $3h_f$ from the edge. In retrospect, this result reinforces the assumption of the preceding chapters that the edge effect associated with the load transfer region is negligible when considering curvature of a film–substrate system with lateral dimensions more than about $50h_f$. Concerning the influence of the modulus ratio k, it is evident from the result that, for a film that is much stiffer (softer) than the substrate so that $k \ll 1$ ($k \gg 1$), the interfacial shear stress has a magnitude that is comparable to σ_m over a region that is large (small) compared to h_f.

4.1.2 Example: An equation governing interfacial shear stress

Consider the physical system consisting of a thin film with a free edge bonded to a relatively thick substrate as depicted in Figure 4.2, and assume that the film deforms as an elastic membrane. For a specified equi-biaxial mismatch stress σ_m, or corresponding mismatch strain $\epsilon_m = \sigma_m/M_f$, determine an integral equation that governs the interfacial shear stress $q(x)$.

Solution:

The condition of overall equilibrium for the portion of the film between the free end and the section at x requires that

$$\int_0^x q(\xi)\, d\xi = t(x), \quad 0 < x < \infty. \tag{4.3}$$

The extensional elastic strain in the film in the y-direction remains as ϵ_m following relaxation due to the plane strain constraint. The extensional strain in the x-direction is

$$\epsilon(x) = \epsilon_m + t(x)/\bar{E}_f h_f - (1 + \nu_f)\epsilon_m = \epsilon_m + \epsilon_f(x) \tag{4.4}$$

where E_f, ν_f are the elastic constants of the isotropic elastic film material, $\bar{E}_f = E_f/(1 - \nu_f^2)$ is the plane strain modulus and $\epsilon_f(x)$ is the elastic strain *relaxation* in the film due to the presence of the free edge. Clearly, $\epsilon_f(x) \equiv 0$ if $t(x)/h_f = \sigma_m = \bar{E}_f(1 + \nu_f)\epsilon_m$ everywhere. The equilibrium equation (4.3) and the constitutive equation (4.4) are combined to obtain

$$\epsilon_f(x) = \frac{1}{h_f \bar{E}_f} \int_0^x q(\xi)\, d\xi - (1 + \nu_f)\epsilon_m, \quad 0 < x < \infty. \tag{4.5}$$

The relationship between the extensional strain along the surface of the substrate $\epsilon_s(x)$ and the shear traction $q(x)$ corresponding to (4.5) is obtained by superposition over the Flamant solution for a concentrated tangential line load acting on the surface of a half-space (Johnson 1985) as

$$\epsilon_s(x) = \frac{2}{\pi \bar{E}_s} \int_0^\infty \frac{q(\xi)}{\xi - x}\, d\xi, \quad 0 < x < \infty. \tag{4.6}$$

Here, $\bar{E}_s$ is the plane strain modulus of the isotropic elastic substrate and the singular integral is defined in the sense of its Cauchy principal value (Muskhelishvili 1953).

The film and substrate are bonded at the interface, and they must deform together during relaxation due to this mutual constraint. The compatibility condition is

$$\epsilon_f(x) = \epsilon_s(x), \quad 0 < x < \infty \tag{4.7}$$

which leads to the integral equation

$$\frac{1}{h_f \bar{E}_f} \int_0^x q(\xi)\, d\xi - \frac{2}{\pi \bar{E}_s} \int_0^\infty \frac{q(\xi)}{\xi - x}\, d\xi = (1 + \nu_f)\epsilon_m\,, \quad 0 < x < \infty \qquad (4.8)$$

for the unknown shear stress $q(x)$. The formulation is completed by noting that $q(x)$ has the properties that

$$q(x) \to 0 \text{ as } x/h_f \to \infty \quad \text{and} \quad \int_0^\infty q(x)\, dx = h_f \sigma_m\,. \qquad (4.9)$$

Thus, a solution of (4.8) having these properties is sought. The form of the equation simplifies considerably by introduction of the nondimensional coordinate $\hat{x} = kx/h_f$ and shear stress $\hat{q}(\hat{x}) = q(x)/k\sigma_m$.

An exact solution of this equation is not known, although equations of this general form have been studied extensively in connection with load transfer in stiffening members attached to thin-walled structures (Koiter 1955). A particularly important feature of the solution of (4.8) is that the shear stress distribution $q(x)$ is square root singular as $x \to 0^+$, as was noted in (4.1). Knowledge of this behavior is invaluable in proceeding with a numerical solution of the integral equation. The interpretation of this square root singular stress distribution will be discussed more completely in connection with crack growth modeling in Section 4.2.

An accurate numerical solution of the integral equation can be obtained by reducing (4.8) to a system of algebraic equations for values of $\hat{q}(\hat{x})$ at a finite number of discrete points by means of Gaussian quadrature formulae (Erdogan and Gupta 1972). The weight function in the quadrature formulae incorporates the known behavior of the solution at its endpoints, and the system of algebraic equations resulting from the integral equation must be augmented by the discrete form of the auxiliary conditions. The numerical solution of (4.8) is illustrated in the nondimensional form of $q(x)/k\sigma_m$ versus kx/h_f in Figure 4.3. Implications of the results shown in Figure 4.3 can be found in Section 4.1.1.

4.1.3 More general descriptions of edge stress

The foregoing analysis of an edge stress singularity based on the idealization of the thin film as a membrane is expected to hold over distances that are large compared to the film thickness. However, over distances from the free edge on the order of the film thickness, the local influence of bending resistance in a thin film should be considered as a possibly significant effect. Thus, in this section, the load transfer between a strained film and its substrate is re-examined for a film that has both extensional and bending resistance to applied loading. After a brief discussion of the main features of a model based on combined extension and bending, other models which involve less *a priori* constraint on deformation fields but which are more difficult to analyze are briefly discussed.

The simplest physical idealization of the film that incorporates bending resistance is that of an elastic plate which admits extensional and bending deformation but no shear deformation; this model is known commonly as a *Kirchhoff plate*. The

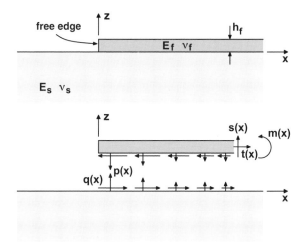

Fig. 4.4. A schematic diagram of a film with the free edge bonded to substrate. The lower portion shows the film and substrate separated, revealing the traction distribution representing the interaction of the film and substrate across the interface.

boundary value problem is illustrated schematically in Figure 4.4, where the film and the substrate have been separated to show the shear stress distribution $q(x)$ and the normal stress distribution $p(x)$ which represent interaction between the film and substrate. Also shown acting on a section of the film at distance x from the edge is an internal extensional force $t(x)$, an internal shear force $s(x)$ and an internal bending moment $m(x)$, all measured per unit thickness in the y-direction. It is important to note that these resultant forces act at a point on the midplane of the film. The objective is to determine $p(x)$ and $q(x)$ in this case for an equi-biaxial mismatch strain ϵ_m in the film with respect to the substrate. Conceptually, the derivation of the equations governing the stress transmitted across the film–substrate interface for this case follows the same steps as for the case of the membrane model. The resulting equations have the form of a coupled pair of integral equations having the same general character as (4.8), and implications of these equations were studied by Shield and Kim (1992). The derivation of the coupled integral equations is left as an exercise.

The traction distributions represented by $p(x)$ and $q(x)$ are expected to be algebraically singular at the film edge $x = 0$. There is a general approach for extracting the strength of such singularities from the governing equations which, in the present case, leads to the behavior (Muskhelishvili 1953)

$$p(x),\ q(x) \sim x^{-\frac{1}{2}+\frac{i}{2\pi}\ln(3-4\nu_s)}. \tag{4.10}$$

The appearance of an imaginary part of the exponent in (4.10) implies that the actual variation of the tractions with position is oscillatory as $x \to 0^+$. This behavior is

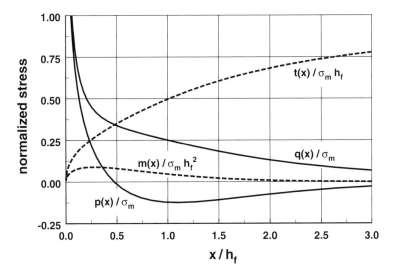

Fig. 4.5. The solid curves show the interface tractions between the film and substrate as determined from the numerical solution of the integral equation (4.93). The dashed curves show the corresponding internal stress resultants in equilibrium with the computed interface tractions.

typical of stress distributions along the interfaces of joined materials near an edge. Several regularization schemes have been proposed for bypassing the non-physical aspects of such behavior, but these tend to be cumbersome and the implications are not readily interpreted. To make the main point of the present discussion, it is sufficient to limit consideration to the case of an elastic substrate that is incompressible, so that $v_s = \frac{1}{2}$ and the imaginary part of the exponent in (4.10) vanishes.

The results of numerical solution of the coupled equations for $p(x)$ and $q(x)$ with $k = 1$ are shown in Figure 4.5, which reveals several noteworthy features. First, the magnitude of the edge singularities of $p(x)$ and $q(x)$ are roughly the same. Therefore, the normal traction is expected to play a more significant role in edge effects than could be anticipated on the basis of the membrane model discussed in the preceding section. It is also seen that the magnitudes of $p(x)$ and $q(x)$ decay to relatively small values compared to σ_m within a distance of 2 or 3 times h_f from the edge of the film. The tensile force $t(x)$ in the film also approaches its remote asymptotic value of $\sigma_m h_f$ within a distance of several times h_f from the edge. At first sight, it might appear from the plot of $m(x)$ in Figure 4.5 that bending effects are not significant, but that is not the case. To see this, it is only necessary to note that the maximum stress in the film due to bending alone, which is $6m(x)/\sigma_m h_f^2$ evaluated at $x \approx 0.3$, is roughly $0.6\sigma_m$ or of the same order of magnitude as the remote tensile stress.

For other values of stiffness ratio k, the distribution of interface stress near the film edge is similar in form to that shown in Figure 4.5 but the amplitudes depend on the

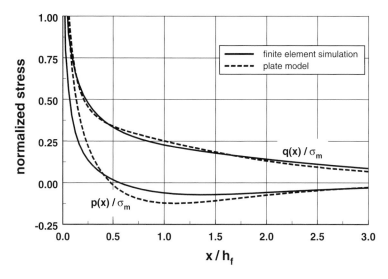

Fig. 4.6. The solid curves show the distribution of interface normal and shear traction near a film edge for the case when the elastic properties of the two materials are identical as determined by means of the numerical finite element method. The corresponding traction distributions based on plate theory and shown in Figure 4.5 are included as dashed curves for purposes of comparison.

value of k. Consistent with the observation drawn from analysis of the membrane film, the load transfer length is very short for a relatively stiff substrate and very long for a relatively compliant substrate.

The extent to which further refinements in the description of edge stress fields can influence the nature of the results is illustrated for the case of a finite element solution of the continuum elasticity equations for the same situation represented by the results in Figure 4.5, that is, an isotropic elastic film bonded to an isotropic incompressible elastic substrate with modulus ratio of $k = 1$. To minimize the influence of remote boundaries in the numerical simulation, a mesh of four-noded quadrilateral elements is defined over the region $-20h_f \leq x \leq 20h_f$, $-20h_f \leq y \leq 0$, with the element dimensions increasing gradually from $h_f/30$ near the film edge to about h_f at the remote boundaries. The film is modeled similarly, with the remote film tension condition enforced by requiring that the edge of the film–substrate system at $x = 20h_f$ is a boundary of reflective symmetry. In effect, the configuration analyzed is that of a film of *finite* extent of $40h_f$ along the substrate surface. At this separation distance, the film edges are expected to have little influence on each other. No elements with special interpolation functions are introduced in an effort to capture the details of the asymptotic singular behavior at the film edge.

The comparison of results obtained from the finite element simulation to those obtained on the basis of the plate model is shown in Figure 4.6. The two sets of results are similar in terms of relative magnitudes absolute magnitudes, decay

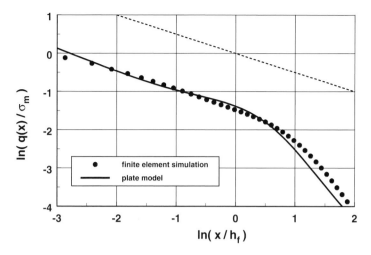

Fig. 4.7. The shear traction distributions shown in Figure 4.6 are graphed in a way that reveals the strength of the assumed algebraic singularity in behavior near the film edge. The behavior seems to agree with $x^{-1/2}$ dependence within the interval $0.1h_\mathrm{f} < x < h_\mathrm{f}$.

characteristics, edge singularity and so on. The strength of the edge singularity is seen to be weaker for the continuum model than for the plate model, a feature that was anticipated in the introduction to this chapter. While such differences should always be kept in mind, their implications for either practical or conceptual concerns are usually of secondary importance.

The nature of the singularity in stress at the interface edge can be pursued somewhat further. If $q(x)$ is proportional to x raised to some power in an interval near the edge $x = 0$, then the slope of the graph of $\ln(q(x)/\sigma_\mathrm{m})$ versus $\ln(x/h_\mathrm{f})$ is equal to that power within the interval. A plot of these logarithmic quantities is shown in Figure 4.7 for both the results of analysis of the plate model and the finite element calculation. For purposes of comparison, a straight dashed line with slope of $-\frac{1}{2}$ is also shown in the same figure. It appears from the numerical results that the interface shear stress in both cases follows an inverse square root dependence on x roughly within the range $0.1 < x/h_\mathrm{f} < 1.0$. This is useful information for the interpretation of some delamination processes.

For all cases discussed up to this point, the load transfer length has been estimated for cases where the substrate extends indefinitely far beyond the edge of the film. What if the substrate has a boundary plane that coincides with the edge of the film, as illustrated in the inset of Figure 4.8? The substrate will appear to be more compliant in this case than if it extends indefinitely far beyond the film edge. There are few analytical techniques available for extracting the behavior of this configuration for which the introduction of an additional free surface is an added complication.

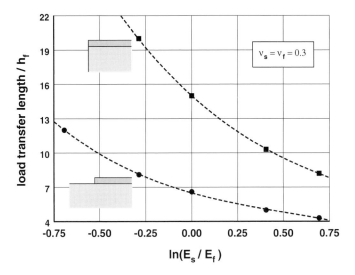

Fig. 4.8. The dependence of load transfer length near the free edge of a film with a remote biaxial mismatch stress σ_m on the modulus ratio of the two materials. The load transfer length is defined as the distance from the free edge of the film at which the internal force resultant has increased to a value equal to 90% of its remote asymptotic value of $\sigma_m h_f$, and this length was determined by means of the numerical finite element method. Results are shown for the case when the substrate extends indefinitely far beyond the free film edge and when the substrate edge coincides with the film edge.

However, the configuration can again be analyzed by means of the finite element method, and results for load transfer length are shown in Figure 4.8. It is evident that this length is substantially larger in the case when the substrate edge coincides with the film edge than in the case when the substrate extends far beyond the free film edge, reflecting the increased compliance of the substrate due to its free surface. Finally, it is noted that the state of stress along the film–substrate interface in the immediate vicinity of the film edge is extremely complicated for the case of differing elastic properties, in general, due to oscillations in the radial variation of stress. However, this effect is quite localized and has little influence on the computed load transfer lengths illustrated.

4.2 Fracture mechanics concepts

The preceding section dealt with the issue of load transfer between a film and its substrate, particularly with the stress concentration in the vicinity of a free edge in a film–substrate system. An important outcome of the study is the estimate of the size of the transition region near the edge of a thin film bonded to a substrate over which the film stress increases in magnitude from its nominally zero value at the edge to a relatively large fraction of its asymptotic value due to constraint of the substrate.

With reference to Figure 4.3, it is also evident that the stress and deformation fields near the film edge lead to the notion of a singular field which is a concept central to the development of linear elastic fracture mechanics. In this section, some general background information on fracture mechanics is summarized. It will be apparent in later sections of this chapter and in Chapter 6 that these concepts provide basic tools for the analysis of delamination and fracture in film–substrate systems.

4.2.1 Energy release rate and the Griffith criterion

Consider a nominally elastic material containing a crack, which may have originated from a region of stress concentration, from a material flaw, or from a free surface, interface or grain boundary. The stress in the material tends to cause the crack to advance while the characteristic strength of the material tends to resist crack growth. To describe this competition, the energy balance approach pioneered by Griffith (1920) is adopted. He recognized that extension of the crack was accompanied by a change in the total potential energy of the system, which is the sum of the elastic strain energy plus the potential energy of any externally applied loading. The advance of a crack also results in the creation of new material surface. Griffith postulated that a certain amount of work per unit area of crack formation must be expended at a microscopic level in creating that area. This work per unit area of cracking is assumed to be a material parameter. In the present context, the work expended at the microscopic level is not included in the continuum potential energy. He also hypothesized that any particular equilibrium state of the material was a *state of incipient crack extension* if the reduction in potential energy due to any small virtual crack advance from that state was equal to the microscopic work of fracture for the new surface area created by that crack extension. By implication, it is understood that the crack cannot extend if the reduction in potential energy associated with any virtual crack advance is less than the material specific fracture energy. The Griffith argument also applies for crack shortening by healing, a phenomenon that has been observed in very clean systems with atomically sharp cracks (Lawn 1993).

The Griffith criterion for crack advance circumvents the need to consider the actual physical mechanism of material separation at the crack edge in detail. It is a reliable concept provided that the size of the region over which fracture energy is absorbed is very small compared to all other relevant physical dimensions. If this is the case, the specific separation energy can be regarded as a material property, independent of geometry and applied loading. The terms fracture energy, work of fracture, energy of separation and energy of fracture are used interchangeably in this discussion. Griffith limited consideration to cases in which the work of fracture is the actual surface energy or energy of cohesion of the material, the cases

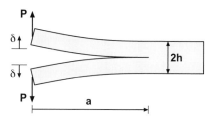

Fig. 4.9. A schematic diagram of a double cantilever crack propagation configuration. The crack of length a advances under the action of the opposed forces P or due to imposed separation δ of the load points.

involving the so-called ideally brittle materials, but the concept has been broadened to include a much wider range of materials in the development of engineering fracture mechanics.

To give the Griffith concept of fracture a concrete structure, which can serve as a framework for quantitative study of thin film systems, consider the configuration of a plate with an edge crack of length a extending along its midplane. As shown in Figure 4.9, this crack is being opened symmetrically either by imposed forces P or by imposed displacements δ. The configuration is two-dimensional and P is the force for unit width in the direction normal to the figure. In any particular case, the imposed quantity will be denoted by a superposed asterisk. The material deforms under plane strain conditions, and all forces and energies are measured per unit depth in the figure, that is, per unit distance along the crack edge. The isotropic elastic material has modulus E and Poisson ratio ν; the plane strain modulus is then $\bar{E} = E/(1 - \nu^2)$. The deformation is adequately described by assuming that each arm of the body deforms as a Kirchhoff plate of length a which is cantilevered at the crack tip. The configuration is commonly referred to as the *double cantilever beam* specimen. The relationship between the end force and the load point displacement is

$$P = \frac{\bar{E}h^3}{4a^3}\delta. \tag{4.11}$$

The corresponding elastic energy per unit depth in each arm is

$$\frac{1}{2}P\delta = \frac{\bar{E}h^3}{8a^3}\delta^2 = \frac{2a^3}{\bar{E}h^3}P^2. \tag{4.12}$$

The specific fracture energy is denoted by Γ, measured as energy expended per unit crack advance per unit distance along the crack edge. For this configuration, what is the value of δ^* necessary to bring the system to a state of incipient fracture under displacement control?

For the case of imposed displacement $\delta = \delta^*$, there is no exchange of energy between the solid and its surroundings during virtual crack advance. Hence, the

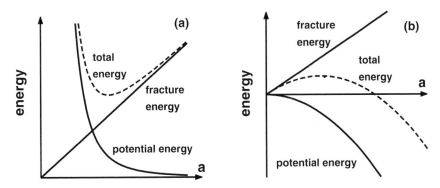

Fig. 4.10. The variation of total potential energy and crack surface fracture energy with crack length is illustrated qualitatively for the configuration depicted in Figure 4.9 for the cases of (a) imposed boundary displacement δ^* and (b) imposed boundary force P^*. The results illustrate the stability of crack advance under increasing displacement in case (a) and instability under increasing force in case (b). Adapted from Freund (1990a).

total potential energy per unit depth $V(a)$ is the elastic energy $\bar{E}h^3\delta^{*2}/4a^3$. The reduction in potential energy per unit crack advance during an infinitesimal virtual crack extension, which is the quantity to be compared to Γ to assess the tendency for crack growth, is

$$\mathcal{G}(a) = -\left.\frac{\partial V}{\partial a}(a)\right|_{\delta^*\ \mathrm{fixed}}. \tag{4.13}$$

This quantity, which is called the *energy release rate*, is of central importance in fracture mechanics; the minus sign accounts for the fact that $\mathcal{G}$ is a measure of the continuum potential energy reduction per unit crack advance per unit depth. Imposition of the Griffith condition $\mathcal{G}(a) = \Gamma$, which is a necessary condition for the onset of crack growth, implies that

$$\delta^* = \delta^*_{\mathrm{cr}} \equiv \left[\frac{4\Gamma a^4}{3\bar{E}h^3}\right]^{1/2}. \tag{4.14}$$

This is the boundary displacement from the stress-free configuration which must be imposed to each arm of the double cantilever specimen to initiate crack growth from an initial crack length of a.

It is instructive to examine the variation of energy with crack length a. As a increases, fracture energy expended increases linearly and potential energy decreases as a^{-3}; these variations are shown schematically in part (a) of Figure 4.10. The sum of the potential energy and fracture energy is stationary at the state of incipient fracture. Furthermore, it is evident that, if $\mathcal{G}(a) = \Gamma$ at a certain crack length, then $\mathcal{G}(a + \Delta a) < \Gamma$ for any increment Δa of crack advance. This implies that δ^* must be increased continually if the crack is to be advanced. Such crack growth is called *stable*. In general, stable crack growth can occur only from states for which

$$\frac{\partial \mathcal{G}}{\partial a}(a) \equiv -\frac{\partial^2 V}{\partial a^2}(a) < 0, \tag{4.15}$$

where the derivatives are calculated at fixed δ^*.

If force is imposed, rather than displacement, what is the value of $P = P^*$ necessary for incipient fracture? For this case, the potential energy is the sum of the elastic energy (4.12) in both arms plus the total external potential energy $-2\delta P^* = -8a^3 P^{*2}/\bar{E}h^2$. It follows that the state of incipient fracture, characterized by $\mathcal{G}(a) = \Gamma$, is defined by the condition

$$P^* = P^*_{cr} \equiv \left[\frac{\bar{E}h^3\Gamma}{12a^2}\right]^{1/2}. \tag{4.16}$$

This is the force that must be applied to each arm of the specimen to induce crack growth. For this case, the potential energy varies with crack length as $-a^2$, and generic graphs of potential energy and fracture energy are shown in part (b) of Figure 4.10. As in part (a) of the figure, the state of incipient fracture is defined by the point at which variation of total energy, potential energy plus fracture energy, is stationary under variations in crack length. In contrast to the case of imposed boundary displacement, however, it is observed that

$$\frac{\partial \mathcal{G}}{\partial a}(a) \equiv -\frac{\partial^2 V}{\partial a^2}(a) > 0 \tag{4.17}$$

at the state of incipient fracture for imposed boundary loading, where the derivatives are calculated with P^* held fixed. This implies that, if $\mathcal{G}(a) = \Gamma$ at some fixed P^*, then $\mathcal{G}(a + \Delta a) > \Gamma$ for any Δa at the same P^*. In other words, the state of incipient fracture is unstable; the crack cannot grow under equilibrium conditions, and inertial or material rate effects are necessarily called into play upon the onset of crack growth (Freund 1990a).

It should be noted that the value of δ corresponding to the critical force P^* in (4.16) is identical to (4.14); likewise, the value of P corresponding to the critical displacement (4.14) is identical to (4.16). In other words, the force–deflection conditions in the states of incipient fracture for applied displacement and imposed force are the same, but the states are fundamentally different because the former is stable while the latter is unstable. A range of responses between these extremes can be described by imposing displacement through compliant links attached to the body of interest, and such links are commonly incorporated to represent the compliance of the loading apparatus. Very stiff links correspond to the above case of imposed displacement while very compliant links correspond to the case of imposed force. Further development of this issue is left as an exercise.

In summary, in order to describe fracture in the context of the Griffith criterion, two ingredients are essential:

- – a representation of the effect of loading, configuration and deformation characteristics of the material through the energy release rate $\mathcal{G}$, and
- – a representation of the resistance of the material to fracture through Γ.

Actual values of Γ cannot be deduced from modeling at this level; such information can be obtained directly only through experiments or it can be estimated on the basis of more fundamental models of the physical processes of material separation. The experimental determination of the fracture resistance of thin film and layered materials is a complex task. Different experimental methods for the determination of Γ will be presented in later sections of this chapter. Values of Γ can vary widely, from about $1\,\mathrm{J\,m^{-2}}$ for separation of atomic planes in brittle materials, for which the strength of the chemical bond dominates the resistance to fracture, to $10^4\,\mathrm{J\,m^{-2}}$ or more in ductile materials, for which plastic flow and ductile tearing primarily determine the energy associated with fracture. For comparison with this energy requirement, consider the energy available in a strained film of thickness $h_{\mathrm{f}} = 1\,\mu\mathrm{m}$, biaxial modulus $M_{\mathrm{f}} = 10^{11}\,\mathrm{N\,m^{-2}}$ and equi-biaxial elastic mismatch strain $\epsilon_{\mathrm{m}} = 0.01$. In such a case, the elastic energy per unit area of interface which is available to drive separation is $M_{\mathrm{f}}\epsilon_{\mathrm{m}}^2 h_{\mathrm{f}} = 10\,\mathrm{J\,m^{-2}}$. Of course, it is unlikely that all of this energy can be used to drive fracture.

The principal merits of the Griffith crack growth condition are evident from the foregoing discussion. When considering the cracking of brittle homogeneous materials, the assumption that the work of separation is a constant appears to be consistent with observed material behavior. For growth of a delamination crack along an interface between dissimilar materials, on the other hand, the work of separation exhibits significant dependence on particular features of the local stress state in the vicinity of the crack edge. Therefore, even within the framework of the Griffith criterion, it becomes necessary at times to take into account certain features of the local crack edge mechanical field. This matter is pursued in Section 4.2.3. Of particular significance is the ratio of the component of shear stress to the component of tensile stress on the prospective crack plane ahead of the crack edge; an appropriate measure of this ratio is defined in Section 4.2.4.

4.2.2 Example: Interface toughness of a laminated composite

A laminated composite is made by bonding two thin strips of steel with a thin film of adhesive epoxy, as shown in Figure 4.11. To characterize the delamination resistance Γ of the epoxy–steel bond, a patch is intentionally left out in the center section to create a crack of total length $2a$. The partially bonded plates are then pulled apart by opposed

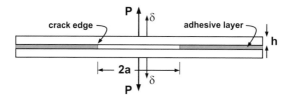

Fig. 4.11. Configuration of bonded plates, showing force per unit width P acting to extend the crack and load point deflection δ. The crack is assumed to extend symmetrically.

forces of magnitude P per unit width of the strips; the load point deflection is denoted by δ. The plane strain modulus of the steel is $\bar{E}$. Assume that the deformation of the strips is two-dimensional and that it can be described in the same way as the double cantilever configuration in Section 4.2.1.

(a) Determine the values of P and δ at which delamination will begin from initial crack length $2a$ in terms of Γ and other system parameters. Note that crack growth consists of the motion of *two* crack edges.
(b) Show that a knowledge of P and δ at the onset of delamination for two different values of a, say a_1 and a_2, is sufficient to estimate Γ from the data. To make the calculation concrete, suppose that $P_1 = 6.74\,\text{N mm}^{-1}$ and $\delta_1 = 1.19\,\text{mm}$ for initiation with $a_1 = 40\,\text{mm}$ and that $P_2 = 7.71\,\text{N mm}^{-1}$ and $\delta_2 = 0.91\,\text{mm}$ for $a_2 = 35\,\text{mm}$. Assume that $\bar{E} = 200\,\text{GPa}$ and $h = 0.97\,\text{mm}$, and obtain a numerical estimate of Γ.

Solution:

(a) The relationship between δ and P at fixed a is $\delta = Pa^3/2\bar{E}h^3$ for this configuration, from elementary plate theory. The potential energy of the system for prescribed loading, say, is $V(a) = -P\delta = -P^2a^3/2\bar{E}h^3$. To satisfy the Griffith criterion $\mathcal{G} = \Gamma$ at each crack edge, it is required that

$$2\mathcal{G} = -\left.\frac{\partial V}{\partial a}(a)\right|_P = \frac{3P^2a^2}{2\bar{E}h^3} = 2\Gamma \qquad (4.18)$$

where the factor 2 appears with $\mathcal{G}$ and Γ to account for both crack edges. It follows immediately that

$$P = \left[\frac{4\Gamma\bar{E}h^3}{3a^2}\right]^{1/2}, \qquad \delta = a^2\left[\frac{\Gamma}{3\bar{E}h^3}\right]^{1/2} \qquad (4.19)$$

at the onset of growth. These two equations together define a path in the plane of P versus δ, usually called the fracture resistance locus or the toughness locus; the path is parametric in crack length a, and is illustrated in Figure 4.12.

(b) The relationship between load and deflection is linear up to the point of fracture initiation for initial crack size $2a_1$, as shown in Figure 4.12, and similarly for the case with initial crack size $2a_2 < 2a_1$. In the latter case, suppose that δ is further increased beyond δ_2 until it equals δ_1. Crack growth takes place with $\mathcal{G} = \Gamma$ during this additional increment in load point displacement and the P versus δ history follows the fracture resistance locus. The area OA_1B under the curve in Figure 4.12 is the external work done in bringing this sample to the state of incipient fracture

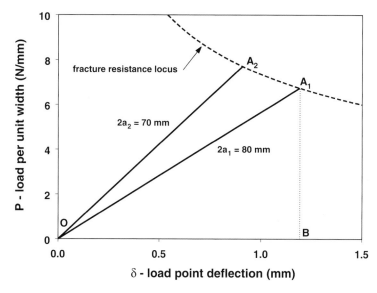

Fig. 4.12. Plane of load P versus deflection δ including the fracture resistance locus for constant Γ. The value of Γ can be estimated from fracture initiation data for two values of initial crack length $2a$. Adapted from Suresh (1998).

at crack length $2a_1$. The area OA_2A_1B under the second curve is the external work done to bring this sample to the same state but from initial crack size $2a_2$. The difference between these two work measures is precisely the energy consumed in advancing the crack ends from total crack length $2a_2$ to $2a_1$, or

$$\text{Area } OA_2A_1B - \text{Area } OA_1B = 2(a_1 - a_2)\Gamma. \tag{4.20}$$

Approximation of the area with a triangle leads to the estimate $\Gamma \approx 293\,\text{J m}^{-2}$. (For comparison, the given 'data' for values of (P_1, δ_1) and (P_2, δ_2) were generated with $\Gamma = 300\,\text{J m}^{-2}$.)

4.2.3 Crack edge stress fields

Descriptions of fracture in materials can be formulated using approaches other than the Griffith energy criterion. The most significant of these is the one based on the concept of *elastic stress intensity factor*. It was recognized by Irwin (1957) that the stress field in the vicinity of the tip or edge of a sharp crack in a nominally elastic solid has universal spatial dependence, in the sense that it is independent of geometry or the nature of loading, and that the magnitudes of the stress components are asymptomatically unbounded as the point of observation approaches the edge of the crack. The scalar amplitude of this universal singular elastic field is the stress intensity factor, and Irwin's fracture condition is the postulate that the state

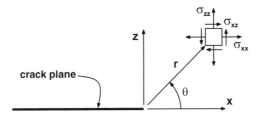

Fig. 4.13. Region near the edge of a crack in which the local field can be described in terms of a two-dimensional stress intensity factor field.

of incipient fracture is reached when the stress intensity factor is increased to a material-specific value termed the fracture toughness.

The stress intensity factor concept is usually discussed in the context of two-dimensional plane strain deformation fields. However, the same characteristics are present near the edge of a three-dimensional crack. To see that this is so, consider a point along the edge at which the curve defining the crack edge has a continuously turning tangent with distance along the edge and which is not on a material boundary other than a crack surface. Then the stress field in any plane that is normal to the crack edge at such a point has the same singular behavior and same spatial dependence as is demonstrated under two-dimensional plane strain conditions. The reason for this outcome is that, very near the crack edge, spatial gradients of fields in the direction parallel to the crack edge tangent are negligibly small compared to the gradients in directions normal to the edge.

Cases for which the crack faces move apart symmetrically with respect to the crack plane without relative sliding are termed mode I or opening mode deformations. The application of a tensile load in a direction normal to the plane of a crack in an isotropic material normally results in mode I deformation. Cases for which the crack faces slide with respect to each other in the direction *normal* to the crack edge without relative opening are termed mode II or in-plane shearing mode deformations. Finally, cases for which the crack faces slide with respect to each other in the direction *parallel* to the crack edge without a relative opening are termed mode III or out-of-plane shearing mode deformations. Although mode III fracture can be a significant factor in situations of both scientific and technological relevance, attention here is limited primarily to the in-plane deformation modes.

Consider a cylindrical coordinate system centered at the edge or tip of a crack whereby the location of any material point in the cracked body can be identified with its radial distance r from the crack tip and angular position θ, measured in the counterclockwise sense about the y-axis that completes the right-handed rectangular coordinate system, with respect to the extension of the crack plane; see Figure 4.13. In this system, the crack faces are at $\theta = \pm\pi$ and $\theta = 0$ is the plane along which the crack extends. The stress field at the edge of a crack in a homogeneous and isotropic elastic solid deformed in mode I under plane strain conditions is

$$\sigma_{xx} = \frac{K_{\mathrm{I}}}{\sqrt{2\pi r}} \Sigma^{\mathrm{I}}_{xx}(\theta) + \sigma^{\mathrm{o}}_{xx} + o(1), \qquad (4.21)$$

as $r \to 0$, where

$$\Sigma^{\mathrm{I}}_{xx}(\theta) = \cos \tfrac{1}{2}\theta \left[1 - \sin \tfrac{1}{2}\theta \sin \tfrac{3}{2}\theta \right], \qquad (4.22)$$

σ^{o}_{xx} is a non-singular stress term, commonly called the T-stress, which need not vanish at the crack tip; the higher order terms vanish as $r \to 0$. The non-singular stress term can have a significant effect on the directional stability of crack growth. The other mode I stress components have the similar representations

$$\sigma_{xz} = \frac{K_{\mathrm{I}}}{\sqrt{2\pi r}} \Sigma^{\mathrm{I}}_{xz} + o(1), \qquad \Sigma^{\mathrm{I}}_{xz}(\theta) = \cos \tfrac{1}{2}\theta \sin \tfrac{1}{2}\theta \cos \tfrac{3}{2}\theta,$$

$$\qquad (4.23)$$

$$\sigma_{zz} = \frac{K_{\mathrm{I}}}{\sqrt{2\pi r}} \Sigma^{\mathrm{I}}_{zz} + o(1), \qquad \Sigma^{\mathrm{I}}_{zz}(\theta) = \sin \tfrac{1}{2}\theta \cos \tfrac{1}{2}\theta \cos \tfrac{3}{2}\theta,$$

except that there is no counterpart to σ^{o}_{xx} for traction free crack faces. In (4.21) and (4.23), the scalar amplitude of the leading term of the asymptotic expansion is the mode I elastic stress intensity factor K_{I}, which is defined formally as

$$K_{\mathrm{I}} = \lim_{r \to 0} \sqrt{2\pi r}\, \sigma_{zz}(r, 0). \qquad (4.24)$$

Similarly, the crack edge singular fields for mode II loading are given by

$$\sigma_{xx} = \frac{K_{\mathrm{II}}}{\sqrt{2\pi r}} \Sigma^{\mathrm{II}}_{xx}, \qquad \Sigma^{\mathrm{II}}_{xx}(\theta) = -\sin \tfrac{1}{2}\theta \left[2 + \cos \tfrac{1}{2}\theta \cos \tfrac{3}{2}\theta \right],$$

$$\sigma_{xz} = \frac{K_{\mathrm{II}}}{\sqrt{2\pi r}} \Sigma^{\mathrm{II}}_{xz}, \qquad \Sigma^{\mathrm{II}}_{xz}(\theta) = \cos \tfrac{1}{2}\theta \left[1 - \sin \tfrac{1}{2}\theta \sin \tfrac{3}{2}\theta \right], \qquad (4.25)$$

$$\sigma_{zz} = \frac{K_{\mathrm{II}}}{\sqrt{2\pi r}} \Sigma^{\mathrm{II}}_{zz}, \qquad \Sigma^{\mathrm{II}}_{zz}(\theta) = \cos \tfrac{1}{2}\theta \left[1 - \sin \tfrac{1}{2}\theta \sin \tfrac{3}{2}\theta \right],$$

and the mode II elastic stress intensity factor is defined as

$$K_{\mathrm{II}} = \lim_{r \to 0} \sqrt{2\pi r}\, \sigma_{xz}(r, 0). \qquad (4.26)$$

The concept of the stress intensity factor, which provides a unique measure of the amplitude of the stress singularity at the edge of a sharp crack, was developed by assuming linear elastic material behavior. Obviously, real engineering materials undergo inelastic deformation or damage in the highly stressed region of the crack edge. Examples include the development of a zone of plastic yielding surrounding the crack edge in a ductile metal and a zone of grain boundary microcracking in a polycrystalline ceramic. As long as the size of this inelastic deformation zone, commonly called the 'plastic zone' or the 'process zone', is small in some sense compared to the length of the crack, the size of the uncracked ligament directly ahead of the crack edge or any other characterizing dimension, the general notion of *small scale yielding* can be invoked (Rice 1968). According to this concept, if the size of the inelastic zone is small compared to any other dimensions of the configuration, then the elastic field surrounding the inelastic zone is still the universal crack tip field which is characterized by the stress intensity factor. In this sense, the stress intensity factor provides a one-parameter representation of the loading imposed on the inelastic zone within which material separation occurs. As a result, a fracture criterion based on attainment of a critical value of stress intensity factor for fracture initiation or sustained crack growth is a credible postulate for understanding fracture of materials and, due to its simplicity, has great appeal. In thin film systems, it is sometimes difficult to justify the use of the small scale yielding hypothesis in any strict sense due to the small geometrical dimensions involved. In such cases, reliance on the stress intensity factor concept becomes a more subjective matter.

The onset of crack advance in a homogeneous material under monotonic loading conditions is characterized by a critical value of the stress intensity factor which may depend on the mode of loading, test temperature and loading rate. The value of the critical stress intensity factor measured under quasi-static plane strain mode I loading conditions is commonly called the *fracture toughness* K_{Ic}. A fracture test specimen of a material susceptible to plastic deformation in the crack edge region is considered to conform to a state of plane strain deformation in the vicinity of the crack edge when the thickness of the test specimen is typically 25 times the maximum extent of the plastic zone at the center of the specimen. For a fixed test environment and loading rate, the fracture initiation toughness K_{Ic} is a material property and is independent of specimen configuration provided that conditions of mode I loading are maintained. Similarly, the plane strain fracture initiation toughness in pure mode II loading is termed K_{IIc}, although advance of a crack in a homogeneous material under purely mode II conditions has not been observed.

An important relationship between stress intensity factors and the energy release rate for planar crack growth under equilibrium conditions was established by Irwin (1960) as

$$\mathcal{G} = \frac{1 - \nu^2}{E} \left(K_I^2 + K_{II}^2 \right) = \frac{1}{\bar{E}} \left(K_I^2 + K_{II}^2 \right). \qquad (4.27)$$

This relationship also provides a connection between the mode I fracture toughness value K_{Ic} for a homogeneous material and the critical value of the energy release rate Γ representing material resistance to fracture according to the Griffith criterion as described in Section 4.2.1. Generalization of the expression for $\mathcal{G}$ when all three crack deformation modes are active is straightforward (Rice 1968).

4.2.4 Phase angle of the local stress state

For any combination of in-plane crack deformation modes, the relative amounts of mode I and mode II loading on the material in the crack edge region are represented by the *phase angle* ψ of the local stress state; it is defined in terms of the stress state as

$$\psi = \lim_{r \to 0} \left\{ \arctan \left[\frac{\sigma_{xz}(r, 0)}{\sigma_{zz}(r, 0)} \right] \right\} = \arctan \left[\frac{K_{\mathrm{II}}}{K_{\mathrm{I}}} \right]. \tag{4.28}$$

The branch of the arctangent function in (4.28) assumed here is such that $-\pi < \psi \le \pi$. Values of ψ in the range $-\pi < \psi < -\pi/2$ or $\pi/2 < \psi < \pi$, which correspond to situations with $K_{\mathrm{I}} < 0$, are troublesome from the physical point of view. A negative mode I stress intensity factor implies that the normal traction on the crack plane ahead of the crack edge is compressive, and that behind the crack front the crack faces tend to displace toward each other. In the usual idealization of an elastic crack, the crack faces both coincide with the same mathematical surface when the solid is free of stress. If the crack faces tend to displace toward each other from such a configuration, they tend to interpenetrate, which must be ruled out on physical grounds. The concept of a negative stress intensity factor is often useful in constructing solutions for crack problems by linear superposition. However, negative values of the total mode I stress intensity factor are ruled out on physical grounds, which implies that the phase angle must be in the range $-\pi/2 \le \psi \le \pi/2$. With $K_{\mathrm{I}} = 0$, the normal stress on the crack plane near the edge can still be compressive, although not singular. This implies that the crack faces will interact in some way, possibly through friction; the possibility of such interaction should be considered in any particular cases for which the assumption of traction free crack faces leads to the conclusion that $K_{\mathrm{I}} \le 0$ (Stringfellow and Freund 1993).

4.2.5 Driving force for interface delamination

If a crack lies in the interface between two materials, with the crack edge separating a part of the interface over which the materials are bonded from a part over which the bond has been broken, then the determination of local fields becomes a complex problem. Dundurs (1969) showed that the solutions to problems in plane strain elastostatics that involve two joined homogeneous and isotropic materials depend

Table 4.1. *Values of the Dundurs parameters D_1 and D_2 for some common substrate and film materials. The entry 'DLC' in this table refers to diamond-like carbon.*

Substrate	Film	E_s (GPa)	ν_s	E_f (GPa)	ν_f	D_1	D_2
Si (100)	Al	130	0.28	70	0.35	−0.28	−0.04
Si (100)	Cu	130	0.28	130	0.34	0.02	0.03
Si (100)	W	130	0.28	85	0.30	−0.20	−0.04
Al$_2$O$_3$	Al	372	0.25	70	0.35	−0.67	−0.14
Al$_2$O$_3$	Au	372	0.25	78	0.44	−0.61	−0.03
Al$_2$O$_3$	Mo	372	0.25	324	0.31	−0.05	0.01
Al$_2$O$_3$	Ni	372	0.25	200	0.31	−0.28	−0.05
SiO$_2$	Cu	71	0.16	130	0.34	0.34	0.14
Steel	DLC	210	0.30	80.8	0.13	−0.47	−0.15
PMMA	Al	3.4	0.30	70	0.35	0.91	0.23

on the four elastic constants involved in a very special way. The elastic fields depend on the material constants only through two independent material parameters defined by

$$
D_1 = \frac{\bar{E}_f - \bar{E}_s}{\bar{E}_f + \bar{E}_s}
$$

$$
D_2 = \frac{1}{4} \left[\frac{\bar{E}_f(1 - \nu_f)(1 - 2\nu_s) - \bar{E}_s(1 - \nu_s)(1 - 2\nu_f)}{\bar{E}_f(1 - \nu_f)(1 - 2\nu_s) + \bar{E}_s(1 - \nu_s)(1 - 2\nu_f)} \right]
$$

(4.29)

where $\bar{E}_f$ and $\bar{E}_s$ are the plane strain moduli of the film and substrate, respectively, and ν_f and ν_s are the corresponding Poisson ratios. The parameters D_1 and D_2 were introduced earlier in (3.98). The labels 'film' and 'substrate' are completely arbitrary in this context, and they can be understood to refer to any joined materials no matter what the configuration might be. The parameter D_1 is usually understood to indicate mismatch in extensional stiffness between the two materials, while D_2 indicates mismatch in volumetric stiffness. Note that $D_1 = 0$ and $D_2 = 0$ when the materials have the same elastic properties. Since the plane strain moduli can only be positive numbers, the parameter D_1 must be in the range $-1 \leq D_1 \leq 1$. For any fixed value of D_1 and Poisson ratios in the range $0 \leq \nu_f, \nu_s \leq \frac{1}{2}$, the parameter D_2 must be in the range $-\frac{1}{4}(1 - D_1) \leq D_2 \leq \frac{1}{4}(1 + D_1)$. Values of the Dundurs parameters for some common substrate–film material combinations are listed in Table 4.1.

When a crack edge lies within either one of the joined materials, the near-edge stress field is still given by the expressions in Section 4.2.3, and there is no particular advantage in restricting the values of D_1 and D_2 in any way. When a crack lies in the interface, however, setting $D_2 = 0$ results in major simplification in the characterization of the crack edge field. As noted in connection with Figure 3.12, any particular value of D_1 with $D_2 = 0$ can be achieved only for certain combinations of ν_s and ν_f. For many of the material combinations shown in Table 4.1, D_2 is indeed much smaller than D_1 and, therefore, this requirement is not an unrealistic restriction. Furthermore, without this restriction, the stress intensity factors for an interface crack are complex-valued, in general. Their interpretation becomes somewhat ambiguous and the utility of the stress intensity factor concept is diminished significantly. It is noted that for Al, Cu, W or Au films on Al_2O_3 substrates, which are materials of interest for applications in microelectronics, D_2 can be set equal to zero for most practical purposes.

With $D_2 = 0$, it is still possible to draw useful conclusions concerning the effect of modulus mismatch through dependence of results on D_1, and this is the situation assumed here. In this case, the definitions of stress intensity factor (4.24) and (4.26) and crack edge phase angle (4.28) given for a crack edge in a homogeneous material are retained for an interface crack. The Irwin relationship between stress intensity factors and the corresponding energy release rate for planar crack growth, which is given by (4.27) for a crack edge in homogeneous material, becomes

$$\mathcal{G} = \frac{\bar{E}_f + \bar{E}_s}{2\bar{E}_f \bar{E}_s} \left(K_I^2 + K_{II}^2\right) = \frac{1}{\bar{E}_s(1 + D_1)} \left(K_I^2 + K_{II}^2\right) \qquad (4.30)$$

for a planar interface crack with $D_2 = 0$. The denominator $\bar{E}_s(1 + D_1)$ has the alternate form $\bar{E}_f(1 - D_1)$. The remarks following (4.28) about physical constraints on achieving states with $K_I < 0$ and the consequent implications for the admissibility of certain ranges of phase angle ψ apply equally well to the case of an interface crack.

As noted above, the work of separation that appears in the Griffith crack growth condition may depend on the phase angle of the local crack edge stress state. Whenever this effect is expected to be significant, as in the case of crack growth along a bimaterial interface, the dependence is emphasized by writing the Griffith condition in the form $\mathcal{G} = \Gamma(\psi)$. An example that plays a central role in delamination buckling is illustrated in (5.20). For a crack in a homogeneous material, where the fracture path is not confined to lie in an interface, crack growth is usually observed to occur in a direction for which $\psi = 0$. In other words, a crack in a homogeneous material selects a path along which it responds to pure tension across the prospective fracture plane.

In many particular cases, there is a direct correspondence between the Irwin and Griffith criteria, as was noted above in connection with the result in (4.27). However, the latter criterion has the distinct advantage that the energy release rate can often be determined, or at least estimated, without the need for a complete solution of the boundary value problem for the stress field in the body. For this reason, it is selected

as the basis for the present discussion. Many of its special features and numerous
extensions of the basic concept will become evident in the sections that follow, in
the course of discussing various issues concerned with delamination and fracture
in thin film configurations.

4.3 Work of fracture

The Griffith criterion, described in Section 4.2.1, was developed as a basis for
considering the onset of crack growth and/or fracture advance in a homogeneous
material by relating the potential energy release rate $\mathcal{G}$ to the microscopic work Γ
required to create the new fracture surfaces. In its original conception, the criterion
was used as a basis for describing the cracking behavior of glass, and the specific
work of fracture was viewed naturally in this case as being twice the surface energy
density of the material. It was expected that the criterion could not be applied in
any situation in which the material exhibited inelastic deformation other than the
formation of the crack.

An important extension was proposed by Irwin (1948) on the basis of a study of
cleavage crack growth in steels, a process invariably accompanied by some amount
of plastic deformation in the material immediately adjacent to the fracture path.
He reasoned that the plastic work dissipation per unit area of interface was also a
characteristic of the separation process, so that the work of fracture could be taken as
the sum of the surface energy density plus this plastic work dissipation per unit area
of crack advance, thus preserving the fundamental structure of the postulated energy
criterion. The expectation was borne out by experimental observations, thereby
expanding the range of applicability of the energy criterion significantly. The same
extension of the Griffith theory was also proposed by Orowan (1955). It does not
matter exactly how the added work is dissipated, as long as it is characteristic of the
separation process. The physical mechanism of separation is determined, to a large
extent, by the nature of the material, but it can also depend on temperature, chemical
environment, rate of loading, rate of cracking and possibly other factors. For the
growth of a crack along a bimaterial interface between two materials, the process can
be viewed in more or less the same way. However, the details of material separation
are complicated by the interaction of two materials which may themselves have
different fracture characteristics.

4.3.1 Characterization of interface separation behavior

In its current state of development, the study of interface fracture has not yet led to a
level of understanding of fundamental processes that can provide a basis for detailed
quantitative description. Indeed, the phenomena encountered are so diverse in type

and complexity that this may never be the case, except perhaps for a few critical material systems. The simplest case involves interface separation as a direct analog of the situation originally considered by Griffith. For example, if crystals of a 'film' material and a 'substrate' material are joined along an interface, then that interface has a characteristic free energy γ_{fs}, as was noted in Section 1.3.3. If the crystals are separated, then the free surface of each has a characteristic surface energy density, γ_f for the film material and γ_s for the substrate material. If the separation of the interface occurs by growth of an interface crack without inelastic deformation in either of the materials involved, then the work of fracture is

$$\Gamma_0 = \gamma_f + \gamma_s - \gamma_{fs} . \tag{4.31}$$

If the two materials are identical, then $\gamma_{fs} = 0$ and $\gamma_f = \gamma_s$, so the original picture conceived by Griffith is recovered. The capability to describe interface fracture quantitatively for a broader range of phenomena and material combinations remains a goal of scientific study. An approach toward describing processes for a fairly wide class of materials that has been found effective is to retain the phenomenological character of fracture mechanics but to add models of behavior at one or two levels of observation, or size scales, between the applied driving force $\mathcal{G}$ and the physical process of material separation. The salient features of this point of view are summarized in this section; the diagrams in Figure 4.14 depict the main features. A similar point of view was found to apply in the case of separation of a cross-linked rubber adhesive from a relatively stiff polymeric substrate (Andrews and Kinloch 1973).

A common representation of the behavior implied by (4.31) within a continuum framework is that of separation of adjacent planes of otherwise elastic materials, with that separation being resisted by a cohesive traction acting across the gap between these planes – the so-called *cohesive zone models* of material behavior. The magnitude of the cohesive traction depends locally on the amount of interface separation δ. In the model developed by Barenblatt (1959), the traction represents atomic interaction, and its magnitude falls to zero beyond some separation distance, say δ_{cr}, on the order of atomic spacing in the materials involved. If the separation of the planes in mode I opening is denoted by δ, which is a function of position on the fracture plane, then the work of fracture is the work done in overcoming this resisting traction or

$$\Gamma_0 = \int_0^{\delta_{cr}} \sigma(\delta) \, d\delta . \tag{4.32}$$

When viewed from a size scale larger than atomic dimensions, the actual shape of the dependence of σ on δ is of only secondary importance. Instead, only some general features of this dependence have significance. For example, the essence of behavior is captured by the area Γ_0 under the traction–displacement curve and the

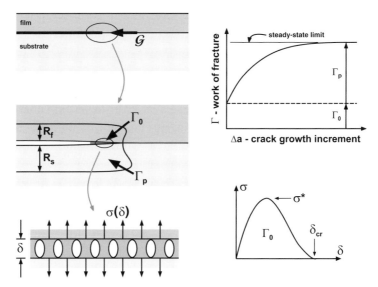

Fig. 4.14. A hierarchical point of view of interface fracture advance, whereby the complex-ities of material separation are lumped into a representative phenomenological cohesive rule that is representative of the system; its essential features are the work per unit area Γ_0 re-quired for separation of the surfaces and the maximum cohesive traction σ^* that arises in the process. The cohesive traction must be imposed by the surrounding film and substrate materials, viewed as elastic–plastic continua. The tendency for significant plastic deforma-tion in either material is determined by the ratio of σ^* to the yield stress of that material. The driving force necessary to effect separation is characterized by an energy release rate $\mathcal{G}$. To sustain crack growth, its value must be large enough to overcome Γ_0 plus plastic dissipation per unit area Γ_p. Adapted from Hutchinson and Evans (2000).

separation displacement δ_{cr} or, equivalently, by the value of Γ_0 and the maximum value of traction σ, say σ^*, within the range $0 < \delta < \delta_{cr}$. The idea that the physical process of material separation can be captured by a cohesive zone is then extended to represent separation mechanisms other than atomic adhesion or cohesion. For example, such a cohesive rule could represent the ductile growth of microvoids to coalescence, the growth of microcracks to eventual linkup, or the rupture of grains or ligaments bridging the opening behind the advancing crack edge. In each case, it is assumed that the behavior within such a planar layer can be represented in terms of the relationship between a traction σ applied on the plane of separation by the surrounding material and an average opening of the faces of the surrounding material across that plane. The layer is assumed to be characterized by two parameters, the work per unit area Γ_0 that must be supplied to reduce the resisting traction to zero and the maximum resisting stress σ^* that is encountered in doing so. The region of the plane of an interface fracture in which separation is described in this way has been termed the *embedded process zone* by Hutchinson and Evans (2000).

Although the embedded process zone has been introduced in terms of symmetric or mode I opening of an interface crack, there is no fundamental impediment to extending the description to include combined tensile and shear cohesive traction resisting the opening and relative sliding of the separating surfaces (Needleman 1990). The net work done per unit area in overcoming opening resistance can be associated with such interaction in a number of ways. Likewise, the ratio of a maximum shear traction to a maximum tensile traction is a measure of the mode mixity of the separation process, as represented by the phase angle in Section 4.2.4. As usual in material characterization, the identification of a suitable description of the embedded process zone for any particular material system must be pursued synergistically by means of conceptual postulates and corresponding experimental confirmation or refutation.

Advancing to a larger size scale, the process zone described above is imagined to be embedded within a region of plastically deforming material, for which the strain magnitudes are sufficiently small to admit a continuum plasticity description for the region. This level is depicted in the left central portion of Figure 4.14. The size of the plastic zone, R_f in the film or R_s in the substrate, is determined largely by the magnitude of the yield stress σ_Y compared to σ^*, the maximum cohesive stress needed to drive the separation process, and by the strain hardening properties of the plastically deforming material. This size tends to increase with increasing values of the ratio σ^*/σ_Y and with decreasing values of the strain hardening rate. In many cases of practical interest, either R_f or R_s will be equal to zero.

The defining characteristics of plastic deformation are that the response at a material point is dependent on the strain history of the point and that stress–strain response is irreversible. As a crack begins to advance along an interface between materials, one or both of which are elastic–plastic, a zone of plastically deforming material begins to take shape, with the extent of plastic deformation depending on the amount of growth. Once the crack has advanced a distance on the order of the extent of the plastically deforming zones, the nature of the plastically deforming regions approaches a steady-state condition, as suggested in Figure 4.14. Thereafter, each active plastic zone leaves behind a layer of plastically deformed but partially unloaded material in its wake. Energy is dissipated through plastic flow in the active plastic zone. If the amount of work dissipated in this way, measured per unit area of the fracture formed, is denoted by Γ_p then the total amount of work per unit area of fracture surface that must be provided to sustain crack growth is the sum of work absorbed in the process zone plus the energy dissipated through plastic deformation, that is, the sum $\Gamma_0 + \Gamma_p$. If the thicknesses of the plastically deformed layers accompanying crack growth are small compared to film thickness or other overall dimensions of the sample, then the value of $\Gamma_p + \Gamma_0$ retains the character of Γ as a material parameter in the Griffith criterion. On the other hand, if the size of either of the plastic regions approaches the film thickness, then Γ_p will depend on sample dimensions. However, Γ_0 may retain its character as being independent of material configuration in either case. Thus, measurement of Γ for a particular delamination process, along with an estimate of Γ_p on the basis of continuum plasticity modeling, still provides an estimate of Γ_0 for the process.

4.3.2 Effects of processing and interface chemistry

The methods used to fabricate multilayered structures, the level of contamination introduced at the interface during processing, the composition of the layers and the interfaces between them, and the chemical and moisture content of the surrounding environment can have a significant effect on interface fracture resistance. From phenomenological observations of interface fracture mechanisms, some general trends can be identified and these are summarized in this section.

When contaminants and segregants weaken the interface, islands of delaminated regions typically form ahead of an advancing interface crack. The coalescence of such weak patches generates crack advance along the interface during which the resistance to crack growth along the interface Γ remains essentially unchanged from the initiation of growth onward. The dependence of interface fracture energy on crack growth is similar to that shown by the dashed line in the diagram in the upper right portion of Figure 4.14. In effect, the interface is too weak to generate a stress σ^* sufficient to induce plastic deformation in the film or substrate, even if either or both of these materials are typically susceptible to plastic flow. Figure 4.15 shows an example of such a failure mechanism at the interface between a Ni–20 at.% Cr alloy and a transparent α-alumina (sapphire) crystal which was produced by liquid phase bonding.

When both the film and substrate materials are resistant to plastic deformation, the toughness of a sharp interface between two dissimilar materials can be very low, typically on the order of 1 J m^{-2}. In such cases, brittle crack growth is likely to occur along the interface with the tip of the crack remaining atomically sharp during extension. The failure mechanism in this case may involve the atomic decohesion process implied by the fracture energy expression in (4.31).

Reaction products, segregants and contaminants present along interfaces can significantly diminish the resistance to interface fracture. Along interfaces between Ti and Al$_2$O$_3$, for example, contaminants lead to the formation of intermetallic phases. The segregation of C is known to be a primary factor in the embrittlement of interfaces between Al$_2$O$_3$ and Au and between Al$_2$O$_3$ and Ni, whereas S segregation reduces the toughness of interfaces between Al$_2$O$_3$ and Ni. Such interface embrittlement is also exacerbated by stress corrosion, particularly that arising from moisture in the environment. When interfaces are produced without contaminants

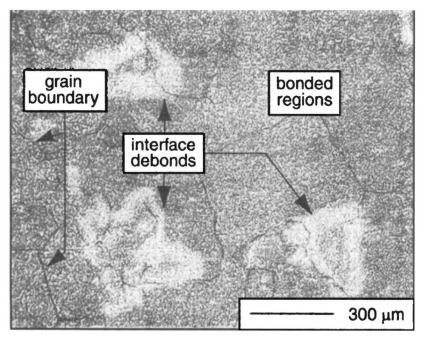

Fig. 4.15. Differential interference contrast optical micrograph of the interface between a 125 μm thick Ni–20 at.% Cr film and a 2.5 mm thick, transparent sapphire layer. This is one of the two similar interfaces in a three-layer structure where the thin metal film was sandwiched symmetrically between two sapphire layers; the interface was produced by liquid phase bonding. The metal film contained a sulfur concentration of up to 200 ppm. The contamination of the interface by S, along with processing by liquid phase bonding, resulted in local debonds which are revealed as the lighter regions. Note that the debonds encompass grain boundaries, and the presence of these debonds leads to an interface fracture energy of only 2–7 J m^{-2}. Reproduced with permission from Gaudette et al. (2000).

and reaction products, high fracture toughness values are realized even when the metallic layer side of the interface is non-epitaxial, highly incoherent and polycrystalline. Such 'clean' interfaces can be produced by employing one of the following two processing strategies:

– Alloy additions are made to the metallic layer side of the interface to 'getter' the contaminants, such as C or OH^{-}, by forming reaction products or precipitates and facilitating unexpurgated metal–oxide bonding. For example, the interface between pure Ni and Al$_2$O$_3$ has a low fracture energy, typically 10 J m^{-2}, due in part to its susceptibility to moisture-induced stress corrosion. However, experiments by Gaudette et al. (1997) reveal that when 20 at.% Cr is added to Ni, Cr-carbides form near the interface thereby gettering C, and that the resulting interface between the Ni–Cr alloy and Al$_2$O$_3$, which cannot be fractured, has a fracture energy Γ well in excess of 100 J m^{-2}.

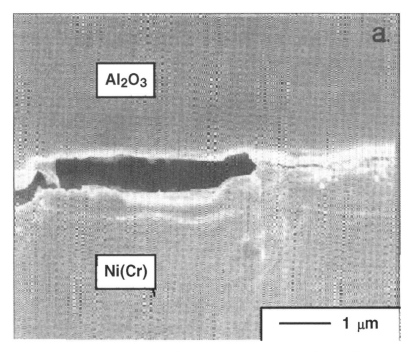

Fig. 4.16. Blunting of a crack along an interface between Ni and Al_2O_3 which is toughened by an order of magnitude by the addition of 20 at.% Cr. Plastic flow in the metal causes the interface crack to blunt. Reproduced with permission from Gaudette et al. (1997).

Figure 4.16 shows an example of a crack along such an interface which accommodates external loading by blunting. In this figure, the residual crack opening δ upon complete unloading from a tensile stress is as much as 0.3 μm. Given the Ni–Cr alloy yield stress of $\sigma_Y = 330$ MPa, this blunting is indicative of the fracture energy $\Gamma \approx 100$ J m^{-2}.

– Excluding contamination by adapting well controlled processing methods also facilitates the production of high toughness interfaces. Consider as an example the interface between γ-Ni(Cr) and α-Al_2O_3. When this interface is produced by solid-state diffusion bonding, the segregation of excess S to the interface is suppressed; experiments by Gaudette et al. (2000) indicate that the resulting 'clean' interface has a very high fracture energy, in excess of 100 J m^{-2}. If, on the other hand, the interface bond between γ-Ni(Cr) and α-Al_2O_3 is produced through a liquid phase at a temperature above the eutectic where excess S is released to the interface region, very low fracture energies in the range 2–7 J m^{-2} result.

The interface fracture energy for a given material system is a strong function of the environment. While moist air embrittles the interface in systems such as

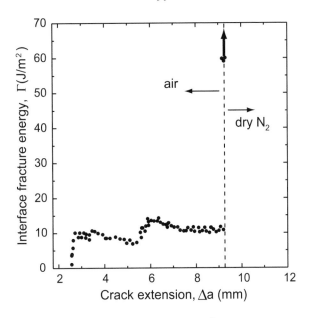

Fig. 4.17. Low interface fracture energy, $\Gamma \approx 10$ J m^{-2}, of a pure Ni–Al$_2$O$_3$ interface in ambient air. Upon switching from moist air to dry nitrogen, the interface fracture energy rises to a value in excess of 100 J m^{-2}. Adapted from Stölken and Evans (1998).

Ni–alumina, possibly due to the combined effects of C segregation at the interface and reaction with water in the environment, resistance to fracture can be significantly enhanced in dry environments such as nitrogen. Figure 4.17 shows an example of a marked increase in the fracture energy of a pure Ni–alumina interface from approximately 10 J m^{-2} to in excess of 100 J m^{-2} when the fracture test environment is switched from moist air (approximately 50–70% relative humidity) to dry N$_2$. Table 4.2 provides a summary of measured interface fracture energies for a variety of substrates, films and test environments.

4.3.3 Effect of local phase angle on fracture energy

In the discussion of interface fracture in Section 4.2.5, it was recognized that an interface fracture energy alone may not be sufficient to characterize the delamination resistance of an interface and that the level of resistance could depend on the relative magnitude of shear traction to normal traction acting on the interface ahead of the delamination. This stress ratio can be represented by the phase angle ψ introduced in Section 4.2.4. Early studies of the influence of phase angle on fracture energy were reported by Trantina (1972), Anderson et al. (1974) and Mulville et al. (1978) for adhesively bonded interfaces between aluminum and epoxy. The results of these studies, as well as those of subsequent experimental studies involving

Table 4.2. *Values of interface fracture toughness at room temperature.*

Substrate[a]	Film	Γ (J m^{-2})	Comment
Al$_2$O$_3$	Al	1	interface contaminated with C; Al film covered with a Ta superlayer[1]
		>100	tough interface; crack blunting[2, 3]
	Au	2	interface contaminated with C[4]
		10	clean interface; moist air test[5, 6]
		250	tested in dry air[4]
	Ni	5–8	interface contaminated with S[7]
		10–40	tested in moist air[7, 8]
		> 200	dry environment; crack blunting[8]
	Ni(Cr)	> 300	moist air test; crack blunting[9]
	Mo	~ 2	generally brittle interface[10]
	Nb	1–20	tested in moist air[10, 11]
	Cu	120–250	tested in moist air[12]
	Al–Cu	5.6	tested with a Ta superlayer[1]
	Ta$_2$N	0.5	deposited layer[13]
Si	W	5.5–9.0	substrate covered with thin SiO$_2$ layer[14]
	Al–Cu	8	tested with W superlayer[15]
SiO$_2$	Cu	2	tested in moist air[16]
		20	interface coated with Cr[10]
	Cu(Cr)	10	tested in moist air[16]
	TiN	10.4±1.3	tested in moist air[17]
Steel	DLC	> 100	interface coated with Cr[10]
Fused silica	untreated epoxy	2.4	moist air test[18]
Soda-lime glass	untreated epoxy	2.0	moist air test[18]

[a] Data sources where further details on experiments and materials can be found: (1) J. A. Schneider et al., *Mater. Res. Soc. Symposium* **522**, p. 347 (1998). (2) J. M. McNaney et al., *Acta Mater.* **44**, p. 4713 (1996). (3) B. J. Dalgleish et al., *Acta Metall. Mater.* **37**, p. 1923 (1989). (4) D. M. Lipkin et al., *Acta Mater.* **37**, p. 4835 (1998). (5) I. Reimanis et al., *Acta Metall. Mater.* **39**, p. 3133 (1991). (6) M. Turner and A. G. Evans, *Acta Mater.* **44**, p. 863 (1996). (7) D. Bonnell and J. Kiely, *Phys. Status Solidi* **166**, p. 7 (1998). (8) J. S. Stölken and A. G. Evans, *Acta Mater.* **66**, p. 5109 (1998). (9) F. A. Gaudette et al., *Acta Mater.* **45**, p. 3503 (1997). (10) A. G. Evans et al., *Acta Mater.* **47**, p. 4093 (1999). (11) H. Ji et al., *Eng. Fract. Mech.* **61**, p. 163 (1998). (12) I. E. Reimanis et al., *J. Amer. Ceram. Soc.* **80**, p. 424 (1997). (13) N. R. Moody et al., *Acta Mater.* **46**, p. 585 (1998). (14) M. D. Kriese et al., *J. Mater. Res.* **14**, p. 3019 (1999). (15) A. A. Volinsky, Ph.D. thesis, University of Minnesota, Minneapolis (2000). (16) A. Bagchi and A. G. Evans, *Interface Sci.* **3**, p. 169 (1996). (17) R. H. Dauskardt et al., *Eng. Fract. Mech.* **61**, p. 141 (1998). (18). J. E. Ritter et al., *J. Mater. Sci.* **33**, p. 4581 (1998).

other model interfaces (Cao and Evans 1989, Charalambides et al. 1989, Wang and Suo 1990, Thouless 1990), implied an apparent increase in work of interface separation with increasing ratio of shear traction to tensile traction on the interface. These investigations required adaptation of a variety of specimen configurations and testing methodologies from traditional fracture mechanics for measuring interface fracture resistance; some of these methods are now widely used in thin film studies and these are described in detail in Section 4.5.

Liechti and Chai (1992) performed systematic experiments to determine the fracture toughness of glass–epoxy interfaces for local stress state phase angle ψ in the range $-60° < \psi < 90°$. They employed an edge-cracked bimaterial specimen made by bonding epoxy and glass layers, each of thickness 12.7 mm; the edge crack was

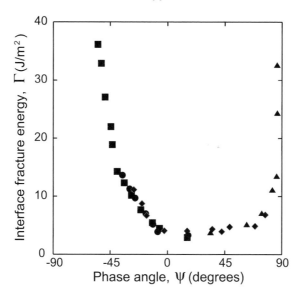

Fig. 4.18. Experimentally determined interface fracture energy Γ as a function of the local stress state phase angle ψ for the epoxy–glass bilayer system. The measurements were made using an edge-cracked bimaterial strip specimen on which prescribed values of normal and shear displacements were imposed. The different symbols represent four different sets of experiments conducted for this material system. Adapted from Liechti and Chai (1992).

introduced by inserting a razor blade along the interface and wedging it open over a length of 75 mm. The specimen was loaded in a specially built biaxial loading device which imposed relative displacement of the clamped top and bottom surfaces of the specimen. This led to mode I opening of the interface crack, and relative shearing displacement at the outer surface of the glass parallel to the interface tending to induce mode II sliding of the interface crack. The elastic moduli of the epoxy and glass layers are 2.03 GPa and 68.95 GPa, respectively, and the corresponding Poisson ratios are 0.37 and 0.2, respectively. The implied values of the Dundurs parameters for this material system are $D_1 = -0.937$ and $D_2 = -0.188$. Liechti and Chai (1992) defined the local stress state phase angle as the arctangent of the ratio of the local shear stress σ_{xz} to the local normal stress σ_{zz}, evaluated at a distance 12.7 mm ahead of the crack edge along the interface.

The measured dependence of interface separation energy Γ on ψ from a series of experiments conducted on the glass–epoxy system is shown in Figure 4.18. The value of Γ is relatively unaffected by the phase angle for $0° < \psi < 45°$, whereas it is strongly dependent on ψ when either $-60° < \psi < 0°$ or $45° < \psi < 90°$; when strong mode II loading is locally imposed at the crack edge, the interface fracture energy is as much as 10 times higher than that for pure local mode I. This increasing work of fracture with increasing relative local shear stress is considered a

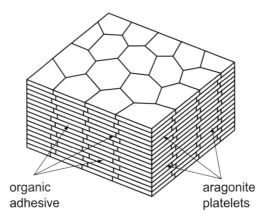

organic
adhesive

aragonite
platelets

Fig. 4.19. Schematic of a cross-sectional view of the correlated interplatelet arrangement of aragonite 'bricks' and interlamellar macromolecular organic 'mortar' material in the architecture of nacre.

consequence of inelastic deformation in the material layer at the interface (Tvergaard and Hutchinson 1993) or frictional contact between the asperities of the crack faces (Stringfellow and Freund 1993) which, in turn, can be influenced by the roughness of the fracture surface. As will be seen in the next chapter, the dependence of interface toughness on the stress state phase angle has important implications for describing the growth of buckle driven delaminations in film–substrate systems.

4.3.4 *Example: Fracture resistance of nacre*

Biomaterials such as teeth, bone and mollusk shell exhibit excellent strength and fracture resistance; their mechanical strength is superior to that of many bulk ceramics and synthetic composites. The fracture resistance of these biological materials is particularly striking in light of the fact that the major constituent phase of the typical composite structure is very brittle. In this example, the structural composition and possible failure mechanisms of nacre, commonly known as 'mother of pearl', are described in order to illustrate how biological multilayer systems may offer inspiration for design of structural materials over multiple length scales.

Nacre, the pearly interior of mollusk shells, is a layered composite with alternating layers of a nanometer-scale biomacromolecular material and crystallographically aligned platelets of aragonite, a mineral form of calcium carbonate ($CaCO_3$) which is very brittle. The organic material occupies less than 5% of the volume of the composite, yet the overall fracture energy of the structure is three orders of magnitude higher than that of bulk $CaCO_3$. Figure 4.19 schematically illustrates the typical cross-section or edge-on view of nacre structure which is commonly found in abalone, nautilus, pearl oyster and blue mussel shells (Currey 1976, Sarikaya et al. 1995). The multilayered composite structure has a 'brick and mortar' arrangement with the hexagon-shaped aragonite bricks, with in-plane dimensions on the order of micrometers and thickness on the order of several hundred nanometers, adhered by an organic 'mortar' which is a composite of proteins and polysaccharides.

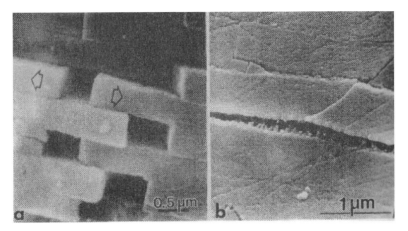

Fig. 4.20. Scanning electron micrographs showing the dominant deformation mechanisms observed in nacre of abalone. (a) Frictional sliding of aragonite platelets in the highly strained region surrounding an indentation. (b) Ligaments of organic material bridging the interplatelet cracks. Reproduced with permission from Sarikaya et al. (1995).

Red abalone has typical shell diameters in excess of 20 cm and a shell thickness in excess of 1.5 cm for specimens that are 10–12 years old. These dimensions facilitate investigations of mechanical properties using standard bend and fracture test geometries whereby the behavior of nacre can be directly compared with that of engineering materials. For example, red abalone which is composed of approximately 95% $CaCO_3$ has a fracture toughness of 4–10 MPa m$^{1/2}$ (Currey 1976, Sarikaya et al. 1995). These values are approximately 20–30 times those of geologically produced monolithic $CaCO_3$. By comparison, polycrystalline Al_2O_3, SiN_4 and ZrO_2 ceramics have fracture toughness values of 3, 4 and 5 MPa m$^{1/2}$, respectively, at room temperature.

Experimental studies of deformation and fracture in nacre indicate that macroscopic inelastic strain is accommodated by massive shearing of the platelets of aragonite and the stretching of the interplatelet organic material that bridges the interface cracks; see Figures 4.20(a) and (b). Figure 4.21(a) is a scanning electron micrograph of polished and plasma-etched abalone nacre which was deformed in compression with the loading axis oriented at 45 degrees to the aragonite platelet interfaces. Note that nanoscale asperities along the interfaces between the aragonite platelets interact during frictional sliding to amplify frictional resistance to shear deformation. The distribution of these asperities on the surfaces of the platelets necessarily engenders dilatation during shear displacement so that adjacent platelets can slide against each other in response to mechanical loading. The result is the formation of interplatelet microcracks.

The nucleation of colonies of microcracks aligned along preferential bands, whose orientation depends on the direction of loading with respect to the arrangement of the aragonite platelets, thus provides a mechanism for energy dissipation and offers an alternative to catastrophic failure by the sudden propagation of a dominant brittle crack. Such dilatation also accommodates the stretching of the organic molecule forming the adhesive between the plates. Atomic force microscope studies of the extension of the organic macromolecules exposed on the surface of freshly cleaved nacre reveal that the protein molecules stretch in discrete jumps as the folded domains pull open (Smith et al. 1999). Figure 4.21(b) is a transmission electron micrograph showing the extension of organic ligaments which bridge

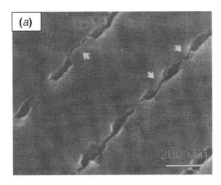
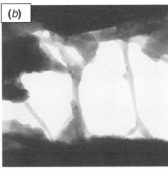

Fig. 4.21. Electron microscope images of the edge-on view of deformed nacre of abalone. (a) Scanning electron micrograph showing contact between aragonite platelets oriented at 45 degrees to the compression axis. Reproduced with permission from Wang et al. (2001). The polished and plasma-etched surface reveals nano-scale asperities which are indicated by the white arrows. (b) Organic ligaments bridging the aragonite platelets, as revealed by the transmission electron micrograph of a cleaved abalone shell. The spacing between the aragonite platelets (the dark regions at the top and bottom of the figure) is approximately 600 nm. The organic adhesive between the tablets had previously been stretched in an atomic force microscope. The organic ligament, originally 30 nm long, elongated twenty times. Reproduced with permission from Smith et al. (1999).

the aragonite platelets. Smith et al. (1999) suggest a modular stretching mechanism whereby a particular molecule extends under local tensile stress; prior to its fracture, another adjacent domain unfolds, absorbing energy in doing so. This sequential deformation of many domains is believed to result in a serrated load–displacement curve which is consistent with the high values of fracture energy exhibited by nacre. Similar mechanisms, involving distributed microcracking and frictional sliding of platelets as energy dissipation processes, and the bridging of the brittle constituent platelets by stretched organic material as a means for toughening, were proposed by Kamat et al. (2000) based on their experimental studies of the fracture resistance of the shell of conch.

4.4 Film delamination due to residual stress

In this section, the phenomenon of *spontaneous* delamination of a strained film from its substrate is examined. The process is termed spontaneous because it occurs without the action of applied loads or any other external cause; it is driven exclusively by the release of elastic energy stored in the film as a result of the mismatch strain. For the time being, delamination is presumed to occur by propagation of a delamination front, or interface crack edge, along the interface between the film and the substrate. The possibility that this advancing fracture will be deflected out of the plane of the interface and into either the film or the substrate will be considered later in this chapter. Furthermore, for the time being, the specific energy Γ, representing the

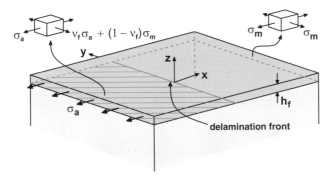

Fig. 4.22. Schematic diagram of the front of a zone of delamination between a film and its substrate and advancing as a line in the x-direction. The stress state prior to delamination is equi-biaxial stress σ_m, and the stress acting in the x-direction is reduced to the level σ_a following delamination.

work per unit area of interface that must be supplied to advance a delamination zone according to the Griffith condition as described in Section 4.2.1, is presumed to be a constant for any particular material system. In other words, Γ is assumed to be independent of the state of stress in the materials or of any other details of the process itself. The validity of this simplifying assumption will be examined subsequently in light of experimental observations.

The physical system that provides the basis for discussion in this section is depicted in Figure 4.22. A film of thickness h_f is bonded to a relatively thick substrate; both materials are presumed to be linearly elastic. When completely bonded, the film carries an equi-biaxial elastic strain of magnitude ϵ_m due to a mismatch of some physical origin with respect to the substrate. The associated biaxial stress magnitude is σ_m and the state of stress far ahead of the delamination front is indicated in the figure. A delamination front is assumed to advance in the x-direction along the interface, resulting in partial relaxation of the elastic strain in the film. The state of stress behind the delamination front is also indicated in the figure, where σ_a is the magnitude of the unrelaxed stress component σ_{xx} acting in the x-direction following delamination. The strain in the direction parallel to the delamination front is locally unaffected by the delamination process, so the stress in the y-direction relaxes to the level $\sigma_{yy} = \nu_f\sigma_a + (1 - \nu_f)\sigma_m$.

It is important to recognize that Figure 4.22 may be interpreted at two different scales of observation. If a delamination front is straight over distances that are very large compared to h_f, then the scale of observation represented by the figure is essentially the entire film and $\sigma_a = 0$ because the edge of the film is stress free. On the other hand, if the delamination front is curved, with radius of curvature large compared to h_f, then the scale of the figure is local to the region of the front; in this case, the front appears to be straight on a local scale and the front curvature is

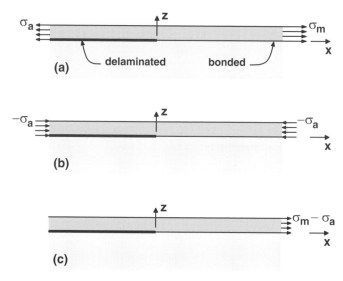

Fig. 4.23. The superposition scheme leading to the conclusion that the state of stress on the interface near the edge of the delamination zone depicted in Figure 4.22 depends only on the difference between stress in the film ahead of and behind the delamination front.

represented through a nonzero value of the unrelaxed stress σ_a. Examples of both situations will be illustrated in this section.

As already noted, the state of stress in the film is essentially homogeneous biaxial tension or compression at distances far ahead and far behind the delamination front compared to h_f. On the other hand, for material points close to the edge of the delamination zone, the state of stress is expected to be very nonuniform and to be highly concentrated. While the details of this stress distribution play little role in the conceptual development of this section, a few observations made on the basis of elementary arguments reveal the full nature of this stress distribution.

A two-dimensional representation of the system in Figure 4.22 is shown in part (a) of Figure 4.23. The remote homogeneous stress component σ_{xx} that acts in the x-direction for $|x/h_f| \gg 1$ is indicated and, due to the fact that $\sigma_a \neq \sigma_m$, some traction must be transmitted across the interface for $x > 0$. Part (b) of Figure 4.23 shows an equilibrium situation with a *uniform* stress $\sigma_{xx} = -\sigma_a$, $\sigma_{yy} = -\nu_f \sigma_a - (1 - \nu_f)\sigma_m$ throughout the film. In this case, no traction is transmitted across the interface and the strain in the film in the y-direction is everywhere $-\epsilon_m$. The result of superimposing the fields of parts (a) and (b) is shown in part (c) of Figure 4.23. The stress state far ahead of the delamination front is $\sigma_{xx} = \sigma_m - \sigma_a$, $\sigma_{yy} = \nu_f(\sigma_m - \sigma_a)$; the stress state far behind the front is $\sigma_{xx} = 0$, $\sigma_{yy} = 0$; the strain in the y-direction in part (c) is everywhere zero. Thus, it is the stress $\sigma_m - \sigma_a$ that must be balanced by the interface traction over $x > 0$. Based on a comparison of part (c) with Figure 4.4, it becomes clear that the interface traction in part (c) of Figure 4.23, and therefore for part (a) as well, is exactly the same as that discussed in Section 4.1 except that the distribution is scaled by the magnitude $\sigma_m - \sigma_a$ rather than by σ_m alone. Furthermore, the deformation in part (c) is plane strain deformation everywhere. Thus, even though this stress distribution does not play a role in the Griffith condition, it is recognized that its principal features are already in hand.

The elastic strain energy density associated with a biaxial state of stress in the film is, in terms of stress components,

$$U = \frac{1}{2E_f}\left[\sigma_{xx}^2 + \sigma_{yy}^2 - 2\nu_f\sigma_{xx}\sigma_{yy}\right]. \tag{4.33}$$

The reduction in potential energy of the system due to advance of the delamination front provides the energetic driving force, or the configurational force, for that process. When the potential energy reduction is expressed per unit area of interface over which delamination occurs, it defines the energy release rate for applied boundary loading which was introduced earlier in this chapter. For the stress states shown in Figure 4.22, the difference in elastic strain energy per unit area of interface between material ahead and behind the front, plus the work done by the applied stress σ_a in a unit advance of the delamination front, yields the energy release rate

$$\boxed{\mathcal{G} = \frac{1 - \nu_f^2}{2E_f}(\sigma_m - \sigma_a)^2 h_f.} \tag{4.34}$$

If the calculation is based on the equivalent stress state in part (c) of Figure 4.23, the driving force may be determined as a change in strain energy per unit area of interface without reference to an external potential energy. The same result is obtained once again.

It is noteworthy that the expression (4.34) does not depend on the properties of the substrate, a feature that could have been anticipated. The assumption of steady-state advance of the delamination front implies that the mechanical fields in the substrate translate without alteration during delamination, and there is no exchange of energy between the substrate and its surroundings. Furthermore, these results are derived under the tacit assumption that the restrictions which underlie the Stoney formula are in place. Within this range of behavior, the ratio of the elastic strain energy in the substrate to the elastic strain energy in the film, both measured per unit area of interface, is approximately $h_f M_f / h_s M_s$. As a result, the energy density itself in the substrate is small compared to the energy change in the film associated with delamination. Consequently, energy variations in the substrate have only a minor influence on the value of $\mathcal{G}$.

4.4.1 A straight delamination front

If the delamination front is straight and the film edge is free then the energy release rate is given in terms of system parameters by (4.34) with $\sigma_a = 0$. According to the Griffith condition, the delamination front will advance or not, depending on whether $\mathcal{G}$ is greater than Γ or less than Γ, the parameter that characterizes the resistance of the interface to separation. The relationship among the system parameters that distinguishes between these two possibilities is

$$
\frac{1 - \nu_f^2}{2 E_f} \sigma_m^2 h_f = \Gamma \quad \Leftrightarrow \quad \frac{(1 + \nu_f) E_f}{2(1 - \nu_f)} \epsilon_m^2 h_f = \Gamma , \tag{4.35}
$$

expressed in terms of mismatch stress or strain, respectively.

For a given material system, the film thickness at which it is first possible to drive a delamination spontaneously is the critical thickness

$$
(h_f)_{cr} = 2 \frac{(1 - \nu_f)\Gamma}{(1 + \nu_f) E_f \epsilon_m^2} = 2 \frac{\bar{E}_f \Gamma}{\sigma_m^2}. \tag{4.36}
$$

The result has several noteworthy features. At this level of modeling, the critical thickness for delamination depends on substrate properties only through their influence on the values of both the mismatch strain ϵ_m and the separation energy Γ. The general form of (4.36) could have been anticipated on the basis of dimensional arguments alone; the analysis provides the numerical factor 2.

Taken at face value, the result (4.35) implies that the film will spontaneously separate from its substrate over the entire area of the interface once a delamination has begun to propagate. This is not necessarily so, however. One reason is that the local value of energy release rate at a place on the delamination front can be influenced by local curvature of the front. This issue is addressed in the next section. A second reason is that the energy release rate for a delamination near a film edge, within a distance of h_f of the edge say, is less than the value for a steadily propagating delamination. Thus, there is an initiation barrier to the onset of delamination at a film edge or at a through-the-thickness cut in the film. While no unambiguous analytical results are available for describing the transition to steady-state as a delamination grows inward from the edge of a film, the magnitude of the effect can be estimated by means of numerical calculations.

A detailed numerical study of this problem specially designed to give accurate results near the edge of the film was carried out by Yu et al. (2001), who showed that the effect is very small for the onset of delamination from a free edge of the film on a flat substrate surface. For example, it was shown that the value of $\mathcal{G}$ was already more than 90 percent of its steady-state value, as given in (4.34), when the

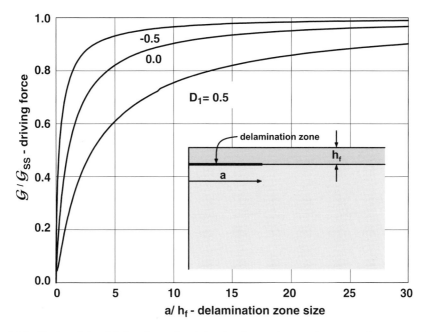

Fig. 4.24. Ratio of the driving force for delamination to its steady-state value approached far from the film edge versus the delamination zone size for three values of D_1. The configuration is shown in the inset. Adapted from Yu et al. (2001).

delamination front has advanced a distance of only $0.05h_f$ from the edge. Thus, the effect of the initiation barrier is minor in this case.

The driving force for delamination of a film from the surface of a substrate in the form of a 90 degree wedge, rather than a flat substrate, was also analyzed by Yu et al. (2001). The free edge of the bonded film in this case was assumed to be at the corner of the wedge; see the inset in Figure 4.24. In contrast to the case of a substrate on a flat surface, this configuration revealed a substantial energetic barrier to initiation of delamination of the film. The ratio of the driving force $\mathcal{G}$ to its steady-state value, labeled $\mathcal{G}_{ss}$ here, is shown in Figure 4.24 for $D_2 = 0$ and $D_1 = 0.5, 0.0$ and -0.5. For the case when the elastic properties of the film and substrate are the same, $\mathcal{G}$ does not increase beyond $0.9\mathcal{G}_{ss}$ until the size of the delamination zone has grown to more than about ten times the film thickness.

4.4.2 Example: Delamination due to thermal strain

Consider a thin film of aluminum, 1 μm in thickness, which is deposited at 220 °C so as to cover partially a (100) Si substrate, which is 300 μm thick and 200 mm in diameter. The thermoelastic properties of the Al film are $E_f = 70$ GPa, $\nu_f = 0.35$, and coefficient of thermal expansion $\alpha_f = 23 \times 10^{-6}$ °C^{-1}. The corresponding properties of the Si substrate are $E_s = 130$ GPa, $\nu_s = 0.28$ and $\alpha_s = 3 \times 10^{-6}\,^\circ\text{C}^{-1}$. The film–substrate system is stress-free at the deposition temperature.

(a) Determine the fracture energy of the interface if the energy release rate due to the mismatch strain of the film with respect to the substrate at a temperature of 20 °C is just enough to propagate a straight delamination front along the interface.
(b) In an attempt to facilitate the deposition of thicker aluminum films, the processing conditions were modified in such a way that the interface separation energy was guaranteed to be at least 5 J m^{-2}. Find the thickness of the aluminum film for which spontaneous delamination of a straight front would occur at the minimum guaranteed interface separation energy.

Solution:

(a) The mismatch strain in the Al film with respect to the substrate at room temperature is

$$\epsilon_m = (\alpha_s - \alpha_f)\Delta T = 4.0 \times 10^{-3}. \tag{4.37}$$

For this mismatch strain, the energy release rate is found from (4.34) to be

$$\mathcal{G} = \frac{1 - \nu_f^2}{2E_f}\sigma_m^2 h_f = 1.1\,\text{J m}^{-2} \tag{4.38}$$

in the absence of any externally applied loads. Since this value of $\mathcal{G}$ is found to be just enough to initiate interface separation, $\Gamma = 1.1$ J m^{-2}.

(b) For $\Gamma = 5$ J m^{-2}, the critical film thickness $(h_f)_{cr}$ at which spontaneous delamination can occur along the interface is found from (4.36) to be

$$(h_f)_{cr} = 2\frac{(1 - \nu_f)\Gamma}{(1 + \nu_f)E_f\epsilon_m^2} = 4.5\,\mu\text{m}. \tag{4.39}$$

4.4.3 An expanding circular delamination front

The same physical system that was considered in the preceding section for the case of a straight delamination front provides the basis for discussion of the effect of front curvature in this section. A film of uniform thickness h_f is bonded to a much thicker substrate over a large area. When fully bonded, the film is under the action of an equi-biaxial stress of σ_m which arises from a mismatch of some physical origin with respect to the substrate. The strength of the interface between the film and the substrate is characterized by an energy of separation of magnitude Γ per unit area of interface. Finally, suppose that the system parameters are such that a straight delamination front would grow spontaneously according to the analysis of the preceding section, that is,

$$\frac{1 - \nu_f}{2E_f}\sigma_m^2 h_f > \Gamma. \tag{4.40}$$

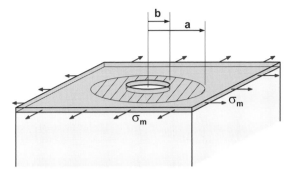

Fig. 4.25. Schematic diagram of a delamination zone expanding outward from the free edge of a circular hole in the film. The expansion is driven by a residual stress in the film.

How is the behavior modified if the delamination front is curved rather than straight?

Suppose that the film completely covers the substrate surface initially. Then, the film is cut through its thickness down to the substrate surface over a circle of radius $b \gg h_f$ as indicated in Figure 4.25. The film material inside the circle is removed and, outside the circle of radius b, a circular delamination front begins to grow outward. When the front is very near the film edge of radius b, the local conditions are very much like those discussed in the preceding section, and the front advances spontaneously.

Consider the situation when the delamination front has advanced to a radius $r = a$ where r is the radial distance measured from the center of the circular cut. Within the annular region $b < r < a$, the film material is completely separated from the substrate, and the edge at $r = b$ is free of traction. The edge at $r = a$ is still connected to the bonded portion of the film across the delamination front so that, in general, it is not free of stress. Suppose that the state of stress in the film at $r = a^-$ is the radial stress $\sigma_a = \sigma_{rr}|_{r=a^-}$; the state of stress is necessarily radial due to the symmetry of the configuration. The use of the notation σ_a for this stress anticipates that it plays exactly the role of σ_a in Figure 4.22. The corresponding radial displacement of the edge of the annular disk at $r = a$, measured from its fully relaxed stress-free configuration, is given by the classical Lamé solution from plane stress elasticity to be

$$u_a = u_r|_{r \to a^-} = \frac{a^3 \sigma_a}{(a^2 - b^2)E_f} \left[(1 + \nu_f)\frac{b^2}{a^2} + (1 - \nu_f) \right]. \qquad (4.41)$$

Because the outer edge of the annular region at $r = a$ is still at its unrelaxed position,

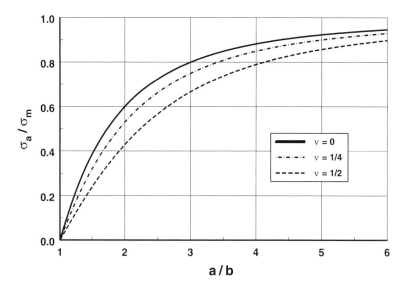

Fig. 4.26. Ratio of the stress σ_a resisting delamination due to front curvature to the mismatch stress σ_m versus the size of the lamination zone for the system depicted in Figure 4.25.

the circumferential strain there is essentially ϵ_m so that

$$\frac{u_a}{a} = \epsilon_m = \frac{(1 - \nu_f)\sigma_m}{E_f} \tag{4.42}$$

if the displacement is to be continuous at $r = a$. Thus, if this radial displacement is eliminated by combining the expressions (4.41) and (4.42), an expression for σ_a as a function of a is obtained as

$$\frac{\sigma_a}{\sigma_m} = \frac{a^2/b^2 - 1}{a^2/b^2 + (1 + \nu_f)/(1 - \nu_f)} . \tag{4.43}$$

It is evident from this result that the stress restraining the film has the value $\sigma_a = 0$ when $a = b$, and that it increases monotonically as a/b increases from the value 1. Note also that σ_a/σ_m asymptotically approaches unity as $a/b \to \infty$. Plots of σ_a/σ_m versus a/b are shown in Figure 4.26 for $\nu_f = 0$, $\frac{1}{4}$ and $\frac{1}{2}$. From the condition (4.40), it is certain that the delamination front would advance some distance from its initial position at $a = b$. From the asymptotic result that $\sigma_a/\sigma_m \to 1$ as $a/b \to \infty$, when considered in light of the general expression for driving force given in (4.34), the delamination front cannot advance to indefinitely large values of a for any nonzero value of Γ because it eventually runs out of driving force.

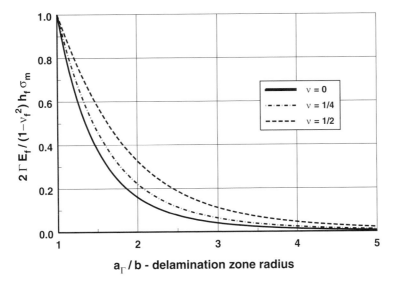

Fig. 4.27. Normalized interface separation energy implied by the extent of the delamination zone.

The value of a at which advance of the delamination front must stop, say a_Γ, is the value at which the Griffith condition

$$\frac{1 - \nu_f^2}{2E_f}(\sigma_m - \sigma_a)^2 h_f = \Gamma \tag{4.44}$$

is first satisfied, with σ_a given by (4.43). For $a < a_\Gamma$, the driving force $\mathcal{G}$ exceeds Γ so that the front continues to advance. On the other hand, the driving force is less than Γ for $a > a_\Gamma$, and hence this range is inaccessible. The resulting relationship can be cast in the nondimensional form

$$\frac{2\Gamma E_f}{(1 - \nu_f^2)h_f\sigma_m^2} = \frac{4}{\left[(1 - \nu_f)a_\Gamma^2/b^2 + (1 + \nu_f)\right]^2}. \tag{4.45}$$

Graphs of normalized separation energy versus a_Γ/b are shown in Figure 4.27 for $\nu_f = 0$, $\frac{1}{4}$ and $\frac{1}{2}$.

Farris and Bauer (1988) introduced an experimental approach to measure the separation energy Γ based on the geometrical configuration under discussion here. Motivated by this approach, the separation energy is expressed as a function of a_Γ in (4.45), rather than the other way around, to anticipate the interpretation of measurements. The material parameters of the film, the film thickness and the mismatch stress are presumably known on the basis of measurements made separately from any delamination experiments. The radius b is controlled in the experiment, and therefore its value is known. Thus, if a_Γ can be observed in an experiment, a value for Γ can be inferred on the basis of (4.45).

The relationship (4.45) between the final radius a_Γ of the delamination zone and the corresponding separation energy Γ was established by adapting some general concepts of fracture mechanics. The same result can be obtained by direct appeal to energy methods. The total elastic energy in the film between the free surface at $r = b$ and some remote circular boundary, say at $r = R > a$, is

$$\mathcal{E} = \pi a \sigma_a u_a h_f + \pi (R^2 - a^2) h_f \sigma_m^2 / M_f. \tag{4.46}$$

If u_a from (4.41) and σ_a from (4.43) are substituted into this expression, equilibrium requires that

$$-\frac{\partial \mathcal{E}}{\partial a}\bigg|_{\sigma_m} = 2\pi a \Gamma, \tag{4.47}$$

which again yields (4.45).

The conclusions of the analysis in this section are independent of the sign of σ_m. From a practical point of view, however, the annular portion of the film would tend to buckle out of plane if $\sigma_m < 0$. This is a mode of deformation that is not included in the analysis, and the conclusions are therefore invalidated if buckling occurs. Film buckling due to compressive σ_m is discussed in the next chapter.

The symmetry of the configuration depicted in Figure 4.25 with a circular delamination zone renders the analysis of delamination simple and transparent. If this symmetry is lost, the situation becomes more complex and the delamination zone shapes which can be observed are surprisingly rich in detail. For example, Choi and Kim (1992) introduced a cut into a polyimide film on a glass substrate and observed the resulting growth of a delamination front from the cut. Because the local conditions varied along the edge of the delamination zone, insight could be gained into the nature of the delamination process in terms of the local stress fields.

4.4.4 Phase angle of the stress concentration field

Under the conditions of film delamination represented by Figure 4.22, the state of stress along the interface immediately in advance of the delamination front can be characterized by the stress intensity factors K_I and K_{II}, provided that the nonlinear process zone is suitably small. In light of the discussion of the superposition scheme represented by Figure 4.23, the loading parameter in the system that gives rise to interfacial stress is $\sigma_m - \sigma_a$. The only length scale in the configuration is the film thickness h_f. Finally, it is noted that both the film and substrate materials are linearly elastic solids with their associated elastic moduli and Poisson ratios. The phase angle of the local stress field can be characterized in a simple way on the basis of these observations.

The stress analysis under consideration is such that both K_I and K_{II} must vary linearly with $\sigma_m - \sigma_a$, and both must vanish when $\sigma_m - \sigma_a = 0$. To render the expressions for stress intensity factors in terms of system parameters dimensionally consistent, each stress intensity factor must be proportional to $(\sigma_m - \sigma_a)\sqrt{h_f}$. The proportionality factor in each case must be dimensionless, it cannot depend on $\sigma_m - \sigma_a$, and it cannot depend on h_f because there is no other length parameter available for forming a dimensionless ratio. Thus, the proportionality factor can depend only on the elastic constants in dimensionless combinations, that is, on the parameters D_1 and D_2 defined in (4.29). The development to follow is restricted to the case when $D_2 = 0$, so that the proportionality factors depend only on D_1, the stiffness ratio.

Suppose that the proportionality factors for K_I and K_{II} are denoted by p_I and p_{II}, respectively, so that $K_I = p_I(\sigma_m - \sigma_a)\sqrt{h_f}$ and similarly for K_{II}. The energy release rate for advance of the delamination front is given by (4.34). If this result is substituted for $\mathcal{G}$ in (4.30), it is evident that the proportionality factors must satisfy

$$\tfrac{1}{2}(1 - D_1) = p_I^2(D_1) + p_{II}^2(D_1) \tag{4.48}$$

identically in D_1. The relationship (4.48) implies the existence of a single function of D_1, say $\omega(D_1)$, such that

$$p_I = \sqrt{\tfrac{1}{2}(1 - D_1)}\, \cos\omega(D_1)\,, \quad p_{II} = \sqrt{\tfrac{1}{2}(1 - D_1)}\, \sin\omega(D_1)\,. \tag{4.49}$$

The value of $\omega(D_1)$ can be determined only through solution of the relevant boundary value problem. A numerical solution of the elasticity problem represented by the configuration depicted in Figure 4.22 has been provided by Suo and Hutchinson (1990). The solution of the corresponding problem when the film is idealized as an elastic plate was described by Shield and Kim (1992) and is determined by numerical solution of the integral equations in (4.93). If p_I and p_{II} are determined from the boundary value problem, then (4.48) can serve as a useful check on the consistency of the results.

Of main interest here is the phase angle ψ of the stress state within the stress concentration field at the edge of the delamination zone, as defined in (4.28). From the foregoing discussion, it is evident that the phase depends only on the nondimensional ratio of material parameter D_1 (for $D_2 = 0$) as defined in (4.29) according to

$$\psi = \arctan\frac{K_{II}}{K_I} = \arctan\frac{p_{II}(D_1)}{p_I(D_1)} = \omega(D_1). \tag{4.50}$$

The dependence of ω on D_1 (for $D_2 = 0$) is shown in Figure 4.28 from Suo and Hutchinson (1990) and from numerical solution of the equations in (4.48). When the film and substrate materials have identical elastic constants, that is, when $D_1 = D_2 = 0$, the elasticity solution yields $\omega \approx 52.1°$ while the solution based

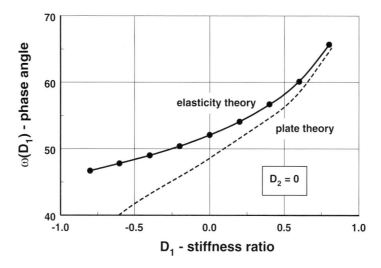

Fig. 4.28. Crack tip phase angle ψ for the configuration shown in Figure 4.22 versus stiffness ratio D_1 for $D_2 = 0$. The dashed line represents an estimate of ψ based on the results for the plate model shown in Figure 4.4.

on the plate idealization yields $\omega \approx 48.6°$. The discrete points in the figure are the numerical results reported by Suo and Hutchinson (1990) and the line through these points represents the expression

$$\omega(D_1) \approx 52.10 + 8.691D_1 + 6.450D_1^2 + 4.893D_1^3, \tag{4.51}$$

which is obtained as a useful approximate fit, with ω expressed in degrees.

In the foregoing discussion, it has been tacitly assumed that $\sigma_m - \sigma_a > 0$. If this is not the case, the phase angle lies in a different quadrant of the K_I, K_{II}-plane in (4.50); for example,

$$\omega|_{(\sigma_m - \sigma_a)<0} = -\pi + \omega|_{(\sigma_m - \sigma_a)>0}, \tag{4.52}$$

where $\omega|_{(\sigma_m - \sigma_a)>0}$ is the quantity graphed in Figure 4.28.

4.4.5 Delamination approaching a film edge

The energetic driving force for film delamination given in (4.34) is based on the presumption of steady advance of a delamination front at the interface between a uniform film and its substrate. In view of the strong interaction between the film and the substrate near the film edge indicated by the results on load transfer length in this region given in Figure 4.8, for example, the driving force for delamination can be expected to depart significantly from the steady-state result (4.34) near a film edge.

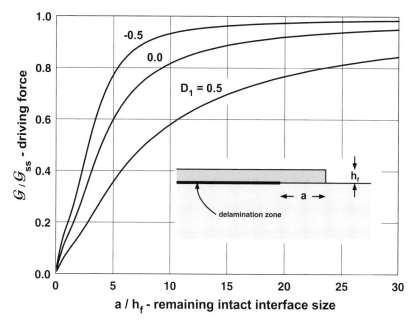

Fig. 4.29. Normalized driving force for a delamination front approaching a film edge versus the size of the remaining intact portion of the interface between the front and the free edge. The configuration is illustrated in the inset, and the response is two-dimensional generalized plane strain deformation. Results are presented for three values of D_1 with $D_2 = 0$. The steady-state value of energy release rate $\mathcal{G}$ is given in (4.34). Adapted from Yu et al. (2001).

This free-edge effect was studied in detail by He et al. (1997) and Yu et al. (2001). A new method for analyzing a fairly wide class of such edge configurations, again based on numerical solution of singular integral equations, was introduced by Yu et al. (2001). They studied both the emergence of delaminations from film edges and the approach of delamination fronts toward film edges.

To examine the change in delamination driving force in the vicinity of the edge, consider the plane strain configuration shown in the inset of Figure 4.29. This diagram represents a straight delamination front approaching the free edge of a film, with the substrate extending far beyond the edge of the film; this arrangement is termed an interior film edge configuration. The width of the portion of the interface that is still intact is denoted by a in this case. When the value of a/h_f is very large compared to unity, the delamination driving force $\mathcal{G}$ has the value given by (4.34), here denoted by $\mathcal{G}_{ss}$ to emphasize its role as a steady-state value. It is reasonable to anticipate that $\mathcal{G}/\mathcal{G}_{ss}$ will vanish as $a/h_f \to 0$. The main point in analyzing this configuration is to understand the nature of the transition between these asymptotic limits and, in particular, to determine the length scale over which this transition occurs.

The results obtained by Yu et al. (2001) for $\mathcal{G}/\mathcal{G}_{ss}$ versus a/h_f with $D_1 = -0.5, 0.0, 0.5$ and $D_2 = 0$ are shown in Figure 4.29. It is evident that the edge effect is quite far reaching in this case. For example, for the case with $D_1 = 0, \mathcal{G}/\mathcal{G}_{ss}$ is greater than 0.8 only for $a/h_f > 10$ and is greater than 0.9 only for $a/h_f > 20$. It was noted in Section 4.4.1 that there is virtually no transition length for initiation of a delamination at an interior film edge, and the result for approach of a delamination toward an interior free edge is in striking contrast to this observation.

The phase angle ψ of the interface stress state at the edge of the delamination front as it approaches the film edge rotates from its positive steady-state value toward 90 degrees as $a/h_f \to 0$. Consequently, as the value of driving force diminishes with diminishing a/h_f, the phase angle ψ becomes larger, implying an increasing work of fracture $\Gamma(\psi)$ for typical material systems showing a phase angle sensitivity in their fracture energies. These effects combined to make an interior film edge resistant to completion of delamination.

4.5 Methods for interface toughness measurement

The measurement of interface fracture toughness is a significantly greater challenge than the measurement of fracture resistance of homogeneous bulk materials. For one thing, the geometries associated with interfaces of practical interest, especially those between thin films and their substrates or adjoining layers, severely constrain the design of test specimens and loading mechanisms. Furthermore, in many situations, the processing of thin film structures in sufficient quantities and dimensions in order to facilitate tests to extract reproducible fracture resistance parameters that are independent of specimen geometry is a severe practical limitation. In addition, mismatch strains in layered structures can induce significant inelastic deformation which, in turn, can restrict choices of experimental methods for determining interface fracture resistance.

Well developed test methods that are commonly employed to measure the fracture properties of homogeneous materials can also be used, with appropriate modifications, for the determination of interface fracture response in layered materials. However, great care must be exercised to account properly for residual stress which can influence the energy release rate $\mathcal{G}$ and the local stress state phase angle ψ in layered structures. Among the various fracture testing techniques available for homogeneous materials, those in which geometric conditions inherently facilitate the stable advance of a crack are preferred for interface toughness measurements. Examples include the double cantilever crack specimen, the symmetric or asymmetric bend crack specimen, and compression test methods involving the Brazilian disk or double cleavage drilled compression specimens. In addition, there exist experimental methods, such as the so-called superlayer test, the peel test and the bulge

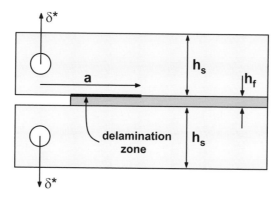

Fig. 4.30. The double cantilever interface crack specimen configuration under imposed diplacement conditions.

or blister test, which have been developed specifically for estimating the interface fracture energies between thin films and their substrates. Each of these methods is briefly described, with the exception of the peel test and the bulge test. The latter two experimental configurations rely on an understanding of deformation within the geometrically nonlinear range for their proper interpretation, and these configurations will be discussed in Chapter 5 once the requisite background is developed.

4.5.1 Double cantilever test configuration

Figure 4.30 schematically shows the adaptation of the double cantilever crack specimen, discussed in Section 4.2.1, for determining the interface fracture properties between a thin film and a surrounding layer of much greater thickness. A debond of length a is introduced at the interface between the thin film and one of the adjoining materials; in many film–substrate systems this debond crack can easily be introduced by depositing a thin layer of carbon between the film and the substrate during sample preparation. The onset and subsequent stable growth of fracture from the initial precrack under imposed boundary displacement is characterized in terms of the energy release rate $\mathcal{G}$. The relationship between $\mathcal{G}$ and the imposed displacement was established in Section 4.2.1. The mismatch in elastic properties between the thin film and the surrounding layers introduces local in-plane shear loading on the crack plane which can alter the crack edge stress field from that of a homogeneous double cantilever specimen. However, the nearly symmetric loading configuration and the much smaller thickness of the film compared to the thickness of the adjoining layers essentially promote mode I fracture.

Considerable plastic deformation can occur if the thin film is made of a ductile metal, and there is no way to separate the actual work of interface separation from the work dissipated in plastic deformation. In any case, if the delamination advances

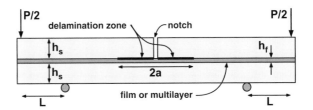

Fig. 4.31. The four-point flexure specimen.

with a more or less steady local stress field, the change in energy that defines the energy release rate is taken to be that given in Section 4.2.1 as

$$\mathcal{G} = \frac{3\bar{E}_s h_s^3}{8a^4}\delta^* = \Gamma. \tag{4.53}$$

Thus, measurement of $\mathcal{G}$ as a function of length a during delamination growth provides an estimate of Γ. As described by Oh et al. (1988) and Evans et al. (1999), the double cantilever specimen has been used to evaluate the fracture energy of interfaces between thin films of such materials as Al and Cu and substrates such as Al$_2$O$_3$ and glass.

4.5.2 Four-point flexure beam test configuration

Figure 4.31 shows an example of a symmetric composite configuration containing two substrates of identical dimensions between which films of much smaller thickness are present; the films may be single layers or multilayers. This entire layered system is subjected to symmetric four-point bend loading with either the bending moment or the end rotation of the specimen midsection being imposed. A central notch is introduced, by indentation or some other technique, through the thickness of the layer(s) that exist above the interface at which the fracture toughness is to be determined. A symmetric delamination precrack connected to the notch is also introduced. Such a debond can be produced along the interface by depositing, during the specimen preparation stage, a layer of carbon over the area in which the delamination precrack is desired.

Within the region of uniform bending moment, the sample is subjected to a uniform bending moment. Consequently, a steady-state value of the energy release rate $\mathcal{G}$ can be used as a measure of the crack driving force provided that the length of the crack lying along the interface is much greater than h_s. This value of $\mathcal{G}$, which is independent of the debond length a, is calculated for the case of applied loading by appeal to elastic beam theory as

$$\mathcal{G} = \frac{21P^2L^2}{16E_s h_s^3}, \tag{4.54}$$

where $PL/2$ is the bending moment per unit width, with $P/2$ being the load per unit width along each loading line and L the spacing between the outer and inner loading lines, h_s is the substrate thickness or the approximate half thickness of the bend specimen, and E_s denotes the elastic modulus of the substrate material (Charalambides et al. 1989). If the depth of the configuration in the direction into the figure is large, then E_s should be replaced by the plane strain modulus $\bar{E}_s$. Again, the deformation of the film material is neglected in this calculation.

The flexure beam is subjected to either a fixed load or a fixed load point displacement. When the load is fixed, the energy release rate is constant and hence the crack advances under steady conditions when it is contained along the interface between the inner loading lines. A displacement transducer is usually attached in the vicinity of the notch to measure the vertical displacement. A plot of the load as a function of the displacement typically provides a positive linear initial slope when there is no extension of the pre-exising debond crack along the interface; the deflection here is merely an outcome of the linear elastic deformation of the loading system. At some critical load, corresponding to the critical value of the energy release rate Γ for the interface, the load–displacement plot exhibits an abrupt change of slope which signifies the onset of crack advance. Slow, time-dependent subcritical crack growth resistance along the interface can also be measured using the flexural beam test specimen by monitoring the change in crack length per unit time as a function of the energy release rate G solely in response to mechanical loading or to environmentally assisted fracture along the interface.

Symmetric three-point or four-point bending of notched and pre-cracked rectangular beams has long been used as a common test method for determining the tensile fracture toughness of homogeneous materials. Asymmetric bending of notched and pre-cracked rectangular beams has also been employed as a means of imposing different combinations of mode I and mode II loading, as described by Suresh et al. (1990) for mixed-mode fracture of ceramic materials. The arrangement shown in Figure 4.31 can also be modified to impose different combinations of local mode I and mode II loads for interface delamination in the layered material. Analyses of the driving force for mixed-mode fracture along the interface between a thin film and a substrate in the four-point flexure specimen have been presented by Charalambides et al. (1989).

An example of the use of the symmetric four-point flexure specimen for the experimental determination of interfacial fracture resistance in a multilayer thin film structure is now considered. Figure 4.32 shows a typical multilayer thin film structure containing an Al–Cu metallization layer and a SiO_2 interlayer dielectric (ILD) film which is used in microelectronic devices. A TiN layer is deposited as a barrier material to inhibit diffusion of Al and Cu during the thermal excursions induced by processing and to enhance adhesion with the ILD. However, debonding

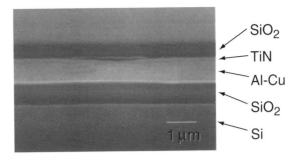

Fig. 4.32. Scanning electron micrograph of the metallization, diffusion barrier and ILD layers in a typical multilayer thin film structure used in microelectronic devices. Note the presence of a delamination crack at the interface between the SiO_2 ILD and the TiN diffusion barrier, the fracture along which can limit the yield and reliability of the device. The top layer is also Si; the two outer layers of Si of the same thickness provide a symmetric multilayer structure within which the thin films are sandwiched. Each Si layer (only a small portion of which is shown in this figure) is approximately 600 μm thick. Reproduced with permission from Dauskardt et al. (1998).

of the ILD–TiN interface and the ensuing delamination of the ILD layer itself can substantially diminish device yield during the fabrication of the microelectronic devices or reduce mechanical integrity during subsequent service. Figure 4.32 shows an example of a delamination along the ILD–TiN interface, which was induced in the vicinity of a fine-scale indentation.

In order to determine the critical fracture initiation toughness Γ as well as the tendency for subcritical crack growth along the interface, Dauskardt et al. (1998) employed the symmetric four-point flexure test geometry. The thin film structure, similar to that shown in Figure 4.32, was fabricated on a Si wafer substrate using standard sputtering methods for the metal and the TiN layers and using plasma-enhanced CVD for the ILD film. In order to create a symmetric multilayer thin film stack, a second Si wafer was diffusion bonded to the top surface of the film stack; a thin layer of Cr was used to enhance adhesion at the bonded surface along with Cu which can be bonded at relatively low temperatures. The copper surfaces were diffusion bonded at 400 °C for 4 h in a vacuum press under 12 MPa pressure.

A topic of particular concern in fracture studies of such multilayer structures is the possibility of energy dissipation due to plastic deformation of the metallic thin film layers. For the particular case of the foregoing multilayer stack comprising very thin metal films, the resulting interface fracture toughness results have been found to be essentially independent of specimen and crack geometry due to the constraints imposed by the surrounding brittle layers. Thus, the energy release rate $\mathcal{G}$ is considered to be an appropriate parameter for characterizing the onset and progression of interface debonding although plastic yielding of the entire metal

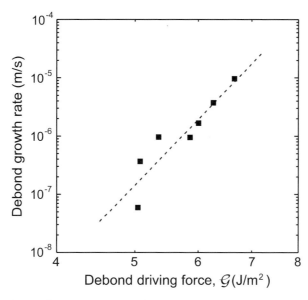

Fig. 4.33. Experimentally measured rate of growth of the delamination along the TiN–SiO$_2$ interface in 45% relative humidity laboratory air environment as a function of the energy release rate $\mathcal{G}$ for the symmetric multilayer structure shown in Figure 4.32 with the thickness of the Al–Cu layer of 0.65 μm. The thin film multilayer structure, sandwiched between two identical Si layers, was tested for interface delamination growth using a symmetric four-point flexure test method, similar to that shown in Figure 4.31. Adapted from Dauskardt et al. (1998).

film may occur at the debond edge. The phase angle for this multilayer flexure specimen is $\psi \sim 43°$. A multitude of tests conducted in this study confirmed that the interface flexure test provides an accurate and reproducible measure of interface fracture energy for the thin film structure under consideration.

Figure 4.33 shows the rate of slow, time-dependent subcritical crack growth, measured using the symmetric four-point flexure specimen geometry, along the TiN–SiO$_2$ interface in the multilayer structure as a function of the energy release rate $\mathcal{G}$. Note that the interface delamination advances subcritically at energy release rate values as small as 60% of the fracture energy of $\Gamma \approx 10.6$ J m^{-2}. As shown in the figure, the debond growth rate has a power-law variation with $\mathcal{G}$. The observed propensity for subcritical fracture of the debond has been ascribed by Dauskardt et al. (1998) to environmentally assisted cracking in the humid air environment, analogous to the stress corrosion cracking of SiO$_2$ glass in moist air.

4.5.3 Compression test specimen configurations

The use of a compressive load applied to a layered structure comprising one or more brittle materials is convenient from an experimental standpoint because such

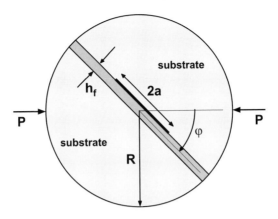

Fig. 4.34. The Brazilian disk interface fracture specimen.

an approach obviates the need for gripping the test specimen to induce tensile stress.
A technique that exploits this feature involves the so-called *Brazilian disk specimen*
configuration which has long been used to evaluate the mixed-mode fracture properties of brittle solids such as ceramic materials (Atkinson et al. 1982, Shetty et al.
1987, Wang and Suo 1990, Huang et al. 1996). The basic features of this method,
adapted for interface fracture studies, are shown schematically in Figure 4.34. In
this configuration, a thin film is processed as a layer that is sandwiched between
two much thicker 'substrate' materials, with the specimen prepared in the shape of
a circular disk of radius R and uniform thickness. An initial precrack or debond of
length $2a$ is introduced artificially along the interface of interest as, for example,
by depositing a thin layer of carbon between the film and the substrate with the
specific purpose of preventing formation of a bond at that interface. A compressive
load P per unit thickness is applied with its line of action inclined at an angle φ
to the interface containing the debond crack, as shown in Figure 4.34. Pure mode I
tensile fracture conditions arise locally at the debond in the homogeneous sample
with a central crack when the loading axis coincides with the crack line, or with
$\varphi = 0$. If the film layer is very thin, then the remotely applied load on the edge of
the delamination zone is pure mode I. Locally, the phase angle is not zero at the
edge of the delamination when $\varphi = 0$, in general. A particularly appealing feature
of this test specimen is that, by changing φ, the local stress state phase angle ψ
can be systematically altered over the entire range $0 \leq \psi \leq 90°$. For a homogeneous material, the crack edge opening and shear stress intensity factors, K_I and
K_{II}, respectively, for the Brazilian disk specimen are of the form

$$K_I = \frac{P p_I}{R} \sqrt{\frac{a}{\pi}}, \qquad K_{II} = \frac{P p_{II}}{R} \sqrt{\frac{a}{\pi}}, \qquad (4.55)$$

where p_I and p_{II} are nondimensional functions of the relative crack size a/R and
the load angle φ. Analyses of stress intensity factors for Brazilian disk specimens

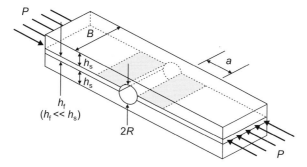

Fig. 4.35. Schematic diagram of the double cleavage drilled compression specimen configuration. The shaded area denotes the interface debond.

of bimaterials were described by O'Dowd et al. (1992). This specimen, like the double cantilever beam and four-point flexure test geometries, is also well suited for assessing the fracture properties of interfaces in materials that can be joined by techniques such as diffusion bonding, welding, explosion cladding or brazing.

The double cleavage drilled compression or DCDC test, which is drawn schematically in Figure 4.35, is another compression test configuration for interface fracture toughness evaluation. This method, first developed by Jansson (1974), typically comprises a rectangular block of a symmetrically layered specimen at the center of which a thin layer of a metal or adhesive is located. One or both of the outerlayers is usually made of a transparent material such as a sapphire single crystal or glass so that the crack can be imaged directly with the aid of a microscope during the interface fracture experiment. The DCDC specimens are usually prepared by diffusion bonding of a thin metal film, such as aluminum, gold or platinum, to sapphire substrates with a {0001} crystallographic orientation. A circular hole is drilled at the center of the rectangular specimen, as shown in Figure 4.35, so as to create a stress concentration from which fracture is nucleated under controlled conditions. Compressive loading parallel to the interfaces in the layered system causes tensile stresses to develop at the poles of the hole where cracks nucleate. These cracks then advance along an interface.

For the particular case where the entire rectangular beam of the DCDC specimen is made of a homogeneous brittle solid with elastic modulus E_s and Poisson ratio ν_s, the energy release rate for fracture is estimated to be

$$\mathcal{G} = \frac{\pi \sigma^2 R}{\bar{E}_s} \left[\frac{h_s}{R} + \left(0.235 \frac{h_s}{R} - 0.259 \right) \frac{a}{R} \right]^{-2} \qquad (4.56)$$

for values of the dimensions within the ranges $h_s/R \leq a/R \leq 15$ and $2 \leq h_s/R \leq 4$; this approximation was obtained as a reasonable fit to the results of a finite element calculation of specimen deformation (Michalske et al. 1993, He et al. 1995). In this expression, h_s is half the thickness of the specimen, R is the radius of the circular hole, a is the length of the symmetric debond emanating from either end of the hole, and $\sigma = P/2h_s$ is the nominal compressive stress due to the applied load P. Since the energy release rate decreases with crack growth at a fixed applied load, crack growth is inherently stable in this test geometry.

If the DCDC specimen is fabricated with the configuration shown in Figure 4.35 but with the thin film sandwiched between two symmetrically configured layers of different materials, then the energy release rate for the interface crack is more complicated than that given in (4.56) because the stress field is no longer symmetric with respect to the crack plane giving rise to mixed-mode applied loading. The local stress intensity factors for the interface crack in the DCDC specimen, as well as the phase angle ψ of the delamination zone edge, can be determined as functions of the Dundurs parameters D_1 and D_2 following the results presented by He et al. (1995) and Suo and Hutchinson (1989).

4.5.4 The superlayer test configuration

As is evident from the foregoing discussion, the application of external loading to drive delamination at the interface between a film and its substrate is a challenging experimental task. In those cases in which the energy stored in the bonded film is not sufficient to drive delamination spontaneously, some additional driving force is essential. Several experimental configurations that rely on externally applied loading have been discussed in the foregoing sections. An approach that provides an alternative to externally applied loading is the use of a strained superlayer. The basic idea is to deposit a second film – the superlayer – on top of a film that is bonded to a relatively thick substrate under conditions in which the elastic energy stored in the superlayer is sufficient to drive delamination of the interface between the film and the substrate. The resulting configuration is called the superlayer test configuration; it is depicted in Figure 4.36.

For example, for metallic films less than 1 μm in thickness and with processing induced residual stress smaller than about 100 MPa, the driving force $\mathcal{G}$ for spontaneous delamination is less than about 0.1 J m^{-2}. However, most interfaces of practical significance require a work of separation Γ that is substantially greater in magnitude than this available driving force. Therefore, it is necessary to elevate the driving force for delamination from that available due to processing alone, but to do so without altering the properties of the film or of the interface being examined. This can be accomplished by means of the superlayer configuration, provided that several conditions are met: (i) the process of depositing the superlayer on the film must not change the microstructure of the film or the properties of the film–substrate interface to be interrogated, (ii) the superlayer film must develop a relatively large

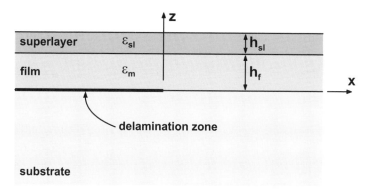

Fig. 4.36. Schematic diagram of a bilayer film on the surface of a relatively thick substrate. The second film can contribute to the driving force for delamination of the first film from the substrate and/or alter the stress state phase angle at the edge of the delamination zone.

residual tensile stress upon deposition in order to drive delamination, and (iii) the interface between the superlayer and the film must remain intact as delamination occurs between the film and the substrate.

Consider the superlayer configuration depicted in Figure 4.36. The equi-biaxial mismatch strain in the film of thickness h_f is ϵ_m and in the superlayer of thickness h_{sl} is ϵ_{sl}. The stored elastic energy per unit area of film–substrate interface is $M_f\epsilon_m^2 + M_{sl}\epsilon_{sl}^2$ prior to delamination. Far behind an advancing straight delamination front, the bilayer consisting of the film and superlayer, now free from the substrate, becomes curved. The amount of curvature depends on the properties of the materials, the magnitudes of the mismatch strains and the thicknesses of the films, and it is determined by the conditions of zero net force and zero bending moment in the detached bilayer. If the delamination growth is assumed to occur under conditions of plane strain deformation, general expressions for the driving force G have been provided by Hutchinson and Suo (1992) and Evans and Hutchinson (1995). For the case when $M_f = M_{sl}$ and $\epsilon_m = 0$, the driving force reduces to

$$G = \tfrac{1}{2}\bar{E}\epsilon_{sl}^2 h_{sl}\left[\frac{h_{sl}^3 c_{30} + h_{sl}^2 h_f c_{21} + h_{sl}h_f^2 c_{12} + h_f^3 c_{03}}{(h_f + h_{sl})^3}\right] \qquad (4.57)$$

where $c_{30} = (1+v)^2$, $c_{21} = 3 + 6v + 2v^2$, $c_{12} = 3 + 6v + 4v^2$ and $c_{03} = 1 + 2v$ in terms of the common value of the Poisson ratio.

Bagchi et al. (1994) have used the superlayer test to determine the separation energies of Cu thin films on silica substrates. In their approach, a thin strip of carbon is first deposited onto the substrate so as to create a debond pre-crack over an area along the interface. The desired thickness h_f of Cu film is then deposited onto the substrate over a region that includes the prepared debond zone. Subsequently, a Cr superlayer of thickness h_{sl} is deposited on top of the first film by electron beam evaporation. When the Cr layer thickness exceeds a certain critical value, spontaneous delamination ensues. Upon delamination, the bilayer film develops a curvature κ which, for the case when $M_f = M_{sl}$ and $\epsilon_m = 0$, depends on the system parameters according to

$$\kappa = \frac{6h_f h_{sl} \nu \epsilon_{sl}}{(h_f + h_{sl})^3}. \tag{4.58}$$

Knowledge of this curvature, along with the values of other system parameters, can be used to extract the interface separation energy Γ using the methods outlined in the earlier sections of this chapter. The superlayer configuration was also used by Zhuk et al. (1998) to study the strength of an interface between a metal film and a polymeric substrate.

An interesting aspect of the bilayer film configuration is the potential for influencing the phase angle ψ of the stress state at the edge of the delamination zone without otherwise altering the interface. Few results of a general nature are available for this configuration. However, the dependence of phase angle on the ratio h_f/h_{sl} for the case when both film layers and the substrate have identical elastic properties and when $\epsilon_m = 0$ was studied by Suo and Hutchinson (1989). Through numerical analysis of the underlying elasticity problem, it was concluded that the phase angle ψ would be zero for the thickness ratio $h_f/h_{sl} = 2.87$.

A rough estimate of the thickness ratio that results in $\psi = 0$ can be obtained on the basis of the following reasoning. For the case considered by Suo and Hutchinson (1989), the elastic mismatch strain across the interface prior to delamination is zero, and the substrate is unstrained. Therefore, the conditions that the extensional strain of the film surface $z = 0^+$ after delamination must still be zero may correspond to a relatively small value of shear stress on the interface at the delamination front; a non-singular value of shear stress $\sigma_{xz} = 0$ corresponds to the condition that $\psi = 0$, provided that the normal stress σ_{zz} is still singular and positive. The methods of Section 2.2 can be used to show that this condition on strain implies that h_f/h_{sl} must have the value 2. While the estimate is crude, it may provide a useful indicator of behavior for arbitrary combinations of material parameters and film thicknesses in cases for which other indicators are not available.

4.6 Film cracking due to residual stress

If a thin film is bonded to a substrate, and if that film is subject to a residual tensile stress as a result of elastic mismatch with the substrate, then the stress can be partially relaxed by formation of cracks in the film. In this section, the behavior of through-the-thickness cracks within the film is considered. First, the behavior of an isolated, fully formed crack is examined and, subsequently, the formation of an array of cracks is considered.

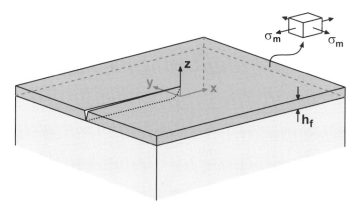

Fig. 4.37. Steady advance of a crack in the x-direction through a thin film. Crack growth is driven by the residual biaxial tensile stress σ_{m} existing prior to cracking.

4.6.1 A surface crack in a film

The phenomenon considered in this section is depicted in Figure 4.37, which shows a through-the-thickness crack advancing in a film bonded to a relatively thick substrate. Initially, the film is subjected to an equi-biaxial tensile stress of magnitude σ_{m}, and this stress is partially relieved by the process of cracking. Indeed, it is the elastic energy stored in the film due to this stress that provides the driving force for the cracking process. The crack is presumed to have been nucleated at a flaw in the material or at a surface damage site; the nucleation process is complex and some of its features are discussed later in this section. For the time being, it is assumed that the crack has expanded to a length in the x-direction which is much greater than h_{f}. The plane of the crack is the xz-plane, and the crack faces with outward normal vectors in the $\pm y$-direction are free of traction. Under these conditions, the ends of the crack no longer interact to any significant degree, and the process of growth of each end can be viewed as a steady-state process as seen by an observer traveling along the film surface with the point at which the crack edge intersects the film surface. The point of view is that even if crack nucleation sites are present in the material, extensive cracking can occur only if cracks which are formed can actually grow into film cracks which are long compared to film thickness. In other words, it is presumed that behavior is dominated by crack growth rather than nucleation.

The cracking process is considered in the context of the Griffith fracture theory. For the time being, the materials are assumed to be isotropic and elastic, and the film and substrate are assumed to have the same elastic properties, that is, $E_{\mathrm{s}} = E_{\mathrm{f}}$ and $\nu_{\mathrm{s}} = \nu_{\mathrm{f}}$. Relaxation of this restriction is considered subsequently. In the configuration shown in Figure 4.37, crack opening is symmetric so the phase angle ψ of the stress state at the crack edge is zero. The energies of separation of the homogeneous film and substrate materials are denoted by Γ_f and Γ_s, respectively.

The basic idea is that the crack will advance if the elastic energy released from the system in the course of that growth equals (or exceeds) the energy of separation of the fracture surface created in the process. The fact that the process is steady-state makes it possible to compute these energy changes at points far ahead and far behind the curved fracture front compared to h_f where the deformation fields are essentially two-dimensional plane strain fields. Tacit in this argument is the assumption that the leading edge of the crack, along with its complicated three-dimensional deformation field, will advance in a self-similar way. This is a reasonable expectation because the environment of the moving crack edge is unchanged as it advances. The process of crack extension is then completely equivalent to conversion of the state of the material far ahead of the advancing crack to the state far behind, a conversion which can be accomplished under two-dimensional plane strain conditions. The steady-state shape of the moving crack edge was determined for the case of a relatively stiff substrate by means of numerical methods by Nakamura and Kamath (1992).

Suppose that the final or steady-state depth of penetration of the crack is a_c which may be smaller than, equal to, or larger than the film thickness h_f. The amount of work per unit crack advance in the x-direction which must be supplied to overcome the resistance of the material is

$$W_c = h_f \Gamma_f \begin{cases} a_c/h_f, & a_c \leq h_f \\ \\ 1 + (a_c/h_f - 1)\,\Gamma_s/\Gamma_f, & a_c > h_f. \end{cases} \tag{4.59}$$

For the fracture to proceed spontaneously, this amount of work must be drawn from the background elastic field.

The calculation of the elastic energy released from the background field under two-dimensional plane strain conditions, say W_m, in terms of the system parameters proceeds in the following way. Suppose a crack moves into an elastic half space in a direction normal to the traction-free surface of the half space. The state of the crack is considered for all depths η between zero and a_c. In the process of growing inward from the surface, the crack negates a normal traction of magnitude σ_m on the crack plane for $0 < \eta < h_f$ but no additional applied traction is relieved for $h_f < \eta$. Suppose that the energy release rate for advance of the crack into the material, say $\mathcal{G}(\eta)$, is known for all η in the range $0 < \eta < a_c$. Then the energy released from the background elastic field per unit distance advance of the steady-state crack in the x-direction in Figure 4.37 is

$$W_m = \int_0^{a_c} \mathcal{G}(\eta)\,d\eta. \tag{4.60}$$

Thus, if the energy release rate $\mathcal{G}(\eta)$ is known for the two-dimensional plane strain

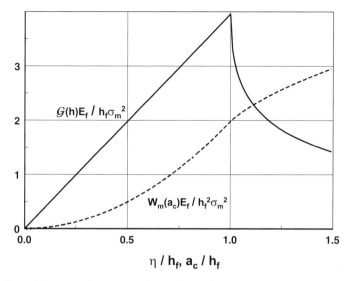

Fig. 4.38. The solid curve shows the driving force $\mathcal{G}$ for insertion of a crack in the thin film as a function of crack depth η. The dashed curve shows the corresponding configurational force W_m defined by (4.60) tending to extend the crack depicted in Figure 4.37 steadily in the x-direction.

state far behind the moving crack edge then W_m can be calculated from it by means of (4.60).

From elastic fracture mechanics (Tada et al. 1985), it is known that $\mathcal{G}(\eta)$ is given approximately by

$$
\mathcal{G}(\eta) = \frac{h_f \sigma_m^2}{\bar{E}_f}
\begin{cases}
\pi c_e^2 \dfrac{\eta}{h_f}, & 0 < \eta \le h_f \\[2ex]
\dfrac{4}{\pi} \dfrac{\eta}{h_f} \left(1.69 - 0.47 \dfrac{h_f}{\eta} + 0.032 \dfrac{h_f^2}{\eta^2} \right) \arcsin^2 \dfrac{h_f}{\eta}, & h_f < \eta
\end{cases}
$$

(4.61)

to within 1% of the results of a detailed numerical solution. The number $c_e = 1.1215$ is a well-established numerical coefficient that arises in describing the crack tip singularity of an edge crack in elastic fracture mechanics (Tada et al. 1985). This function $\mathcal{G}(\eta)$ is plotted in Figure 4.38. The energy release rate increases linearly from zero, as it must on dimensional grounds alone, as the crack depth increases from $\eta = 0$ toward $\eta = h_f$. As the crack depth increases beyond $\eta = h_f$, the crack is forming in a region of the body that is not stressed due to mismatch. Consequently, the energy release rate diminishes with further increase in crack depth.

For the case when $\eta \gg h_f$, the net forces of magnitude $\sigma_m h_f$ which tend to open the crack are essentially concentrated forces acting at the corners of right angle wedges. The energy release rate has been shown to be

$$\mathcal{G}(\eta) \approx \frac{h_f \sigma_m^2}{\bar{E}_f} \frac{4\pi}{\pi^2 - 4} \frac{h_f}{\eta} \tag{4.62}$$

for $\eta \gg h_f$ by Freund (1978), independently of the approximate result given in (4.61), which provides another check on the quality of the approximation.

The result of calculating W_m according to (4.60) is also shown in Figure 4.38 as the dashed curve. The general features of the dependence of W_m on a_c are readily anticipated from the properties of $\mathcal{G}(\eta)$ due to the relationship $W_m'(a_c) = \mathcal{G}(a_c)$. The value of W_m increases parabolically with crack depth as the depth increases from $a_c = 0$ to $a_c = h_f$. Thereafter, the value of W_m continues to increase with a_c beyond $a_c = h_f$ but at a diminishing rate.

Whether or not a through-the-thickness crack will form in the film is determined by a comparison of the energy supply, represented by W_m, to the energy required for fracture, represented by W_c. According to the Griffith condition, a crack cannot grow if $W_m < W_c$ for all admissible values of a_c. If cracking is confined to the film, then the most easily achieved crack depth is $a_c = h_f$. To see this, it is only necessary to observe that W_m increases quadratically with a_c/h_f while W_c increases linearly. Thus, if $W_m = W_c$ for some $a_c < h_f$, then $W_m > W_c$ for any crack depth η in the range $a_c < \eta < h_f$. In other words, there is always a surplus of driving force which will tend to grow the crack deeper. The crack depth in the film at which the condition $W_m = W_c$ is *first* satisfied, without a surplus of driving force, is $a_c = h_f$. The critical thickness h_f for film cracking at the given level of mismatch σ_m which is implied by $W_m = W_c$ is then

$$(h_f)_{cr} = \frac{2}{\pi c_e^2} \frac{\Gamma_f \bar{E}_f}{\sigma_m^2} . \tag{4.63}$$

It is important to recall that this result applies only when the elastic moduli of the film and the substrate are the same.

Suppose that conditions are favorable for formation of a crack in the film. Can the crack grow to depths greater than h_f, that is, can it penetrate into the substrate? To pursue the answer to this question, consider the schematic diagram of W_m and W_c versus a_c/h_f in Figure 4.39. The graphs have been drawn so that $W_m = W_c$ with $a_c/h_f = 1$; in other words, conditions are such that $h_f = (h_f)_{cr}$. The change in slope of W_c versus a_c/h_f represents the change in fracture resistance Γ_s of the substrate material with respect to the resistance Γ_f in the film. The change in slope of W_m versus a_c/h_f, on the other hand, reflects the fact that the crack surface traction which is relieved by crack growth extends through the film but not into the substrate. With

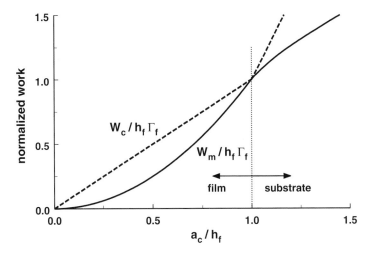

Fig. 4.39. The solid curve shows the driving force W_m tending to extend a film crack versus the depth of penetration of that crack. The dashed curve shows the material resistance to extension as the function of depth, drawn in this case for $\Gamma_s > 2\Gamma_f$.

this understanding, the answer to the question posed above is clear; it hinges on the relative slopes of the two graphs at $a_c/h_f \to 1^+$. The crack *will not* tend to penetrate into the substrate if $dW_m/da_c \leq dW_c/da_c$ there. Alternatively, this condition can be stated as $\mathcal{G}(h_f) \leq \Gamma_s$. Using the explicit expressions in (4.59) and (4.61), the condition can be expressed in terms of system parameters as

$$\frac{\pi c_e^2 h_f \sigma_m^2}{\bar{E}_f} \leq \Gamma_s. \tag{4.64}$$

But, from (4.63), the left side of the inequality is $2\Gamma_f$ so it becomes simply

$$2\Gamma_f \leq \Gamma_s. \tag{4.65}$$

Thus, the crack will not tend to penetrate into the substrate as long as the fracture resistance of the substrate material is at least twice as large as the resistance of the film material. This condition is predicated on the assumption that h_f is just equal to the critical value given in (4.63). If, in fact, $h_f > (h_f)_{cr}$ then crack penetration into the substrate is more likely. The details can be pursued by scaling the solid curve in Figure 4.39 upward but leaving the dashed curve unchanged. The depth of penetration will be defined by the rightmost intersection point of the two curves.

There are other inferences to be drawn from the sketch in Figure 4.39. For example, if (4.65) is not satisfied, that is, if the fracture resistance of the substrate is less than twice the fracture resistance of the film, then the crack can penetrate into the substrate even though the condition (4.63) may not be satisfied. With reference to Figure 4.39, for example, this will be the case if $W_m < W_c$ at $a_c/h_f = 1$ but $W_m = W_c$ for some $a_c/h_f > 1$ where the two curves are tangent to each other. This depth is a stable position of the crack and it arises at the smallest value of h_f for fixed σ_m at which crack growth can occur in this case.

When the elastic properties of the film and the substrate materials differ significantly, the behavior can be altered from that for the case of equal properties. The qualitative influence of modulus difference can be anticipated from simple cases with extreme modulus differences, all other things being equal. For example, suppose that a crack approaches the film–substrate interface from within the film and that the substrate stiffness is much greater than that of the film. Under such conditions, the substrate constrains the film from giving up its stored elastic energy to drive crack growth more than it does in the case of equal stiffnesses. Thus, crack formation in the film is more difficult in the case of a relatively stiff substrate. Similarly, if the substrate material is relatively compliant compared to the film material, then it permits relaxation of residual stress over a relatively large volume of film material. This implies that more stored energy can be extracted to drive the crack than is available in the case of equal stiffnesses, so crack formation in the film is less difficult in the case of a relatively compliant substrate than in the case of equal stiffnesses.

A quantitative basis for these ideas has been provided by Beuth (1992) for the case of steady-state advance of a surface crack in an isotropic elastic film bonded to an isotropic elastic substrate. Under circumstances of steady-state crack propagation, the conditions on system parameters that are necessary for growth can be expressed in terms of the states of plane strain deformation which exist far ahead of and far behind the advancing crack segment; see Figure 4.37. The four material parameters ν_f, E_f, ν_s, E_s enter these conditions only through two dimensionless combinations, the Dundurs parameters D_1 and D_2 which were defined in (3.98) and reintroduced in (4.29). To illustrate the effect of modulus difference between the film and substrate, the two cases with $D_1 = -0.6$ and 0.6 are compared to the equal modulus case of $D_1 = 0$, all with $D_2 = D_1/4$. These cases correspond to $\bar{E}_s = 4\bar{E}_f$ and $\bar{E}_s = \bar{E}_f/4$, both with $\nu_f = \nu_s = \frac{1}{2}$, for example.

Attention is again focused on the plane strain deformation field far behind the steadily advancing portion of the crack edge. The dependence of energy release rate $\mathcal{G}(\eta)$ on the crack depth η from the free surface is illustrated in Figure 4.38 for $D_1 = D_2 = 0$. The deviation from this behavior for $D_1 \neq 0$ can be anticipated. For $\eta/h_f \ll 1$, the effect of the substrate is minimal, so the initial slope will be unchanged for any value of D_1. As η increases across the thickness of the film, a compliant (stiff) substrate will enhance (retard) relaxation of stored energy in the film, so that $\mathcal{G}(\eta)$ will increase more (less) rapidly with η than for the case

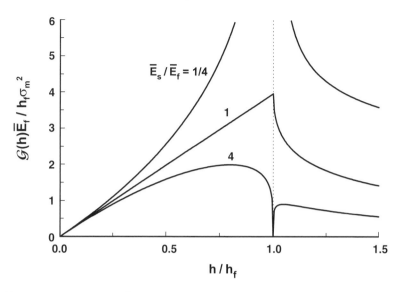

Fig. 4.40. The driving force $\mathcal{G}$ for insertion of a crack in a film–substrate system under plane strain conditions for $D_2 = 0$ and $\bar{E}_s/\bar{E}_f = 1/4$, 1 and 4. The middle curve reproduces the result shown by a solid curve in Figure 4.38. Numerical data provided by Vijay Shenoy, Indian Institute of Science, Bangalore (1999).

when $D_1 = 0$. The behavior for the particular cases when $D_1 = 0.6$ ($\bar{E}_s = \bar{E}_f/4$) and $D_1 = -0.6$ ($\bar{E}_s = 4\bar{E}_f$) are shown in Figure 4.40, as obtained by numerical simulation. The energy release rate is unbounded as $\eta \to h_f$ in the former case and it is zero as $\eta \to h_f$ in the latter case. For either $\eta > h_f$ or $\eta < h_f$, the stress field is square root singular at the crack edge for any combination of material parameters. When $\eta = h_f$, however, the singularity in the stress field at the crack edge differs from being square root singular for dissimilar materials (Zak and Williams 1963), being stronger or weaker in a way consistent with the results of Figure 4.40.

The strong influence of the interface persists as the crack edge emerges into the substrate, as is evident in Figure 4.40. As the crack edge moves further into the substrate, the energy release rate has the asymptotic behavior given in (4.62). This behavior is independent of the properties of the film. Thus, the large differences observed in Figure 4.40 for $\eta/h_f > 1$ are due primarily to the fact that energy release rate $\mathcal{G}(\eta)$ is scaled by the film modulus $\bar{E}_f$ in the normalization rather than any physical effect.

The dependence of recovered elastic energy W_m per unit length as the propagating segment of the crack advances along the film is indicated in Figure 4.41 for the same cases as were illustrated in Figure 4.40. While quantitative conclusions of various kinds can be drawn for any particular system from this type of analysis,

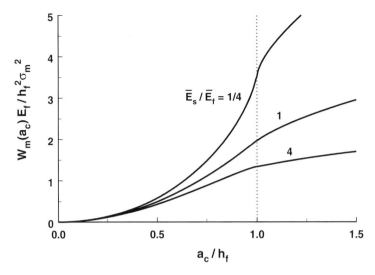

Fig. 4.41. The driving force W_m for extension of a crack along the film, as defined by (4.60), calculated from the results in Figure 4.40 versus final crack depth a_c for the case when the film and substrate have different elastic constants.

only some general qualitative observations are included here. It is evident from Figure 4.41 that a decrease in substrate stiffness from the level of the film stiffness (or, more properly, an increase in D_1 from $D_1 = 0$) will tend to enhance growth of a surface crack. Furthermore, if such a crack grows, it will likely penetrate some distance into the substrate. On the other hand, an increase in the stiffness of the substrate with respect to the film will tend to make crack growth more difficult at a given level of mismatch, and it will make penetration of the crack through the full thickness of the film or into the substrate less likely than for the case of equal properties.

4.6.2 A tunnel crack in a buried layer

Suppose that a film of thickness $2h_f$ is buried between two relatively thick layers of material, which will again be called the substrate material, and that the film carries a residual equi-biaxial mismatch tensile stress of magnitude σ_m with respect to the substrate material. A *tunnel crack* is a crack that forms on a plane which is perpendicular to the plane interfaces between the film and its confining layers and which is large in extent in a direction parallel to the interface compared to h_f; see Figure 4.42. The crack edges, which are separated by the distance $2a_c$ and which are parallel to the interfaces, are stationary far behind the advancing edge. The crack grows by advance of the portion of crack edge which threads across the thickness of the film. If this portion is far from the surface of the solid (for a crack growing in

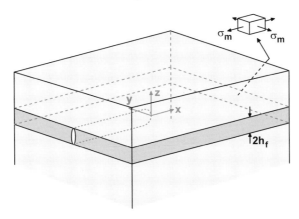

Fig. 4.42. Schematic diagram of steady advance of a tunnel crack in a strained layer embedded between two relatively thick substrates. Crack growth is driven by residual stress in the buried layer.

from an edge) or is far from the other advancing segment at the remote end (for an interior crack) then the advancing part of the crack responds to a spatially uniform environment and the process is steady-state as viewed by an observer traveling with that part. The analysis of the process closely follows that developed for an isolated crack in a surface film which was outlined in Section 4.6.1. For the time being, the two materials are assumed to have identical elastic properties.

The steady-state width of the tunnel crack is $2a_c$ which may be less than, equal to or greater than $2h_f$. The amount of work per unit crack advance in the x-direction which must be supplied to overcome material resistance is

$$W_c = 2h_f \Gamma_f \begin{cases} a_c/h_f, & a_c \leq h_f \\ 1 + (a_c/h_f - 1)\Gamma_s/\Gamma_f, & a_c > h_f. \end{cases} \tag{4.66}$$

For the fracture to proceed, this amount of work must be drawn from the background elastic field.

The energy release rate $\mathcal{G}(\eta)$ at total crack width 2η for each crack edge far behind the advancing portion of the tunnel crack, measured per unit length along the edge, is

$$\mathcal{G}(\eta) = \frac{\pi h_f \sigma_m^2}{\bar{E}_f} \begin{cases} \dfrac{|\eta|}{h_f}, & 0 < |\eta| \leq h_f \\ \dfrac{2}{\pi}\sqrt{\dfrac{|\eta|}{h_f}}\,\arcsin^2\dfrac{h_f}{|\eta|}, & h_f < |\eta|. \end{cases} \tag{4.67}$$

It follows immediately that W_m, the total elastic energy released per unit advance of the propagating portion of the crack, is

$$W_{\mathrm{m}} \;=\; 2 \int_0^{a_{\mathrm{c}}} \mathcal{G}(\eta)\, d\eta$$

(4.68)

$$= \; \frac{8\sigma_{\mathrm{m}}^2 h_{\mathrm{f}}^2}{\pi \bar{E}_{\mathrm{f}}} \left\{ \begin{array}{ll} \dfrac{\pi^2 a_{\mathrm{c}}^2}{8 h_{\mathrm{f}}^2}\,, & a_{\mathrm{c}} \le h_{\mathrm{f}} \\[2ex] \left[\dfrac{a_{\mathrm{c}}^2}{h_{\mathrm{f}}^2} - 1\right]^{1/2} \arcsin\dfrac{h_{\mathrm{f}}}{a_{\mathrm{c}}} + \dfrac{a_{\mathrm{c}}^2}{2 h_{\mathrm{f}}^2}\arcsin^2\dfrac{h_{\mathrm{f}}}{a_{\mathrm{c}}} + \ln\dfrac{a_{\mathrm{c}}}{h_{\mathrm{f}}}\,, & a_{\mathrm{c}} > h_{\mathrm{f}}. \end{array} \right.$$

Note that this result applies only when the elastic properties of the film and the substrate are the same. The variation of W_{m} and W_{c} with crack depth a_{c} resembles that shown in Figure 4.39 for the case of a surface crack. A necessary condition for a tunnel crack to form when $\Gamma_{\mathrm{s}} \ge 2\Gamma_{\mathrm{f}}$ at mismatch stress σ_{m} is

$$(h_{\mathrm{f}})_{\mathrm{cr}} = \frac{2}{\pi}\frac{\Gamma_{\mathrm{f}}\bar{E}_{\mathrm{f}}}{\sigma_{\mathrm{m}}^2}$$

(4.69)

and this crack will form across the full thickness of the layer, that is, with $a_{\mathrm{c}} = h_{\mathrm{f}}$.

4.6.3 An array of cracks

The formation of a through-the-thickness crack in a film subjected to a residual or applied tensile stress relieves that stress in the film material at points adjacent to the crack path. At points in the film at some distance from the crack path, the stress remains unrelaxed due to the constraint of the substrate. Consequently, a long crack that is parallel to the first formed crack can also form. Indeed, an array of parallel cracks over the entire film surface is likely, and the point of the discussion in this section is to provide an estimate of the dependence of the spacing between cracks in such an array on the film thickness h_{f} and the mismatch stress σ_{m}. The discussion is limited to the case when the equi-biaxial mismatch stress is uniform throughout the film, the elastic properties of the film and substrate are nominally the same, and $h_{\mathrm{s}}/h_{\mathrm{f}}$ is sufficiently large so that the behavior is insensitive to the substrate thickness h_{s}. Furthermore, it is assumed that the cracks grow through the thickness of the film to the depth h_{f}, but that they do not penetrate into the substrate. There is no fundamental barrier to relaxation of these limitations, but the relatively simple system is sufficiently rich in physical detail to reveal the principal features of behavior.

The simplest conception of the phenomenon of crack array formation is depicted in Figure 4.43. This system has all the features of the crack formation process illustrated in Figure 4.37, except that a periodic array of cracks is formed instead of only a single isolated crack. The crack-to-crack spacing is λ, and the array extends

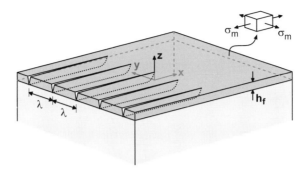

Fig. 4.43. A periodic array of cracks advancing steadily through a thin film, driven by residual stress in the film.

indefinitely in the positive and negative y-directions. A particular feature of this situation is that the propagating segment of each crack is identical to that of every other crack and that these segments advance in unison in the x-direction. This feature will be reconsidered below.

The process of advance of the propagating segments of the cracks can be understood in terms of an energy balance. Some energy is absorbed in the process of forming the fractures as they grow and, at the same time, internal stress is relieved in the film which reduces the stored elastic energy as the cracks grow. For steady advance of the propagating segments of the cracks, the energy change per unit crack advance is completely equivalent to the energy change due to removal of a slice of material of unit thickness in the x-direction from far ahead of the advancing crack segments, introduction of an edge crack array in this slice under plane strain conditions, and insertion of this crack slice back into the system at some section far behind the advancing crack segments. This reasoning makes it clear that the energy balance can be enforced through a two-dimensional plane strain deformation process.

The underlying two-dimensional configuration is evident in Figure 4.43 on a section for some fixed negative value of x for which $|x/h_f| \gg 1$. The energy per period of length λ that is consumed as each propagating crack segment advances a unit distance is $W_c = h_f \Gamma_f$ where Γ_f is the specific fracture energy of the film material. The initial elastic energy per period per unit crack advance is $\sigma_m^2 h_f \lambda / M_f$. This stored energy is partially relieved to drive the crack growth process. The energy released per period per unit crack advance necessarily has the form

$$W_m = \int_0^{h_f} \mathcal{G}(\eta) \, d\eta = f(\lambda/h_f) \sigma_m^2 h_f^2 / \bar{E}_f \qquad (4.70)$$

on dimensional grounds alone, where $\mathcal{G}(\eta)$ is the energy release rate for each crack advancing through the film under plane strain conditions. The nondimensional factor $f(\lambda/h_f)$ must be determined by means of stress analysis. It can be expected to be

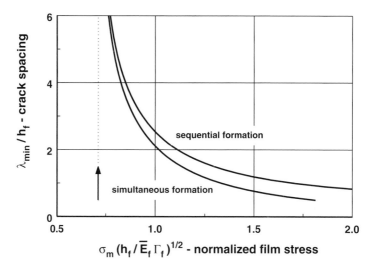

Fig. 4.44. The minimum spacing possible for an array of cracks formed simultaneously, as in Figure 4.43, or sequentially, as in Figure 4.45, versus residual stress in the film. The arrow identifies the stress at which cracking first becomes possible as defined by the condition (4.63).

an increasing function of its argument and to have the asymptotic behavior that $f(\lambda/h_f) \to 0.63\pi$ as $\lambda/h_f \to \infty$, consistent with (4.61).

The energy release rate per crack for growth of a periodic array of edge cracks normal to the free edge of a solid under plane strain conditions is given in graphical form by Tada et al. (1985). This result can be fit to any desired degree of accuracy by use of elementary functions, and then integrated to obtain W_m according to (4.70), which yields the total energy released by each crack in the array as it grows from the free surface of the film to the film–substrate interface at depth $a_c = h_f$. In particular, the function $f(\lambda/h_f)$ can be determined. An approximate form of this function that provides a compromise between accuracy and simplicity is

$$f(\lambda/h_f) \approx 0.63\pi \left(1 - e^{-\lambda/3h_f}\right) . \qquad (4.71)$$

Once this function is known, the condition for minimum crack spacing $\lambda_{\min}$ is determined by $W_m = W_c$. If the spacing falls below $\lambda_{\min}$, then $W_m < W_c$ or there is insufficient energy available to drive each of the fractures without external influence. Thus, $\lambda_{\min}$ must satisfy the condition

$$f(\lambda_{\min}/h_f) = \frac{\bar{E}_f \Gamma_f}{\sigma_m^2 h_f} . \qquad (4.72)$$

When the approximation (4.71) is used, this equation can be inverted to yield the dependence of $\lambda_{\min}/h_f$ on the normalized film characteristic stress $\sigma_m \sqrt{h_f/\bar{E}_f \Gamma_f}$, and the result is shown in Figure 4.44 where it is labeled as simultaneous formation.

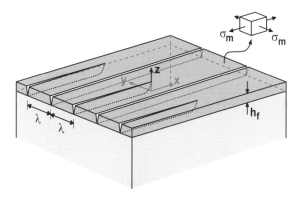

Fig. 4.45. A periodic array of film cracks formed sequentially. First, an array with spacing 2λ is formed. Then, a second array, with its members midway between adjacent cracks of the first formed set, is inserted.

As might be expected, the dependence has a vertical asymptote at $\sigma_{\mathrm{m}}\sqrt{h_{\mathrm{f}}/\bar{E}_{\mathrm{f}}\Gamma_{\mathrm{f}}} =$ 0.71 which is the value corresponding to the isolated crack limit obtained when $\lambda_{\min}/h_{\mathrm{f}} \to \infty$ in (4.63). As $\sigma_{\mathrm{m}}\sqrt{h_{\mathrm{f}}/\bar{E}_{\mathrm{f}}\Gamma_{\mathrm{f}}}$ is increased above the threshold value, the minimum spacing decreases monotonically.

Among the various simplifications that were incorporated in the foregoing discussion of the process of formation of a distributed array of surface cracks, perhaps the most extreme is that the propagating segments of these cracks advance in unison along the film. It is likely that cracks which are nucleated from the most severe defects form first and grow to lengths which are large compared to h_{f}. As the mismatch is increased through continued temperature change or increased external load, cracks form from the less severe defects. Whether or not these cracks can then grow into additional long surface cracks depends on their proximity to previously formed cracks.

To consider this sequential nature of crack formation, the condition that the propagating segments of the forming cracks advance in unison is relaxed. Instead, it is assumed that a periodic array of cracks first forms in the film at spacing 2λ; the mechanics of formation of this initial array is not an issue in this discussion. After these cracks are fully formed and their propagating segments have advanced into remote regions, it is assumed that a second set of cracks is formed, with each member of this set being located midway between two cracks of the earlier formed set. The propagating segments of the second set are again assumed to advance more or less in unison, as illustrated in Figure 4.45. It is the mechanics of formation of this second set that is of interest here. The final state of the system is again a periodic array of surface cracks with spacing λ, but the second set of cracks must form under conditions somewhat different from those prevailing when the first set was formed.

Nonetheless, the energetics of formation can again be expressed in terms of the function $f(\cdot)$ which was introduced earlier in this section. The main result follows from comparisons of the plane strain deformation states which exist (i) prior to formation of any cracks, (ii) following formation of the first set at spacing 2λ but prior to formation of the second set, and (iii) following formation of the second set of cracks to produce a periodic array at spacing λ.

The elastic energy density per unit area of interface stored in the film prior to formation of any cracks is $U_0 = \sigma_m^2 h_f / M_f$, as before. Then, following formation of the first set of cracks but prior to formation of the second set, the energy density per unit area of interface is reduced to

$$U_1 = \frac{\sigma_m^2 h_f}{M_f} - f(2\lambda/h_f)\frac{\sigma_m^2 h_f^2}{2\bar{E}_f \lambda} . \tag{4.73}$$

Likewise, if both sets of cracks are formed simultaneously, the energy density per unit area of interface would be reduced to

$$U_2 = \frac{\sigma_m^2 h_f}{M_f} - f(\lambda/h_f)\frac{\sigma_m^2 h_f^2}{\bar{E}_f \lambda} . \tag{4.74}$$

Thus, the energy release per unit advance of *each* crack formed in the second set is

$$W_m = 2\lambda\,(U_1 - U_2) = \frac{\sigma_m^2 h_f^2}{\bar{E}_f}\left[2f(\lambda/h_f) - f(2\lambda/h_f)\right] . \tag{4.75}$$

The smallest value of λ for which formation of the second set of cracks is energetically possible, say λ_{min}, is established by the condition $W_m = W_c = h_f \Gamma_f$. For a value of λ smaller than λ_{min}, the relationship $W_m < W_c$ implies that spontaneous crack formation is not possible. The condition $W_m = W_c$ in this case implies that

$$2f(\lambda_{min}/h_f) - f(2\lambda_{min}/h_f) = \frac{\bar{E}_f \Gamma_f}{\sigma_m^2 h_f} . \tag{4.76}$$

This equation can also be inverted to yield the dependence of λ_{min}/h_f on $\sigma_m\sqrt{h_f/\bar{E}_f \Gamma_f}$, and the result is shown in Figure 4.44 where it is labeled as sequential formation. The present assumption on sequential formation of film cracks, which is thought to be more realistic than that of simultaneous formation, predicts a substantially larger minimum spacing than was found for the case of simultaneous formation. Experimental data supporting the sequential formation conjecture as being more realistic are reported by Thouless et al. (1992) and by Delannay and Warren (1991). Among situations of this type, perhaps the most realistic would be to establish threshold conditions for insertion of the last crack into a periodic array. The implications of this approach have not yet been examined for cracks, but the corresponding situation for dislocations is discussed in Chapter 6.

4.6.4 Example: Cracking of an epitaxial film

An $In_{0.25}Ga_{0.75}As$ film is grown epitaxially on a thick InP substrate, resulting in a state of equi-biaxial tension. The mismatch strain can be estimated to be $\epsilon_m = 0.02$ from the lattice parameters shown in Figure 1.17 and the rule of mixtures method of inferring lattice mismatch of alloys implied by (1.15). Assume the film material to be isotropic with elastic modulus $E_f = 76.8$ GPa and Poisson ratio $\nu_f = 0.32$. Furthermore, assume the moduli of the substrate to have the same values, although the elastic stiffness of the substrate is actually only about 75% that of the film. The fracture energy of the film material is $\Gamma_f = 1.6\,\mathrm{J\,m^{-2}}$.

(a) Determine the value of film thickness h_f at which through-the-thickness cracks can be expected to appear in the film during growth.
(b) If the cracks do not penetrate into the substrate, estimate the crack spacing when the film thickness h_f has increased to twice the thickness determined in part (a).
(c) Discuss the likelihood of crack penetration into the substrate.

Solution:

(a) The mismatch stress is $\sigma_m = M_f \epsilon_m = 2.26$ GPa in this case. The critical thickness for onset of film cracking can be estimated from (4.63) to be

$$(h_f)_{cr} = \frac{2}{\pi c_e^2} \frac{1.6 \times 76.8 \times 10^9}{2.26^2 \times (1 - 0.32^2) \times 10^{18}} = 13.6\,\mathrm{nm}. \tag{4.77}$$

(b) An expression for the minimum spacing of film cracks, assuming that they do not penetrate into the substrate, is given in (4.76). Adopting the approximation for the function $f(\,)$ provided in (4.71), the minimum crack spacing when the film thickness is twice the critical thickness is estimated from (4.76) to be

$$\lambda_{min} = -6(h_f)_{cr} \ln\left(1 - \sqrt{\bar{E}_f \Gamma_f / 1.26\pi \sigma_m^2 (h_f)_{cr}}\right) = 99.9\,\mathrm{nm} \tag{4.78}$$

for $h_f = 2(h_f)_{cr}$.

(c) For this system, the fracture energies of the film and substrate materials are very similar. Consequently, the condition $\Gamma_s \geq 2\Gamma_f$ for confinement of the fractures to the film material given in (4.65) is not met and it is likely that the cracks will penetrate into the substrate immediately upon formation. As a result of crack penetration into the substrate, the actual crack spacing likely will be less than that anticipated in part (b). This expectation is borne out by the experimental observations of Wu and Weatherly (1999).

4.7 Crack deflection at an interface

In Section 4.4, conditions for delamination growth at a film–substrate interface were considered under the tacit assumption that the fracture was naturally confined to propagate in the interface. Similarly, in Section 4.6, conditions for crack growth through the thickness of the strained film in a direction normal to the film–substrate interface, and possibly into the substrate, were considered. Again, it was tacitly assumed that the fracture plane was always perpendicular to the interface. In reality, a crack growing along an interface may prefer to kink out of the interface and into the interior of one of the materials, or a crack approaching an interface from one of the joined materials may prefer to deflect into the interface rather than to continue across it. Thus, the circumstances which admit this kind of crack deflection behavior are examined in this section for conditions of plane strain deformation.

While the details of any particular situation can become complicated, the strategy is simple and fundamental. The state of a planar crack edge at any position is represented by the current value of energy release rate $\mathcal{G}$, which is the configurational force tending to drive the crack forward, and by the local phase angle ψ, which characterizes the degree of shear loading compared to tensile loading on the extension of the crack plane ahead of the crack edge. The material resistance to crack advance in this state is represented by the local work of fracture $\Gamma(\psi)$, where the possibility that this material specific quantity depends on the local phase angle is made explicit. From any such state, it is now imagined that the crack can sample its prospects for further growth in *any* direction, including the possibility of continued straight ahead planar growth. This sampling is done for any possible direction of growth by sending out a virtual crack extension, or kink, which is very short compared to any other dimensions in the configuration. Let ω_k denote the angle of deflection of this kink from the plane of the main crack; $\omega_k = 0$ corresponds to straight ahead growth. The sign convention adopted for ω_k is such that a crack will tend to kink with $\omega_k > 0$ when both $K_I > 0$ and $K_{II} > 0$. Associated with the leading edge of this kink is an energy release rate or driving force $\mathcal{G}_k$ and a local stress state phase angle ψ_k. The material into which the crack kinks resists the formation of this kink by requiring that the work of fracture per unit area must be at least $\Gamma_k(\psi_k)$, where again this work of fracture may depend on the local phase angle. The critical condition for crack advance is then defined by the smallest value of applied loading (represented by ϵ_m, σ_m or some other parameter) for which $\mathcal{G}_k = \Gamma_k(\psi_k)$ at some ω_k, with $\mathcal{G}_k < \Gamma_k(\psi_k)$ for all other values of ω_k. If ω_k turns out to be zero then straight ahead growth is preferred; otherwise, crack deflection to a direction defined by $\omega_k \neq 0$ is preferred. To summarize, continued growth of the crack in its own plane occurs with $\mathcal{G} = \Gamma(\psi)$ while $\mathcal{G}_k < \Gamma_k(\psi_k)$ for all possible ω_k. On the other hand, growth of an oblique kink in a direction ω_k^{max} occurs with $\mathcal{G}_k^{max} = \Gamma_k(\psi_k)$ while $\mathcal{G} < \Gamma(\psi)$ where

$$\mathcal{G}_k^{max} = \max_{-\pi < \omega_k < \pi} \mathcal{G}_k \qquad (4.79)$$

at some fixed load level and ω_k^{max} is the direction in which $\mathcal{G}_k$ achieves its maximum value. This implies that the outcome of competition between straight ahead cracking and kink crack formation depends on whether the ratio $\mathcal{G}/\mathcal{G}_k^{\max}$ is greater than or less than the fracture resistance ratio $\Gamma(\psi)/\Gamma_k(\psi_k)$, respectively.

This general strategy is next applied to consider the possibility of a delamination crack on a bimaterial interface deflecting into one or the other of the joined materials and the possibility of a crack approaching a bimaterial interface deflecting into the interface rather than arresting or continuing across the interface. This development is limited to the case of joined isotropic elastic materials with the material parameters restricted by $D_2 = 0$ where the material parameter D_2 is defined by (4.29). This avoids the ambiguity that arises from the oscillatory stress concentration field when $D_2 \neq 0$.

4.7.1 Crack deflection out of an interface

Consider first the situation in which the main crack lies in the interface between the film and substrate under two-dimensional conditions, as illustrated in Figure 4.22. The equi-biaxial stress due to mismatch which exists far ahead of the delamination front is σ_m and the stress remaining in the film on a section far behind the delamination front with normal in the direction of delamination advance is σ_a. As was illustrated in Figure 4.23, the stress fields in the film and substrate in this configuration differ from the stress fields in a plane strain configuration by, at most, spatially uniform equilibrium stress fields, and the interfacial stress distributions are identical in the two configurations. The plane strain configuration that is equivalent (in this sense) is that illustrated in part (c) of Figure 4.23 where the loading factor is $\sigma_m - \sigma_a$. The superposition of the equilibrium plane strain field with uniform stress $\sigma_{xx} = -(\sigma_m - \sigma_a)$ in the film and zero stress in the substrate reduces the configuration to that illustrated in part (a) of Figure 4.46. In this configuration, there is no residual stress far ahead of the delamination front, and the delamination process is driven by the resultant force per unit depth of magnitude $(\sigma_m - \sigma_a)h_f$ acting far behind the delamination front.

For the time being, suppose that $\sigma_m > \sigma_a > 0$ which corresponds to a film with residual tensile stress. For this case, it was noted in Section 4.4 that the phase angle of the stress concentration field at the edge of a delamination is approximately $\psi = 52.1°$ and that $K_I > 0$. Because the phase of a delamination crack with this type of loading is constant, the interfacial fracture energy will be written simply as Γ, as before, thus suppressing any dependence of this quantity on ψ under these special conditions.

For the loading present in the film delamination process, a state of large tensile stress is generated in a direction from the edge of the delamination zone corresponding to $\omega_k > 0$, and the tendency for formation of an alternative crack path is toward kink formation into the substrate. The fracture resistance of the substrate, a

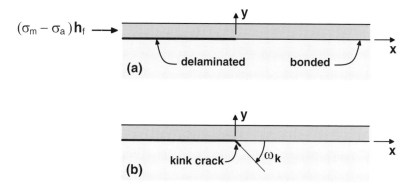

Fig. 4.46. Part (a) shows a delamination crack propagating along the film–substrate interface. In part (b), the possibility that the crack edge deflects out of the interface is considered, with the new direction of growth being inclined at an angle ω_k to the interface plane.

homogeneous material, will be denoted by the specific fracture energy $\Gamma_k(\psi_k) = \Gamma_s$ for any ψ_k. For the cases of interest, it is found that $\psi_k \approx 0$ or the stress state at the edge of the kink crack is essentially mode I. The critical condition for a delamination to deflect out of the interface and into the substrate is then

$$\frac{\mathcal{G}_k^{\max}}{\mathcal{G}} > \frac{\Gamma_s}{\Gamma} \tag{4.80}$$

and, by implication, such crack kinking is unlikely if the inequality is reversed.

The fracture mechanics parameters for an elastic interface crack deflecting out of the interface into one of the joined materials have been determined numerically by He and Hutchinson (1989b). From their results, it can be shown that the energy release rate for the kink segment in Figure 4.46, which is small in length compared to h_f, depends on the kink angle ω_k as shown in Figure 4.47 for $\psi \approx 52.1°$. From this plot, it is evident that

$$\frac{\mathcal{G}_k^{\max}}{\mathcal{G}} \approx 1.76\,, \quad \omega_k^{\max} \approx 62° \tag{4.81}$$

where $\omega_k^{\max}$ is the kink angle at which $\mathcal{G}_k$ takes on its maximum value and $\mathcal{G}$ is the energy release rate which prevailed prior to formation of the out-of-plane kink crack. It was also demonstrated by He and Hutchinson (1989b) that $\psi_k \approx 0$ at $\omega_k = \omega_k^{\max}$, so the direction of kink formation selected on the basis of a maximum energy release rate condition essentially corresponds with the direction selected on the basis of a requirement of purely mode I crack growth in a homogeneous material. The immediate implication of the result in (4.81) is that a delamination crack formed between a film in tension and its unstressed substrate will be confined

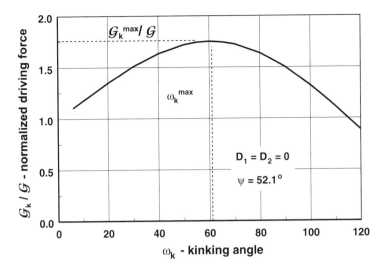

Fig. 4.47. The ratio of driving force for the kink crack in Figure 4.46(b) versus kink angle. $\mathcal{G}_k$ is normalized by the value of energy release rate $\mathcal{G}$ for continued growth of the delamination crack in the interface.

to the interface as long as $\Gamma_s/\Gamma > 1.76$, whereas crack deflection into the substrate is likely if $\Gamma_s/\Gamma < 1.76$.

The results described in the foregoing discussion can be combined with the basic delamination criterion of Section 4.4 to produce the interface crack deflection behavior diagram shown in Figure 4.48. The material and geometrical parameters of the system are combined into two key nondimensional ratios, $(\sigma_m - \sigma_a)^2 h_f/2\bar{E}_f\Gamma$ and Γ_s/Γ, and the plane with the values of these ratios varying along rectangular axes can be divided into zones of behavior as illustrated. In labeling one zone as 'substrate cracking only', it is presumed that the substrate crack can be formed from some interfacial flaw that did not grow as a delamination crack or perhaps that the substrate crack can be formed at the film edge. Finally, recall that systems are being considered for which $\sigma_m > \sigma_a > 0$.

The behavior map shown in Figure 4.48 was based on numerical results obtained for the particular case when $D_1 = D_2 = 0$. The crack kinking boundary value problem was also analyzed by He and Hutchinson (1989b) for several nonzero values of D_1. To illustrate the influence of variations in the value of D_1 on system response, the lines representing the critical condition $\mathcal{G}_k^{max}/\mathcal{G} = \Gamma_s/\Gamma$ for $D_1 = 0.5$ and -0.5 have been included in Figure 4.48; for these two values of the stiffness ratio, the phase angles of the edge stress concentration are seen from Figure 4.28 to be approximately $\psi \approx 59°$ and $\psi \approx 48°$, respectively. It is evident from these limited results that $D_1 < 0$ (substrate stiffer than film) suppresses crack deflection into the substrate whereas $D_1 > 0$ enhances such crack deflection, all other things being equal.

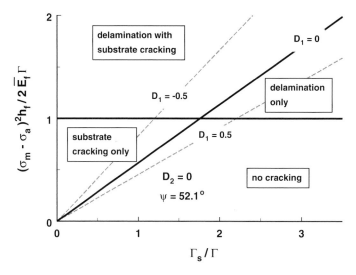

Fig. 4.48. The plane spanned by two nondimensional combinations of system parameters, with $(\sigma_{\mathrm{m}} - \sigma_{\mathrm{a}})^2 h_{\mathrm{f}}/2\bar{E}_{\mathrm{f}}\Gamma$ representing crack driving force and $\Gamma_{\mathrm{s}}/\Gamma$ representing substrate fracture resistance, with both measures normalized by the same interface separation energy. Based on the developments in this chapter, the plane can be divided into regions in which no cracking is possible, only substrate fracture is possible, only interface delamination is possible, and either interface or substrate fracture is possible.

4.7.2 Crack deflection into an interface

In Section 4.4, the conditions that are necessary for a strained thin film to separate spontaneously from its substrate by growth of a delamination crack were considered and, in the preceding subsection, the possibility of that delamination crack kinking into the substrate was considered. Similarly, the conditions that are necessary for formation of a through-the-thickness crack to form in a strained film in tension were considered in Section 4.6. In this subsection, the possibility of such a crack either advancing into the substrate or kinking into the interface between the film and substrate is considered. Indeed, the possibility of advance of such a crack into the substrate when the film and substrate materials have identical elastic properties was addressed in (4.64) and (4.65), but the circumstances are generalized here to include both a difference in properties between the materials and the competition between advance of the crack into the substrate versus kinking into the interface. The strategy for determining the outcome of the competition which was outlined in the introduction is followed in this discussion.

Suppose that a crack extends through the thickness h_{f} of an isotropic elastic film from its free surface to the interface with an isotropic elastic substrate under plane strain conditions. The crack is presumed to have formed in order to partially relieve the residual equi-biaxial tensile stress σ_{m} in the film, so the crack opening is symmetric with respect to the crack plane. The substrate thickness is assumed to

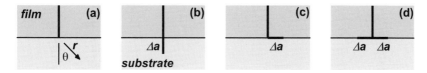

Fig. 4.49. A through-the-thickness crack in a film is illustrated in part (a). Parts (b), (c) and (d) depict the possibilities that the crack will extend a small distance Δa into the substrate, in one direction along the interface, and in both directions along the interface, respectively.

be much greater than h_f. The crack edge lies in the interface, and the material in the vicinity of the edge is depicted in part (a) of Figure 4.49. The elastic near-tip stress field in the neighborhood of the crack edge is singular, but not necessarily square root singular for general combinations of material parameters D_1 and D_2. As noted above, attention here is restricted to the case of $D_2 = 0$ but D_1 is arbitrary for the time being. Under these conditions, the local stress field in terms of the polar coordinates indicated in part (a) of Figure 4.49 has the universal form (Zak and Williams 1963)

$$\sigma_{ij}(r, \theta) = \frac{k_o}{(2\pi r)^{\lambda_o}} \Sigma_{ij}^o(\theta) \qquad (4.82)$$

plus less dominant terms, where Σ_{ij}^o is a known function of angular coordinates (but its explicit form is inconsequential for present purposes), λ_o is an exponent in the range $0 < \lambda_o < 1$, and k_o is a stress intensity factor which is independent of r and θ and which has physical dimensions of force $\times$ (length)$^{\lambda_o-2}$. For any value of D_1, the exponent λ_o is determined by the equation

$$\frac{\cos \lambda_o \pi}{1 - 4\lambda_o + 2\lambda_o^2} + D_1 = 0. \qquad (4.83)$$

A plot of λ_o versus D_1 implied by (4.83) is shown in Figure 4.50, where it can be seen that the singularity is stronger than (weaker than) square root if the plane strain modulus of the film material is greater than (less than) the plane strain modulus of the substrate material.

The main crack is now imagined to advance from its position in the interface in one of three possible ways; the three ways are represented by parts (b), (c) and (d) of Figure 4.49. The crack samples its prospects for advance by extending in its own plane a distance Δa into the substrate beyond the interface (part (b) of Figure 4.49) or by extending along the interface either as a single kink of length Δa (part (c) of Figure 4.49) or as a double kink each of which is of length Δa (part (d) of Figure 4.49). In each case, it is assumed that the sampling length Δa is small compared to h_f.

In the case of advance of the crack into the substrate, the configuration remains symmetric with respect to the crack plane and the crack advances in mode I. The

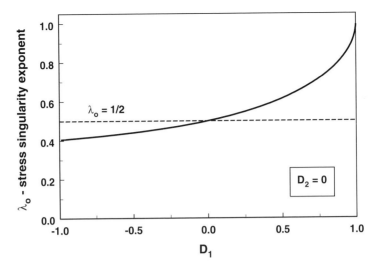

Fig. 4.50. The strength of the singularity in stress, as defined in (4.82) and (4.83), versus the stiffness ratio D_1.

magnitude of the stress relieved by crack advance is set by k_o so, due to requirements of dimensional consistency, the mode I stress intensity factor of the added growth segment must depend on k_o and Δa according to

$$K_{\mathrm{I}} = C_{\mathrm{I}}\, k_o\, \Delta a^{1/2-\lambda_o} \qquad (4.84)$$

where C_{I} is a dimensionless function of D_1 which must be determined from the solution of the relevant boundary value problem. From (4.24), it follows that the corresponding energy release rate or crack driving force is

$$\mathcal{G} = \frac{K_{\mathrm{I}}^2}{\bar{E}_{\mathrm{s}}} = \frac{C_{\mathrm{I}}^2\, k_o^2\, \Delta a^{1-2\lambda_o}}{\bar{E}_{\mathrm{s}}}. \qquad (4.85)$$

For a crack that advances as a kink in the interface, either a single kink or a symmetric double kink, as depicted in parts (c) or (d), respectively, of Figure 4.49, the local field near the edge of a kink has a nonzero phase angle ψ, in general. The state of opening of the kink cracks is some combination of mode I and mode II in both cases. The respective stress intensity factors of the kink cracks will be denoted by $K_{\mathrm{Ik_n}}$ and $K_{\mathrm{IIk_n}}$ where $n = 1$ for a single kink and $n = 2$ for a double kink configuration. The mode II stress intensity factors differ in algebraic sign but not in magnitude for the two crack edges in the case of double kinking. For either single or double kinking, the stress intensity factors are necessarily proportional to k_o due to the linearity of the underlying boundary value problem. Furthermore, the requirements of consistency of physical dimensions dictate that the expressions for the stress intensity factors in terms of other variables must have the form

$$K_{\mathrm{Ik_n}} = C_{\mathrm{Ik_n}} k_o\, \Delta a^{1/2-\lambda_o}\,, \qquad K_{\mathrm{IIk_n}} = C_{\mathrm{IIk_n}} k_o\, \Delta a^{1/2-\lambda_o} \qquad (4.86)$$

where C_{Ik_n} and C_{IIk_n} are dimensionless quantities depending on D_1; these coefficients must be determined from the solutions of the pertinent boundary value problems. The energy release rate corresponding to the stress intensity factor expressions in (4.86) is, from (4.30),

$$\mathcal{G}_{k_n} = \frac{K_{Ik_n}^2 + K_{IIk_n}^2}{\bar{E}_s(1 + D_1)} = \frac{k_o^2 \, \Delta a^{1-2\lambda_o} \left(C_{Ik_n}^2 + C_{IIk_n}^2 \right)}{\bar{E}_s(1 + D_1)}. \tag{4.87}$$

The quantity $\mathcal{G}_{k_2}$ is the energy release rate for one of the kinks in the case of double kinking. From (4.28), the stress concentration field phase angle corresponding to (4.86) is

$$\psi_{k_n} = \arctan\left[\frac{K_{IIk_n}}{K_{Ik_n}}\right] = \arctan\left[\frac{C_{IIk_n}}{C_{Ik_n}}\right] \tag{4.88}$$

where ψ_{k_2} is the phase angle for the right going kink in part (d) of Figure 4.49.

Note that both the energy release rate $\mathcal{G}$ for crack penetration into the substrate given by (4.85) and the energy release rate $\mathcal{G}_{k_n}$ for single or double interface kink formation depend on the increment of crack advance Δa in the same way. Consequently, the ratio of $\mathcal{G}_{k_n}$ to $\mathcal{G}$, which specifies the relative energy release for kink crack formation of some length Δa to substrate penetration to the same depth, is independent of Δa. This ratio, which is given by

$$\frac{\mathcal{G}_{k_n}}{\mathcal{G}} = \frac{C_{Ik_n}^2 + C_{IIk_n}^2}{(1 + D_1)C_I^2}, \tag{4.89}$$

depends solely on the stiffness ratio D_1 for $D_2 = 0$. The dependence of the parameters C_I, C_{Ik_n} and C_{IIk_n} has been determined by He and Hutchinson (1989b) through numerical solution of the relevant elastic crack boundary value problems, and values are tabulated by He et al. (1994). The ratio $\mathcal{G}_{k_n}/\mathcal{G}$ versus D_1 implied by these results is shown in Figure 4.51 for both single and double interface kink formation, and the crack edge phase angles for single and double kinking are shown in Figure 4.52; recall that $\psi = 0$ for substrate penetration by straight ahead crack growth.

Several useful observations follow immediately from these results. For one thing, the two curves in Figure 4.51 are nearly identical over the full range of values of D_1. Consequently, the tendencies toward single and double kinking may be considered to be essentially the same for any practical purposes. Secondly, the graphs of $\mathcal{G}_{k_n}/\mathcal{G}$ versus D_1 are quite flat over the range $-0.5 < D_1 < 0.5$ which implies that a stiffness difference between the film and substrate materials does not have a large influence on the competition between extension of a film crack into the substrate versus extension as a single or double kink along the interface for stiffness ratios in this range. Yet another useful observation from Figure 4.52 is that the stress concentration phase angles for the cases of single and double kinking depend on D_1

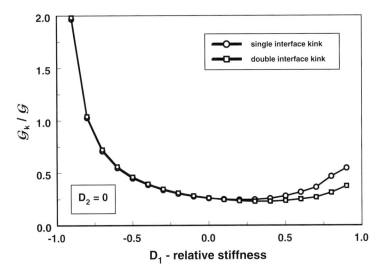

Fig. 4.51. Ratio of the energy release rate for incremental crack extension along the interface, as in part (c) or (d) of Figure 4.49, to the energy release rate for incremental crack extension into the substrate, as in part (b) of Figure 4.49, versus the stiffness ratio D_1. Note that the difference between single and double kinking into the interface is minor

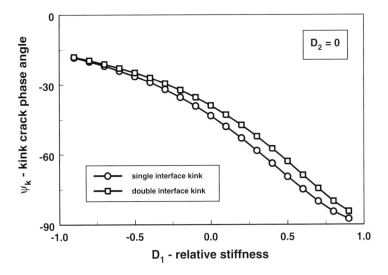

Fig. 4.52. Phase angle of a right going interface kink crack, as illustrated in parts (c) and (d) of Figure 4.49, versus the stiffness ratio D_1. Note that the difference between single and double kinking into the interface is minor.

in much the same way. Thus, even in cases when the interface fracture resistance $\Gamma(\psi_{k_n})$ depends on the local phase angle, the resistance to a single kink or each of a pair of double kinks can be viewed as being about the same for any practical purposes.

Finally, the outcome of the competition between crack penetration into the substrate versus interface kink crack formation is determined by comparing $\mathcal{G}_{k_n}/\mathcal{G}$ to the fracture resistance ratio $\Gamma(\psi)/\Gamma_s$ where $\Gamma(\psi)$ is the fracture resistance of the interface at the appropriate phase angle and Γ_s is the fracture resistance of the substrate material to mode I crack growth. For example, at $D_1 = 0$, the ratio is approximately 0.26. Thus, a film crack will likely penetrate into the substrate upon extension if $0.26\Gamma_s < \Gamma$ but will likely deflect into the interface if $0.26\Gamma_s > \Gamma$. The fracture resistance Γ is that at a phase angle of roughly $\psi_{k_n} \approx -40°$.

For any specified value of D_1, the results of the foregoing discussion can be combined with the basic film cracking criterion of Section 4.6 to produce a cracking behavior diagram. Such a diagram is illustrated here for the case when $D_1 = 0$, that is, when the film and substrate elastic properties are the same. The film of thickness h_f is presumed to carry an initial equi-biaxial mismatch tensile stress σ_m. The fracture resistance of the film material, the substrate material and the interface are denoted by Γ_f, Γ_s and Γ where the last is understood to be the value at the appropriate phase angle. The material and geometrical parameters are combined into three nondimensional groups, namely, $\sigma_m^2 h_f/\bar{E}_f\Gamma_f$, Γ/Γ_f and Γ/Γ_s. The plane with rectangular coordinates of $\sigma_m^2 h_f/\bar{E}_f\Gamma_f$ versus Γ/Γ_f can then be divided into regions of behavior representing film cracking and interface delamination or debonding. The use of Γ/Γ_s as a parameter in this plane allows the further reduction of this plane into regions concerned with crack penetration into the substrate, as well. The behavior map for $\Gamma/\Gamma_s < 0.26$ is fundamentally different from that for $\Gamma/\Gamma_s > 0.26$, so the map is illustrated for each of these two cases separately in the upper and lower parts, respectively, of Figure 4.53.

The horizontal dividing line in the behavior diagram follows directly from (4.63). The inclined dividing line separating the region in which film cracking is possible from the region in which interface debonding also becomes possible is determined by combining two pieces of information. For $D_1 = 0$, the energy release rate for the film crack when it meets the interface is $\pi c_e^2 \sigma_m^2 h_f/\bar{E}_f$ from (4.61), and the driving force varies continuously with crack length as the crack penetrates into the substrate; see Figure 4.38. Then, from Figure 4.51 with $D_1 = 0$, it is seen that $\mathcal{G}_{k_n} = 0.26\mathcal{G}$. Thus, if Γ is the fracture resistance of the interface, the critical condition dividing those circumstances that admit crack kinking into the substrate from those that do not is

$$\frac{\sigma_m^2 h_f}{\bar{E}_f\Gamma_f} = \frac{1}{0.26\pi c_e^2}\frac{\Gamma}{\Gamma_f} = 1.03\frac{\Gamma}{\Gamma_f} \tag{4.90}$$

which is the equation of the dividing line. Finally, the line dividing those circumstances that admit penetration of the film crack into the substrate from those that do

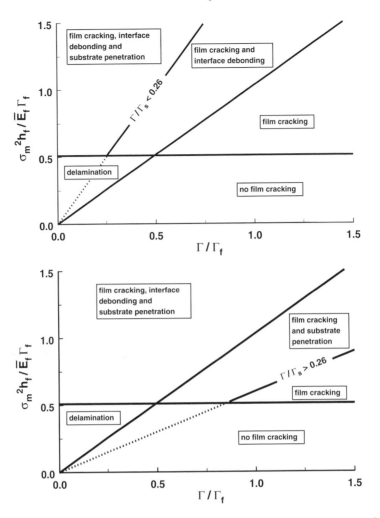

Fig. 4.53. Plane of the dimensionless groups of system parameters, in the form of $\sigma_m^2 h_f / \bar{E}_f \Gamma_f$ versus Γ / Γ_f, divided into ranges of fracture behavior based on conclusions formed in this chapter. The upper (lower) diagram applies for the case in which $\Gamma / \Gamma_s < 0.26$ (> 0.26).

not follows from (4.64) in the case of equality, that is,

$$\frac{\sigma_m^2 h_f}{\bar{E}_f \Gamma_f} = \frac{1}{\pi c_e^2} \frac{\Gamma_s}{\Gamma_f} = \frac{1}{\pi c_e^2} \frac{\Gamma_s}{\Gamma} \frac{\Gamma}{\Gamma_f}. \tag{4.91}$$

This line falls above or below the line defined by (4.90), depending on whether $\Gamma / \Gamma_s < 0.26$ or > 0.26, respectively. These are the two possibilities illustrated in the upper and lower parts of Figure 4.53 for arbitrary choices of the ratio Γ / Γ_s in the appropriate ranges.

The mechanics of the growth of cracks approaching interfaces is a topic of broad interest for layered coatings that are used for thermal, environmental and tribological protection of surfaces. Experimental observations of the effects of mismatch in material properties across interfaces on the growth of cracks subjected to monotonic and cyclic loading have been considered by Suresh (1998) and Suresh et al. (1993).

4.8 Exercises

1. The double cantilever beam (DCB) is a widely used fracture test specimen for both brittle and ductile materials. Consider the DCB specimen schematically shown in Figure 4.9. From the discussion presented in Section 4.2.1, answer the following questions.
   (a) Calculate the energy release rate $\mathcal{G}$ for the double cantilever specimen subjected to a tensile opening load P.
   (b) Derive an expression for the mode I stress intensity factor K_I in terms of the applied load, the specimen geometry and elastic properties of the double cantilever specimen.
   (c) Discuss how calibrations of $\mathcal{G}$ and K_I for the double cantilever specimen may be obtained experimentally using measurements of compliance changes during crack growth.

2. Two polymeric beams are glued together using an epoxy adhesive as shown in Figure 4.54. The air bubbles trapped in the adhesive during the joining process lead to the formation of several flat, circular, disk-like cracks in the adhesive; these cracks are aligned parallel to the interface between the two plastic beams. The dominant flaw is usually a semi-circular (thumb-nail) crack located along the interface in the adhesive near the outer surface of the composite beam. The average radius of such a crack is $a = 1.2$ mm. The dimensions indicated in Figure 4.54 are: $L = 1.7$ m, $W = 10$ cm, and $B = 7.5$ cm. If the fracture toughness of the epoxy is 0.8 MPa m$^{1/2}$, calculate the maximum load P which the beam can support, assuming elastic deformation. For the semi-circular crack, the mode I stress intensity factor is $K_I \approx 0.71\sigma_{max}\sqrt{\pi a}$ where σ_{max} is the maximum opening stress at the outer fiber of the composite beam.

3. A single crystal thin film of a SiGe alloy, with 80 atomic percent Si and the rest Ge, is grown epitaxially at room temperature on a $\{100\}$ silicon substrate which is $600\ \mu$m thick. The mean lattice parameter of the alloy is 0.5476 nm while that of Si is 0.5431 nm. The biaxial moduli of the film and substrate materials are 173 GPa and 180.5 GPa, respectively, while the corresponding Poisson ratios of the two materials are 0.25 and 0.27, respectively. If the fracture energy of the interface between the film and the substrate, which is a $\{100\}$ plane of the two materials, is 1 J m^{-2}, and if spontaneous delamination occurs along the interface at the film edge due to the epitaxial mismatch strain in the film, determine the thickness of the film.

4. In the discussion of the double cantilever configuration in Section 4.2.1 and Section 4.2.2, it was assumed that the load or displacement was imposed directly on the sample. In an actual experiment, loading must be imposed through an apparatus that has an elastic compliance of its own. Re-examine the double cantilever configuration depicted in Figure 4.9 assuming that the loading apparatus is represented by a pair of springs, each of compliance C^M, in series with the sample, so that the

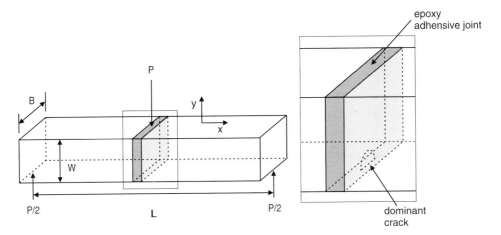

Fig. 4.54. The composite beam, formed by gluing two polymeric beams together end-to-end, is loaded in three-point bending. A semi-circular surface crack in the glue that joins the two beams is shown in the enlarged view of the glue joint on the right. Adapted from Suresh (1998).

relationship between the applied loading P and the load point deflection δ_p is

$$\delta_p = \left(C^M + \frac{4a^3}{\bar{E}h^3} \right) P \qquad (4.92)$$

instead of (4.11). The compliance of the loading apparatus cannot affect the load at which fracture begins but it can alter the stability of crack advance. Determine the value of C^M separating the range of stable growth from the range of unstable growth in terms of crack length a and other system parameters.

5. In Section 4.4.3, the growth of a delamination zone along the interface between a stressed film and substrate was considered. It was found that a delamination zone with a circular front growing outward from the free edge of the circular hole would eventually be arrested. Consider the complementary case of a circular delamination front that is contracting in size. A film of thickness h_f on a relatively thick substrate, with residual biaxial stress σ_m, is cut through its thickness down to the substrate over a circle of radius $b \gg h_f$. The film material outside the circle is removed, and a circular delamination front begins to grow inward from the free edge. Assume that the work of fracture Γ is uniform over the interface, and determine whether or not the delamination front can be arrested before the circular patch is fully delaminated from the substrate.

6. In Section 4.4.5, it was observed that the driving force of an interface delamination approaching a film edge is influenced significantly at distances from the edge that are large compared to film thickness. Study this phenomenon quantitatively by adopting the membrane film model for the case when the elastic properties of the film and substrate are the same.

   (a) Assuming that the film edge is at $x = a$ and that the delamination zone occupies the region $x < 0$, derive an integral equation governing the interfacial shear stress $q(x)$ in the interval $0 < x < a$.

(b) Solve the integral equation numerically to determine the strength of the singularity in $q(x)$ as $a/h_f \rightarrow 0^+$ relative to the result in (4.1).

(c) Estimate $\mathcal{G}_{ss}/\mathcal{G}$ versus a/h_f for this case and compare the result to the data graphed in Figure 4.29.

7. In the discussion of Section 4.6.1, steady advance of a crack along a film due to residual tension in the film was considered. It was assumed from the outset that the crack depth would be equal to film thickness, and it was subsequently verified that this would indeed be the case provided that $\Gamma_s \geq 2\Gamma_f$. Consider the particular case in which $\Gamma_s = \Gamma_f$, and estimate the depth of penetration of the steady crack into the substrate, assuming that any crack kinking is suppressed.

8. Assuming that deformation can be adequately described by means of elementary beam theory, derive the expression for energy release rate for the four-point flexure beam test configuration given in (4.54).

9. Derive the pair of coupled integral equations which govern the residual interfacial traction components $p(x)$ and $q(x)$ for a plate with a free edge bonded to a substrate, as illustrated in Figure 4.4. The state of residual stress arises from a mismatch strain ϵ_m in the plate with respect to the substrate. The derivation is carried out by enforcing equilibrium of the isotropic elastic materials in the absence of external loading and continuity of deformation across the interface during relaxation of the edge stress. Show that the results can be put in the form (cf. Shield and Kim 1992)

$$\frac{1}{\pi} \int_0^\infty \frac{p(\xi)}{\xi - x} d\xi - \frac{1 - 2v_s}{2(1 - v_s)} q(x) =$$

$$-\frac{3k}{h_f^3} \int_0^x x[(x - 2\xi) p(\xi) - h_f q(\xi)] d\xi + \frac{3k}{h_f^3} \int_x^\infty \xi[\xi p(\xi) + h_f q(\xi)] d\xi$$

$$\frac{1}{\pi} \int_0^\infty \frac{q(\xi)}{\xi - x} d\xi + \frac{1 - 2v_s}{2(1 - v_s)} p(x) =$$

$$-\frac{k}{2}\sigma_m + \frac{2k}{h_f} \int_0^x q(\xi) d\xi + \frac{3k}{h_f^2} \int_0^x (\xi - x) p(\xi) d\xi. \qquad (4.93)$$

5

Film buckling, bulging and peeling

Interface delamination and film fracture induced by residual stress in the thin film were the focus of discussion in the preceding chapter where models were developed within the framework of linear elastic fracture mechanics. Such analyses did not account for the actual separation of the film from its substrate. However, there are delamination processes for which transverse deflection of the film away from the substrate becomes an important consideration in a variety of practical applications. Examples include:

- transverse buckling instability of a film in compression, as in the case of a ceramic thermal barrier coating or a diamond-like carbon wear resistant coating on a metallic substrate, as illustrated in Figure 5.1,
- the bulging of a segment of a thin film away from the substrate under the influence of an applied pressure, which is a deformation mode of interest for the assessment of residual stress, interface fracture energy and mechanical properties in film–substrate systems and MEMS structures, and
- the forcible peeling of a film from its substrate as a means of evaluating the adhesion energy of the interface between the film and the substrate.

This chapter deals with the transverse or out-of-plane deflection of a thin film, and it includes quantitative descriptions of the phenomena associated with the buckling, bulging or peeling of a film from its substrate. A common thread throughout the discussion is that system behavior extends beyond the range of geometrically linear deformation. Consequently, aspects of finite or nonlinear deformation must be incorporated to capture essential features of behavior. The progression of delamination associated with transverse deflection is modeled by invoking the principles of linear elastic fracture mechanics in the immediate vicinity of the delamination front where local deformation is typically within the range of geometrically linear response (Hutchinson 1996). Features of the models developed are compared with available experimental observations wherever appropriate.

Fig. 5.1. A cross-sectional micrograph of an electron-beam physical vapor deposited yttria-stabilized zirconia film which is partially delaminated from the nickel-base superalloy substrate. This ceramic layer and the interlayers comprising the bondcoat and thermally grown oxide serve as the thermal-barrier coating system in gas turbine engines, as described in Section 1.2.4. Reproduced with permission from Padture et al. (2002).

5.1 Buckling of a strip of uniform width

Consider a thin elastic film which is bonded to a much thicker substrate. Let the film be subjected to an equi-biaxial mismatch stress of magnitude σ_m which, for the present discussion, is assumed to be compressive, that is, $\sigma_m < 0$. The elastic strain energy density in the film is $\sigma_m^2(1 - \nu_f)/E_f$. Suppose that some portion of the interface is not actually bonded, so that the film is free to deflect laterally away from the substrate over this portion. Whether or not the film actually does deflect spontaneously is determined by the energetics of the situation. Upon deflection, the lengths of material line elements along the midplane of the film increase, in general, implying a reduction in strain energy associated with extensional deformation. However, lateral deflection is also accompanied by bending deformation which leads to some increase in strain energy. If the consequent net change in energy is negative, the film undergoes spontaneous lateral deflection over the debonded region, that is, the film *buckles*. This propensity for buckling is suppressed if the net change in energy is positive. The critical state for onset of buckling or bifurcation of equilibrium states is determined by the neutral condition which discriminates between these two ranges of behavior.

The foregoing qualitative picture of film buckling makes no mention of energy changes in the portion of the film outside the buckle zone, which remains bonded to the substrate; it also does not address the energy change in the substrate itself. The reason for the tacit neglect of such effects is that the detached portion of the film is usually regarded as being very compliant compared to the still-bonded portion of the film or compared to the substrate. Consequently, the edge of the buckled portion of the film at the boundary of the detached region can be viewed as being rigidly clamped along this boundary. With this assumed constraint, the buckled portion of the film nonetheless exerts forces and moments on the still-attached portions of the film. The changes in magnitude of these interaction forces and moments due to buckling are determined by the post-buckling configuration of the film. If the lateral extent of the buckled region is not very large compared to h_f or if the configuration

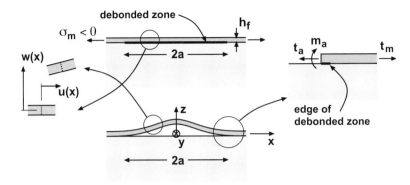

Fig. 5.2. Schematic showing the two-dimensional profile of a straight-sided buckle with the associated nomenclature.

at the edge of the buckle region is not actually very stiff, the edge compliance can possibly influence the condition for buckling, as well as the post-buckling response, as shown by Yu and Hutchinson (2002).

An important consequence of lateral buckling of a portion of the film is that it induces a driving force for expansion of the size of the buckled region. Prior to buckling, the traction transmitted across the film–substrate interface is zero even for points in the still-bonded portion of the interface, as was discussed in Chapter 2. After buckling, the stress state in the film is no longer spatially uniform adjacent to the buckle; a traction is necessarily induced on the interface to maintain equilibrium. Furthermore, the transition in stress state from the unrelaxed condition in the still-bonded region to the relaxed condition within the buckled region occurs quite abruptly across the boundary of the buckled area of the interface. This leads to stress concentrations on the interface at the edge of the buckled region. Such stress concentrations can influence the subsequent growth in size of a buckled region or a change in its shape due to ongoing delamination.

In order to examine the phenomenon of film buckling and possible subsequent delamination, this section begins with a description of the simplest case of buckling under plane strain conditions. This is followed, in subsequent sections, by analysis of circular buckles, secondary buckling phenomena and buckling of a fully bonded film.

5.1.1 Post-buckling response

Consider a film of thickness h_f on the plane surface of a relatively thick substrate. A rectangular coordinate system is introduced with the xy-plane coinciding with the film midplane and the z-direction coinciding with the outward normal direction to the substrate as shown in Figure 5.2. Prior to formation of a buckle, the stress in the film is the uniform equi-biaxial mismatch stress $\sigma_m < 0$. For purposes of discussion in this section, it is assumed that the interface is not bonded over the strip $-a < x < a$, $-\infty < y < \infty$. The origin of coordinates is on the film midplane prior to lateral deflection.

Suppose that a buckle forms in the film with a two-dimensional profile as sketched in Figure 5.2. The buckle-induced displacement of a point located on the midplane of the film depends only on its pre-buckling position x; the components of displacement in the x- and z-directions are $u(x)$ and $w(x)$, respectively. Once the buckle forms, the normal force $t_x(x)$ per unit length in the y-direction acting on any cross-section $x = \text{constant}$ is reduced in magnitude from its initial value of $t_m = \sigma_m h_f < 0$. The reduction in force at the edge of the buckled region is $t_a - t_m$, where $t_a = t_x(a)$ is a positive quantity. The parameter t_a is introduced because only this *change* in internal force in the film affects the stress acting on the interface; a spatially uniform compressive force can always be added to any state of stress without affecting the traction acting on the interface, as indicated in the introduction to Chapter 2. A bending moment $m_x(x)$ acting on the film cross-section $x = \text{constant}$ and measured per unit length in the y-direction is also induced by film buckling. The deformation of the film associated with buckle formation can be determined by means of elastic plate theory under the assumption that the edges of the buckle are clamped at $x = \pm a$. The stress resultants $t_x(x)$ and $m_x(x)$ can be calculated from the deformation. These quantities can be used, in turn, as input into a fracture mechanics analysis to assess whether or not the buckle can drive further delamination of the film.

In the particular configuration being discussed in this section, all deformation fields over the midplane are independent of y and are represented in terms of the coordinate x as illustrated in Figure 5.2. As noted by Fung (1965), equilibrium deformation within the buckle can be described by means of the nonlinear von Kármán equations for thin plates. Such an analysis is predicated upon several assumptions and observations, as listed below.

- As is the case with most plate theories, the governing equations are expressed in terms of fields defined over the midplane surface of the undeformed plate or over the uniformly compressed midplane of the plate prior to buckling. Deformation at material points away from the midplane is described in terms of these fields.
- The through-the-thickness variation of fields is assumed to be consistent with the plane stress approximation and elementary bending theory. In particular, a kinematic assumption is adopted whereby material lines that are initially straight and perpendicular to the midplane of the plate or film remain so during deformation, and the influence of tractions acting on planes parallel to the midplane is assumed to be negligibly small.

– The central idea of the von Kármán theory is to account for the fact that the cumulative effect of small midplane curvature over distances large compared to the thickness of the film can result in relatively large rotation of the midplane, which is represented here by $w'(x)$. This rotation makes a potentially significant contribution to the extensional strain of the midplane, approximately $\frac{1}{2}w'(x)^2$, which may be of the same order of magnitude as the direct extensional strain $u'(x)$. This geometrically nonlinear term is therefore retained in the strain–displacement relationship, but the material response continues to follow the linear Hooke's law. The same line of reasoning was followed in the analysis of geometrically nonlinear deformation of thin films on substrates in Section 2.5.

If these features are incorporated into the model, the strain–displacement relations take the form

$$\epsilon_{xx}(x, z) = \epsilon_{\mathrm{m}} + u'(x) + \tfrac{1}{2}w'(x)^2 - zw''(x), \quad \epsilon_{yy}(x, z) = \epsilon_{\mathrm{m}}, \qquad (5.1)$$

where $z = 0$ coincides with the undeflected midplane of the film. The in-plane normal force in the film is related to the displacement components through

$$t_x(x) = \bar{E}_{\mathrm{f}} h_{\mathrm{f}} [\epsilon_{\mathrm{m}}(1 + \nu_{\mathrm{f}}) + u'(x) + \tfrac{1}{2}w'(x)^2], \qquad (5.2)$$

where, as before, the symbols E_{f}, ν_{f}, $\bar{E}_{\mathrm{f}}$ and h_{f} denote the elastic modulus, Poisson ratio, plane strain modulus and thickness of the film, respectively. Note that $t_x(x) = \sigma_{\mathrm{m}} h_{\mathrm{f}} = t_{\mathrm{m}}$ when $u(x) = 0$ and $w(x) = 0$. Similarly, the resultant bending moment acting on a cross-section identified by its x-coordinate is

$$m_x(x) = \frac{\bar{E}_{\mathrm{f}} h_{\mathrm{f}}^3}{12} w''(x). \qquad (5.3)$$

The elastic strain energy density $\frac{1}{2}\bar{E}_{\mathrm{f}}[\epsilon_{xx}^2 + \epsilon_{yy}^2 + 2\nu_{\mathrm{f}}\epsilon_{xx}\epsilon_{yy}]$ can be expressed in terms of the initial state and the relaxation displacement components $u(x)$ and $w(x)$. The total potential energy of the film, which is equal to the elastic energy in the absence of externally applied loads, can be obtained by integration of the strain energy density over the volume of film material per unit depth in the y-direction within the buckled region. The integration on z can be carried out explicitly. The principle of stationary potential energy then implies that, for the film to be in equilibrium, the elastic deflections $u(x)$ and $w(x)$ must satisfy the differential equations

$$[u'(x) + \tfrac{1}{2}w'(x)^2]' = 0,$$

$$\qquad (5.4)$$

$$w''''(x) - \frac{12}{\bar{E}_{\mathrm{f}} h_{\mathrm{f}}^3} w''(x) t_x(x) = 0$$

within the region $-a < x < a$, where the prime denotes differentiation with respect to x. The first equilibrium equation $(5.4)_1$ implies that $t_x(x)$ as given by (5.2) is a constant, and its value is therefore necessarily equal to $t_x(a) = t_a$. The second equilibrium equation $(5.4)_2$ then takes the form

$$w''''(x) - \frac{12 \, t_a}{\bar{E}_f h_f^3} w''(x) = 0, \quad -a < x < a. \tag{5.5}$$

Even though the deformation is in the geometrically nonlinear range, the differential equation governing $w(x)$ is linear. This is a fortuitous outcome that follows naturally from the nonlinear von Kármán plate theory. A solution of (5.5) is sought subject to the boundary conditions

$$w(\pm a) = 0 \quad \text{and} \quad w'(\pm a) = 0, \tag{5.6}$$

which imply that the edges of the buckle are clamped, as discussed above. Once $w(x)$ is known, the in-plane displacement $u(x)$ can be found from $(5.4)_1$ and the boundary condition $u(\pm a) = 0$. The differential equation in (5.5) and the boundary conditions (5.6) have the form of an eigenvalue problem for a one-dimensional continuous system where the variable t_a represents the eigenvalue. The eigenvalue of least magnitude for which a nontrivial solution exists is

$$t_a = -\frac{\pi^2 \bar{E}_f h_f}{12} \left(\frac{h_f}{a} \right)^2. \tag{5.7}$$

No positive eigenvalues exist in this case. The compressive force in the film cannot be increased by buckle formation, so that $\Delta t_a \equiv t_a - t_m > 0$. The expression (5.7) provides a relationship between the size a of the debonded region and the compressive force in the film following buckle formation. The constraint that Δt_a is necessarily positive implies that the *smallest* debond size that will result in buckle formation in a film with compressive stress σ_m is

$$\boxed{a_m = \frac{\pi h_f}{2} \sqrt{\frac{\bar{E}_f}{3 |\sigma_m|}}.} \tag{5.8}$$

Alternatively, the compressive stress level required to induce a buckle in a film with debonded zone of size a_m is

$$\boxed{\sigma_m = -\frac{\pi^2 \bar{E}_f}{12} \left(\frac{h_f}{a_m} \right)^2,} \tag{5.9}$$

which is the expression obtained by solving (5.8) for σ_m. For a debonded zone of size a_m, a stress smaller in magnitude than $|\sigma_m|$ in (5.9) does not produce a buckle. Likewise, for a compressive stress of magnitude $|\sigma_m|$, a debonded zone smaller than a_m in (5.8) does not produce a buckle. The film force t_a can be expressed in terms of a_m and σ_m, with the result that

$$\Delta t_a = t_a - t_m = |t_m| \left(1 - \frac{a_m^2}{a^2} \right). \tag{5.10}$$

The shape of the buckle prescribed by the differential equation (5.5) is

$$w(x) = \frac{1}{2} w_0 \left[1 + \cos \frac{\pi x}{a} \right], \tag{5.11}$$

where $w_0 = w(0)$. The task remaining is to establish a relationship between the force reduction Δt_a and the deflection w_0. This relationship follows from the constitutive equation (5.2) and the differential equation $(5.4)_1$. The strain in the y-direction is unchanged by buckling. The force change has the spatially uniform value Δt_a over $-a < x < a$, as noted above. Integration of $(5.4)_1$ over $-a < x < a$, with the constraint that $u(\pm a) = 0$ and the explicit expression (5.11) for $w(x)$, yields

$$w_0^2 = \frac{16a^2}{\pi^2} \frac{\Delta t_a}{h_f \bar{E}_f}. \tag{5.12}$$

This result, which completes the analysis of the post-buckling response of the configuration, enables determination of the edge bending moment $m_x(a) = m_a$ from (5.3) as

$$m_a = \frac{\bar{E}_f h_f^3}{12} w''(a) = \frac{|t_m| h_f}{\sqrt{3}} \frac{a_m^2}{a^2} \sqrt{\frac{a^2}{a_m^2} - 1}. \tag{5.13}$$

Graphs of the normalized edge stress resultant force $(t_a - t_m)/|t_m|$ and moment $m_a/|t_m| h_f$ versus a/a_m are shown in Figure 5.3. It is observed that the bending moment m_a increases from zero at $a = a_m$ to a maximum value, and then it decreases as a/a_m becomes large. The compressive stress reduction Δt_a, on the other hand, increases monotonically from zero at $a = a_m$, becoming asymptotic to $\Delta t_a/|t_m| = 1$ as a/a_m becomes large, implying that the compressive force t_a is almost completely relaxed once a has increased to about $3a_m$. The stress resultant ratio $m_a/h_f(t_a - t_m)$ is also shown by the dashed line in Figure 5.3. This ratio has a strong dependence on buckle size, implying that the phase angle ψ of the state of stress on the interface adjacent to the buckle varies significantly as the buckle grows in size due to delamination; the phase angle is defined in (4.28). This variation of the phase angle will be taken into account in the discussion of buckle driven delamination in the next subsection.

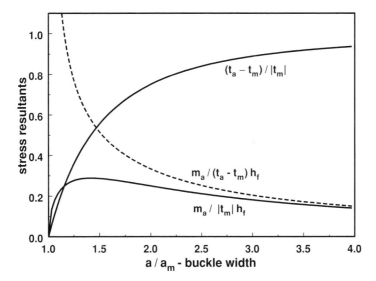

Fig. 5.3. Variation of the normalized edge stress resultant force $(t_a - t_m)/|t_m|$ and moment $m_a/|t_m|h_f$ (solid lines), and the normalized edge moment $m_a/h_f(t_a - t_m)$ (dashed line) as functions of the normalized debond size a/a_m.

The process of buckle driven delamination can be described in either of two ways. One point of view is that the mismatch stress in the film is the independent variable. As the stress is increased, it induces buckling from debonded zones, followed by delamination. The goal then is to determine the size of the delamination as a function of the film stress according to some delamination criterion. An alternate point of view is that the size of the zone of delamination is the independent variable. For a fixed level of film stress, the goal is to determine the size of buckles that form and the final size that can be achieved by buckle driven delamination. The latter point of view is pursued next.

5.1.2 Driving force for growth of delamination

Conditions governing the expansion of the zone of delamination within the buckled region of the film through the propagation of its edges can be determined from the deformation field. Such an analysis requires determination of the energy release rate $\mathcal{G}$ at the edge of the buckle and the phase angle ψ of the stress state. According to the Griffith condition introduced in Section 4.2.1, the edges will move with $\mathcal{G}$ equal to the work of separation $\Gamma(\psi)$.

Without loss of generality, it is assumed that any delamination takes place symmetrically at both edges of the buckle; if only one edge moves, there is a compensating shift in the reference point from which a is measured and the expressions obtained remain valid. The total elastic energy in one-half of a buckle of total width

$2a$ is expressed generally in terms of the stress resultants as

$$W(a) = \frac{1}{2\bar{E}_f h_f} \int_0^a \left[t_x(x)^2 + \frac{1 - \nu_f}{1 + \nu_f} t_m^2 + \frac{12}{h_f^2} m_x(x)^2 \right] dx. \tag{5.14}$$

This is the elastic energy per unit length along the crest of the buckle. The resultant force has the spatially uniform value t_a which is given in terms of a in (5.7). The resultant moment is determined by the post-buckling shape (5.11), which is also a known function of a. Substitution of these fields into (5.14) and evaluation of the integral leads to the expression

$$W(a) = \frac{t_m^2}{2\bar{E}_f h_f} \left[\frac{1 - \nu_f}{1 + \nu_f} a + 2\frac{a_m^2}{a} - \frac{a_m^4}{a^3} \right]. \tag{5.15}$$

When the edge of the buckle at $x = a$ is advanced a unit distance, the associated reduction in the elastic energy in one-half of the buckle is $-W'(a)$. In addition, as a delamination front advances along the interface, the area of the still-bonded portion of the film is diminished. This reduction in area provides an additional contribution to the energy release rate equal to the initial elastic energy in the film per unit area of interface, namely, $(1 - \nu_f)t_m^2/E_f h_f$. Thus, the total energy release rate for each edge of the buckle is

$$\mathcal{G}(a) = \frac{(1 - \nu_f)t_m^2}{E_f h_f} - W'(a). \tag{5.16}$$

From (4.34) and the discussion in Section 4.4.1, the plane strain energy release rate for advance of a straight delamination front from a free edge is

$$\mathcal{G}_m = \frac{t_m^2 \left(1 - \nu_f^2\right)}{2 E_f h_f}. \tag{5.17}$$

Combining (5.15) through (5.17) and noting that $t_m = \sigma_m h_f$, the total driving force for each edge of the delamination is written as

$$\boxed{\mathcal{G}(a) = \mathcal{G}_m \left(1 - \frac{a_m^2}{a^2} \right) \left(1 + 3\frac{a_m^2}{a^2} \right), \qquad \mathcal{G}_m = \frac{\sigma_m^2 \left(1 - \nu_f^2\right) h_f}{2 E_f}.} \tag{5.18}$$

Since buckling is precluded for $a < a_m$, the range of relevance of the expression in (5.18) is $a > a_m$. A plot of $\mathcal{G}(a)/\mathcal{G}_m$ versus a/a_m is shown in Figure 5.4. It is observed that $\mathcal{G}(a)$ increases from zero at $a = a_m$. It overshoots the value of $\mathcal{G} = \mathcal{G}_m$, reaching a maximum value of $\mathcal{G} = 4\mathcal{G}_m/3$ at $a = a_m\sqrt{3}$, and then gradually approaches $\mathcal{G} = \mathcal{G}_m$ asymptotically from above as a/a_m becomes large compared to unity.

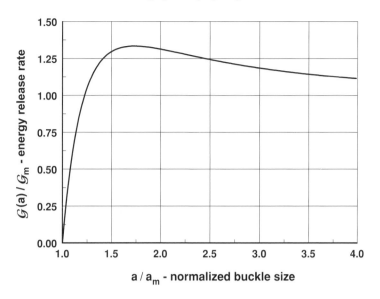

Fig. 5.4. A plot of the total energy release rate for each edge of a straight-sided buckle $\mathcal{G}(a)/\mathcal{G}_\mathrm{m}$ versus the normalized buckle size a/a_m.

5.1.3 Phase angle of local stress state at interface

For any particular material system, the delamination resistance of the interface is characterized by a fracture energy $\Gamma(\psi)$ per unit area which depends on the phase angle ψ of the stress state acting on the interface immediately in advance of the delamination front, in general. This concept was discussed in Chapter 4. The delamination is postulated to advance according to the condition that

$$\mathcal{G}(a) = \Gamma(\psi). \tag{5.19}$$

As noted in the preceding chapter, available data for such processes suggest a phenomenological expression for fracture energy of the form

$$\Gamma(\psi) = \Gamma_\mathrm{Ic}\left[1 + \tan^2 \eta_\mathrm{c}\psi\right], \tag{5.20}$$

where Γ_Ic gives the separation resistance under a purely opening-mode local stress state and η_c is a parameter in the range $0 \leq \eta_\mathrm{c} \leq 1$; this latter parameter represents the influence of phase angle on delamination resistance. When $\eta_\mathrm{c} = 0$ the interface toughness is independent of the phase of stress state, whereas when $\eta_\mathrm{c} = 1$ the interface toughness is independent of the shear traction on the interface ahead of the delamination zone. In the latter case, $\mathcal{G} = \Gamma$ reduces to $K_\mathrm{I}^2/\bar{E}_\mathrm{s}(1 + D_1) = \Gamma_\mathrm{Ic}$ for $D_2 = 0$ where, as noted in Chapter 4, E_s is the elastic modulus of the substrate and D_1 and D_2 are Dundurs parameters. A value of η_c of roughly 0.7 or 0.8 gives a representation of resistance consistent with the data reported in Chapter 4.

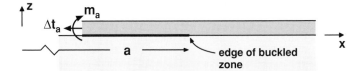

Fig. 5.5. Schematic showing the combination of pure bending and extension at a distance behind the delamination front that is on the order of the film thickness.

Determination of the phase angle ψ requires consideration of the local stress state at the edge of the buckled zone. Invoking the superposition argument depicted in Figure 4.23, it follows that the singular stress distribution on the interface is established by loading that acts at some distance behind the edge, as shown in Figure 5.5. This distance need only be on the order of the film thickness h_f. The loading parameters $\Delta t_a = t_a - t_m$ and m_a have the interpretations prescribed in (5.7) and (5.13), and the state of deformation in this reduced configuration is plane strain.

At a distance behind the delamination front on the order of h_f, the state of deformation is essentially a combination of pure bending and pure extension. With Δt_a and m_a of fixed magnitude and applied at some material cross-section behind the front, advance of the delamination front in Figure 5.5 results in a reduction in potential energy of the system. This reduction per unit distance is represented by the energy release rate

$$\mathcal{G} = \frac{1}{2\bar{E}_f h_f} \left[\Delta t_a^2 + 12 \frac{m_a^2}{h_f^2} \right]. \tag{5.21}$$

At the edge of the delamination zone, the applied loading locally induces stress distributions with edge singularities in the normal and shear tractions on the film–substrate interface. The strengths of these singularities are again denoted by K_I and K_{II}. Each stress intensity factor must be linear in both Δt_a and m_a for small deformations and linear elastic material response. Furthermore, in this reduced problem, h_f is the only significant physical length. It follows that K_I and K_{II} must each have the form

$$K_I = \alpha_{It} \frac{\Delta t_a}{h_f^{1/2}} + \alpha_{Im} \frac{m_a}{h_f^{3/2}},$$

$$\tag{5.22}$$

$$K_{II} = \alpha_{IIt} \frac{\Delta t_a}{h_f^{1/2}} + \alpha_{IIm} \frac{m_a}{h_f^{3/2}},$$

where the unknown coefficients are dimensionless. Based on the results in Section 4.1, both α_{It} and α_{IIt} can be expected to be negative. On physical grounds, it can also be anticipated that $\alpha_{Im} > 0$ and $\alpha_{IIm} < 0$.

In terms of the stress intensity factors K_I and K_{II}, the energy release rate is given by

$$\mathcal{G} = \frac{1}{\bar{E}_f(1 - D_1)} \left(K_I^2 + K_{II}^2 \right) \tag{5.23}$$

according to (4.2). But (5.21) and (5.23) represent one and the same physical quantity. For these representations to be consistent, the unknown coefficients in (5.22) must satisfy the relationships

$$\alpha_{It}\alpha_{Im} + \alpha_{IIt}\alpha_{IIm} = 0,$$

$$\alpha_{It}^2 + \alpha_{IIt}^2 = \tfrac{1}{2}(1 - D_1), \quad \alpha_{Im}^2 + \alpha_{IIm}^2 = \tfrac{1}{2}(1 - D_1). \tag{5.24}$$

All three relationships can be satisfied by introducing a single parameter ω such that

$$\alpha_{It} = -\sqrt{\tfrac{1}{2}(1 - D_1)} \cos \omega, \qquad \alpha_{IIt} = -\sqrt{\tfrac{1}{2}(1 - D_1)} \sin \omega,$$

$$\alpha_{Im} = \sqrt{\tfrac{1}{2}(1 - D_1)} \sin \omega, \qquad \alpha_{IIm} = -\sqrt{\tfrac{1}{2}(1 - D_1)} \cos \omega, \tag{5.25}$$

where the signs have been chosen so that the coefficients have the characteristics that were anticipated when ω falls within the range $0 < \omega < \pi/2$. The boundary value problem illustrated in Figure 5.5 must be solved to determine the dependence of the parameter ω on D_1; an analysis carried out for this purpose, based on numerical solution of a singular integral equation, has been reported by Thouless et al. (1987). However, based on the discussion in Section 4.4 which involves the same physical system but with $m_a = 0$, it is clear that ω is precisely the quantity shown numerically in Figure 4.28. With that understanding, the stress intensity factors are completely known for the case of delamination buckling for any buckled zone size a.

Of particular interest is the phase angle ψ of the stress state at the edge of the straight-sided buckle. In light of the results obtained in (5.22) and (5.25), the definition of phase angle (4.28) leads to the result that, in the case of the straight-sided buckle,

$$\psi(a) = \arctan \left[\frac{-\Delta t_a h_f \sin \omega - 2\sqrt{3}\, m_a \cos \omega}{-\Delta t_a h_f \cos \omega + 2\sqrt{3}\, m_a \sin \omega} \right], \tag{5.26}$$

where $\omega = \omega(D_1)$ is the function plotted in Figure 4.28; recall that it is assumed that $D_2 = 0$ throughout this discussion. The algebraic signs of the various contributions to phase angle indicate the quadrant of the phase angle. For example, positive values of both Δt_a and m_a induce negative values of stress intensity factor K_{II},

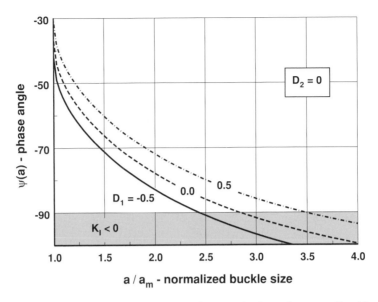

Fig. 5.6. The dependence of local stress state phase angle ψ on the normalized buckle size a/a_{m} for $D_1 = 0$ ($\bar{E}_{\mathrm{f}} = \bar{E}_{\mathrm{s}}$), $D_1 = 0.5$ ($\bar{E}_{\mathrm{f}} = 3\bar{E}_{\mathrm{s}}$) and $D_1 = -0.5$ ($\bar{E}_{\mathrm{f}} = \bar{E}_{\mathrm{s}}/3$) for $D_2 = 0$. The shaded region represents values of the phase angle for which the mode I stress intensity factor is negative.

which implies a phase angle in the third or fourth quadrant of the phase angle plane. Similarly, a positive value of m_{a} induces a positive value of stress intensity factor K_{I} whereas a positive value of Δt_{a} induces a negative K_{I}. On physical grounds, the net value of K_{I} cannot be negative, a constraint that limits the range of applicability of the solution being developed here. Thus, values of ψ in the fourth quadrant of the phase plane are anticipated in the present case, that is, $-90° \leq \psi \leq 0$.

The stress resultants Δt_{a} and m_{a} in (5.26) are both known explicitly as functions of buckle size a. Furthermore, values of angle $\omega(D_1)$ are known from Figure 4.28 for any combination of isotropic materials with $D_2 = 0$. The dependence of phase angle ψ on buckle size a is illustrated in Figure 5.6 for $D_1 = 0$ ($\bar{E}_{\mathrm{f}} = \bar{E}_{\mathrm{s}}$), $D_1 = 0.5$ ($\bar{E}_{\mathrm{f}} = 3\bar{E}_{\mathrm{s}}$) and $D_1 = -0.5$ ($\bar{E}_{\mathrm{f}} = \bar{E}_{\mathrm{s}}/3$). From Figure 4.28, it is seen that $\omega(-0.5) \approx 48°$, $\omega(0) \approx 52°$ and $\omega(0.5) \approx 59°$. As was anticipated from the dependence of the stress resultant ratio on delamination zone size shown in Figure 5.6, the phase angle $\psi(a)$ varies significantly with buckle size a, decreasing from a value of about $-35°$ when $a = a_{\mathrm{m}}$ to about $-90°$ when a has roughly doubled or tripled its initial size, depending on the value of the material parameter D_1. The region of the plot in Figure 5.6 for $\psi \leq -90°$ is shaded to emphasize that this region is inaccessible because negative values of K_{I} must be ruled out on physical grounds.

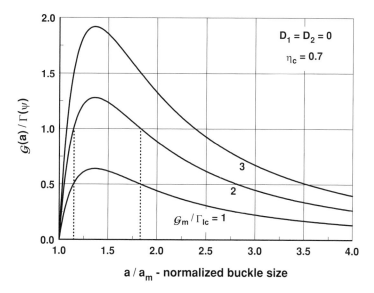

Fig. 5.7. Normalized driving force $\mathcal{G}(a)/\Gamma(\psi)$ plotted as a function of the normalized buckle size for $\mathcal{G}_m/\Gamma_{Ic} = 1$, 2 and 3, with the assumption that $D_1 = D_2 = 0$ and that $\eta_c = 0.7$.

The important role of edge stress phase angle can now be illustrated by combining the results obtained in (5.18), (5.20) and (5.26) in the form of plots of $\mathcal{G}(a)/\Gamma(\psi)$ versus a for several values of $\mathcal{G}_m/\Gamma_{Ic}$ in Figure 5.7. For purposes of illustration, only the case with $D_1 = 0$ and $\eta_c = 0.7$ is represented in the figure. As seen from the graphs, the delamination condition is not satisfied for any flaw size if $\mathcal{G}_m/\Gamma_{Ic} = 1$. On the other hand, the curve for $\mathcal{G}_m/\Gamma_{Ic} = 2$ crosses $\mathcal{G}/\Gamma(\psi) = 1$ at two values of a/a_m. The interpretation of the curve is as follows. The ratio $\mathcal{G}_m/\Gamma_{Ic} = 2$ sets the level of mismatch stress relative to interface strength for the system. At this stress level, any strip debonded zone that is wider than $2a_m$ results in the formation of a straight-sided buckle. Any buckle produced at a debonded zone of width $a > 1.15a_m$ results in further delamination of the film from the substrate. Finally, if a buckle broadens by advance of the delamination front at its edges, it can grow no larger than the size $a = 1.83a_m$.

5.1.4 Limitations for elastic–plastic materials

The analysis of the thin film buckling phenomenon discussed in this section is based on the assumptions that the material deformation is everywhere within the elastic range, and that inelastic deformation associated with the delamination front at the edge of the buckled zone is sufficiently localized to fall within the range of applicability of linear elastic fracture mechanics.

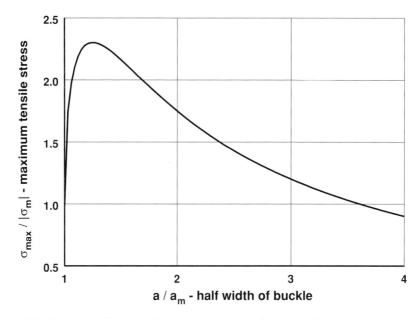

Fig. 5.8. Maximum tensile stress in absolute value at the edge of the buckled zone versus normalized size of the buckled zone. The maximum tensile stress normalized by $|\sigma_m|$ is defined in (5.27).

The restriction that the overall deformation should be elastic can be examined by considering the state of stress at the edge $x = a$ of the buckled zone. The membrane force resultant has the value $t_a < 0$ throughout the zone $-a < x < a$. On the other hand, the bending moment $m_x(x)$ varies throughout the zone, but it is readily confirmed that it does so between the limits of $-m_a$ and m_a. The tensile stress σ_{xx} that is largest in absolute value for this state of combined tension and bending is

$$\frac{\sigma_{max}}{|\sigma_m|} = \frac{|t_a|}{|t_m|} + 6\frac{m_a}{h_f|t_m|}, \tag{5.27}$$

where the stress has been normalized by the absolute value of the initial compressive stress in the film. A compressive stress of this magnitude arises on the face of the film away from the substrate at $x = \pm a$. The quantity σ_{max} is the stress magnitude that is to be compared to a plastic yield stress, say σ_Y, or perhaps to some other limiting stress magnitude.

The normalized maximum stress defined in (5.27) is plotted versus buckled zone size a/a_m in Figure 5.8. This stress has the value unity when $a = a_m$, and it increases rapidly to a local maximum value of about 2.3 when $a \approx 1.25a_m$. For larger values of a/a_m, the value of σ_{max} decreases monotonically, approaching zero asymptotically as $a/a_m \rightarrow \infty$. The stress component σ_{yy} is also compressive but smaller in magnitude than σ_{xx}, and the component $\sigma_{zz} = 0$ by the Kirchhoff hypothesis. Consequently, the maximum shear stress at this point is $\frac{1}{2}\sigma_{max}$. According to a maximum shear stress yield criterion, this is the quantity to be compared to the shear stress necessary to produce plastic flow, say $\frac{1}{2}\sigma_Y$ according to the Tresca criterion, where σ_Y is a tensile yield stress for an elastic–plastic film material. The strip buckling process can occur without general plastic deformation provided that

$$2.3|\sigma_m| \leq \sigma_Y. \tag{5.28}$$

A rough estimate of the limits on the range of applicability of elastic fracture mechanics for the present problem can also be established. For a tensile crack in an elastic solid, with the local field described in terms of an elastic stress intensity factor K, the effective stress on the crack plane ahead of the crack tip reaches the value of σ_Y at the distance from the edge of roughly K^2/σ_Y^2 (Rice 1968). This quantity, which has the dimensions of length, can serve as a rough estimate of the size of the inelastic zone around the crack edge. If the crack, or delamination in the present case, is at the stage of incipient growth, then $\mathcal{G} = K^2/\bar{E}_f = \Gamma$. The concepts of elastic fracture mechanics are applicable provided that the inelastic zone size is much smaller than the characteristic physical dimensions of the elastic solid containing the crack. In the case of delamination buckling, the film thickness h_f is the length parameter to which a comparison should be made. Thus, the requirement that

$$\boxed{\frac{\bar{E}_f \Gamma}{\sigma_Y^2} \ll h_f} \tag{5.29}$$

provides a restriction on parameters which must be enforced if film behavior is to be assessed within the framework of elastic fracture mechanics.

5.2 Buckling of a circular patch

The case of a circular buckle can be described in the same way as that of the straight-sided buckle, and it also involves essentially the same physical system of a film of uniform thickness h_f which initially is subjected to an equi-biaxial compressive stress σ_m. This stress is partially relaxed over a circular region of radius a as the film deflects away from the substrate; the film remains bonded to the substrate over the remainder of the interface. If the size of the buckled region increases by advance of the delamination front, it is assumed that the shape of the delamination region remains circular. It is not essential to assume that the shape also remains concentric but it is convenient to do so. The analysis proceeds by determining the post-buckling equilibrium configuration from which the energy release rate associated with expansion of the buckled zone can be calculated. The energy release rate is then compared to a measure of interface delamination resistance in order to gain an understanding of the mechanics of delamination behavior. The analysis is complicated significantly, however, by the fact that the equilibrium equations governing the post-buckling shape, corresponding to (5.5), are not linear. This difference necessitates the use of numerical methods for solution of the equilibrium equations. Two approaches to the problem have been developed in the literature. In one approach, the field equations are reduced to ordinary differential equations with the radial coordinate as the independent variable. These equations are not sufficiently transparent to reveal general aspects of behavior, and numerical methods must be used to extract useful information. Alternatively, the numerical finite element method provides a direct approach to the study of behavior.

An examination of the behavior of the system in the vicinity of the edge of the delamination zone in the context of fracture mechanics shows that the energy release rate is determined by the limiting values of the resultant extensional force and resultant bending moment in the buckled region of the film in the limit as the edge of the buckled region is approached. This line of reasoning is developed in this section, largely following the summary provided by Hutchinson and Suo (1992). Early contributions to this problem were made by Thompson and Hunt (1973), Yin (1985) and Chai (1990), among others. Notation introduced in the discussion of the straight-sided buckle is retained to the extent possible.

5.2.1 Post-buckling response

Consider an axially symmetric buckle that forms on a circular region of the interface of radius a. The depiction of the two-dimensional system in Figure 5.2 applies here as well with only minor modifications. The displacement of a point on the midplane of the film due to buckle formation depends only on its radial distance r from the axis of symmetry of the configuration. The nonzero components of displacement of a point on the film midplane are $u(r)$ in the radial direction and $w(r)$ in the transverse or z-direction; the displacement in the circumferential direction vanishes. These displacement components are zero in the initially flat but uniformly stressed configuration of the material. Upon formation of the buckle, the normal force $t_r(r)$ in the radial direction, measured per unit length in the circumferential direction, acting on any cross-section $r =$ constant is reduced in magnitude from its initial value $t_m = \sigma_m h_f < 0$. The reduction in force at the edge of the buckle is $\Delta t_a = t_a - t_m > 0$, where $t_a = t_r(a)$. Prior to buckle formation, the bending moment is everywhere zero in the film, but not so afterward. The resultant bending moment $m_r(r)$ per unit length in the circumferential direction acts on any cross-section

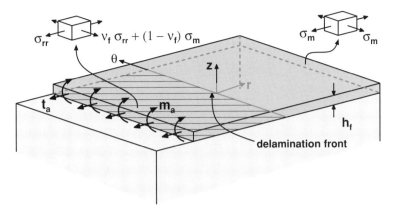

Fig. 5.9. Schematic representation of the edge force and bending moment for an axisymmetric buckle which forms on a circular region along the film–substrate interface.

$r = \text{constant}$. The edge bending moment is $m_{\mathrm{a}} = m_r(a)$, as shown in Figure 5.9. The kinematic boundary conditions at the edge of the film buckle are

$$u(a) = 0, \quad w(a) = 0, \quad w'(a) = 0, \tag{5.30}$$

and, at the center of the buckle,

$$u(0) = 0, \quad w'(0) = 0 \tag{5.31}$$

due to symmetry and continuity.

One way to proceed with analysis of the circular buckle is to adopt the von Kármán plate theory for small strain and moderate rotation. In this case, the strain–displacement relations are

$$\epsilon_{rr}(r) = \epsilon_{\mathrm{m}} + u'(r) + \tfrac{1}{2}w'(r)^2 - zw''(r), \quad \epsilon_{\theta\theta} = \epsilon_{\mathrm{m}} + \frac{1}{r}u(r) - \frac{z}{r}w''(r). \tag{5.32}$$

With expressions for strain in hand, the elastic energy density can be written in terms of $u(r)$ and $w(r)$. Integration of the strain energy density over the volume of the film $0 < r < a, 0 < \theta < 2\pi, -h_{\mathrm{f}}/2 < z < h_{\mathrm{f}}/2$ yields an expression for the total potential energy of the buckle as a functional of $u(r), w(r)$ over $0 < r < a$. The Euler equations which determine the particular functions for which the potential energy is stationary are the equilibrium equations for the configuration in the form of a pair of coupled nonlinear ordinary differential equations for $u(r)$ and $w(r)$. Additional natural boundary conditions follow from application of the variational principle. The differential equations and the boundary conditions can be solved numerically, as was done by Thompson and Hunt (1973), for example. Alternatively, the equilibrium configuration can be studied within the framework of finite deformation elasticity theory by means of the finite element method. The latter option is employed here, and specific numerical results are discussed after the mechanics of energy release at the edge of the buckle zone is considered.

The particular problem that has been solved approximately by means of the finite element method is the post-buckling response of a circular thin plate which is constrained against transverse deflection and midplane rotation along its outer edge, that is, $w(r) = 0$, $w'(r) = 0$ and $u(r) = 0$ at $r = a$. In the formulation adopted, the material response is linear elastic, but the magnitude of the deflection of the film midplane is unrestricted. The post-buckling response is determined by first deflecting the film laterally by means of an artificial pressure. A mismatch stress σ_m is then introduced by increasing the temperature, with a nonzero coefficient of linear thermal expansion assigned to the material, while the artificial pressure is gradually relaxed to zero. The resulting equilibrium solution defines a post-buckling shape for a given buckle radius a and mismatch stress σ_m. A range of post-buckling solutions can be obtained by finding equilibrium solutions as the mismatch stress σ_m is gradually increased or reduced by changing the temperature imposed on the plate. For each level of σ_m, the corresponding values of $w(0)$, t_a and m_a can be determined.

The procedure is very accurate until the mismatch extensional stress resultant t_m becomes very close to the classical buckling load of the circular film or plate, that is, at an extensional stress resultant of t_b. In other words, for $a \gg h_f$, the critical value of stress resultant t_b, or the equivalent mismatch stress $\sigma_b = t_b/h_f$, for buckling of a debond or blister of radius a is determined by recourse to the results based on nonlinear plate theory for buckling of a clamped circular plate subjected to an in-plane radial compressive loading along its perimeter. The critical load of the lowest axially symmetric buckling mode is

$$t_b = \sigma_b h_f = -1.2235 \frac{\bar{E}_f h_f^3}{a^2}. \tag{5.33}$$

For a fixed value of the Poisson ratio, all results that are important for present purposes can be expressed in terms of the ratio of t_m to the buckling load for the current buckle radius or, equivalently, in terms of the ratio of the current buckle radius a to the smallest radius that is unstable at the in-plane normal stress resultant t_m. The latter option is followed here. The results are not very sensitive to changes in the value of the Poisson ratio ν_f, and all numerical results included have been obtained for $\nu_f = \frac{1}{4}$. An asymptotically exact relationship between the compressive stress in the undeflected film and the transverse deflection $w(0)$ of the film along the axis of symmetry following buckling was obtained by Hutchinson and Suo (1992). In terms of the critical buckling stress resultant t_b given in (5.33), this asymptotic

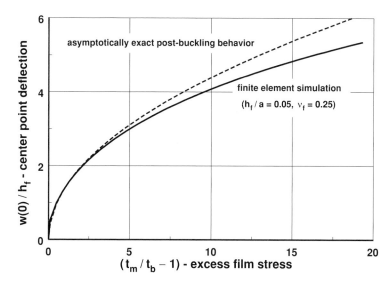

Fig. 5.10. Normalized transverse deflection along the axis of symmetry following buckling versus the normalized value of excess compressive stress in the film. The figure shows the asymptotic prediction of post-buckling response (dashed line) and finite element simulation (solid line) for $h_f/a = 0.05$ and $v_f = 0.25$.

relationship is

$$\frac{w(0)}{h_f} = \sqrt{\frac{t_m/t_b - 1}{c_v}}, \qquad c_v = 0.2473 \left[(1 + v_f) + 0.2231(1 - v_f^2) \right]. \quad (5.34)$$

Note that the parameter c_v depends only on the Poisson ratio of the film. The ratio $w(0)/h_f$ was also extracted from a finite element simulation of the buckling process for an elastic film with $h_f/a = 0.05$ and $v_f = 0.25$, and the result is illustrated by the solid curve in Figure 5.10. The asymptotic result (5.34) has been evaluated for the same parameters, and the result is shown as the dashed curve in Figure 5.10. The numerical results are in excellent agreement with the asymptotic result where the latter is valid, namely, for $0 \le (t_m/t_b - 1) \ll 1$, and it diverges only moderately for much larger values of the excess film stress.

The smallest circular zone of delamination which is unstable in the presence of a mismatch force $t_m < 0$ is denoted by a_m and is given by

$$a_m = 1.106 h_f \sqrt{\frac{\bar{E}_f}{|\sigma_m|}}. \quad (5.35)$$

This is the analog for a circular buckle of the quantity identified by the same symbol for the straight-sided buckle in (5.8). For a buckle of this radius at the compressive stress level σ_m, the edge stress resultants are $t_a = t_m$ and $m_a = 0$. For any larger

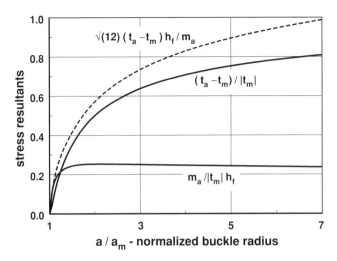

Fig. 5.11. Variation of the normalized edge stress resultant force $(t_a - t_m)/|t_m|$ and moment $m_a/|t_m|h_f$ (solid lines), and the normalized edge moment $\sqrt{(12)}\,\Delta t_a h_f/m_a$ (dashed line) as functions of the normalized debond size a/a_m for the circular buckle.

buckle $a > a_m$, the edge force t_a is less in magnitude than t_m and the edge moment m_a is greater than zero. Numerical results for the normalized quantities $\Delta t_a/|t_m|$ and $m_a/|t_m|h_f$ versus a/a_m are shown in Figure 5.11. It is seen from the figure that $\Delta t_a = t_a - t_m$ increases gradually from zero as a increases from a_m at fixed t_m, and that it approaches unity as a becomes much larger than a_m. The edge bending moment, on the other hand, increases from zero at $a = a_m$ to a maximum value at about $a/a_m = 2$, whereafter it decays very slowly in value with increasing values of a/a_m. The dependence of the ratio $\Delta t_a h_f/m_a$, which determines the magnitude of the phase angle of the local interface stress state, is also shown in the same figure.

5.2.2 *Example: Temperature change for buckling of a debond*

A 1 μm thick diamond-like carbon film is deposited at 500 °C on a Ti alloy substrate. The film, with elastic modulus $E_f = 500$ GPa and Poisson ratio $\nu_f = 0.2$, is essentially free of internal stress at the deposition temperature. When cooled to the temperature 20 °C, however, an equi-biaxial compressive mismatch stress of 5 GPa is expected to exist in the film as a consequence of thermal mismatch with the substrate. An unbonded circular patch, 30 μm in diameter, developed at the film–substrate interface during film deposition. Determine whether the film buckles upon cooling to 20 °C and if so, determine the temperature at which buckling begins.

Solution:

Using the notation introduced in Section 5.2.1, the radius of the circular region of debond at the interface $a = 15\,\mu$m, the film thickness $h_f = 1\ \mu$m, and the film biaxial modulus

$\bar{E}_f = E_f/(1 - v_f^2) = 521$ GPa. A temperature change $\Delta T = 480\,°C$ from the stress-free deposition temperature to room temperature, produces a compressive mismatch stress $\sigma_m = -5$ GPa. The mismatch stress necessary to cause the film to buckle is given by (5.33) as

$$\sigma_b = -1.2235\,\frac{\bar{E}_f h_f^2}{a^2} = -2.83\,\text{GPa}. \qquad (5.36)$$

Since this compressive mismatch stress is smaller than that in the film upon cool down to room temperature, it is seen that the film buckles during cooling from the deposition temperature. Noting that the compressive thermal mismatch stress in the film varies linearly between 500 °C and 20 °C, from an initial value of zero to a final value of 5 GPa, the film mismatch stress which would cause the film to buckle upon a decrease in temperature ΔT_b from the deposition temperature is

$$\sigma_b = \sigma_m \frac{\Delta T_b}{\Delta T} \qquad \Rightarrow \qquad \Delta T_b = 272\,°C. \qquad (5.37)$$

Thus, the temperature at which buckling begins is $T_b = (500 - 272)\,°C = 228\,°C$.

5.2.3 *Driving force for delamination*

Consider a portion of the edge of the delamination zone that is close to the edge compared to a, as depicted in Figure 5.9. At this level of observation, the edge is essentially straight and the state of deformation is generalized plane strain. Far ahead of the delamination front (on this scale), the stress state is uniform biaxial compression without bending. Far behind the front, but still close to the edge, the stress resultants are the normal force t_a and the bending moment m_a which were introduced above.

The superposition argument outlined in Figure 5.12 shows that the energy release rate associated with advance of the delamination front due to this loading depends only on m_a and on the extensional force *reduction* $\Delta t_a = t_a - t_m$. The stress state in (B) is a uniform equi-biaxial stress of magnitude $-t_m$; the state (C) is the superposition of states (A) and (B). Because there is no energy release associated with advance of the front in (B), the energy release rates for (A) and (C) are identical. This can be seen by considering the portion of the film between the loaded end in part (C) of Figure 5.12 and the delamination front. As the front advances, its complicated nonuniform stress field advances self-similarly because the loading on that region remains unchanged. A consequence is that the energy of this region remains unchanged. Thus, the energy change which represents the energy release rate can be calculated from the spatially uniform fields near the loaded end of the film segment behind the front. The elastic energy due to advance of the front a unit distance increases, while the external potential energy of the loading decreases. The anticipated spatial uniformity of the stress fields near the loaded end suggests the use of linear elastic plate theory as a sufficiently accurate model. The state of

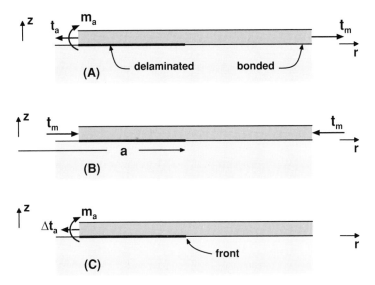

Fig. 5.12. (A) Schematic showing a uniform equi-biaxial force t_m ahead of the circular delamination front and an extensional force t_a and bending moment m_a behind the front. (B) represents a state of uniform equi-biaxial compression. The superposition of (A) and (B) leads to the stress state shown in (C).

deformation in part (C) is essentially plane strain. In terms of the stress resultants t_a and m_a, the elastic energy per unit length in the r-direction (per unit length in the circumferential direction) is $(\Delta t_a^2/h_f + 12m_a^2/h_f^3)/2\bar{E}_f$. Advance of the delamination front a unit distance in the r-direction with fixed loads Δt_a and m_a acting on the remote material boundary results in an increase of elastic energy by that amount, but a decrease in external potential energy of twice that amount. Thus, an analysis on this basis leads to

$$\mathcal{G}(a) = \frac{1}{2\bar{E}_f}\left(\frac{\Delta t_a^2}{h_f} + 12\frac{m_a^2}{h_f^3}\right) \tag{5.38}$$

as the energy release rate for this configuration, which is identical to (5.21). The main inference to be drawn from this result is that the energy release rate for advance of the delamination front is determined by the edge loads $\Delta t_a = t_a - t_m$ and m_a acting on the buckle. These loads must be determined from the numerical simulation of post-buckling behavior, as described in the preceding section.

The dependence of energy release rate on a/a_m implied by these stress resultants is determined from the numerical results according to (5.38) and is shown in Figure 5.13. The energy release rate is normalized by $\mathcal{G}_m$ which is the energy release rate for advance of a delamination zone with both $t_a = 0$ and $m_a = 0$ as determined in (5.18). The dependence of energy release rate on Poisson's ratio is weak.

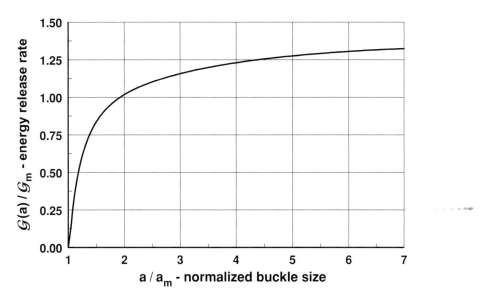

Fig. 5.13. Normalized energy release rate $\mathcal{G}(a)/\mathcal{G}_\mathrm{m}$ plotted as a function of the normalized buckle size a/a_m for the circular delamination.

The ratio of stress resultants $h_\mathrm{f}\Delta t_\mathrm{a}/m_\mathrm{a}$ has been included in Figure 5.11. This ratio determines the phase angle ψ of the interfacial stress state at the edge of the delamination zone. The fairly strong dependence of the stress resultant ratio on a implies that the phase angle of the edge stress changes as the buckle expands, similar to the behavior that was observed for the case of the straight-sided buckle. The dependence of phase angle $\psi(a)$ on buckle size implied by the ratio of stress resultants is shown in Figure 5.14 for three combinations of elastic properties of the film and substrate materials. With ν_f fixed at the value $\frac{1}{4}$, the three cases are $\bar{E}_\mathrm{f} = \bar{E}_\mathrm{s}$ $(D_1 = 0)$, $\bar{E}_\mathrm{f} = 3\bar{E}_\mathrm{s}$ $(D_1 = 0.5)$ and $\bar{E}_\mathrm{f} = \bar{E}_\mathrm{s}/3$ $(D_1 = -0.5)$, all with $D_2 = 0$. The corresponding values of ω in (5.26) are approximately $52°$, $48°$, and $59°$, respectively.

For the case of a straight-sided buckle, the ratio of the delamination driving force $\mathcal{G}(a)$ to the phase-angle-dependent delamination resistance $\Gamma(\psi)$ was shown in Figure 5.7 for a particular choice of the parameter η_c introduced in (5.20) and for three values of the system parameter $\mathcal{G}_\mathrm{m}/\Gamma_\mathrm{Ic}$. A similar representation could be produced for the case of a circular buckle. An alternate representation in terms of a *mode-adjusted driving force*, which is somewhat more readily interpreted, was introduced by Hutchinson et al. (1992) in their discussion of the circular buckle. In the present context, the mode-adjusted driving force for delamination is defined as the nondimensional quantity $[\mathcal{G}(a)/\mathcal{G}_\mathrm{m}][\Gamma_\mathrm{Ic}/\Gamma(\psi)]$. The dependence of this mode-adjusted driving force on buckle size a/a_m for fixed σ_m is shown as the solid

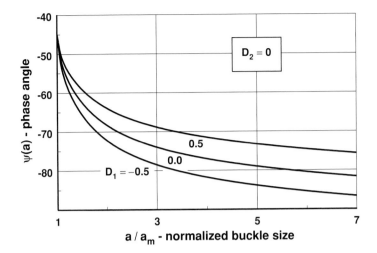

Fig. 5.14. Variation of phase angle $\psi(a)$ with normalized circular buckle size a/a_m for three different values of the Dundurs parameter D_1 (–0.5, 0 and 0.5), all with $D_2 = 0$ and $v_f = \frac{1}{4}$.

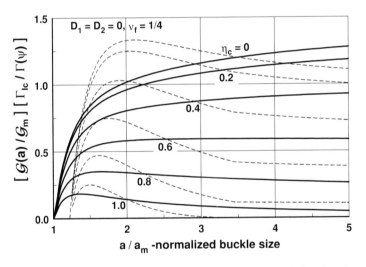

Fig. 5.15. Mode-adjusted driving force for circular delamination, defined as the nondimensional quantity $[\mathcal{G}(a)/\mathcal{G}_m][\Gamma_{\mathrm{Ic}}/\Gamma(\psi)]$, versus the normalized circular buckle size a/a_m for fixed film mismatch stress σ_m for six values of η_c (solid curves), for the particular case where the film and substrate have no elastic mismatch and where $v_f = \frac{1}{4}$. The dashed curves show the corresponding results for the straight-sided buckle.

curves in Figure 5.15 for six values of the parameter η_c and for the particular case when the film and substrate have identical elastic properties. A given level of film stress is most effective in causing delamination when $\eta_c = 0$, and the effectiveness diminishes as η_c increases. This ordering of the curves of mode-adjusted driving force versus buckle size could have been anticipated by recalling that η_c is a measure of the relative importance of shear effects in driving delamination; an increasing η_c corresponds to decreasing importance of local shear stress on delamination.

The set of dashed curves included in Figure 5.15 shows the dependence of the mode-adjusted driving force on buckle size for a straight-sided buckle. Thus, the figure permits a direct comparison of the mode adjusted driving force for the circular and straight-sided buckles for a given set of system parameters, namely, the elastic constants, σ_m, h_f, Γ_{Ic} and η_c. For the straight-sided buckle, the point of zero driving force is shifted slightly to the right of the corresponding point for a circular buckle because the value of a_m in the former case is approximately 0.82 times its value in the latter case. Also, a slight kink is evident in the dashed curves at a value of a/a_m of roughly 3.5. This kink corresponds to the amount of buckle advance at which the buckle edge phase angle ψ falls below $-90°$, an eventuality that must be ruled out on physical grounds. Therefore, for a/a_m greater than roughly 3.5, the value of ψ is held at $-90°$. The comparison of the two cases suggests that, in the early stages of advance of buckle delamination, the driving force for the straight-sided buckle is larger than for a circular buckle. For larger amounts of growth of the delamination zone, on the other hand, the relative magnitudes of the driving forces are reversed for all values of η_c. This feature may be a contributing factor in creating the irregular shapes of delamination buckling patterns commonly observed (Hutchinson et al. 1992).

5.2.4 Example: Buckling of an oxide film

Oxide films of different thicknesses were thermally grown at a temperature of 1200 °C on a relatively thick metal alloy substrate. Upon cooling to room temperature (20 °C), the oxide films were found to develop a large equi-biaxial residual compressive stress, primarily as a consequence of its thermal contraction mismatch with the substrate. The elastic modulus and Poisson ratio for the film are $E_f = 400\,\text{GPa}$ and $\nu_f = 0.25$, respectively, at room temperature, and the coefficients of thermal expansion for the film and the substrate are $\alpha_f = 8 \times 10^{-6}\,°\text{C}^{-1}$ and $\alpha_s = 14 \times 10^{-6}\,°\text{C}^{-1}$, respectively.

(a) Determine the size of the smallest edge debond at the interface between the film and the substrate which would form a straight-sided buckle when the oxide–alloy system is cooled to room temperature in the series of experiments in which the film thickness was 5 μm.

(b) Experiments on interface fracture toughness, such as those described in Section 4.5, indicate that the mode I separation energy for the interface between the oxide and the alloy is approximately $\Gamma_{Ic} = 28\,\text{J}\,\text{m}^{-2}$ for $h_f = 5\,\mu$m, and that the value of the parameter η_c in (5.20) is 0.7. Assuming that the elastic mismatch between the film and the substrate does not have a noticeable influence on the growth of the delamination, describe if and how a straight buckle produced at the film edge would propagate.

(c) In another set of experiments, circular imperfections were found along the film–substrate interface for a film thickness of 5 μm. If complete debonding occurs along the film–substrate interface over the area of such imperfections, determine the smallest imperfection size which would cause spontaneous deflection of the film away from the substrate.

Solution:

(a) The equi-biaxial mismatch stress in the film upon cooling from the oxidation temperature of 1200 °C to 20 °C is

$$\sigma_{m} = \frac{E_f}{1 - \nu_f} \, (\alpha_s - \alpha_f) \, \Delta T = -3.776 \text{ GPa.} \tag{5.39}$$

For $h_f = 5$ μm, the width of the smallest interface delamination at the edge of the film which would buckle under the influence of the compressive mismatch stress in the film is found from (5.8) to be $a_m = 45.6$ μm.

(b) Figure 5.7 shows the variation of $\mathcal{G}(a)/\Gamma(\psi)$, the ratio of driving force for delamination to the phase-angle-dependent separation energy, as a function of a/a_m for $D_1 = D_2 = 0$ and $\eta_c = 0.7$, parameters which represent the circumstances assumed here for the oxide–alloy system under consideration. From (5.17), the plane strain energy release rate for the advance of a straight-sided edge delamination front in the oxide is found to be

$$\mathcal{G}_m = \frac{\sigma_m^2 h_f \left(1 - \nu_f^2\right)}{2 E_f} = 83.5 \text{ J m}^{-2}, \tag{5.40}$$

from which it is seen that $\mathcal{G}_m / \Gamma_{Ic} \approx 3$. Referring to the result plotted in Figure 5.7, it is seen that the $\mathcal{G}_m / \Gamma_{Ic} = 3$ curve crosses the $\mathcal{G}(a)/\Gamma(\psi) = 1$ line at $a/a_m \approx 1.2$ and again at $a/a_m \approx 2.37$. Thus, if the debond zone width a is greater than $1.2 a_m \approx 55$ μm, then delamination of the film from the substrate would continue to progress until $a \approx 2.37 a_m = 108$ μm.

(c) The critical radius of smallest imperfection which induces spontaneous buckling of a circular debond is found from (5.35) to be $a_m = 58.8$ μm.

5.3 Secondary buckling

The elastic strain energy in a film is reduced from its value for an initially flat film by the spontaneous formation of a buckle. This reduction results from relaxation of some, but not necessarily all, stress components in the film stress components. For example, in the case of the straight-sided buckle depicted in Figure 5.2, the compressive stress in the film acting in the direction normal to the edge of the delamination zone, that is, the component σ_{xx}, can be reduced in magnitude substantially by buckle formation and subsequent buckle growth. On the other hand, the stress component σ_{yy} which acts in the direction parallel to the sides of the buckle is relaxed in magnitude only indirectly through the Poisson effect; the strain component ϵ_{yy} remains unchanged from its initial value of ϵ_m as a result of primary buckle formation and the stress component σ_{yy} remains relatively large. Thus, it is plausible that the compressive stress acting along the length of the straight-sided buckle following primary buckling can eventually induce *secondary buckling* in this direction, thereby disrupting the translational invariance of the configuration in the y-direction.

The case of a circular buckle is similar, although it is somewhat more complicated to analyze due to its axially symmetric configuration. Upon formation of a circular buckle, the radial stress is certainly reduced in magnitude from its initial value of $|\sigma_m|$. Near the edge of the circular delamination zone, the film is constrained against relaxation of strain in the circumferential direction by the portion of the film that is still bonded to the substrate. Therefore, the circumferential compressive stress is still large in magnitude near the edge of the delamination zone following formation of a circular buckle. This large compressive stress has the potential for inducing secondary buckling in the circumferential direction, thereby disrupting the rotational invariance of the circular buckle configuration. Near the center of a well-developed axially symmetric circular buckle, however, both the radial and circumferential stress components are substantially relaxed; indeed, both these stress components may become positive as a result of the three-dimensional effects. Therefore, a secondary buckling configuration that might be expected for a circular buckle is a configuration with wrinkling of the film shape in the circumferential direction near the edge of the zone of delamination, but with little or no wrinkling of the film surface near the center of the zone of delamination. An example of this kind of secondary buckling configuration for a circular delamination is illustrated in the next section.

In either case, the process of primary buckle formation can be viewed as a bifurcation of equilibrium states for a given delamination zone of size a and increasing loading parameter $|\sigma_m|$. The buckled configuration represents a stable branch of equilibrium states following bifurcation. In the same way, the formation of a secondary buckle is a second generation bifurcation of equilibrium states, with the more irregular film shape representing the stable branch of equilibrium states and the symmetric primary buckling configuration representing the unstable branch following this bifurcation. Such phenomena have not yet been studied systematically within the framework of buckle driven delamination of compressed thin films but a framework for doing so has been proposed by Jensen (1993) and Audoly (1999). Some indication of secondary buckling behavior can be seen by examining cases of secondary post-buckling configurations for a straight-sided delamination zone. Secondary buckling modes are illustrated in Figure 5.16 in the form of level curves of transverse film deflection $w(x, y)$ normalized by the film thickness h_f over a portion of the undeflected film plane. The undeflected compressed film extends over the strip $-a \leq x \leq a$, $-\infty < y < \infty$, and the primary buckling shape is a crest of uniform profile in the y-direction. Secondary buckling modes were sought that

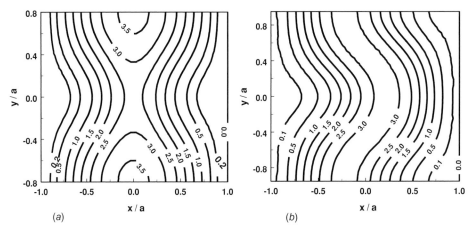

Fig. 5.16. Post-buckling configurations plotted for a stress level that is approximately nine times the primary buckling stress, or 1.5 times the secondary buckling stress. These figures show contours of constant values of transverse deflection $w(x, y)$ of the film, normalized by the film thickness h_f, over a portion of the undeflected film plane. Part (a) shows the mode that is symmetric in the x-direction for fixed y, whereas part (b) shows the most easily achieved mode in the y-direction for fixed x.

are periodic in the y-direction. Figure 5.16(a) shows what appears to be the most easily achieved mode that is symmetric in the x-direction for fixed y, which has a wavelength in the y-direction of approximately $1.6a$. Similarly, Figure 5.16(b) shows the most easily achieved mode that is asymmetric in the x-direction for fixed y; this mode has a wavelength of approximately $1.9a$ in the y-direction. For a fixed value of a, *both* of these modes arise for a film stress magnitude that is roughly six times the stress required to form the primary straight-sided buckle, that is, about $6\pi^2 \bar{E}_f (h_f/a)^2$. Both Figures 5.16(a) and (b) are post-buckling configurations plotted for a stress level that is approximately nine times the primary buckling stress, or 1.5 times the secondary buckling stress. Alternatively, if σ_m is the stress level at which a straight-sided buckle is formed for a strip width a_m, as given in (5.9), and if the delamination zone subsequently broadens to a width of a with the stress level held fixed, then these secondary buckling modes become possible once a has increased to approximately $2.5a_m$.

Perhaps the most noteworthy aspect of the secondary buckling phenomenon illustrated here is that it is accompanied by an energy release rate at the edge of the zone of delamination that varies with position in the y-direction. This provides a mechanism by which a straight-sided delamination zone may develop a wavy shape through buckle driven delamination, or a circular delamination zone may develop a lobed shape. The consequences of secondary buckling in thin films have not been studied systematically.

5.4 Experimental observations

A rich variety of experimental observations has been reported in the literature on the buckling of compressively stressed thin films on substrates and on the development of secondary buckling phenomena which lead to configurations with reproducible characteristic shapes. Examples of such observations are presented in this section.

5.4.1 Edge delamination

A series of experiments was reported by Thouless et al. (1992) who compared observations on buckle driven delaminations of thin films under plane strain compression with the analysis of the phenomenon presented in Section 5.1.2. The experiments were conducted with sheets of mica of thickness in the range $30\,\mu\text{m} \leq h_f \leq 120\,\mu\text{m}$ bonded to a steel substrate. For this material system, the values of the Dundurs parameters are roughly $D_1 = 0.09$ and $D_2 = 0.06$. A debonded zone of predetermined width could be introduced artificially. The substrate thickness was about 6 mm and the sample width was 25 mm, large enough to maintain the plane strain constraint. The mismatch stress was generated by subjecting the composite system to loading in the four-point bending arrangement, resulting in pure bending in the region of the debonded zone. In the context of the present discussion, this loading arrangement provided control over the level of the compressive stress $|t_m|$ in the film. The width of the debonded zone increased as the load was increased beyond the level required to initiate the process. Based on the assumption of an interface delamination resistance of the form (5.20), a value of the parameter η_c of about 0.7 was deduced for this system as long as the delamination was actually open, that is, as long as $\psi > -90°$. In some cases, it was observed that the delamination continued to advance with $\psi = -90°$, implying that it did so exclusively in mode II deformation and that there was contact between the film and substrate surfaces within some portion of the delamination zone.

5.4.2 Initially circular delamination

In a series of experiments, Ogawa et al. (1986) demonstrated the configurational instabilities in focused ion beam sputtered and magnetron sputtered molybdenum films on glass substrates where the nucleation and progression of film delamination and buckling were documented using Nomarski interference contrast microscopy. The delamination and buckling of SiC coatings on Si substrates were studied experimentally by Argon et al. (1989) in an investigation of the intrinsic toughness of interfaces. They deposited amorphous hydrogenated thin films of SiC, 0.1–1.0 μm in thickness, on (100) Si single crystal substrates, 250 μm in thickness and 25.4 mm in diameter, using a plasma-assisted chemical vapor deposition (PACVD) process.

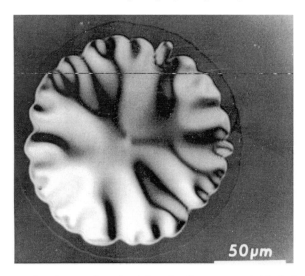

Fig. 5.17. An example of secondary buckling leading to a wrinkled shape in a circular blister in a SiC thin film on a (100) Si substrate, as revealed by Nomarski interference contrast microscopy. Reproduced with permission from Argon et al. (1989).

The as-deposited SiC films were in a state of residual compression, with the magnitude of the film mismatch stress increasing from about 700 MPa for the thinnest film to approximately 2 GPa for the thickest film. When the coating thickness was in excess of a critical value, spontaneous circular delamination was observed to initiate from interface flaws whose origin was linked to occasional local contaminants along the interface. The circular delaminations along the interface were essentially non-propagating defects when the SiC film thickness was much less than 1 μm. This is consistent with the expectations predicated upon the analyses presented in Section 5.2. However, when the film thickness exceeded 1 μm, complex wrinkles developed along the perimeter of the expanding circular delaminations, consistent with the mechanisms discussed qualitatively in Section 5.3. Figure 5.17 shows an image of such a blister taken in Nomarski interference contrast microscopy. The dark outer ring that is about 5–7 μm beyond the outer edge of the buckle was ascribed by Argon et al. (1989) to the relative radial slippage between the film and the substrate.

Hutchinson et al. (1992) performed a systematic experimental study of the growth and configurational stability of initially circular delaminations of films in equibiaxial compression. Their material system comprised a thin film of mica which was bonded to a relatively thick aluminum substrate by means of a thermoplastic resin. Mica sheets, with in-plane dimensions of 150 mm × 150 mm and 14–130 μm in thickness, were chosen as the film material because of their availability in large quantities and in a variety of uniform thicknesses. The aluminum substrate offered a high coefficient of thermal expansion and good thermal conductivity. Thus, by

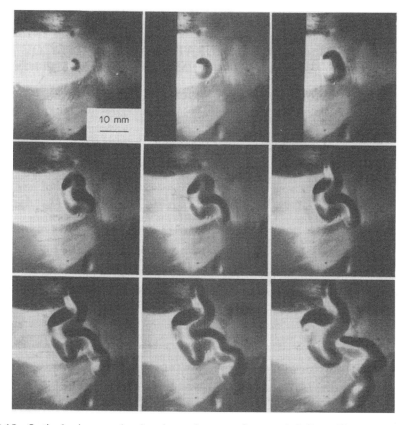

Fig. 5.18. Optical micrographs showing axisymmetric growth followed by telephone cord instability in response to an increase in mismatch stress in a mica film bonded to an aluminum substrate. Reproduced with permission from Hutchinson et al. (1992).

varying the temperature of the film–substrate system, the magnitude of the mismatch stress in the film could be controlled by exploiting the difference in thermal expansion coefficients between mica and aluminum. Residual stress in the film was estimated by monitoring substrate curvature changes during thermal excursion using optical interferometry, and the reference temperature below which the mismatch stress developed was estimated to be $52 \pm 2\,°C$. At room temperature, the equi-biaxial compressive stress in the mica film was approximately 100 MPa. Values of the Dundurs parameters for the mica–aluminum system are $D_1 = 0.4$ and $D_2 = 0.1$. For these values, the corresponding value of the parameter ω defined in (4.50) is essentially the same as that shown in Figure 4.28 for $D_1 = 0$ and $D_2 = 0$. This provides a rationale for applying the results for a homogeneous system to the mica–aluminum bilayer.

Controlled circular delaminations of different diameters were introduced at the interface between the mica and the resin by using a screw that was threaded through the substrate, from the back face of the substrate to the interface. Experiments conducted with different values of film stress σ_m and different thicknesses of mica film (which also gave rise to different values of interface fracture energy Γ) revealed that the growth of delamination occurred in a relatively narrow range of values of $\sigma_m^2 h_f / \bar{E}_f \Gamma$.

There exists no driving force for an initial interface delamination if the film does not undergo buckling. Once buckling occurs, the interface delamination begins to advance when the energy release rate exceeds the interface fracture energy which depends on the local stress state phase angle. The shape of the buckle driven delamination is also a strong function of the energy release rate. Figure 5.18 comprises a sequence of optical micrographs which reveal the growth of an initially circular interfacial debond under the influence of an increasing film stress σ_m which was generated by lowering the temperature of the mica–aluminum system. Note that the blister retains a near-circular shape at lower values of σ_m; the circular symmetry is lost in the left middle micrograph after which a so-called telephone cord buckle configuration develops. The term telephone cord buckle has been adopted generally to describe film buckles that are essentially straight over distances many times the width of the buckles, but that consist of curly segments with dimensions on the scale of the width of the buckles that are periodically arrayed along the length. It was suggested that a possible origin of nonaxisymmetric shapes is the instability of the expanding circular delamination front to noncircular perturbations. The fact that the mode-adjusted crack driving force decreases with increasing radius of curvature of the circular blister, as shown in Figure 5.15 for η_c in the range 0.7 to 1.0, beyond a certain critical radius would indicate that a local radius of curvature which is smaller than the average radius of curvature would lead to a higher local driving force when the blister size far exceeds the critical radius. In addition, note that the relative significance of mode II in influencing local stress state phase angle increases less precipitously for the circular blister than for the straight-sided blister, as seen in a comparison of Figure 5.14 with Figure 5.6. Consequently, perturbations with a smaller local curvature or shapes that lead to enhanced waviness are favored beyond a certain delamination radius because a curved crack front is subject to a higher driving force than a straight-sided one. Such arguments provide a rationale for the observed increase in the number of lobes in the shape of an initially circular delamination with an increase in the equi-biaxial compressive stress in the film (Hutchinson et al. 1992). The unstable delamination phenomena described here for film–substrate systems have also been observed in connection with interfacial debonding in laminated composite materials (Whitcomb 1986).

An example of the progressive growth, as a function of time, of wavy ridge patterns in a compressively stressed film is provided in Figure 5.19 from the work of Ogawa et al. (1986). This optical micrograph shows the advance of a telephone cord

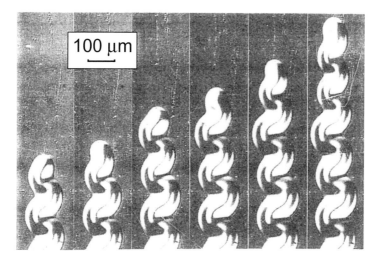

Fig. 5.19. A sequence of optical micrographs showing the advance of a telephone cord instability in a magnetron sputtered molybdenum thin film on a relatively thick glass substrate. The six images shown here reveal the progressive growth of the wavy ridge pattern in the film at 0, 180, 512, 692, 872 and 1,052 s after the film–substrate system was removed from the sputtering chamber and placed in an optical microscope. Reproduced with permission from Ogawa et al. (1986).

instability in a magnetron sputtered, 900 ± 90 nm thick molybdenum film on a relatively thick glass substrate. The images shown in this figure were photographed immediately after the film–substrate system was removed from the sputtering chamber and placed in an optical microscope for observation. The equi-biaxial compressive stress in the film was estimated to be approximately 2.2 GPa. Note that the wavy ridge pattern, approximately 100 μm wide, advances in a straight line, with the length of the telephone cord pattern increasing by an amount equal to its wavelength every 300 s or so.

5.4.3 Effects of imperfections on buckling delamination

Axisymmetric buckling of a thin film on a flat substrate surface becomes possible when there exists an interface separation with minimum radius a_m as given in (5.35). Analysis of this phenomenon is predicated on the tacit assumption that the film is nominally uniform in thickness and that the interface is nominally flat; in other words, it is assumed that no large scale imperfections, inclusions or other defects exist along the interface between the film and the substrate. For typical values of elastic moduli and film mismatch stress that are representative of thermal barrier and wear resistant coatings on metallic and brittle substrates, such homogeneous nucleation of a buckle requires the existence of interface flaws whose minimum size

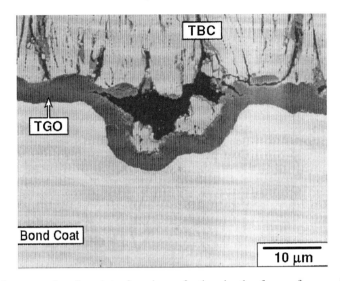

Fig. 5.20. An example of an interface imperfection in the form of an undulation in a thermal-barrier coating (TBC) system comprising a Pt aluminized bondcoat on a Ni-based superalloy substrate. A thermally grown oxide (TGO) layer is formed between the bondcoat and the zirconia TBC. Thermal cycling leads to interfacial fracture and cracking in the TBC in the region of high local tensile stress adjacent to the TGO layer. Such cracking and interfacial delamination ultimately lead to the onset of large scale buckling. Reproduced with permission from Evans et al. (2001).

is as large as $20h_f$. In fact, experimental observations reveal that buckling can occur for interface flaw sizes that are much smaller than $20h_f$ (Christensen et al. 1997, He et al. 1998). Figure 5.20 shows an example of an interface imperfection in the form of an undulation or wrinkle along the interface between a Pt aluminized bondcoat and a thermally grown oxide (TGO) layer of α-Al$_2$O$_3$. The bondcoat is processed by depositing a 3–10 μm thick layer of Pt on a Ni-based superalloy substrate and then aluminizing it by chemical vapor deposition. Aluminum from the bondcoat in conjunction with the ingress of oxygen from the top layer of thermal-barrier coating (TBC), which is typically a cubic/tetragonal phase zircon containing yttrium in solid solution, leads to the formation of α-Al$_2$O$_3$. This *in situ* buildup of the oxide results in a high compressive stress in the TGO layer, typically on the order of 3–6 GPa.

Analysis of the possible role of imperfections in influencing buckling-driven delamination indicates that the presence of imperfections along the film–substrate interface could markedly alter the conditions for the nucleation and growth of the buckle. The nucleation of delamination at imperfections located along an interface is generally considered to be associated with a peak value of the energy release rate $\mathcal{G}_{max}$ (Evans et al. 1998, He et al. 1998). In the absence of buckling, small flaws are nucleated at sites of imperfections along interfaces near regions of highest local tensile stress with the result that the flaws so induced pop in along the interface and then arrest. Concomitantly, the energy release rate, which attains a peak value at the onset of delamination nucleation at imperfections, subsequently declines and reaches the interface fracture energy Γ when the delamination arrests. If there exists an interface flaw whose radius is larger than a_m, as given in (5.35), the film buckles. As a result, the energy release rate rises again and approaches a value with magnitude comparable to $\mathcal{G}_m$ as the delamination propagates; see Figure 5.13. These opposing trends during the nucleation and growth of delamination suggest a minimum in the value of the energy release rate, say $\mathcal{G}_{min}$. If this minimum value approaches the work of separation Γ of the interface, the buckle continues to advance.

Hutchinson et al. (2000) have performed finite element simulations of energy release rates for interface imperfections in the form of axisymmetric undulations of different wavelengths and amplitudes. Their results suggest that the variation of $\mathcal{G}_{min}/\mathcal{G}$ with a_0/a_m is independent of the residual strain $\sigma_m/\bar{E}_f$, where a_0 is the radius of the interface wrinkle or imperfection. When the size of the axisymmetric interface flaw is in the range $0.12a_m \leq a_0 \leq 0.6a_m$, the minimum value of the energy release rate for buckle propagation is of the form

$$\mathcal{G}_{min} \approx 0.4 \left(\frac{a_0}{a_m} \right) \mathcal{G}_m. \tag{5.41}$$

When $\mathcal{G}_{min}$ is equated with the interface fracture energy at the appropriate local stress state phase angle ψ, a critical wavelength or radius for the interface imperfection $a_0 = a_{cr}$, can be extracted by combining (5.41) and (5.17) to yield

$$a_0^{cr} \approx 11 \left(\frac{\bar{E}_f \Gamma}{\sigma_m^2} \right) \sqrt{\frac{\bar{E}_f}{\sigma_m}}. \tag{5.42}$$

This result implies that when the imperfection wavelength is smaller than a_{cr}, the energy release rate is not sufficiently high to induce buckling. On the other hand, when the imperfection wavelength is larger than a_0^{cr}, a buckle forms, provided that there is a sufficiently large interface separation, and it evolves into a telephone cord-like shape.

It is noted that a_{cr} as given by (5.42) is independent of the film thickness h_f, which is counter to experimental evidence. Hutchinson et al. (2000) circumvent this problem by postulating that $\mathcal{G}_{max} \sim \mathcal{G}_m = 2\Gamma$ gives the critical film thickness h_f^{cr} at which the imperfections lead to the nucleation of delamination as

$$h_f^{cr} \approx \frac{4\bar{E}_f \Gamma}{\sigma_m^2}, \tag{5.43}$$

which has a weak dependence on a_{cr}/h_f. When this result is combined with (5.42), it is seen that $a_{cr}/h_f^{cr} \approx 2.7\sqrt{\bar{E}_f/\sigma_m}$. Therefore, it can be argued that a delamination will not nucleate for circumstances in which both $h_f < h_f^{cr}$ and $a_0 < a_0^{cr}$ prevail. Furthermore, any delamination comparable to the undulation size a_0 will arrest before instability sets in. On the other hand, when $h_f < h_f^{cr}$ and $a_0 > a_0^{cr}$ are both satisfied, an interface delamination will initiate and propagate in an unstable manner. In summary, the inception and advance of buckle driven delaminations in the presence of interface imperfections can be regarded as being governed by a nucleation process which occurs at a critical film thickness and by unstable buckling which occurs at a critical wavelength of the imperfection.

5.4.4 Example: Buckling instability of carbon films

The occurrence of buckling instability is a topic of particular interest in the case of amorphous carbon films. These films, commonly referred to as diamond-like carbon (DLC), exhibit many desirable properties similar to those of diamond: optical transparency, wear resistance, high modulus to density ratio, low coefficient of friction during contact with most materials, high electrical resistivity, high thermal conductivity and chemical inertness. These attributes underlie the interest in their use in a variety of engineering applications which include: tribological coatings for transmission gears and head drums in videocassette recorders; overcoat layers for speaker diaphragms and surface acoustic wave devices; and protective coatings for magnetic storage media and infared optical windows. A common method of depositing DLC films onto substrates entails ion beam deposition or capacitively coupled RF glow discharge in a gas such as CH_4 or C_6H_6. These films can be formed with high residual compressive stresses, typically 1–6 GPa, depending on the processing methods employed. The range is due, in part, to the large concentration of hydrogen incorporated into the film during deposition.

Figure 5.21, taken from the work of Gille and Rau (1984), shows an example of the onset and progression of buckling of a 200-nm thick ion-beam-deposited carbon layer on a relatively thick quartz glass substrate. The intrinsic compressive stress in the film is estimated to be 3–6 GPa, and the energy of the interface between the film and the substrate is 4–7 J m^{-2}. Irregularly buckled regions, which are seen in the lower half of Figure 5.21, initiate at the film edge. As the wrinkled region advances away from the edge towards the center of the specimen, a remarkably regular telephone cord buckle develops. These undulations exhibit clear periodicity with the amplitude of the waviness typically in the range 1–2 μm and with the mean spacing between the peaks in the range 10–20 μm. The moisture content of the humid air environment is also known to have an influence on the propagation of the buckle in carbon films.

Figure 5.22(a) shows an example of stress relief pattern in a compressively stressed DLC film which was deposited in an RF glow discharge in a hydrocarbon-containing gas onto a Si substrate. Buckling is initiated at the film edge or at an interior defect, upon exposure to atmospheric air. Upon growth away from the film edge or the defect site, the buckle develops into the shape of a 'telephone cord'. Figure 5.22(a) shows a sinusoidal stress relief pattern which populates a significant area of the film. The intersection of the buckle with a crack causes a break in the telephone cord pattern. Figure 5.22(b) shows another example of a telephone cord buckle in a DLC film whose origin can be traced to the presence of an imperfection which was present on the surface of the soda lime glass substrate prior to film deposition. The imperfections in this system were composed of silica and Ti silicate phases. Experiments by Moon et al. (2002) reveal that when the energy release rate for

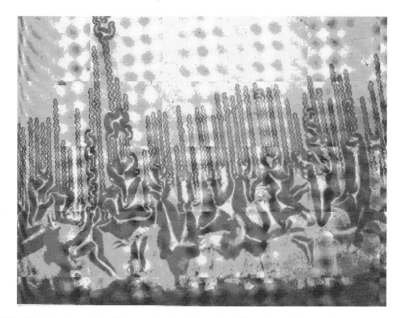

Fig. 5.21. Nucleation of buckles and wrinkles at the film edge followed by the development of telephone cord instability in a 200-nm thick carbon film deposited on a glass substrate. Reproduced with permission from Gille and Rau (1984).

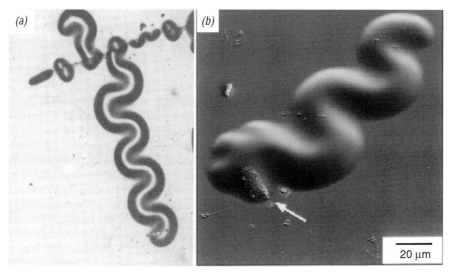

Fig. 5.22. Examples of telephone cord buckles in DLC films. (a) A telephone cord buckle, initiated at the film edge, intersected by a crack. Reproduced with permission from Nir (1984). (b) A telephone cord buckle whose origin (indicated by the arrow) is identified to be an imperfection on the surface of the glass substrate. Reproduced with permission from Moon et al. (2002).

the initial separation of the film from the substrate at the site of the imperfection exceeds the interface fracture energy, suitably representing the appropriate local stress state phase angle, the buckle becomes unstable and propagates in the shape of a telephone cord.

5.5 Film buckling without delamination

In the discussion of film buckling in Sections 5.1 and 5.2, it was tacitly assumed that a portion of the film was able to buckle laterally because it had become detached from the substrate over some portion of the interface. However, there are situations in which film buckling can occur without delamination. For example, this phenomenon can arise if the film stress is very large and/or if the substrate stiffness is small compared to the film stiffness. To investigate the range of parameters for which this can occur, consider the possibility of plane strain buckling of a compressed film of thickness h_f and plane strain modulus $\bar{E}_f$ which is bonded to a relatively thick substrate with plane strain modulus $\bar{E}_s$. Prior to buckling, the film is subjected to an equi-biaxial stress of magnitude $\sigma_m < 0$.

For present purposes, it is assumed that, upon buckle formation, the transverse deflection of the film midplane is equal to the normal surface displacement of the substrate, a reasonable approximation in light of the thinness of the film. Furthermore, it is assumed that the in-plane displacement component of the film midplane from its initial uniformly stressed configuration is zero. This assumption is also reasonable for a thin film, provided that the deflections are not large. Buckling theories of this kind have been considered in the context of geological folding by Biot (1965), layered structural panels by Allen (1969), and material surfaces with negative surface energy or surface stress by Andreussi and Gurtin (1977).

5.5.1 Soft elastic substrate

The differential equation that governs the transverse deflection $w(x)$ of the film midplane is again (5.5), modified to account for the resistance to transverse deflection exerted by the substrate. This resistance is represented by a pressure $p(x)$ over the midplane of the film. The result is the ordinary differential equation

$$w''''(x) - \frac{12t_m}{\bar{E}_f h_f^3} w''(x) = -\frac{12}{\bar{E}_f h_f^3} p(x), \tag{5.44}$$

where $t_m < 0$ is the membrane stress resultant in the film. The dependence of $p(x)$ on $w(x)$ is dictated by the response of the substrate. In general, if a normal tensile traction $p(x)$ is imposed on the surface of the substrate, then the normal displacement $w(x)$ of the substrate surface can be expressed in terms of $p(x)$ by

$$w'(x) = \frac{2}{\pi \bar{E}_s} \int_{-\infty}^{\infty} \frac{p(\xi)}{\xi - x} d\xi \tag{5.45}$$

when the shear traction on the surface vanishes everywhere, analogous to (4.6) (Johnson 1985). The stability of the undeflected film configuration is considered next on the basis of (5.44) and (5.45).

Only the question of initial bifurcation is considered here; post-buckling behavior is not examined. The undeflected configuration with $w(x) = 0$ and $p(x) = 0$ is always a solution. The objective is to find a range of parameters for which a nontrivial solution can be found. A spatially periodic deflection

$$w(x) = W_m \cos \frac{2\pi x}{\lambda} \tag{5.46}$$

with wavelength λ and amplitude W_m is sought. If a solution can be found with $W_m \neq 0$ for some value of t_m, then that stress resultant value defines a buckling stress level for the film. If $w(x)$ varies with x according to (5.46), then $p(x)$ must also have the form

$$p(x) = P_m \cos \frac{2\pi x}{\lambda}. \tag{5.47}$$

If (5.46) and (5.47) are substituted into (5.45) and if the integral is evaluated, it is found that $\pi W_m / \lambda = P_m / \bar{E}_s$. Then, substitution of this result and (5.46) into (5.44) yields a relationship between stress and wavelength in the form of

$$\left(\frac{2\pi}{\lambda}\right)^3 + \frac{12 t_m}{\bar{E}_f h_f^3} \left(\frac{2\pi}{\lambda}\right) + \frac{6 \bar{E}_s}{\bar{E}_f h_f^3} = 0. \tag{5.48}$$

It is readily established that the smallest magnitude of $|t_m|$ for which a positive real solution of this cubic equation for wavelength can be found is

$$|t_m| = \left(\frac{3}{8}\right)^{2/3} h_f \bar{E}_f^{1/3} \bar{E}_s^{2/3}. \tag{5.49}$$

This value is therefore the critical stress for onset of plane strain wrinkling of the film on the substrate surface. The corresponding wavelength is

$$\frac{\lambda}{h_f} = \pi \left(\frac{8 \bar{E}_f}{3 \bar{E}_s}\right)^{1/3}. \tag{5.50}$$

If the substrate properties are such that $\bar{E}_s = 10^{-3} \bar{E}_f$, for example, then the critical stress magnitude is $\sigma_m \approx -0.0052 \bar{E}_f$ from (5.49), and the corresponding wavelength is $\lambda = 43.6 h_f$.

Behavior of this type was studied experimentally by Bowden et al. (1998) who found a wide range of buckling patterns in gold films deposited on elastomeric

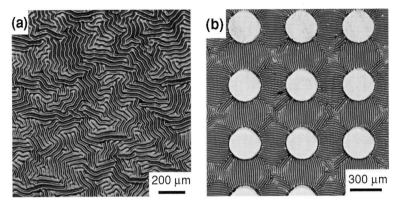

Fig. 5.23. Optical micrographs showing film buckling patterns in 50-nm thick gold films on relatively thick PDMS (polydimethylsiloxane) substrates, and with an intermediate layer of titanium or chromium a few nm thick to promote adhesion. The wavelengths in the buckling patterns varied between 20 and 50 μm, with depths of a few μm. For a uniform film, the disordered pattern shown on the left developed, whereas the introduction of free edges to the film by substrate patterning prior to deposition led to ordered buckling patterns as illustrated on the right. Reproduced with permission from Bowden et al. (1998).

polymer substrates. In those cases in which the film was apparently uniform over a large area of the substrate surface, it did not buckle into a one-dimensional array of long ridges and valleys. Instead, the buckling patterns consisted of a disordered array of small regions, each of which was essentially a one-dimensional array of ridges and valleys. An example of such a pattern is shown in part (a) of Figure 5.23. The reason for the disorder is unclear, and perhaps a detailed study of post-buckling behavior would provide clues to its existence. Much more regular arrays of wrinkles were found when the polymer substrate was patterned prior to deposition of the metal film with bas-relief features in the form of elevated ridges or elevated mesas of square or circular shape. These features interrupted the continuity of the film, and essentially provided stress-free boundaries in the interior of the film. In these cases, the wrinkles were aligned normally to the film boundary at the elevated features, presumably because stress sufficiently large to induce buckling could be maintained only in directions parallel to free edges. Characteristic patterns were formed as these families of wrinkles merged with those emanating from adjacent surface features. An example with circular gaps in the film is shown in part (b) of Figure 5.23.

The partial differential equation that generalizes (5.44) for the case of two-dimensional buckling patterns is

$$\frac{\bar{E}_\mathrm{f} h_\mathrm{f}^3}{12} \nabla^4 w - t_{xx} \frac{\partial^2 w}{\partial x^2} - t_{yy} \frac{\partial^2 w}{\partial y^2} - 2 t_{xy} \frac{\partial^2 w}{\partial x \partial y} = -p, \qquad (5.51)$$

where all fields are functions of both x and y, and the nonuniform membrane stress resultants t_{xx}, t_{yy} and t_{xy} appearing in (5.51) represent the equilibrium plane stress field prior to buckling that satisfies the appropriate boundary conditions.

5.5.2 Viscous substrate

If the substrate material supporting the compressed film is a linear viscous material, instead of an elastic material as in the preceding discussion, then the film deflection $w(x, t)$ is time-dependent and the interfacial stress $p(x)$ appearing in (5.44) depends on the transverse velocity $\partial w/\partial t$ of the film. If the viscous substrate material is assumed to be incompressible, then (5.45) is replaced by

$$\frac{\partial^2 w}{\partial x \partial t}(x, t) = \frac{4}{9\pi \eta_s} \int_{-\infty}^{\infty} \frac{p(\xi, t)}{\xi - x} d\xi , \qquad (5.52)$$

where η_s is the viscosity of the substrate material.

Analysis of buckling behavior for spatially periodic film deflections proceeds as in Section 5.5.1, except that the buckling amplitude $W_m(t)$ appearing in (5.46) is time-dependent. The resulting equation governing $W_m(t)$ implied by (5.44) in this case takes the form

$$\dot{W}_m(t) = \left[\frac{12|t_m|}{27\eta_s} - \frac{\bar{E}_f h_f^3}{27\eta_s} \left(\frac{2\pi}{\lambda} \right)^2 \right] \left(\frac{2\pi}{\lambda} \right) W_m(t) , \qquad (5.53)$$

where the superposed dot represents differentiation with respect to time. It is evident that the film can undergo lateral buckling for any value of the initial compressive stress $|t_m|$ if the wavelength λ is large enough so that the term in square brackets in (5.53) becomes negative. Of greater interest is the wavelength, say λ_{gr}, for which the amplitude $W_m(t)$ will grow most rapidly for a given value of $|t_m|$. This wavelength is readily found to be

$$\lambda_{gr} = \pi h_f \sqrt{\frac{\bar{E}_f h_f}{|t_m|}} = \frac{\pi h_f}{\sqrt{(1 + \nu_f)|\epsilon_m|}}. \qquad (5.54)$$

For a linear viscous material, this fastest-growing wavelength is independent of the viscosity of the substrate, as is evident from (5.53). The rate of growth itself, which is determined by substituting (5.54) into (5.53), is inversely proportional to the viscosity but is independent of the film thickness. The corresponding results for the case of a viscous substrate of finite thickness were presented by Sridhar et al. (2001).

The configuration of an elastic film bonded to a viscous substrate is of interest in semiconductor-on-insulator configurations in microelectronics. The interest stems from the prospect of relaxing undesirable stress in the film by activating viscous flow in the amorphous substrate during a short duration, high temperature annealing cycle. The viscous flow in the substrate, however, can also induce film buckling as was verified in the experiments reported by Hobart et al. (2000). Compressively

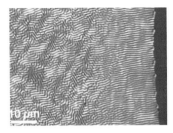 

Fig. 5.24. The interference contrast (Nomarski) microscopy image on the left shows a plan view of the buckling pattern for a $Si_{0.7}Ge_{0.3}$ film, approximately 30 nm in thickness, on a borophosphorosilicate glass (BPSG) substrate following high temperature annealing. The scanning electron microscopy image on the right shows a cross-section of the sample, revealing the fully formed buckling pattern on its BPSG substrate. Reproduced with permission from Hobart et al. (2000).

strained $Si_{0.7}Ge_{0.3}$ films were bonded to substrates consisting of either SiO_2 or silicate glass doped with boron and phosphorus to reduce its viscosity. The films were deposited epitaxially on Si substrates and then transferred to the glass substrates without a reduction in mismatch stress. For the relatively low viscosity borophosphorosilicate glass (BPSG), buckling was found to occur at a temperature of about $800\,°C$. Representative images are shown in Figure 5.24. The plan view image of a buckling pattern on the left was obtained by interference contrast (Nomarski) microscopy; it shows the same type of behavior as was illustrated in Figure 5.23, wherein the buckling pattern is regular near the free film edge on the right but has a random patchiness far from the free edge. The side view, obtained by scanning electron microscopy, clearly illustrates the periodicity of the buckling pattern. Interestingly, Hobart et al. (2000) were able to suppress buckling by patterning the stressed films into smaller area patches. Through further development of the model (Liang et al. 2002), it was established that buckling could be suppressed as a result of stress relief associated with in-plane expansion, as outlined in Section 6.7.3, for a film with lateral dimensions 10–15 times the critical wavelength (5.54).

5.5.3 Example: Buckling wavelength for a glass substrate

A $Si_{0.7}Ge_{0.3}$ film that is 35 nm in thickness is transferred from its Si substrate onto the surface of a borophosphorosilicate glass without reduction in mismatch strain. The structure is then heated to the temperature of $800\,°C$. Determine the wavelength of the buckling pattern that is expected to develop due to relaxation of compressive stress and the characteristic time for its formation. Base the estimate on the assumed properties $E_f = 124\,\text{GPa}$, $\nu_f = 0.26$ and $\eta_s = 10^7\,\text{Pa s}$.

Solution:

The mismatch strain between the SiGe film and its Si substrate is approximately $\epsilon_m = -0.012$. From (5.54), the ratio of the wavelength of the fastest growing buckling mode to the film thickness is

$$\frac{\lambda_{gr}}{h_f} = \frac{\pi}{\sqrt{1.26 \times 0.012}} = 25.6 \,. \tag{5.55}$$

For $h_f = 35$ nm, this wavelength is roughly $1\,\mu$m. From the differential equation (5.53), the characteristic time for this wavelength to develop is

$$\frac{27\eta_s}{16 M_f \sqrt{1 + \nu_f} \, |\epsilon_m|^{3/2}} \approx 0.1 \, \text{s} \,. \tag{5.56}$$

5.6 Pressurized bulge of uniform width

In this section, the case of an initially flat film that is not attached to its substrate over the strip $-a \leq x \leq a$, $-\infty < y < \infty$ is considered. The film is deflected away from its substrate over this region by applying a pressure p to the face of the film toward the substrate. As a result of the invariance of the configuration under translation in the y-direction, this deflection will be independent of y. The straight-sided bulge configuration is perhaps of less practical significance than the corresponding circular bulge configuration. However, the fact that the response of the film to the applied pressure can be described in a fairly transparent way at various levels of approximation makes it a useful device for introducing ideas that are important in the study of all pressurized bulge configurations. The straight-sided configuration is depicted in Figure 5.25, where it is intimated that the pressure is applied hydrostatically by means of a fluid. The notation introduced in the discussion of buckling delamination in Section 5.1 is followed to the extent possible. This discussion of a pressurized bulge configuration is similar to that for a buckled configuration, the main difference being that the deflection results from external loading in the latter case whereas it occurs spontaneously in the former case. An equi-biaxial residual stress σ_m, which may be tensile or compressive but of magnitude below the buckling stress in the latter case, may also be present. However, consideration of this possibility is postponed until the discussion of the circular pressurized blister is taken up in Section 5.7.

5.6.1 Small deflection bending response

The simplest level of approximation for small deflection of the film is geometrically linear elastic plate theory. In effect, it is presumed that the pressure is resisted exclusively by bending resistance of the film in this case, and the bending stress is not coupled to any membrane stress which may be present. In this case, the

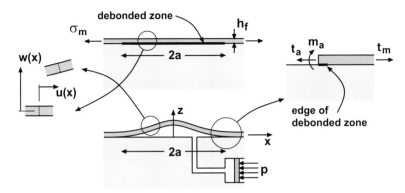

Fig. 5.25. Schematic representation of a straight-sided debond where an externally imposed hydrostatic pressure causes the film to bulge.

transverse deflection $w(x)$ is known from linear elastic plate theory to be

$$\frac{w(x)}{h_{\mathrm{f}}} = \frac{1}{2} \frac{p a^4}{\bar{E}_{\mathrm{f}} h_{\mathrm{f}}^4} \left(1 - \frac{x^2}{a^2} \right)^2 \tag{5.57}$$

for any pressure p. In particular, the deflection of the center of the bulge, say $w(0) = w_0$, is related to the pressure through the linear relationship

$$\frac{w_0}{h_{\mathrm{f}}} = \frac{1}{2} \frac{p a^4}{\bar{E}_{\mathrm{f}} h_{\mathrm{f}}^4}. \tag{5.58}$$

The small deflection approximation is usually thought to be accurate for a center point deflection up to roughly $w_0 \approx h_{\mathrm{f}}$. Therefore, the pressure at which $w_0 = h_{\mathrm{f}}$ is identified as the reference pressure

$$p_r = 2 \bar{E}_{\mathrm{f}} \frac{h_{\mathrm{f}}^4}{a^4} \quad \Rightarrow \quad \frac{w_0}{h_{\mathrm{f}}} = \frac{p}{p_r} \tag{5.59}$$

for purposes of comparison among the various results obtained by adopting different levels of approximation.

The configurational driving force for delamination at the edge of the pressurized zone can be calculated directly from expressions for the total potential energy of the bulge configuration or by using the general formula (5.38). If the latter option is adopted, then $\Delta t_{\mathrm{a}} = 0$ within this level of approximation and $m_{\mathrm{a}} = \frac{1}{12} \bar{E}_{\mathrm{f}} h_{\mathrm{f}}^3 w''(a) = \frac{1}{3} p a^2$, so that

$$\mathcal{G}(a) = \mathcal{G}_{\mathrm{b}}(a) = \frac{2}{3} \frac{p^2 a^4}{\bar{E}_{\mathrm{f}} h_{\mathrm{f}}^3}. \tag{5.60}$$

The corresponding phase angle ψ for the stress state on the interface immediately beyond the pressurized zone is given by (5.26) as $\psi = -45°$. Even if $t_m \neq 0$, the energy release rate is given by $\mathcal{G}_b(a)$ because bending and membrane stresses are not coupled at this level of approximation.

5.6.2 Large deflection response

If the magnitude of the center point deflection of the film w_0 increases to values on the order of film thickness h_f, then the potential arises for generation of significant membrane stress in the film due to transverse deflection (in addition to any residual membrane stress which may be present in the film prior to deflection). As in the case of film buckling, the von Kármán plate theory provides a useful and effective framework for describing response with center point deflection w_0 of magnitude equal to several times the film thickness. In the present case, the von Kármán equations reduce to the pair of ordinary differential equations

$$w''''(x) - \frac{12t_0}{\bar{E}_f h_f^3} w''(x) = \frac{12p}{\bar{E}_f h_f^3},$$

(5.61)

$$u'(x) + \frac{1}{2} w'(x)^2 = \frac{t_0}{\bar{E}_f h_f}$$

for the transverse and in-plane deflection components of the film midplane as defined in Figure 5.25. The solutions of these equations are subject to the boundary conditions

$$w(\pm a) = 0, \quad w'(\pm a) = 0, \quad u(\pm a) = \pm a(1 - \nu_f)\sigma_m/E_f. \quad (5.62)$$

The parameter t_0 is the spatially uniform membrane tension in the film; its value prior to application of the pressure p is $\sigma_m h_f$ where σ_m is the uniform residual stress. The six boundary conditions in (5.62) yield values for the five integration constants which arise from solution of the differential equations and for the membrane stress t_0 following application of the pressure p.

For the time being, consider the case of no initial residual stress, that is, $\sigma_m = 0$. For this case, the relationship between the membrane stress t_0 in the film and the applied pressure p that is implied by (5.61) and (5.62) is given by

$$\frac{p}{p_r} = \frac{\tau^2 \sinh \sqrt{\tau}}{12 \left[(24 + 4\tau)\cosh 2\sqrt{\tau} - 18\sqrt{\tau} \sinh 2\sqrt{\tau} - 24 - 16\tau\right]^{1/2}}, \quad (5.63)$$

where p_r is the reference pressure defined in (5.59) and $\tau = 24t_0 h_f/p_r a^2$ is the normalized membrane tension in the film. The result (5.63) is illustrated by the solid curve in Figure 5.26. The transverse deflection of the pressurized bulge implied

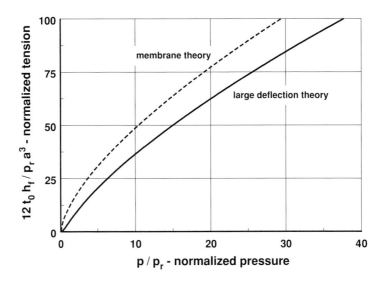

Fig. 5.26. Normalized membrane tension as a function of normalized hydrostatic pressure imposed on the debond region from the predictions of membrane theory and large deflection theory.

by the solution of the deflection differential equations (5.61) and the boundary conditions (5.62) is

$$\frac{w_0}{h_f} = 12\frac{p}{p_r}\left[\frac{\sqrt{\tau} - 2\tanh\frac{1}{2}\sqrt{\tau}}{\tau^{3/2}}\right]. \tag{5.64}$$

When (5.64) is considered together with (5.63), the two relationships provide the dependence of w_0 on pressure p parametrically in t_0. This result, which is based on bending theory with finite rotations of material elements, is shown graphically by the solid line in Figure 5.27. The corresponding result from small deflection bending theory given in (5.58) is also shown in Figure 5.27 by the dashed line. In the limit as $\tau \to 0$, the result in (5.64) reduces to the small deflection result given in (5.59). Note that nonlinear deformation effects have a strong influence on response for $w_0/h_f \geq 0.5$.

The energy release rate $\mathcal{G}(a)$ for spread of the zone of application of the pressure is again given by (5.38). In the present instance, the membrane stress at the edge $x = a$ of the pressurized region is $\Delta t_a = t_a - t_m = t_0$ and the bending moment at the edge is determined from the solution of the deflection equation (5.61) to be

$$m_a = \frac{1}{12}\bar{E}_f h_f^3 w''(a) = pa^2\left[\frac{\sqrt{\tau}\coth\sqrt{\tau} - 1}{\tau}\right]. \tag{5.65}$$

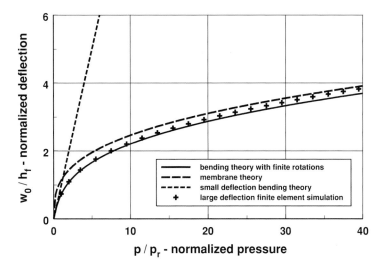

Fig. 5.27. A comparison of different approaches to estimate the normalized deflection of the pressurized bulge as a function of the normalized pressure.

The normalized energy release rate is then

$$
\frac{\mathcal{G}(a)}{\mathcal{G}_b(a)} = \frac{\tau^2}{8}\frac{p_r}{12p} + 9\left[\frac{\sqrt{\tau}\coth\sqrt{\tau}-1}{\tau}\right]^2 , \tag{5.66}
$$

where $\mathcal{G}_b(a)$ is the energy release rate for small deflection bending theory given in (5.60). With p/p_r given in terms of τ in (5.63), the expression (5.66) can be viewed as providing $\mathcal{G}/\mathcal{G}_b$ versus p/p_r parametrically in τ. This relationship is shown graphically by the solid line in Figure 5.28 for pressure in the range $0 \le p/p_r \le 30$. It is evident from the graph that $\mathcal{G}(a)/\mathcal{G}_b(a) \to 1$ as $p/p_r \to 0$, so the result (5.66) based on the von Kármán finite deformation theory is consistent with the elementary bending result when the pressure is very small. However, for any particular value of the half width a, the dependence of $\mathcal{G}/\mathcal{G}_b$ on p/p_r for small values of the pressure is found by noting that $\mathcal{G}/\mathcal{G}_b \sim 1 - 13\tau/105$ for small values of τ and by recalling that $(p/p_r)^2 \sim 35\tau^2/256$ for small values of τ. Together, these two asymptotic expressions imply that $\mathcal{G}/\mathcal{G}_b \sim 1 - 0.91(p/p_r)^2$. This result reveals the significant fact that the energy release rate falls off dramatically from its value based on linear plate theory alone as the pressure increases within the range where small deflection bending theory is generally regarded as being valid.

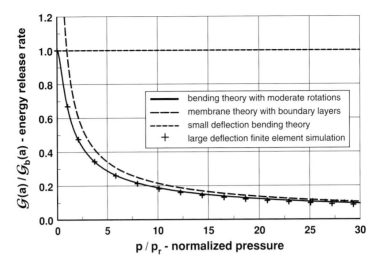

Fig. 5.28. The energy release rate $\mathcal{G}(a)$ for the spread of the debond zone over which hydrostatic pressure is applied, normalized by the energy release rate for small deflection $\mathcal{G}_b(a)$, as a function of the normalized pressure p/p_r.

5.6.3 Membrane response

From the foregoing examination of deflection of a pressurized strip of a thin film, it is evident that the pressure is resisted less by bending stress and more by membrane stress as the pressure p increases beyond the reference pressure p_r defined in (5.59). In terms of midpoint deflection w_0, membrane effects come into play when $w_0 \approx \frac{1}{2}h_f$ and they tend to dominate response for $w_0 > h_f$. Within this range, the full span of the film within the bulge may be viewed as being divided into two regimes of behavior, as illustrated in Figure 5.29. Over most of the span, the response to the pressure is assumed to be exclusively due to membrane stress in the film, and bending effects are ignored. In terms of the differential equations which govern the deflection response given in (5.61), behavior is determined by

$$-t_0 w''(x) = p, \qquad u'(x) + \frac{1}{2}w'(x)^2 = 2t_0 h_f^3 / p_r a^4 \qquad (5.67)$$

over $-a < x < a$. The first of these equations is recognized as a version of the Laplace–Thompson equation involving curvature, membrane tension and pressure. The solution of (5.67) must satisfy boundary conditions $w(\pm a) = 0$ and $u(\pm a) = 0$. The additional condition $w'(\pm a) = 0$ which was considered in (5.62) cannot be enforced here because of the restriction to membrane effects. The matching of the solution of (5.67) to the constraint of the film supports at $x = \pm a$ must rely on a boundary layer treatment.

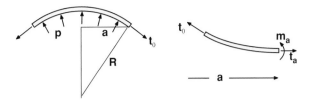

Fig. 5.29. Schematic diagram of a section of a straight-sided pressurized bulge for large deflections, with the main part of the film exhibiting membrane behavior as illustrated on the left. The diagram on the right represents a boundary layer region near the edge of the bulge in which bending effects must be taken into account to enforce the boundary conditions.

Within the range of behavior described by (5.67), the center point deflection of the bulge w_0 is found to depend on pressure according to

$$\frac{w_0}{h_{\mathrm{f}}} = \frac{1}{2}\left(\frac{12p}{p_r}\right)^{1/3}.$$ (5.68)

This behavior is also shown in Figure 5.27 by means of the long-dash curve, along with the corresponding results for small deflection bending theory and for large deflection theory. The increasing stiffness dp/dw_0 with increasing pressure exhibited by the large deflection response is due to the more significant role of membrane stress at larger pressures. As expected, this trend is also shown by the membrane response being considered in this section. For very small deflection, the behavior (5.68) based on membrane effects alone has unrealistically small stiffness, corresponding to the large slope in the graph of w_0 versus p shown in Figure 5.27 for this case. The membrane stress t_0 is related to the pressure according to

$$\frac{t_0}{p_r h_{\mathrm{f}}} = \frac{1}{12}\frac{a^2}{h_{\mathrm{f}}^2}\left(\frac{12p}{p_r}\right)^{2/3},$$ (5.69)

and this behavior is compared to the corresponding result for large deflection theory in Figure 5.26, where it is seen that the distribution for membrane behavior implies a membrane stress approximately 15–20% larger than for large deflection theory.

The deflection calculated on the basis of membrane theory is slightly greater than that for large deflection theory, at least up to the largest pressure level included in Figure 5.27. This difference is most likely due to the fact that the film is more constrained kinematically in plate theory due to the additional condition $w'(\pm a) = 0$ than it is when only membrane resistance is taken into account. To pursue the issue of the influence of edge conditions on response in the large pressure range, the film response near $x = a$ is examined as a boundary layer phenomenon. This approach makes it possible to understand the transition from local bending behavior satisfying the edge condition $w'(\pm a) = 0$ to membrane behavior, and to proceed to a calculation of energy release rate for advance of the delamination zone on the same basis as for the other models of response of the pressurized bulge.

Within a region of extent small compared to a near the edges of the bulge at $x = \pm a$, it is expected that gradients in deformation fields are relatively large. On this basis, only the gradient terms in (5.61) are maintained, that is, transverse deflection near $x = a$ is assumed to be governed by

$$w''''(x) - \frac{2}{a^2}\left(\frac{12p}{p_r}\right)^{2/3} w''(x) = 0 \qquad (5.70)$$

for $x < a$, where the coefficient of the second term has been expressed in terms of the uniform membrane tension (5.69) deduced from the foregoing membrane analysis. A solution of (5.70) is sought subject to the edge conditions $w(a) = 0$, $w'(a) = 0$. At points of the film far from $x = a$, the solution is required to match the limiting behavior of the membrane solution as $x \to a^-$. Thus, the additional conditions imposed on the solution of (5.70) are

$$w'(x) \to -\frac{h_f}{a}\left(\frac{12p}{p_r}\right)^{1/3}, \qquad w''(x) \to 0 \quad \text{as} \quad x \to -\infty. \qquad (5.71)$$

The resulting unique solution exhibits two features of particular interest, namely, the curvature $w''(a) = \sqrt{2}\,(h_f/a^2)(12p/p_r)^{2/3}$ at the edge of the bulge, from which the edge bending moment can be calculated according to $m_a = \bar{E}_f h_f^3 w''(a)/12$, and the curvature at adjacent points within the boundary layer region relative to the value $w''(a)$. This latter result is

$$\frac{w''(x)}{w''(a)} = e^{-\sqrt{2}(12p/p_r)^{1/3}(1-x/a)} \qquad (5.72)$$

for $x < a$. This expression provides a basis for estimating the thickness or extent of the boundary layer region. Suppose that the boundary layer thickness is estimated to be the distance from $x = a$ at which the ratio (5.72) has decayed to e^{-1} times its boundary value. This criterion leads to the conclusion that the boundary layer thickness is $a/10$ when p/p_r is approximately equal to 30, and the size of the boundary layer diminishes gradually as p/p_r increases to larger values. Thus, the boundary layer thickness is a significant fraction of a for smaller values of pressure which may account for the differences in response between membrane theory and large deflection theory seen in Figure 5.27. Aspects of boundary layer behavior in this context have been considered by Gioia and Ortiz (1997) and by Jensen (1998).

A main objective in developing the boundary layer solution is to determine the bending moment at the edge of the bulge, which is needed to determine the driving force for further delamination of the film. The bending moment is determined from the solution to be

$$\frac{m_a}{p_r h_f^2} = \frac{1}{12\sqrt{2}} \frac{a^2}{h_f^2} \left(\frac{12p}{p_r}\right)^{2/3}. \tag{5.73}$$

With this result, and the membrane stress $\Delta t_a = t_a = t_0$ given in (5.69), the energy release rate can be calculated according to (5.38) for the membrane deformation model. The result, normalized by the corresponding energy release rate $\mathcal{G}_b(a)$ for small deflection bending theory given in (5.60), is

$$\boxed{\frac{\mathcal{G}(a)}{\mathcal{G}_b(a)} = \frac{21}{4} \left(\frac{p_r}{12p}\right)^{2/3}.} \tag{5.74}$$

This result is also shown in Figure 5.28 along with results obtained on the basis of other models of deformation of the same pressurized bulge configuration.

5.6.4 Mechanics of delamination

In the foregoing sections, the energy release rate $\mathcal{G}(a)$ for symmetric increase in width a of the strip bulge was determined for film response described by elementary bending theory in (5.60), for large deflection theory with coupling between bending and membrane stress in (5.66), and for membrane theory with edge boundary layers in (5.74). The focus in each case was on a comparison of the dependence of $\mathcal{G}(a)$ on pressure ratio p/p_r among the various levels of description. In order to consider film delamination due to the application of the pressure, it is necessary to know the driving force as a function of bulge width a, under conditions in which either the pressure p acting on the film or the volume V of the pressurizing medium is controlled. This volume (per unit length along the strip) is defined as the volume swept out by the film in undergoing transverse displacement, that is,

$$V \approx \int_{-a}^{a} w(x)\, dx \tag{5.75}$$

for small displacement $u(x)$. Volume can be controlled if the compressibility of the pressurizing medium is negligible over the pressure range of interest.

For the case of fixed applied pressure p forming the bulge, the driving force given in (5.60) for small deflection bending theory can be written as a function of a in the nondimensional form

$$\frac{\mathcal{G}(a)}{h_f p} = \frac{4}{3} \frac{p a^4}{2\bar{E}_f h_f^4}. \tag{5.76}$$

This result is shown graphically by the dot-dash curve in Figure 5.30. The corresponding expression for large deflection theory is given parametrically in (5.66) and

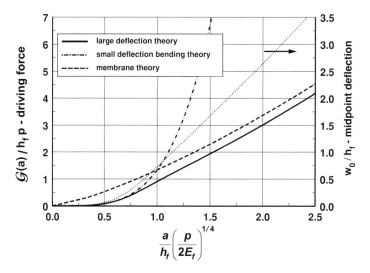

Fig. 5.30. Predicted variation of the normalized driving force $\mathcal{G}/(h_f p)$, using (5.76) based on small deflection bending theory, (5.66) and (5.63) based on large deflection theory, and (5.74) and (5.63) based on membrane theory, as a function of the normalized bulge width $(a/h_f)(p/2\bar{E}_f)^{1/4}$. The dotted line represents the normalized midpoint deflection w_0/h_f which is plotted as a function of the normalized bulge width using the results from (5.64) and (5.63).

(5.63), with parameter τ; likewise, the equivalent boundary layer result based on membrane theory is given parametrically in (5.74) and (5.63). Both of these cases can be expressed in terms of the same normalized variables as were used in (5.76), and the results are shown graphically in Figure 5.30. As was already noted above, the midpoint deflection w_0/h_f versus $(a/h_f)(p/2\bar{E}_f)^{1/4}$, which is given parametrically by (5.64) and (5.63), is included as the dotted curve in the figure for reference. As long as the midpoint deflection is less than about $h_f/2$, the descriptions based on the two bending theories are indistinguishable; membrane theory is not expected to be valid within this region. For larger deflections, the trends of descriptions based on membrane theory and large deflection theory agree very well but the actual values of energy release rate show a systematic difference. If the two descriptions are compared at a given level of driving force in the large deflection range, then the two values of a differ by a distance that is roughly equal to the width of the boundary layer introduced with the membrane theory to match the constraints at the bulge edges. This boundary layer width is not negligibly small compared to a until $(a/h_f)(p/2\bar{E}_f)^{1/4}$ becomes very large, well beyond the range of parameters represented in Figure 5.30. This provides a limitation on the use of membrane theory within a range of behavior of practical interest. To some extent, this limitation is offset by the simplicity of the approach based on membrane theory.

If delamination at the edges of the bulging strip is driven by controlling the volume V of the pressurizing fluid between the film and its substrate, the dependence of driving force $\mathcal{G}(a)$ on width a is quite different from that shown in Figure 5.30. For the case of small deflections, $w(x)$ is given explicitly in (5.57) in terms of pressure p and various system parameters. Therefore, the relationship between volume V as defined in (5.75) and pressure can be established as

$$\frac{V}{ah_f} = \frac{16}{15}\frac{p}{p_r} \qquad (5.77)$$

in a convenient normalized form. In terms of the same variables, the driving force for edge delamination can be expressed as a function of width a at fixed volume V from (5.60) as

$$\frac{\mathcal{G}_b(a)}{p_r h_f} = \frac{4}{3}\left(\frac{p}{p_r}\right)^2 = \frac{75}{64}\left(\frac{V}{ah_f}\right)^2. \qquad (5.78)$$

Derivation of the corresponding result for large deflection response is left as an exercise.

It was noted above that the phase angle of the interface stress at the edge of the delamination zone is always $-45°$ when it is calculated according to small deflection bending theory with $t_m = 0$. The incorporation of effects of finite deflection through (5.61) leads to a nonzero membrane stress t_0 even when $t_m = 0$. The nondimensional ratio of membrane stress t_a to bending moment m_a at the edge of the delamination zone is found from the solution of these equations to be

$$\frac{t_a h_f}{m_a} = \frac{1}{24}\frac{p_r}{p}\left[\frac{\tau^2}{\sqrt{\tau}\coth\sqrt{\tau} - 1}\right]. \qquad (5.79)$$

This result is shown in Figure 5.31. It is evident from the figure that the membrane stress t_a takes on relatively large values for moderate levels of applied pressure. In contrast, this stress component is identically zero according to small deflection bending theory. The dependence of membrane stress on pressure implies a corresponding dependence of phase angle ψ of the local stress field at the edge of the delamination zone at $x = a$ on pressure, as implied by (5.26). The crack edge phase angle ψ is sensitive to the level of pressure at fixed bulge size a or, equivalently, the phase angle is sensitive to the size a of the bulge at a fixed level of pressure p. The stress resultant ratio based on membrane theory with a boundary layer transition zone is also shown in Figure 5.31 as the horizontal dashed line. The results based on both large deflection analysis and on the large deflection finite element simulation approach the membrane behavior as p/p_r becomes very large, but the approach is gradual.

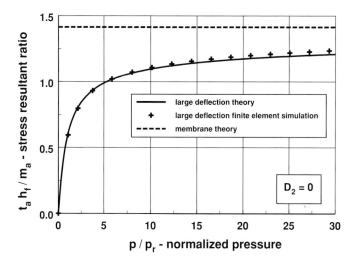

Fig. 5.31. The variation of membrane tension in the film versus pressure applied within the bulge region. The solid curve represents the result of large deflection analysis given in (5.79), and the discrete points follow from a finite element simulation of the bulge. Both of these results approach the high-pressure asymptote represented by membrane theory, shown as the dashed line, but only very gradually.

5.7 Circular pressurized bulge

The case of a circular pressurized bulge in a thin film is considered in this section. The film is bonded to its substrate everywhere except over a circular portion of the interface of radius a, and the film is deflected away from the substrate by means of a pressure p applied to the face of the film adjacent to the substrate. The deformation in the film resulting from application of the pressure is assumed to be axially symmetric. The radial and transverse components of displacement of a point on the film midplane at distance r from the axis of symmetry are denoted by $u(r)$ and $w(r)$, respectively. The periphery of the film is assumed to be constrained from translation or rotation due to the relative rigidity of the substrate, so that $u(a) = 0$, $w(a) = 0$ and $w'(a) = 0$. The deformation may also be subject to symmetry conditions as $r \to 0^+$. The configuration is similar to that depicted in Figure 5.25, with only minor modifications to that figure being necessary to account for axial symmetry.

In the sections that follow, the pressurized film is described successively in the frameworks of small deflection bending response, elementary membrane response, and large deflection response with coupled bending and tension. The development takes advantage of experience gained with the straight-sided pressurized strip wherever possible. One obvious difference is that the finite deflection response for a circular bulge cannot be determined analytically, unlike the case of a strip bulge, so numerical methods are invoked at the outset. A second difference is that the influence of an equi-biaxial residual stress σ_m in the film prior to pressurization is considered here. It is assumed from the outset that, if this residual stress is compressive, its magnitude is not large enough to induce film buckling.

5.7.1 Small deflection bending response

Historically, the pressurized bulge configuration was analyzed on the basis of the assumption that the applied pressure is resisted entirely by bending stiffness of the film. If any membrane stress is present, it is not coupled to the bending stress within the range of small deflection behavior. On this basis, elementary elastic plate theory leads to the expression

$$\frac{w(r)}{h_f} = \frac{3pa^4}{16\bar{E}_f h_f^4} \left(1 - \frac{r^2}{a^2}\right)^2 \tag{5.80}$$

for axially symmetric deflection. As in the case of a plane strain strip bulge, the value of pressure at which $w(0) = h_f$ is identified as a *reference pressure* for purposes of comparison of the results obtained by different approaches. This pressure and the corresponding relationship between the center point deflection w_0 and the applied pressure p are then

$$p_r = \frac{16}{3}\bar{E}_f \frac{h_f^4}{a^4} \quad \Rightarrow \quad \frac{w_0}{h_f} = \frac{p}{p_r}. \tag{5.81}$$

Although the same notation that was used in the case of the strip bulge is adopted here for a circular bulge, it should be evident from the context which expressions are applicable in any particular case.

The driving force for expanding the size of the circular bulge through interface delamination can be deduced from the general expression in (5.38). The magnitude of the bending moment at the periphery of the circular bulge is $m_a = \frac{1}{12}\bar{E}_f^3 w''(a) = \frac{1}{8}pa^2$ in the present case, so that

$$\mathcal{G}(a) = \mathcal{G}_b(a) = \frac{3}{32}\frac{p^2 a^4}{\bar{E}_f h_f^3}. \tag{5.82}$$

Again, the phase angle of the delamination edge stress field is $\psi = -45°$, no matter whether membrane stress is present or not.

5.7.2 Membrane response

One conclusion to be drawn from the foregoing analysis of bulge formation under plane strain conditions in Section 5.6 is that membrane effects become relatively

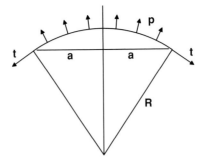

Fig. 5.32. Cross-sectional diagram of a circular patch of film of initial radius a deforming into a spherical cap of radius R under the action of an applied pressure p.

more significant as the film deflection increases beyond h_f in magnitude. On this basis, it is reasonable to expect that the asymptotic behavior of w_0/h_f for pressures that are large compared to the reference pressure can be deduced on the assumption of membrane behavior alone, which is the line of reasoning pursued in this section. This is achieved first on the basis of a fairly crude description of film deformation and the implications of more realistic assumptions on deformation are discussed subsequently.

Suppose that the circular film patch to be pressurized extends over $0 < r \leq a$. It is assumed from the outset that the state of stress in the film is an equi-biaxial membrane tension with isotropic force-per-unit-length stress resultant $t_r = t_\theta = t$. For linear elastic material response, the stress resultant t is related to the equi-biaxial strain of magnitude $\epsilon_r = \epsilon_\theta = \epsilon$ according to

$$t = \frac{E_f h_f}{1 - \nu_f} \epsilon. \tag{5.83}$$

It is immediately clear that the assumption of a uniform biaxial extensional strain in the film, which implies that the film deforms into a spherical cap, violates a condition of constraint at $r = a$ whereby the circumferential extensional strain due to deflection should be exactly zero. Nonetheless, the simplicity of the relationship between w_0 and p obtained on the basis of this model justifies its examination. The edge condition $w'(a) = 0$ does not play a role in any membrane model because the bending moment necessary to enforce this condition cannot be supported by a membrane.

With reference to the cross-sectional diagram in Figure 5.32, overall equilibrium of the film requires that the axial resultant force due to membrane tension t acting along the circumference of length $2\pi a$ is balanced by the resultant force of the pressure p acting on the area πa^2 to yield

$$p = 2t/R, \tag{5.84}$$

where R is the radius of the spherical cap. This radius is related to the biaxial extensional strain ϵ in the film by the geometrical compatibility condition

$$\epsilon = \frac{1}{a}\left[R\sin^{-1}\left(\frac{a}{R}\right) - a\right] \approx \frac{a^2}{6R^2} \tag{5.85}$$

to lowest order in powers of $a/R \ll 1$. Likewise, it is evident from the diagram that the center point deflection w_0 is related to R, and thus to strain ϵ, according to

$$w_0 = R - \sqrt{R^2 - a^2} \approx \frac{a^2}{2R} \tag{5.86}$$

to lowest order in powers of a/R.

The four relationships (5.83) through (5.86) involve the five parameters t, ϵ, R, p and w_0. Elimination of t, ϵ and R leaves the single relationship

$$\frac{w_0}{h_{\mathrm{f}}} = \left[\frac{2}{1 + \nu_{\mathrm{f}}}\frac{p}{p_r}\right]^{1/3} \tag{5.87}$$

between center point deflection w_0 and applied pressure p, where p_r is the reference pressure defined in (5.81). It is immediately apparent that the response of the film represented by (5.87) is much stiffer than the small deflection response of (5.81).

As noted above, the assumed deformation on which (5.87) is based is not kinematically admissible. An estimate of membrane response of the circular pressurized bulge which is, in fact, kinematically admissible can be obtained with only slightly more effort by appeal to energy arguments. To illustrate the approach, assume a two-parameter, axially symmetric deformation field for the bulge of the form

$$u(r) = u_1\frac{r}{a}\left(1 - \frac{r}{a}\right), \quad w(r) = w_0\left(1 - \frac{r^2}{a^2}\right), \tag{5.88}$$

where the two parameters u_1 and w_0 are to be determined by the requirement that the total potential energy functional must be stationary under variations of these parameters, in the sense of variational calculus. Note that the radial curvature $w''(r)$ and the circumferential curvature $w'(r)/r$ implied by (5.88) are still uniform, so the deformed shape is still spherical as long as the center point deflection is small compared to a. The radial and circumferential membrane strain fields implied by (5.88) are

$$\epsilon_r(r) = \frac{u_1}{a}\left(1 - 2\frac{r}{a}\right) + 2\frac{w_0^2 r^2}{a^4}, \quad \epsilon_\theta(r) = \frac{u_1}{a}\left(1 - \frac{r}{a}\right) \tag{5.89}$$

to a level of approximation consistent with the von Kármán theory on the deformation of the film midplane. This assumed deformation satisfies the kinematic boundary conditions $w(a) = 0$ and $u(a) = 0$, the symmetry conditions $u(0^+) = 0$

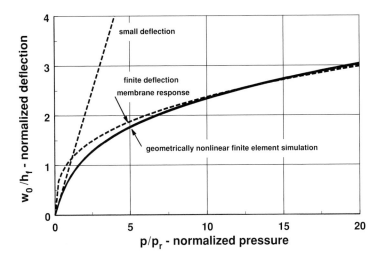

Fig. 5.33. Center point deflection of a thin film due to action of a uniform pressure p. The deflection is normalized by film thickness and the pressure is normalized by the reference pressure defined in (5.81).

and $w'(0^+) = 0$, and the compatibility condition $\epsilon_r(0) = \epsilon_\theta(0)$. The elastic strain energy per unit area of the film midplane is

$$U(r) = \frac{1}{2}\bar{E}_f h_f \left[\epsilon_r(r)^2 + 2\nu_f \epsilon_r(r)\epsilon_\theta(r) + \epsilon_\theta(r)^2\right]. \tag{5.90}$$

The total potential energy for deflection of moderate amplitude which incorporates effects of membrane stress is then

$$V(u_1, w_0) = \pi \bar{E}_f h_f \int_0^a U(r)\, r\, dr - 2\pi p \int_0^a w(r)\, dr. \tag{5.91}$$

The equilibrium condition of stationary potential energy requires that values of the parameters u_1 and w_0 must be chosen as those which satisfy the equations $\partial V/\partial u_1 = 0$ and $\partial V/\partial w_0 = 0$. The resulting expression for w_0 in terms of pressure and system parameters is

$$\frac{w_0}{h_f} = \left[\frac{50}{23 + 18\nu_f - 3\nu_f^2}\frac{p}{p_r}\right]^{1/3}, \tag{5.92}$$

where the reference pressure p_r is defined in (5.81). The expression has the same form as does the cruder result in (5.87), but the dimensionless coefficient depending only on the Poisson ratio ν_f is slightly larger here than in the former result. The relationship (5.92) is plotted in Figure 5.33, along with the corresponding result (5.81) based on linear bending theory.

5.7.3 *Large deflection response*

The small deflection bending analysis and the membrane deformation analysis provide limiting behaviors for small and large values of pressure ratio p/p_r. To describe the transition in behavior between these limiting behaviors, a coupled bending-membrane analysis is required. However, even for a configuration as simple as the axially symmetric circular bulge, the von Kármán equations representing the coupling between bending effects and membrane effects for moderately large rotations of the film midplane cannot be solved analytically. The situation was analyzed by Jensen (1991) through numerical solution of the von Kármán equations. Here, with the benefit of a complete analysis of the plane strain pressurized bulge in Section 5.6, the film deflection for the circular bulge is estimated directly by means of the numerical finite element method. The formulation adopted presumes small elastic strain with strain components varying linearly through the film thickness. The possibility of shear deformation, which is excluded in the von Kármán thin plate theory, is admitted but it is not a significant effect for thin films, say for $h_f/a \leq 1/20$ for a bulge of radius a. The formulation is valid for arbitrary midplane deflections, and the pressure loading is imposed on the film surface in the deformed configuration. The transverse deflection of the midplane is again denoted by $w(r)$, and this deflection is subject to the symmetry condition $w'(0^+) = 0$ and the boundary conditions $w(a) = 0$, $w'(a) = 0$. Calculations have been carried out for $v_f = 1/4$. Because the deformation is less constrained for this computational model than it is for either the small deflection bending theory or the von Kármán theory for moderately large deflections, it is expected that the computed response will be more compliant than either of the plate theories within those ranges of behavior where the respective theories are valid, but only slightly so.

The computed result for transverse deflection $w_0 = w(0)$ of the center point of the film versus applied pressure p is shown in Figure 5.33 for two values of aspect ratio, $a/h_f = 20$ and 100; however, the graphs are virtually indistinguishable. It is evident from the figure that the response deviates significantly from linearity for center point deflection greater than about $w_0 = 0.5h_f$. The main reason for this deviation is that membrane stresses, which are uncoupled from the applied pressure in the range of linear response, come into play beyond $w_0 = 0.5h_f$ to significantly stiffen the response. For an applied pressure of $10p_r$, for example, the center point deflection is approximately $2.3h_f$ when membrane stresses are taken into account but is $10h_f$ when they are ignored. The coupled response shows very little sensitivity to variations in film aspect ratio for values larger than about $a/h_f = 20$.

To consider delamination of the interface between the film and the substrate due to application of the pressure, it is useful to know the dependence of the elastic energy release rate $\mathcal{G}(a)$ per unit length along the periphery of the bulge on system

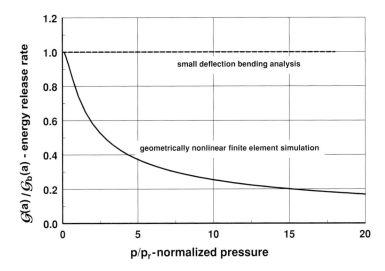

Fig. 5.34. The elastic energy release rate for axially symmetric expansion of the circular pressurized bulge. The energy release rate is normalized by the corresponding result for bending deformation only defined in (5.82) and pressure is normalized by the reference pressure defined in (5.81).

parameters, particularly a and p. The ratio $\mathcal{G}(a)/\mathcal{G}_b(a)$, where $\mathcal{G}_b(a)$ is defined in (5.82) along the basis of small deflection bending response, can be determined from the radial stress resultant $t_r(a) = t_a$ and the bending stress resultant $m_r(a) = m_a$ along $r = a$. The expression for $\mathcal{G}(a)$ is given in (5.38). The result of evaluating the ratio $\mathcal{G}/\mathcal{G}_b$ as a function of normalized pressure p/p_r is shown in Figure 5.34. The result is similar to the corresponding result for the plane strain bulge shown in Figure 5.28. The corresponding stress resultant ratio $t_a h_f/m_a$ is shown in Figure 5.35. The increasing ratio of tensile stress resultant to bending stress resultant as the pressure increases reflects the increasing role of membrane stresses compared to bending stresses in resisting the applied pressure as the pressure increases. A description of film response based entirely on generation of membrane stresses will be considered subsequently.

5.7.4 The influence of residual stress

Suppose that a residual equi-biaxial membrane stress σ_m is present in the film prior to application of the pressure p. The buckling stress for this configuration is $\sigma_b = -1.2235 \bar{E}_f h_f^2/a^2$, and it is assumed from the outset that $\sigma_m > \sigma_b$ so that the undeflected configuration of the film is stable against buckling deformation. The purpose of this section is to examine the influence of σ_m on the transverse deflection of the film once the pressure p is applied.

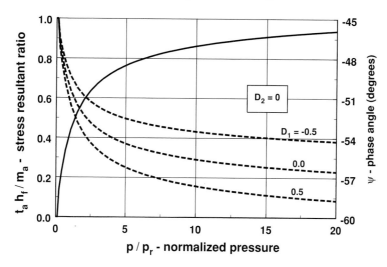

Fig. 5.35. The ratio of membrane stress resultant to bending stress resultant at $r = a$ versus normalized applied pressure (solid line) and the dependence of phase angle on pressure (dashed lines).

For $p/p_r \ll 1$, the linear relationship (5.81) between w_0 and p is unchanged because, in this regime of behavior, membrane and bending effects are not coupled to each other. For $p/p_r \gg 1$, the membrane analysis leading to (5.87) is unchanged by the presence of the residual stress except that the compatibility equation (5.85) must be replaced by

$$\epsilon = \frac{a^2}{6R^2} + \frac{1 - \nu_f}{E_f}\sigma_m \tag{5.93}$$

in order to properly account for the elastic strain associated with the initial stress σ_m prior to transverse deflection. The modified relationship between w_0 and p which follows from incorporation of this additional feature is

$$\frac{1 + \nu_f}{2}\left(\frac{w_0}{h_f}\right)^3 + \frac{3a^2\sigma_m}{4h_f^2\bar{E}_f}\frac{w_0}{h_f} = \frac{p}{p_r}. \tag{5.94}$$

Thus, it appears that the role of the residual stress σ_m for large p/p_r is to stiffen or soften the response of the bulging film to applied pressure compared to its behavior in the absence of residual stress, depending on whether $\sigma_m > 0$ or $\sigma_m < 0$, respectively.

For the residual stress σ_m to have a significant influence on the response, the second term on the left side of (5.94) must be at least of magnitude unity when $w_0 \approx h_f$. In other words, the magnitude of σ_m should be roughly $\bar{E}_f h_f^2/a^2$ or greater if the residual stress is to influence the film response. It is noted that, when σ_m is compressive, this magnitude is roughly equal to the buckling stress for the circular patch of film.

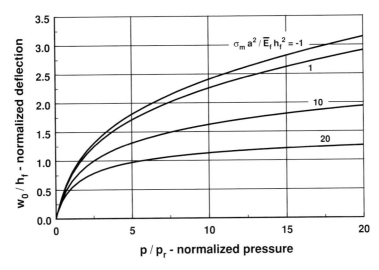

Fig. 5.36. The center point deflection of a film bulge due to applied pressure for several values of normalized initial residual membrane stress in the film.

With the foregoing qualitative estimate as guidance, the relationship between w_0 and p has been computed numerically by means of the finite element method for geometrically nonlinear response. The results for several values of nondimensional parameter $\sigma_m a^2 / \bar{E}_f h_f^2$ are shown in Figure 5.36. The results are insensitive to the aspect ratio a/h_f as long as $a/h_f > 20$. A study of the sensitivity of the response of a pressurized bulge to initial stress and other system parameters was reported by Small and Nix (1992).

5.7.5 *Delamination mechanics*

Suppose that the size a of the bulge can increase by interface delamination once the pressure has been increased to a value large enough so that the condition (5.19) is satisfied, where the dependence of delamination energy Γ on local stress phase angle ψ is again assumed to be as given in (5.20). Thus,

$$\mathcal{G}(a) = \Gamma_{Ic}\left[1 + \tan^2(\eta_c\psi)\right] \tag{5.95}$$

where Γ_{Ic} is the fracture resistance when $\psi = 0$ and η_c is a second failure parameter which reflects the influence of phase angle on resistance, as discussed following (5.20). The dependence of $\mathcal{G}$ on p/p_r is given in Figure 5.34 and the dependence of ψ on p/p_r is given in Figure 5.35 for $D_2 = 0$ and several values of stiffness ratio D_1, reflecting the influence of the substrate modulus. The variation of ψ is quite small over the considered range of p/p_r, so the influence of phase angle in enforcing (5.95) is expected to be weak.

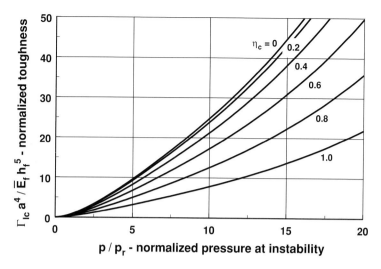

Fig. 5.37. Normalized delamination energy versus normalized pressure at which the edge of the bulge becomes unstable against further increase in pressure.

The pressurized bulge is unstable for the delamination condition (5.95) under circumstances of fixed pressure p. In other words, if (5.95) is to be satisfied continuously for *increasing* a then the value of p must *decrease* in a prescribed way. This implies that, for an ideal system satisfying (5.95) with initial bulge radius a, the bulge size remains fixed at its initial value as the pressure is increased to the level at which (5.95) is first satisfied. The slightest increase in pressure beyond that level causes the bulge to expand unstably, provided that the pressure is sustained. Thus, for a given set of system parameters $\bar{E}_f$, h_f and ν_f, the level of pressure at instability is related to Γ_{Ic}, the interface toughness parameter. This pressure has been termed the *blow-off* pressure by Gent and Lewandowski (1987). The dependence of the normalized toughness $\Gamma_{Ic} a^4 / \bar{E}_f h_f^5$ on normalized pressure p/p_r at instability for any given a can be deduced directly from the data of Figure 5.34 and Figure 5.35, both of which presume $\nu_f = 1/4$. The result is shown in Figure 5.37 for values of η_c spanning its full range. The experiments reported by Gent and Lewandowski (1987) showed a slight super-linear dependence of toughness on critical pressure, consistent with the result shown in Figure 5.37.

The qualitative aspects of behavior of a pressurized bulge in a film are rendered most transparent when considered at the level of the membrane idealization. For example, for the most rudimentary membrane model depicted in Figure 5.32, it follows that the elastic strain energy for any given level of applied pressure is approximately

$$V_{el} = \frac{\pi}{12} \left[\frac{3 p^4 a^{10}}{M_f h_f} \right]^{1/3}, \tag{5.96}$$

where $M_f = E_f/(1 - v_f)$ is the biaxial elastic modulus of the film. The displaced volume v of the film due to deflection is approximately $v = \pi w_0 a^2/2$ in terms of center point deflection, so the external potential energy at fixed applied pressure is

$$V_{\text{ext}} = -pv = -\frac{\pi}{4}\left[\frac{3p^4 a^{10}}{M_f h_f}\right]^{1/3}. \tag{5.97}$$

The energy released per unit area of interface during axially symmetric expansion of the bulge at fixed applied pressure is

$$\mathcal{G} = -\frac{1}{2\pi a}\frac{\partial V}{\partial a}\bigg|_p = \frac{5}{18}\left[\frac{v p^4 a^4}{M_f h_f}\right]^{1/3}, \tag{5.98}$$

where V is the sum of the elastic strain energy and external potential energy.

If the energy release rate $\mathcal{G}$ is then assumed to be equal to a specific fracture energy Γ at the point of instability of the configuration under variations in the position of the delamination front, say when $p = p_{cr}$, it follows that Γ is proportional to $p_{cr}^{4/3}$. This kind of behavior was demonstrated by Gent and Lewandowski (1987). This behavior is consistent with that illustrated in Figure 5.37, at least for some fixed value of the product $\eta_c \psi$. In adopting any membrane model of the delamination process, the possible dependence of delamination behavior on the phase angle of the stress state at the edge of the bulge cannot be addressed directly. A boundary layer approach whereby a transition in behavior of the film from membrane response to coupled membrane-bending response in a narrow zone around the periphery of the film has been considered by Jensen (1998). A similar boundary layer approach has been considered in connection with buckling in Section 5.1 above.

Control of the displaced volume v of the bulging film, rather than of the pressure itself, offers the possibility of observing stable advance of the delamination front. In this case, the rate of potential energy release per unit area of interface during axially symmetric advance of the delamination front is defined as

$$\mathcal{G} = -\frac{1}{2\pi a}\frac{\partial V_{\text{el}}}{\partial a}\bigg|_v = \frac{20\pi^6}{9}\frac{M_f h_f w_0^4}{a^4}. \tag{5.99}$$

The rightmost term in the continued equality (5.99) is based on the same membrane idealization that was used to get the result in (5.98). The final result in (5.99) is expressed in terms of w_0 rather than v because the deflection is perhaps the most readily measured kinematic variable in observing the process. In any case, (5.99) shows that $\mathcal{G}$ decreases rapidly with increasing a at fixed volume v, implying a stably advancing delamination front.

5.7.6 Bulge test configurations

The bulge test commonly refers to the configuration in which a free-standing window of a thin film is pressurized from one side so as to deflect it. In such an experiment, similar to that schematically shown in Figure 5.25, the value of the out-of-plane deflection of the film is monitored as a function of the applied pressure p. From the analysis presented in Sections 5.6 and 5.7, it is evident that the deflection of the film at a given pressure is influenced by the size and shape of interface delamination, the thickness and material properties of the film, the film mismatch stress and the fracture energy of the interface between the film and the substrate. Consequently, different test configurations involving the pressurized bulge geometry have been employed to determine experimentally the elastic modulus of the film, the film mismatch stress or the interface fracture energy (Hinkley 1983, Mehregany et al. 1987, Storåkers 1988, Jensen 1991, Vlassak and Nix 1992, Hutchinson and Suo 1992, Osterberg and Senturia 1997, Jensen 1998). The configuration is alternately known as the blister test, but no distinction is made between these descriptors here.

For a pressurized bulge of uniform width, the dependence of film deflection w_0 as a function of the applied pressure p is shown in Figure 5.27 based on small deflection bending theory (5.59), large deflection bending theory (5.64), membrane theory (5.68), and finite element simulations which account for large deflection. Similar results are presented in Figure 5.33 for the circular bulge on the basis of small deflection theory (5.81), large deflection membrane response (5.92), and finite element simulations incorporating large deflection. These results can be used in conjunction with the experimental results of w_0 versus p from the bulge test to determine the elastic modulus of the film if the film thickness h_f, film Poisson ratio ν_f and the initial debond size a are known.

If a mismatch stress σ_m acts in the film in the undeflected configuration, presumably engendered either during film deposition or as a consequence of thermal excursions, it is evident from (5.94) and Figure 5.36 that the film bulging response will be altered by this stress provided that it is comparable to or greater than the quantity $\bar{E}_f h_f^2 / a^2$. If the elastic properties of the film and the initial debond size are known, the film mismatch stress can be assessed by matching the experimentally determined variation of w_0 with p from the bulge test with the result given by (5.94).

In addition to providing a means of measuring material properties or the initial stress, the bulge test configuration offers a common experimental technique for probing the fracture energy of the interface between a film and a substrate. As noted earlier, if the pressure applied along an exposed region of a thin film reaches a value at which the energy release rate attains the critical magnitude given by (5.95), the debond edge begins to propagate in an unstable manner. Experimental observations of this critical condition for debond growth and instability can be combined with

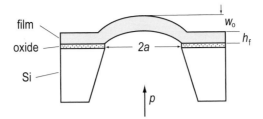

Fig. 5.38. Schematic diagram showing the geometry of the bulge and blister configurations created using anisotropic etching of silicon employing standard MEMS processing methods.

the results for the delamination energy plotted in Figure 5.37 as a function of the pressure at instability for a pressurized circular blister to infer the interface fracture energy Γ_{Ic}.

Figure 5.38 shows an example of the use of the bulge test for the determination of the film elastic modulus, mismatch stress or adhesion energy using silicon micromachining processes that are commonly employed in the fabrication of MEMS structures (Peterson 1982, Mehregany et al. 1997). This configuration, formed by recourse to conventional anisotropic etching of a silicon substrate, is achieved through several steps: (a) an etch stop layer is first formed by depositing boron from a high temperature solid source, (b) a thermal oxide is grown on the Si single crystal substrate at high temperature, (c) a test site pattern is defined by opening a window on the back side of the oxide using standard photolithography techniques (see Section 1.5.1) while the front side of the oxide is covered with a photoresist, and a silicon diaphragm is formed, (d) a polyimide thin film is spin cast on to the oxide layer, and (e) the diaphragm holding the film is etched away in an SF_6 plasma to obtain a free-standing polyimide membrane, as shown in Figure 5.38. An example of the use of bulge experiments involving such suspended polyimide membranes of square shape to determine the elastic properties of the film, the residual stress in the film and energy of adhesion of the film with the substrate can be found in Mehregany et al. (1987).

5.8 Example: MEMS capacitive transducer

An effective configuration for a MEMS transducer intended to convert acoustic signals into electric signals is a parallel plate capacitor (Scheeper et al. 1994). One plate of the capacitor is rigid, or nearly so, and the other plate is sufficiently flexible to respond to acoustic pressures in the audible range. The two plates are closely spaced with an air gap g between them, and the rigid plate is typically perforated with many holes to avoid pressure buildup in the gap. A fixed bias voltage V_b is maintained between the plates; the physical dimensions of voltage are (force×length/charge). As a pressure is applied to the flexible film, the gap changes size

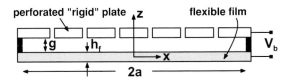

Fig. 5.39. Schematic diagram of a parallel plate capacitor in which one plate is essentially rigid whereas the other plate is a deformable film. The configuration has potential for function as an acoustic microphone. A characteristic feature of the configuration is that, for any voltage greater than a particular critical voltage, the system has no stable equilibrium configurations.

and a current is induced in order to maintain a fixed bias voltage. Most fabrication processing methods for capacitive elements for applications in acoustics are based on planar silicon technologies developed for microelectronic applications (Tien 1998).

The plates of a capacitor are oppositely charged, so they exert an attractive electrostatic force on each other. The force per unit area of parallel capacitor plates separated by vacuum at bias voltage V_b is

$$p_c = \frac{e_0 V_b^2}{2g^2}, \tag{5.100}$$

where e_0 is the electric permittivity of free space and edge effects have been neglected. The physical dimensions of permittivity are charge^2/(force×length^2). This pressure acts to reduce the gap dimension from its initial value g. Upon imposition of a bias voltage V_b between the plates of the capacitor which are not constrained against deflection by a dielectric material in the gap, the compliant plate deflects toward the rigid plate due to electrostatic attraction. This is a strongly nonlinear attractive force with the potential for inducing an instability at sufficiently large V_b. Nonlinearity in the elastic response of the solid can be a significant stabilizing influence.

Consider a configuration with the essential characteristics described, as depicted in Figure 5.39. Suppose that the deformable film is square with in-plane dimensions of $2a \times 2a$ and thickness h_f. If the film deforms with local displacement $w(x, y)$ in the z-direction, the gap is locally reduced to $g - w(x, y)$, and the upward pressure p_c is increased accordingly. The film is assumed to support an initial or residual stress $\sigma_m = \bar{E}_f(1 + \nu_f)\epsilon_m$ when $w(x, y) \equiv 0$, and the plane strain modulus of the film material is $\bar{E}_f$. Finally, the edges of the film at $x = \pm a$ and $y = \pm a$ are constrained against displacement and rotation. The thickness of the deformable plate is very small, with the ratio h_f/a being less than 0.01 and possibly as small as 0.001. In such a case, resistance to deformation can arise from both membrane effects and bending effects. Thus, the general framework for modeling this capacitor configuration has already been established previously in Section 5.6. However, this is a film bulge configuration that is more complex in its behavior than were those considered in Sections 5.6 and 5.7 due to the way in which the loading 'pressure' depends on the deformation.

Before pursuing a detailed calculation on the basis of the model, consider the system qualitatively. If the film is deflected by application of a uniform pressure p so that its midpoint displacement w_0 is larger than the thickness h_f, the relationship between it and p has the form shown schematically in Figure 5.40. This response curve can be expanded or contracted in the w_0 direction by decreasing or increasing, respectively, the residual tension. To obtain a second input curve for comparison, consider the pressure p_c in (5.100) with g

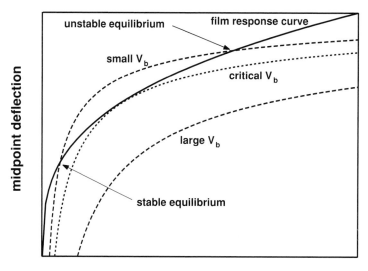

Fig. 5.40. Schematic diagram showing the nature of the behavior of a flexible thin film capacitor. The film response curve (solid line) represents midpoint deflection due to uniform pressure in a case for which membrane effects are significant. The dashed curves represent a pressure on the film due to electrostatic attraction of the capacitor plates, in one case due to a small bias voltage and in another case due to a large bias voltage. Bias voltage is small or large according to whether or not there are equilibrium configurations represented by intersection between the input pressure curves and the response curve. A critical voltage, called the pull-in voltage, exists which is the value that defines the transition from one kind of behavior to the other.

replaced by $g - w_0$, which implies that

$$w_0 \sim g - V_b \sqrt{\frac{e_0}{p_c}}. \tag{5.101}$$

Graphs of this dependence of w_0 on pressure p_c are also shown schematically in Figure 5.40 as dashed curves, one for a 'small' value of V_b and another for a 'large' value of V_b. The interpretation of small and large will have to await a quantitative analysis of the configuration.

The two intersection points of the response curve of the film represented by the solid line and the input curve for small V_b represent equilibrium configurations. Furthermore, it is evident that the equilibrium configuration for the lower value of w_0 is stable. To see this, imagine the system is at this state. Then, an external stimulus increases w_0 slightly from its equilibrium value. The response curve shows that the pressure needed to maintain the new position has also increased, whereas the input curve shows that the available pressure cannot be increased enough to provide what is needed for continued equilibrium. Consequently, when the external stimulus is removed, the system returns to its equilibrium position. By probing both equilibrium positions in both possible directions of departure, it is concluded that the equilibrium position for the smaller w_0 is stable and that for larger w_0 is unstable. For the case of large V_b, there are no intersection points of the response curve with the input curve, implying that the gap has been closed catastrophically as the voltage is increased,

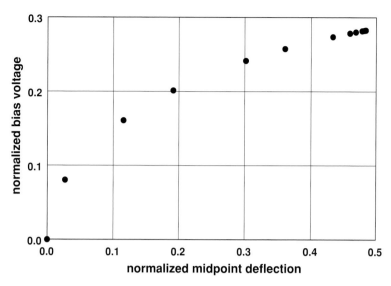

Fig. 5.41. Dependence of normalized bias voltage on normalized midpoint deflection for a square capacitor film obtained by means of a finite element simulation. The calculation was based on the following values of the system characterizing parameters: $\epsilon_m = 0.001$, $a/h_f = 250$ and $g/h_f = 3$. The pull-in voltage is defined by the largest value of bias voltage for which an equilibrium equation could be established.

without arriving at an equilibrium position. The discriminating value of V_b between the two types of behavior, corresponding to an input curve that is tangent to the response curve at a single point, defines the critical voltage or *pull-in voltage* for instability. Knowledge of the instability voltage is an important design consideration for such devices (Osterberg and Senturia 1997). The goal here is to estimate this pull-in voltage for representative values of system parameters by direct appeal to the numerical finite element method.

Results of a calculation for a device characterized by $\epsilon_m = 0.001$, $a/h_f = 250$ and $g/h_f = 3$ are shown in Figure 5.41. The data are shown in the form of normalized bias voltage $\hat{V}_b$ versus normalized midpoint deflection α/g of the square film, where the normalized bias voltage is defined in terms of the applied bias voltage and other system parameters by $V_b^2 a^2 e_0 / h_f^4 \bar{E}_f = \hat{V}_b^2$. No equilibrium configurations could be found for values of normalized bias voltage greater than 0.283, so this value defines the pull-in voltage. It is noted that the variation of $\hat{V}_b$ with midpoint deflection appears to have a slightly rising slope at the highest value of $\hat{V}_b$ for which an equilibrium configuration could be found. The dependence of bias voltage on volume under the deflected film or on device capacitance, either of which would seem to be a more reliable variable due to its global nature, showed the same general behavior. In any case, the slope is slight and the absence of an equilibrium configuration is perhaps a stronger indicator of the pull-in phenomenon than is the vanishing of the slope of this response curve.

To provide a measure of residual stress σ_m, which is necessary to estimate pull-in voltage, the fabrication process usually involves deposition of a film of the device plate material, commonly polysilicon, onto a solid Si substrate. This structure can then be observed by means of wafer curvature techniques, for example, to infer a value of residual stress in the film layer. Among the system parameters that influence functional characteristics, the

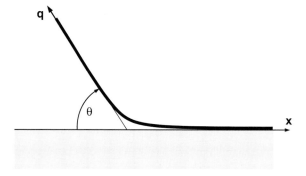

Fig. 5.42. Steady-state peeling of a film from a substrate to which it has been bonded by the action of peel force q in the direction θ.

residual stress is the most difficult to assess properly. On the other hand, control of its value provides an additional means for controlling device characteristics.

5.9 Film peeling

Steady film peeling in the peel test configuration has been adopted as a method to measure the strength of a bonded interface between a film and its substrate or to measure the strength of an adhesive layer between a film and a substrate. The configuration, which is essentially two-dimensional, is depicted in Figure 5.42 for the situation in which the point of separation advances steadily in the x-direction. Far ahead of the line of separation, the film is bonded to the substrate, and it may support a stress due to elastic mismatch. The delamination process is driven by a peel force q measured per unit length along the line of separation, acting on the detached portion of the film far behind the line of separation. The peel force acts in a direction inclined to the substrate surface by an angle θ as shown. If the substrate stiffness $\bar{E}_s$ is similar to or greater than the film stiffness, then the substrate compliance does not significantly influence the peeling process. In this case, the substrate can be assumed to be rigid for any practical purpose.

5.9.1 The driving force for delamination

Consider steady advance in the x-direction of the two-dimensional profile shown schematically in Figure 5.42. If the line of separation advances a unit distance, then the point of application of the peel force also moves a unit distance in the x-direction plus a unit distance in the direction defined by the angle θ. The inner product of the force with this displacement implies that the work done by the force per unit advance of the separation line is

$$W_q = q(1 - \cos\theta) \qquad (5.102)$$

per unit length of separation line. This result is based on the tacit assumption of an inextensible film.

Suppose next that extensibility of the film is taken into account. If the bonded film supports a uniform equi-biaxial mismatch stress σ_m then the elastic energy density per unit area of interface far ahead of the line of separation compared to h_f is $W_+ = \sigma_m^2 h_f / M_f = t_m^2 / h_f M_f$. For points far behind the separation point, the nonzero stress components are $\sigma_{xx} = q/h_f$ and $\sigma_{yy} = \nu_f q/h_f + (1 - \nu_f)\sigma_m$; these values are based on the constraint that the out-of-plane extension ϵ_{yy} is maintained at the value $\epsilon_m = \sigma_m / M_f$. The elastic energy per unit area of film midplane far behind the separation point is then

$$W_- = \left[(1 - \nu_f^2)q^2 + (1 - \nu_f)^2 t_m^2\right] / 2E_f h_f. \qquad (5.103)$$

If the film midplane is extensible, there is an additional contribution to W_q that is subtle in nature but of the same order as W_+ and W_-. If the separation line moves a unit distance with respect to the substrate in the x-direction then the point of application of q also moves a unit distance in the x-direction plus some distance in the direction specified by θ. Because of the change in strain in the film from $\epsilon_{xx}^+ = \epsilon_m$ to $\epsilon_{xx}^- = q/\bar{E}_f h_f - \nu_f(1 - \nu_f)\sigma_m / E_f$, the point of application of q displaces in the θ direction a distance $1 - \epsilon_{xx}^+ + \epsilon_{xx}^-$. With this effect taken into account, the driving force $\mathcal{G}$ for peeling can be extracted. This quantity is defined as the excess of the work done by q plus the elastic energy density before peeling over the elastic energy density remaining in the film after peeling, and it is given by the expression

$$\mathcal{G} = W_q + W_+ - W_- = q(1 - \cos\theta) + \frac{1}{2\bar{E}_f h_f}(t_m - q)^2. \qquad (5.104)$$

Note that the first term on the right side of (5.104) vanishes as $\theta \to 0$ and that the expression reduces to (4.34), as it should. When $\theta \sim \pi/2$, on the other hand, the second term on the right side of (5.104) is typically negligible compared to the first term.

5.9.2 Mechanics of delamination

The driving force per unit length of separation line for peeling is given in (5.104). If the interface resistance to peeling can be characterized by a work of fracture Γ then

$$\mathcal{G} = \Gamma, \qquad (5.105)$$

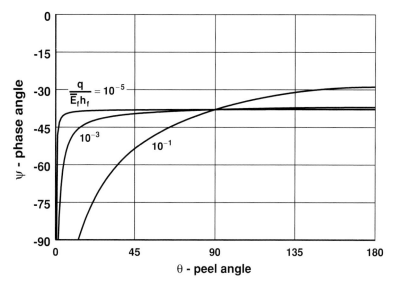

Fig. 5.43. Phase angle ψ of interface traction ahead of an advancing separation line versus direction θ of application of the peel force q for several values of $q/\bar{E}_{\mathrm{f}}h_{\mathrm{f}}$.

and the value of Γ can be inferred from measurements of $\mathcal{G}$ for a given material system. In view of the possible sensitivity of interface fracture resistance to the phase angle ψ of the local stress state acting on the interface adjacent to the zone of delamination, it is important to know the phase angle that arises in peeling. This issue was addressed by Thouless and Jensen (1992), who demonstrated that the phase angle can be determined in the following way.

The local conditions at the edge of the line of separation can again be represented by the situation depicted in Figure 5.2 where t_{a} and m_{a} are the stress resultants at the edge of the detached region. At this level of observation, the driving force for delamination has been given by (5.21) where $\Delta t_{\mathrm{a}} = t_{\mathrm{a}} - t_{\mathrm{m}}$. Equilibrium of the detached portion of the film implies that $t_{\mathrm{a}} = q \cos \theta$ so that

$$\Delta t_{\mathrm{a}} = q \cos \theta - t_{\mathrm{m}}. \tag{5.106}$$

The edge bending moment m_{a} can then be found in terms of q, θ and system parameters by noting that the local form of energy release rate in (5.21) must be identical to the form (5.104). This identification leads to the result that

$$m_{\mathrm{a}} = \frac{h_{\mathrm{f}}}{\sqrt{12}} \left[2q \left(\bar{E}_{\mathrm{f}}h_{\mathrm{f}} - t_{\mathrm{m}} \right) (1 - \cos \theta) + q^2 \sin^2 \theta \right]^{\frac{1}{2}}. \tag{5.107}$$

An expression for phase angle in terms of Δt_{a} and m_{a} is given in (5.26) where the parameter ω depends on the relative stiffnesses of the substrate and the film according to the numerical results plotted in Figure 4.28. When the film and substrate have identical elastic properties, $D_1 = D_2 = 0$ and the parameter ω has the value $52.1°$. To illustrate the sensitivity of ψ to changes in angle θ for a specified level of peel force q, consider the case of identical elastic properties for the film and substrate and no initial mismatch stress, that is, $\sigma_m = 0$. The results of evaluating ψ versus θ for several values of $q/\bar{E}_f h_f$ are shown in Figure 5.43. The most noteworthy feature of the result is that ψ is relatively insensitive to angle θ when the film and substrate properties are similar. For values of tensile peeling stress q/h_f less than about $0.001\,\bar{E}_f$, the phase angle is approximately $-38°$ for angles θ.

5.10 Exercises

1. Suppose that a straight sided film bulge is deformed under conditions for which the volume V per unit length between the bulged film and its substrate is being controlled. For a film with fixed bulge width a in the geometrically nonlinear deformation range, the pressure producing the bulge is related to the normalized membrane tension τ in the film by (5.63) according to the von Karman equations (5.61).

    (a) By integrating (5.61)$_1$ and using the definition of bulge volume V given in (5.75), determine the normalized volume V/ah_f in terms of normalized film tension τ.
    (b) An expression for the delamination driving force $\mathcal{G}$ is given in (5.66). Express $\mathcal{G}(a)/p_r h_f$ in terms of film tension τ, and combine the result with the result of part (a) to plot $\mathcal{G}(a)/p_r h_f$ versus V/ah_f parametrically in τ.
    (c) Expand the expressions for both normalized driving force and bulge volume to lowest order in τ, and verify that the results are consistent with the expression in (5.78) which was based on small deflection theory.

2. In deriving a relationship between midpoint deflection of a pressurized membrane bulge and the applied pressure given in (5.92), the influence of initial stress was neglected. Determine how the relationship should be modified to account for an initial equi-biaxial tensile stress in the film.

3. Consider the situation of cool down from deposition temperature of a diamond–like carbon film on a Ti substrate with a circular interface debond zone as described in Section 5.2.2. It was illustrated in the example that the debond zone, $30\,\mu m$ in diameter, will buckle before the film cools to room temperature. If the separation resistance of the interface is $\Gamma = 10\,\text{J/m}^2$, determine whether or not the buckle region will begin to expand in size prior to reaching room temperature. If expansion does occur, estimate the final size of the circular buckle.

4. The expression (5.12) provides a relationship between the post-buckling equilibrium midpoint deflection of a straight-sided buckle and the corresponding stress drop within the buckle region. Derive this result by solving the ordinary differential equations (5.4) subject to the boundary conditions on $w(\pm a)$, $w'(\pm a)$ and $u(\pm a)$ stated in the text below the differential equations.

5. Consider the formation of straight-sided buckles due to temperature change under plane strain conditions in a film–substrate system for which the interface separation

energy of a buckle depends on the phase angle ψ of the stress state at the edges. Suppose the elastic modulus and Poisson ratio of both the film and substrate materials are 10^{11} N/m^2 and 0.25, respectively. The thickness of the film is $10\,\mu$m and the substrate is comparatively very thick. The interface separation energy is given in terms of the phase angle by $\Gamma(\psi) = 7.0[1 + \tan^2(0.7\psi)]$ J/m^2 where ψ is expressed in radians. It is observed that, among all debonded zones at the interface, only those greater in half–width than $a_{\mathrm{m}} = 125\,\mu$m developed into buckles during the temperature change. On the basis of this observation, determine the final width of those buckles that formed and expanded during the temperature change.

6. Suppose the peel test is applied to measure the energy of separation between a metal film of thickness h_{f} and its substrate due to an adhesive layer for the case when the peel force acts in the direction $\theta = \pi/2$. There is no residual stress in the adhered film. The tensile yield stress of the film material is denoted by σ_{Y} and the elastic modulus is E_{f}.

   (a) Assume a stress distribution in the film at the point of separation from the substrate that is consistent with elementary bending theory, and determine the smallest peel force magnitude q that will induce plastic flow in the film at the point of separation in terms of the thickness and material properties. Base the estimate on the maximum shear stress plastic yield condition.

   (b) Determine the largest adhesive separation energy Γ that can be measured in this way without inducing plastic deformation in the film. Evaluate the result numerically for a case with $h_{\mathrm{f}} = 4\,\mu$m, $E_{\mathrm{f}} = 100$ GPa and $\sigma_{\mathrm{Y}} = 60$ MPa.

   (c) Discuss the influence of plastic deformation in the film during the peel test, if it does occur, on the value of separation energy that is inferred from a measurement of the peel force (cf. Kim and Aravas 1988).

6

Dislocation formation in epitaxial systems

The generation of film stress as a natural consequence of epitaxial constraint arising from mismatch between the lattice parameters of a thin film and its substrate was discussed in Section 1.7, and elastic strain induced in layered epitaxial materials was analyzed in Section 3.4. The stress associated with the elastic strain acts as a driving force for formation and growth of structural defects. An image of dislocations that have formed at the interface between SiGe and Si in this way appears as Figure 1.16 and a second image of dislocations that have been formed at the interface between CdTe and GaAs appears as Figure 1.18. The presence of such misfit dislocations, in turn, can have a detrimental effect on the performance of strained layer material systems. Lattice strain in SiGe materials alters the electronic bandgap, a behavior that can be exploited to good advantage in some circumstances. Strain can also alter band edge alignment, thereby converting an indirect bandgap material into a direct bandgap material and rendering it a candidate material for optoelectronic applications. In cases in which lattice strain is induced to achieve some functional objectives, the formation of strain-relieving dislocations can be particularly deleterious as noted in Section 7.3.2. Even for cases in which it is not essential that the strain be maintained, the presence of dislocations can have an adverse effect on electronic performance of semiconductor materials, serving as easy diffusion paths for dopants or as recombination centers to diminish carrier density in devices (Mahajan 2000). The means of controlling structural defects induced by misfit strain and of interpreting their consequences is, therefore, a central concern in the fabrication and characterization of strained layer materials.

This chapter is devoted to a discussion of situations in which the defect of concern is a crystal dislocation. Particular attention is focused on the conditions necessary for formation of interface dislocations which partially relieve the elastic strain in the film, and on cases for which the strained layer is a single crystal or a polycrystal with grain size in the plane of the film–substrate interface that is large compared to film thickness. More general questions of dislocation interaction and strain relaxation are addressed in Chapter 7.

6.1 Dislocation mechanics concepts

The discussion is based on the continuum theory of elastic dislocations and it proceeds in the following manner. First, the work that must be expended to form an elastic dislocation with a particular configuration is examined in a general setting, and the variation of this work with change in the position of the dislocation is interpreted as a configurational force acting on the dislocation. This notion of force leads naturally to the question of equilibrium of dislocation configurations and the stability of those configurations. The ideas are then applied in subsequent sections to determine the smallest thickness of a uniformly strained layer on a substrate, termed the critical thickness, at which it first becomes energetically possible for a dislocation to form spontaneously in the layer. Attention is then shifted to determining the conditions for the advance of the threading segment of the strain-relieving glide dislocation through the strained film. Experimental observations of threading dislocation motion in constrained thin films are then presented to substantiate the mechanistic basis for the theoretical results. The influence on critical thickness of multiple layering, grading of mismatch strain, nonplanarity and other factors is also examined. Some applications of these concepts are then described with particular reference to the strategies for mismatch strain control in the heteroepitaxial structures used in optoelectronic devices. Conditions for the formation of strain-relieving dislocations in quantum wires are also examined as functions of wire geometry and constraint of the surrounding matrix material.

6.1.1 Dislocation equilibrium and stability

The main competing effects in the formation of a dislocation in a thin strained layer are illustrated schematically in Figure 6.1. For purposes of this discussion, a block of elastic material is strained by means of a very stiff loading apparatus, and the loaded boundary is subsequently held fixed to preclude the possibility of energy exchange between the solid and its surroundings. The elastic energy stored in the crystal due to this loading can be reduced by the advance of a glide dislocation through the crystal. It is presumed that there is indeed a traction on the glide plane due to the loading. This effect leads to the concept of an applied *configurational force* on the dislocation due to the elastic field, defined as the reduction in mechanical energy of the solid per unit distance advance of the dislocation and denoted by F_{appl} in the figure (Eshelby 1951).

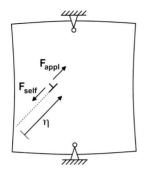 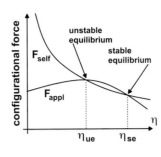

Fig. 6.1. A schematic diagram of a glide dislocation in a strained elastic solid, showing the configurational forces acting on the dislocation, F_{appl} due to the background elastic field and F_{self} due to the proximity to the free surface.

On the other hand, the fact that this dislocation motion occurs in the proximity of a free surface implies the presence of an opposing effect. In forming a dislocation, an elastic material appears to be more compliant near a free surface than far from it. In other words, less work is required to form a dislocation of given characteristics near a free surface than far from it. The observation that the energy of a dislocation can be reduced by moving it closer to a free surface, that is, by reducing its distance η to the free surface along the glide direction, suggests the presence of a configurational force tending to pull the dislocation out of the crystal through the free surface. This force, which is denoted by F_{self} in the figure, is commonly called the *image force*; the name derives from superposition of the field of a dislocation with a mirror image field with respect to the free surface in order to satisfy boundary conditions, a strategy that is useful in the analysis of some very simple configurations. Because this force arises solely from the equilibrium field of the dislocation itself, in the absence of any other loading on the body, it is called the *self-force* acting on the dislocation.

The dislocation is in equilibrium when the sum of all forces acting on it vanishes, as suggested on the right side of Figure 6.1 where two equilibrium positions are identified. These equilibrium positions, however, differ in a fundamental aspect. When the dislocation is displaced a small distance along the glide plane in either direction along its glide plane from its equilibrium position $\eta = \eta_{\mathrm{se}}$, it comes under the action of a force that tends to restore it to this equilibrium position. Consequently, this constitutes a stable equilibrium position. For the case $\eta = \eta_{\mathrm{ue}}$, however, a small change in location of the dislocation from the equilibrium position induces a force which tends to push the dislocation further from that position. Consequently, the latter is an unstable equilibrium position.

Two conditions must be satisfied before a strain-relieving dislocation is formed in a closed mechanical system which cannot exchange energy with its surroundings.

First, a stable equilibrium configuration must be accessible, but this feature alone is not a sufficient condition. In addition to an accessible equilibrium configuration, it is necessary that the work done in forming the dislocation must be provided by the background elastic field as the strain is relaxed in the material. This latter requirement, in the context of the schematic diagram of the configurational force in Figure 6.1, implies that in order for a dislocation to form at $\eta = \eta_{se}$, the area under the plot of F_{appl} versus η over the interval $0 \leq \eta \leq \eta_{se}$ must be equal to the area under the plot of F_{self} versus η over the same interval. For a proper quantitative analysis of this problem, a sign convention for force should be adopted so that all forces are positive in the direction of increasing η, and this is done in the sections that follow.

A significant portion of this chapter is devoted to obtaining estimates of the applied and self forces acting on dislocations, along with the associated energy changes, in layered systems of general practical interest. The discussion is based almost entirely on the elastic theory of dislocations. It is perhaps ironic that a phenomenon which derives solely from the discreteness of the crystal lattice is described by a continuum theory. However, what is known about the behavior of dislocations in systems of interest here stems almost exclusively from elastic analysis. For one thing, dislocations interact with boundaries and with each other through long range effects which can be described adequately by means of elasticity theory. For this to be a valid approach, it is necessary to assume that the dislocation core structure in the crystal is autonomous, in the sense that it is unaffected by changes in the mechanical environment of the dislocation. If this is so, the fact that it does not influence changes in the mechanical state renders it of secondary importance even though the actual core structure is unknown. Even with an autonomous core structure, the dislocation core energy can come into play in situations in which the dislocation line length changes. This effect is not taken into account here, but the modifications needed to do so are relatively minor and self-evident.

Finally, the elastic theory of dislocations is encumbered with the notion of a core cutoff radius, a stratagem made necessary by the existence of a singular stress field at the core of the dislocation line. This singularity renders the elastic strain energy density non-integrable in the usual sense of continuum mechanics. To circumvent this difficulty, material within some radial distance, say r_0, of the dislocation line is presumed to be removed and the surface of the tunnel of circular cross-section created by material removal is subjected to traction such that no relaxation occurs in the remaining material. By the argument of autonomy, this traction at each section of the tunnel is given by the stress field of a straight dislocation in an unbound solid with its line orientation and Burgers vector coinciding with the local line orientation and Burgers vector of the dislocation under consideration.

The selection of a particular value for r_0 is arbitrary, although several criteria for it have been proposed. Among these are: the radius at which an elastic strain

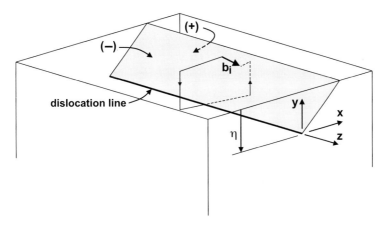

Fig. 6.2. A long straight dislocation in an elastic solid is parallel to a traction-free surface and at a distance η from it. The dislocation is formed by an offset of one side of the shaded plane with respect to the other side defined by the Burgers vector b_i.

component first becomes large in magnitude upon approach to the dislocation line; a radius equal to some fixed multiple of the lattice parameter or Burgers vector length; or a radius chosen so that the elastic energy of the material remaining outside the cutoff radius equals the energy computed for the complete dislocation by some more fundamental approach (Beltz and Freund 1994). The last criterion, which often leads to the smallest estimate for r_0, is preferred if a choice is necessary. It also has the feature that core energy, which is presumably included in any fundamental approach, would then be incorporated indirectly through the choice of cutoff radius. In any case, if the description of any phenomenon hinges on a particular choice of r_0 for the existence of that phenomenon, then any conclusions drawn should be viewed with caution.

6.1.2 Elastic field of a dislocation near a free surface

Consider an isotropic elastic material with a traction-free planar boundary. Other boundaries of the solid are sufficiently remote so that they are inconsequential for present purposes. Points in the material are located by means of an xyz rectangular coordinate system, oriented so that $y = \eta$ is the plane boundary in the reference configuration, and the material occupies the region $y \leq \eta$, as shown in the diagram in Figure 6.2. The condition that the surface is free of traction is enforced by requiring the stress components σ_{xy}, σ_{yy} and σ_{yz} to vanish at all points on the surface.†

---

† Note that the orientation of the xyz-coordinate axes with respect to the free surface is not the same here as it was in preceding chapters. Conventions have been adopted largely for historical reasons, and quantities that appear in the course of this discussion that have counterparts in preceding chapters are entirely consistent.

Suppose that the material is distorted by a long, straight dislocation line lying within the solid parallel to the plane surface. For definiteness, suppose that the dislocation line coincides with the z-axis; the plane of displacement discontinuity is a plane containing this line. The sense of the Burgers vector is determined with respect to a choice of positive direction of traversal along the dislocation line, here chosen to be the z-direction. For any given plane of displacement discontinuity (shown shaded in the figure), a material loop is imagined to exist in the solid. This loop is closed prior to formation of the dislocation, and it surrounds the eventual dislocation line. The loop is formed by moving in the positive sense around the line according to a right-hand rule, beginning on the $(-)$ side of the plane and ending on the $(+)$ side. The formation of the dislocation introduces a discontinuity in this material loop so that it is no longer closed. If the elastic displacement field resulting from the formation of the dislocation is represented by u_i, then the Burgers vector b_i is defined by the difference in values of u_i for adjacent points on the loop on opposite sides of the plane of discontinuity as

$$b_i = u_i^{(+)} - u_i^{(-)}. \tag{6.1}$$

The Burgers vector of the dislocation has components b_x, b_y and b_z in the coordinate directions. As components of a vector, these parameters may take on either positive or negative values. The *edge component* of the Burgers vector has length $\sqrt{b_x^2 + b_y^2}$ and the *screw component* has length $|b_z|$.

The material plane on which the offset characterized by the Burgers vector is imposed to form the dislocation is unspecified for the time being. The elastic stress field of the dislocation is independent of the orientation of this plane but, in a sense to be discussed subsequently, the energy of the dislocation does have a weak dependence on the choice of this plane. The equilibrium stress field of the dislocation is independent of z due to the invariance of the system under translation in the z-direction.

The six independent components of the symmetric stress matrix are recorded here for later reference. The stress field arising from the edge component of the Burgers vector is conveniently represented in terms of the Airy stress function

$$A(x, y) = \frac{\mu}{2\pi(1-v)} \left[\tfrac{1}{2}\left(xb_y - yb_x\right) \ln\left(\frac{x^2 + y^2}{x^2 + (2\eta - y)^2}\right) \right.$$
$$\left. + \frac{2b_y\eta x(\eta - y) + b_x\eta[x^2 + 2y\eta - y^2]}{x^2 + (2\eta - y)^2} \right], \tag{6.2}$$

where μ is the elastic shear modulus and v is the Poisson ratio of the isotropic elastic material. The stress components themselves are then derived from the Airy function according to

$$\sigma_{xx}(x, y) = \frac{\partial^2 A}{\partial y^2}(x, y), \qquad \sigma_{xy}(x, y) = -\frac{\partial^2 A}{\partial x \partial y}(x, y),$$

$$\sigma_{yy}(x, y) = \frac{\partial^2 A}{\partial x^2}(x, y), \qquad \sigma_{zz}(x, y) = v\left[\sigma_{xx}(x, y) + \sigma_{yy}(x, y)\right]. \quad (6.3)$$

The explicit expressions for these stress components are readily obtained for purposes of calculation, but they are too complicated to be of intrinsic interest for the present discussion.

The remaining stress components are associated with the screw component of the Burgers vector, and the fields are

$$\sigma_{xz}(x, y) = -\frac{\mu b_z}{2\pi}\left[\frac{y}{x^2 + y^2} + \frac{2\eta - y}{x^2 + (2\eta - y)^2}\right],$$

$$\sigma_{yz}(x, y) = \frac{\mu b_z}{2\pi}\left[\frac{x}{x^2 + y^2} - \frac{x}{x^2 + (2\eta - y)^2}\right]. \qquad (6.4)$$

It can be confirmed by direct calculation that (6.3) and (6.4) together represent an equilibrium stress field, and that the boundary conditions $\sigma_{xy}(x, \eta) = 0, \sigma_{yy}(x, \eta) = 0$ and $\sigma_{yz}(x, \eta) = 0$ are satisfied by this stress field.

Next, the energy of the dislocation near a free surface is considered. This energy is normally understood to be equal to the work that must be done to form the dislocation in the absence of any other stresses in the material. Equivalently, it can also be envisioned as the elastic energy stored in the solid as a consequence of the formation of the dislocation. Because all fields are uniform in the z-direction in the present case, the energy per unit length in the z-direction is considered. For purposes of discussion, it is assumed that the dislocation is formed by glide on a specific crystallographic plane. For any glide dislocation, the inner product of the Burgers vector with the glide plane normal is zero. For example, the solid might be a cubic crystal with the free surface being the (010) plane and the glide plane being the (111) plane. For this plane in a face-centered cubic or diamond cubic material in the orientation identified, the most likely directions for the Burgers vector are the crystallographic directions $[10\bar{1}]$, $[1\bar{1}0]$ and $[01\bar{1}]$. Clearly, each of these directions lies in the glide plane and each is orthogonal to the glide plane normal direction. All members of the $\{111\}$ family of planes are expected to respond in the same way under the same local resolved shear stress.

To calculate the energy, consider the situation depicted in Figure 6.3. The material is initially stress-free. Then, the material is imagined to be cut along the glide plane from the surface to the dislocation line. In addition, a core of material of radius $r_0 \ll \eta$ and centered on the eventual position of the dislocation is removed. Then, displacements are imposed on the surfaces created by these cuts to produce a relative

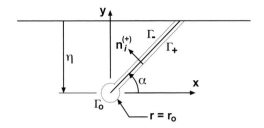

Fig. 6.3. To form the dislocation, the material is cut along the dotted line consisting of the parts Γ_+, Γ_- and Γ_0 and appropriate displacements are imposed on the surfaces created. The energy of the dislocation equals the work required to form it.

offset b_i of the $(+)$ side of the glide plane cut with respect to the $(-)$ side and in a displacement on $r = r_0$ which is that of a dislocation with the same Burgers vector in an unbounded solid.

According to the theory of elasticity, the energy stored in the material as a result of these operations, per unit length in the z-direction, is

$$W_d(\eta) = \int_\Gamma \tfrac{1}{2} T_i u_i \, dl, \tag{6.5}$$

where Γ_+, Γ_- and Γ_0 together comprise the boundary Γ of the region created by the cuts in the material as shown in Figure 6.3, l is the arc length along this boundary, u_i is the displacement imposed on that boundary, and T_i is the traction required to maintain the imposed displacement. Traction on a surface with outward normal n_i is related to the limiting value of stress σ_{ij} on the boundary through the relationship $T_i = \sigma_{ij} n_j$, known commonly as the Cauchy stress principle. The tractions acting on Γ_+ and Γ_- are equal in magnitude but opposite in direction, and the displacements satisfy (6.1), so these contributions can be combined to yield

$$W_d(\eta) = \int_{r_0}^{\eta \csc \alpha} \tfrac{1}{2} b_i \sigma_{ij} n_j^{(+)} \, dl + \int_0^{2\pi} \tfrac{1}{2} \left[\sigma_{ij}^{(\infty)} n_j^c u_i^{(\infty)} \right]_{r=r_0} r_0 \, d\theta, \tag{6.6}$$

where $\alpha = \arctan(b_y/b_x)$, $n_i^{(+)}$ is the outward unit normal on Γ_+, n_i^c is the outward unit normal on Γ_0 and the superscript (∞) identifies fields of the dislocation in an unbounded solid. Evaluation of the integrals in (6.6), in light of (6.3) and (6.4), yields

$$W_d(\eta) = \frac{\mu[b_x^2 + b_y^2 + (1-\nu)b_z^2]}{4\pi(1-\nu)} \ln \frac{2\eta}{r_0} + \frac{\mu[b_x^2 + b_y^2]}{8\pi(1-\nu)} \left[\cos 2\alpha - \frac{(1-2\nu)}{2(1-\nu)} \right].$$

$$\tag{6.7}$$

For most purposes, the logarithmic term in (6.7) is the dominant contribution so

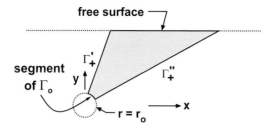

Fig. 6.4. A planar region of material (shaded area) adjacent to the dislocation core which is bounded by the free surface, by two choices of Γ_+ that are labelled Γ'_+ and Γ''_+, and by a segment of Γ_o.

that

$$W_d(\eta) \approx \frac{\mu[b_x^2 + b_y^2 + (1 - \nu)b_z^2]}{4\pi(1 - \nu)} \ln \frac{2\eta}{r_o}, \tag{6.8}$$

which provides a very good approximation to W_d for $\eta \gg r_o$. However, the remaining terms which arise naturally in the derivation are included in (6.7) for completeness.

Note that the work that must be done to form the long, straight dislocation depends on the orientation of the plane on which the offset b_i is enforced, provided that there is an edge component to b_i. The second integral in (6.6) is obviously independent of α, and hence the dependence of $W_d(\eta)$ on α stems from the first integral. Because the Burgers vector b_i can be taken outside the integral sign, this integral has the simple interpretation of an inner product of $\frac{1}{2}b_i$ with the total force acting on Γ_+. Thus, the dependence of $W_d(\eta)$ on α that is reflected in (6.7) must arise through this force (Freund 1994).

Figure 6.4 shows a region of the plane bounded by two choices of Γ_+, say Γ'_+ and Γ''_+, by a segment of Γ_o, and by a segment of the free surface. The total force acting on this region must vanish since the material in this region is in equilibrium. There is no force contribution from the free surface segment. However, the segment of Γ_o carries a finite and nonzero force, no matter how small r_o becomes, due to the strength of the stress singularity in the elastic field of the dislocation. It is then evident that the force acting on each of Γ'_+ and Γ''_+ must be different from the force on the other, and this difference accounts for the α dependence which emerged in (6.7).

The configurational force acting on a dislocation due to its proximity to a free surface was introduced in Section 6.1.1 on the basis of a qualitative argument. In light of the expression in (6.7) for the work required to form a dislocation W_d as a function of distance η from the free surface, a mathematical expression for the self-force $F_d(\eta)$ is readily obtained in terms of system parameters. The work that must be added to the system to move the dislocation from distance η to the infinitesimally greater distance $\eta + d\eta$ is $W_d(\eta + d\eta) - W_d(\eta) \approx W_d'(\eta) \, d\eta$. This work must be done to overcome the effect of the configurational force, or self-force, acting on the dislocation due to its own elastic field. Thus,

$$F_d(\eta) = -W_d'(\eta) = -\frac{\mu[b_x^2 + b_y^2 + (1 - \nu)b_z^2]}{4\pi(1 - \nu)\eta} \tag{6.9}$$

is the self-force tending to move the dislocation *away from* the free surface. The minus sign in (6.9) indicates that the force actually tends to pull the dislocation toward the free surface. This force is represented schematically in Figure 6.1. For a material for which $\mu = 50 \times 10^9 \, \text{N m}^{-2}$, $\nu = 0.25$, Burgers vector length $b = \sqrt{b_i b_i} = 0.5 \, \text{nm}$ and $r_o = \frac{1}{2}b$, the energy per unit length of a dislocation at a distance $\eta = 100 b$ from the free surface is roughly $W_d \approx 8 \times 10^{-9} \, \text{J m}^{-1}$. The magnitude of the force tending to draw the dislocation toward the free surface is roughly $|F_d| \approx 0.03 \, \text{N m}^{-1}$.

Finally, it is evident from the definition of the force in (6.9) that it acts on the dislocation in the direction of the free surface normal vector. The force acting in a direction parallel to the free surface is zero due to the invariance of the configuration under translation of the dislocation in any direction parallel to the surface. The component of the configurational force acting along the glide plane normal to the dislocation line is $F_d \sin \alpha$. If the motion of the dislocation is restricted to its glide plane, then the component of the force normal to the glide plane is workless and, therefore, inconsequential.

6.2 Critical thickness of a strained epitaxial film

Consider a single crystal film of thickness h_f grown epitaxially on a substrate of thickness $h_s \gg h_f$. For this case, the evolution of substrate curvature for arbitrary combinations of substrate and film moduli M_s and M_f, respectively, was analyzed in Section 2.2.1. If the modulus ratio M_f/M_s is of order unity then, to lowest order in the thickness ratio h_f/h_s, the curvature of the substrate as given in (2.19) is zero, the elastic strain in the substrate is zero, and the mismatch ϵ_m between the two materials is completely accommodated by elastic strain in the film. It follows that the uniform state of strain in the film is

$$\epsilon_{xx} = \epsilon_{zz} = \epsilon_m, \qquad \epsilon_{yy} = -\frac{2\nu_f}{1 - \nu_f}\epsilon_m, \tag{6.10}$$

where ν_f is the Poisson ratio of the film material. These strain components are referred to an xyz rectangular coordinate system oriented so that the y-direction is the outward normal direction to the free surface of the film, and the origin of coordinates is in the film–substrate interface. For the case of a relatively thick substrate, the parameter h_s does not enter the calculation. The film thickness h_f is then the only characteristic dimension in the calculation, other than the length of the Burgers vector.

For many film–substrate systems, the elastic properties of the two materials are similar and it is assumed for the present that $M_s \approx M_f = 2\mu_f(1 + \nu_f)/(1 - \nu_f)$. If the difference in elastic constants is taken into account in the critical thickness calculation, only minor additional insights are gained at the expense of substantially greater complexity. Once the critical thickness condition for like properties is in hand, the influence of modulus difference will be examined more closely within the framework of a model problem.

The spatially uniform state of stress in the perfectly coherent film is

$$\sigma_{xx} = \sigma_{zz} = \sigma_m = M_f \epsilon_m, \quad \sigma_{yy} = 0. \tag{6.11}$$

This is the stress field that is available to do work as a dislocation is formed in the film. The corresponding elastic energy in the film per unit area of the interface is $\sigma_m \epsilon_m h_f = M_f \epsilon_m^2 h$. For a film of thickness $h_f = 50$ nm, a biaxial modulus of $M_f = 10^{11}$ N m^{-2} and a mismatch strain of $\epsilon_m = 0.01$, the elastic energy in the film per unit area of interface is 0.5 J m^{-2}. For comparison to the work required to form a dislocation, consider the energy per unit length in a strip of film of width h_f which, in this case, is 2.5×10^{-8} J m^{-2}. This energy per unit length is of the same order of magnitude as the estimate of the energy per unit length of a dislocation in the same film. While such a comparison is informative, it is not fully appropriate for purposes of establishing the necessary condition for formation of a dislocation. The proper statement of this condition is developed next.

6.2.1 The critical thickness criterion

Suppose that a long, straight dislocation is introduced into the material. The dislocation line is parallel to the free surface of the film and at a distance η from it. The coordinate system is oriented so that the z-axis is parallel to the dislocation line, as shown in Figure 6.2. The dislocation is assumed to have the same properties as were adopted for the analysis in Section 6.1.2. To calculate the work done by the background stress field (6.11) in forming the dislocation, the material is imagined to be cut along the glide plane from the free surface to the dislocation line. Then, displacements are imposed on the surfaces created by this cut to produce a relative offset b_i of the $(+)$ side of the glide plane cut with respect to the $(-)$ side. For a glide dislocation, the inner product $b_i n_i^{(+)}$ vanishes. The stress field due to the mismatch strain is, by itself, an equilibrium field and it is unaffected by the formation of the dislocation. Thus, with reference to Figure 6.3, the work done by this field in forming the dislocation is

$$
W_{\mathrm{m}}(\eta) = \int_{(h_{\mathrm{f}}-\eta)\csc\alpha}^{h_{\mathrm{f}}\csc\alpha} b_i \sigma_{ij} n_j^{(+)} \, dl. \tag{6.12}
$$

If the inner products implied by the integrand are formed, and if the fact that the mismatch stress is zero throughout the substrate is taken into account, then

$$
W_{\mathrm{m}}(\eta) = \begin{cases} -b_x \sigma_{\mathrm{m}} \eta & \text{for} \quad 0 < \eta \le h_{\mathrm{f}}, \\ -b_x \sigma_{\mathrm{m}} h_{\mathrm{f}} & \text{for} \quad h_{\mathrm{f}} < \eta < \infty. \end{cases} \tag{6.13}
$$

The identity $n_x^{(+)} \csc\alpha = -1$ for $0 < \alpha < \pi$ has been used in obtaining the result. Note that $W_{\mathrm{m}}(0) = 0$, that $W_{\mathrm{m}}(\eta)$ varies linearly with η if the dislocation is formed in the film and that it is independent of η if the dislocation is formed in the substrate. The corresponding configurational force acting on the dislocation in the direction of increasing η is then

$$
F_{\mathrm{m}}(\eta) = \begin{cases} b_x \sigma_{\mathrm{m}} & \text{for} \quad 0 < \eta < h_{\mathrm{f}}, \\ 0 & \text{for} \quad h_{\mathrm{f}} < \eta < \infty. \end{cases} \tag{6.14}
$$

This force tends to move a strain-relieving dislocation in the film away from the free surface and toward the film–substrate interface. The force is discontinuous at $\eta = h_{\mathrm{f}}$ and is everywhere zero in the substrate.

With estimates of both the work W_{d} required to form a dislocation in this system and the work W_{m} that can be recovered from the background mismatch field in forming the dislocation, the condition for the formation of the dislocation *spontaneously* in the system can be determined. Evidently, if h_{f} is small enough then the elastic energy extracted from the material in forming a dislocation is inadequate to compensate for the energy of formation of the dislocation, and dislocation formation is precluded. On the other hand, if h_{f} is large enough then any amount of elastic energy can be released in forming a dislocation. The thickness that distinguishes one type of behavior from the other is called the *critical thickness* for the system. It is defined as the smallest value of h_{f}, say h_{cr}, for which $W_{\mathrm{d}}(\eta) + W_{\mathrm{m}}(\eta) = 0$ for any possible value of $\eta > 0$.

Representative graphs of the normalized work measures $W_{\mathrm{d}}(\eta)/\mu_{\mathrm{f}}b^2$, $W_{\mathrm{m}}(\eta)/\mu_{\mathrm{f}}b^2$ and $[W_{\mathrm{d}}(\eta)+W_{\mathrm{m}}(\eta)]/\mu_{\mathrm{f}}b^2$ versus η/b are shown in Figure 6.5 for the particular case when $h_{\mathrm{f}} = h_{\mathrm{cr}}$. From this illustration, it is clear that the optimum choice of η is $\eta = h_{\mathrm{f}}$, so the critical thickness condition becomes

$$
W_{\mathrm{d}}(h_{\mathrm{cr}}) + W_{\mathrm{m}}(h_{\mathrm{cr}}) = 0. \tag{6.15}
$$

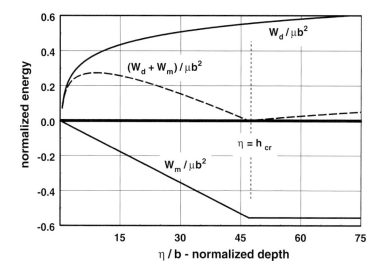

Fig. 6.5. Representative graphs of $W_d(\eta)/\mu_f b^2$ from (6.8) and $W_m(\eta)/\mu_f b^2$ from (6.13) versus η/b for a case where the strained film thickness h_f is exactly the critical thickness h_{cr}. The dashed curve represents the sum of these two energies, and the dislocation position $\eta = h_{cr}$ is evidently the only possible equilibrium position in the closed system. For purposes of illustration, the plot is drawn for $\nu_f = 1/4$, $r_0 = b/2$, $b_z = 0$, $b_x = b_y$ and $\epsilon_m = 0.005$.

From (6.7) and (6.13), this condition is

$$\frac{[b_x^2 + b_y^2 + (1 - \nu_f)b_z^2]}{8\pi(1 + \nu_f)b_x h_{cr}} \ln \frac{2h_{cr}}{r_0} + \frac{[b_x^2 + b_y^2]}{16\pi(1 + \nu_f)b_x h_{cr}} \left[\cos 2\alpha - \frac{(1 - 2\nu_f)}{2(1 - \nu_f)} \right] = \epsilon_m .$$

(6.16)

On physical grounds, only strain-relieving dislocations are possible. This notion is reflected in (6.16) by the requirement that ϵ_m and b_x must have the same algebraic sign for positive roots of h_{cr} to exist.

In cases for which $h_{cr} \gg b$, which is the situation for many film–substrate systems of practical interest, the left side of (6.16) is dominated by the logarithmic term. As a consequence, the critical thickness condition is often written in the simpler form

$$\boxed{\frac{[b_x^2 + b_y^2 + (1 - \nu_f)b_z^2]}{8\pi(1 + \nu_f)b_x h_{cr}} \ln \frac{2h_{cr}}{r_0} \approx \epsilon_m, \qquad h_{cr} \gg b.}$$

(6.17)

Broad implications of this relationship were discussed by Matthews and Blakeslee (1974) and it has become known commonly as the Matthews–Blakeslee condition.

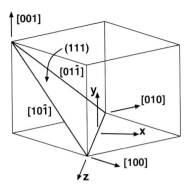

Fig. 6.6. Orientation of crystallographic axes for a cubic crystal relative to the underlying *xyz* coordinate axes for dislocation formation by slip on the (111) plane. The interface between the film and the substrate is the (001) plane.

6.2.2 Dependence of critical thickness on mismatch strain

To examine the relationship between mismatch strain and critical thickness implied by the condition (6.16), consider a thin film of a cubic material for which the free surface is the (001) plane and for which the slip plane is the (111) plane illustrated in Figure 6.6. If a dislocation were to form spontaneously on the slip plane in this strained epitaxial film, its line would be oriented along the $[1\bar{1}0]$ direction, and the slip direction would be parallel to either the $[10\bar{1}]$ direction or the $[01\bar{1}]$ direction; see Figure 6.6. The x, y and z coordinate directions are then $[110]$, $[001]$ and $[1\bar{1}0]$, respectively, in crystallographic directions.

The Burgers vector in either of the two slip directions depends on the sign of ϵ_m. If $\epsilon_m > 0$, the Burgers vector of length b is

$$b_i = b\left\{\tfrac{1}{2}, -\tfrac{1}{\sqrt{2}}, \pm\tfrac{1}{2}\right\}; \tag{6.18}$$

results for $b_z > 0$ and $b_z < 0$ are identical. The angle between the dislocation line and the x-axis, as defined in Figure 6.3, is $\alpha = \arccos(-1/\sqrt{3}) \approx 125.3°$. If $\epsilon_m < 0$, the direction of the Burgers vector must be reversed, but the calculation is otherwise identical.

For this case, evaluation of the critical thickness condition (6.16) yields the results shown in Figure 6.7 in the form of plots of $\ln(h_{cr}/b)$ versus $|\epsilon_m|$ for $r_o = \tfrac{1}{2}b$, b and $2b$. The choice of cutoff radius has a systematic influence on the predicted values of critical thickness, although the existence of a critical thickness and the qualitative aspects of the prediction do not depend on any particular value of r_o. The basis for choosing a value of r_o must be provided from another source, either more fundamental calculations in which this arbitrariness is precluded or experimental data. The limited input from either source generally supports a choice from among the smaller values of r_o, say $\tfrac{1}{2}b$.

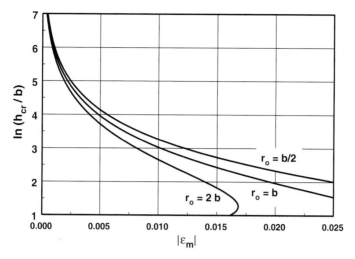

Fig. 6.7. Normalized critical thickness $\ln(h_{cr}/b)$ versus mismatch strain magnitude $|\epsilon_m|$ for three values of core cutoff radius $r_o = \tfrac{1}{2}b,\ b,\ 2b$ according to (6.16).

The discussion here is restricted, for the most part, to cases of isotropic elastic response of both the film and substrate materials. However, the influences of two of the most important consequences of anisotropy can be taken into account in a relatively straightforward way. One principal influence of anisotropy concerns the level of resolved shear stress due to the mismatch stress on the potential glide planes for dislocation motion. This influence enters geometrically through the relative components of the Burgers vector, and it is illustrated in Section 6.2.5 for the case of cubic materials. The second principal influence concerns the relative magnitudes of elastic moduli. The mismatch strain is fixed by lattice dimensions and it does not depend on elastic constants. The mismatch stress, on the other hand, does depend on elastic constants as indicated in (6.11) for isotropic materials. For some anisotropic film orientations, it is possible to estimate the mismatch stress for equibiaxial mismatch strain by adopting an effective biaxial modulus, such as the value $M_{(001)}$ given in (3.45) for a cubic material system with a {001} interface or the value $M_{(111)}$ given in (3.51) for a cubic material system with a {111} interface. The elastic constants also enter the expression for the energy of the self-stress field of the dislocation, which is given in (6.8) for isotropic material response. An approach to energy of the self-stress field for anisotropic materials introduced by Hirth and Lothe (1982) leads to an expression of the form

$$W_d(\eta) \approx \frac{K_f[b_x^2 + b_y^2 + b_z^2]}{4\pi} \ln \frac{2\eta}{r_o}, \qquad (6.19)$$

in place of (6.8), where K_f is a scalar factor which has physical dimensions of elastic modulus and which is determined completely by the components of the stiffness matrix c_{ij} of the film as introduced in Section 3.1. The critical thickness condition then takes the form (6.17) with the factor $1/2(1 + \nu_f)$ replaced by K_f/M_f. The influence of elastic anisotropy is discussed for the case of cubic materials by Shintani and Fujita (1994) who make use of the image force theorem of Barnett and Lothe (1974) for anisotropic materials.

6.2.3 Example: Critical thickness of a SiGe film on Si(001)

A single crystal alloy thin film with composition $Si_{0.85}Ge_{0.15}$ is grown on an initially flat Si(001) substrate which is 0.5 mm thick. The lattice parameter of Si at room temperature is $a_{Si} = 0.5431$ nm, while that of Ge is $a_{Ge} = 0.5656$ nm. Any dislocations formed as a consequence of epitaxial mismatch between the film and the substrate are known to be $60°$ dislocations with Burgers vectors in the family represented by (6.18). Assume that the biaxial moduli of Si(001) and Ge(001) crystals are $M_{Si(001)} = 180.5$ GPa and $M_{Ge(001)} = 142$ GPa, respectively, that the Poisson ratio of the film is $\nu_f \approx 0.25$, that $r_o = b/2$, and that $b \approx 0.4$ nm. Curvature measurements are made continuously using the multibeam optical stress sensor method (see Section 2.3.2) so as to monitor the evolution of internal stress during film deposition. Estimate the substrate curvature at which misfit dislocations are first able to form at the interface between the film and the substrate.

Solution:

Using (1.14), the mean lattice parameter of the $Si_{0.85}Ge_{0.15}$ alloy at room temperature is approximated as

$$a_{SiGe} \approx 0.85a_{Si} + 0.15a_{Ge} = 0.5465 \text{ nm}. \tag{6.20}$$

The equi-biaxial mismatch strain in the film at room temperature is calculated using (1.15) to be

$$\epsilon_m = \frac{a_{Si} - a_{SiGe}}{a_{SiGe}} = -0.0062. \tag{6.21}$$

The normalized critical thickness h_{cr}/b can be computed from the condition (6.16) assuming $r_o = b/2$ for the given conditions and for the dislocation Burgers vector whose components are listed in (6.18). Alternatively, it is readily noted from Figure 6.8 that $\ln(h_{cr}/b) \approx 3.9$ for $|\epsilon_m| = 0.0062$. With $b \approx 0.4$ nm, the critical thickness is estimated to be $h_{cr} \approx 19.8$ nm.

Using (2.12) and (2.13) with $M_{SiGe} \approx 0.85M_{Si} + 0.15M_{Ge} = 174.7$ GPa, the curvature at which misfit dislocations are first able to form at the interface is found to be

$$\kappa = \frac{6\epsilon_m M_{SiGe}h_{cr}}{M_{Si}h_{Si}^2} = -2.85 \times 10^{-3} \text{ m}^{-1}. \tag{6.22}$$

The negative curvature signifies that the substrate flexes with a convex shape on the film side, consistent with the existence of a compressive mismatch strain in the film. The formation of the first few misfit dislocations is not detectable by curvature measurements, as noted in Section 6.2.4.

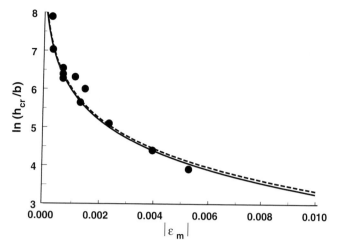

Fig. 6.8. Experimentally observed critical thickness versus mismatch strain for the SiGe/Si(100) system is shown as discrete points. The solid curve represents the predicted critical thickness condition (6.16) for this material system and the dashed curve is the pre-dicted condition simplified by retaining only the logarithmic term as in (6.17), both for $\nu_f = 0.25$ and $r_o = \frac{1}{2}b$. Adapted from Houghton (1991).

6.2.4 Experimental results for critical thickness

Critical thickness experiments are very challenging because, ideally, it is neces-sary to detect the *first* misfit dislocation that appears within a relatively large area of interface. Observations based on strain measurement, either monitoring sub-strate curvature by optical methods or direct measurement of lattice strain by x-ray diffraction, are not able to detect the onset of strain relaxation, mainly because these techniques lack the necessary resolution. In early studies of strained layer epitaxy, this inability to detect the earliest stages of strained relaxation led to a distinction between the 'theoretical critical thickness', as defined in Section 6.2.1, and an 'ex-perimental critical thickness' representing the extent of relaxation detectable by a particular technique for measuring average strain change in a sample.

   An effective technique for examining the critical thickness concept was developed by Tuppen and Gibbings (1989). The technique was based on detecting the *threading* segment at the end of an interface misfit dislocation, where the dislocation line turns from the interface toward the free surface of the film and threads through the film thickness to the surface; threading dislocations will be considered in the context of critical thickness in Section 6.3. The technique, which involves chemical etching, made it possible to detect threading dislocations at a density as low as one per square centimeter of interface. This technique was applied by Houghton (1991) to determine the thickness of a Si_xGe_{1-x} film on a Si(100) substrate for $0 < x < 0.15$. The normalized data are presented in Figure 6.8. The solid curve included in the

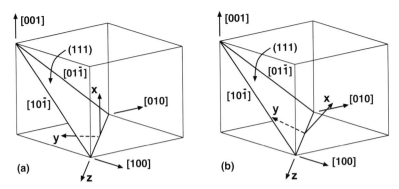

Fig. 6.9. Relative orientation of crystallographic axes for a cubic crystal and the underlying xyz coordinate axes for dislocation formation by slip on a {111} plane. The interface between the film and the substrate is (a) the $(\bar{1}\bar{1}0)$ plane and (b) the $(\bar{1}\bar{1}1)$ plane.

figure is the graph of the critical thickness condition for this material system as given in (6.16), and the dashed curve represents the condition modified by retaining only the logarithmic term on the left side of (6.16), as in (6.17). In both cases, the cutoff radius was chosen to be $r_o = \frac{1}{2}b$. It was verified in the experiments that the dislocations that were observed were virtually all $60°$ dislocations with Burgers vector in the family represented by (6.18). The data provide strong support for the validity of the critical thickness condition.

6.2.5 *Example: Influence of crystallographic orientation on* h_{cr}

The nature of the critical thickness condition (6.16) was illustrated in Figure 6.7 for the particular case in which the film–substrate interface was the (001) crystallographic surface of the epitaxial cubic film. Other orientations of the material lead to different predictions for critical thickness versus mismatch strain. Two alternate orientations for slip on a {111} plane of the cubic crystal are identified in Figure 6.9. In (a), the interface is the $(\bar{1}\bar{1}0)$ plane in crystallographic coordinates, whereas in (b) the interface is the $(\bar{1}\bar{1}1)$ plane. Show how the variation of the normalized critical thickness h_{cr}/b as a function of the film mismatch strain is affected as the orientation of the film is changed from that given in Figure 6.6 to that indicated in either Figure 6.9(a) or (b).

Solution:

For the $(\bar{1}\bar{1}0)$ oriented film shown in Figure 6.9(a), the Burgers vector is

$$b_i = b\{\tfrac{1}{\sqrt{2}}, \tfrac{1}{2}, \pm\tfrac{1}{2}\} \tag{6.23}$$

and $\alpha = \arccos(\sqrt{\tfrac{2}{3}}) \approx 35.3°$. For the $(\bar{1}\bar{1}1)$ interface in Figure 6.9(b), on the other hand, the Burgers vector is

$$b_i = b\{\tfrac{1}{\sqrt{6}}, \tfrac{1}{\sqrt{6}}, \tfrac{2}{\sqrt{6}}\} \tag{6.24}$$

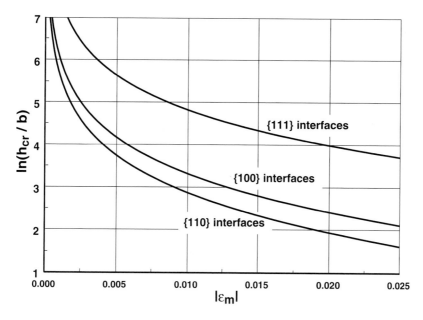

Fig. 6.10. A comparison of the critical thickness (6.17) for three different crystallographic orientations of the epitaxial interface, with $v_f = 0.25$ and $r_o = b/2$. Although the diagrams in Figure 6.9 identify particular crystallographic glide planes for dislocation formation, the results obtained apply for families of glide planes and, therefore, for the families of interface orientations indicated.

and $\alpha = \arccos(\frac{1}{3}) \approx 70.5°$. The corresponding critical thickness values, computed from (6.17), are shown in Figure 6.10 along with the result for the $\{100\}$ interface, all for $r_o = \frac{1}{2}b$. The crystallographic orientation has a strong influence on critical thickness, as seen in Figure 6.10. This geometrical effect is a consequence of variations in the shear traction, due to the mismatch stress field (6.11), acting on the glide plane in the direction of the Burgers vector.

It has been observed experimentally that the formation of Shockley partial dislocations separated by stacking faults is much more common in diamond cubic epitaxial films in some crystallographic orientations than it is in other orientations (Kvam and Hull 1993). The discriminating aspect between the two types of behavior seems to be related to the fact that dislocation glide in a $\langle 110 \rangle$ direction on a $\{111\}$ plane in such a material actually occurs as an ordered sequence of two glide increments which together sum to a complete Burgers vector (Hirth and Lothe 1982). For example, the Burgers displacement $\frac{1}{2}[\bar{1}01]$ would be executed as an increment $\frac{1}{6}[\bar{2}11]$ followed by an increment $\frac{1}{6}[\bar{1}\bar{1}2]$; the order of the increments cannot be reversed. In some crystallographic orientations, the resolved shear stress due to the background mismatch elastic field will be larger in the direction of the first increment whereas, in other cases, it will be larger in the direction of the second increment. If this resolved shear stress is larger in the direction of the first increment, then formation of a pair of partial dislocations is favored; the energy of the stacking fault between the partial dislocations contributes to the driving force on both the leading and trailing partial dislocations. On the other hand, if the resolved shear stress in the direction of the second increment is larger, then there is a natural tendency for the two partial dislocations to move together as a single or complete dislocation.

6.3 The isolated threading dislocation

In Section 6.2, a critical thickness condition was established for a uniformly strained single crystal film on a substrate. This condition discriminates between circumstances under which a system is susceptible to elastic strain relaxation by formation of interface misfit dislocations and those circumstances under which such relaxation is precluded on energetic grounds. The condition is obtained by comparing two states of the system, one with a uniformly strained and fully coherent film, and the other with a partially relaxed film and a completely formed interface misfit dislocation. More precisely, the condition is obtained by calculating the work which must be done on the system to convert it from the first state to the second. If the result is that positive work must be done then the dislocation cannot form spontaneously whereas, if the work required is negative, there is an energetic driving force for the spontaneous formation of a dislocation.

When considered in this framework, the resulting critical thickness condition is independent of any physical mechanism by which the interface misfit dislocation is actually formed in the system. This approach to critical thickness is an outgrowth of the pioneering work of Frank and van der Merwe (1949) and the criterion was formulated in the way described here by van der Merwe and van der Berg (1972). The most commonly observed mechanism for formation of misfit dislocations is through glide of a threading dislocation. In this section, the notion of critical thickness is re-examined on the basis of this mechanism.

6.3.1 Condition for advance of a threading dislocation

The physical system to be considered is depicted in Figure 6.11. A uniformly strained layer of thickness h_f is epitaxially bonded to a commensurate substrate of relatively large thickness. The equi-biaxial mismatch strain is ϵ_m. A representative glide plane, which is shared by the film and the substrate, is shown shaded in the figure. A dislocation is assumed to exist on this glide plane in the presence of the background mismatch strain field. The dislocation line extends from the free surface of the film to the interface; this portion spans the film thickness and it is called the *threading* segment. The dislocation line then continues along the interface for a distance large compared to h_f; this is the *interface misfit* segment. As the threading segment of the dislocation, or simply the threading dislocation, advances or recedes along the glide plane, it leaves behind an ever longer or shorter

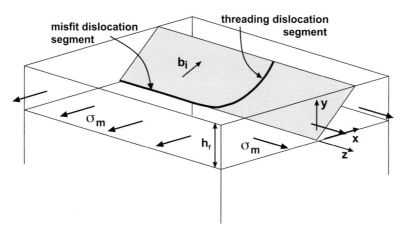

Fig. 6.11. Threading dislocation on a glide plane in a film subject to a uniform mismatch stress σ_m. As the threading dislocation glides in the z-direction, it leaves behind an increasing length of interface misfit dislocation.

length, respectively, of misfit dislocation. Thus, the critical thickness condition for formation of strain-relieving misfit dislocations can also be phrased in terms of the tendency for the threading dislocation to advance or recede for given mismatch strain ϵ_m, Burgers vector b_i, film thickness h_f and elastic properties μ_f and ν_f. The basic idea behind this approach was introduced by Matthews et al. (1970) and it was exploited more fully by Matthews and Blakeslee (1974).

It is worthwhile to consider briefly the implication of the invariance of the system, in the absence of the dislocation, under translation in the z-direction. This invariance immediately implies that the environment for the threading segment at some location in the z-direction along the glide plane is identical to the environment at any other location. In light of this observation, it is reasonable to assume that the threading dislocation glides in a self-similar way, or with fixed shape, as it moves along the glide plane. If this is so, then there are no energy changes associated with the *shape* of the threading dislocation during its advance or retreat. This is an important observation because this shape is difficult to specify with any certainty, and it is generally unknown at the level of modeling developed here.

For the arrangement shown in Figure 6.11, there is a configurational force acting on the threading dislocation, in general, and the main contributions to this driving force have already been identified in the discussion of Section 6.2. These contributions are readily identified in the present context. There is an equilibrium stress field in the body due to the presence of the dislocation, aside from the mismatch stress field; this is the self-stress field. There is always a tendency for the threading segment to retreat under the action of this stress field, thereby reducing the energy of the system. Likewise, there is an equilibrium stress field in the body due to the mismatch strain, aside from the dislocation stress field. If the Burgers vector orientation is such that this is a strain-relieving dislocation, then there is always a tendency for the threading segment to advance along its glide plane under the action of this stress field, thereby reducing the elastic energy associated with the mismatch strain.

These two tendencies compete with each other. When the film is very thin, there is a strong configurational force pulling the dislocation toward the free surface while the force available to advance the threading dislocation is relatively small. As a consequence, the net tendency is for retreat of the threading segment. On the other hand, when the film thickness is relatively large, the free surface force is diminished but the force due to the mismatch stress field in the film is increased. Consequently, if the film thickness is sufficiently large, the net tendency is for advance of the threading dislocation. The discriminating film thickness again identifies a *critical thickness*, but this time on the basis of a *mechanism* for formation of a strain-relieving misfit dislocation. What is the relationship of this critical thickness to that identified through (6.16) in Section 6.2? On the basis of a relatively straightforward argument that invokes the cutting and pasting of material to alter the length of the misfit dislocation, it can be established that the two approaches to determining the critical thickness lead to *identical* results.

Suppose that the threading dislocation in Figure 6.11 advances a unit distance in the z-direction without a change in shape. This results in a change in the total energy of the system, in general. Let this energy change be denoted by $-G(h_f)$ at film thickness h_f. Defined in this way, $G(h_f)$ is the *configurational force* on the threading dislocation acting in the z-direction. If $G(h_f)$ is positive (negative), then the energy of the system decreases (increases) as the threading dislocation advances. In either case, it leaves behind an additional length of misfit dislocation. As noted above, the critical thickness condition expressed in terms of the behavior of a threading dislocation is

$$G(h_{cr}) = 0. \tag{6.25}$$

Next, suppose that the same advance of the threading dislocation is accomplished in a different way. At some distance ahead of the threading dislocation that is very large compared to h_f, a slab of material of unit thickness is imagined to be removed from the system by making cuts on two parallel planes which are perpendicular to the z-axis and which are separated by a unit distance. Tractions are maintained on the cut surfaces so that the two-dimensional state of deformation is maintained. The gap created by removal of the slab is then closed and the faces are rejoined without affecting the elastic fields. A dislocation is introduced into the slab under two-dimensional conditions; this dislocation has precisely the properties of the interface misfit dislocation. A cut is then made in the body along a plane perpendicular to the z-axis at some distance behind the threading dislocation that is very large compared to h_f. The cut is opened a unit distance, the dislocated slab is inserted, and the joined faces are imagined to be rebonded.

The result of this imaginary procedure is *indistinguishable* from the result of simply advancing the dislocation a unit distance in the z-direction. The change in energy in the latter case has already been denoted by $-G(h_f)$. But the change in energy in the former case is already known. It is precisely the sum of the work terms established in Section 6.2 representing the work per unit length $W_d(h_f)$ required to form a dislocation under two-dimensional conditions and the work per unit length $W_m(h_f)$ done by the mismatch stress field in forming the dislocation. The expressions for these two contributions appear in (6.7) and (6.13), respectively, and it follows that

$$-G(h_f) = W_d(h_f) + W_m(h_f). \tag{6.26}$$

An immediate consequence of this fundamental result is that the critical thickness condition based on an energy comparison obtained in Section 6.2 is *identical* to that based on consideration of configurational forces acting on the threading dislocation, so that (6.16) results in either case (Freund 1987).

A second consequence of (6.26) is that it provides an unambiguous expression for the force acting on a threading dislocation at any thickness h_f at or beyond the critical thickness h_{cr} (Freund and Hull 1992). This is the element needed in a specification of the nonequilibrium behavior of the threading dislocation in the form of a kinetic relation between the rate of advance of the threading dislocation along its glide plane and this driving force, a relationship considered in Section 6.3.3. The expression is

$$G(h_f) = bh_f M_f \left[\epsilon_m \frac{b_x}{b} - \frac{[b_x^2 + b_y^2 + (1 - \nu_f)b_z^2]}{8\pi(1 + \nu_f)bh_f} \ln \frac{4h_f}{b} \right] \tag{6.27}$$

where M_f is the biaxial elastic modulus of the material; only the logarithmic term in (6.7) has been retained and $r_o = \frac{1}{2}b$. The ratio $G(h)n_x^{(+)}/bh_f$ is sometimes called the *excess stress* acting on the threading dislocation (Dodson and Tsao 1987, Tsao 1993). The excess stress defined in this way is representative of the resolved shear stress due to the mismatch stress field for the slip system of the dislocation, but it is not actually a measure of stress on any surface. A surface plot of the normalized quantity $G(h_f)/bh_f M_f$ over the plane of ϵ_m and h_f/b is shown in Figure 4.12 for the material system depicted in Figure 6.6. The height of the surface above the plane $G = 0$ represents the driving force acting on a threading dislocation at the corresponding coordinates $(\epsilon_m, h_f/b)$ in that plane. Negative driving forces $G < 0$ are not relevant to the consideration of critical thickness and, consequently, the surface is not extended to negative values of G in the figure.

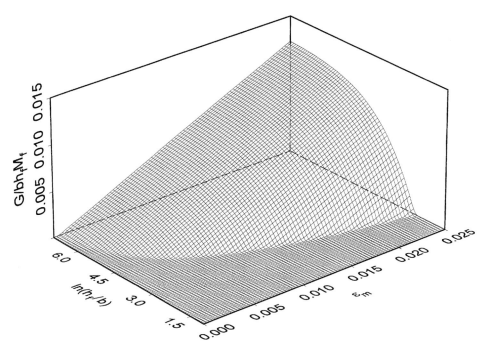

Fig. 6.12. Normalized driving force on threading dislocation versus mismatch strain ϵ_m and normalized film thickness h_f/b for the cubic material system represented in Figure 6.6. The surface has been truncated at $G = 0$ so that negative values are not shown. The kink in the surface, at the boundary between the flat and rising portions, is identical to the critical thickness condition shown in Figure 6.7 for $r_o = \frac{1}{2}b$. Adapted from Freund (1990a).

Fig. 6.13. Cross-sectional TEM image of a thin Al film epitaxially deposited on a single crystal α-Al$_2$O$_3$ substrate. A threading dislocation, indicated by the horizontal arrow, leaves behind a misfit dislocation segment at some distance from the interface. This misfit dislocation segment was found to interact with dislocations formed previously at the film–substrate interface. Reproduced with permission from Dehm et al. (2002a).

Figure 6.13 is a cross-sectional TEM micrograph which provides an example of a threading dislocation in an epitaxial aluminum film on a relatively thick Al$_2$O$_3$ substrate. In this case, a 350 nm thick epitaxial Al film was magnetron sputtered on a clean (0001) surface of α-Al$_2$O$_3$. Dislocation nucleation at the film–substrate interface and motion through the film were observed by cross-sectional TEM specimens (Dehm et al. 2002a). Two beam diffraction conditions were used with the $\mathbf{g}_{111}$ diffraction vector oriented normally to the film–substrate interface and close to the $\langle 11\bar{2} \rangle$ or $\langle 01\bar{1} \rangle$ Al zone axes. The micrograph shows a threading dislocation, indicated by the horizontal arrow, which deposited a 60 degree dislocation segment nearly parallel to the interface. *In situ* TEM observations revealed that localized heating of the specimen with a focused electron beam resulted in the motion of the dislocation through the Al film on the $(\bar{1}11)$ Al plane. The *in situ* TEM experiments also indicated that the threading dislocation advancing through the film left behind a dislocation segment parallel to the Al–Al$_2$O$_3$ interface, as indicated in Figure 6.11. The standoff of the misfit dislocation from the interface is likely due, in part, to the fact that the elastic modulus of the substrate material is approximately five times that of the film material: see Table 2.3. The issue of modulus difference is addressed in Section 6.5.2. Details of similar observations are discussed in the context of film plasticity in Chapter 7.

6.3.2 Limitations of the critical thickness condition

The critical thickness theory is a fundamentally sound, self-consistent mechanistic theory that has played a central role in understanding strain relaxation in heterostructures (Matthews and Blakeslee 1974, Matthews and Blakeslee 1975). The situation in which the critical thickness approach is most directly applicable is when dislocations, which already exist in the substrate and which terminate at the substrate surface, are simply continued into the epitaxial film during growth. Such a situation is illustrated by the TEM image in Figure 6.14. A pre-existing dislocation in a Si substrate was continued into a SiGe film as it was epitaxially deposited onto the substrate surface. Once the film thickness reached the level of the critical thickness, the portion of the dislocation in the film began to glide in the direction corresponding to strain relief according to its Burgers vector. The interface misfit dislocation segment left behind by the propagating threading segment as it propagated is evident in the figure. However, there are certain assumptions that underlie the theory that must be recognized in applying the criterion to assess observations. Several of these are briefly noted here.

The criterion refers to the conditions that prevail when it first becomes possible for *any* threading dislocation to advance. The ability to detect the onset of relaxation depends on the resolution of the technique used to observe dislocation motion (Fritz 1987, Gourley et al. 1988). X-ray diffraction methods detect average strain over an area on the order of a square millimeter, and a detectable change in average strain occurs only after extensive dislocation activity at thicknesses well in excess of the true critical thickness. Consequently, such methods are ineffective in efforts to actually observe the onset of dislocation glide in strained layers.

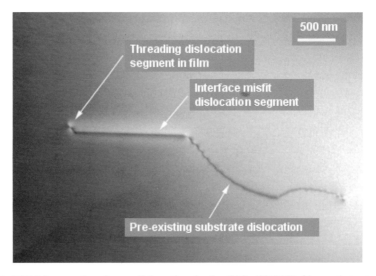

Fig. 6.14. TEM image showing a dislocation in the SiGe/Si(100) film–substrate material system under circumstances in which the dislocation in the film is a continuation of a pre-existing dislocation in the substrate. The film has been grown to thickness beyond the critical thickness. Consequently, the segment of the dislocation in the film began to propagate in the direction corresponding to strain relief, leaving behind the interface misfit dislocation segment and the continuation of the dislocation into the substrate. Image provided by E.A. Stach, Lawrence Berkeley National Laboratory.

The statement of the criterion is based on the *a priori* assumption that a threading dislocation exists, and it addresses only the *tendency* for that threading segment to advance or recede. It has nothing to say about the source of such threading segments. From time to time, it has been observed that films can be grown to thicknesses in excess of the critical thickness under certain circumstances, and it has been argued that such observations imply that the critical thickness theory is incorrect. This is not the case, however, because the premises of the critical thickness theory are not met in such cases. In particular, if no threading dislocation is present then the question of whether or not such a dislocation will glide cannot be considered. The important question of dislocation nucleation is discussed briefly in Section 6.8.

The tendency for an existing threading segment to advance or recede does not imply that it will actually do so in any particular way. A kinetic model of the glide process is required to consider this aspect, including effects that cannot be taken into account in a continuum model. Configurations for which there is a tendency for strain relaxation to occur, but for which kinetic barriers prevent it from occurring, usually due to low temperature, are often called *metastable* configurations. Commonly, the term metastable as applied to a material structure describes a state of equilibrium that is stable under 'small' amplitude fluctuations in its configuration but not stable under 'large' amplitude fluctuations. The kinetic barriers associated with a metastable state are activation energies characterizing the fluctuation amplitudes needed for a system to depart from a metastable equilibrium state into the realm of nonequilibrium configurations, through which it searches out another equilibrium state at lower energy, assuming that the transition proceeds spontaneously.

The model also presumes that the threading dislocation exists in a layer that is otherwise spatially uniform. However, the behavior of a dislocation can be strongly influenced by other dislocations present, either on parallel or intersecting glide planes. Dislocation interactions will be considered in Chapter 7. Behavior can also be influenced by geometrical features in the film. For example, patterning can result in an array of trenches with free surfaces within the film or small mesas on which the film material is deposited. The motion of a dislocation can also be influenced by fluctuations in surface topography of the film.

At the level of the above discussion, the critical thickness condition $G = 0$ is based on elastic continuum concepts only. This overlooks a variety of effects that may bear on the process in some cases. Among these are the glide resistance due to the Peirls stress of the material (Matthews 1975), the role of surface or interface energy (Cammarata and Sieradzki 1989), the resistance due to creation or annihilation of a surface ledge in the wake of the threading dislocation (Matthews 1975), the presence of a deposition flux during dislocation motion, and the possibility of dissociation of the dislocations into Shockley partial dislocations (Alexander 1986). Thus, the critical thickness condition could be restated as $G = \mathcal{R}$ where $\mathcal{R}$ is a resisting force representing any or all of these additional effects.

6.3.3 Threading dislocation under nonequilibrium conditions

Early observations of elastic strain relaxation during growth of epitaxial layers led to paradoxical results. An attempt to interpret the observations on the basis of the critical thickness theory in its most elementary form suggested that, once the thickness of a film exceeded the critical thickness, the final elastic strain of the film should be determined by the thickness of the film alone, independent of the original, or fully coherent, mismatch strain. This is implied by the result in (6.27), which states that the mean elastic strain predicted by the equilibrium condition $G(h_f) = 0$ is completely determined by h_f beyond critical thickness, no matter what the value of ϵ_m. However, it was found that the post-growth elastic strain as measured by x-ray diffraction methods did indeed vary with the initial elastic mismatch strain, and it did so in different ways for different film thicknesses (Bean et al. 1984). As a consequence, the critical thickness theory came under question, and various alternate models were proposed to replace it. However, further study of the problem has revealed the relaxation process to be much richer in physical phenomena than anticipated, with the critical thickness theory revealing only part of the story.

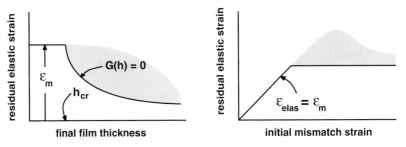

Fig. 6.15. Schematic representation of the dependence of final residual elastic strain, or unrelaxed elastic strain, on final film thickness at some fixed level of mismatch (left) and on initial mismatch strain at some fixed level of final film thickness (right) according to equilibrium theory. The shaded regions suggest where observations would fall if kinetic constraints on relaxation were taken into account.

In the interpretation of the observations, it had been tacitly assumed that dislocations would appear fully formed whenever it was energetically possible for them to exist, but this is not the case. Instead, the rate of dislocation nucleation in materials with low initial dislocation densities and the rate of dislocation motion in semiconductor materials emerged as important features. Roughly, the time for a threading dislocation to glide a distance equal to the lateral extent of a film at a typical growth temperature under the action of the mismatch strain is often of the same order of magnitude as the growth time itself. Consequently, the variability observed in the relaxation data could be associated with the kinetics of nucleation and glide of dislocations. This realization led to the use of post-growth annealing experiments to study relaxation (Fiory et al. 1984). In this approach, a strained layer material is grown beyond critical thickness with minimal relaxation, due to low growth temperatures or other controls, and then cooled. In the resulting structures, any threading dislocation which exists is still subjected to a positive driving force but it is kinetically restrained from relaxing the elastic strain in the film. The resulting structures have been labeled as metastable, as noted above, an appropriate term if thermal fluctuation is understood to be the external stimulus by which stability is probed. In any case, under well controlled conditions, the previously deposited film is heated and observed, sometimes simultaneously and sometimes alternately, to study the mechanisms of strain relaxation (Hull and Bean 1989, Noble et al. 1989, Houghton 1991). The results of these experiments, which yield behavior suggested by the shaded regions in Figure 6.15, have been of central importance in resolving some of the paradoxes. The annealing or temperature cycling experiments are also relevant to industrial fabrication technology where a device incorporating a strained layer heterostructure may be subjected to several temperature excursions in a manufacturing sequence.

Virtually all observations of dislocation glide in bulk covalent crystals at stress levels above some modest threshold level on the order of 10^3 N m^{-2} reveal that the normal glide velocity on a particular slip system of a given material varies with resolved shear stress on the glide plane $\tau_{\text{res}} = \sigma_{ij} n_j^{(+)} b_i / b$ and absolute temperature T according to an Arrhenius relationship of the form

$$v = v_0 \left(\frac{\tau_{\text{res}}}{\mu} \right)^m e^{-Q_0/kT} \qquad (6.28)$$

where v_0 is a material constant with dimensions of speed, Q_0 is an activation energy and k is the Boltzmann constant (Alexander 1986). The activation energy is most commonly assumed to be constant but, in some cases, its potential dependence on stress level is taken into account explicitly. For Si and Ge, the exponent m appears to be slightly temperature dependent, but it is usually assumed to be a constant in the range $1 \le m \le 2$. The velocity factor v_0 is roughly 10^5 m s^{-1} for undoped Si and 10^6 m s^{-1} for undoped Ge. It follows immediately that the speed of self-similar advance of an isolated threading dislocation is (Tsao and Dodson 1988, Tsao et al. 1987, Freund and Hull 1992)

$$v_{\text{td}} = v_0 \left(\frac{G(h_{\text{f}}) n_x^{(+)}}{b h_{\text{f}} M_{\text{f}}} \right)^m e^{-Q_0/kT} . \qquad (6.29)$$

The stress measure enclosed within parentheses is defined so that (6.29) reduces to (6.28) in the limit as the film becomes very thick. It is likely that glide is accomplished by the thermally activated motion of kinks along the threading segment (Tuppen and Gibbings 1990, Hull et al. 1991). A free surface or bimaterial interface may offer a better site for kink formation than interior points along the dislocation line, and this is an influence on relaxation that cannot be addressed directly by approaches based on continuum mechanics.

The values of activation energy Q_0 appearing in (6.29) have been determined experimentally. Tuppen and Gibbings (1990) examined dislocations originating from roughness at the film edges or from scratches artificially introduced on the surfaces of the films. This was done by annealing MBE grown films with very low dislocation densities, and then measuring dislocation propagation distances from their sources, that is, the distances that were traversed during the annealing step, by means of a defect selective chemical etching (Tuppen and Gibbings 1989). Relaxation in buried strained films was also examined. The same material system was subsequently studied by Nix et al. (1990). The results of *in situ* high voltage transmission electron microscopy measurements of threading dislocation velocities in Si$_{1-x}$Ge$_x$/Si(100) films as reported by Nix et al. (1990) are shown in Figure 6.16, interpreted according to the relationship (6.29). To obtain these data, an isolated threading dislocation was found in the film at low temperature, and the motion of this dislocation was then monitored as the sample was slowly heated to successively higher temperatures, up to a maximum of 675 °C. The prospect of following this

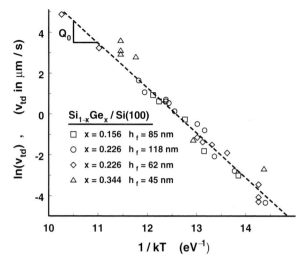

Fig. 6.16. Glide speed of an isolated threading dislocation in a SiGe/Si(100) film versus inverse temperature from *in situ* transmission electron microscope observations. The slope of the line fitted to the data provides an estimate of the activation energy according to (6.29). Adapted from Nix et al. (1990).

approach for still higher temperatures was precluded by the fact that dislocations moved too rapidly to be tracked, so the data in this range were obtained by chance observations over a fixed area. When plotted in the form of $\ln v_{td}$ versus $1/kT$, the slope of the best-fit line provides a direct estimate of the activation energy Q_0. The inferred value of Q_0 is about 2.2 eV. Similar values have been obtained by other means. The data in Figure 6.16 cover a speed range from about 5 nm s^{-1} to 150 μm s^{-1}.

The issue of glide of the threading dislocation segment according to (6.29) was pursued further in experiments reported by Stach et al. (1998). For films with atomically clean surfaces, they reported an activation energy depending on the Ge atomic fraction x in the Si$_{1-x}$Ge$_x$/Si(001) alloy films as $Q_0 = (2.2 - 0.6 x)$ eV. They also found that dislocation velocities were significantly larger during annealing of samples that had native oxide present on the film surfaces. This effect was attributed to stress-assisted dislocation kink nucleation at the interface between the oxide and the strained layer, with interface roughness resulting in local stress concentrations (Stach and Hull 2001).

6.4 Layered and graded films

The necessary condition established in Sections 6.2 and 6.3 for dislocation formation in a uniformly strained film can readily be extended to cases where the mismatch strain varies in an arbitrary way through the thickness of the film, provided that the material is still elastically homogeneous. With reference to Figure 6.11, the mismatch strain is a function of the y coordinate and, in this case, it is written as $\epsilon_m(y)$. The mismatch is independent of the x and z coordinates, thus preserving the important property of invariance under translation in the xz-plane. The fact that the stress components σ_{xx} and σ_{zz} in the background mismatch field vary with y does not disturb the equilibrium state of the background field, even if that variation is discontinuous in the y-direction.

Examples of material systems within this category include a uniformly strained layer covered or *capped* with an unstrained layer of uniform thickness, a film in which the mismatch with respect to the substrate varies continuously across the thickness due to compositional gradation, and a film that comprises a periodic arrangement of many thin layers of alternating mismatch strain, a so-called strained *superlattice* structure. Conditions for dislocation formation in such structures are examined in this section.

The restriction to cases of elastic homogeneity implies that the work of forming a misfit dislocation parallel to the free surface and at a distance η from it is given by (6.2) in all cases. This result is independent of the mismatch strain distribution. The work of the background mismatch stress field as the dislocation is formed is still defined by (6.12). Because the stress components now vary with position along the glide plane, however, the expression for $W_m(\eta)$ given in (6.13) must be replaced by

$$W_m(\eta) = \int_{(h_f-\eta)\csc\alpha}^{h_f \csc\alpha} b_i \sigma_{ij} n_j^{(+)}\, dl = -2b_x \mu_f \frac{1+\nu_f}{1-\nu_f} \int_{h_f-\eta}^{h_f} \epsilon_m(y)\, dy \qquad (6.30)$$

for isotropic material response.

As before, the critical thickness h_{cr} is defined as the smallest value of h_f for which any position η exists such that

$$W_d(\eta) + W_m(\eta) = 0. \qquad (6.31)$$

To be of physical significance, the position η must also be a stable equilibrium position. In the present context, this is understood to mean that any perturbation $\Delta\eta$ in position η from the value satisfying (6.31) must result in an increase in energy of the system, that is,

$$W_d(\eta + \Delta\eta) + W_m(\eta + \Delta\eta) > 0 \qquad (6.32)$$

for any small $\Delta\eta \neq 0$. Elastic fields do not vary continuously across interfaces in all cases and, consequently, the stability condition cannot always be stated in the more familiar but weaker form involving the derivatives of these work functions evaluated at η, that is, as $W_d''(\eta) + W_m''(\eta) > 0$.

6.4.1 Uniform strained layer capped by an unstrained layer

Suppose that a strained layer with uniform mismatch strain ϵ_m is deposited on a substrate to a thickness h_{sl} that exceeds the critical thickness h_{cr} for the layer itself. Then, an unstrained layer of uniform thickness h_{ul} is deposited on the surface of the strained layer. For example, suppose a SiGe alloy film is deposited epitaxially on a Si substrate to a thickness beyond its critical thickness, and then a Si capping layer is deposited on the surface of the alloy layer. The total thickness of the composite film is $h_f = h_{sl} + h_{ul}$. The case of a Si capping layer is the simplest case that can be considered because, in the absence of a dislocation within a strained layer, the epitaxial capping layer is free of mismatch stress; this particular system was studied experimentally by Nix et al. (1990). The case of a capping layer that is not matched to the substrate can be handled in essentially the same way, but with slightly greater complexity in the details.

The capping layer tends to stabilize the configuration against dislocation formation. The basis for this expectation is that the formation of the dislocation at the film–substrate interface requires distortion of the material over the full depth h_f. On the other hand, the background mismatch stress field does work as the dislocation is formed only over a fraction of the total thickness $h_{sl} < h_f$. To what extent can the capping layer stabilize the structure against dislocation formation?

The most advantageous position for formation of a dislocation in this structure is at the film–substrate interface, that is, for $\eta = h_f$. The critical condition $W_d(h_f) + W_m(h_f) = 0$ then takes the form

$$\frac{[b_x^2 + b_y^2 + (1 - \nu_f)b_z^2]}{8\pi(1 + \nu_f)b_x} \ln \frac{4(h_{sl} + h_{ul})}{b} = h_{sl}\epsilon_m, \qquad (6.33)$$

where the value of cutoff radius $r_o = \frac{1}{2}b$ has been incorporated and only the logarithmic term in W_d is retained as in (6.17). If the thickness of the strained layer and the capping layer are normalized by the critical thickness h_{cr} of the strained layer alone according to

$$p = h_{sl}/h_{cr}, \quad q = h_{ul}/h_{cr}, \qquad (6.34)$$

then (6.33) reduces to

$$q = \left(\frac{4h_{cr}}{b}\right)^{p-1} - p. \qquad (6.35)$$

This result defines the value of q, representing the thickness of the capping layer, which is required to stabilize the composite film against formation of a misfit dislocation. This value is a function of h_{cr}/b and p, the parameters that completely characterize the strained layer. The quantity q given by (6.35) is plotted versus p for several values of h_{cr}/b in Figure 6.17. For example, suppose that the critical

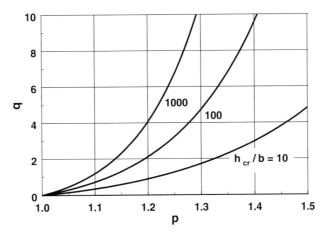

Fig. 6.17. A plot of normalized capping layer thickness q versus normalized strained layer thickness p needed to stabilize the composite film against interface misfit dislocation formation according to (6.35).

thickness of the strained layer for a particular material system is $100\,b$. If a film is grown to a thickness of $120\,b$, then deposition of an unstrained capping layer of thickness $200\,b$ will stabilize the system against formation of misfit dislocations at the film–substrate interface.

The idea of stabilization of an epitaxial structure by subsequent overgrowth has significant practical implications. The prospect of inserting material structures that are unstable or only weakly metastable into applications for which long life is required is undesirable. However, if material structures that are unstable as fabricated can be stabilized by overgrowth then the risk of material failure is diminished significantly. This issue plays a role in several quantum wire configurations, as discussed in Section 6.6, and it focuses additional emphasis on the processing of small epitaxial material structures.

While the expression in (6.35) represents the answer to the question posed, it is an incomplete answer in the following sense. Formation of a misfit dislocation at the film–substrate interface is indeed a possible strain-relief mechanism for the case of a composite film consisting of a strained layer and an unstrained capping layer. However, it may not be the preferred mechanism in all cases. Because the configuration involves two interfaces, there is a possibility that strain relief can occur by means of a pair of misfit dislocations, one at the film–substrate interface and one at the interface between the strained layer and the capping layer. Configurations giving rise to pairs of dislocations in this way have been studied experimentally by Tuppen and Gibbings (1990) and Nix et al. (1990). Such a mechanism has the advantage that it does not require glide through the unstrained capping layer in order

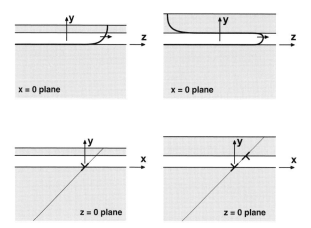

Fig. 6.18. Schematic diagram illustrating strain relaxation mechanisms by which the glide of a threading dislocation leads to a single misfit dislocation (left) and a misfit dislocation dipole (right), depending on the relative thicknesses of the unshaded strained layer and the unstrained capping layer.

to relax elastic strain. On the other hand, the dislocation pair will tend to annihilate each other, a tendency that must be overcome by the influence of the background mismatch field in keeping them separated. The circumstances that lead to formation of a double misfit dislocation configuration are examined next.

It is helpful to consider the two-misfit-dislocation mechanism in the framework of glide of a threading dislocation. If the capping layer is very thin, then the config-uration is similar to that depicted in Figure 6.11. In particular, the configuration of the dislocation line on the projection of the glide plane onto the plane $x = 0$ has the features illustrated in the upper left side of Figure 6.18. The glide of the threading segment within the strained layer, driven by the mismatch stress field there, drags along the short segment in the capping layer. As the unstrained capping layer be-comes thicker, the burden of deforming the capping layer material without recovery of any stored elastic energy becomes greater. Eventually, due to this burden, the threading segment in the capping layer falls behind the segment in the strained layer, as illustrated in the upper right side of Figure 6.18. Intrinsic lattice resistance to glide facilitates this separation. A material section of the configuration on the upper right side of Figure 6.18 is shown in the lower right portion of Figure 6.18. This section plane is perpendicular to the z-axis at some point behind the threading segment in the strained layer but ahead of the threading segment in the capping layer. This is precisely the two-dislocation mechanism introduced above, with one misfit dislocation at each of the two interfaces involved.

For the case of the strain relaxation by formation of a dislocation dipole, the energy of formation of the dislocations must be modified to account for the fact that they are formed together. There are several equivalent ways in which this calculation can be carried out. Perhaps the most direct approach is to proceed as in (6.6) with the stress field representing the superposition of the equilibrium fields of the two dislocations with equal but opposite Burgers vectors separately, and the range of integration being the slipped portion of the glide plane between them. An approach that is less direct but which takes advantage of results already obtained is to recognize that the total energy of the dislocation pair is the sum of the energies of the two dislocations, each computed in the absence of the other as was done in (6.8), plus the energy of interaction between the two dislocations. Following the latter option and using the form of self-energy in (6.8), the net energy of the dislocations is

$$
\begin{aligned}
W_{\mathrm{d}} &= \frac{\mu_{\mathrm{f}}[b_x^2 + b_y^2 + (1 - \nu_{\mathrm{f}})]}{4\pi(1 - \nu_{\mathrm{f}})} \ln \frac{4h_{\mathrm{f}}h_{\mathrm{ul}}}{r_{\mathrm{o}}^2} - \int_{r_{\mathrm{o}}}^{h_{\mathrm{sl}}\,\csc\alpha} b_i \sigma_{ij} n_j^{(+)}\, dl \\
&= \frac{\mu_{\mathrm{f}}[b_x^2 + b_y^2 + (1 - \nu_{\mathrm{f}})b_z^2]}{2\pi(1 - \nu_{\mathrm{f}})} \ln \frac{2h_{\mathrm{sl}}(h_{\mathrm{f}}h_{\mathrm{ul}})^{1/2}}{r_{\mathrm{o}}[h_{\mathrm{sl}}^2 + 4h_{\mathrm{f}}h_{\mathrm{ul}} \sin^2\alpha]^{1/2}},
\end{aligned}
\tag{6.36}
$$

where $\sigma_{ij}(x, y)$ is the stress field given in Section 6.1.2.

If the critical thickness of the strained layer is again denoted by h_{cr}, and if the normalized thicknesses of the strained layer and the unstrained capping layer are represented by p and q as defined in (6.34), then the condition $W_{\mathrm{d}} + W_{\mathrm{m}} = 0$ takes the form

$$
\frac{p^2 q(p + q)}{p^2 + 4q(p + q)\sin^2\alpha} = \left(\frac{4h_{\mathrm{cr}}}{b}\right)^{p-2}.
\tag{6.37}
$$

For given values of h_{cr}/b and p, the solutions q of this equation define the critical state. For values of q smaller than the admissible roots of (6.37), the two-dislocation mechanism can operate whereas, for larger values of q, it is energetically prohibited from operating. Values of q that satisfy (6.37) for several choices of h_{cr}/b are given in Figure 6.19 for a range of values of p. It is evident from (6.37) that all curves intersect at $p = 2$, and that each curve has a vertical asymptote at a value of p which satisfies $p^2 = 4(4h_{\mathrm{cr}}/b)^{p-2} \sin^2\alpha$. This implies values of p only slightly larger than 2 for cases of practical interest. Each vertical asymptote defines the critical thickness of a buried strained layer at a given level of mismatch as a multiple of the critical thickness of a strained surface layer of the same material at the same level of mismatch. Usually, the term *buried layer* applies to a strained layer of thickness h_{f} that is positioned between substrates, both of which are much greater in thickness than h_{f}. In this case, h_{f} is the only length parameter of significance in examining the stability of strained epitaxial layer against dislocation formation.

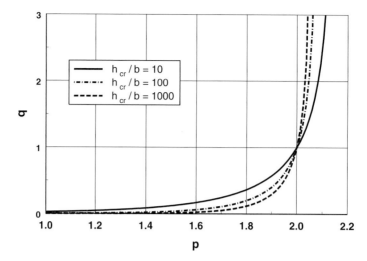

Fig. 6.19. Normalized capping layer thickness q versus the normalized strained layer thickness p for which the composite film is stable against formation of a misfit dislocation dipole according to (6.37). Stable values are those above each curve for given values of p and h_{cr}/b. Each curve has a vertical asymptote representing a strained layer thickness beyond which a capping cannot stabilize the configuration under any circumstances.

6.4.2 Strained layer superlattice

In this case, the film consists of a periodic arrangement of many layers, as described in Section 2.4. Suppose that the superlattice is composed of alternating homogeneous layers of materials a and b of thicknesses Δh_a and Δh_b, respectively. The materials in the film each have a mismatch with respect to the substrate; these are denoted by ϵ_{ma} and ϵ_{mb}, respectively. The period of the structure is λ and the total number of periods in the film is N_λ; the total thickness is then $h_f = \lambda N_\lambda$. Differences in the elastic properties of the two materials with respect to each other and with respect to the substrate are neglected.

The energy of formation of a misfit dislocation at distance η from the free surface is again given by (6.7) or (6.8). The elastic energy extracted from the system as the dislocation is formed or, equivalently, the work done by the background mismatch stress field as the dislocation is formed, is obtained by means of a straightforward generalization of (6.13). If the misfit dislocation is formed at the interface between the superlattice film and the substrate, then the result is

$$W_m(h_f) = -2b_x \mu_f \frac{1 + \nu_f}{1 - \nu_f} N_\lambda \left(\epsilon_{ma} \Delta h_a + \epsilon_{mb} \Delta h_b \right). \qquad (6.38)$$

If an effective mismatch is defined by the linear weighting rule as

$$\epsilon_{m,eff} = N_\lambda \left(\epsilon_{ma} \frac{\Delta h_a}{h_f} + \epsilon_{mb} \frac{\Delta h_b}{h_f} \right), \qquad (6.39)$$

then the critical thickness condition $W_d(h_f) + W_m(h_f) = 0$ for the total thickness h_f of the superlattice at which misfit dislocation formation at the film–substrate interface becomes possible reduces to (6.17) with the quantity $\epsilon_{m,eff}$ taking the place of mismatch strain ϵ_m. Thus, the results of Section 6.2 and Section 6.3 apply for the case of a superlattice film without modification. It is readily confirmed that $\eta = h_f$ is the most favorable location for formation of a misfit dislocation in this configuration under the assumed constraints on material properties.

6.4.3 Compositionally graded film

In this section, a film with a linear variation in mismatch strain, from $\epsilon_m = 0$ at the interface $y = 0$ to $\epsilon_m = \epsilon_{top}$ at the free surface $y = h_f$, is considered. In this case,

$$\epsilon_m(y) = \epsilon_{top} \frac{y}{h_f}. \tag{6.40}$$

Any variation of the elastic properties of the system in the y-direction is neglected in this development.

The energy of formation of the dislocation, which is independent of the distribution of mismatch, is given by (6.8). On the other hand, the work done by the background mismatch stress field is completely determined by the mismatch strain distribution according to (6.30). With the distribution of mismatch as specified in (6.40), this work is

$$W_m(\eta) = \begin{cases} -\mu_f b_x \dfrac{1 + \nu_f}{1 - \nu_f} \epsilon_{top} \eta \left(2 - \dfrac{\eta}{h_f}\right), & 0 < \eta \le h_f, \\[4mm] -\mu_f b_x \dfrac{1 + \nu_f}{1 - \nu_f} \epsilon_{top} h_f, & h_f < \eta < \infty. \end{cases} \tag{6.41}$$

As before, the critical thickness h_{cr} of the film is defined as the smallest value of h_f for which a stable equilibrium position η for the dislocation can be found such that $W_d(\eta) + W_m(\eta) = 0$.

As an illustration, consider the case of a cubic material, oriented so that the interface is a $\{001\}$ plane, with dislocation glide on a $\{111\}$ plane. The Burgers vector is given in (6.18). Graphs of $W_d(\eta)/\mu_f b^2$, $W_m(\eta)/\mu_f b^2$ and $[W_d(\eta) + W_m(\eta)]/\mu_f b^2$ are shown in Figure 6.20 for $\epsilon_{top} = 0.015$, $h_f/b = 40$ and $\nu_f = 0.25$. From the graphs, it is evident that this thickness is approximately equal to the critical thickness for the specified mismatch strain, that is, $h_{cr} \approx 40b$. It first becomes energetically possible for a dislocation to form when the local minimum in $W_d(\eta) + W_m(\eta)$ just touches the η-axis from the positive side, which first occurs at $\eta_{eq} \approx 35b$. From the fact that $W_d'(\eta_{eq}) + W_m'(\eta_{eq}) = 0$, this is an equilibrium position. Furthermore, the property that $W_d''(\eta_{eq}) + W_m''(\eta_{eq}) > 0$ implies that the equilibrium position is stable.

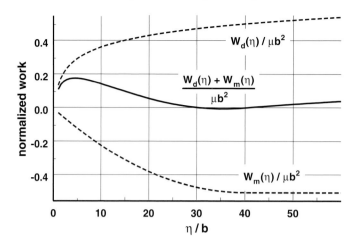

Fig. 6.20. Normalized energy of formation $W_d/\mu_f b^2$ of a dislocation and normalized work recovered from mismatch field $W_m/\mu_f b^2$ versus normalized depth of formation of the dislocation η/b for the case of a linear gradation of mismatch with surface mismatch $\epsilon_{top} = 0.015$, film thickness $h_f = 40b$ and Poisson ratio $\nu_f = 0.25$. For this case, the critical thickness is approximately $h_{cr} \approx 40b$.

6.5 Model system based on the screw dislocation

The analysis in Sections 6.2 and 6.3 provides a complete continuum elastic description of relaxation of an equi-biaxial mismatch strain by formation of a dislocation of arbitrary Burgers vector on a glide plane of arbitrary orientation. In considering extensions of this analysis to include the effects of elastic modulus difference between the film and the substrate or of dislocation interactions, the analysis becomes complicated and unwieldy. For purposes of illustrating important ideas in strain relaxation without this encumbrance, it is both convenient and instructive to consider the simpler case of strain relaxation by screw dislocations (Head 1953, Nix 1998). The purpose of the present section is to reproduce the critical thickness argument in the context of strain relaxation by a screw dislocation in order to illustrate that the results are qualitatively identical and quantitatively similar to the corresponding results for the case of general Burgers vector. The screw dislocation model is then extended to explore additional features of strain relaxation phenomena in Sections 6.5.2, 7.1 and 7.2. Sources providing the details for more realistic configurations are cited in those instances for which they are available.

6.5.1 Critical thickness condition for the model system

Upon introduction of a screw dislocation at the interface between a strained layer and its substrate, it becomes apparent that this dislocation will not relax any background elastic mismatch strain if the material is isotropic and if this strain is equi-biaxial

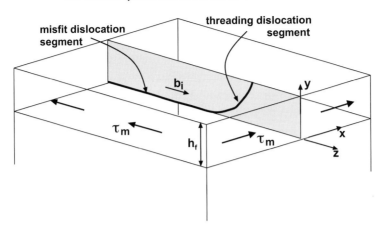

Fig. 6.21. Schematic illustration of a model system where a screw dislocation with a Burgers vector along the z-axis constitutes the interface misfit segment and the glide plane, along which the threading dislocation with mixed edge–screw character moves, is oriented perpendicularly to the film–substrate interface.

extension or compression. Therefore, to develop a model which parallels that given in Section 6.2, the mismatch of the film of thickness h_f with respect to the relatively thick substrate is necessarily a *shear* strain. The model is represented schematically in Figure 6.21, where notation from Section 6.2 is retained to the extent possible. The elastic strain in the film due to mismatch is

$$\epsilon_{xz} = \tfrac{1}{2}\gamma_m, \tag{6.42}$$

where γ_m is the spatially uniform shear strain required to render the film compatible with the substrate. The corresponding stress due to mismatch is

$$\sigma_{xz} = \tau_m = \mu_f \gamma_m, \tag{6.43}$$

which is indicated by the surface tractions in Figure 6.21. All other stress components are zero in the background mismatch field. The interface misfit segment of the strain-relieving dislocation, which is a screw dislocation, lies in the interface $y = 0$ along the negative z-axis; its Burgers vector is in the z-direction and is denoted by $b = b_z$. The threading segment of the dislocation has mixed screw–edge character. Since the orientation of the glide plane is immaterial in this case, it is chosen to be perpendicular to the interface for convenience, that is, $\alpha = 90°$.

If the energy comparison of Section 6.2 is repeated for this configuration, it is found that the work per unit length required to create the interface misfit screw dislocation (far behind the threading segment) is

$$W_d(h) = \frac{\mu_f b^2}{4\pi} \ln \frac{2h_f}{r_0}, \tag{6.44}$$

and the work per unit length done by the background mismatch elastic stress field (far behind the threading segment) in creating this dislocation is

$$W_{\mathrm{m}}(h) = -\tau_{\mathrm{m}} b h_{\mathrm{f}}. \tag{6.45}$$

It follows immediately that the critical thickness condition implied by these results is

$$\frac{b}{4\pi h_{\mathrm{cr}}} \ln \frac{2h_{\mathrm{cr}}}{r_{\mathrm{o}}} = \gamma_{\mathrm{m}}, \tag{6.46}$$

where r_{o} is again the core cutoff radius. Note that (6.46) is identical in form to the general result (6.16) but it is much simpler in detail. The driving force on the threading segment of the dislocation for any value of h_{f}, analogous to (6.27), is

$$G(h_{\mathrm{f}}) = bh\mu_{\mathrm{f}} \left[\gamma_{\mathrm{m}} - \frac{b}{4\pi h_{\mathrm{f}}} \ln \frac{4h_{\mathrm{f}}}{b} \right], \tag{6.47}$$

where the value $r_{\mathrm{o}} = \frac{1}{2}b$ of the core cutoff radius has been incorporated.

6.5.2 The influence of film–substrate modulus difference

The general behavior of a long straight dislocation parallel to an interface between two isotropic elastic materials is well established. A configurational force acts on the dislocation due to its proximity to the interface. The direction of the force is normal to the interface, a direct consequence of the invariance of the configuration under translation in any direction within the interface. The magnitude of the force varies with the inverse of the distance from the interface and, finally, whether the force is attractive or repulsive depends on the relative magnitudes of the elastic constants of the two materials. For a screw dislocation, the force exerted by the interface on the dislocation is attractive if the dislocation is in the material with the larger shear modulus, and it is repulsive otherwise. For an edge dislocation, the dependence of the sign of the force on elastic properties is more intricate but commonly the force tends to act in the direction from the more stiff to the less stiff material.

In the discussion of critical thickness of a strained layer on a substrate in Sections 6.2 and 6.5, the elastic properties of the film and substrate were assumed to be *identical*. Thus, the interface between the film and substrate played no role in the analysis other than defining the range over which the background mismatch stress was operative and, as a consequence of energy considerations, the misfit dislocations which were formed at or beyond critical thickness were positioned on the interface. It is no longer possible for the misfit dislocations to reside precisely on the interface if the film and substrate materials have elastic properties that differ significantly. To gain some insight into the influence of this effect on the critical thickness condition, the conditions for formation of a screw dislocation in a strained film on a substrate are re-examined in this section for the case of a film and substrate with different elastic stiffnesses.

Before undertaking the full critical thickness calculation, a relatively simple preliminary calculation can shed some light on the effect. Consider a screw dislocation near the interface between a strained film and a substrate, and suppose that the force on the dislocation due to the free surface of the film is negligible. This might be so if the distance of the dislocation from the interface, which can be specified by its y-coordinate, is much smaller than h_f, for example. If the Burgers vector of the dislocation is such that it is strain relieving, the background mismatch field exerts a force $-\tau_m b$ in the y-direction per unit length on the dislocation if the dislocation is in the film ($y > 0$) and no force at all if the dislocation is in the substrate ($y < 0$). The negative sign indicates that the force acts *toward* the interface if the dislocation is in the film. The configurational force per unit length exerted by the bimaterial interface on the dislocation, on the other hand, is

$$F_b = -\frac{\mu_f b k_\mu}{4\pi |y|}, \quad k_\mu = \frac{\mu_f - \mu_s}{\mu_f + \mu_s}, \tag{6.48}$$

where the shear moduli in the film and substrate are denoted by μ_f and μ_s, respectively (Hirth and Lothe 1982). The parameter k_μ is clearly the counterpart for antiplane shear deformation of the Dundurs parameters defined in (3.98) for plane strain deformation. The force acts in the positive (negative) y-direction if $\mu_s > \mu_f$ ($\mu_s < \mu_f$). It follows immediately that, if $\mu_s > \mu_f$, there is a stable equilibrium position for the dislocation within the film at a distance

$$y_{eq} = -\frac{\mu_f b k_\mu}{4\pi \tau_m} \tag{6.49}$$

from the interface. If $\mu_s < \mu_f$, on the other hand, there is no stable equilibrium position for the dislocation at any value of y, although the net force (6.48) acting on the dislocation falls off very rapidly with distance into the substrate. The purpose of going through the critical thickness calculation is to determine how these preliminary conclusions are modified by the presence of the free surface at $y = h_f$.

The complete solution for the case of a screw dislocation in a configuration consisting of a layer of uniform thickness bonded to a substrate has been provided by Chou (1966). The critical thickness analysis proceeds by calculating the work $W_d(\eta)$ that must be done to insert a dislocation into the material at a distance η from the interface, and then comparing the result to the work $W_m(\eta)$ that is done by the background mismatch stress field as the dislocation is formed. As in Section 6.2, the critical thickness is defined as the smallest value of h_f for which *any* value of η can be found for which $W_d(\eta) + W_m(\eta) = 0$. When $\mu_f \neq \mu_s$, there is no reason to expect that $\eta = h_f$, as was the case for $\mu_f = \mu_s$. The relationship between the shear moduli of the film and the substrate is specified by the parameter k_μ defined in (6.48). Although the solution provided by Chou (1966) is valid for the full range of elastic constants, it is given in the form of an infinite series in powers of k_μ. Only terms up to and including those that are *linear* in k_μ are retained here. This means that the present result is valid only for $|k_\mu| \ll 1$. If this is understood to imply that $|k_\mu| \le 0.1$, then the results are expected to be accurate for

$$\frac{9}{11} \le \frac{\mu_f}{\mu_s} \le \frac{11}{9}. \tag{6.50}$$

To this level of accuracy, the work of formation of the dislocation is

$$W_d(\eta) = \frac{\mu_f b^2}{4\pi} \begin{cases} \ln \dfrac{2\eta}{r_o} + k_\mu \ln \dfrac{h_f^2 - \eta^2}{h_f^2}, & r_o < \eta < h_f - r_o, \\[4mm] \ln \dfrac{2\eta}{r_o} - 2k_\mu \ln \dfrac{2\eta(\eta - h_f)^{1/2}}{r_o(\eta + h_f)^{1/2}}, & h_f + r_o < \eta < \infty. \end{cases} \tag{6.51}$$

When $k_\mu = 0$, it is evident that this result reduces to the appropriate limit for a homogeneous material given in (6.44). Furthermore, if $\mu_s > \mu_f$ ($\mu_s < \mu_f$), then the work required to form a dislocation anywhere in the film is increased (decreased) from the corresponding result for a homogeneous material, as might be expected. Note that the terms that are proportional to k_μ in (6.51) are logarithmically singular at the interface $\eta = h$. Thus, no significance is attached to values of these terms within the range $h_f - r_o < \eta < h_f + r_o$.

The work done by the background mismatch field in forming a dislocation at the distance η from the free surface of the film is

$$W_m(\eta) = \begin{cases} -\mu_f \gamma_m b \eta, & \eta \le h_f, \\[3mm] -\mu_f \gamma_m b h_f, & \eta > h_f. \end{cases} \tag{6.52}$$

For the dislocation to be strain relieving, it must satisfy the condition that $\gamma_m b > 0$. There is no loss of generality in assuming that $\gamma_m > 0$ and $b > 0$ separately; if the signs are actually opposite to these in any particular situation, then γ_m and b can be understood as magnitudes.

The influence of modulus difference on the behavior of dislocations in a thin strained film epitaxially bonded to a substrate was considered in a series of observations reported by Wu and Weatherly (2001). They deposited $In_{1-x}Ga_xAs_{1-y}P_y$ films onto InP(100) substrates at 480 °C. Through cross-sectional transmission electron microscopy, it was observed that strain relaxation occurred by twinning in the case of $In_{0.5}Ga_{0.5}As_{0.5}P_{0.5}$ films, for which the mismatch strain was approximately 0.02. In this case, the elastic modulus of the film is approximately 1.3 times the elastic modulus of the substrate. It was established that the leading partial dislocation of the twin structure was often found to be well within the substrate. While there is a force on the leading partial due to the stacking fault energy that tends to pull it back toward the interface, the relative compliance of the substrate was apparently sufficient to draw the dislocation away from the interface and deeper into the substrate. Wu and Weatherly (2001) carried out an approximate dislocation image force analysis, similar to that reported by Chou (1966) for the case of a screw dislocation, to estimate the magnitude of the force pushing the dislocation further into the more compliant substrate. An exact, but quite intricate, analysis of this system for an edge dislocation has been reported by Weeks et al. (1968).

It can be readily established from (6.49)–(6.52) that the elastic modulus mismatch between the film and the substrate can strongly influence the propensity for formation of misfit dislocations for a given mismatch strain as well as the preferred location for dislocation formation. These effects of modulus mismatch are illustrated quantitatively in the following numerical example.

6.5.3 Example: Modulus difference and dislocation formation

Consider the model system for an epitaxial thin film on a substrate depicted in Figure 6.21 with a shear strain $\gamma_m = 0.01$.

(a) Consider first an elastically homogeneous system with $\mu_f = \mu_s$. Determine the critical film thickness at which it becomes unstable against formation of an interface misfit dislocation. Assume that $\alpha = 90°$ and $r_o = \frac{1}{2}b$.

(b) Now consider the two different cases of elastic modulus difference with $\mu_s = \frac{9}{11}\mu_f$ and $\mu_s = \frac{11}{9}\mu_f$. Discuss how the elastic modulus difference influences the propensity for the formation of the misfit dislocation and the location at which such dislocation formation preferentially occurs.

Solution:

(a) For the elastically homogeneous film–substrate system with $\mu_s = \mu_f$, the critical thickness is found from (6.46) to be $h_{cr} \approx 40.5b$. In Figure 6.22, a graph of $W_d(\eta) + W_m(\eta)$ is shown for $\mu_s = \mu_f$ and $h_f = 40.5b$ (the solid curve), from which it is clear that $W_d(\eta) + W_m(\eta)$ first becomes zero at $\eta = h_f$ when h_f reaches the value h_{cr}.

(b) The variations of $(W_d + W_m)/\mu_f b^2$ versus η/b for $\mu_s = \frac{9}{11}\mu_f$ and $\frac{11}{9}\mu_f$ are also plotted in Figure 6.22. By comparison with the case for which $\mu_s = \mu_f$, the graph for $\mu_s < \mu_f$ passes well *below* the axis defining $W_d + W_m = 0$. This means that a critical thickness for dislocation formation was reached at a value of h_f *smaller* than h_{cr}, and that the relatively greater elastic compliance of the substrate assists in the formation of dislocations. Furthermore, the preferred location of a misfit dislocation is in the substrate at the distance from the interface coinciding with the minimum in the graph of $W_d(\eta) + W_m(\eta)$. This minimum is rather shallow, and the gradual rise in the curve beyond the minimum is due solely to the influence of the free surface of the film. In contrast, the graph for $\mu_s > \mu_f$ passes well *above*

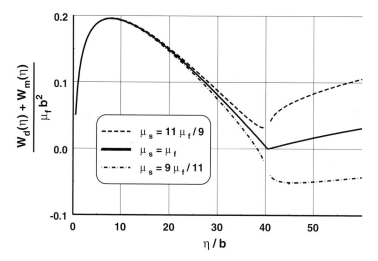

Fig. 6.22. Variation of $W_d + W_m$ normalized by $\mu_f b^2$ as a function of η/b for the three cases: $\mu_s = \frac{9}{11}\mu_f$, $\mu_s = \mu_f$ and $\mu_s = \frac{11}{9}\mu_f$ for $\gamma_m = 0.01$, $\alpha = 90°$ and $r_o = \frac{1}{2}b$.

the axis defining $W_d + W_m = 0$. This means that film thickness is still below the critical thickness for this case, and that the relatively greater stiffness of the substrate retards the formation of dislocations. If the thickness were to increase until the graph just touches the $W_m + W_d = 0$ axis, it would be found that it does so at a location within the film at a small distance from the interface. This distance was anticipated in the result (6.49) above, which provides a good approximation to its magnitude.

6.6 Nonplanar epitaxial systems

The epitaxial material structures considered in the preceding sections were two-dimensional or planar structures, according to the categorization introduced in Section 1.1. In this section, attention is redirected to nonplanar epitaxial structures. Perhaps the simplest among these is a one-dimensional structure known commonly as a *quantum wire*. An epitaxial quantum wire structure is fabricated so that the spatial extent of the deposited material, which will continue to be identified as the film material throughout the discussion, is very large in one direction compared to its extent within a cross-section perpendicular to that direction. Because of the higher degree of confinement of an epitaxial wire, compared to a large area planar quantum well, the strain levels that can be achieved before the structure becomes unstable against the formation of strain-relieving dislocations are substantially greater than those for planar structures. This will be demonstrated for particular structures in this section.

As the name implies, such structures are of interest in microelectronics and op-toelectronics in situations for which the cross-sectional dimensions are sufficiently small so that electronic behavior is dominated by quantum mechanical effects. The combined influences of two-dimensional quantum confinement of charge carriers and of elastic strain due to the constraint of epitaxy result in unique opportunities for tailoring the electronic or optical properties of lasers, optical modulators and other devices. The extra degree of charge carrier confinement, over and above that afforded by planar structures, gives rise to a narrowing of the distribution of density of states which results in sharper optical spectral features. In addition, the range of elastic strain states accessible in the wire configuration allows for significant modifications of band structure of the material. An interesting and technologically important issue is whether or not electronic characteristics of the structure can be interpreted within the framework of electronic band theory. The theory presumes wave transport in a periodic medium, and the influence of uniform strain on band structure has been studied extensively (Luttinger and Kohn (1955), O'Reilly (1989) and Yu et al. (1992), for example). Models have also been developed to consider the influence of strain that is spatially nonuniform on a scale that is very long compared to atomic dimensions (Yang et al. (1997) and Johnson et al. (1998), for example). However, robust predictive methodologies for characterizing electronic behavior of very small nonuniformly strained material structures, particularly those incorporating alloys, are not yet available.

In planar epitaxial systems for which in-plane dimensions are large compared to layer thickness, the background equilibrium stress field due to mismatch, if any, is laterally uniform. This feature leads to significant simplification in the calculation of the work done by the mismatch stress field during formation of a dislocation, introduced as W_m in Section 6.2. In the case of a *nonplanar* epitaxial structure with lattice mismatch, assessment of the stability of the system against formation of strain-relieving dislocations is conceptually unchanged from the planar case but it is practically more difficult to extract quantitative estimates of instability conditions. For any particular configuration, the goal in considering the onset of elastic strain relaxation is still to establish circumstances under which $W_m + W_d$ first approaches zero through positive values, as in Section 6.2.1. The task is complicated by the fact that the background mismatch stress distribution is no longer spatially uniform, as it is in the planar case. Consequently, the calculation of W_m for any particular prospective dislocation becomes more difficult and the location that is most favorable for dislocation formation is not always self-evident, as it was in the planar configurations.

In this section, two particular configurations are discussed. Both are two-dimensional configurations in which the mismatch stress field results in a state of generalized plane strain deformation. The first is a long misfitting inclusion with rectangular cross-section which is embedded within and is epitaxially bonded to a surrounding medium. All other boundaries are sufficiently remote so as to be inconsequential. This configuration was analyzed by Gosling and Willis (1995) for a periodic array of inclusions. The second configuration is a long misfitting inclusion which is epitaxially deposited in a surface groove in a second material. Both configurations are within a class identified above as quantum wire configurations. The first of these configurations is by far the simpler to analyze due to the rectangular geometry and the absence of free surfaces, and a critical condition for dislocation formation will be obtained by constructing the exact stress field in the material. On the other hand, the second configuration with a nonplanar material structure and a free boundary is cumbersome by means of standard stress analysis, and a computational strategy for estimating the critical conditions is introduced. Various two-dimensional configurations of periodic quantum wire arrays were considered by Gosling (1996) within the framework of anisotropic elastic material response.

6.6.1 A buried strained quantum wire

The two-dimensional configuration to be considered is depicted in the upper portion (a) of Figure 6.23. A long slender elastic inclusion occupies the region $-\frac{1}{2}w < x < \frac{1}{2}w, 0 < y < h$ within an otherwise unbounded matrix of an elastic material with nominally the same properties as the inclusion. The mismatch strain of the quantum wire inclusion with respect to the surrounding material is assumed to be a pure volumetric strain represented in terms of the linear mismatch ϵ_m as $\epsilon_{ij}^m = \epsilon_m \delta_{ij}$. The formation of this material structure can be imagined to be the result of the following steps, starting with the two materials completely separated. The inclusion is subjected to an artificial uniform normal surface traction $\sigma_m = 2\mu_f(1 + \nu_f)\epsilon_m/(1 - 2\nu_f)$ where the combination of parameters $2\mu_f(1 + \nu_f)/3(1 - 2\nu_f)$ is the elastic bulk modulus. Under these conditions, the elastic strain in the inclusion is precisely ϵ_{ij}^m and the inclusion fits perfectly into the rectangular cavity with zero stress everywhere in the matrix material. This is the state represented in the lower left portion (b) of Figure 6.23, with the tractions acting on the surface of the inclusion but not on the surrounding matrix. The interface is next bonded to enforce the constraint of epitaxy. Finally, the artificial normal traction σ_m on the interface is removed from the lateral surface of the inclusion, as represented in the lower right portion (c) of Figure 6.23.

At this point, the inclusion strain is partially relaxed under generalized plane strain conditions and a stress distribution has been generated in the surrounding elastic material. The elastic strain field in the inclusion is quite different from the mismatch strain ϵ_m at this point, emphasizing the distinction between elastic strain

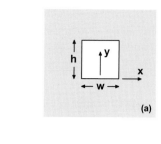

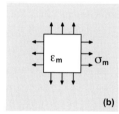 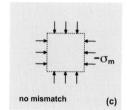

Fig. 6.23. The two-dimensional configuration of a buried strained quantum wire (upper figure (a)), where a long slender elastic inclusion of the quantum wire is embedded in an otherwise unperturbed matrix of an elastic material. In the lower left figure (b), the inclusion is subject to an imaginary uniform normal traction σ_m at its surface in such a manner that it fits perfectly into the rectangular cavity of the surrounding matrix without inducing any stress in the matrix material. The lower right figure (c) represents the situation where the artificial normal traction σ_m on the inclusion surface is removed, whereby the strain in the inclusion is partially relaxed and the surrounding elastic matrix becomes stressed.

as a field descriptor and mismatch ϵ_m as a system parameter. Since the equi-triaxial volumetric strain necessary to fit the inclusion into the cavity induces no shear stress in the material, it is irrelevant in considering the work done in forming a glide dislocation. It is only the stress arising from relaxation of the artificial surface tractions in (c) that gives rise to shear stress which can drive dislocation formation by glide. A necessary condition on h_f, w, ϵ_m and Burgers vector b_i for formation of a strain-relieving dislocation is considered next.

An elegant general framework exists for the description of elastic fields in solids with misfitting inclusions (Eshelby 1957). However, the configuration under consideration here is sufficiently simple so that the stress distribution can be determined directly in a relatively straightforward way.

Suppose that the equilibrium stress field resulting from the relaxation step shown in part (c) of Figure 6.23 is denoted by $\sigma_{ij}^r(x, y)$. An expression for this stress field can be written immediately by appeal to the representation theorem for stress in terms of the elastic Green's function for a concentrated force under plane strain conditions. Suppose a stress field $\Sigma_{ijk}(x, y)$ can be found which, for fixed k, is the stress field in the plane due to a concentrated force of unit magnitude applied at the origin $x = 0$, $y = 0$ and acting in the k-th direction. This singular solution is known, so the solution for any distribution of concentrated forces can be constructed by superposition. The components of $4\pi(1 - \nu_f)(x^2 + y^2)^2 \Sigma_{ijk}(x, y)$ are

$$
\begin{bmatrix}
-x\left[(3 - 2\nu_f)x^2 + (1 - 2\nu_f)y^2\right] & -y\left[(3 - 2\nu_f)x^2 + (1 - 2\nu_f)y^2\right] \\[2ex]
-y\left[(3 - 2\nu_f)x^2 + (1 - 2\nu_f)y^2\right] & x\left[(1 - 2\nu_f)x^2 - (1 + 2\nu_f)y^2\right]
\end{bmatrix}
\tag{6.53}
$$

for $k = x$, and

$$
\begin{bmatrix}
-y\left[(1 + 2\nu_f)x^2 - (1 - 2\nu_f)y^2\right] & -x\left[(1 - 2\nu_f)x^2 + (3 - 2\nu_f)y^2\right] \\[2ex]
-x\left[(1 - 2\nu_f)x^2 + (3 - 2\nu_f)y^2\right] & -y\left[(1 - 2\nu_f)x^2 + (3 - 2\nu_f)y^2\right]
\end{bmatrix}
\tag{6.54}
$$

for $k = y$. The concentrated force distribution is that shown in part (c) of Figure 6.23 where the magnitude of each line of concentrated force is $2\mu_f(1 + \nu_f)\epsilon_m/(1 - 2\nu_f)$. Thus,

$$
\sigma_{ij}^r(x, y) = -2\mu_f\epsilon_m \frac{1 + \nu_f}{1 - 2\nu_f} \int_S \Sigma_{ijk}(x - \xi_s, y - \eta_s) n_k(s)\, ds,
\tag{6.55}
$$

where S is the boundary curve between the inclusion and the matrix, s is arc length along S, (ξ_s, η_s) is a point on S represented by s and $n_k(s)$ is the outward unit normal vector to S at s. The integral can be evaluated in terms of elementary functions, but the result is too complicated to warrant its inclusion here for general values of the parameters.

The orientation of the glide plane on which the dislocation dipole is formed and the location of the dislocations depends on the geometrical configuration and crystallographic orientation of the material. To make the discussion more concrete, consider the case when the material has cubic crystallographic structure and the preferred slip planes are the {111} planes. Suppose that the xz-plane coincides with a {100} crystallographic plane, and the length of the quantum wire or, equivalently, the z-axis, is aligned with a ⟨110⟩ direction. If the generic slip plane indicated in Figure 6.24 is a {111} plane lying along the length of the inclusion, then $\alpha = 54.7°$ or $125.3°$. Although the formation of dislocation lines on the {111} planes which cut across the cross-section cannot be ruled out, such dislocations would result in limited amounts of elastic strain relaxation compared to those lines which run along the length of the inclusion and, therefore, such dislocations are not considered here. The dislocation dipole is imagined to be formed by glide of threading dislocations in opposite directions along the length of the inclusion, leaving behind the dipole pair as they move apart, as illustrated in Figure 6.18 for dislocation formation in a buried strained layer. In general, the dislocations may be formed inside the inclusion, outside the inclusion or spanning the interface.

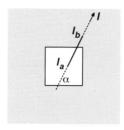

Fig. 6.24. A buried strained quantum wire with a slip plane oriented at an angle α to the x-axis. The glide of a threading dislocation leaves behind a dislocation dipole pair with l_a and l_b denoting the positions of the dislocations. l denotes distance along the slip plane.

Consider formation of $60°$ dislocations on the glide plane with orientation α as depicted in Figure 6.24. The work done by the stress field $\sigma_{ij}^r(x, y)$ in forming a dislocation dipole on a part of a plane with normal $n_i^{(+)}$ and Burgers vector b_i is

$$W_m = \int_{l_a}^{l_b} b_i \sigma_{ij}^r n_j^{(+)} \, dl \,, \tag{6.56}$$

analogous to (6.12), where l is distance along the dashed glide plane measured in the sense shown in the figure from some arbitrary origin. From (6.36), the work that must be done to form this dipole in the absence of any other stress field is

$$W_d = \frac{\mu_f[b_x^2 + b_y^2 + (1 - \nu_f)b_z^2]}{2\pi(1 - \nu_f)} \ln \frac{l_b - l_a}{r_o}. \tag{6.57}$$

The remaining task is to locate the positions of the dislocations, represented symbolically by l_a and l_b in (6.56) and (6.57), for which it first becomes possible to have $W_m + W_d = 0$.

In terms of α, the Burgers vector in (6.56) is

$$b_i = b \left\{ \tfrac{\sqrt{3}}{2} \cos \alpha, \tfrac{\sqrt{3}}{2} \sin \alpha, \tfrac{1}{2} \right\} \tag{6.58}$$

for $\epsilon_m > 0$ and the slip plane normal is

$$n_i^{(+)} = \{- \sin \alpha, \cos \alpha, 0\}. \tag{6.59}$$

If $\epsilon_m < 0$, then the sign of the Burgers vector must also be reversed but the calculation is otherwise unaffected.

Suppose that dislocation formation is restricted to the inclusion itself. Then the preferred location for the dipole in the inclusion becomes clear from examination of the integrand of (6.56). A contour plot of the non-dimensional quantity $W_m^* = (1 - \nu_f)b_i \sigma_{ij}^r(x, y)n_j^{(+)}/2\mu_f(1 + \nu_f)\epsilon_m b_x$ for an inclusion with a square cross-section,

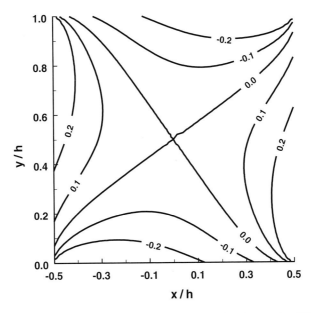

Fig. 6.25. Contours of constant values of the nondimensional quantity $W_{\mathrm{m}}^* = (1 - \nu_{\mathrm{f}}) \times b_i \sigma_{ij}^{\mathrm{r}}(x, y) n_j^{(+)}/2\mu_{\mathrm{f}}(1 + \nu_{\mathrm{f}})\epsilon_{\mathrm{m}} b_x$ for an inclusion with a square cross-section ($w = h$ and $\alpha = 54.7°$).

that is, with $w = h$ and $\alpha = 54.7°$, is shown in Figure 6.25. In fact, $-W_{\mathrm{m}}^*$ represents the density per unit length along the slip plane of work done by the background mismatch field; regions in which W_{m}^* is negative and large in magnitude contribute most to dislocation formation, whereas regions in which W_{m}^* is positive act to oppose dislocation formation. On the basis of this intuitive argument, it is clear that the glide plane should pass through the geometric center of the inclusion at $x = 0$, $y = \frac{1}{2}h$ and that the dislocation pair should be symmetrically positioned on the glide plane with respect to the center. Finally, in light of the expression for W_{d} in (6.57), the dislocations should be as far apart as possible; that is, they should lie in the interfaces at $y = 0$ and $y = h$.

For any value of $w \geq h \cot \alpha$, W_{m} has the form

$$W_{\mathrm{m}} = -2\mu_{\mathrm{f}}\frac{(1 + \nu_{\mathrm{f}})}{(1 - \nu_{\mathrm{f}})}\epsilon_{\mathrm{m}} b_x h I(w/h, \alpha), \qquad (6.60)$$

where $I(w/h, \alpha)$ is a nondimensional factor depending only on the aspect ratio w/h and the angle α. It can be verified by direct calculation that $I(w/h, \alpha) \to 1$ as $w/h \to \infty$. Consequently, the correct result is obtained in the limit as the inclusion shape approaches that of a uniform buried layer. The condition necessary for the formation of a strain-relieving dislocation dipole, obtained from $W_{\mathrm{m}} + W_{\mathrm{d}} = 0$,

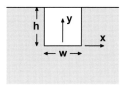

Fig. 6.26. A strained quantum wire structure formed in the patterned surface of a substrate.

is then

$$\frac{[b_x^2 + b_y^2 + (1 - v_f)b_z^2]}{4\pi(1 + v_f)b_x h} \ln \frac{h \csc \alpha}{r_o} = \epsilon_m I(w/h, \alpha).$$

$$(6.61)$$

For $\alpha = \arccos(1/\sqrt{3}) \approx 0.955$, the factor $I(w/h, \alpha)$ is given approximately by

$$I(w/h, 0.955) \approx 0.91 - 0.84\frac{h}{w} + 0.0029\frac{w}{h}$$

$$(6.62)$$

for $1 < w/h < 20$. The finite width of the rectangular wire has a very strong effect on the strain needed to drive dislocation formation at a given value of h. For example, $I(1, 0.955) \approx 0.1$ which implies that the strain required for dislocation formation in a square $h \times h$ wire is an order of magnitude larger than for a uniformly strained layer of the thickness h with a free surface, and four times larger for a buried layer of that same thickness. This result illustrates the point made above that the range of elastic strains in quantum wire configurations of certain cross-sectional dimensions is substantially larger than strain levels at which a uniform thin film of comparable thickness loses stability against formation of strain-relieving dislocations.

6.6.2 *Effect of a free surface on quantum wire stability*

While the results obtained in the preceding section for a buried quantum wire are useful and constructive, a more critical stage in the formation of such a material structure is when the wire is being deposited onto the patterned surface of the substrate. Such a configuration is illustrated in Figure 6.26 for a wire of rectangular cross-section. The configuration is identical to that in Figure 6.23 except that the plane $y = h$ is a free surface, and it may be viewed as an intermediate configuration encountered in the process of fabricating a buried wire. The presence of the free surface makes this structure more susceptible to strain relaxation by dislocation formation than is the corresponding buried wire.

Based on the study of the buried wire in Section 6.6.1, it is clear that the critical thickness condition in this case will again have the form (6.61) where $I(w/h, \alpha)$ is a dimensionless factor depending only on the aspect ratio w/h and the angle α. As before, the factor $I(\cdot)$ incorporates the complexity of the geometry of the structure.

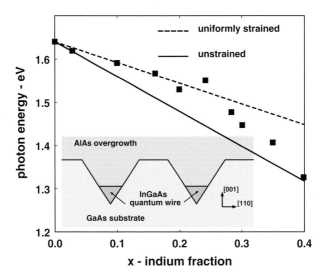

Fig. 6.27. The discrete points, adapted from Arakawa et al. (1993), show the measured photoluminescence peak energy versus In fraction x for the quantum wires illustrated in the inset. The solid (dashed) line provides a theoretical estimate of peak energy for unstrained (strained) $In_x Ga_{1-x} As$.

The same general approach can be followed for more complex configurations. The most significant additional difficulty encountered is that neither the stress analysis problem for the background mismatch field nor that for the work of formation of the dislocation is mathematically tractable in most cases. Consequently, both W_m and W_d must be determined by numerical calculation. A situation of this kind of practical interest is described next.

A structure that has been studied experimentally by Arakawa et al. (1993) is illustrated schematically in the inset to Figure 6.27. Quantum wires of $In_x Ga_{1-x} As$ were deposited between the $(1\bar{1}1)$ and $(\bar{1}11)$ side walls of an array of [110]-oriented V-grooves formed by lithographic patterning of a (001) GaAs substrate. Once deposited, the wires were covered by a thick overgrowth of AlGaAs, a material that is lattice matched with the GaAs substrate. The width of the wires was estimated to be of the order of 10 nm and the spacing between the wires was a little more than 100 nm. Photoluminescence spectra were obtained, yielding the emission energy peak as a function of the In composition x. The values determined from the experiments are represented by the discrete points in Figure 6.27. The solid and dashed lines give the results of calculations by Arakawa et al. (1993) based on a one-band quantum mechanical model. The solid line gives results for an unstrained material, and the dashed line gives the equivalent results for a strained material. At low In fractions the experimental values agree closely with those calculated with the strain effect included. As the In fraction is increased, the peak position tends

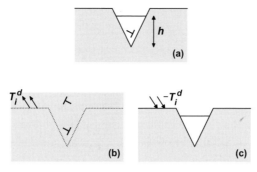

Fig. 6.28. Part (a) shows a schematic illustration of a quantum wire in a V-groove formed by patterning a $\{100\}$ surface of a cubic crystal substrate, with the glide dislocation on a $\{111\}$ crystallographic plane within the structure. The conditions for formation of such a dislocation are considered in terms of a superposition of the stress fields of the configurations depicted in parts (b) and (c). The superposition scheme is described in the text.

to move toward the value for an unstrained material and the peak width at half its maximum value was observed to increase, strongly suggesting the occurrence of mechanical degradation and consequent strain relaxation.

Here we discuss the stability of the quantum wire structure at that point in its formation when the top surface of the wire is free. For purposes of this discussion, it is assumed that the spacing between the wires and the depth of the V-grooves are both large enough for it to be a good approximation to model each wire as being isolated and in a deep groove. Furthermore, the apex of the groove is taken as being sharp and the wire cross-section is assumed to be triangular with a flat top surface. The possible formation of thin layers of material up the side walls of the groove is ignored. This idealized configuration is illustrated in part (a) of Figure 6.28. It is assumed that the strain tends to relax via the formation of a $60°$ dislocation along the wire, having a $\{111\}$ glide plane and $\frac{1}{2}\langle 110 \rangle$ Burgers vector. The formation of the dislocation is driven by the elastic energy arising through lattice mismatch ϵ_m. The expectation is that for a given h there will be a critical value of ϵ_m that divides the regimes which are stable and unstable against dislocation formation. The aim is to define this critical relationship between h and ϵ_m by adapting the approach that was introduced in the preceding section.

The task is to calculate W_d, the work required to form a dislocation at any position on any $\{111\}$ plane within the wire, and W_m, the work done by the mismatch stress field as that dislocation is formed. The critical state is then defined as being the most conservative combination of ϵ_m and h for which $W_d + W_m = 0$. Both W_d and W_m can be computed effectively by means of the finite element method. This numerical method cannot accurately represent the singular field of an elastic dislocation, but the discrete dislocation can be incorporated analytically by the superposition scheme in Figure 6.28. Suppose that the goal is to compute W_d for the position of the dislocation in part (a) of the figure. In part (b), the entire elastic plane is considered with a dislocation in the assigned location and a second dislocation with equal but opposite Burgers vector at a nearby location *outside* the region of the solid being considered. Problem (b) can be simply analyzed, and the solution has already been used in Section 6.6. The second dislocation is not included to enforce any particular boundary condition. Instead, its inclusion significantly improves the accuracy of the numerical solution of the complementary problem discussed next.

The pair of dislocations in problem (b) induces a traction distribution T_i^d along the surface which coincides in position with the free surface in the physical problem at hand. Thus, the complementary boundary value problem is one concerned with the actual configuration of the solid, as shown in part (c) of Figure 6.28, but with no dislocation present. Instead, the surface of the solid is subjected to a traction distribution $-T_i^d$. This boundary value problem must be solved numerically, and it is here where the significance of the second dislocation becomes evident. The numerical procedure is necessarily effected only over a finite portion of the solid. The second dislocation ensures that the stress components decay in amplitude sufficiently rapidly as $(x^2 + y^2)^{1/2}$ becomes large with respect to h. Consequently, the calculation can be carried out over a region of size $-10h \leq x \leq 10h$, $-10h \leq y \leq 10h$ with confidence that the stress resultant on the outer boundary will be very small indeed and that the results in the vicinity of the wire will be accurate. Superposition of the two solutions (b) and (c) provides the solution for a dislocation in the solid in (a) with a free surface. W_d can then be calculated according to (6.6) where σ_{ij} is the *net* stress obtained from superposition of the two solutions.

This approach has been applied by Freund and Gosling (1995) in order to determine critical conditions for dislocation formation in the structure illustrated in part (a) of Figure 6.28. Intuitively, the dislocation will most likely form along the interface on the right side of the wire in the orientation illustrated in the figure. Calculations were performed for a range of positions of the dislocation along this interface in order to search out the location of greatest susceptibility; both W_d and W_m were calculated for each of these positions. For given dimensions, the smallest value of mismatch strain ϵ_m for which dislocation formation becomes possible is plotted against the normalized height h/b of the wire in Figure 6.29. The parameter range within which degradation of structures occurred in the experiments is shown by the shaded rectangle in the figure. For conditions on the left and bottom sides of the rectangle, no degradation was detected whereas, for conditions on the right and top sides, degradation was observed. The results of the critical thickness calculation are consistent with this observation.

These results can be compared to the critical mismatch strain results for a buried wire of height h and for a uniformly strained layer of thickness h. It is found that the

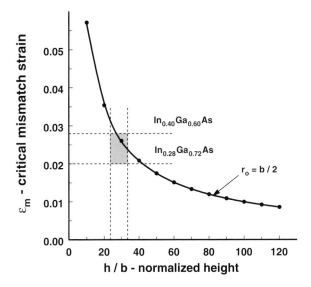

Fig. 6.29. The critical mismatch strain ϵ_m, plotted as a function of normalized quantum wire height h/b, for core cutoff radius $r_o = b/2$, as determined by the computational procedure based on Figure 6.28; adapted from Freund and Gosling (1995). The shaded area represents the range of values of mismatch strain and wire size within which degradation of the optical characteristics was observed to occur. Adapted from Arakawa et al. (1993).

critical mismatch strain for the quantum wire is approximately 20% of the critical mismatch for the corresponding buried wire and about 250% of the critical mismatch strain for a uniform layer. Therefore, much of the advantage seen for the case of a buried or completely confined wire in Section 6.6.2 is lost when the wire is in a configuration with a free surface. This observation emphasizes the importance of preventing dislocation nucleation in the structures at this critical juncture in their fabrication; once they become fully confined, their stability against strain relaxation by dislocation formation is enhanced significantly.

6.7 The influence of substrate compliance

The film thicknesses at which many strained epitaxial thin film materials begin to degrade in quality due to dislocation formation are too small to permit their use in electronic device development. For very thick substrates of the kind assumed in preceding sections in this chapter, any lattice mismatch between the film and the substrate must be accommodated by deformation of the film. This situation can be alleviated to some degree if the thickness of the substrate is of the same order of magnitude as the thickness of the film deposited on it. In this case, accommodation of a mismatch in lattice parameter can be shared between the film and substrate materials, thereby decreasing the elastic strain in the film and extending the range

Fig. 6.30. A film–substrate bilayer system which is constrained against bending by a relatively thick handle wafer. However, the structure is free to extend or contract in its own plane.

of thickness over which a particular film–substrate system is stable against formation of strain-relieving dislocations. The substrates that have been developed for this stress management function in epitaxial systems are called *compliant substrates*.

The development of a compliant substrate presents a challenging fabrication problem. The basic idea is to provide a substrate of good quality which has a lateral extent on the order of centimeters and thickness on the order of tens of nanometers. A free-standing structure of this kind is very fragile. A more promising configuration is that in which the very thin compliant substrate is *weakly* bonded to a substantial *handle wafer* which constrains it to remain flat but permits it to extend or contract in its own plane. The goal of this section is to estimate the degree to which the critical thickness for dislocation formation can be increased by means of substrate compliance of this kind.

6.7.1 A critical thickness estimate

Consider the substrate–film system shown in Figure 6.30. The bilayer is constrained against bending, presumably by a relatively thick handle wafer, but it is unconstrained against extension or contraction in the plane of the interface. The isotropic elastic materials of the film and the substrate have the same shear modulus μ and Poisson ratio ν. The common biaxial modulus is then $M = 2\mu(1+\nu)/(1-\nu)$. The elastic biaxial extensional strains in the substrate and film are denoted by ϵ_s and ϵ_f, respectively. The mismatch in lattice parameter is denoted by ϵ_m.

The compatibility of deformation of the layers is assured if

$$\epsilon_f - \epsilon_m = \epsilon_s, \tag{6.63}$$

and the equilibrium condition of zero net force on any internal plane perpendicular to the interface requires that

$$M\epsilon_f h_f + M\epsilon_s h_s = 0. \tag{6.64}$$

These two conditions determine the elastic strain in each layer prior to formation of dislocations as

$$\epsilon_f = \epsilon_m \frac{h_s}{h_s + h_f}, \qquad \epsilon_s = -\epsilon_m \frac{h_f}{h_s + h_f}. \tag{6.65}$$

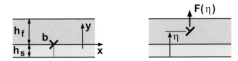

Fig. 6.31. A dislocation with Burgers vector of magnitude b and components $\{b_x, b_y, b_z\}$ which is introduced at the film–substrate interface from within the substrate.

Note that this result is consistent with the more familiar result for a thick substrate, that is, $\epsilon_s \to 0$ and $\epsilon_f \to \epsilon_m$ as $h_f/h_s \to 0$.

Suppose now that a dislocation with Burgers vector of magnitude b and components $\{b_x, b_y, b_z\}$ is introduced at the film–substrate interface from within the substrate, as depicted in Figure 6.31. If the net work which must be done to form such a dislocation is negative (positive), then the dislocation can (cannot) form spontaneously. Thus, the discriminating condition of zero net work provides the critical thickness condition, as discussed in Section 6.2. To be more specific, the work of tractions acting on the plane, shown as a dashed line in Figure 6.31, during formation of the dislocation is estimated. The work of the initial stress, corresponding to the strain in (6.65), is $W_m = -M\epsilon_s h_s b_x$ where it has been tacitly assumed that the direction of the Burgers vector is such that $\epsilon_m b_x > 0$, that is, the dislocation relieves background strain.

Next, the self-energy of the dislocation, which is the work required to create the dislocation in the bilayer in the absence of any other loading, must be determined. A mathematical solution of general applicability for a dislocation in a plate with free surfaces is not available. Thus, an estimate of this self-energy is made for a general Burgers vector in order to derive an approximate critical thickness condition. The quality of the estimate is examined in Section 6.7.4. It is well known that a dislocation parallel to a flat free surface in an otherwise unbounded elastic material is subject to an image force which is inversely proportional to the distance of the dislocation line from the free surface. This result is illustrated in Section 6.1, and the configurational force on the dislocation is termed the *half-space image force*. The estimate here is based on the assumption that the total force on a dislocation in the bilayer is the sum of the half-space image forces acting on the dislocation due to the two free surfaces separately. That is, the total image force on a dislocation in the bilayer at a distance η from the free surface of the substrate, and acting in a direction of increasing η as shown in Figure 6.31, is estimated to be

$$F(\eta) = \frac{\mu[b_x^2 + b_y^2 + (1-\nu)b_z^2]}{4\pi(1-\nu)} \left[\frac{1}{h_s + h_f - \eta} - \frac{1}{\eta} \right]. \qquad (6.66)$$

The self-energy of the dislocation at the interface is then the total work which must be done in overcoming this image force as the dislocation is moved from the surface of the substrate to the interface, which is

$$W_d = -\int_{r_0}^{h_s} F(\eta)\, d\eta = \frac{\mu[b_x^2 + b_y^2 + (1-v)b_z^2]}{4\pi(1-v)} \ln \frac{h_s h_f}{r_0(h_s + h_f)}, \qquad (6.67)$$

where r_0 is the core cutoff radius.

The external work done in creating the dislocation in the bilayer initially strained according to (6.65) is $W_d + W_m$. This work is positive if the thicknesses of the layers are sufficiently small, but it can become negative for larger thicknesses. The discriminating condition of zero net work defines the smallest thickness of the film for fixed thickness of the substrate (or vice versa) at which dislocation formation becomes energetically possible. Thus, the *critical thickness condition* based on this reasoning is (Freund and Nix 1996)

$$\boxed{\epsilon_m = \frac{[b_x^2 + b_y^2 + (1-v)b_z^2]}{8\pi(1+v)b_x} \frac{h_s + h_f}{h_s h_f} \ln \frac{h_s h_f}{r_0(h_s + h_f)}.} \qquad (6.68)$$

Several noteworthy features of (6.68) are immediately evident. For one thing, it is symmetric in h_s and h_f, a feature that might have been anticipated in advance. Because there is no fundamental distinction between the film and the substrate, their roles should be interchangeable. A second observation is that, if the condition is examined in the limit of a thin film on a much thicker substrate, then the result obtained in Section 6.2 is recovered, namely,

$$\epsilon_m \to \frac{[b_x^2 + b_y^2 + (1-v)b_z^2]}{8\pi(1+v)b_x} \frac{1}{h_f} \ln \frac{h_f}{r_0} \quad \text{as} \quad \frac{h_s}{h_f} \to \infty. \qquad (6.69)$$

_____

6.7.2 Example: Critical thickness for a compliant substrate

A thin film of SiGe, a cubic material, is epitaxially grown on a Si(100) substrate. Show how the substrate thickness influences the film mismatch strain at which dislocation formation first becomes energetically possible in the film–substrate bilayer as a function of the film thickness. Assume that the system relaxes by the formation of $60°$ dislocations as described in Section 6.2.2, $v = 0.25$ and $r_0 = \frac{1}{2}b$. For this material combination, the mismatch $\epsilon_m < 0$ is negative. Compare the critical conditions for dislocation formation with those that prevail for a thin film on a relatively thick substrate.

Solution:

If $\epsilon_m < 0$, then a possible choice of Burgers vector is $b_x = -b/2, b_y = b/\sqrt{2}, b_z = b/2$. For this case, the critical thickness condition (6.68) for the compliant substrate is plotted as graphs of the magnitude of mismatch strain at which dislocations begin to form in the bilayer versus film thickness. These plots are shown in Figure 6.32 for a range of values of substrate thickness. The critical thickness condition on h_f for a very thick substrate is also

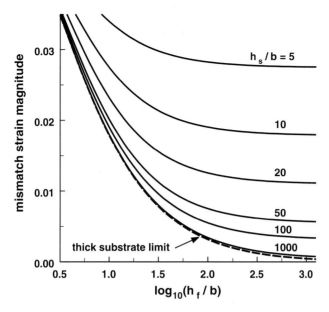

Fig. 6.32. Plots of the magnitude of mismatch strain at which dislocations begin to nucleate in the film–substrate bilayer as a function of the film thickness for different values of substrate thickness. These results were obtained for a cubic system which relaxes by the formation of $60°$ dislocations. The Burgers vector is chosen for a negative mismatch strain in the film.

shown as a dashed line for the case of $h_f/h_s \to 0$, $h_f \neq 0$. It is obvious that if h_f is less than the thin film critical thickness as determined from (6.69), then dislocations should not form no matter how thick the substrate might be. Furthermore, it may still be possible to grow a sub-critical strained film of thickness in excess of the thick substrate critical thickness on a sufficiently thin substrate before the bilayer critical thickness condition (6.68) is violated. For example, if the mismatch strain is 0.01, then the thick substrate critical thickness is approximately $20\,b$. But if the substrate has a thickness of only $h_s = 40\,b$, say, then the compliance of this substrate implies that a film can be grown up to a thickness of about $60\,b$ before the bilayer critical thickness condition (6.68) is exceeded. It is emphasized that this result is based on the assumption of ideal compliance. The degree to which this ideal situation can be approached in practice is uncertain.

6.7.3 Misfit strain relaxation due to a viscous underlayer

The concept of compliant substrate as introduced in Section 6.7.1 is based on the idea of relaxation of misfit strain in a film deposited epitaxially on a thin substrate through viscous flow in a layer between the thin compliant substrate and its support. Such structures have been fabricated by several methods, motivated by the potential advantages of compliance and by the success of the silicon-on-insulator device configurations as a means of electronic isolation of active elements. One approach

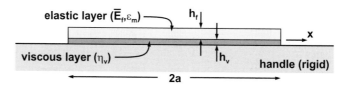

Fig. 6.33. Schematic diagram of a thin elastic film bonded to a thin viscous layer which, in turn, is bonded to a thick substrate. Strain in the layer is relaxed by flow in the viscous layer which progresses in a diffusive manner from the edges of the film inward toward its center.

is to implant oxygen ions into a Si wafer at some depth, with the implant energy chosen to result in a distribution of oxygen within a narrow range of depth centered a few hundred nanometers from the surface. Upon high-temperature annealing, damage caused to the lattice by passage of high energy ions is repaired and the implanted oxygen reacts with the Si to form a SiO_2 layer buried below a crystalline Si layer at the wafer surface. In another approach, the surface of a Si wafer is oxidized to a certain depth. This wafer is then bonded to a second Si wafer with the oxidized surface at the bonded interface. One of the wafers is then etched away from the back side to an etch stop, again leaving the configuration of a thin crystalline layer above the buried oxide layer. In either case, the final configuration is a thin crystalline Si layer a few tens of nanometers thick bonded to a SiO_2 layer of comparable thickness which is, in turn, bonded to a thick handle wafer. The idea behind the compliant substrate concept is that a thin film can be grown epitaxially on the thin crystalline layer at high temperature. If strain is generated due to lattice mismatch, then the strain can be partially relaxed through viscous flow in the oxide layer, thereby diminishing the likelihood of dislocation formation in the layer. The purpose in this section is to obtain an estimate of the time required for such relaxation to occur in terms of the relevant system parameters. The comparison of this relaxation time to the time elapsed in other stages of processing is an important consideration in assessing the potential integrity of material structure.

A simple version of the material configuration of interest is depicted in Figure 6.33. A thin elastic layer with plane strain modulus $\bar{E}_f$ and thickness h_f is bonded to a thin viscous layer with viscosity η_v and thickness h_v. The opposite face of the viscous layer is bonded to a relatively thick base which is assumed to be rigid for purposes of this discussion. The lateral extent of both materials is $2a$ as shown, and the deformation is assumed to occur under two-dimensional plane strain conditions. If the uniform strain ϵ_m exists in the elastic film at time $t = 0$, how much time must elapse before substantial relaxation occurs at the center of the film? As a result of the edges of the film being free of stress, diffusive relaxation of the film stress progresses inward from each free edge toward the center. For this reason, the stress level at the center of the film will be larger than the stress level remaining anywhere else along its length. The corresponding strain relaxation problem with $a/h_{\rm f} \to \infty$ was considered by Elsasser (1969), in the context of deformation of the earth's crust, in order to estimate the rate of stress diffusion inward from an isolated edge.

The configuration of practical interest is slightly more complicated than that depicted in Figure 6.33, and the elastic layer consists of both a film and a substrate with a mismatch in lattice parameter. In that case, the film is initially strained and the compliant substrate is initially stress free. Under these conditions, the long-time strain states in these components once the structure is fully relaxed are those given in (6.65). The aspect of principal interest here is the characteristic time for that relaxation process, and that time is precisely the time estimated in this section, as long as the moduli of the film and substrate materials are comparable in magnitude.

The gradient of film force $h_{\rm f}\sigma(x, t)$ along the length of the elastic film is balanced by the shear stress $q(x, t)$ acting on the interface between the film and the viscous layer, or

$$\frac{\partial \sigma}{\partial x} h_{\rm f} = q \,. \tag{6.70}$$

For simple shear deformation in the viscous layer, the shear stress $q(x, t)$ is proportional to the shearing rate in the layer, which is the speed of the top surface with respect to the bottom surface in a direction parallel to the interface, or

$$q = \frac{\eta_{\rm v}}{h_{\rm v}} \frac{\partial u}{\partial t} \tag{6.71}$$

where $u(x, t)$ is the displacement of the cross-section in the elastic film from its position x in the unrelaxed state of the film. Finally, the extensional strain $\epsilon(x, t) = \sigma(x, t)/\bar{E}_{\rm f}$ in the elastic film is related to the displacement u according to

$$\epsilon = \epsilon_{\rm m} + \frac{\partial u}{\partial x} \,. \tag{6.72}$$

Elimination of u and q from among (6.70), (6.71) and (6.72) yields the single differential equation for $\epsilon(x, t)$ in the form

$$\frac{\partial^2 \epsilon}{\partial x^2} - \frac{\eta_{\rm v}}{\bar{E}_{\rm f} h_{\rm f} h_{\rm v}} \frac{\partial \epsilon}{\partial t} = 0 \tag{6.73}$$

with its boundary and initial conditions

$$\epsilon(\pm a, t) = 0 \quad \text{and} \quad \epsilon(x, 0) = \epsilon_{\rm m} \,, \tag{6.74}$$

respectively.

The boundary value problem represented by (6.73) and (6.74) is an elementary diffusion problem in a finite domain, and a solution is readily obtained in the form of an infinite series by means of separation of variables. Plots of the distribution of

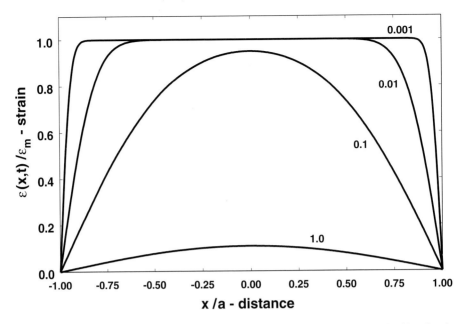

Fig. 6.34. Unrelaxed strain $\epsilon(x, t)/\epsilon_m$ versus distance x/a along the elastic film for four values of normalized time $t\bar{E}_f h_f h_v/a^2 \eta_v$ as determined from solution of (6.73) and (6.74). Unpublished results obtained in collaboration with W. D. Nix, Stanford University.

normalized strain $\epsilon(x, t)/\epsilon_m$ as a function of normalized distance x/a along the film for several values of normalized time $t\bar{E}_f h_f h_v/a^2 \eta_v$ are shown in Figure 6.34. From the plots, it is evident that relaxation begins at the center of the film at normalized time of approximately 0.1, and it becomes fully relaxed at a normalized time of approximately 1.0. The behavior of the stress relaxation solution implies that a characteristic time for the process is

$$t_{char} = \frac{a^2 \eta_v}{\bar{E} h_f h_v}. \tag{6.75}$$

This result makes clear the important influence of configuration size, as represented by a, on the relaxation process. For example, suppose that $a = 1\,\mathrm{cm}$, $\bar{E}_f = 10^{11}\,\mathrm{N/m^2}$ and $h_f = h_v = 100\,\mathrm{nm}$. The borosilicate glass at $800\,^\circ\mathrm{C}$ might have a viscosity as low as $\eta_v = 10^7\,\mathrm{N\,s\,m^{-2}}$, in which case the characteristic time is roughly $10^6\,\mathrm{s}$ or about 10 days. A small metastable material structure held at elevated temperature for that length of time would lose its defining features. At the $a = 100\,\mu\mathrm{m}$ length scale, on the other hand, the stress relief mechanism can be effective.

6.7.4 Force on a dislocation in a layer

In order to examine the mechanics of dislocation formation in a compliant film–substrate system, the energy of a dislocation at an arbitrary position in this structure

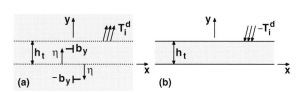

Fig. 6.35. Two edge dislocations in an unbounded elastic plane. (a) Schematic showing the traction distribution, represented by $T_i^{\rm d}$, on the planes $y = 0$ and $y = h_{\rm t}$ which is induced by the dislocations. (b) Complementary boundary value problem which can be solved by the finite element method.

in the absence of any other stress field is required. This energy is equivalent to the configurational force on the dislocation as a function of position through the thickness of the composite layer. In order to determine an approximate form for the critical thickness condition in Section 6.7.1, an ad hoc assumption on the variation of this force was made in (6.66). While the assumed variation of the force is asymptotically correct near either free surface and it has the obvious virtue of simplicity, the quality of the approximation is not evident. Thus, in this section, the variation of this force with position is examined in greater detail.

For the case of a screw dislocation in the layer with free surfaces, a complete solution of the boundary value problem is known (Eshelby 1979). For the notation established in Figure 6.31, the force on a screw dislocation is

$$\bar{F}_z = \frac{2\pi h_{\rm t}}{\mu b_z^2} F(\eta) = \frac{\pi}{2} \tan \frac{\pi \left(\eta - \frac{1}{2}h_{\rm t}\right)}{h_{\rm t}}, \qquad (6.76)$$

where $h_{\rm t} = h_{\rm s} + h_{\rm f}$. The approximation of the magnitude of this force as given in (6.66) is asymptotically exact as $\eta \to 0$ or $\eta \to h_{\rm t}$, and it is a good approximation for any practical purposes over the full range $0 < \eta < h_{\rm t}$.

For the edge component of the Burgers vector, a simple closed form expression for the force $F(\eta)$ comparable to (6.76) is not available. However, an accurate result can be obtained by a numerical simulation based on the finite element method. As in Section 6.6.2, the result is obtained by relying on linear superposition so that the singular field of the dislocation is taken into account analytically and a non-singular boundary value problem is solved numerically. Convergence of the numerical solution is greatly facilitated by including the field of a second dislocation in the superposition scheme as depicted in Figure 6.35, where the origin of coordinates is chosen arbitrarily to lie on the lower face of the layer. The two dislocations in an unbounded elastic plane shown in part (a) of Figure 6.35 induce a traction distribution represented by $T_i^{\rm d}$ on the planes $y = 0$ and $y = h_{\rm t}$. The displacement field diminishes with distance as $(x^2 + y^2)^{-1/2}$ for $(x^2 + y^2)^{1/2} \gg h_{\rm t}$ due to the presence of the second dislocation. The force on the dislocation within the layer in part (a) alone is $F_{\rm a}(\eta) = -\mu b_y^2 / 4\pi (1 - \nu)\eta$.

An accurate numerical solution of the complementary boundary value problem shown in part (b) of Figure 6.35 can be obtained by means of the finite element method. The contribution to the total force on the dislocation from this solution is $F_b(\eta) = b_y \sigma_{xy}(0, \eta)$ where σ_{xy} is the numerically computed shear stress component at the position of the dislocation in part (a). On the basis of dimensional considerations, it is evident that σ_{xy} has a factor μb_y. To see this, note that the traction T_i^d found in part (a) must be linear in b_y and must have the physical dimensions of stress. It follows that each component of T_i^d must have the form of a product of $\mu b_y / h_t$ and a dimensionless factor. Part (b) is a traction boundary value problem in elasticity. In this case, the stress field magnitude at any point in the material scales with the magnitude of the applied traction, but it is otherwise independent of the elastic moduli. It follows that the stress component σ_{xy} from which force on the dislocation is determined has the form of $\mu b_y / h_t$ times a dimensionless factor.

The force on the dislocation has been computed at a large number of discrete points in the range $0 < \eta < \frac{1}{2} h_t$, and then at a small number of points in the range $\frac{1}{2} h_t < \eta < h_t$ to confirm that the result is anti-symmetric with respect to the midplane of the layer. The computed results were then approximated by means of a least squares fit to an analytic form which represents a compromise between simplicity and accuracy. A fit that agrees with all computed data to within 5% is

$$\bar{F}_y = \frac{2\pi(1-v)h_t}{\mu b_y^2} F(\eta) = \frac{(\eta_t - \frac{1}{2})}{\eta_t(1-\eta_t)} + 9.0(\eta_t - \frac{1}{2})\eta_t(1-\eta_t) \qquad (6.77)$$

for $b_x = b_z = 0$, where $\eta_t = \eta / h_t$, and

$$\bar{F}_x = \frac{2\pi(1-v)h_t}{\mu b_x^2} F(\eta) = \frac{(\eta_t - \frac{1}{2})}{\eta_t(1-\eta_t)} - 31.5(\eta_t - \frac{1}{2})\eta_t(1-\eta_t) \qquad (6.78)$$

for $b_y = b_z = 0$.

Graphs of the expressions (6.76), (6.77) and (6.78) for force on a dislocation in a traction free layer for the three components of the Burgers vector are shown in Figure 6.36. A predictable feature of this graph is that the force is zero at the midplane position, as it must be due to symmetry. Therefore, this is an equilibrium position for all three cases. While this equilibrium position is unstable for $b_y \neq 0$ and $b_z \neq 0$, it is surprising to find that this position is a stable equilibrium position for $b_x \neq 0$. Apparently, this is due to the bending stress field which arises in the layer when the dislocation is formed.

Comparison of the more accurate results for force to the ad hoc form assumed in (6.66) for the choice of Burgers vector $b_x = -\frac{1}{2}b$, $b_y = \frac{1}{\sqrt{2}}b$, $b_z = \frac{1}{2}b$ as for a $60°$ dislocation in a material with $\{100\}$ surfaces, confirms that the ad hoc approximation is reliable for this case. For other crystallographic orientations of the

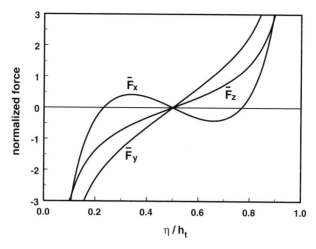

Fig. 6.36. Graphs of the configurational force on the dislocation in a layer associated with each component of the Burgers vector. The expressions for force are given in (6.76), (6.77) and (6.78), where the subscript indicates the nonzero component of Burgers vector in each case.

surfaces of the layer, say $\{111\}$ planes or $\{110\}$ planes, the ad hoc approximation should be applied with caution.

6.8 Dislocation nucleation

The discussion of dislocation formation in the foregoing sections follows the point of view of global stability of a small-scale strained material structure against dislocation formation. For the various configurations considered, relationships among system parameters – elastic moduli, geometrical shape and crystallographic features, for example – that are necessary for insertion of a dislocation into the material structure were established. It was recognized in the course of obtaining such results that they are not sufficient to ensure that dislocations will indeed form once the critical conditions are met. The free energy path in configuration space from a crystallographically 'perfect' material structure to a dislocated material structure does not always proceed along a path of steadily decreasing free energy. Instead, the appearance of a dislocation implies only that the end state has lower free energy than the beginning state. Before a dislocation can relieve strain by motion on its glide plane, it must be nucleated somehow. If there is a significant energetic barrier to nucleation in a particular system, then the process of strain relaxation in that system is controlled by dislocation nucleation and the conditions necessary for dislocation glide take on a less significant role.

The purpose in this section is to consider the question of dislocation nucleation in small-scale epitaxial material structures. This is an issue with many facets and with few clear, unambiguous conclusions. At the simplest level, the question concerns the activation energy for nucleation of the first dislocation to appear in a small structure and comparison of that energy to the available energy, usually in the form of background thermal energy. Such a question is addressed in Section 6.8.1. At a significantly higher level of complexity, all aspects of dislocation behavior are considered together to draw conclusions on ensemble behavior that affects overall strain relaxation. This might include consideration of nucleation statistics, dislocation glide kinetics, dislocation interactions or reactions, and multiplication mechanisms; some illustrations are considered in Chapter 7 but conclusions are largely qualitative. However, such approaches seem to represent the path to be followed in order to reach quantitative or predictive understanding of the connections between growth/processing conditions and the appearance of macroscopically detectable strain relaxation.

These issues have enormous practical ramifications for producing high quality material structures. The most relevant point of view usually depends on the materials involved, the geometrical configuration of the material structure of interest, the steps involved in processing, and the eventual functional environment of the structure. For example, once certain strained quantum structures – planar wells, wires or dots – have been formed on their substrates, they are overgrown epitaxially with a material that is lattice matched to the substrate; a quantum wire illustration was discussed in Section 6.6.2. Circumstances are commonly such that the structures are unstable against dislocation formation when deposited, but are stabilized by the overgrowth. Consequently, the time and temperature sensitivities of dislocation nucleation during the intermediate processing phase are of importance. As another example, consider the situation in which the amount of elastic strain relaxation in a thin film structure is not of paramount importance, but rather the number density of threading dislocations terminating at the free surface in the final structure is important. It is obvious that a certain level of strain relaxation, represented by the net length of interface misfit dislocation created per unit area, can be achieved by having had many threading dislocations within that area each glide a small distance or by having had a small number each glide a large distance. Thus, situations are desirable for which the likelihood of dislocation nucleation is diminished, the likelihood of an increase in dislocation glide speed at a certain level of stress is increased, and the likelihood of impeding dislocation motion upon interactions with other dislocations is diminished. Such effects can be influenced by temperature, strain gradients, growth surface morphology and other factors (Fitzgerald 1995).

In the sections that follow, illustrations of homogeneous and nonhomogeneous dislocation nucleation are presented. The former case implies dislocation formation in a material system that is otherwise spatially uniform; nucleation is equally likely at all locations. Nonhomogeneous nucleation, on the other hand, implies that spatial

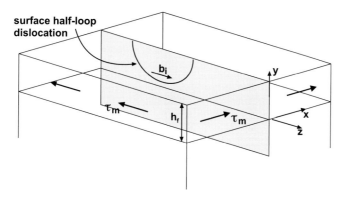

Fig. 6.37. A semi-circular dislocation loop at a free surface with Burgers vector parallel to the surface. The dislocation loop is under the influence of a shear stress τ_m.

nonuniformity arising through configuration, material structure or material defects renders certain sites in the structure far more susceptible to dislocation nucleation than other sites.

6.8.1 Spontaneous formation of a surface dislocation loop

In the case of antiplane shear deformation, the behavior of a semi-circular dislocation loop at a free surface with Burgers vector parallel to the surface under the action of a mismatch shear stress τ_m, as depicted in Figure 6.37, is identical to the behavior of a circular loop in an unbounded solid due to the same shear stress. The traction on the boundary vanishes due to symmetry in the latter case. According to Hirth and Lothe (1982), the energy of formation of such a surface half loop of radius R is

$$W_d(R) = \frac{2-\nu}{8(1-\nu)}\mu b^2 R \ln \frac{4R}{e^2 r_o} \qquad (6.79)$$

for elastic constants μ and ν, Burgers vector of magnitude b parallel to the free surface, and cutoff radius r_o; the quantity e is the natural logarithm base. The value of cutoff radius is assumed to be $\frac{1}{2}b$ throughout this discussion.

The corresponding work done against the background elastic deformation field due to mismatch is

$$W_m(R) = -\frac{1}{2}\tau_m b\pi R^2, \qquad (6.80)$$

where the minus sign reflects the fact that the slip occurs in a sense for which energy is drawn from the elastic energy reservoir due to mismatch in forming the dislocation.

The total work done in forming the dislocation is then

$$W(R) = W_d(R) + W_m(R). \qquad (6.81)$$

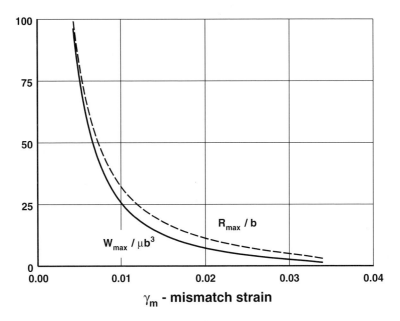

Fig. 6.38. Variation of R_{max}/b and $W_{\mathrm{max}}/\mu b^3$, from (6.82) and (6.83), respectively, as a function of the mismatch strain γ_{m} for $\nu = \frac{1}{4}$ and $r_{\mathrm{o}} = \frac{1}{2}b$.

For very small values of R/b, the behavior of the function is dominated by $W_{\mathrm{d}}(R)$, and $W(R)$ increases from $W(0) = 0$ with increasing R/b. On the other hand, for large values of R/b, the behavior is dominated by $W_{\mathrm{m}}(R)$ as $W(R)$ decreases to ever more negative values. The function assumes a maximum value at some intermediate value of R; this maximum value represents the activation energy that must be supplied through random thermal fluctuations in order to nucleate a dislocation in the form of a surface half loop. The value of R at which this maximum occurs, say R_{max}, satisfies the equation $W'(R_{\mathrm{max}}) = 0$ which implies that

$$\frac{2-\nu}{8(1-\nu)} \ln \frac{8 R_{\mathrm{max}}}{eb} = \pi \gamma_{\mathrm{m}} \frac{R_{\mathrm{max}}}{b} \tag{6.82}$$

for $r_{\mathrm{o}} = \frac{1}{2}b$. Aside from numerical factors, this equation is identical to the critical thickness condition (6.46). The corresponding activation energy is $W_{\mathrm{max}} = W(R_{\mathrm{max}})$ or

$$W_{\mathrm{max}} = \mu b^3 \left(\pi \gamma_{\mathrm{m}} \frac{R_{\mathrm{max}}^2}{b^2} + \frac{2-\nu}{4(1-\nu)} \frac{R_{\mathrm{max}}}{b} \right). \tag{6.83}$$

Graphs of R_{max}/b and $W_{\mathrm{max}}/\mu b^3$ versus γ_{m} are shown in Figure 6.38 for $\nu = \frac{1}{4}$. Based on modeling of this kind, it appears that spontaneous or thermally activated nucleation of the surface dislocation loops in strained films is unlikely (Fitzgerald et al. 1989). An activation energy of $10 \mu b^3$ implies that $W_{\max} \approx 10^{-16}$ J. The corresponding loop radius is roughly $15b$, so the number of atoms which must displace in coordination to form this loop is on the order of 100. The associated thermal energy is roughly $100kT$ or about 10^{-18} J at a typical film growth temperature. Consequently, the likelihood of homogeneous surface loop nucleation is remote.

The activation energy estimates obtained above can be lowered somewhat if it is recognized that thermal fluctuations are needed only to create an unstable stacking fault in the material, rather than a fully formed dislocation (Beltz and Freund 1993). This would result in a reduction of the estimates of activation energy by perhaps a factor of two, but surely not by the two orders of magnitude difference between activation energy and background thermal energy noted above. It is more likely that heterogeneous dislocation nucleation sources are required to produce the dislocation densities observed in real material systems.

6.8.2 Dislocation nucleation in a perfect crystal

The issue of dislocation nucleation in an initially perfect crystal has been studied experimentally by Gouldstone et al. (2001) who used the soap bubble raft as a model system for metal crystals. In their study, a two-dimensional, defect-free closed-packed array of soap bubbles was subjected to indentation in the plane of the raft in an attempt to determine the conditions for the nucleation of dislocations in the highly strained region of indentation. The bubble raft represents an analog of atoms in equilibrium positions in a face-centered cubic crystal, and the interaction among bubbles provides a simple description of the interatomic potential of Cu (Bragg and Nye 1947). When the edge of the raft in contact with the rounded tip of the indenter is 'atomically' smooth, dislocation nucleation is observed at the interior point at which the shear stress is the largest (Johnson 1985). However, when asperities with width and height of a few 'atomic' radii are introduced at the indented surface, dislocation nucleation occurs at the peaks of the asperities where they contact the indenter which has tip radius that is much larger than the asperity dimension, as shown in Figure 6.39(a). When the asperity size is large compared to the atomic dimension but smaller than the indenter tip radius, dislocation nucleation is observed at the reentrant corners of the base of the asperities due to the stress concentration there, as shown in Figure 6.39(b). When the asperity size is comparable to the indenter tip radius, the stress level at the base of the asperities is reduced geometrically. Consequently, dislocation nucleation appears to occur at an interior point, similar to the case of an atomically smooth surface subjected to indentation: see Figure 6.39(c).

On the basis of this analogy, Gouldstone et al. (2001) have studied the possibility of nucleation of dislocations in the interior of face-centered cubic metal single

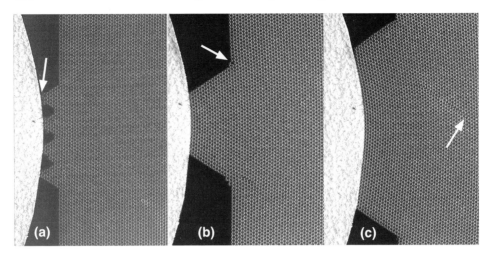

Fig. 6.39. Images of dislocation nucleation in a bubble raft model of a single crystal subjected to surface indentation. In the case of the smallest surface roughness (a), dislocation nucleation occurs at the peaks of the asperities where they contact the indenter. At the intermediate scale (b), dislocation nucleation occurs at the reentrant corners at the bases of the asperities. Finally, for the largest scale asperity (c), the stress level near the stress concentrations has been reduced geometrically, and dislocations nucleated at an interior point where the shear stress is the largest. Each bubble in this raft is 1 mm in diameter and it represents an atom which is approximately 0.3 nm in diameter. Reproduced with permission from Gouldstone et al. (2001).

crystals subjected to indentation. Figure 6.40 is a plot of the indentation load P versus the depth of penetration of the indenter h into a single crystal of pure aluminum along the $\langle 133 \rangle$ crystallographic direction. During this load-controlled indentation, elastic deformation of the crystal is observed until a critical load at which the indenter abruptly sinks into the material, giving rise to a discontinuity or displacement 'burst' in the $P - h$ curve. Post-indentation transmission electron microscopy observations of $\{111\}$-textured polycrystalline Cu films by Gouldstone et al. (2000) revealed that defect nucleation occurs in the highly strained region of indentation through the emission of dislocations. The maximum shear stress induced in the crystal interior by the applied indentation load at the onset of the first displacement burst is on the order of several GPa, or the theoretical shear strength of aluminum, suggesting the possibility of nucleation at the point interior to the crystal. Similar trends have also been noted for Al, Cu, Au and Ir single crystals of several different crystallographic orientations which were subjected to ultra-fine scale indentation (Corcoran et al. 1997, Kiely and Houston 1998, Suresh et al. 1999, Gouldstone et al. 2000, Minor et al. 2001). These results are somewhat surprising, given the expectation that steps, ledges and reentrant corners of asperities at the indented surface would easily provide sites for the heterogeneous nucleation of dislocations. However, the observations shown in Figure 6.39 suggest that nucleation of dislocations

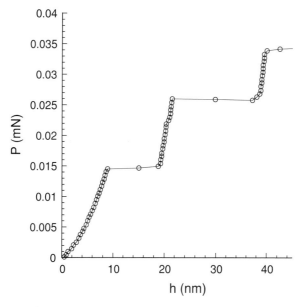

Fig. 6.40. Variation of indentation load *P* as a function of depth of penetration *h* of a three-sided diamond pyramid indenter into the {133} surface of single crystal aluminum. Adapted from Gouldstone et al. (2000).

in the crystal interior is possible when the scale of surface roughness is comparable to the tip radius of the indenter. Because ultra-fine scale depth-sensing instrumented indentation, such as that used to produce the result plotted in Figure 6.40, is commonly performed with a pyramid-shaped diamond indenter with a tip radius on the order of 50 nm, nucleation of dislocations may be possible even in the presence of surface asperities with dimensions larger than 50 nm. Results such as those plotted in Figure 6.40 also indicate that the elastic properties of thin films are unaffected by size scale since the rising portion of the *P* versus *h* curve between the displacement bursts at indenter penetration depths as small as tens of nms can be predicted from knowledge of the elastic properties of the film material obtained by means of specimens of macroscopic dimensions (Gouldstone et al. 2000).

To examine the mechanics of dislocation formation in the interior of a crystal, Li et al. (2002) have adopted a criterion for material stability introduced by Hill (1962) and Rice (1976) for the study of strain localization in nonlinear elastic materials. The basic idea is to examine the deformation field throughout a crystal during an application of indentation loads to its surface and to identify the interior point and the load level at which the material first becomes unstable. In this context, instability is implied by vanishing of the incremental stiffness for any homogeneous deformation, vanishing of the propagation speed of any incremental plane acceleration wave or loss of ellipticity of the elasticity tensor; these criteria are basically equivalent. Physically, the instability is manifested in the deformation field by the formation of the stationary slip band and with a certain orientation and a certain polarization of displacement discontinuity across it.

In a continuum analysis, Li et al. (2002) adopted an approach whereby the non-linear elastic response of the material was based on the known homogeneous deformation response of the single crystal of interest. The stress field throughout the performing elastic crystal is then determined incrementally as the indentation force is applied, and the stability of response at each material point is monitored by calculating the quantity

$$\Lambda\left(\mathbf{k},\ \mathbf{w}\right) = \left(c_{ijkl}^{0} w_i w_k + \sigma_{jl}^{0}\right) k_j k_l > 0, \tag{6.84}$$

where σ_{ij}^{0} is the local value of the stress tensor at the current level of indentation load, c_{ijkl}^{0} is the local value of the incremental elasticity tensor, k_i is the propagation vector of the test acceleration wave and w_i is the displacement polarization direction of the acceleration wave. Material response is locally stable as long as $\Lambda > 0$ for all possible orientations of k_i and w_i. The response becomes locally unstable once $\Lambda = 0$ for any possible combination of k_i and w_i.

Physically, the onset of instability at the point is interpreted as the formation of a crystal defect there. It should be noted that local instability does not imply global instability of the configuration. If the angle between the vectors $\mathbf{k}$ and $\mathbf{w}$ established in this way is larger than $60°$, then the nucleated defect is expected to be a dislocation with slip plane normal vector $\mathbf{k}$ at Burgers vector approximately along $\mathbf{w}$. On the other hand, if the angle between $\mathbf{k}$ and $\mathbf{w}$ at instability is less than $30°$, the nucleated defect would likely be a microcrack or a void.

Li et al. (2002) have implemented this defect nucleation criterion in a molecular dynamics simulation of an fcc crystal undergoing surface indentation for purposes of comparison to the continuum elasticity results. It was found that the vectors $\mathbf{w}$ and $\mathbf{k}$ which render $\Lambda = 0$ agree well with the slip direction and slip plane normal, respectively, determined in the molecular dynamics simulations. These simulations also provide insights into the mechanisms by which the first discontinuity during indentation triggers nucleation of dislocations at GPa-level local stresses and into the dislocation interaction processes by which subsequent discontinuities occur at MPa-level local stresses. Although the foregoing discussion has been presented in the context of indentation, the criterion for defect formation is expected to be applicable for a wide variety of deformation conditions.

6.8.3 Effect of a stress concentrator on nucleation

The circumstances of formation of a dislocation loop at the surface of a strained film considered in Section 6.8.1 presumed no conditions to make dislocation nucleation more likely at one location than at another. This is the essential characteristic of homogeneous nucleation. In reality, epitaxial films are never geometrically perfect. Irregularities can take the form of surface or interface features (Jesson et al. 1993), precipitate particles due to chemical impurities (Perovic et al. 1989), residual oxide or other debris on the growth surface (Tuppen et al. 1989), grown-in stacking faults (Eaglesham et al. 1989), and so on. Any such irregularity has the potential to elevate the stress field in the material locally, thereby making dislocation nucleation more likely in the vicinity of the irregularity than elsewhere. Nucleation under circumstances of threading dislocations leaving behind interface misfit dislocations as they glide away from debris particles from which they were nucleated has been modeled by Zhang and Yang (1993).

The range of circumstances that have the potential for giving rise to nonhomogeneous nucleation is enormous, and models intended to assess the effectiveness of mechanisms quantitatively can become very complex. Here, only a relatively simple model of nonhomogeneous nucleation is considered. Study of this model leads to a basis for some conclusions concerning dislocation nucleation of general applicability for strained films. The features of the model are depicted in Figure 6.41; for the most part, these features are retained from the material system represented in Figure 6.37 and considered in the preceding section. The main difference is that the free surface of the film is no longer flat. Instead, the surface shape is perturbed by a cylindrical groove that is semi-circular in cross-section and of radius a. The value of a is assumed to be smaller than the film thickness h_f. The groove provides a site of potential local stress concentration. Retaining the simplicity of shear loading exploited in Section 6.8.1, the shear stress component σ_{xz} on the glide plane extending into the film from the base of the groove along the plane $x = 0$ varies as

$$\sigma_{xz} = \tau_m \left(1 + \frac{a^2}{r^2} \right), \quad r \equiv h_f - y \geq a . \tag{6.85}$$

Far from the groove, the film stress has the magnitude τ_m, while the magnitude is $2\tau_m$ at the root of the groove at $r = a$. In other words, the stress concentration or amplification factor for the groove is 2. It follows immediately that dislocation nucleation is more likely from the base of the groove than from the surface of the uniform film. The enhanced propensity for nucleation can be examined quantitatively, following the discussion of Freund et al. (1989).

Suppose that the semi-circular glide dislocation loop of radius R and Burgers offset b parallel to the groove axis emerges from the base of the groove into the film as depicted in Figure 6.41. The free surface condition is somewhat different here from that for the case of homogeneous nucleation, but the expression for self-energy of formation W_d given in (6.79) is again adopted for convenience.

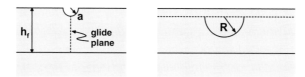

Fig. 6.41. Schematic diagram of dislocation nucleation from the base of a cylindrical groove in the surface of a strained film. The radius of the groove of semi-circular cross-section is a and the radius of the semi-circular dislocation line is R.

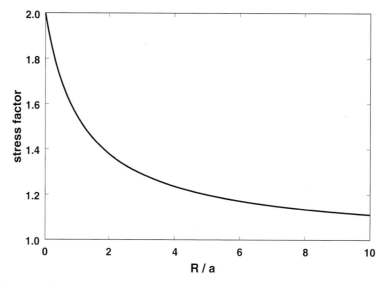

Fig. 6.42. Plot of the stress amplification factor enclosed in square brackets in (6.86). The stress magnitude is effectively doubled by the presence of the groove when R/a is small, but the effect is diminished as R/a becomes larger.

The work done against the background elastic deformation field, equivalent to W_m in (6.80) for homogeneous nucleation, is more involved in this case because the stress field varies nonuniformly over the glide plane prior to dislocation formation, as indicated in (6.85). This work of formation is readily determined to be

$$W_m(R) = -\frac{1}{2}\tau_m b\pi R^2 \left[1 + \frac{2}{\pi}\frac{a^2}{R^2}\int_0^\pi \int_0^{R/a} \frac{1}{(1+s\sin\phi)^2}s\,ds\,d\phi \right]. \quad (6.86)$$

The integral expression in (6.86) can be evaluated in terms of elementary functions, but the result does not render the behavior transparent. Instead, the dependence of the nondimensional quantity enclosed in square brackets in (6.86) on R/a is illustrated graphically in Figure 6.42. This factor has the value 2 for $R/a \to 0$ and it decreases monotonically as R/a increases. As was anticipated by the stress concentration factor of 2, the effect of the groove is to double the work done by the background elastic field in forming the dislocation, compared to the case of homogeneous nucleation, when R/a is small. The magnitude of the effect is diminished, however, as R/a becomes larger. The result leads immediately to the following general observation. Qualitatively, if the loop can be expanded to its unstable equilibrium size within the region of stress concentration of the notch, then the presence of the groove has a strong effect on activation energy. However, if the equilibrium size of the loop is large compared to the size of the groove, then the presence of the groove has little effect on the process of dislocation loop nucleation.

A number of additional trends can be noted. For example, if the stress concentration factor is larger than 2 – it is approximately 3 for a circular groove in a tension field – then the site is more effective as a source of dislocations. If the configuration of the stress concentrator is a notch with a very high curvature of the notch surface at its root, then the stress concentration factor can be very large compared to 2, but the spatial extent of the localized stress field is significantly reduced from that of the circular stress concentrator. For the case of a planar crack, which is the ultimate sharp notch, the issue of dislocation nucleation has been modeled by Rice and Thomson (1973) and Rice and Beltz (1994). Similar techniques have been adapted for the study of dislocation nucleation at the edge of an epitaxial island (Johnson and Freund 1997).

6.9 Exercises

1. A long straight edge dislocation with Burgers vector in the $x-$direction lies along the $z-$axis in an unbounded elastic solid, resulting in a state of plane strain deformation.

   (a) Suppose that an identical edge dislocation with Burgers vector in the $x-$direction is formed along the line $x = \xi$, $y = 0$. Calculate the work done by the stress field of the first dislocation in forming the second dislocation; this work is the *energy of interaction* between the two dislocations, say W_{int}.
   (b) Evaluate the configurational force $-\partial W_{\text{int}}/\partial \xi$ acting on the second dislocation due to its proximity to the first dislocation. Is this force attractive or repulsive?
   (c) Discuss how the configurational force would be affected by a change in sign of the Burgers vector of either dislocation or a change in sign of the Burgers vector of both dislocations.

2. The method of images is a technique for constructing the elastic fields for dislocations, as well as other singularities of various types, in bounded regions by means of superposition of dislocation fields for unbounded regions. The method is applicable in cases for which the boundary conditions are satisfied as a result of symmetry in the superimposed fields.

   (a) The stress components for a straight screw dislocation in an elastic half-space are given in (6.4) for the case in which the dislocation line is parallel to the free surface and at a distance η from it. Show that the stress field is the superposition of the fields of two parallel screw dislocations in an unbounded solid, one along the line $x = 0$, $y = 0$ with Burgers vector b_z

and the other along $x = 0$, $y = 2\eta$ with Burgers vector $-b_z$. Confirm that the superimposed stress fields yield a stress free boundary along $y = \eta$.

(b) For the case of an elastic plate of thickness h_t, the configurational force acting on the dislocation due to its proximity to the free surfaces of the plate is given in (6.76). Show that this expression can be obtained by means of a construction involving an infinite series of image dislocations. Verify that the stress field obtained by this construction satisfies the conditions of vanishing traction on the plate surfaces.

3. A new experimental technique for substrate curvature measurement is proposed by an inventor. It is claimed that this high precision technique is capable of detecting the critical curvature at which misfit dislocations nucleate in a film–substrate system by monitoring continuously the changes in curvature during the epitaxial growth of the film on the substrate and by identifying very small discontinuities in the curvature versus film thickness plot associated with the nucleation of dislocations. The inventor attempts to demonstrate the accuracy of the method by identifying the critical value of substrate curvature during the deposition of a cubic epitaxial film on a 0.7 mm thick Si(100) substrate. The film mismatch strain is –0.4%, the biaxial modulus of the film is 96% of that for the substrate, and the Poisson ratio of the film is 0.25. The Burgers vector magnitude for misfit dislocations in this system is approximately 0.5 nm, and the dislocation core cut-off radius $r_0 \approx \frac{1}{2} b$. The inventor claims that the radius of curvature of the substrate is 880 m when the critical thickness condition is reached, that the substrate is concave on the face away from the bonded film, and that the critical film thickness for dislocation formation is within 10% of the value that can be expected from the theory developed in Section 6.2. Are the inventor's claims valid?

4. The lattice parameter of a film material is 0.6% larger than that of a substrate material. In one set of experiments, a strained buried quantum wire, similar to the arrangement shown in Figure 6.23, is patterned inside the substrate material. The rectangular wire, with $w/h = 5$, is epitaxially bonded to its surrounding matrix. In another experiment, a continuous layer of the same film material, of thickness h, is epitaxially grown on the same substrate material. In both sets of experiments, the free surface is parallel to the $\{100\}$ crystallographic plane of the cubic film and substrate materials, and the preferred slip plane is the $\{111\}$ crystallographic plane, which is oriented at an angle of $54.7°$ to the free surface. Is the mismatch strain required to nucleate the misfit dislocation in the first experiment higher, lower or the same as that in the second experiment? If it is different, by how much?

5. In discussing the influence of crystallographic orientation on the stability of a strained epitaxial film against dislocation formation in Section 6.2.5, it was observed that, in some orientations, there is a natural tendency for dislocations to dissociate into partial dislocations. Consider the case of glide of a dislocation with net Burgers vector $\frac{1}{2}b[\bar{1}01]$ and line orientation $[\bar{1}01]$ on the plane (111).

(a) If the complete dislocation separates into partial dislocations with Burgers vectors $\frac{1}{6}b_{p1}[\bar{2}11]$ and $\frac{1}{6}b_{p2}[\bar{1}\bar{1}2]$, determine the magnitudes b_{p1}/b and b_{p2}/b.

(b) Suppose that a long, straight dislocation splits into parallel partial dislocations in an unbounded crystal that is otherwise free of stress. If the energy per unit area of the stacking fault between the partial dislocations is γ_{st}, determine the equilibrium separation distance between the partials at which the repulsion due to elastic interaction is balanced by the attraction due to the stacking fault energy.

(c) Discuss whether a strained layer is more or less stable against formation of partial dislocations than against formation of complete dislocations.

6. A thin film is deposited epitaxially onto a substrate with a linear compositional gradient in mismatch strain as indicated in (6.40). Determine the relationship between the value of mismatch strain ϵ_{top} at the free surface of the film and the largest value of thickness h_f of the film for which the configuration is stable against misfit dislocation formation.

7. The strain profiles shown in Figure 6.34 are based on the solution of the partial differential equation (6.72) and its boundary conditions (6.74). Determine the solution of this equation by means of the method of separation of variables. Examine the behavior of the midpoint strain $\epsilon(0, t)/\epsilon_m$ as a function of time implied by the solution, and confirm the result for the characteristic relaxation time given in (6.75).

8. Spontaneous formation of a semicircular dislocation loop was analyzed in Section 6.8. Consider a thin film on a substrate for which the mismatch strain is ϵ_m.

(a) In some situations, it may be assumed that the nucleation rate for dislocation loops is of the form

$$\dot{N} \sim \frac{A_{dn}}{\Omega} \exp\left[-\mathcal{E}_{act}/kT\right], \qquad (6.87)$$

where $\dot{N}$ denotes the number of dislocation nucleation events per unit volume per unit time, A_{dn} is a constant with units of inverse time, $\mathcal{E}_{act}$ is the activation energy for nucleation, Ω is the atomic volume, k is the Boltzmann constant and T is absolute temperature. Derive an expression for the approximate dependence of the dislocation nucleation rate on the mismatch strain and temperature.

(b) A change in film deposition process is introduced as a consequence of which the film processing temperature is changed from $500\,°C$ to $700\,°C$. Approximately what relative change in the film mismatch strain should occur, in response to this process change, in order that the dislocation nucleation rate remains fixed.

7

Dislocation interactions and strain relaxation

The issue of dislocation formation in a strained epitaxial heterostructure was the focus of attention in the preceding chapter. Residual stress was assumed to originate from the combination of a mismatch in lattice parameters between the materials involved and the constraint of epitaxy. The discussion in Chapter 6 led to results in the form of minimal conditions which must be met by a material system, represented by a geometrical configuration and material parameters, for dislocation formation to be possible. Once the values of system parameters are beyond the point of fulfilling such minimal conditions, dislocations begin to form, propagate and interact. The ensemble behavior is usually termed strain relaxation.

There are several practical aspects of strain relaxation that originate from the small size scales involved, the relatively low dislocation densities that are observed, and the fact that kinetic processes occur on a timescale comparable to growth or processing timescales. Can significant strain relaxation be suppressed? Can threading dislocation densities be controlled? Under what conditions can ensemble dislocation behavior be captured by a continuum plasticity representation? How do length scales associated with geometrical configuration and microstructure, such as film thickness and grain size, respectively, influence the process of strain relaxation?

Progress toward resolving such questions is summarized in this chapter. The discussion begins with the issue of fundamental dislocation interaction phenomena and nonequilibrium behavior of interacting dislocations. Attention is then shifted from consideration of films with low dislocation density to the modeling of inelastic deformation of thin films with relatively high densities of dislocations. For this purpose, constitutive models for time-independent and time-dependent deformation of thin films are examined by appeal to continuum plasticity theory. Overall features of material behavior captured by such theories are then compared with available experimental results and observations, and some general trends pertaining to the size dependence of inelastic deformation in thin films are extracted. Commonly

464

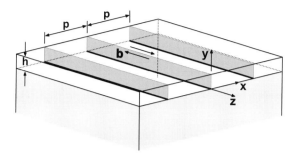

Fig. 7.1. Schematic representation of a periodic array of parallel misfit dislocations at the interface between a strained layer of thickness h and its substrate. The assumed glide planes are shown shaded. The mismatch strain is a shear strain $\gamma_{xz} = \gamma_m$, the common Burgers vector of length b lies along the dislocation line, and the dislocation spacing is p.

used experimental methods for probing the inelastic properties of thin films are described along with a brief summary of the advantages and limitations of each method.

7.1 Interaction of parallel misfit dislocations

Consider a strained thin film on a relatively thick substrate, and suppose that the film thickness is beyond the critical thickness as defined in Chapter 6. If a misfit dislocation is formed at the film–substrate interface by glide on some crystallographic plane, it does so by drawing elastic energy from the background elastic mismatch strain field and, as a result, strain is partially relaxed in the vicinity of the dislocation line. At some distance from the dislocation line, large compared to film thickness, the elastic strain remains essentially unrelaxed from its initial value based on lattice mismatch. Consequently, it is likely that other misfit dislocations will form at some distance away on parallel glide planes. The question addressed in this section concerns the spacing of such dislocations for a given set of system parameters. To provide quantitative answers, attention is limited to periodic arrays of equally spaced dislocations. Furthermore, to avoid having the main ideas obscured by algebraic complexity, the discussion is based largely on the case of residual shear strain γ_m in the film and Burgers vector orientations appropriate for screw dislocation models. Results for more general Burgers vectors are summarized and original literature sources are cited.

The physical system studied is depicted in Figure 7.1. A thin film of thickness h is epitaxially bonded to a relatively thick substrate, and the lateral extent of the interface is assumed to be very large compared to h. The lattice mismatch between the film and substrate materials is represented by the mismatch shear strain γ_m. Prior to formation of any dislocations in the film, the elastic strain in the film is uniform and is given by $\epsilon_{xz} = \frac{1}{2}\gamma_m$, $\gamma_{yz} = 0$. Dislocation formation occurs at the expense of the energy stored in this strain field. The critical thickness condition for dislocation formation is given in (6.46) for general cutoff radius r_0, and it is restated here for the particular value $r_0 = \frac{1}{2}b$ as

$$\frac{b}{4\pi h_{cr}} \ln \frac{4h_{cr}}{b} = \gamma_m, \qquad (7.1)$$

where b is the magnitude of the Burgers vector. The sense of the Burgers vector must be consistent with a reduction of elastic energy in the film due to dislocation formation. In other words, γ_m and b must have the same algebraic sign in (7.1). Without loss of generality, it is assumed that both $\gamma_m > 0$ and $b > 0$ for the remainder of this discussion. It is implied in Figure 7.1 that the film thickness is such that $h > h_{cr}$, and that a periodic array of identical dislocations has been formed at the film–substrate interface. The spacing between adjacent dislocations is p. Given that such an array exists, what is the smallest value for p based on overall work and energy considerations? This question is addressed in several ways in the subsections that follow.

7.1.1 Spacing based on mean strain

A rough estimate of the minimum equilibrium spacing of the dislocations in a periodic array can be obtained by basing the calculation on the mean elastic strain in the film, rather than on the spatially nonuniform elastic strain field that actually exists. While the strain field due to the dislocations in the film in Figure 7.1 is highly nonuniform, the average displacement gradient per period in the array is necessarily b/p. This is the mean value of the shear strain $2\epsilon_{xz}$ due to dislocation formation and it represents a *reduction* in average value of elastic strain from γ_m to $(\gamma_m - b/p)$. If it is assumed that the current film thickness h is the critical thickness for this reduced background elastic strain, then (7.1) is reinterpreted, in accordance with the mean field consideration, as

$$\frac{b}{4\pi h} \ln \frac{4h}{b} = \gamma_m - \frac{b}{p}. \qquad (7.2)$$

This provides a relationship between the film thickness h and the dislocation spacing p for all $h > h_{cr}$. As $h/b \to \infty$, the result implies that the dislocation spacing approaches b/γ_m or, equivalently, that all strain is relieved in the film. The behavior implied by (7.2) is plotted in Figure 7.2 in the form of $\ln(p/b)$ versus $\ln(h/b)$. Clearly, the critical thickness condition is recovered from (7.2) as $p/b \to \infty$. As h/b becomes very large, (7.2) implies that all elastic strain is eventually relaxed for a relatively thick film. Figure 7.2 also shows estimates of equilibrium spacing of misfit dislocations obtained by recourse to other lines of reasoning, which are described in the following paragraphs.

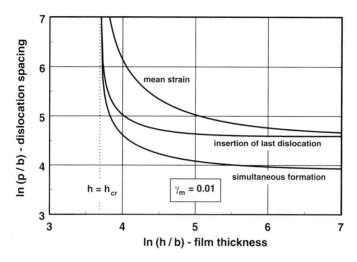

Fig. 7.2. Plots of equilibrium spacing $\ln(p/b)$ of a periodic array of misfit dislocations versus film thickness $\ln(h/b)$ for a mismatch strain of $\gamma_m = 0.01$. The result based on mean strain is given in (7.2), on simultaneous dislocation formation in (7.6), and on insertion of the last dislocation in the array in (7.12).

7.1.2 Spacing for simultaneous formation of dislocations

An estimate of the minimum value of p can also be obtained by following the approach adopted for calculation of the critical thickness condition in Section 6.2. According to this approach, it is necessary to compare the work required to form each dislocation in the periodic array in the absence of any other stress field to the work done by the background mismatch stress field in forming that dislocation. These two work quantities will again be denoted by W_d and W_m, respectively, and they are measured per unit length of dislocation line per period. As before, W_d is the dislocation self-energy and W_m is the interaction energy between the dislocation field and the background mismatch strain field. The self-energy in this case must be understood to be equal to the work required to form a dislocation in a cell of width p subjected to periodic boundary conditions.

The value of $W_m = -\tau_m bh$ is unchanged from that in (6.45). The background stress which does work as each dislocation forms is still τ_m and the displacement offset as a result of dislocation formation is still b.

The calculation of W_d is a bit more involved, but is again based on the fundamental expression

$$W_d = -\int_{\frac{1}{2}b}^{h} \frac{1}{2}b\sigma_{xz}(0, y)\,dy,\qquad(7.3)$$

which is a particular case of (6.6) for the problem at hand, with the cutoff radius chosen as $r_0 = \frac{1}{2}b$ (see Section 6.1). The stress distribution $\sigma_{xz}(0, y)$ on the glide plane of the dislocation forming at $x = 0$ that appears in (7.3) can be obtained by superposition from the elastic field of a single dislocation as given in (6.4), such that

$$
\begin{aligned}
\sigma_{xz}(0, y) &= -\frac{\mu b}{2\pi} \sum_{n=-\infty}^{\infty} \left[\frac{y}{y^2 + n^2 p^2} - \frac{y - 2h}{(y - 2h)^2 + n^2 p^2} \right] \\
&= -\frac{\mu b}{2\pi} \left[\frac{\pi}{p} \coth \frac{\pi y}{p} - \frac{\pi}{p} \coth \frac{\pi(y - 2h)}{p} \right].
\end{aligned} \tag{7.4}
$$

Evaluation of the integral in (7.3) yields

$$
W_{\mathrm{d}}(h, p) = \frac{\mu b^2}{4\pi} \ln \left(\frac{2p}{\pi b} \sinh \frac{2\pi h}{p} \right). \tag{7.5}
$$

As anticipated, this result reduces to (6.44) as $p/b \to \infty$.

For a given value of h, the minimum value that p can have for spontaneous dislocation formation is again established by the condition that $W_{\mathrm{d}} + W_{\mathrm{m}} = 0$ or

$$
\frac{b}{4\pi h} \ln \left(\frac{2p}{\pi b} \sinh \frac{2\pi h}{p} \right) = \gamma_{\mathrm{m}}. \tag{7.6}
$$

This condition also reduces to the critical thickness condition (7.1) as $p/b \to \infty$.

The behavior implied by (7.6) is also plotted in Figure 7.2 in the form of $\ln(p/b)$ versus $\ln(h/b)$ and it is labeled 'simultaneous formation'. The reason for this label becomes evident if the process of formation of the dislocation array is considered to be a consequence of glide of threading dislocations. With this point of view, the arguments outlined in Section 6.3 again imply that

$$
G(h, p) = -[W_{\mathrm{d}}(h, p) + W_{\mathrm{m}}(h, p)] \tag{7.7}
$$

is the configurational force or driving force on each threading dislocation in the array as that dislocation glides, leaving behind its corresponding interface misfit dislocation. Furthermore, because all dislocations are treated in exactly the same manner, it follows that all threading dislocations move through the film in perfect unison when the result is viewed from the mechanistic point of view. This is a particularly unrealistic aspect of this approach, not only because such a degree of coordination is unlikely but also because the configuration with all threading dislocations positioned along the line is probably unstable. It is likely that the free energy of the system could be reduced simply by letting the threading dislocations break rank and move in a less coordinated way. Furthermore, as h/b becomes large, (7.6) implies that $p \to b/2\gamma_{\mathrm{m}}$. If this were so, then the final elastic strain after all dislocations were formed would actually be *opposite* in sense to γ_{m}, or the film would be over-relaxed. These unrealistic aspects of (7.6) suggest an alternative and more realistic point of view for estimating p based on the notion of sequential formation of the dislocations in the array.

7.1.3 Spacing based on insertion of the last dislocation

Consider, once again, the formation of the periodic array of parallel misfit dislocations shown in Figure 7.1. In the present instance, however, the dislocations do not all form simultaneously. Instead, it is assumed that all dislocations have formed with spacing p except for one member of the array. This leaves a gap of width $2p$ in the array which may accommodate one more misfit dislocation. What is the smallest value of p for which the final dislocation can be inserted spontaneously? In the discussion to follow, it is assumed without loss of generality that the last dislocation formed is the one at $x = 0$ in the array.

In providing an answer to the question, three independent equilibrium stress fields must be considered, rather than the two fields considered in similar cases up to this point. These equilibrium fields are the background mismatch field characterized by τ_m or γ_m, the net field of all previously formed dislocations, and the self-stress field of the last dislocation being inserted. Thus, in analogy with (6.46), the configurational driving force on the threading segment of the last dislocation being inserted incorporates three contributions,

$$G(h, p) = -[W_d(h, p) + W_m(h, p) + W_a(h, p)],\qquad(7.8)$$

where W_d is again the self-energy contribution given in (6.44) and W_m is the interaction energy between the background field and the dislocation as it forms, as given in (6.45). The remaining contribution W_a on the right side of (7.8) is the work of interaction between the dislocations formed previously and the one just being inserted. This work of interaction is

$$W_a(h, p) = -\int_0^h b\sigma_{xz}^{(a)}(0, y)\,dy,\qquad(7.9)$$

where $\sigma_{xz}^{(a)}(0, y)$ is the stress distribution on the prospective glide plane $x = 0$ of the last formed dislocation in the array. This stress distribution is readily constructed by subtracting the field of a single dislocation at $x = 0$, $y = 0$ from the field (7.4) of the complete array to obtain

$$\sigma_{xz}^{(a)}(0, y) = -\frac{\mu b}{2\pi}\left[\frac{\pi}{p}\coth\frac{\pi y}{p} - \frac{1}{y} - \frac{\pi}{p}\coth\frac{\pi(y - 2h)}{p} + \frac{1}{y - 2h}\right].\qquad(7.10)$$

Evaluation of the integral expression in (7.9) yields

$$W_a(h, p) = \frac{\mu b^2}{2\pi}\ln\left(\frac{p}{2\pi h}\sinh\frac{2\pi h}{p}\right)\qquad(7.11)$$

for this interaction energy.

All three contributions to the driving force G defined in (7.8) are now known in terms of h and p. If the available energy supply is more than adequate to drive the formation of the last dislocation in the array, then conditions are such that $G > 0$. On the other hand, if the energy supply is inadequate, $G < 0$ and any threading dislocation inserted will retract rather than advance. The critical or discriminating condition between these two types of behavior is defined by $G = 0$ or

$$\frac{b}{4\pi h} \ln \frac{4h}{b} + \frac{b}{2\pi h} \ln \left(\frac{p}{2\pi h} \sinh \frac{2\pi h}{p} \right) = \gamma_{\mathrm{m}}. \tag{7.12}$$

For a given value of h, the value of p that satisfies this equation yields a minimum possible spacing for which the last dislocation can be inserted into the periodic array. The behavior implied by (7.12) is plotted in the form of $\ln(p/b)$ versus $\ln(h/b)$ for $\gamma_{\mathrm{m}} = 0.01$ in Figure 7.2.

This result can be compared directly with similar results obtained from consideration of mean strain relaxed by the dislocations and of the simultaneous formation of all members of the dislocation array. Examination of the expression (7.12) reveals that the mean stress approximation is accurate only for $p/h \ll 1$. In terms of the results plotted in Figure 7.2, this is the case in a region roughly to the right and below the straight line $\ln(p/b) \approx \ln(h/b) - 2$. Comparison of the condition for insertion of the last dislocation in a periodic array in Figure 7.2 with the corresponding result for simultaneous formation adds support to the suggestion that the premise of the latter point of view is unrealistic and that the result is flawed. This is an issue requiring further study from the mechanistic point of view.

7.2 Interaction of intersecting misfit dislocations

The preceding section was concerned with the interaction of dislocations that form on parallel glide planes in a strained film. These interactions arise through the elastic fields of the dislocations, particularly through the influence of the elastic field of a misfit dislocation on those strain-relieving dislocations that form subsequently. The same general issue is again addressed in this section, but with a focus on the interaction of dislocations that form on *intersecting* glide planes. This is a more complex interaction to investigate, mainly because the feature of translational invariance of the configuration in one direction along the interface is lost. Nonetheless, the general nature of the interaction can be understood. The discussion is motivated by considering the behavior of a thin film of a cubic material that is epitaxially bonded to a cubic substrate with the shared {100} surface as the interface. An equi-biaxial strain in the film is presumed to exist due to a mismatch in film and substrate lattice parameters. A pyramid of {111} planes on which dislocation glide can occur in the film is shown in Figure 7.3. The cube edge directions are [100] and [010] in the interface and the direction normal to the interface is [001]. The possible orientations

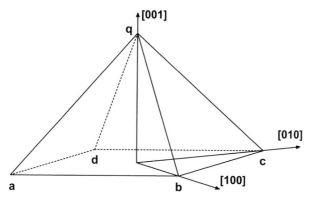

Fig. 7.3. Pyramid of {111} planes in a cubic film material epitaxially bonded to a cubic substrate with a (001) interface. The planes represent potential glide planes for strain-relieving dislocations in the film, and the edges of the pyramid extending from the apex identify the directions of possible Burgers vectors of 60° dislocations. The edges of the cube cell are identified by the triad of crystallographic vectors.

of Burgers vectors of the glide dislocations are the four edges qa, qb, qc and qd extending from the apex of the pyramid, and the lines of interface misfit dislocations lie along the $\langle 110 \rangle$ directions in the interface.

Suppose that an interface misfit dislocation has been formed by glide of a threading dislocation and that this misfit lies along the line bc, that is, along $[\bar{1}10]$. It could have been formed by glide on the plane qbc with Burgers vector along either qb or qc, or by glide on a plane parallel to qad with Burgers vector along either qa or qd. The direction of the Burgers vector must be consistent with relaxation of the background mismatch strain. Subsequently, a threading dislocation approaches this misfit dislocation, leaving behind an interface misfit dislocation with its line orthogonal to the one in its path which was formed earlier, say along the line ad in the $[1\bar{1}0]$ direction. The threading dislocation glides on the plane qab with Burgers vector along one of the edges qa or qb. Again, the sense of the Burgers vector must be consistent with relaxation of the mismatch strain.

When the threading dislocation is far from the misfit dislocation in its path, its driving force depends only on the background mismatch stress field and geometrical parameters. More precisely, it depends on the shear stress acting on its glide plane due to the background stress field. As the threading dislocation approaches the misfit dislocation in its path, this shear stress is altered by the stress field of the misfit dislocation and this effect, in turn, alters the driving force. This complex interaction process has been discussed in detail by Freund (1990b). The misfit dislocation can have any of four Burgers vectors, as noted above, and the approaching threading dislocation can have either of two possible Burgers vectors so that there are eight possible combinations to be considered. In some cases, the dislocations may react with each other, say if their Burgers vectors are colinear or if they split into certain combinations of partial dislocations. A complete study of the interaction would require determination of the ever-changing shape of the threading segment as it travels through the nonuniform stress field of the misfit dislocation, as well as determination of any deflection of the misfit dislocation out of the interface. Progress in this direction on the basis of detailed computational models has been reported by Schwarz and Tersoff (1996). That blocking does indeed occur has been demonstrated experimentally. The possibility of blocking was suggested on the basis of transmission electron microscopy images by Hull et al. (1989) and clear evidence in the form of direct observation has been provided by Stach et al. (1998). A transmission electron microscopy image from this work is shown in Figure 7.4; the phenomenon of blocking of the threading dislocation by an interface misfit dislocation is evident in the image.

7.2.1 Blocking of a threading dislocation

The issue of blocking of a threading dislocation is investigated here on the basis of screw dislocation models, following the general approach outlined by Freund (1990c) for a buried layer and by Freund (1990b) for interaction of 60 degree dislocations in an equi-biaxial strain field in a film with a free surface. The model system is illustrated in Figure 7.5. A film of thickness h is bonded to a relatively thick substrate. The film supports a mismatch shear strain $\epsilon_{xz}^m = \frac{1}{2}\gamma_m = \frac{1}{2}\tau_m/\mu$. An interface misfit dislocation with its line oriented parallel to the x-axis is assumed to exist from the outset. Then, a threading dislocation on a perpendicular glide plane is considered as it approaches the misfit dislocation. In the figure, the glide plane of the threading dislocation lies in the yz-plane; the misfit dislocation it leaves behind as it advances lies along the z-axis.

When the threading segment is still far from the misfit dislocation in its path compared to h, its behavior is that of an isolated threading dislocation in the strained film. As it approaches the misfit dislocation, however, the shear stress acting on its glide plane is altered from the field consisting only of its self-stress and the uniform background mismatch stress. The alteration in stress field is readily determined from (6.4) by replacing the coordinate x with the coordinate z in the expressions for stress to be consistent with the orientation shown in Figure 7.5. Thus, the additional stress acting on the glide plane due to the misfit dislocation is

$$\sigma_{xz}^{(md)}(y, z) = -\frac{\mu b}{2\pi}\left[\frac{y}{y^2 + z^2} - \frac{y - 2h}{(y - 2h)^2 + z^2}\right]. \qquad (7.13)$$

The shear stress acting on this glide plane due to the mismatch field is $\sigma_{xz} = \tau_m$, $\sigma_{xy} = 0$. The additional shear stress due to the stress-relieving misfit dislocation

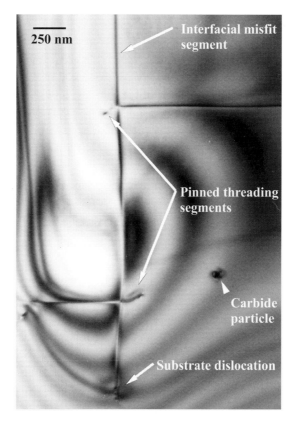

Fig. 7.4. Transmission electron microscopy image of threading dislocations in a SiGe/Si(001) material system that have been blocked by an interface misfit dislocation. The image was produced by means of a high-resolution TEM equipped with *in-situ* CVD capabilities at 600 °C. It shows two threading dislocations, the upper of which had been traveling from right to left and the lower from left to right, that have been blocked by the same interface misfit segment. Reproduced with permission from Stach (2000).

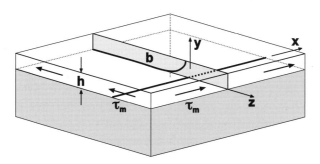

Fig. 7.5. Schematic diagram of a threading dislocation gliding on the yz-plane approaching an interface misfit dislocation (parallel to the x-axis) lying in its path.

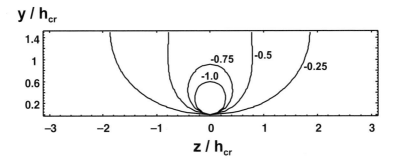

Fig. 7.6. Contours of constant shear stress $\sigma_{xz}^{(\mathrm{md})}/\tau_\mathrm{m}$ due to the pre-existing misfit dislocation acting on the glide plane of the threading dislocation for $\gamma_\mathrm{m} = 0.01$ and $h = 1.5h_\mathrm{cr}$.

is illustrated in Figure 7.6 by means of level curves of normalized shear stress $\sigma_{xz}/\tau_\mathrm{m}$ on the glide plane for the case when $h = 1.5h_\mathrm{cr}$ for this particular system. The value of critical thickness for this system is defined in (6.46). Because this additional stress is opposite in sign to the mismatch stress, it tends to *retard* the progress of the threading dislocation. The total work of interaction that must be provided to move the threading dislocation past the misfit dislocation, to a distance far beyond it compared to h, is

$$-\int_0^h \int_{-\infty}^{\infty} b\,\sigma_{xz}^{(\mathrm{md})}(y, z)\,dy\,dz = \mu b^2 h\,. \tag{7.14}$$

This is the total work that must be done to overcome the retarding force as the threading dislocation moves past the misfit dislocation, but the result provides no direct information on how this work is distributed over the path taken by the threading dislocation. However, it is evident that the stress field of the misfit dislocation acts to some degree to retard the progress of the threading dislocation.

Various strategies can be adopted to arrive at an estimate of the retarding effect. For example, it is evident from Figure 7.6 that the stress field of the misfit dislocation completely negates the background stress field over some portion of the glide plane; this is the region bounded by the contour labeled -1.0. Suppose that the average stress is τ_* over the remaining gap of width h_* on the glide plane between the boundary of this region and the free surface. Then, a plausible criterion for identifying the conditions that discriminate between the range of parameters for which blocking occurs and the range for which it is unlikely to occur is that h_* must be the critical thickness for an effective strain $\gamma_* = \tau_*/\mu$ according to (7.17). It is readily shown that $\sigma_{xz}^{(\mathrm{md})}(y, 0) = -\tau_\mathrm{m}$ for

$$y = y_* = h - \sqrt{h^2 - bh/\pi\gamma_\mathrm{m}} = h - h_*\,. \tag{7.15}$$

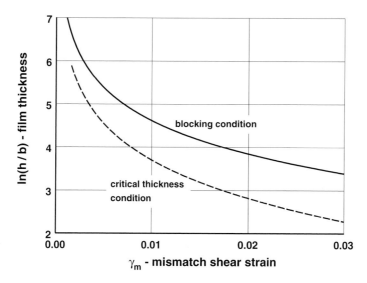

Fig. 7.7. Curve in the plane of film thickness $\ln(h/b)$ versus mismatch strain γ_m showing the criterion described by (7.17) for blocking of a threading dislocation due to a misfit dislocation in its path. Also shown by the dashed curve is the critical thickness criterion for the same system as given in (7.13). Blocking is predicted for states below and to the left of the solid curve, whereas the blocking mechanism should not be effective for states above and to the right of the solid curve.

The average strain remaining within this gap is

$$
\begin{aligned}
\gamma_* &= \gamma_m + \frac{1}{\mu h_*} \int_{y_*}^{h} \sigma_{xz}^{(md)}(y,0)\, dy \\
&= \gamma_m - \frac{b}{2\pi h \sqrt{1 - b/\pi h \gamma_m}} \ln \left(\frac{1 + \sqrt{1 - b/\pi h \gamma_m}}{1 - \sqrt{1 - b/\pi h \gamma_m}} \right).
\end{aligned}
\tag{7.16}
$$

If h_* is then required to be the critical thickness corresponding to the effective mismatch strain γ_*, that is, if

$$
\frac{b}{4\pi h_*} \ln \frac{4h_*}{b} = \gamma_*,
\tag{7.17}
$$

it is found that h and γ_m are related as shown in Figure 7.7. This plot is constructed by finding roots h of (7.17) numerically for many choices of γ_m within the range of interest.

This criterion can be interpreted in the following way. For thin film growth with steadily increasing h at fixed mismatch γ_m, glide of threading dislocation segments is not possible until h increases to a value of h_{cr}. For thicknesses greater than h_{cr} but still below the thickness given by the blocking condition, any threading dislocations that have been formed are able to glide, leaving behind ever increasing lengths of

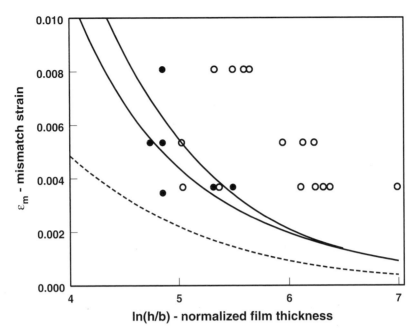

Fig. 7.8. Summary of *in-situ* transmission electron microscopy observations of encounters between advancing threading dislocation segments and misfit interface dislocation segments lying across their paths on a graph of elastic film strain due to misfit versus film thickness normalized by the Burgers vector magnitude. Open symbols show situations in which blocking was not observed, whereas filled symbols indicate situations in which blocking was observed to occur. Examples of encounters that resulted in blocking are shown in Figure 7.4. The observations were made on SiGe/Si(001) film–substrate systems at 600 °C. Adapted from Stach (1998).

interface misfit dislocations and thereby contributing to relaxation of the background elastic strain. However, if the path of a threading dislocation were crossed by a misfit dislocation formed earlier in the relaxation process, its progress would be arrested, as shown in Figure 7.4, if the thickness did not exceed the blocking thickness. However, as the film thickness h is increased further to bring the state of the system to a point above the solid curve in Figure 7.7, a threading dislocation is able to bypass a misfit dislocation in its path. The range of stress and film thickness for which blocking occurs was determined experimentally by Stach et al. (1998) for the $Si_{1-x}Ge_x$/Si(001) material system, and these experimental results are reproduced in Figure 7.8. The curves appearing in this figure represent results based on the same conceptual model as was developed here but with crystallographically correct Burgers vectors rather than the screw dislocation Burgers vector. An interesting aspect of this interaction reported by Stach et al. (1998) is that, once a threading dislocation is actually stopped for some time by a misfit dislocation in its path, it is unlikely to resume its motion even after the film thickness has been increased by further deposition to a value beyond which blocking should not occur. The reason for this behavior is not clear.

An alternate interpretation of the blocking criterion has been given by Gillard et al. (1994) on the basis of strain relaxation experiments, which were also conducted with the $Si_{1-x}Ge_x/Si(001)$ material system. In these experiments, films that were essentially dislocation-free were grown at a relatively low temperature to thicknesses well beyond the critical thickness for various choices of mismatch strain, as determined by Ge content of the film alloy. The materials were then annealed at a temperature well above the growth temperature. Strain relaxation was monitored at elevated temperature by means of the substrate curvature method, and it was allowed to proceed until no further relaxation was detectable. It was hypothesized that the magnitude of residual elastic strain in the film after the cessation of relaxation and the current value of film thickness could be identified with points on the blocking curve. The reasoning was simply that if any threading dislocation could continue to propagate, then relaxation would also continue as observed in the experiments. On the other hand, once the background elastic strain fell to a magnitude that was too small to drive threading dislocations past misfit dislocations, all threading dislocations would be arrested and stress relaxation would stop. The experimental observations were consistent with this hypothesis. The data in Figure 7.8 were obtained in effect by varying thickness at fixed elastic strain, that is, by a sequence of states in Figure 7.7 along a line parallel to the thickness axis. The experiments reported by Gillard et al. (1994) are complementary, in the sense that a sequence of states along the line parallel to the strain axis were interrogated. The identification of the necessary condition for blocking in terms of thickness and elastic strain was similar in the two approaches.

7.2.2 Intersecting arrays of misfit dislocations

The foregoing results in Section 7.1 on interaction of parallel misfit dislocations and in the preceding subsection on intersecting misfit dislocations are now combined to obtain a result of potential use in describing overall elastic strain relaxation in a strained film. The model is again based on the idea of critical conditions for advance of a single threading dislocation through the film, as illustrated schematically in Figure 7.5. However, in the present case, it is assumed that a complete periodic array of misfit dislocations, each parallel to the x-axis and equally spaced at intervals p in the z-direction, has already been formed. Also, it is assumed that a second array of interface misfit dislocations, each parallel to the z-axis and equally spaced at intervals of p in the x-direction, also exists; this array is complete except for a single missing member. This last misfit dislocation is to be inserted and, without loss of generality, the glide plane of this last dislocation is taken to lie in the plane

$x = 0$. The mismatch strain to be relieved is a shear strain $\epsilon_{xz}^{\mathrm{m}} = \gamma_{\mathrm{m}}/2$ and $\epsilon_{yz} = 0$, and the Burgers vector of each dislocation has length b and orientation parallel to the interface.

Consider a threading dislocation advancing in the film along the plane $x = 0$, thereby depositing a misfit dislocation along the line $x = 0$, $y = 0$ as it moves. What is the driving force G on this dislocation? This driving force varies with distance traveled by the threading dislocation in the z-direction, but it does so periodically with spatial period p. Therefore, while determination of the driving force G itself at all points along the path of the threading dislocation is an unsolved problem, it is possible to calculate the *average* driving force per period, say

$$\bar{G}(h, p) = \frac{1}{p} \int_{z_0}^{z_0+p} G(h, p, z)\, dz, \tag{7.18}$$

which is independent of the choice of z_0.

There are contributions to $\bar{G}(h, p)$ from several sources. Among these are the self-energy per unit length W_{d} of the misfit dislocation being formed and the work of interaction W_{m} with the background mismatch strain field; these contributions are defined in (6.44) and (6.45), respectively. There is also the work of interaction W_{a} with all misfit dislocations that are parallel to the one being inserted; this interaction energy is given in (7.11). Finally, there is the work of interaction between the dislocation being formed and the misfit dislocation array with its members lying across the path of the advancing threading dislocation. The shear stress on the plane $x = 0$ in the film, say $\sigma_{xz}^{(\mathrm{ia})}(y, z)$, due to the array of misfit dislocations parallel to the x-axis and intersecting the path of the threading dislocation can be constructed by superposition based on (6.4) to yield

$$\sigma_{xz}^{(\mathrm{ia})}(y, z) = \mu b \frac{\sinh \frac{2\pi h}{p} \left(\cosh \frac{2\pi h}{p} - \cos \frac{2\pi z}{p} \cosh \frac{2\pi(h-y)}{p}\right)}{\left(\cos \frac{2\pi z}{p} - \cosh \frac{2\pi y}{p}\right)\left(\cos \frac{2\pi z}{p} - \cosh \frac{2\pi(2h-y)}{p}\right)}. \tag{7.19}$$

The average interaction energy due to the intersecting array, say W_{ia}, has the value

$$W_{\mathrm{ia}} = -\frac{1}{p} \int_{-p/2}^{p/2} \int_0^h b\, \sigma_{xz}^{(\mathrm{ia})}(y, z)\, dy\, dz = \frac{\mu b^2 h}{p}, \tag{7.20}$$

where the value $z_0 = -p/2$ in (7.18) has been chosen for convenience.

If the four contributions to $\bar{G}(h, p)$ identified above are taken into account, then

$$\bar{G}(h, p) = \mu b h \left[\gamma_{\mathrm{m}} - \frac{b}{4\pi h} \ln \frac{4h}{b} - \frac{b}{2\pi h} \ln \left(\frac{p}{2\pi h} \sinh \frac{2\pi h}{p} \right) - \frac{b}{p} \right]. \tag{7.21}$$

This is the period-averaged driving force on the threading dislocation as it leaves

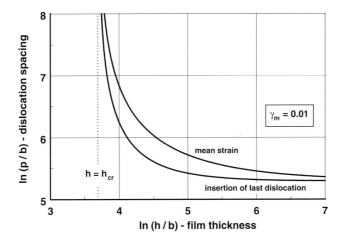

Fig. 7.9. Illustration of the results established as critical conditions for formation of two arrays of periodic interface misfit dislocations, one array orthogonal to the other, at the interface between a strained epitaxial film and its substrate. The lower curve is a plot of the result (7.21) for insertion of the last dislocation necessary to complete one of the arrays, the other being already complete, and the upper curve is the equivalent result (7.22) based only on mean strain measures and the critical thickness condition for insertion of an isolated misfit dislocation. The graphs are based on the screw dislocation model with a mismatch strain $\gamma_m = 0.01$.

behind the misfit dislocation that completes a doubly periodic array with spacing p and film thickness h. The equation $\bar{G}(h, p) = 0$ provides a lower bound on possible dislocation spacing for fixed values of h and γ_m for which the last dislocation can be inserted. The relationship between p/b and h/b defined by this condition is shown in Figure 7.9 for the particular choice in $\gamma_m = 0.01$. Asymptotic behavior as $p/b \to \infty$ is again the critical thickness condition (7.1). As $h/b \to \infty$, on the other hand, this critical condition implies that $p/b \to 2/\gamma_m$, which is the spacing for the doubly periodic array for which the elastic strain γ_m is completely relaxed, on the average.

A mean strain approach, as described in Section 7.1, to this same relaxation problem would lead to a condition of the form

$$\frac{b}{4\pi h} \ln \frac{4h}{b} = \gamma_m - \frac{2b}{p} \tag{7.22}$$

for the minimum spacing p of a doubly periodic array. This result is also shown in Figure 7.9 for purposes of comparison. The mean strain result (7.22) shares asymptotic behavior with (7.21), but the transition between asymptotes is somewhat sharper in the latter case.

7.3 Strain relaxation due to dislocation formation

How the motion of individual dislocations influences bulk strain relaxation is not fully understood at this time, but some progress has been made (Tsao et al. 1987, Dodson and Tsao 1987, Gosling et al. 1994). Motivated by results obtained in Chapter 6 on dislocation formation and in this chapter on dislocation interactions, a procedure for obtaining a closed set of evolution equations that describes elastic strain relaxation in a strained film is outlined here. Numerous trade-offs between physical detail and mathematical simplicity must be considered and, as is the case with all constitutive theories, questions of validity and consistency of any particular description obtained can only be addressed through appeal to experiments. In this section, the procedure for constructing a complete relaxation model is outlined by integrating the essential elements on which it is based. This is followed by a summary of experimental observations of relaxation processes which are designed to achieve structures with specific features; in one case, the goal is to minimize relaxation whereas, in a second case, the goal is to maximize relaxation but with as few remaining threading dislocations as possible.

7.3.1 Construction of a relaxation model

The matter is considered first within the framework of the model based on relaxation of elastic shear strain by the formation of screw dislocations. This avoids a certain degree of complexity without abandoning essential physical features. Consider a small rectangle or patch of interface between a strained film and its relatively thick substrate. The geometrical configuration of the film–substrate system is illustrated in Figure 7.5, along with a convenient rectangular coordinate system. Suppose that the extent of the patch is Δx in the x-direction and Δz in the z-direction. Furthermore, suppose that, at a certain instant of time, the elastic strain in the film is partially relaxed by periodic arrays of dislocations, one aligned with the x-axis and another aligned with the z-axis, both of period p. The direction of the Burgers vector of both arrays is such that a positive background elastic strain γ_m is partially relaxed by dislocation formation; the magnitude of the Burgers vector is b. The total shear strain in the film during relaxation is fixed at γ_m by symmetry; this is the elastic strain before any dislocations form. On the average, the inelastic strain due to the dislocation arrays is $\gamma_d = 2b/p$; this relationship gives the microscale quantity b/p a macroscopic interpretation. The *residual elastic strain* γ_r in the film following formation of these arrays is

$$\gamma_r = \gamma_m - \gamma_d = \gamma_m - 2b/p. \tag{7.23}$$

Consider once again the film attached to the same patch of interface of dimensions Δx by Δz. Suppose that the edges of the patch parallel to the z-direction are held

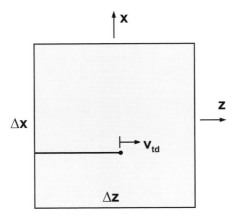

Fig. 7.10. Schematic diagram of a single threading dislocation (represented by the circular dot) traveling across the width of a typical patch of film at speed v_{td}, leaving behind an interface misfit dislocation as it goes.

fixed for the moment, and imagine the consequences of moving a single threading dislocation at speed v_{td} across the width Δz of the patch, as depicted schematically in plan view in Figure 7.10. The net change in residual elastic strain is a reduction, that is, the inelastic strain due to formation of the misfit dislocation is

$$\Delta \gamma_d = \frac{b}{\Delta x} \tag{7.24}$$

once the threading dislocation has moved completely across the patch. The transit time of the dislocation across the patch is $\Delta t = \Delta z / v_{td}$, so the average strain rate due to dislocation formation during transit is approximately

$$\frac{\Delta \gamma_d}{\Delta t} = \frac{b v_{td}}{\Delta x \, \Delta z}. \tag{7.25}$$

Any misfit dislocations already present do not contribute to strain rate.

Suppose next that, instead of just one dislocation, many threading dislocations glide across the patch, all moving at approximately the same speed. Assume also that there are n threading dislocations per unit area of interface at the present instant, so that the total number of threading dislocations contributing to strain relaxation within the patch is $n \Delta x \Delta z$. Each makes essentially the same contribution to inelastic strain rate. Consequently, the total instantaneous inelastic strain rate due to glide of the threading dislocations is

$$\dot{\gamma}_d = n b v_{td}, \tag{7.26}$$

which is the classical Orowan equation for the configuration at hand (Orowan 1940). It provides a macroscopic interpretation for the microscopic parameter v_{td} through the homogenization parameter n.

The average threading dislocation speed v_{td} can be expressed in terms of the average driving force $\bar{G}(h, p)$ acting on it, which was defined in (7.21) as

$$v_{td} = v_0 \left(\frac{\bar{G}(h, p)}{\mu h b} \right)^m e^{-Q_0/kT},\tag{7.27}$$

which is the form of (6.29) for the case when the glide plane of the threading dislocation is perpendicular to the interface and the relevant elastic modulus is the shear modulus μ. The parameter p can be eliminated from (7.27) in favor of γ_d by appeal to (7.23), yielding

$$\dot{\gamma}_d(t) = n(t)bv_0 \left(\frac{\bar{G}\,[h(t), 2/\gamma_d(t)]}{\mu b h(t)} \right)^m e^{-Q_0/kT}.\tag{7.28}$$

It is recognized here that γ_d, as well as density n, thickness h and temperature T, may vary with time t. Since the thickness and temperature are normally controlled parameters in a relaxation process, the histories $h(t)$ and $T(t)$ can be assumed to be known for many circumstances. However, $n(t)$ is an internal variable that cannot be controlled. The density evolves during a relaxation process in a way that has not yet been captured in a concise and generally applicable representation. This evolution includes contributions resulting from nucleation, multiplication, annihilation and blocking. For most purposes, it is probably adequate to postulate that $n(t)$ evolves according to a rate equation of the form

$$\dot{n}(t) = \dot{n}_1 N_1 \left[n(t), \gamma_d(t), h(t), T(t) \right],\tag{7.29}$$

where $N_1(\cdot)$ is some dimensionless function of its arguments and $\dot{n}_1$ is a material parameter with dimensions of number per unit area per unit time. An alternate and simpler form for an evolution equation for $n(t)$ is

$$n(t) = n_0 N_0 \left[\gamma_d(t), h(t), T(t) \right],\tag{7.30}$$

where $N_0(\cdot)$ is some other dimensionless function of its arguments and n_0 is a material parameter with dimensions of number per unit area representing a reference density of threading dislocations.

Suppose that $n(t)$ is scaled by some reference threading dislocation density n_r. Furthermore, attention is limited for the time being to cases of relaxation at constant temperature. Then (7.28) can be rewritten as

$$\dot{\gamma}_d(t) = \frac{1}{t_\gamma} \left(\frac{\bar{G}(t)}{\mu b h(t)} \right)^m \frac{n(t)}{n_r},\tag{7.31}$$

where $t_\gamma = e^{-Q_0/kT}/n_r b v_0$ is a characteristic time for the process. Similarly, (7.29) can be rewritten as

$$\frac{\dot{n}(t)}{n_r} = \frac{1}{t_n} N_1(t),\tag{7.32}$$

where $t_n = n_r/\dot{n}_1$ is a second characteristic time. Then, if a nondimensional time variable s is introduced as $s = t/\sqrt{t_\gamma t_n}$, the two differential equations (7.31) and (7.32) reduce to the nondimensional forms

$$\gamma_d'(s) = \left(\frac{\bar{G}(s)}{\mu b h(s)}\right)^m \beta(s), \quad \beta'(s) = N_1(s), \tag{7.33}$$

where

$$\beta(s) = \frac{n(t)}{n_r}\sqrt{\frac{t_n}{t_\gamma}} \tag{7.34}$$

and both $\bar{G}(s)$ and $N_1(s)$ depend on s through the arguments identified in their definitions in (7.21) and (7.29), respectively. The equations (7.33) are universal in the sense that all material parameters have been suppressed through scaling of variables. Solution of the pair of ordinary differential equations requires specification of initial conditions; the choices

$$\gamma_d(0) = 0, \quad \beta(0) = 0 \tag{7.35}$$

are suitable for most cases, provided that $N_1(0) \neq 0$.

As a specific form for $N_1(s)$ in (7.33), assume that

$$N_1(s) = \sqrt{\gamma_m^2 + \gamma_d(s)^2}. \tag{7.36}$$

This form is motivated by the idea that the dislocation formation rate is dominated by nucleation and that it scales with the level of background elastic strain γ_m when $\gamma_d \approx 0$. Once the value of γ_d due to density of misfit dislocations becomes comparable in magnitude to γ_m, which implies that many intersections of misfit dislocations have been formed, the dislocation formation rate may involve both nucleation and multiplication. Under such conditions, the rate is enhanced by the increasing magnitude of γ_d. The parameter t_n has dimensions of time and it represents a timescale for dislocation formation. A certain amount of relaxation can be achieved by having many threading dislocations with each moving for a short time (large n_r and/or small t_n) or having a few threading dislocations with each moving for a long time (small n_r and/or large t_n). A numerical solution of the pair of differential equations (7.33) is shown in Figure 7.11 for the case when $\gamma_m = 0.01$. If a different value is chosen for γ_m, the same curve results due to the universal form of the differential equations but the timescale is altered. This result has the general form of elastic strain relaxation with time that was observed in the experiments by Gillard et al. (1994).

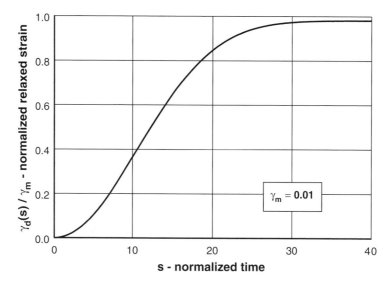

Fig. 7.11. Strain relaxation history determined by numerical solution of the differential equations (7.33) with the particular form (7.36) to represent the rate of increase of threading dislocations.

7.3.2 Example: Dislocation control in semiconductor films

The stress associated with the elastic strain induced in heteroepitaxial material systems with lattice mismatch provides a driving force for the formation and growth of strain-relieving dislocations. In some device applications, it is essential that the induced strain be maintained because of its influence on functional characteristics. For cases involving significant mismatch strain, the critical thickness condition for dislocation formation discussed in Chapter 6 imposes a severe constraint on feature dimensions that can be achieved without loss of stability of the strained configuration. Consequently, effective approaches that lead to stabilization of strained heteroepitaxial structures against dislocation formation and/or growth are of great practical interest. One such approach is based on the compliant substrate concept introduced in Section 6.7. Progress in the use of growth area patterning, another promising approach, is described in this section. In other device applications, it is desirable to relax as much of the mismatch strain as possible, and to do so in such a way that the relaxed material provides a high quality growth surface for subsequent deposition. Progress in this direction is also described in this section.

Both of the cases described are concerned with the epitaxial deposition of a SiGe alloy on a Si substrate. The choice of substrate is influenced by the widespread use of Si as the bulk substrate material in complementary metal–oxide semiconductor (CMOS) applications. SiGe alloys are attractive materials for film growth for several reasons, in addition to their basic compatibility with Si substrates. Since Si and Ge are completely miscible over the full range of solid solution compositions, alloy crystals of very high quality can be grown at moderate temperatures, and dopants are readily introduced. Retention of mismatch strain is critical for microelectronic applications because of the way in which the strain alters electronic characteristics of the material. Perhaps the most significant changes are the decrease in electronic bandgap with an increase in Ge content at fixed strain, the decrease in electronic bandgap with increase in magnitude of biaxial strain of either sign at fixed composition, and alteration of band alignment at SiGe/Si interfaces.

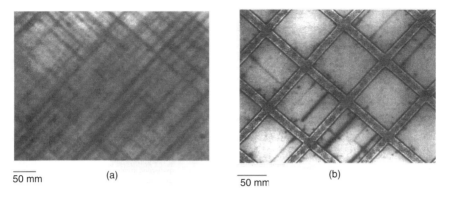

50 mm   (a)   (b)
50 mm

Fig. 7.12. Electron-beam-induced current (EBIC) images showing the plan view of (a) GeSi deposited on unpatterned Si substrates and (b) GeSi deposited on patterned Si substrates. The figures reveal changes in misfit dislocation densities in a $Ge_{0.19}Si_{0.81}$ alloy on a (001) Si substrate. Adapted from Fitzgerald et al. (1991) and Fitzgerald (1995). Reproduced with permission from E.A. Fitzgerald, Massachusetts Institute of Technology.

Specific device configurations in which curtailment of misfit dislocation formation enhances performance are the heterojunction bipolar transistor and the modulation doped field effect transistor. In order to suppress misfit dislocation formation, device designers can tailor SiGe film thicknesses to those below the critical value h_{cr} at which misfit dislocations nucleate and/or modulate mismatch strains by recourse to alloying. Further gains in performance can be achieved by creating highly strained, metastable layers of SiGe by purposely introducing kinetic barriers that serve to limit the extent of misfit dislocations (Fitzgerald 1995). This is accomplished through the introduction of patterns on the substrate prior to deposition. Following deposition, the pattern features restrict the range of influence of active dislocation sources. Figure 7.12(a) is an electron-beam-induced current (EBIC) image of a $Si_{0.81}Ge_{0.19}$ alloy grown on an unpatterned Si substrate. Misfit dislocations, which appear as dark lines in the image, are seen to be aligned along two $\langle 110 \rangle$ directions. Figure 7.12(b) is an EBIC micrograph of the specimen in which the Si substrate was patterned and etched to form a square grid – approximately 50–70 μm in length/width and 2 μm deep with edges along the $\langle 110 \rangle$ directions – before the deposition of the SiGe film by MBE at 550 °C. Significant reduction in misfit dislocation density is evident in the case of the patterned GeSi/Si system, compared to the case of the sample with a spatially uniform growth surface. The range of influence of any active nucleation source is confined by the grid features, particularly the trenches which serve to arrest the progress of advancing threading dislocations. Substrate patterning thus suppresses relaxation of misfit strains by limiting the length of the misfit dislocations. Such patterning methods employing selective epitaxy inside oxide windows have been used to fabricate metastable GeSi diodes.

Some applications, such as optoelectronic devices which emit at 1.3 μm and 1.55 μm wavelengths for fiber-optic telecommunications, require relaxed epilayers for optimal performance. For example, epitaxial growth of relaxed GaAs or GeSi on Si substrates is required for high transconductance transistors and optoelectronic devices. Relaxed InGaAs on GaAs substrate facilitates the integration of 1.3 μm wavelength optoelectronic devices with GaAs electronics. The ability to produce relaxed InGaP layers on GaP substrates allows the fabrication of transparent substrate light-emitting diodes (LEDs). Similarly, GeSi alloy epilayers

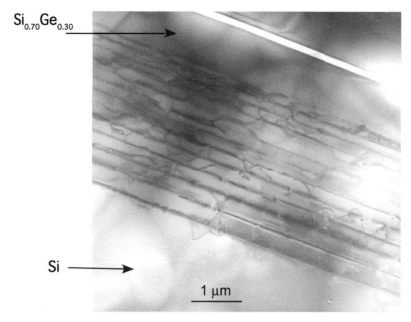

Si$_{0.70}$Ge$_{0.30}$

Si

1 μm

Fig. 7.13. Cross-sectional TEM image showing a dislocation-free Si$_{0.70}$Ge$_{0.30}$ cap layer which is grown on a graded SiGe buffer layer. The thick, bright white line running from the top to the right of the figure is a quantum well light-emitting diode. Adapted from Fitzgerald (1995). Reproduced with permission from E. A. Fitzgerald, Massachusetts Institute of Technology.

need to be relaxed to the bulk state for infrared detectors and devices requiring biaxial tensile layers. Such relaxation of mismatch strains necessitates the introduction of misfit dislocations. The objective then is to control the density and kinetics of threading dislocations in the relaxed epilayers so that certain wavelength emitting quantum wells could be protected and the device performance enhanced.

A different situation – one in which it is desirable to relax all mismatch strain in a SiGe/Si epitaxial film structure – arises in seeking a means to integrate direct bandgap III–V compound semiconductors into Si-based systems in order to exploit the optical properties of the direct bandgap materials. A fully relaxed SiGe film for which the free surface has high Ge content provides a growth surface with lattice parameter comparable to that of GaAs. If the relaxed SiGe film also has a low density of threading dislocations, it provides a substrate surface for deposition of high quality films of GaAs or other III–V semiconductor films.

The stress relaxation of the SiGe film proceeds by formation of misfit dislocations which are left behind as threading dislocations propagate throughout the film. The goal is to force all of the threading dislocations to the edges of the wafer so that they cannot be extended into the layer to be deposited subsequently. To achieve this end, it is desirable to grow a low-mismatch layer at a high temperature in such a manner that it is relaxed without a high threading dislocation density. This relaxed layer could then be used to grow another layer on top of it with the same step-wise increase in lattice constant. This process could be repeated to generate a gradual change in lattice constant in a multilayer structure. In the limit, a linearly graded composition results where a virtual buffer layer of

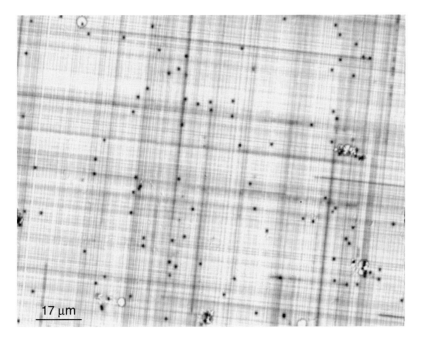

17 μm

Fig. 7.14. EBIC image showing the plan view of the cap layer where a low density of threading dislocations, revealed as black dots, is seen. Adapted from Fitzgerald (1995) and Fitzgerald et al. (1991, 1999). Reproduced with permission from E. A. Fitzgerald, Massachusetts Institute of Technology.

an alloy (such as GeSi or InGaAs) with varying composition in the film thickness direction is created on a bulk substrate. Proper grading of composition facilitates the suppression of rapid dislocation nucleation, and the enhancement of dislocation glide. Thus, proper compositional grading leads to relaxed heteroepitaxial layers with low threading dislocation densities.

Figure 7.13 is a cross-sectional TEM micrograph of a SiGe alloy whose composition is graded from the Si substrate to the cap layer of $Si_{0.70}Ge_{0.30}$ over a distance of approximately 0.25 μm. Note the slow grading of composition at approximately 10% Ge μm^{-1}. The cap layer is a $Si_{0.70}Ge_{0.30}$ alloy with a uniform thickness of 1.5 μm. Note that the cap layer with 30% Ge has no dislocations as a result of the introduction of the graded buffer layer. Had the $Si_{0.70}Ge_{0.30}$ cap layer been grown directly on the Si substrate, a high rate of dislocation nucleation would have resulted in a high density of threading dislocations. Figure 7.14 is a plan-view EBIC image of the relaxed $Si_{0.70}Ge_{0.30}$ layer of the same system shown in Figure 7.13. The black dots in Figure 7.14 are individual threading dislocations. The threading dislocation density is of the order of 10^{10} m^{-2}. This is some three orders of magnitude smaller than that expected for a uniform composition $Si_{0.70}Ge_{0.30}$ alloy grown directly on Si.

7.4 Continuum analysis of ideally plastic films

The issue of elastic strain relaxation has been discussed up to this point within the framework of formation, growth and interaction of glide dislocations. While such an approach provides considerable physical insight, the prospect of developing quantitative models of experimentally observed phenomena in material systems exhibiting high dislocation densities in this way is remote. The benefits of such modeling must be seen in the trends in behavior that are predicted, in the sensitivity of predicted response to values of parameters incorporated to characterize the models, and in the admissible forms of constitutive equations implied. The most likely circumstances for capturing important aspects of physical behavior with dislocation models are probably in the study of films with low dislocation densities, such as epitaxial semiconductor films. In this section, the task of modeling inelastic behavior of films is addressed for the opposite extreme of relatively high densities of dislocations, typical of metal films. In such cases, the tools of continuum plasticity theory can be adopted. These tools include constitutive equations that serve as models of the behavior of materials. The underlying phenomena are too complicated to represent in any complete or exact sense. Instead, constitutive equations involving a few parameters, values of which must be determined by experiments, are used as approximate representations of these phenomena for practical purposes. The transition between discrete dislocation modeling and continuum plasticity modeling remains a work in progress, and some aspects will be addressed later in this chapter.

Continuum descriptions of the evolution of stress, substrate curvature and plastic deformation also provide a useful and broad framework with which experimental observations of strain relaxation in thin films on substrates can be assessed. As shown in later sections of this chapter, such analyses provide some tools with which the plastic deformation response can be experimentally determined by recourse to methods such as substrate curvature measurement, x-ray diffraction or indentation.

7.4.1 *Plastic deformation of a bilayer*

The issue of substrate curvature in an elastic bilayer system with arbitrary layer thicknesses was considered in Section 2.2.1, where results were presented for the variation of equi-biaxial stress through the thickness of isotropic film–substrate systems for any ratio of thicknesses and any ratio of elastic moduli. In this section, extensions of these results into the realm of plastic response are pursued whereby useful insights into the conditions governing the onset and spread of plastic flow in the bilayer can be extracted.

If one or both of the materials are modeled as being elastic–ideally plastic, then the methods adopted for study of elastic response can be extended to establish conditions for onset of plastic deformation and the progression of plastic flow through the thickness of the system. This line of reasoning is pursued in this section for the case when stress in the film–substrate system is a consequence of mismatch in thermal expansion properties between the two materials involved; the typical system comprises a metal film with a relatively high coefficient of thermal expansion bonded to an insulating substrate with the relatively low coefficient of thermal expansion. As shown by Suresh et al. (1994), certain specific values of temperature can be identified with the onset of plastic deformation and with transitions in the overall response arising from plastic deformation.

The discussion here follows directly from that in Section 2.2. A film of thickness h_f is bonded to a substrate of thickness h_s, with no restrictions on the thickness ratio. The stress and deformation fields are referred to a cylindrical coordinate system with polar coordinates in the plane of the system and with the z-direction normal to the interface; the origin of coordinates lies in the substrate midplane. The equi-biaxial stress components are referred to polar coordinates, but they could equally well be expressed in rectangular coordinates. As long as the response is in the range of geometrically linear behavior, the common shape of the film and substrate in plan view is immaterial. As in Section 2.2, the biaxial elastic moduli of the film and substrate are M_f and M_s, respectively, and the corresponding coefficients of linear thermal expansion are α_f and α_s, respectively.

The film material is assumed to be elastic–ideally plastic for the time being; the influence of strain hardening in the course of plastic strain accumulation will be considered subsequently. The substrate material is taken to be elastic, although the analysis can be generalized to account for the onset of plastic deformation in both the film and the substrate. For plastically incompressible material response, the condition on stress for plastic yield due to equi-biaxial tension σ acting on a material element is identical to the condition under uniaxial compression of the same magnitude. Consequently, the conditions for plastic yielding under equi-biaxial stress conditions can be expressed in terms of a uniaxial yield or flow stress. As will be seen subsequently, however, the overall plastic response under biaxial deformation is different from that for uniaxial deformation. It is assumed that, at any temperature T, there is a level of stress denoted by $\sigma_Y(T) > 0$ such that the material responds elastically when the equi-biaxial stress σ is such that $|\sigma| < \sigma_Y(T)$ and that plastic flow may occur when $|\sigma| = \sigma_Y(T)$. The range $|\sigma| > \sigma_Y(T)$ is inaccessible. In the case of equality, whether or not plastic flow does occur depends on whether or not an increment in plastic strain, due to an increment in temperature in this case, corresponds to a positive or negative plastic dissipation of mechanical work in the material; only increments in plastic strain resulting in positive dissipation are admissible. Although the behavior of the system can be given a more formal mathematical description, the response in the present case is evident without doing so. When referring specifically to the film material, the flow stress is denoted by σ_{Y_f}.

Some state of the film–substrate system is identified as the initial state. The temperature in this state is the reference temperature T_o, and T represents the current temperature. All components of stress and plastic strain are presumed to be equal to zero in the reference state; minor modifications are required to consider any other initial state. The reference substrate curvature is also zero at T_o. A change in temperature from the reference state is denoted as $\bar{T} = T - T_o$ in all subsequent discussion.

The in-plane extensional strain and curvature in the film–substrate system arise as a consequence of thermal expansion mismatch between the film and substrate materials during temperature excursion from the reference temperature. Conditions are assumed to be such that the temperature is uniform throughout the film and substrate at all times during thermal cycling. Some particular temperatures at which distinct transitions occur in the elastoplastic deformation of the film material are first identified by adopting the assumption that, over the range of temperature, the properties of the film and substrate materials remain essentially unchanged; effects of temperature dependence of plastic yield or flow behavior on the evolution of film stress and substrate curvature are examined subsequently.

The variations of elastic equi-biaxial stress and in-plane extensional strain through the thickness of a film–substrate system of arbitrary layer thickness were described in Section 2.2 for the case of arbitrary mismatch strain ϵ_m; a distribution of the in-plane extensional strain through the thickness of a film–substrate system is illustrated in Figure 2.5. Combining the results of (2.19) through (2.21) with (2.31), the variation of radial and circumferential stress components σ_{rr} and $\sigma_{\theta\theta}$, respectively, through the layer thickness during thermal excursion from the reference temperature prior to plastic yielding is

$$\sigma_{rr}(z) = \sigma_{\theta\theta}(z) = \begin{cases} M_s\left(\epsilon_o - \kappa z\right) & \text{for } -\tfrac{1}{2}h_s < z < \tfrac{1}{2}h_s, \\ \\ M_f\left(\epsilon_o - \kappa z + \epsilon_m\right) & \text{for } \tfrac{1}{2}h_s < z < \tfrac{1}{2}h_s + h_f \end{cases} \tag{7.37}$$

where the curvature κ of the substrate midplane is given in (2.19), the in-plane extensional strain ϵ_o of the substrate midplane is given in (2.20), and the mismatch strain is

$$\epsilon_m = \left(\alpha_s - \alpha_f\right)\bar{T} \tag{7.38}$$

in this case.

In terms of the thickness ratio $\eta = h_f/h_s$ and the modulus ratio $m = M_f/M_s$, the curvature κ and midplane extensional strain ϵ_o appearing in (7.37) are

$$\kappa = \frac{6\epsilon_m}{h_s}\left[\frac{\eta m}{1 + 4\eta m + 6\eta^2 m + 4\eta^3 m + \eta^4 m^2}\right], \tag{7.39}$$

$$\epsilon_o = -\epsilon_m\left[\frac{m\eta(1 + \eta^3 m)}{1 + 4\eta m + 6\eta^2 m + 4\eta^3 m + \eta^4 m^2}\right]. \tag{7.40}$$

It follows from the through-the-thickness profile of biaxial stress given in (7.37) that the location of the stress of largest magnitude in the film is at the film–substrate interface $z = \frac{1}{2}h_s$. The sign of the stress there is the same as the sign of ϵ_m. It follows that the change in temperature at which the stress in the film first satisfies the plastic yield condition, say $\bar{T}_Y$, is given by

$$\bar{T}_Y(\alpha_s - \alpha_f) = \pm\frac{\sigma_{Y_f}}{M_f}\left[\frac{1 + 4\eta m + 6\eta^2 m + 4\eta^3 m + \eta^4 m^2}{1 + 3\eta^2 m + 4\eta^3 m}\right]. \tag{7.41}$$

The upper sign on the right side applies if $\epsilon_m > 0$, whereas the lower sign applies if $\epsilon_m < 0$. This is consistent with the intuitive notion that the film will yield in tension (compression) if the mismatch strain is positive (negative). The substrate curvature at first yield, say κ_Y, is given by

$$\kappa_Y = \pm\frac{6\sigma_{Y_f}}{h_s M_f}\left[\frac{\eta m(1 + \eta)}{1 + 3\eta^2 m + 4\eta^3 m}\right] \tag{7.42}$$

where the sign is chosen as in (7.41). In other words, if the film yields plastically in tension, then the substrate curvature is positive, or concave to the film side, whereas the curvature is negative if the film first yields in compression.

With further increase in the magnitude of temperature difference $\bar{T}$, the edge of the zone of plastically deforming film material advances from the interface to the free surface, whereupon the entire film is deforming within the plastic range. The temperature difference at which the plastic region extends completely across the film thickness is denoted by $\bar{T}_{pl}$. Upon reaching this temperature difference, the distribution of equi-biaxial stress in the film is uniform and is given by

$$\sigma_{rr}(z) = \sigma_{\theta\theta}(z) = \pm\sigma_{Y_f} \tag{7.43}$$

throughout $\frac{1}{2}h_s < z < \frac{1}{2}h_s + h_f$. In the substrate, the stress distribution varies linearly across the thickness according to

$$\sigma_{rr}(z) = \sigma_{\theta\theta}(z) = \mp\sigma_{Y_f}\left[\left(\frac{z}{h_s} + \frac{1}{2}\right)\eta(4 + 3\eta) + \left(\frac{z}{h_s} - \frac{1}{2}\right)\eta(2 + 3\eta)\right] \tag{7.44}$$

for $-\frac{1}{2}h_s < z < \frac{1}{2}h_s$ where the sign is again chosen in a manner consistent with the convention established in (7.41). This stress distribution does not change with further increase in the magnitude of $\bar{T}$ beyond $\bar{T}_{pl}$.

The corresponding extensional strain and curvature of the substrate midplane are

$$\epsilon_0 = \alpha_s \bar{T} \mp \frac{\sigma_{Y_f}}{M_f} m \eta (4 + 3\eta), \tag{7.45}$$

$$\kappa_{pl} = \pm \frac{6\sigma_{Y_f}}{h_s M_f} m \eta (1 + \eta), \tag{7.46}$$

respectively. For the assumptions invoked in the results leading up to this point, that is, for elastic–ideally plastic response of the film material with no dependence of flow stress on temperature, the substrate curvature κ_{pl} is a limiting value which remains unchanged during further increase in the magnitude of temperature differ- ence. However, the extensional strain of the midplane as given in (7.45) continues to vary linearly with increasing magnitude of temperature $\bar{T}$, albeit at a different rate with respect to temperature than that given by the elastic solution (7.40). This is solely a thermoelastic response, and it is accommodated by ongoing plastic flow at constant stress in the film.

Combining (7.37) and (7.38) with (7.46), the temperature change from the refer- ence state beyond which the curvature of the bilayer remains unaltered with further increase in the magnitude of temperature change can be written in nondimensional form as

$$\bar{T}_{pl} (\alpha_s - \alpha_f) = \pm \frac{\sigma_{Y_f}}{M_f} \left[1 + m \eta (4 + 9\eta + 6\eta^2) \right]. \tag{7.47}$$

Figure 7.15 schematically shows the transitions in the onset and spread of plastic yielding in a bilayer comprising an elastic–ideally plastic film on an elastic substrate, in response to an increase in temperature from a reference state, presuming that $\alpha_s < \alpha_f$ and that $\bar{T} > 0$. The temperature increase $\bar{T}_Y$ at which plastic yielding commences in the film and the temperature change $\bar{T}_{pl}$ at which the entire film material first becomes fully plastic are marked on this figure. Note that the substrate curvature varies linearly with temperature when both the film and substrate are elastic, that is, for $\bar{T} < \bar{T}_Y$; beyond this point, the slope of the curvature versus temperature plot becomes nonlinear. As described in Section 7.7.3, this transition point can be used to extract the plastic yield strength of a film material from a thermal cycling experiment using any of the substrate curvature measurement methods described in Section 2.3. Once the film becomes fully plastic at $\bar{T}_{pl}$, the substrate curvature attains a value κ_{pl} which remains unaltered with further change in temperature. This transition point can also be determined experimentally from the development of a plateau in the substrate curvature versus temperature plot, provided that the response of film material is captured by an elastic–perfectly plastic idealization with a temperature-independent yield strength over the range of temperature of

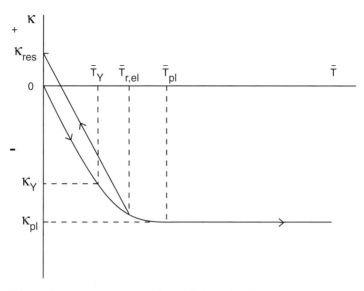

Fig. 7.15. Schematic representation of the variation of substrate curvature κ as a function of the temperature change $\bar{T}$ for an arbitrary bilayer for which $\alpha_{\rm s} < \alpha_{\rm f}$. It is assumed that the film material is an elastic–ideally plastic solid whose properties do not vary with temperature. The onset of plastic yielding at $\bar{T} = \bar{T}_{\rm Y}$ and of complete yielding at $\bar{T} = \bar{T}_{\rm pl}$ of the film material and the corresponding substrate curvatures, $\kappa_{\rm Y}$ and $\kappa_{\rm pl}$, respectively, are marked in the figure. Note that reversing the temperature from $\bar{T} < \bar{T}_{\rm r,el}$ to $\bar{T} = 0$ produces only elastic unloading with residual curvature $\kappa_{\rm res}$.

interest. If $\alpha_{\rm s} > \alpha_{\rm f}$ and/or if the temperature is decreased from the reference value, the behavior is as shown in the Figure 7.15, but with the curves reflected with respect to one or both of the coordinate axes.

The development of thermal mismatch strain in the bilayer is accommodated by substrate bending, plastic yielding in the film, or the in-plane extensional strain $\epsilon_{\rm o}$. Once the film material becomes fully plastic at $\bar{T}_{\rm pl}$, the first two of these three avenues of strain relief are exhausted, and any further change in thermal mismatch strain due to a change in temperature is accommodated solely through a change in $\epsilon_{\rm o}$, as indicated by (7.45).

Conditions governing the onset of reverse yielding during thermal cycling can also be identified for the present problem. As the temperature is raised beyond $\bar{T}_{\rm Y}$, there exists another critical temperature, say $\bar{T} = \bar{T}_{\rm r,el}$, such that unloading at or prior to this temperature leads only to elastic deformation without reversed plastic flow in the film. If the tensile and compressive yield strengths of the elastic–ideally plastic film are equal, elastic unloading is ensured when the temperature is reversed from a value smaller than

$$\bar{T}_{\rm r,el} = 2\bar{T}_{\rm Y} \,. \tag{7.48}$$

The temperature $\bar{T}_{\mathrm{r,el}}$ is also indicated in Figure 7.15. If the bilayer is subjected to thermal cycles between the temperature limits $\bar{T} = 0$ and $\bar{T} = \bar{T}_{\mathrm{r,el}}$, there is no net accumulation of plastic strain with increasing number of thermal cycles, and the system is said to have undergone *elastic shakedown*. The relative magnitudes of $\bar{T}_{\mathrm{Y}}$, $\bar{T}_{\mathrm{r,el}}$ and $\bar{T}_{\mathrm{pl}}$ are strongly influenced by the elastic modulus ratio m and the thickness ratio η for the film–substrate system.

If the substrate is also an elastic–ideally plastic material with yield strength magnitude $\sigma_{\mathrm{Y_s}}$, it follows from (7.37) through (7.40) that plastic yielding will always commence at the interface within the film material and that there will be no yielding in the substrate material if

$$\frac{\sigma_{\mathrm{Y_f}}}{\sigma_{\mathrm{Y_s}}} < \frac{1 + 3\eta^2 m + 4\eta^3 m}{\eta(4 + 3\eta + \eta^3 m)} \,. \tag{7.49}$$

Note that the magnitude of stress in the substrate is higher after full plastic yielding of the film than at the onset of yielding in the film. It can also be easily shown from (7.44) that, if the substrate were also to be an elastic–ideally plastic metal with equal yield strength magnitude $\sigma_{\mathrm{Y_s}}$ in tension and compression, it would remain in the elastic range along its entire thickness if

$$\frac{\sigma_{\mathrm{Y_s}}}{\sigma_{\mathrm{Y_f}}} > \eta(4 + 3\eta) \,. \tag{7.50}$$

7.4.2 *Thin film subjected to temperature cycling*

If the film bonded to a substrate is thin enough compared to the substrate, say for $\eta = h_{\mathrm{f}}/h_{\mathrm{s}} \leq 0.025$, the state of stress within the film material is essentially uniform throughout its thickness and the deformation response does not depend on the deformation of the substrate (see Chapter 2). These are the essential features of the thin film approximations underlying derivation of the Stoney formula, and they can be invoked here to great advantage. Along with the property of translational invariance of the film–substrate structure in directions parallel to the interface, these features imply that the total in-plane biaxial strain at any point in the film, consisting of elastic strain plus plastic strain plus thermal strain, does not change during temperature cycling. This observation enables a useful picture of the film response during excursions in ambient temperature. Circumstances at the film edges will be discussed subsequently.

For a start, the critical temperatures for the onset and spread of plasticity in the thin film can easily be determined from the results obtained in the preceding section for the bilayer of arbitrary layer thickness. For a thin film on a relatively thick substrate, consider first the situation where the film is elastic–ideally plastic with equal magnitudes of yield strength in tension and compression which do not vary

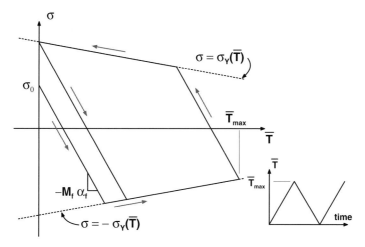

Fig. 7.16. Stress–temperature history for a thin film of elastic–ideally plastic material on a relatively thick substrate during a temperature excursion. The tensile yield stress of the film material depends on temperature according to $\sigma_Y(\bar{T})$. After one temperature cycle, the response adopts a cyclic behavior for fixed temperature limits.

with temperture. In this case, the temperature $\bar{T}_Y$ at which plastic yielding first occurs in the film and the temperature $\bar{T}_{pl}$ at which plastic flow has spread throughout the film thickness differ only in second-order terms involving the small thickness ratio $\eta = h_f/h_s$. Consequently, these two temperatures which define transitions in behavior are virtually identical for thin films. Similarly, the corresponding values of substrate curvature, κ_Y at first yield and κ_{pl} for fully plastic response, are also identical to first-order in the small ratio $\eta = h_f/h_s$, so that they are also indistinguishable for thin films. Thus, for thin elastic–ideally plastic films on relatively thick substrates

$$\bar{T}_Y = \bar{T}_{pl} = \frac{\sigma_{Y_f}}{M_f(\alpha_s - \alpha_f)}, \quad \kappa_Y = \kappa_{pl} = \frac{6\sigma_{Y_f}h_f}{M_s h_s^2}. \tag{7.51}$$

Now consider the system comprising the thin elastic–ideally plastic film on a relatively thick elastic substrate which is subjected to a temperature history as shown in the inset of Figure 7.16. Starting from the reference temperature $\bar{T} = T - T_0 = 0$, the temperature is first increased to $\bar{T}_{max}$, then decreased to zero, and then again increased to $\bar{T}_{max}$. This history is plotted against 'time' but, because no aspect of the response is sensitive to rate of deformation, the detailed shape or 'waveform' of the curve of temperature versus time is unimportant except for the general property that it varies monotonically from zero to T_{max}, decreases from T_{max} to zero, and again increases to T_{max}.

The stress response of the film is constructed in Figure 7.16. The stress–temperature response is confined to the range between the dashed flow stress–temperature yield curves $|\sigma| = \sigma_Y(\bar{T})$ in this plane; if not on a yield locus, the response must follow a straight line with slope $-M_f\alpha_f$ for $|\sigma| < \sigma_Y(\bar{T})$. Starting from the point where the stress magnitude equals some arbitrary initial stress $\sigma = \sigma_0$ and $\bar{T} = 0$, the equi-biaxial stress decreases elastically along a straight line until the lower stress–temperature yield locus is reached. Thereafter, the stress increases gradually due to the assumed thermal softening of the material during heating until the temperature $\bar{T}_{max}$ is reached at the stress level $\sigma_Y(\bar{T}_{max})$. As the temperature is reduced from $\bar{T}_{max}$, the material undergoes a stress reversal with its magnitude following a straight line with slope $-M_f\alpha_f$ for $|\sigma| < \sigma_Y(\bar{T})$ and passing through zero to tensile values. Upon reaching the upper stress–temperature yield locus, extensional in-plane plastic strain begins, and it continues until the temperature is reduced to zero at a stress level of $\sigma_Y(0)$. If the temperature is again increased from zero to $\bar{T}_{max}$, the response follows an elastic line to the lower yield locus, and then moves along that yield locus until the point at which $\bar{T} = \bar{T}_{max}$ is reached. For further cycling of the temperature between the limits zero and $\bar{T}_{max}$, the response follows the closed loop established in Figure 7.16.

Once the stress history is determined or, equivalently, once the dependence of stress on temperature of a prescribed temperature history is determined, the dependence of curvature on time or on temperature follows immediately from the Stoney formula (2.7) for thin films. Qualitatively, the stress versus temperature history depicted in Figure 7.16 captures the essential features of observations such as those reported by Doerner et al. (1986) and Shen and Suresh (1995a) for Al films on the Si substrates. Stress variations based on more complex constitutive behavior are considered in the following sections.

7.5 Strain-hardening response of thin films

In this section, the description of plastic response as it applies to the behavior of thin metal films is broadened significantly. The ideas of a range of elastic behavior and of a condition for plastic yielding were adopted in the preceding section in order to discuss plastic response under thermal cycling in the most idealized circumstances. Here, the basic tenets of rate-independent plasticity theory are briefly reviewed so that subsequent discussion can be placed in the broader setting of plastic yield conditions, history dependence of plastic deformation and flow rules relating plastic strain to the prevailing state of stress in the material. The topic of rate-dependent plastic deformation will be addressed in the following section in the same vein.

No single choice of constitutive relation for a material can be expected to provide an accurate description of behavior in all circumstances. On the other hand, some structure of constitutive representation of the material must be adopted as a basis for comparison and correlation of data and, ultimately, to make predictions of material behavior for prescribed circumstances. A few of the unifying concepts in constitutive theory for rate-independent deformation of solids, primarily metals, that exhibit permanent deformation are summarized. Once a constitutive structure is established, the concepts are specialized to the states of stress and deformation typical of thin metal films.

The first of the concepts of broad applicability is the existence of a *yield surface* in stress space; this same quantity is known by a variety of names, including yield locus, the elastic limit surface and yield function. The function defining the surface is denoted by ϕ and its role is to identify the boundary of the elastic range in a space spanned by the stress components σ_{ij}. In the most general circumstances to be considered here, this function takes the form

$$\phi(\sigma_{ij}, \xi) \equiv \Phi(\sigma_{ij} - \alpha_{ij}(\xi)) - \beta(\xi), \qquad (7.52)$$

where $\Phi(\cdot)$ is a prescribed function which defines the shape of the yield surface in stress space. The scalar ξ is an internal variable that represents history of deformation in some way; the most common choices of history variables are measures of accumulated plastic strain and dissipated plastic work per unit volume at the material point. In some cases, more than one such internal variable representing history of deformation is found to be useful. The quantity α_{ij} is a reference stress, sometimes called a *back stress*, that identifies the location of the yield surface in the stress space. Finally, β is a history dependent quantity that specifies the size of the yield surface in stress space. The condition $\phi(\sigma_{ij}, \xi) = 0$ defines the yield surface for any particular value of ξ. In general, the plastic strain rate vanishes for any stress state for which $\phi < 0$, and stress states for which $\phi > 0$ are inaccessible. If $\phi = 0$ for a particular state of stress, the plastic strain rate may be nonzero but it is not necessarily so.

A concept that has been central to the development of relationships between plastic strain rate and current state of stress, which are the *flow rules* of plasticity, underlies the postulate of a *maximum plastic resistance*. This postulate can be stated in the following way. Consider an elastic–plastic material under circumstances in which the state of stress σ_{ij} satisfies the yield condition $\phi(\sigma_{ij}, \xi) = 0$. In geometrical terms, σ_{ij} is a point on the surface $\phi(\sigma_{ij}, \xi) = 0$ in stress space. Let σ_{ij}^* be *any* stress that is distinct from σ_{ij}, being either inside or on the yield surface. According to this postulate, the rate of plastic strain $\dot{\epsilon}_{ij}^{\mathrm{p}}$ will always adopt values for which the actual rate of plastic dissipation $\sigma_{ij}\dot{\epsilon}_{ij}^{\mathrm{p}}$ is equal to or greater than the rate of plastic dissipation $\sigma_{ij}^*\dot{\epsilon}_{ij}^{\mathrm{p}}$ for any other possible stress state, or

$$\sigma_{ij}\dot{\epsilon}_{ij}^{\mathrm{p}} \geq \sigma_{ij}^*\dot{\epsilon}_{ij}^{\mathrm{p}}. \qquad (7.53)$$

Immediate consequences of the postulate are that the yield function defines a convex surface in stress space† and that the plastic strain rate $\dot{\epsilon}_{ij}^{\mathrm{p}}$ is normal to the yield surface at any point at which the surface is smooth. Generalization of the feature of normality to surfaces with an edge or an apex is straightforward. The feature of normality implies that

$$\dot{\epsilon}_{ij}^{\mathrm{p}} = \Lambda \frac{\partial \phi}{\partial \sigma_{ij}}, \qquad (7.54)$$

where Λ is a scalar function of stress σ_{ij}, stress rate $\dot{\sigma}_{ij}$ and history ξ. The relation (7.54) is usually called the *associated flow rule*, reflecting the connection between the yield function and the plastic strain rate implied by a particular stress on the yield surface.

To render the response rate-independent, the quantity Λ must be linear in $\dot{\sigma}_{ij}$. In approximate terms, any stress rate that is directed from the yield surface into the elastic range of response, or that is tangent to the yield surface with $\dot{\sigma}_{ij} \, \partial \phi / \partial \sigma_{ij} = 0$, induces no plastic deformation. This implies that only the component of $\dot{\sigma}_{ij}$ acting in the direction of the outward yield surface normal contributes to plastic strain. It follows that (7.54) must have the form

$$\dot{\epsilon}_{ij}^{\mathrm{p}} = \frac{1}{H} \frac{\partial \phi}{\partial \sigma_{ij}} \frac{\partial \phi}{\partial \sigma_{kl}} \dot{\sigma}_{kl}, \qquad (7.55)$$

where H is a scalar function of plastic deformation history which must be specified in a way that renders it consistent with the yield condition. The relationship applies as long as $\phi = 0$ and $\dot{\sigma}_{ij} \, \partial \phi / \partial \sigma_{ij} > 0$; otherwise, the plastic strain rate is zero. Once the yield function ϕ and a hardening function H are chosen, the plastic response is fully specified. Two particular idealized hardening rules – isotropic hardening and kinematic hardening, both with and without temperature-dependent yield characteristics – are considered here. In the former case, the yield surface expands in stress space in a self-similar fashion without translation whereas, in the latter case, the yield surface translates in stress space without change in shape or size. In terms of the general form of the yield function in (7.52), isotropic hardening is the special case with $\alpha_{ij} \equiv 0$ and kinematic hardening is the special case with β = constant. Combinations of these two effects can be considered in order to capture various subtleties in plastic behavior. The case of isotropic hardening is examined next.

---

† A surface in stress space is said to be convex if, for any two states of stress $\sigma_{ij}^{(a)}$ and $\sigma_{ij}^{(b)}$ either inside or on the yield surface, all states of stress $\sigma_{ij} = \lambda \sigma_{ij}^{(a)} + (1 - \lambda) \sigma_{ij}^{(b)}$ for $0 \le \lambda \le 1$ are also either inside or on the yield surface.

7.5.1 Isotropic hardening

Assume that plastic response of the material is isotropic. This is usually a reasonable assumption when considering initial yielding of a material or when ongoing plastic deformation occurs without load reversals or stress cycling. The condition of isotropy implies that the yield function can depend on stress only through the principal stresses or, equivalently, through the scalar invariants of stress. For metals deforming plastically by crystallographic slip under common circumstances, response is insensitive to mean normal stress or pressure, the first stress invariant. A stress measure that has been central to the description of isotropic plastic response of metals under combined stress loading is the *Mises stress* (Von Mises 1913). This is a scalar *equivalent stress* denoted by σ_M and defined in terms of the values of principal stresses σ_1, σ_2, σ_3 at a point as

$$\sigma_M = \sqrt{\tfrac{1}{2}[(\sigma_1 - \sigma_2)^2 + (\sigma_2 - \sigma_3)^2 + (\sigma_3 - \sigma_1)^2]}. \qquad (7.56)$$

This stress measure has the properties that it is invariant under coordinate transformations, it is insensitive to changes in hydrostatic stress, it has the same value under reversal of sign of all stress components, and its value is equal to the magnitude of the tensile or compressive stress for the special case of uniaxial state of stress. From its definition in terms of differences of principal stress values, it is seen that σ_M is a measure of the tendency for shear of any state of stress. Because inelastic deformation, particularly deformation due to dislocation glide or twinning, is essentially a shearing response, the usefulness of σ_M as a driving force for such a response is not surprising. An equivalent definition of Mises stress in terms of deviatoric stress components s_{ij} is

$$s_{ij} = \sigma_{ij} - \tfrac{1}{3}\sigma_{kk}\delta_{ij} \quad \Rightarrow \quad \sigma_M = \sqrt{\tfrac{3}{2}s_{ij}s_{ij}}. \qquad (7.57)$$

If the plastic response remains essentially isotropic, then there must be an equivalent plastic strain measure, having the form of some function of the principal values of the plastic strain rate tensor, such that the Mises stress times the equivalent plastic strain rate is the local rate of plastic work per unit volume of material. In terms of the principal plastic strain rates $\dot{\epsilon}_1^P$, $\dot{\epsilon}_2^P$ and $\dot{\epsilon}_3^P$, this equivalent plastic strain rate is defined as

$$\dot{\epsilon}_M^P = \sqrt{\tfrac{2}{9}[(\dot{\epsilon}_1^P - \dot{\epsilon}_2^P)^2 + (\dot{\epsilon}_2^P - \dot{\epsilon}_3^P)^2 + (\dot{\epsilon}_3^P - \dot{\epsilon}_1^P)^2]}. \qquad (7.58)$$

The subscript 'M' is included to indicate that this plastic strain rate measure is work-conjugate to σ_M, so that $\sigma_M \dot{\epsilon}_M^P$ is the plastic work rate per unit volume of material.

The most commonly adopted isotropic yield function is the von Mises yield function which depends on stress only through the equivalent stress σ_M. If the accumulated equivalent plastic strain ϵ_M^P, which is a non-negative measure of the history of plastic deformation, is adopted as a particular choice of the history parameter ξ, then

$$\phi(\sigma_{ij}, \xi) \equiv \sigma_M - \beta(\epsilon_M^P). \tag{7.59}$$

In this case, the associated flow rule (7.55) takes the form

$$\dot{\epsilon}_{ij}^P = \frac{3}{2} \frac{1}{H(\epsilon_M^P)} \frac{\dot{\sigma}_M}{\sigma_M} s_{ij}. \tag{7.60}$$

If the magnitude of each side of this expression is formed, it follows that

$$\dot{\epsilon}_M^P = \frac{\dot{\sigma}_M}{H(\epsilon_M^P)}, \tag{7.61}$$

so that H emerges as the tangent modulus of the σ_M versus ϵ_M^P relationship during plastic flow according to an isotropic hardening rule. Note that the condition $\phi(\sigma_{ij}, \xi) = 0$ must be satisfied identically in time, so that $\dot{\sigma}_M = \beta'(\epsilon_M^P)\dot{\epsilon}_M^P$ where the prime denotes differentiation of β with respect to its argument. It follows that the functions β and H must be related by

$$\beta'(\epsilon_M^P) \equiv H(\epsilon_M^P) \tag{7.62}$$

for consistency. In view of (7.61), either of these functions of effective plastic strain can be determined from a single experimental stress–strain curve for monotonic loading, and the result completes specification of the flow rule (7.60) for any stress history and isotropic hardening.

Consider some simple cases in which the directions of principal stress and deformation are aligned with the directions of xyz coordinate axes. For the special case of plastic straining due to a uniaxial tensile stress, say $\sigma_{xx} = \sigma$ with all other stress components being zero, the Mises stress is $\sigma_M = |\sigma|$. For a plastically isotropic response, the plastic strain rate in the x-direction is denoted by $\dot{\epsilon}_{xx}^P = \dot{\epsilon}^P$ where $\dot{\epsilon}^P$ has the same sign as σ. The other two principal plastic strain rates are then equal to each other with values $\dot{\epsilon}_{yy}^P = \dot{\epsilon}_{zz}^P = -\frac{1}{2}\dot{\epsilon}^P$ for deformation without volume change. The equivalent plastic strain rate is $\dot{\epsilon}_M^P = |\dot{\epsilon}^P|$. The product $\sigma_M \dot{\epsilon}_M^P = \sigma \dot{\epsilon}^P$ is clearly the rate of plastic work per unit volume in this case.

Another special state of stress of particular relevance for the study of thin films is equi-biaxial tension, say $\sigma_{xx} = \sigma_{zz} = \sigma$ with all other stress components being equal to zero in value. The Mises stress in this case is again $\sigma_M = |\sigma|$. The nonzero components of plastic strain rate in this case are $\dot{\epsilon}_{xx}^P = \dot{\epsilon}_{zz}^P = \dot{\epsilon}^P$, $\dot{\epsilon}_{yy}^P = -2\dot{\epsilon}^P$ for plastic response that is isotropic and incompressible. It follows that $\dot{\epsilon}_M^P = 2|\dot{\epsilon}^P|$. Consequently, the rate of plastic work per unit volume in this case is $\sigma_M \dot{\epsilon}_M^P = 2\sigma \dot{\epsilon}^P$, as expected. For the case of equi-biaxial stress and strain, the relationship between plastic strain rate and stress rate given in (7.60) becomes

$$\frac{\dot{\sigma}}{\dot{\epsilon}^{\text{p}}} = 2H(\epsilon_{\text{M}}^{\text{p}}) \qquad (7.63)$$

where $\epsilon_{\text{M}}^{\text{p}}$ is twice as large as the equivalent strain under uniaxial conditions, as was noted above. It is noteworthy that the hardening rate under biaxial loading, as represented by (7.63), is significantly larger than for the same material under uniaxial loading at the given level of stress.

7.5.2 Example: Temperature cycling with isotropic hardening

A thin film of rate-independent elastic–plastic material is subjected to temperature cycling between some reference absolute temperature T_0 and a higher temperature $T_0 + \bar{T}$. The film material exhibits isotropic strain hardening, and the quantity β in the yield function (7.59) has been determined to be

$$\beta(\epsilon_M^{\text{p}}) = \sigma_{\text{Y}_{\text{f}}} \left[1 + \frac{\epsilon_{\text{M}}^{\text{p}}}{\sigma_{\text{Y}_{\text{f}}}} E_{\text{f}} \right]^{N} \qquad (7.64)$$

from a uniaxial stress experiment, where $\sigma_{\text{Y}_{\text{f}}}$ is the initial tensile yield stress, E_{f} is the elastic modulus and $N = 0.2$. This is an example of *power-law hardening* behavior. The temperature is cycled sufficiently slowly so that the temperature distribution can be assumed to be uniform throughout the material. The initial stress in the film is zero in this case and $\nu_{\text{f}} = 0.3$.

(a) If $\sigma_{\text{Y}_{\text{f}}}$ and other parameters do not depend on temperature, determine the history of normalized equi-biaxial stress $\sigma/\sigma_{\text{Y}_{\text{f}}}$ versus temperature for the range $0 \le E_{\text{f}}(\alpha_{\text{f}} - \alpha_{\text{s}})\bar{T}/\sigma_{\text{Y}_{\text{f}}} \le 2$ in the film.
(b) Suppose that the quantity $\sigma_{\text{Y}_{\text{f}}}$ is the initial yield stress at temperature $T_0 = 2\sigma_{\text{Y}_{\text{f}}}/(\alpha_{\text{f}} - \alpha_{\text{s}})E_{\text{f}}$, and that the initial yield stress at a higher temperature $\bar{T} > 0$ is $\sigma_{\text{Y}_{\text{f}}}T_0/(T_0 + \bar{T})$ for the range $0 \le \bar{T} \le T_0$. Determine the history of normalized equi-biaxial stress in the film in terms of normalized temperature within this temperature range.

Solution:

(a) For a thin film on a relatively thick substrate, the film strain is not altered significantly by deformation of the substrate. Consequently, it remains constant during temperature cycling. The contributions to the total equi-biaxial strain in this case are the elastic strain $\sigma(t)/M_{\text{f}}$, the thermal strain $(\alpha_{\text{f}} - \alpha_{\text{t}})\bar{T}(t)$ and the plastic strain $\epsilon^{\text{p}}(t)$. The time rate of change of total strain, which is the sum of these three contributions, is zero or

$$\frac{\dot{\sigma}(t)}{M_{\text{f}}} + (\alpha_{\text{f}} - \alpha_{\text{s}}) \dot{\bar{T}}(t) + \dot{\epsilon}^{\text{p}}(t) = 0, \qquad (7.65)$$

where the plastic strain rate is given in terms of stress by (7.63). The initial condition is $\sigma(0) = 0$. This ordinary differential equation is readily recast in terms of the normalized stress and normalized temperature identified above, and the resulting differential equation is integrated numerically. The temperature $\bar{T}$ is cycled between zero

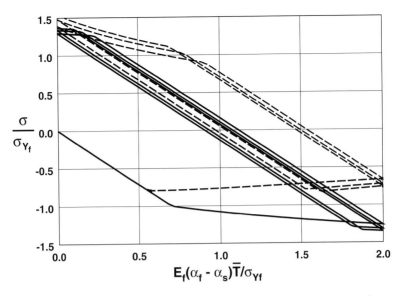

Fig. 7.17. Variation of normalized equi-biaxial film stress as a function of normalized temperature, for the example considered in Section 7.5.2 where the film material is modeled as an isotropically hardening solid. The solid lines denote the response obtained from the numerical integration of (7.65) for the first three thermal cycles where it is assumed that the material properties do not vary with temperature over the range considered. The dashed lines denote the corresponding behavior for the case where the thin film plastic response is taken to be temperature-dependent.

and its maximum value, but the results are insensitive to its actual time dependence. For present purposes, the form

$$\bar{T}(t) = \frac{2\sigma_{Y_f}}{(\alpha_f - \alpha_s)E_f} \sin^2(3\pi t), \quad 0 \leq t \leq 1 \tag{7.66}$$

is found to be convenient. The result of the numerical integration for the first three cycles is shown by the solid curve in Figure 7.17.

(b) If the plastic response of the material is temperature-dependent as prescribed, the procedure for finding stress is essentially unaltered. The only modification required from part (a) is that some coefficients in the ordinary differential equation to be integrated for stress are now time dependent. The result of numerical integration in this case for the first three cycles is shown by the dashed curve in Figure 7.17.

7.5.3 Kinematic hardening

According to the assumption of isotropic strain hardening of an elastic–plastic material, the yield surface expands or contracts in a geometrically self-similar manner in stress space, as discussed in Section 7.5.1. In the idealization of strain hardening commonly known as kinematic hardening, the yield surface retains its shape and size during ongoing plastic straining but the surface *translates* in stress space in a prescribed way. Kinematic hardening reflects a feature that is common to most metal thin films undergoing plastic deformation.

Suppose that plastic deformation first proceeds in a certain direction of plastic strain rate due to a certain state of stress. Then, let the direction of stress be reversed to the point of causing plastic straining in the opposite direction. If the magnitude of the stress at which reversed yielding first occurs is significantly less than the magnitude of the stress that caused forward yielding, the ensuing asymmetry of plastic yielding upon reversal of stress direction is called the Bauschinger effect (Bauschinger 1886). As reviewed by Suresh (1998), the origins of the Bauschinger effect in bulk metals and alloys are related to changes in the internal stress systems arising from the changes in dislocation substructure, from the interaction of dislocations with other dislocations, particles or interfaces, and from the dissolution of sub-grain boundaries during reverse loading. Structural obstacles opposing the motion of dislocations generate a back stress during forward deformation; this back stress assists the motion of dislocations upon load reversal, thereby leading to the asymmetry.

Contributions to strain hardening in a metallic material during forward deformation may arise from a variety of mechanisms, including intrinsic hardening effects such as solid solution strengthening and glide resistance of grain boundaries or interfaces; interaction of mobile dislocations with intersecting or forest dislocations in the film; and a back stress arising through dislocation interactions at barriers to glide. The first two of these three effects must be overcome in essentially the same way for forward straining as well as for reversed straining. However, the last of these three factors aids reverse deformation rather than opposing it, thereby engendering the Bauschinger effect.

For a thin metallic film on a thick substrate, deformation in the absence of any diffusive processes occurs by dislocation glide or mechanical twinning. The likely scenario is that native oxide or the passivation layer and the substrate serve as primary obstacles to shear deformation. Since the thickness of the film is small, the spread of plastic deformation by cross slip or dislocation climb is restricted. Instead, dislocations would simply glide back and forth on the planes on which they were formed (Shen et al. 1998). This process then provides a possible mechanism for the generation of back stress during cyclic variations in stress through dislocation interactions, and a mechanistic rationale for the Bauschinger effect reflected in kinematic hardening response in thin metal films on elastic substrates. This back stress can give rise to a size effect in plastic deformation of thin films (Kraft et al. 2002).

The essential features of this asymmetry are represented by a yield function that has the form (7.52) with β being constant, independent of deformation history. The location of the yield surface $\phi = 0$ in stress space for any given straining history represented by ξ is specified by the symmetric tensor $\alpha_{ij}(\xi)$. If plastic deformation occurs with zero volume change, that is, if $\epsilon_{kk}^{\mathrm{p}} = 0$ identically in time, then α_{ij} is a deviatoric tensor or $\alpha_{kk} = 0$. In this case, (7.52) takes the form

$$\dot{\epsilon}_{ij}^{\mathrm{p}} = \frac{1}{H} \frac{\partial \Phi}{\partial s_{ij}} \frac{\partial \Phi}{\partial s_{kl}} \dot{s}_{kl}, \tag{7.67}$$

where H is again a scalar function of deformation history.

Many special forms of the dependence of α_{ij} on plastic straining history have been proposed in the literature (Martin 1975). To make the discussion more concrete, suppose that the yield function (7.52) is given by

$$\Phi \equiv \sqrt{\tfrac{3}{2}(s_{ij} - \alpha_{ij})(s_{ij} - \alpha_{ij})}\,, \quad \alpha_{ij} \equiv \gamma_{\mathrm{p}}(\epsilon_M^{\mathrm{p}})\epsilon_{ij}^{\mathrm{p}}, \tag{7.68}$$

where γ_{p} is a scalar function of accumulated equivalent plastic strain. This definition of α_{ij} assumes that the 'origin' of the current yield locus in stress space is displaced from its initial position in the direction of current plastic strain. In this case, the associated flow rule (7.67) takes the form

$$\dot{\epsilon}_{ij}^{\mathrm{p}} = \frac{3}{2H\beta^2}(s_{ij} - \gamma_{\mathrm{p}}\epsilon_{ij}^{\mathrm{p}})(s_{kl} - \gamma_{\mathrm{p}}\epsilon_{kl}^{\mathrm{p}})\dot{s}_{kl}\,. \tag{7.69}$$

For uniaxial deformation with extensional plastic strain ϵ^{p} in the direction of tensile stress σ, this reduces to

$$\dot{\epsilon}^{\mathrm{p}} = \frac{1}{H}\dot{\sigma}\,. \tag{7.70}$$

Thus, H again serves the role of uniaxial incremental strain-hardening modulus, as in (7.61). Finally, the consistency condition in the case of kinematic hardening is

$$H(\epsilon_M^{\mathrm{p}}) = \gamma_{\mathrm{p}}'(\epsilon_M^{\mathrm{p}})\epsilon_M^{\mathrm{p}} + \gamma_{\mathrm{p}}(\epsilon_M^{\mathrm{p}})\,, \tag{7.71}$$

analogous to the corresponding result (7.62) for isotropic hardening.

Suppose that a thin film of elastic–plastic material bonded to an elastic substrate is subjected to temperature cycling, as in the example discussed in Section 7.5.2. Furthermore, suppose that the film material has precisely the same uniaxial stress–strain response under monotonic loading as does the material considered there. However, in the present context, the film is assumed to strain harden according to the kinematic hardening idealization. Under these circumstances of temperature cycling, with the conditions described in Section 7.5.2, the stress versus temperature history that is obtained by numerical integration of the differential equation (7.65) is shown in Figure 7.18. For the first half cycle of temperature variation, the behavior is identical to that illustrated in Figure 7.17. Thereafter, the two hardening idealizations lead to significantly different responses, even though the two materials have identical uniaxial response under monotonic loading.

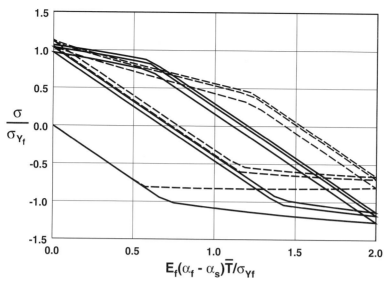

Fig. 7.18. Variation of normalized equi-biaxial film stress as a function of normalized temperature, using conditions specified in the example in Section 7.5.2. The film material is now modeled as a kinematically hardening solid. The solid lines denote the response obtained from the numerical integration of (7.65) for the first three thermal cycles where it is assumed that the material properties do not vary with temperature, over the range considered. The dashed lines denote the corresponding behavior for the case where the thin film plastic response is taken to be temperature-dependent.

A more effective or versatile model for describing plastic response of rate independent materials under complex loading histories is to adopt a yield function which combines isotropic and kinematic hardening. The general form of such a function is shown in (7.52), and the basic idea is to consider both nonzero α_{ij} and a value of β that depends on deformation history. Any such model must be calibrated by appeal to an experimental data set that is sufficiently rich in stress histories, both magnitude range and deviation from straight line or proportional loading/unloading in stress space, to deduce particular forms for the parameters involved.

7.5.4 Proportional stress history

In some circumstances of plastic deformation under multi-axial states of stress of practical interest, neither the principal stress directions nor the ratios of principal stresses themselves change over the course of time. Stress histories with this characteristic are called *proportional stressing* or *proportional loading* histories. In such cases, the state of stress can be represented by the value of a single time-dependent amplitude, the so-called stress factor or load factor, and the constant stress ratios. An elastic–plastic thin film sustaining an equi-biaxial stress is an example of a case of proportional stressing. In this instance, the stress factor is the time-dependent magnitude of the equi-biaxial stress σ and the corresponding plastic strain rate $\dot{\epsilon}^{\mathrm{p}}$ is given by (7.63) for monotonic loading with $\sigma \geq \sigma_{\mathrm{Y}}$ and either isotropic hardening, kinematic hardening or some combination of the two forms of hardening. Given the dependence of the hardening parameter on accumulated plastic strain, that equation can be integrated to yield a relationship between σ and ϵ^{p} at constant temperature.

A particular choice of the function H that has found utility in the study of thin films is

$$H(\xi) = Cn\xi^{n-1} \tag{7.72}$$

where $C > 0$ is a material constant with dimensions of stress and n is a dimensionless number in the range $0 < n < 1$. The implied relationship between uniaxial stress and plastic strain for this hardening function is the power-law hardening expression

$$\sigma = C(\epsilon^{\mathrm{p}})^n \quad \text{for} \quad \sigma \geq \sigma_{\mathrm{Y}}. \tag{7.73}$$

For the case of the same material subjected to equi-biaxial stress,

$$\sigma = C(2\epsilon^{\mathrm{p}})^n \quad \text{for} \quad \sigma \geq \sigma_{\mathrm{Y}} \tag{7.74}$$

with the same material parameters as in (7.73). For this description of hardening, the stress level is 2^n higher for a given level of plastic strain during biaxial stressing than during uniaxial stressing.

Another explicit relationship between equi-biaxial stress σ and its corresponding plastic strain ϵ^{p} that has been found useful in describing thin film behavior is a variant of power-law hardening in which current stress is related to total strain, rather than only plastic strain, according to

$$\sigma = C(\epsilon^{\mathrm{e}} + \epsilon^{\mathrm{p}})^n \quad \text{for} \quad \sigma \geq \sigma_{\mathrm{Y}}, \tag{7.75}$$

where C and n are again assumed to be material parameters. Although it is somewhat more awkward to cast this form of the relationship between stress and plastic strain in the form of the rate equation (7.63), it has the advantage that C can be determined in terms of the biaxial elastic modulus M_{f} and the yield stress $\sigma_{\mathrm{Y_f}}$ of the film material. This is accomplished by recognizing that $\epsilon^{\mathrm{e}} = \sigma_{\mathrm{Y_f}}/M_{\mathrm{f}}$ and $\epsilon^{\mathrm{p}} = 0$ when $\sigma = \sigma_{\mathrm{Y_f}}$ during initial loading, from which it follows that $C = M_{\mathrm{f}}^n \sigma_{\mathrm{Y_f}}^{1-n}$.

Observations of plastic response of metal thin films deforming plastically under conditions of proportional stressing with occasional stress reversals suggest that the monotonic loading stress–strain relation (7.75) can also be used to describe response under stress reversal (Shen et al. 1998). Suppose that a film is deformed under forward loading according to (7.75) until some state of stress σ_* and total strain $\epsilon_* = \epsilon_*^{\mathrm{e}} + \epsilon_*^{\mathrm{p}}$ is achieved. Upon stress reversal, within the constraint of proportional stressing, the film deforms elastically according to $\sigma_* - \sigma = M_{\mathrm{f}}(\epsilon_*^{\mathrm{e}} - \epsilon^{\mathrm{e}})$ as σ

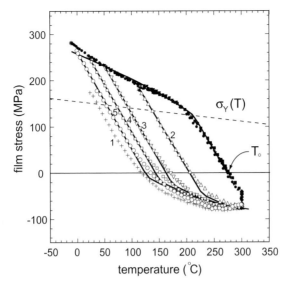

Fig. 7.19. Experimentally determined stress versus temperature hysteresis data for a 1 μm thick Al film deposited on a relatively thick elastic substrate. The specimen was first heated from room temperature to 300 °C (the data point set marked '1'), held at that temperature for 30 min, and then subsequently cooled to a minimum temperature before being heated again to 300 °C. This minimum temperature was chosen to be 110, 50, 20 and −10 °C for the four thermal cycles, the heating portions of which are denoted by the numbers 2, 3, 4 and 5, respectively. The specimen was held at 300 °C for 30 min during each thermal cycle. The solid curves in Figure 7.19 show the response for elastic and plastic deformation implied by (7.75) and (7.76). T_{o} denotes the stress-free reference temperature. Experimental data provided by Y. J. Choi, Massachusetts Institute of Technology (2002).

decreases from σ_* until $\sigma_* - \sigma = 2\sigma_{Y_f}$. For further reduction in σ, the plastic response can be approximated by

$$\sigma_* - \sigma = 2M_{\mathrm{f}}^n \sigma_{Y_{\mathrm{f}}}^{1-n} \left[\tfrac{1}{2}(\epsilon_*^{\mathrm{e}} - \epsilon^{\mathrm{e}}) + \tfrac{1}{2}(\epsilon_*^{\mathrm{p}} - \epsilon^{\mathrm{p}}) \right], \qquad (7.76)$$

where the values of material parameters are the same as those for forward loading. Any temperature dependence of plastic flow arises through the temperature dependence of $\sigma_{Y_{\mathrm{f}}}$. This description presumes that the extent of the elastic range at the prevailing temperature is $2\sigma_{Y_{\mathrm{f}}}$. Aside from the kinematic translation of the plastic range along the stress axis during plastic deformation, dependence of the response on plastic deformation history is wiped out upon each load reversal.

The stress–temperature histories implied by the constitutive equations for kinematic hardening under proportional stressing given in (7.75) and (7.76) are compared with experimental observations of stress–temperature variations in thin Al films subjected to thermal cycling in Figure 7.19. The discrete data points in this figure show the experimentally determined film stress versus temperature variation for several thermal cycles for a 1 μm thick Al–1 wt% Si film which was deposited at room temperature onto a 525 μm thick (100) Si which had a diameter of 10 cm. The specimen was first heated from room temperature to 300 °C, held at that temperature for 30 min, and then cooled to a minimum temperature before being heated again to 300 °C. This minimum temperature was chosen to be 110, 50, 20 and −10 °C for the four thermal cycles for which experimental results are plotted in Figure 7.19. A heating rate of 10 °C min^{-1} and a cooling rate of 6 °C min^{-1} were used; during each cycle the specimen was held at 300 °C for 30 min so as to relax the film stress. The substrate curvature was continuously monitored during thermal cycling by recourse to the scanning laser method; film stress was then determined from the substrate curvature using the Stoney formula (2.7). The first heating cycle was applied to exhaust the tendency for grain growth in the film material, and comparison between observations and the model stress–strain relations was considered only after completion of this half cycle. Thus, the comparison begins with the film at the stress-free reference temperature of approximately $T_0 = 275$ °C. The solid curves in Figure 7.19 show the predicted constitutive response for elastic and plastic deformation using (7.75) and (7.76). The dashed line in this figure represents the assumed variation of tensile yield stress of the film as a function of temperature, which was approximated by the linear relationship

$$\sigma_{\mathrm{Yf}} = \sigma^* \left(1 - \frac{T}{T^*} \right), \tag{7.77}$$

where T is the prevailing absolute temperature at any instant, σ^* is a reference yield strength at absolute zero, and temperature T^* is chosen so that $-\sigma^*/T^*$ approximates the gradient of yield stress with respect to temperature in the range of interest (Shen et al. 1998). A single choice of values for the material constants with $n = 0.4$, $\sigma^* = 190\,\mathrm{MPa}$ and $T^* = 1373\,\mathrm{K}$ is seen to match the observed film stress variation with temperature during all four thermal cycles.

7.6 Models based on plastic rate equations

The range of material behavior considered next is broadened significantly by appeal to the notion of a plastic rate equation as a model for any possible physical mechanism of deformation that may be operative. The ideas will be developed for general states of stress, but will be applied primarily for the case of thin films in equi-biaxial tension. Constitutive relationships that serve as models for inelastic response of materials for a wide variety of physical mechanisms of deformation have been compiled by Frost and Ashby (1982). These constitutive equations are represented as scalar equations expressing the inelastic equivalent strain rate $\sqrt{3}\dot{\epsilon}_{\mathrm{M}}^{\mathrm{p}}$ in terms of the effective stress $\sigma_{\mathrm{M}}/\sqrt{3}$ and temperature T. These strain rate and stress measures are denoted by $\dot{\gamma}$ and σ_{s} by Frost and Ashby (1982), and the rate equations representing models of material behavior all take the form

$$\dot{\gamma} = f_{re}(\sigma_s, T) \quad \Leftrightarrow \quad \dot{\epsilon}_M^p = \frac{1}{\sqrt{3}} f_{re}\left(\sigma_M/\sqrt{3}, T\right), \tag{7.78}$$

where the function f_{re} includes all the material parameters and microstructural features that are needed to describe behavior. When written in this form, it is tacitly assumed that microstructural features and values of material parameters remain unchanged as plastic straining proceeds at a material point under the action of applied stress; in other words, a 'steady state' of deformation is deemed to occur during which there is no net change in key microstructural features such as dislocations, grain boundaries, etc. Because (7.78) is a first-order differential equation, an initial condition of some kind is required if it is to be integrated over time to examine deformation history.

For a general state of stress and deformation at a material point, how are individual components of plastic strain rate related to stress components in this framework? An answer is provided through the work of Rice (1970) on the general structure of stress–strain relations for time-dependent plastic deformation. In the present setting, it is most conveniently expressed in terms of deviatoric stress components s_{ij} defined in terms of stress σ_{ij} in (7.57). Rice (1970) has shown that the components of plastic strain rate $\dot{\epsilon}_{ij}^p$ have the representation

$$\dot{\epsilon}_{ij}^p = \Lambda s_{ij}, \tag{7.79}$$

where Λ may depend on stress through the value of σ_M corresponding to s_{ij}, on plastic strain history, and possibly on other parameters. For (7.79) to be consistent with the rate equation (7.78), Λ must have the value

$$\Lambda = \frac{\sqrt{2}}{3} \frac{f_{re}\left(\sigma_M/\sqrt{3}, T\right)}{\sigma_M}. \tag{7.80}$$

Thus, given the state of stress and the temperature, the current rate of plastic straining is specified through (7.79) and (7.80).

Some implications of these models of material behavior based on plastic rate equations are considered in the following subsections. One of the practical advantages in using plastic rate equations for this purpose is that they are presumed to be in effect over all possible values of equivalent stress and temperature. Whether or not the plastic strain corresponding to any particular physical mechanism of plastic response actually contributes in a significant way to the total plastic strain rate is determined by the instantaneous magnitudes of stress and temperature. This feature distinguishes the present approach from the classical plasticity theories of structural mechanics which are based on the existence of a well-defined yield locus in stress space bounding a region of stress states for which the response is strictly elastic.

Such a case was considered in Section 7.5. This distinction will be illustrated in the representative calculations that follow. For example, if rate equations admitting the possibility of response according to several different mechanisms are included in the expression for total plastic strain rate, the particular mechanism dominating response may change over the course of a process as the magnitudes of stress and/or temperature change.

7.6.1 Thermally activated dislocation glide past obstacles

Consider once again the specific phenomenon treated in Section 7.5, namely, a thin film bonded to a relatively thick substrate subject to a periodic fluctuation in temperature that varies over time between some initial relative temperature $\bar{T} = T - T_0$, where T_0 is a reference temperature, and a higher temperature $\bar{T} = \bar{T}_{\max}$. The material is again assumed to be isotropic and to exhibit elastic–plastic behavior in this temperature range. The initial state at the time $t = 0$ is characterized by an initial value of the biaxial stress $\sigma(t)$, and all components of plastic strain are assumed to be zero initially.

The observation that the total strain in the film in any direction along the interface does not change significantly during temperature cycling leads immediately to an ordinary differential equation for the time history of equi-biaxial stress $\sigma(t)$ in the film during temperature cycling. The fact that the total strain remains unchanged implies that the total strain rate is zero. The rate of change of in-plane elastic biaxial strain is $\dot{\epsilon}^e(t) = \dot{\sigma}(t)/M_f$ where any temperature sensitivity of M_f is neglected. The stress-free biaxial thermal strain in the film in any direction is $\alpha_f T(t)$ so that the rate of change of thermal strain is $\dot{\epsilon}^T(t) = \alpha_f \dot{T}(t)$ in terms of the prescribed temperature history. This expression ignores thermal expansion of the substrate, an effect that is easily taken into account simply by replacing α_f with the relative coefficient of thermal expansion $(\alpha_f - \alpha_s)$. Finally, the in-plane biaxial plastic strain rate $\dot{\epsilon}^p(t)$ is expressed in terms of stress and temperature by means of one or more plastic rate equations.

In the present discussion, it will be assumed that the plastic response is dominated by the thermally activated glide of crystal dislocations past discrete obstacles in the lattice. With reference to (7.78), the plastic rate equation proposed by Frost and Ashby (1982) has the form

$$f_{re}(\sigma_s, T) = \dot{\gamma}_0 \exp\left[-\frac{\Delta F}{k(T_0 + \bar{T}(t))} \left(1 - \frac{\sigma_s}{\bar{\tau}}\right) \right] \qquad (7.81)$$

where $\dot{\gamma}_0$ is a strain rate material parameter with physical dimensions of time^{-1}, ΔF is a material parameter with physical dimensions of energy representing the activation energy of the barrier to dislocation motion, $\bar{\tau}$ is a material parameter with physical dimensions of stress representing the applied stress level necessary to push dislocations past obstacles at very low temperature, k is the Boltzmann constant, and T_0 is the absolute reference temperature. Note that $T_0 + \bar{T}(t)$ is then the absolute temperature of the material.

As noted in Section 7.6, the effective stress and equivalent biaxial strain for equibiaxial tension of an isotropic film are $\sigma_M = |\sigma|$ and $\dot{\epsilon}_M^p = 2|\dot{\epsilon}^p|$. The condition that the total rate of strain parallel to the interface is zero implies that $\dot{\epsilon}^e(t) + \dot{\epsilon}^p(t) + \dot{\epsilon}^T(t) = 0$ or

$$\frac{\dot{\sigma}(t)}{M_f} + \frac{\text{sgn}[\sigma(t)]}{2\sqrt{3}} \dot{\gamma}_0 \exp\left[-\frac{\Delta F}{k(T_0 + \bar{T}(t))}\left(1 - \frac{|\sigma(t)|}{\sqrt{3}\,\bar{\tau}}\right)\right] + \alpha_f \dot{T}(t) = 0. \quad (7.82)$$

For a prescribed history $\bar{T}(t)$ of temperature for $t > 0$, relative to an initial absolute reference temperature T_0, and an initial value of stress $\sigma(0) = \sigma_0$, (7.82) is an ordinary differential equation for stress history. The signum function $\text{sgn}[\sigma]$ introduced in (7.82) has value $-1, 0$ or $+1$ according to whether its argument $\sigma(t)$ is negative, zero or positive, respectively.

As an illustration, suppose that a periodic relative temperature history

$$\bar{T}(t) = \bar{T}_{\max} \sin^2\left(\frac{\pi t}{t_0}\right) \quad (7.83)$$

with time period t_0 is imposed on a copper film bonded to a relatively thick substrate. The values of $T_{\max} = 400\,°\text{C}$ and $t_0 = 1\,\text{h}$ are assumed for the calculation. The thermoelastic properties of copper are $E_f = 1.15 \times 10^{11}\,\text{N m}^{-2}$, $\nu_f = 0.31$ and $\alpha_f = 17 \times 10^{-6}\,°\text{C}^{-1}$. The values of the parameters in the plastic rate equation as given by Frost and Ashby (1982) are $\dot{\gamma}_0 = 10^6\,\text{s}^{-1}$, $\Delta F = 3.5 \times 10^{-19}\,\text{J}$ and $\bar{\tau} = 2.4 \times 10^8\,\text{N m}^{-2}$. Finally, it is assumed that the reference temperature is approximately room temperature, or $T_0 = 300\,\text{K}$.

The result of determining an approximate solution of the ordinary differential equation (7.82) by numerical integration is shown in Figure 7.20. The figure is a parametric plot of stress history versus temperature history over several cycles of temperature, with the time t as the coordinating parameter between the histories. The stress–temperature loop shown remains essentially unchanged after the first temperature cycle.

It may be surprising that the computed response shown in Figure 7.20 has essentially the same form as the schematic diagram of behavior for an elastic–ideally plastic response in Figure 7.16, in view of the fact that the former is based on a rate equation while the latter is a case of strictly rate-independent behavior. However, that this might be a reasonable outcome can be seen through a comparison of the terms in the differential equation (7.82). Once plastic straining begins, the rate of elastic straining becomes relatively small, that is, the second and third terms in

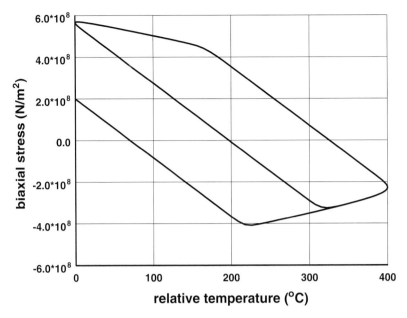

Fig. 7.20. The stress–temperature history for a thin film of elastic–plastic material on a relatively thick substrate during a temperature excursion is shown. The dependence of plastic strain rate on stress is prescribed by the plastic rate equation (7.81) and is based on the assumption that resistance to plastic deformation is due to thermally activated motion of dislocations past obstacles. After one temperature cycle, the response adopts a repeated cyclic behavior for fixed temperature limits.

the differential equation balance each other with only a modest contribution from the first term. The magnitude of the thermal strain rate is essentially $\alpha_f T_{\max}/2t_0$. Therefore, stress necessarily varies with temperature in such a way that

$$-\frac{\Delta F}{k(T_0 + T(t))}\left(1 - \frac{|\sigma(t)|}{\sqrt{3}\bar{\tau}}\right) = \ln\frac{\alpha_f T_{\max}\sqrt{3}}{t_0\dot{\gamma}_0} \qquad (7.84)$$

during plastic straining. This is the equation of a straight line in the plane of σ versus T as reflected in the computed behavior in Figure 7.20 and as is shown for the case of ideally plastic response in Figure 7.16.

7.6.2 Influence of grain boundary diffusion

The graph of film stress versus temperature in Figure 7.20 is based on the *a priori* assumption that all inelastic deformation in the film is due to obstacle-controlled glide of dislocations. Whether or not this is actually the case in any particular situation can be established only through appeal to experimental data. Perhaps a more rational approach to associating response characteristics with physical mechanisms is to include contributions to the total plastic strain rate for *all* mechanisms that might be active, and then to discard any of those contributions that do not influence the response for the relevant ranges of stress and temperature. The purpose in this subsection is less ambitious than incorporation of all possible rate equations. Here, only two possible mechanisms will be included to illustrate how the net response can be dominated by different mechanisms over different regions of the stress–temperature plane. The issue will be pursued by adding the possibility of inelastic straining due to mass transport along grain boundaries in the film microstructure as a result of gradients in grain boundary chemical potential.

The phenomenon of inelastic straining at constant stress due to mass transport along grain boundaries is known commonly as Coble creep (Frost and Ashby 1982). The driving force for this thermally activated process is a gradient in normal stress acting on the grain boundaries. In the case of a thin film with columnar grain structure, the grain boundary surfaces are predominantly normal to the film–substrate interface. For an equi-biaxial state of stress, there are no significant gradients of stress on the grain boundaries. However, if the surface of the film is free, mass can readily flow over the free surface at elevated temperatures to be drawn into or pushed out of grain boundaries in response to differences between the surface and grain boundary chemical potentials. This mass transport mechanism can contribute to overall inelastic straining in response to applied stress.

The process of combined surface diffusion and grain boundary diffusion was discussed in detail by Thouless (1993) who developed a plastic rate equation to describe the process. In the present notation, this rate equation has the form

$$\dot{\epsilon}^{\text{p}}_{\text{M}} = \frac{6\Omega\sigma_{\text{m}}}{kTdh_{\text{f}}^2} D_{0\text{b}} \exp\left[-\frac{Q_{\text{b}}}{kT}\right] \tag{7.85}$$

where Ω is the atomic volume of the material, $D_{0\text{b}}$ is the pre-exponential factor of the diffusion coefficient, Q_{b} is the activation energy for grain boundary diffusion, and d is the mean grain size of the columnar structure as observed in a plane parallel to the film–substrate interface. Other parameters retain the definitions established earlier in this section. It follows that the differential equation governing the time history of film stress $\sigma(t)$ is identical to that given in (7.82), but now with the term

$$\frac{3\Omega\sigma(t)}{dh_{\text{f}}^2 k(T_0 + \bar{T}(t))} D_{0\text{b}} \exp\left[-\frac{Q_{\text{b}}}{k(T_0 + \bar{T}(t))}\right] \tag{7.86}$$

included on the left side. The values of the parameters appropriate for copper are $\Omega = 1.18 \times 10^{-29}\,\text{m}^3$, $D_{0\text{b}} = 5.0 \times 10^{-15}\,\text{m}^3\,\text{s}^{-1}$ and $Q_{\text{b}} = 1.728 \times 10^{-19}\,\text{J}$.

For columnar grain structure, the value of d typically falls within the range $0.1 \leq d/h_{\text{f}} \leq 10$. Approximate solutions of the differential equation for stress

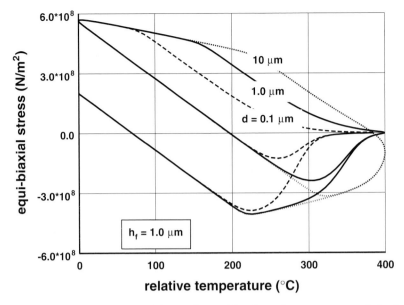

Fig. 7.21. The stress–temperature history for a thin film of copper on a relatively thick substrate during a temperature excursion is shown. The dependence of plastic strain rate on stress includes (7.81) to represent obstacle-controlled glide of dislocations and (7.86) to represent grain boundary diffusion. Results are shown for film thickness of 1 μm and grain sizes of 0.1 μm, 1.0 μm and 10 μm. In each case, after one temperature cycle, the response adopts a repeated cyclic behavior for fixed temperature limits.

history have been determined by numerical integration with initial stress $\sigma_0 = 2 \times 10^8$ N m^{-2}, film thickness of $h_f = 1\,\mu$m and grain sizes of $d = 0.1\,\mu$m, $1\,\mu$m and $10\,\mu$m. The temperature history is again given by (7.83). The results are shown in Figure 7.21. The stress–temperature material response for the largest of the three grain sizes represented by the results in this figure is dominated by dislocation glide, and the influence of grain boundary diffusion on overall response is relatively minor. On the other hand, for the smallest of the three grain sizes, the response is altered in a dramatic way by the role of grain boundary diffusion on stress relaxation at the higher temperatures in the temperature cycling range. Loading rate and bulk parameters are identical in the two cases, and this difference in response is solely a consequence of the difference in grain size of the film.

By combining continuum analysis of inelastic deformation, such as that described in Section 7.4, with the mechanistic models of the foregoing subsections, Shen and Suresh (1996) have identified the variation of equi-biaxial film stress through the thickness of the film during steady-state creep deformation. An appealing feature of such an approach is that the evolution of substrate curvature, spatial variation of residual stress and relative dominance of different strain relaxation mechanisms can be identified for a given film–substrate system subjected to a known thermal history.

The results reveal that different deformation mechanisms can become dominant at different through-thickness locations of a bilayer of arbitrary layer thickness at any particular stage of thermal cycling. The principal drawback of this approach, however, is that transient effects commonly associated with the strain relaxation phenomena are not captured in the analysis.

7.7 Structure evolution during thermal excursion

Strain relaxation in thin films was examined in the preceding sections by considering two extreme situations: epitaxial films with a very low density of dislocations where the interactions between dislocations can be used to rationalize the overall strain relaxation process, and films with a very high density of dislocations where phenomenological constitutive models are used to relate experimental observations to continuum level analysis of time-independent or time-dependent inelastic deformation. As stated earlier in this chapter, the simplifying assumptions invoked in each of these approaches facilitate the derivation of analytical results that provide quantitative understanding of the overall response of thin films. It is commonly found, however, that many thin film materials of technological interest exhibit microstructural evolution processes which can lead to observed behavior that is markedly different from that presumed in these models. Some examples of these processes involving structural changes are examined in this section.

7.7.1 Experimental observation of grain structure evolution

To keep the discussion consistent with the analysis presented in Section 7.4, attention is focused on the evolution of film microstructure in response to mismatch strain generated by thermal excursion. Figure 7.22(a) shows an example of a cross-sectional TEM view of a 0.63-μm thick, unpassivated Al film which was sputter-deposited onto a 620 μm thick, oxidized (100) Si substrate; the film has an average grain size of approximately 0.39 μm in the as-deposited condition. Note that the Al film surface is actually coated with a layer of native oxide, approximately 5 nm in thickness. This initial microstructure reveals grains and sub-grains or dislocation cells separated by boundaries and dislocation networks of irregular shapes. Regions within the interior of the grains are populated with randomly distributed dislocations as well as piled-up dislocations. Two key features of this as-deposited film microstructure are that the average size of the grains and dislocation cells is smaller than the film thickness, and that the grains are irregular in shape, suggesting that the microstructure is not in a low energy state. Figure 7.22(b) is a cross-sectional TEM image of a similarly produced unpassivated Al film, 1-μm thick, on a similar Si substrate; this image was obtained after the film–substrate system was subjected

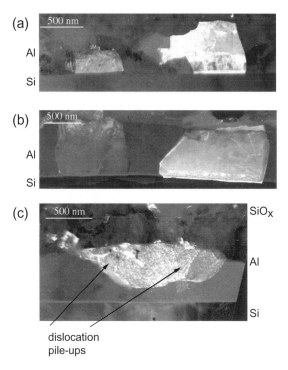

(a)

Al

Si

(b)

Al

Si

(c)

SiO_x

Al

Si

dislocation
pile-ups

Fig. 7.22. Cross-sectional TEM images of the grain and dislocation structures in (a) an unpassivated 0.63-μm thick aluminum film after sputter deposition, (b) an unpassivated 1-μm thick aluminum film after three thermal cycles between 20 and 450 °C, and (c) a 1-μm) thick aluminum film capped with a 1-μm thick SiO_x passivation layer after 20 thermal cycles between 20 and 450 °C, all on a Si substrate. Reproduced with permission from Legros et al. (2002).

to three thermal cycles between 20 and 450 °C. The first heating cycle was found to result in significant grain growth as well as alignment and straightening of grain boundaries in directions essentially normal to the film–substrate interface. The grain size of the film in Figure 7.22(b) is comparable to the film thickness, and thermal cycling has also resulted in grain interiors that are essentially free of dislocations. This image, along with the *in-situ* TEM studies by Legros et al. (2002) of structure evolution during temperature excursion, reveals that the first few thermal cycles result in grain growth, disentanglement of low angle grain boundaries, and absorption of dislocations into the metal–oxide and metal–substrate interfaces. Grain growth is also accompanied by preferential texturing of the film which tends to eliminate crystallographic orientations in the grain with the largest strain misfit with the substrate. Grain growth is complete when all the grain boundaries have reoriented to become perpendicular to the interface resulting in large prismatic grains. Once this saturation state is reached, no further changes in the microstructure occur with continued thermal cycling.

When the aluminum film is passivated with an oxide capping layer, the evolution of microstructure in response to temperature excursion shows distinct differences from that seen for an unpassivated film, even though the initial grain structure and dislocation structure are very similar for the two cases. Figure 7.22(c) is a cross-sectional TEM image of the 1-μm thick Al film on the Si substrate which was capped with a 1-μm thick SiO_x passivation layer at 300 °C by means of low-pressure chemical vapor deposition (LPCVD) and subsequently subjected to 20 thermal cycles between 20 and 450 °C. The presence of the capping layer prolongs the grain growth phase of microstructural evolution well beyond the first thermal cycle, and it results in a noticeable anisotropy in the shape in the grains, with the extent of grain growth in the film thickness direction being relatively smaller than in the in-plane direction. For the particular thermal excursion imposed in this case, several thermal cycles were necessary prior to cessation of grain growth. These experimental observations are also consistent with the substrate curvature measurements of Shen and Suresh (1995b) who found that the presence of a passivation layer strongly influenced the number of thermal cycles needed to reach a steady-state stress–temperature hysteresis loop for the Al film on Si substrate. As seen from Figure 7.22(c), passivation also leads to dislocation pile up at grain boundaries.

7.7.2 Experimental observation of threading dislocations

As discussed in Chapter 6 and in Sections 7.1 through 7.3 of this chapter, the propensity for threading dislocations to advance or recede provides a mechanistic basis for understanding elastic strain relaxation in thin films. If the threading dislocation is isolated from other defects, the underlying issue is the critical thickness condition. On the other hand, for circumstances well beyond that of critical thickness for dislocation formation, the interaction of the threading dislocation with its surroundings – other threading dislocations, misfit dislocations in its path, grain boundaries and so on – underlies macroscopic relaxation of elastic strain. *In-situ* observation of the formation and motion of threading dislocations in non-epitaxial metal films during the development of mismatch strain is a challenging task owing to the difficulties in the preparation of thin foil specimens of the layered structure for observation in a TEM; if the mismatch strain derives from lattice mismatch and the constraint of epitaxy in a single crystal film then the task is simplified considerably. Challenges arise in mounting the foil in a specimen holder in which a controlled mismatch strain can be imposed and in imaging the nucleation and motion of dislocations during such straining. *In-situ* TEM studies on the motion of dislocations in metal films during the simultaneous imposition of a mismatch strain have been reported by Kuan and Murakami (1982), Venkatraman et al. (1990) and Jawarani et al. (1997); these deal primarily with plan-view observation of the film–substrate system, the situation in which the motion of a threading dislocation extending through the thickness of the thin film and the interaction of such a dislocation with the film–substrate interface or other defects cannot be documented during the experiment. An advantage of plan-view observations is that the state of stress in the film may be the same as or similar to the state of stress in the film prior to preparation of the TEM sample.

In an attempt to view the through-the-thickness behavior of threading dislocations in a thin metal sample, Keller-Flaig et al. (1999), Legros et al. (2001) and Legros et al. (2002) introduced a wedge-shaped cross-sectional TEM specimen resembling a passivated thin metal film on a Si substrate. Mismatch strains were induced by subjecting the specimen to thermal excursion inside the TEM, and simultaneously performing *in-situ* observations of threading dislocation motion. The material chosen for investigation was a SiO_x-passivated polycrystalline aluminum film on a (100) Si substrate. The 1-μm thick aluminum layer was sputter-deposited on a (100) Si wafer, 620 μm in thickness and 10 cm in diameter. Prior to film deposition, an 18-nm thick amorphous oxide layer was formed on the Si wafer surface as a diffusion barrier between Al and Si. The aluminum film was subsequently capped with a 0.5-μm thick SiO_x passivation layer which was introduced by LPCVD at a temperature of 300 °C.

Cross-sectional TEM foils were made by removing rectangular specimens, approximately 0.5 mm × 2.0 mm in size, from the passivated Al films. Then two wafer pieces were glued together and the resulting sample was polished in a tripod sample holder to form a wedge shape with a mirror finish and an electron transparent tip. Thermal mismatch strains could be induced in a controlled manner by cycling the foil between 20 °C and 450 °C at a heating or cooling rate of 10–40 °C min^{-1} in a high temperature specimen holder. The complex geometry of the layer structure as well as the state of stress in the wedge-shaped specimen differs markedly from the equi-biaxial state of stress in a continuous thin film.

Grain growth during the first two thermal cycles results in the formation of a columnar grain structure with individual grains spanning the thickness of the film; the grains themselves are relatively free of dislocations. The dislocation density decreases with each successive thermal cycle and the remaining grain boundaries migrate until they are oriented in the direction perpendicular to both the film–substrate and the film–oxide interfaces. Figures 7.23(a)–(d) show a sequence of images captured during the third thermal cycle; the motion of threading dislocations induced by thermal mismatch strain is evident in this sequence. Figure 7.24(a) schematically shows the thermal cycle in which the points a, b, c and d correspond to the instants during the cooling sequence at which the images in Figures 7.23 identified by the matching letter were captured. Figure 7.24(b) is a schematic illustration of the progression of dislocation 1 through the Al grain, and it serves as an aid in interpreting the electron micrographs of Figures 7.23(a)–(d). Two threading

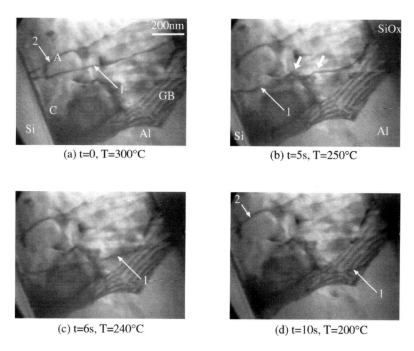

(a) t=0, T=300°C

(b) t=5s, T=250°C

(c) t=6s, T=240°C

(d) t=10s, T=200°C

Fig. 7.23. The transgranular motion of threading dislocation in a 1-μm thick polycrystalline aluminum film which is passivated with a 0.5-μm thick silicon oxide layer. The TEM images in Figures (a)–(d) show the positions of two threading dislocations, marked 1 and 2, during the cooling phase of a thermal cycle. Dislocation 2 is immobile during this entire sequence, whereas dislocation 1 traverses through the grain. The short, thick arrows in (b) denote two points at which a segment of dislocation 1 is pinned before breaking away. The small curvature of the dislocation about the pinning points indicates that they serve as only weak obstacles to dislocation motion. Reproduced with permission from Legros et al. (2002).

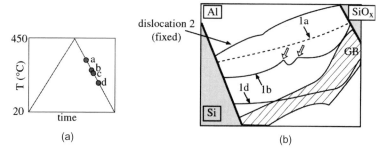

(a)

(b)

Fig. 7.24. (a) Schematic of the thermal cycle with indications of the instants, a, b, c and d, at which the images in Figures 7.23(a)–(d), respectively, are captured. (b) Schematic of the movement of threading dislocation 1 during this sequence of imaging in Figures 7.23(a)–(d), where the position corresponding to Figure (c) has been omitted for clarity. The shaded region in (b) denotes the fringes along the grain boundaries. Adapted from Legros et al. (2002).

dislocations marked 1 and 2 are visible in these figures; both are nearly perpendicular to the Al–Si interface. Dislocation 2 is immobile during this sequence and can be used as a reference. Dislocation 1 traverses a distance of several hundred nms within the Al grain without encountering any obstacles. The two pinning points indicated by the thicker arrows in Figure 7.23(b) are easily overcome, as is seen in Figure 7.23(c) as dislocation 1 continues to traverse the grain. The dislocation is ultimately arrested at the grain boundary in Figure 7.23(d). The role of threading dislocations in influencing the plastic yield properties of thin films will be considered in later sections of this chapter.

7.7.3 Strain relaxation mechanisms during temperature cycling

It is evident from the discussion presented in Section 7.4 that observation of substrate curvature, interpreted as mean stress in a film bonded to an elastic substrate, during temperature cycling provides a convenient framework with which the inelastic strain relaxation characteristics of thin film materials can be inferred. Indeed, substrate curvature measurement made in the course of temperature cycling is among the most common experimental methods employed to determine the elastoplastic response of thin films. In addition to the scientific interest, such observations also mimic excursions in temperature experienced in processing steps such as film deposition, passivation and material characterization, or in service functions.

The various curvature measurement methods described in Section 2.3 provide convenient tools with which the film stress can be assessed by using the approaches outlined in Chapters 2 and 3 and Section 7.4. However, these approaches are often predicated on assumed microstructural conditions which may differ markedly from real behavior in some material systems. Microstructural evolution and competition between strain relaxation mechanisms can influence the evolution of substrate curvature during thermal excursions and the manner in which such effects are strongly affected by the geometry and material properties of the particular thin film–substrate system being considered. The following points are discussed in order in the remainder of the section, primarily in the context of polycrystalline metal films:

- competing effects of plastic strain hardening and thermal softening,
- grain growth in the film during the first temperature cycles,
- hillock formation on the film surface,
- formation of voids or cracks during the tensile portion of cycles, and
- stabilization of cyclic response beyond the grain growth stage.

The thin film, modeled as an elastic–ideally plastic material, was discussed in Section 7.4, where it was shown that a closed loop of film stress (or, equivalently, elastic substrate curvature) versus temperature was established after the first thermal cycle. When the film–substrate system is first heated from the reference state, initial elastic deformation leads to a linear variation of curvature with temperature. Once the film begins to undergo plastic flow at $\bar{T}_Y$, substrate curvature does not change with further increase in temperature if the yield strength of the ideally plastic film does not vary with temperature. If the film material exhibits strain hardening and no appreciable change in yield strength with temperature, over the range of temperature of interest, then the magnitude of substrate curvature increases with further increase in temperature beyond the point of film yield, as illustrated in Figure 7.17. Since the yield strength of the film decreases with increasing temperature, as illustrated in Figures 7.16 and 7.17, the magnitude of film stress and substrate curvature may increase, decrease or remain the same with further increase in temperature depending on the relative dominance of the increase in film flow stress due to strain hardening and the decrease in film flow stress due to rising temperature. These different possibilities arise even in the complete absence of any change in the grain structure or dislocation structure within the film.

As described in Section 7.7.1, metallic thin films may undergo grain growth during the first heating cycle. The densification of the film resulting from such grain growth and the ensuing net reduction in the grain boundary area and dislocation density partially relieves the compressive stress in the film, thereby reducing the magnitude of change in substrate curvature measured with respect to that at the reference state. The early thermal cycles also lead to reorientation of grain boundaries, as illustrated in Figure 7.22, so as to create the least strain mismatch with the substrate. These changes in both grain size and texture can also alter the yield strength, the strain hardening response, the temperature dependence of yield strength, and even the elastic or plastic anisotropy of the film. Consequently, microstructural changes induced by the first heating phase can influence the evolution of film stress and substrate curvature in a manner that is not taken into account in the methods presented in Section 7.4.

The evolution of structural damage in the thin film during the first heating cycle can introduce additional complications in the estimation of film stress. The equibiaxial compressive mismatch stress generated in a metal film on Si or SiO_2, owing to its larger thermal expansion coefficient, causes the film material to extrude out of plane through the free surface. As a result, the film material can 'squeeze out' through any defects or cracks in its protective oxide layer or passivation layer, producing a mound of extruded material known commonly as a *hillock*. Figure 7.25 shows a focused-ion-beam (FIB) image of the cross-section of a hillock formed in a blanket thin film on a Si substrate following an annealing treatment at 450 °C. This figure also shows the elongated structure of the grains within the extruded region of the hillock. Hillocks represent permanent damage to the film material which is not relieved upon temperature reversal.

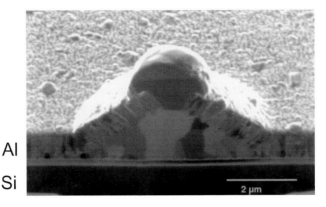

Fig. 7.25. Cross-sectional focused-ion-beam image of a hillock penetrating through the surface of a 1-μm thick Al film on a relatively thick (100) Si substrate. Reproduced with permission from Kim et al. (2000).

Fig. 7.26. An SEM image of a whisker formed in an Al film extruding through the surface of a TiN capping layer. Reproduced with permission from Blech et al. (1975).

While short time exposure to elevated temperature usually results in hillock formation, prolonged high temperature annealing can sometimes result in the formation of whiskers. The whiskers may nucleate at grain boundaries, and hillocks or inside individual grains. Figure 7.26 is a laboratory simulation of whisker extrusion in a capped metal film. This SEM image shows a whisker formed in an aluminum film on a silicon substrate, where the film is covered with a TiN capping layer. The whisker extrudes through the TiN layer, causing fracture in the capping layer.

Upon cooling from a maximum temperature, the film undergoes elastic unloading which is followed by plastic flow in tension; see Figure 7.16. Temperature dependence of yield stress, strain hardening behavior, and any changes in the microstructure of the thin film influence the film stress and substrate curvature during the first cooling phase of the thermal cycle. If the film composition, processing and environmental conditions during thermal cycling are such that weakly bonded grain boundaries, internal oxide particles, or inclusions exist in the thin film, the tensile stresses generated upon cooling from the maximum temperature can lead to the formation and growth of voids and cracks at internal interfaces; these irreversible damage processes, in turn, reduce the film stress and the substrate curvature.

Changes in microstructure as a result of grain growth and hillock formation usually stabilize after the first complete thermal cycle, and the history of film stress versus temperature attains a steady-state hysteresis loop which does not vary significantly from cycle to cycle thereafter. Figure 7.27 shows an example of the evolution of film stress as a function of temperature for an unpassivated aluminum thin film on a relatively thick Si substrate which was subjected to three thermal cycles between 25 and 450 °C. Shen and Suresh (1995b) obtained the results in this figure by converting the experimentally determined substrate curvature values to film stress using the Stoney formula (7.51). During the first heating phase, the metallic film yields at a temperature of approximately 100 °C, following which a gradual reduction in film stress is observed with increasing temperature in response to grain growth and hillock formation. When cooling commences from the maximum temperature of 450 °C, elastic unloading is followed by a rapid rise in film stress with decreasing temperature due to the combined effects of temperature dependence of yield strength and strain hardening. Note that the film stress versus temperature plot is essentially the same for the second and third thermal cycles. These observations based on substrate curvature measurements are consistent with the microstructural changes seen from the TEM images, such as those presented in Figure 7.22(a) and (b).

For a metallic thin film, diffusive mechanistic processes generally become more dominant compared to dislocation interactions as the temperature is raised. In unpassivated films, atoms are transported preferentially along grain boundaries and along the free surface of the film. The attendant shear on the interface can be equilibrated by interface slip which contributes to strain relaxation; this behavior is illustrated in Figure 7.21. With a passivating layer present on the film, it is necessary that diffusion occurs along the interface between the film and the substrate and between the film and the passivation layer in order for an analogous strain relaxation process to take place. However, even thin passivation layers curtail diffusion and/or interfacial slip. In addition, thermal cycling produces continual changes in microstructure over

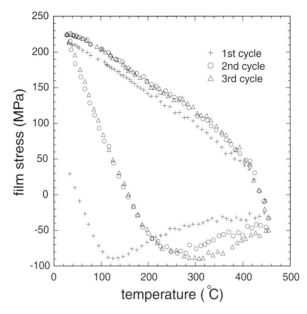

Fig. 7.27. Film stress plotted as a function of temperature for a 1-μm thick, unpassivated Al film on a relatively thick Si substrate which was subjected to three thermal cycles between 25 and 450 °C at a constant heating and cooling rate of 10 °C. Adapted from Shen and Suresh (1995b).

far greater numbers of thermal cycles in passivated films than in unpassivated films, as indicated in Figure 7.22.

Whether continuum descriptions of elastoplastic deformation or diffusion-dominated effects capture the stress versus temperature history or, equivalently, the substrate curvature versus temperature history, during thermal cycling strongly depends on the specific materials involved and their constraint conditions. For example, the experimental results shown in Figures 7.19 and 7.27 for unpassivated Al films generally follow trends anticipated on the basis of continuum elastoplastic analyses, such as those presented in Section 7.4 because of the relative dominance of time-independent plastic deformation over diffusion. However, over the same range of temperature excursion, other metallic materials, such as unpassivated pure Ag, Au or Cu films, on Si substrates may exhibit time-dependent strain relaxation processes involving atomic diffusion along the grain interiors, grain boundaries, interfaces between the film and the substrate or oxide/passivation layer, or along the free surface of the film (Thouless 1993, Vinci et al. 1995, Keller et al. 1999, Gao et al. 1999, Weiss et al. 2001, Kobrinsky and Thompson 2000). Such competing phenomena can be taken into account in describing behavior by appeal to the rate equation approach described in Section 7.6. These diffusive processes, which are

functions of temperature as well as the rate of thermal excursion and the waveform of the thermal cycle, are also anticipated to be strongly influenced by the film
thickness itself. In addition, the presence of a passivation layer plays an important
role in influencing microstructural evolution and dislocation mobility; as indicated
by the TEM images in Figure 7.22. Experimental studies by Vinci et al. (1995),
Shen and Suresh (1995b) and Kobrinsky and Thompson (2000) of stress evolution
in Al, Cu and Ag thin films undergoing temperature excursions show that the presence of a capping layer influences both time-independent and time-dependent strain
relaxation processes through mechanisms that are not yet well understood.

The role of constrained grain boundary diffusion in promoting dislocation plasticity in a polycrystalline metal film has been investigated by Gao et al. (1999). They
consider an initially uniformly strained thin film in which the grain boundaries are
oriented perpendicularly to the film–substrate interface; the interface does not accommodate any sliding or diffusion. Atomic diffusion is postulated to result in the
relaxation of the normal traction along the grain boundary which is equivalent to
the insertion of a wedge of material at the boundary. The edge of the grain boundary
wedge serves as a source of stress concentration, analogous to the tip of a crack,
thereby enhancing dislocation plasticity in the film and facilitating nucleation of
dislocations near the tip of the wedge, as shown schematically in Figure 7.28(a).
The issue of grain boundary diffusion and its influence on stress relaxation in a
polycrystalline thin film on a substrate will be taken up for further discussion in
Section 9.5.1 where the competition between different nonequilibrium processes
will also be addressed.

Some experimental evidence for the process of constrained grain boundary diffusion has been obtained by Dehm et al. (2002b) who observed the nucleation of
parallel glide dislocations, such as those sketched in Figure 7.28(a), by recourse to
in-situ TEM studies of strain relaxation during thermal cycling. The material chosen
was a predominantly {111} textured polycrystalline Cu film, 200 nm in thickness, on
a relatively thick Si substrate. Substrate curvature measurements revealed an equi-
biaxial compressive mismatch stress of 260 MPa in the film at 500 °C, whereas
a maximum tensile stress of 640 MPa was estimated to develop when the film–
substrate system was cooled to 40 °C. Figure 7.28(b) is a plan-view TEM image
of a series of ten dislocations nucleated within a single grain of the Cu film during
cooling from 500 °C to 40 °C. All the dislocations nucleate near the bottom right
triple junction of the grain, and the leading dislocation subsequently traverses the
entire length of the grain as the dislocations glide toward the left of the grain. Several
observations can be made about the dislocation structure found in Figure 7.28(b).
(i) Given the texture of the film and grain dimensions, it is evident from the *in-situ*
TEM observation that the dislocations glide on a {111} plane of the fcc metal thin
film that is oriented parallel to the film surface. Such an observation runs counter to

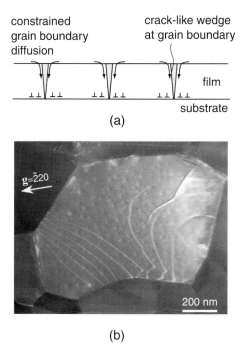

(a)

(b)

Fig. 7.28. (a) Schematic illustration of the formation of crack-like grain boundary wedges due to constrained diffusion and the emission of dislocations due to the enhanced stress concentration at the edge of the wedge. Adapted from Gao et al. (1999). (b) Plan-view, weak-beam TEM image of the dislocation structure produced inside a single grain in a 200-nm thick polycrystalline Cu film which was cooled from 500 °C to 40 °C. Reproduced with permission from Balk et al. (2001).

the general expectation that crystallographic planes oriented parallel to the plane of the film would not operate as slip planes because the film is in a state of equi-biaxial tension during the cooling stage. (ii) Such parallel glide appears to have replaced the possible occurrence of threading dislocation motion within the film. (iii) The first six dislocations on the left side of the grain exhibit very little curvature indicating the absence of any significant shear stress in this region of the grain. (iv) The four dislocations on the right of the grain appear to avoid the top grain boundary and to curve instead, possibly as a consequence of the rotation of the adjacent grain boundary there by as much as 17°. (v) The increasing spacing seen between successively generated dislocations is indicative of dislocation pile up and of the influence that the newly generated dislocations exercise on the glide of the ones nucleated earlier. Collectively, these observations point to the existence of a highly inhomogeneous state of stress within the grain, and suggest that diffusion at the grain boundaries may be responsible for the dislocation structures seen during thermal cycling.

7.8 Size-dependence of plastic yielding in thin films

The foregoing discussion of strain relaxation phenomena in thin films reveals that, unlike bulk metals and alloys, metal thin films exhibit plastic deformation characteristics that are influenced by the physical dimensions of the structure and its constraint conditions. In addition, characteristic dimensions associated with microstructural features such as grains and dislocation cell boundaries may also influence the plastic properties of the film. Quantitative understanding of the dependence of yield strength in thin films on various physical and microstructural length scales is not yet established. The task of doing so is particularly challenging due to the complex interplay among different mechanisms of deformation. Additional challenges arise as a consequence of the dependence of overall plastic response on material-specific effects such as those introduced by composition, crystallographic texture, geometrical configuration, constraint conditions and processing history. Furthermore, it is a challenging experimental task to determine the true material response *in situ* during the application of thermomechanical loads to the film in a manner that circumvents the incorporation of experimental artifacts into the interpretation of intrinsic deformation mechanisms. The effects of film thickness and microstructural parameters on the overall yield behavior of metal films are discussed in this section.

7.8.1 Observation of plastic response

The dependence of thin film yield stress on geometrical and microstructural length scales has been investigated extensively. Values of flow stress of polycrystalline thin films of different pure metals and alloys as functions of film thickness, grain size and texture are listed in Table 7.1. These data are based on experiments conducted at a single fixed temperature using such techniques as direct tension loading, microbeam bending or x-ray diffraction. The data in Table 7.2 are based on experiments in which the stress was imposed by temperature cycling of a thin film bonded to nominally elastic substrate and was measured by means of simultaneous observation of substrate curvature. The experimental techniques applied in obtaining these data are discussed in Section 7.9.

Examples of some trends are illustrated in Figures 7.29 and 7.30. Venkatraman and Bravman (1992) estimated the yield strength of pure Al and Al–0.5% Cu thin films on Si substrates by performing substrate curvature measurements during the repeated application of a thermal cycle. A particularly appealing feature of their experiments is that the film thickness was systematically altered without changing the grain size. This was accomplished by the repeated growth and dissolution of a barrier anodic oxide which was grown uniformly on the film. Substrate curvature

Table 7.1. *Representative values of the flow stress of polycrystalline thin films of metals and alloys as a function of film thickness (h_f) and average grain size (g.s.). Data obtained from microtensile tests, microbeam deflection and x-ray diffraction measurements.*

Materials and experimental conditions[a]	h_f (μm)	g.s. (μm)	Flow stress (MPa)
Cu film on 125-μm thick polyimide substrate at $\sim$ 20 °C; plastic strain $\sim$ 0.5%; {111} texture film; microtensile tests.[1]	1.0	0.18	481
		0.6	320–355
		1.0	268–306
		1.2	268
Same as above; but with {100} texture film.[1]	1.0	0.6	166–178
		1.0	137
		1.2	76
Al–0.5 wt% Cu film on 13-μm thick polyimide substrate at $\sim$ 20 °C; plastic strain $\sim$ 0.2%.[2]	0.2	0.14	529
	0.5	0.2	457
	0.9–1.1	0.3	450
	1.6–1.7	0.5	354–373
Pb film on 250-μm thick Si(100) substrate coated with 60-nm thick Si_3N_4 layer; x-ray and TEM at −173 °C after cooling from $\sim$ 20 °C; most grains with {111} along surface.[3]	0.1	4.5	395–454
	0.4		181
	0.6		70–79
	1.1		60
Free-standing thin film of 99.99% pure Ag with strong {111} texture; microtensile test at $\sim$ 20 °C at strain rate of 1.1×10^{-5} s^{-1}; 0.2% offset yield strength.[4]	5.8	5.8	$\sim$270
Free-standing thin film of 99.99% pure Cu with strong {111} texture; same test conditions as above.[5]	4.3	$\sim$4.3	$\sim$300
Pure Ag films magnetron-sputtered in ultra-high vacuum (6×10^{-7} Pa) at a rate of 77 nm s^{-1} on micromachined SiO_2 cantilever beams 2.83 μm thick; microbeam deflection tests at $\sim$ 20 °C with a nanoindenter tip; grain sizes with standard deviation of 41–66 μm; film residual stress in the range 26–43 MPa.[6]	0.2	0.9	300±7
	0.4	0.9	290±8
	0.6	0.8	270±7
	0.8	0.9	250±5
	1.0	1.0	210±10
	1.5	1.0	76

[a] References to various data quoted in this table pertain to information taken from the following sources where further details on experiments and materials can be found: (1) M. Hommel and O. Kraft, *Acta Mater.*, vol. 49, p. 3935 (2001). (2) F. Macionzyk and W. Brückner, *J. Appl. Phys.*, vol. 86, p. 4922 (1999). (3) T. S. Kuan and M. Murakami, *Metall. Trans. A*, vol. 13A, p. 383 (1982). (4) H. Huang and F. Spaepen, *Acta Mater.*, vol. 48, p. 3261 (2000). (5) R.-M. Keller et al., *J. Mater. Res.*, vol. 13, p. 1307 (1998). (6) R. Schwaiger and O. Kraft, *Acta Mater.*, vol. 51 (2003) 195–206.

measurements leading to estimates of film stress versus temperature were obtained after successively removing 0.1 μm of Al film, thereby facilitating the separation of the contributions to flow strength of the material from film thickness and grain size. The results reveal that the flow stress of the film varies inversely with film thickness at a fixed grain size, as shown in Figure 7.29.

Table 7.2. *Representative values of the flow stress of polycrystalline thin films of metals and alloys as a function of film thickness (h_f) and average grain size (g.s.). Data obtained from substrate curvature measurements.*

Materials and experimental conditions[a]	h_f (μm)	g.s. (μm)	Flow stress (MPa)
Magnetron-sputtered 99.999% pure Al film on Si(100) substrate; data obtained at room temperature after cooling from 450 °C; biaxial tensile yield strength.[1]	0.21	0.3	393
	0.26	-	344
	0.42	-	323
	0.59	-	266
	1.09	1.4	181
Pure Al film on Si(100) substrate; successive etching of film to vary h_f; data gathered at 60 °C after cooling from 460 °C.[2]	0.48	> 10	170
	0.65		124
	1.04		79
	1.5		50
	0.25	1.9	363
	0.55		188
	0.96		117
	1.03–1.52		130–134
	0.08	0.9	533
	0.48		272
	0.68		224
	0.90		189–196
Magnetron-sputtered pure Cu film on oxidized, 530 μm-thick Si substrate; obtained at room temperature after cooling from 600 °C.[3]	0.6	1.0	410±24
	1.0	1.3	280±16
Same as above, but with Cu films passivated with 50 nm-thick Si_3N_4.[3]	0.45	0.35	745±42
	0.60	0.83	585±33
	1.0	1.3	455±26
E-beam deposited pure Ag film on 290 μm-thick, oxidized Si(100); data obtained at –50 °C after cooling from 500 °C.[4]	0.21		267
	0.47–0.49		308–317
	0.72		277
	0.97		197
	1.21		163
Same as above, but with Ag films passivated with 50–70 nm-thick SiO_x layer.[4]	0.21		572
	0.35		477
	0.47		369
	1.21		216

[a] References to various data quoted in this table pertain to information taken from the following sources where further details on experiments and materials can be found: (1) M. F. Doerner et al., *J. Mater. Res.*, vol. 1, p. 845 (1986). (2) R. Venkatraman and J. C. Bravman, *J. Mater. Res.*, vol. 7, p. 2040 (1992). (3) R.-M. Keller et al., *J. Mater. Res.*, vol. 13, p. 1307 (1998). (4) M. J. Kobrinsky and C. V. Thompson, *J. Appl. Phys.*, vol. 73, p. 2429 (1998).

The results presented in Tables 7.1 and 7.2 and in Figure 7.29 also reveal that the flow strength of a polycrystalline thin film is several times larger than that exhibited by a bulk material of the same composition and grain size. For example, a 1 μm thick continuous film of pure Al on a relatively thick Si substrate, with a grain size comparable to film thickness, has a tensile flow stress of approximately 200 MPa at room temperature, whereas the tensile flow stress of a bulk specimen of Al of the same purity and grain size is approximately 70 MPa.

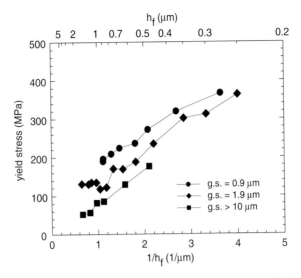

Fig. 7.29. Experimentally determined variation of the effect of film thickness on the tensile yield stress of polycrystalline Al thin films on Si substrates at 60 °C. The data are shown for three different grain sizes. Adapted from Venkatraman and Bravman (1992).

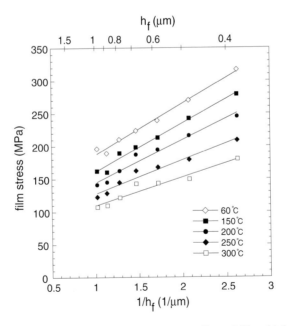

Fig. 7.30. Experimentally determined variation of the effect of film thickness and temperature on the tensile yield stress of polycrystalline Al thin films on Si substrates. The data are shown for a grain size of 0.9 μm. Adapted from Venkatraman and Bravman (1992).

Figure 7.30 is a plot of the tensile yield stress as a function of the film thickness at several different temperatures between 60 and 300 °C for the pure Al film from the same set of curvature experiments conducted by Venkatraman and Bravman (1992). The initial film thickness and grain size in these experiments were 1.0 μm and 0.9 μm, respectively, and the film thickness was progressively reduced, after each set of experiments at a fixed temperature, to explore the effect of film thickness at an essentially constant grain size on the plastic response of the film. The results clearly indicate that tensile yield stress varies inversely with film thickness at a given temperature, and with temperature at a given film thickness. Similar results were also obtained by Venkatraman and Bravman (1992) for the compressive yield behavior of the Al thin film. At elevated temperatures, noticeable differences were observed between the flow characteristics of the film in tension and compression.

The experimental results shown in Tables 7.1 and 7.2, as well as in Figures 7.29 and 7.30, indicate that the yield stress of metallic thin films varies inversely with film thickness h_f. Such observations cannot be rationalized by recourse to conventional continuum plasticity theories which are devoid of intrinsic length scales. Trends in modeling the plastic response of thin films are discussed briefly in the next section.

7.8.2 Models for size-dependent plastic flow

Efforts to develop models of plastic deformation of thin films and other small structures have typically followed one of two main approaches, both starting from well-established models of material behavior. In one approach, the point of departure is the behavior of a single threading dislocation in a strained crystalline layer, as described in detail in Section 6.3. In this approach, critical conditions are established by a balance between the energy needed to alter a dislocation configuration in a certain way and the external work that can be provided by the system in effecting that alteration. The conceptual approach can be extended to include arrays of dislocations as in Sections 7.1 and 7.2, configurations with multiple interfaces as in Section 6.4 and glide kinetics as in Section 7.3 to consider time-dependent phenomena. On this basis, Chaudhari (1979) concluded that residual inelastic strain, and therefore the current value of yield stress, in a crystalline metal film varied with the film thickness as h_f^{-1}. The early literature on this approach is reviewed by Murakami et al. (1982), Nix (1989) and Arzt (1998). The general approach was further extended by Nix (1998) in a discussion of plastic yield and strain hardening in a periodic multilayer film incorporating alternating layers of ductile material and elastic material. This study considered intersecting dislocations as discussed in Section 7.2, and a fairly high strain hardening rate was deduced. In virtually all examples of this type, the value of yield stress at any state of relaxation depends on the geometrical dimensions of the material system in some way.

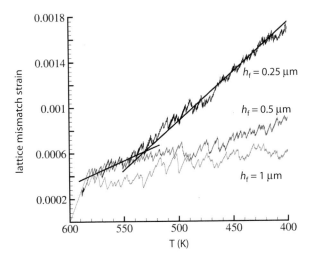

Fig. 7.31. Numerical prediction of the average lattice mismatch strain as a function of the change in temperature from a stress-free temperature for three different thicknesses of an aluminum film on an elastic substrate. Adapted from Nicola et al. (2003).

For even a relatively small number of interacting dislocations which have independent degrees of freedom in a thin film system, the behavior becomes very complex. In order to deal with this complexity, Nicola et al. (2003) reformulated the problem of dislocation formation and interaction in a strained single crystal thin film to allow for determination of its features by numerical methods. A single crystal thin film on an elastic substrate was considered to deform under plane strain conditions. Edge dislocations could exist on predetermined slip systems, and film stress was introduced as a consequence of thermal mismatch between the film and the substrate during temperature change. Dislocations were generated in the initially dislocation-free film through a random distribution of sources which, in the context of the two-dimensional analysis, produced edge dislocation dipoles whenever the Peach–Koehler force at the source site exceeded a critical value. The elastic fields of the dislocations were determined by using the analytical stress fields for a dislocation in an unbounded elastic solid, and then relying on the finite element method to determine an additional non-singular stress-field in order to satisfy boundary conditions; this superposition approach is outlined in Section 6.6.2.

Figure 7.31 shows the numerically predicted variation of the average value of extensional strain in the film in the direction parallel to the interface due to thermal mismatch with respect to the substrate as a function of temperature change from the initial stress-free temperature. The unrelaxed mismatch strain, which is a measure of the average film stress, is markedly higher for the thinnest film than for thicker films, indicating a size effect on the plastic strain hardening response. Nicola et al.

(2003) showed that the effect of film thickness on film yield behavior is chiefly a consequence of the development of a boundary layer due to the pile-up of disloca- tions at the film–substrate interface. The width of the boundary layer varies with the orientation of the slip system and is independent of the film thickness. When the film thickness is below a certain critical value, the back stress induced by the dislocations piled-up at the interface causes a reduction in dislocation nucleation, thereby contributing further to the hardening process. In very thin films, this role of back stress in suppressing nucleation occurs earlier in the relaxation process.

The origin of such a size effect in plastic response can be understood in the following way. First, recall that constitutive behavior as represented by classical plasticity theory is based on the presumption of simple material behavior, that is, mechanical behavior is fully characterized by response under states of homogeneous deformation. Against that background, consider deformation of a single crystal, large in size compared to atomic dimensions, due to dislocation motion on a single slip system. Plastic strain measured from an arbitrary initial configuration is a measure of the number of dislocations that have passed entirely through the crystal, either from one side to the other or from an interior source outward through the boundary. This number, along with the length of the Burgers vector and the geometry of the slip system, define a homogeneous simple shear strain as a starting point for a continuum description of the process. In arriving at this description, there is no reference to dislocations remaining in the sample but only to those having passed completely through it. If the sample includes a random distribution of dislocations at the outset that does not change statistically during plastic straining then the argument is unaffected and a homogeneous deformation can still be identified. On the other hand, if a Burgers circuit with dimensions on the scale of the sample size leads to a net offset that changes in the course of plastic deformation, then the interpretation as a homogeneous deformation becomes ambiguous.

The need to depart from the foregoing ideas underlies the development of strain gradient plasticity theories. The reasoning is as follows. There are some cases of plastic deformation in which a certain density of dislocations *must* be present for reasons of geometric compatibility; these are the so-called geometrically necessary dislocations introduced by Cottrell (1964) and discussed by Ashby (1970). If some net density of dislocations is geometrically necessary in a particular plastically deforming crystal, then the deformation of a small sample of that material cannot be represented as being locally homogeneous. Instead, the deformation of a small sample also depends on the density of geometrically necessary dislocations which is reflected macroscopically through a plastic strain gradient. These ideas, which are discussed only loosely here, have been given a mathematical structure by Fleck et al. (1994) and Fleck and Hutchinson (1997). The structure of the theory will not be discussed further here.

The results presented in Tables 7.1 and 7.2 and in Figure 7.29 indicate that, in addition to the film thickness, the grain size of thin films has a strong influence on yield stress. As an example, consider Figure 7.29 which reveals that the yield stress of aluminum thin films varies inversely with grain size at a fixed film thickness. In order to provide a mechanistic basis for the effect of grain size on film yield stress, Thompson (1993) suggested that the deposition of segments of a surface-nucleated dislocation loop (such as that described in Section 6.8.1) at grain boundaries leads to a contribution to film yield stress that scales with the inverse of the grain size. Such a variation, however, is not generally observed in polycrystalline thin films. For example, Keller et al. (1999) found that the flow stress of polycrystalline Cu films scales with the inverse square root of the grain size, as envisioned in the classical Hall–Petch type models for strengthening from grain refinement. Choi and Suresh (2002) invoked the assumption that relaxation of mismatch strain in the thin film is accommodated by the introduction of dislocation loops whose population, dimensions and interaction are determined by the film thickness and grain structure. By accounting for the interactions among such an assumed population of dislocation loops constrained by the grains in the thin film, they rationalized the experimentally documented combined effects of film thickness and grain size on the yield properties of polycrystalline thin films. A drawback of all of these mechanistic models of polycrystalline films is that the dislocation configurations postulated to be generated during plastic deformation have not been experimentally verified. Consequently, no satisfactory quantitative model is currently available for predicting the plastic deformation response of a polycrystalline thin film in terms of its structure.

7.8.3 *Influence of a weak film–substrate interface*

As has been noted, plastic deformation of a metal film consists largely of driving dislocations through the film toward the film–substrate interface. TEM observations of the interfaces of plastically deformed films show that interface dislocations are present in some cases but not in others. It was also reported that interface dislocations disappear during TEM imaging (Müllner and Arzt 1998). In virtually all cases, the substrate material is not deformed plastically, so the difference between these observations must be a consequence of variations in interface quality.

An interface that is fully or partially coherent will likely preserve the structure of nearby dislocations. In particular, the lattice distortion near the dislocation line that results in the contrast seen in TEM images will be preserved. If the interface is incoherent, on the other hand, its intrinsic resistance against slip can be much smaller than the flow strength of the film. Macroscopically, no traction is transmitted across the interface for a uniform film, and therefore the background film stress does not test the strength of the interface. In the case of a weak interface, the stress field of a dislocation near the interface, or perhaps of a group of similar dislocations, will result in a traction on the interface with the potential for inducing slip. The slip, in turn, softens the response detected by the dislocations, compared to the response for a perfectly bonded interface; in other words, the interface begins to take on the character of a free surface. This reduction in stiffness of response, in turn, leads to an attractive force on the dislocations. Consequently, a dislocation is spontaneously drawn toward a weak interface, which increases the shear stress induced on the interface, which then results in additional slip, and so on. These circumstances imply that the equilibrium configuration is unstable. The distance from the interface over which the effect is dominant is roughly $\mu b / 2\pi \tau_0$ where τ_0 is the shear stress required to induce slipping on the interface. For $\mu = 100\,\text{GPa}$, $b = 0.3\,\text{nm}$ and $\tau_0 = 100\,\text{MPa}$, this distance is about 50 nm.

The interaction between a dislocation and a weak interface is inherently nonlinear because the slipping zone at first expands as a dislocation is attracted to the interface. As the dislocation moves closer, the slipping zone necessarily contracts, leaving behind a slipped but no longer slipping portion. This nonlinear interaction has been described within the framework of elastic dislocation theory by Hurtado and Freund (1999). The main consequence of the interaction is that the dislocations near weak interfaces are essentially drawn into the interface as slipping progresses. As a result, the large lattice distortions associated with dislocation cores give way to small lattice distortions over a relatively large part of the interface; the strain contrast of the core region is no longer seen in TEM observations, giving the impression that the dislocations have vanished. This is a possible explanation for the fact that interface misfit dislocations are not seen in some systems. Complete absorption of a dislocation by an interface requires minor atomic rearrangement at the interface, but this is quite possible, especially at elevated temperature.

7.9 Methods to determine plastic response of films

Experimental techniques most commonly used to probe the plastic properties of thin film materials involve direct tensile loading of either a free-standing film or a film deposited onto a deformable substrate material, microbeam bending of films on substrates, substrate curvature measurement or instrumented depth-sensing nanoindentation. Salient features of these methods, as well as specific examples of the adaptation of these methods for the study of mechanical properties in thin films, are briefly addressed in the following subsections.

7.9.1 Tensile testing of thin films

The determination of stress–strain response of thin films, typically a micrometer or so in thickness, is a challenging experimental task in view of the difficulties associated with gripping the specimen and with imposing and monitoring *in situ* the mechanical load and the specimen strain, while ensuring that measurements are not often skewed by effects such as out-of-plane bending or twisting of the specimen or by the deformation of the loading frame itself. In addition, producing a wrinkle-free, free-standing film of uniform thickness with sufficiently large in-plane dimensions to facilitate direct load application and *in-situ* strain measurement is a difficult task in itself. In the application of conventional tensile testing methods developed for bulk materials, such as those adopted by Sharpe (1982), Baral et al. (1985) and Ruud et al. (1993), or the bulge test described in Chapter 5, it is difficult to measure the strain in the film accurately and reliably during the application of a mechanical stress. To overcome these limitations, a variety of *in-situ* strain measurement methods have been adapted for use with mechanical tests; among these are the interferometric strain gauge method (Sharpe 1982), diffraction techniques (Schadler and Noyan 1995, Hommel et al. 1999, Huang and Spaepen 2000) and speckle interferometry (Read 1998). Two examples of such approaches are presented here.

Huang and Spaepen (2000) conducted uniaxial tensile tests on free-standing poly-crystalline thin films of pure Ag, Cu and Al and multilayer films of Ag–Cu. In each case, the film was deposited by electron-beam evaporation on a glass substrate which was appropriately masked to produce a dog-bone-shaped specimen with a gauge length of 10 mm, gauge width of 3.1 mm and a thickness in the sub-micrometer to several micrometers range. Micro-lithographic methods were then used to deposit a thin two-dimensional polymeric grid on the surface of the film, whereby an unstrained pattern of photoresist material was created in the form of a square array of islands with 0.6 μm side width and 10 μm spacing. The polymeric layer has a much lower elastic modulus than the film material and forms unconnected islands; consequently, its influence on the interpretation of measured applied force in terms of stress in the film material is negligible; this was ascertained by varying the thickness of the photoresist grids which was found to have no effect on the measured film response. The films were carefully peeled from the glass substrate for uniaxial loading in a microtensile testing machine. The grips of this device were equipped with a mechanism to monitor the imposed force through a load cell. A focused He–Ne laser beam, directed onto the film surface, was diffracted by the grid pattern in the photoresist; the in-plane longitudinal and transverse strains in the film are proportional to the displacement of the scattered laser spots in the respective directions. This strain measurement, with a resolution of 0.002%, in conjunction with the force measurement in the microtensile tester with a stress resolution capability of 0.1 MPa, provide plots of the stress–strain curves for the thin films.

Macionczyk and Brückner (1999) and Hommel and Kraft (2001), on the other hand, studied the tensile stress–strain response of polycrystalline thin films deposited on compliant substrates. In the latter study on Cu films, x-ray diffraction during the tensile test facilitated examination of the mechanical response as well as the evolution of dislocation density and the crystallographic texture of the grains. A 125-μm thick polyimide foil, capable of undergoing tensile strains as large as 0.03, was used as the deformable substrate. Copper thin films, with thickness in the range from 0.4 to 3.2 μm, were sputter-deposited onto the substrate through a mask over the gauge section of the dog-bone-shaped specimens. The films were subsequently annealed in vacuum at a base pressure of 3×10^{-4} Pa for 60 min at 200 °C. The specimens, with gauge section dimensions of 20 mm by 6 mm, were loaded in uniaxial tension in a specially built electromechanical testing machine which was housed inside an x-ray goniometer. The specimen was incrementally strained in steps of 10 μm displacement; at each step, the stress in the film was estimated by performing x-ray diffraction measurements for a duration of approximately 30 min. While x-ray diffraction provided the film stress, the overall strain in the specimen was measured with the use of a non-contact laser extensometer that recorded the instantaneous spacing between two reference points on the specimen. Hommel and Kraft (2001) used {331} planes in the {111}-oriented grains and {420} planes in the {100}-oriented grains to monitor the position of the Bragg peak in order to estimate the film stress. The interplanar spacings of the {331} and {420} planes were then calculated and the in-plane stress components along and across the tensile direction were calculated for the appropriate elastic constants of Cu using the methods described in Section 3.6. In addition to providing estimates of the microstress in the film, the x-ray analysis provides a measure of the dislocation density which is related to the width of the measured x-ray intensity peaks. Experimental results obtained by Huang and Spaepen (2000) and Hommel and Kraft (2001) on the tensile stress–strain response of several different thin film materials were examined in Section 7.8.1.

The challenges inherent in the measurement of stress–strain response of thin film materials by means of direct tensile testing are commonly more than offset by the distinct advantage that properties characterizing deformation resistance of the material in the plastic range can be determined under *isothermal* conditions for a relatively simple state of stress on the specimen. However, these techniques are not readily amenable to modifications that can accommodate uniaxial compression, simple shear stress, equi-biaxial stress or more general states of stress on the specimen. As a result, it is difficult to draw conclusions concerning the dependence of plastic response on stress path history. It is noted that results for some cyclic tension–compression experiments were reported by Hommel et al. (1999).

7.9.2 Microbeam deflection method

Isothermal mechanical tests can also be performed on thin films and layered materials using microbeam deflection methods introduced by Weihs et al. (1988) and Baker and Nix (1994). In this approach, precise micromachining techniques, such

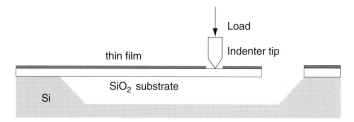

Fig. 7.32. Schematic of the microbeam deflection experimental setup. Adapted from Schwaiger and Kraft (2003).

as those described in Chapter 1, are employed to fabricate small cantilever beam structures. A transverse deflection is then imposed on these beams by mechanical means over distances on the order of micrometers. Figure 7.32 shows an example of such a system studied by Schwaiger and Kraft (2003) who used an anisotropically etched, oxidized Si substrate to produce a silicon oxide cantilever beam, approximately 100 μm long, tens of μm wide, and several μm thick. Thin films of Ag, 200 nm to approximately 1 μm in thickness, were then magnetron-sputtered on the cantilever beam under ultra-high vacuum. The cantilever beam specimen was placed inside a nanoindentation system so that the wedge-shaped tip of the diamond indenter could be used to deflect the beam a distance on the order of a μm at the load point, a location approximately 30 μm from the free end of the beam. Components of stress and strain induced in the film due to this imposed deflection were analyzed by means of finite element simulations, and the appropriate stress–strain characteristics of the film were extracted from these results; examples of such results on the yield properties of Ag thin films were provided in Section 7.8.1.

A particularly appealing feature of the test system shown in Figure 7.32 is that it makes it possible to perform cyclic deformation experiments on thin films on substrates in order to explore systematically the effects of cyclic frequency, waveform, mean stress and film thickness on the evolution of damage and failure. For example, Schwaiger and Kraft (2003) have superimposed a sinusoidal force oscillation at 45 Hz on a static force signal to investigate the fatigue response of Ag films over millions of imposed stress cycles. Progression of fatigue damage is quantitatively assessed by recording the change in beam stiffness with increasing numbers of fatigue cycles; this measurement is made using the lock-in amplifier in the nanoindentation setup which provides a continuous measure of the dynamic stiffness. Figure 7.33(a) shows the variation of the bilayer beam stiffness as a function of the number of fatigue cycles at room temperature for 0.6 and 1.5 μm thick Ag films on the silicon oxide substrate. The imposed tensile stress amplitudes for the two cases were 50 and 54 MPa, respectively, and the corresponding tensile mean stress values were 255 MPa and 350 MPa, respectively.

Microscopic examination of the fatigued films revealed that the change in the bilayer stiffness correlated well with the onset of damage in the film in the form of voids, surface extrusions and microcracks. The onset of distributed fatigue damage in the thicker film led to a precipitous drop in the bilayer stiffness at approximately 8×10^5 stress cycles; the thinner Ag film was capable of sustaining cyclic stresses without any apparent damage. Figure 7.33(b) is a plot of the stress amplitude necessary to generate a noticeable decrease in beam stiffness within 3.8×10^6 fatigue cycles as a function of the thickness of the Ag films. These stress amplitudes represent maximum values acting along the beam axis at the surface of the film near the fixed end of the cantilever beam. The dashed line in Figure 7.33(b) represents a demarkation line between conditions that led to this reduction in stiffness (open symbols) and conditions for which no stiffness change was detected. This figure shows that a 200 nm thick Ag film required twice the amplitude of stress to undergo the same amount of stiffness change as a 1000 nm thick film. An increase in mean stress at a fixed stress amplitude leads to a reduction in the number of cycles needed for a stiffness loss to occur; however, the critical stress amplitude for damage evolution is unaffected by the mean stress. Figure 7.33(c) is a micrograph of a cross-section of the Ag film on the silicon oxide substrate which was produced by focused-ion-beam (FIB) milling and imaged at a tilt angle of 45 degrees. The evolution of surface extrusion and voids in response to cyclic stressing is evident in this micrograph. Although the mechanisms by which fatigue damage evolution in thin films differs from that in bulk metals and alloys (Suresh 1998) are not fully understood at this time, the microbeam deflection method provides a convenient experimental tool with which the cyclic deformation and failure characteristics of thin films can be studied. Similar fatigue experiments have also been reported for Cu thin films on polyimide substrates where cyclic tensile tests were conducted in an electromechanical testing machine using the methods described in Section 7.9.1.

7.9.3 Example: Thin film undergoing plane strain extension

For the case of thin films being deformed plastically under the action of equi-biaxial states of stress, the stress history follows a straight line path in stress space. In some of the experimental methods for study of thin film plasticity that have been described in this section, the in-plane stress components are not equal, in general, and the trajectory in stress space during deformation is not a straight line. Consider a film material being deformed under plane strain extension in the x-direction; the plane strain constraint is enforced in the z-direction. The elastic modulus is E_f and the Poisson ratio is $\nu_f = 0.3$. The initial yield locus for the material in terms of tensile yield stress σ_{Y_f} is the Mises condition $\sigma_M = \sigma_{Y_f}$. The material undergoes isotropic linear hardening beyond yield, with $d\epsilon_M^p/d\sigma_M^p = 10/E_f$ in a tension test which provides the value for H. Determine the trajectory in stress space for an elastic–plastic film constrained to deform in plane strain extension.

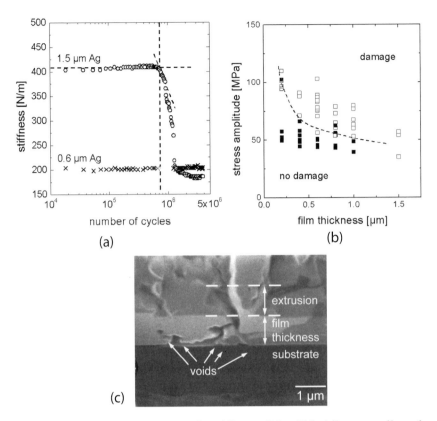

(a)   (b)

(c)   1 μm

Fig. 7.33. (a) Effect of cyclic stress on the stiffness of Ag–SiO$_2$ bilayer cantilever beams
for Ag film thickness of 0.6 μm and 1.5 μm. (b) Fatigue damage map showing the tensile
stress amplitude necessary to induce a change in the stiffness of the bilayer cantilever beam
within 3.8×10^6 fatigue cycles as a function of the thickness of the Ag films. (c) FIB image,
shown at a 45 degree tilt, of the cross-section of the Ag film on the SiO$_2$ substrate showing
the evolution of cyclic damage in the form of voids and surface extrusions. Reproduced
with permission from Schwaiger and Kraft (2003).

Solution:

Initially, the response is within the elastic range of behavior. In this range, the plane
strain constraint $\epsilon_{zz} = 0$ is sufficient to determine the stress trajectory. From Hooke's
law,

$$\epsilon_{zz} = \frac{1}{E_\mathrm{f}}\left(\sigma_{zz} - \nu_\mathrm{f}\sigma_{xx}\right) \tag{7.87}$$

which implies that the stress state moves out along the line $\sigma_{zz} = \nu_\mathrm{f}\sigma_{xx}$ in stress space from
the origin until the line intersects the yield locus. The portion of the stress trajectory within
the elastic range is shown by the line with slope ν_f in Figure 7.34.

Once plastic flow begins, the total strain rate is the sum of the elastic and plastic strain
rates. The plastic strain rate is given in terms of stress by (7.60), and the expressions for

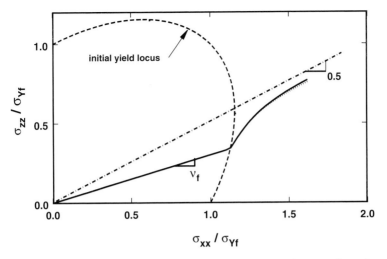

Fig. 7.34. The trajectory in stress space resulting from plane strain extension of an elastic–plastic film. The film material is assumed to undergo isotropic linear hardening, as described in the text.

total strain rate are

$$\dot{\epsilon}_{xx} = \frac{1}{E_f} (\dot{\sigma}_{xx} - \nu_f \dot{\sigma}_{zz}) + \frac{10}{E_f} \frac{\dot{\sigma}_M}{\sigma_M} \left(\sigma_{xx} - \tfrac{1}{2}\sigma_{zz} \right) \tag{7.88}$$

$$\dot{\epsilon}_{zz} = \frac{1}{E_f} (\dot{\sigma}_{zz} - \nu_f \dot{\sigma}_{xx}) + \frac{10}{E_f} \frac{\dot{\sigma}_M}{\sigma_M} \left(\sigma_{zz} - \tfrac{1}{2}\sigma_{xx} \right) . \tag{7.89}$$

The plane strain constraint still requires that $\dot{\epsilon}_{zz} = 0$ and the imposed extensional strain requires that $\dot{\epsilon}_{xx} = 5\sigma_{Y_f}/E_f$. Therefore, the left-hand sides of both equations are known. The equations (7.88) and (7.89) then provide a pair of first-order ordinary differential equations to be integrated for σ_{xx} and σ_{zz} as functions of time. The initial values are determined from the intersection of the straight line trajectory in the elastic range with the yield locus. Note that E_f is a common factor in each term of the differential equations, so the stress trajectory does not depend on elastic modulus. Furthermore, if both stress components are normalized with respect to the yield stress σ_{Y_f}, then it also drops out of the formulation.

Integration of the differential equations from first yield for an elapsed time of $t = 1$ yields the result shown in Figure 7.34. The dashed curve is the initial yield locus. Note that the stress trajectory deviates sharply from linearity once yielding begins, even though the state of deformation is always uniaxial plane strain extension.

As the magnitude of plastic strain becomes large in the material compared to the elastic strain, the stress trajectory approaches the dot–dashed line with slope 0.5. This is precisely the trajectory that would be predicted if elastic strains would be neglected entirely from the outset. For comparison, the corresponding results for kinematic hardening are shown in the graph by means of a dotted curve, barely visible under the solid curve.

7.9.4 Substrate curvature method

The ease with which the film mismatch stress can be estimated from the change in curvature of a relatively thick substrate subjected to thermal excursion has led to the widespread use of the substrate curvature method for the study of mechanical properties. As reviewed in Section 2.3, substrate curvature can be continuously monitored with high precision using a variety of commercially available instruments.

Advantages in the use of the substrate curvature method for the extraction of film properties can be summarized as follows.

- The film stress can be determined from the Stoney formula and the measured substrate curvature without knowledge of the thermal and mechanical properties of the film material, as indicated by (2.7). If both the film and substrate deform elastically during thermal excursion, the slope of the stress versus temperature plot, such as that shown in Figure 7.16, has a magnitude of $M_f\alpha_f$. If either of these material properties is known through other means, the other can be determined from this slope. Since the stress-free strain in the film is induced by thermal expansion mismatch with respect to the substrate, the curvature method obviates the need for mechanically loading the specimen, thereby circumventing the complexities associated with the design, fabrication and gripping of the specimen.
- The onset of plastic deformation in the thin film causes a distinct change in the slope of stress versus temperature curve during both heating and cooling. This transition point is commonly used to assess the magnitude of the biaxial yield strength of metallic thin films in both tension and compression in different temperature regimes.
- It was shown in Section 7.7.3 that the curvature versus temperature or the stress versus temperature hysteresis loops for an unpassivated thin film generally attain a saturated state after the first thermal cycle. Since the subsequent thermal cycles are not affected by the prior thermal loading history (in the absence of any transient damage processes such as void growth or crack advance or interfacial delamination), the method offers considerable experimental flexibility in exploring a wide variety of mechanical responses (such as curvature evolution with different heating and cooling rates, or with different hold periods at a fixed temperature, tension–compression asymmetry of plastic flow at different temperatures, etc.), and loading histories (such as thermal cycling between different temperature limits as, for example, shown in Figure 7.19) using a single specimen.
- Since substrate curvature can be accurately measured, even in the presence of rigid body motion or vibration, by use of the multi-beam optical stress sensor or the coherent gradient sensor techniques described in Chapter 2,

the method is readily amenable for *in-situ* stress measurement during film deposition and passivation, or for use as a quality control tool to monitor film stress and properties.

- By comparing experimentally determined stress versus temperature hysteresis loops with simple constitutive models of plastic deformation, approximate estimates of plastic properties such as the film yield strength in tension or compression, extent of strain hardening and the temperature-dependence of flow stress can be inferred for some thin film materials (see, for example, Figures 7.16 and 7.19).

Several examples were presented in Section 7.4 on the use of the substrate curvature method for the determination of plastic properties of thin films.

The curvature method also has some inherent limitations. Since the method entails determination of plastic response through the imposition of a temperature change or phase transformation, which is known to alter the plastic properties of metals, care should be exercised in the interpretation of strain relaxation phenomena from curvature measurements. There are also no direct ways to isolate the individual contributions to overall curvature evolution seen experimentally from such factors as plastic yielding, strain hardening, diffusional creep, or microstructural changes, without recourse to other independent experimental and observational tools.

7.9.5 Instrumented nanoindentation

Nanoindentation experiments entail use of a mechanical probe in the form of a three-sided or four-sided pyramid, commonly referred to as the Berkovich or Vickers indenter, respectively. The indenter penetrates the specimen surface under either load or displacement rate control, and the corresponding variation of the load as a function of displacement is continuously recorded during both the loading and unloading phases of indentation. Figure 7.35 shows schematic illustrations of two different types of instrumented nanoindenters. In Figure 7.35(a), the depicted configuration facilitates the application of sub-micronewton indentation loads on a surface by appropriately modulating the current in the coil which is surrounded by a magnet; the indenter tip is located at the end of a load train which originates at the coil. This arrangement is connected to the frame through suspension springs, with the entire assembly being capable of vertical translation by a controlled amount. The capacitance displacement gauge shown in this figure provides a record of the depth of penetration of the indenter tip into the surface, to a resolution on the order of a tenth of a nm. A small oscillation in the applied force, produced by the superposition of an alternating current which provides a frequency of load oscillation of 10 to 100 Hz, is also used in this arrangement so that the resulting oscillations in displacement provide a continuous measurement of contact stiffness during the entire indentation

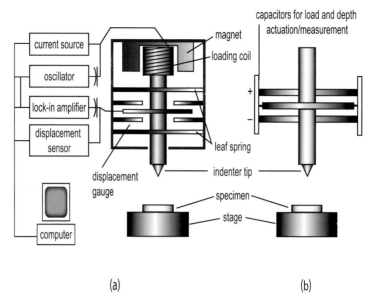

Fig. 7.35. Schematic illustrations of two different designs for an instrumented nanoindenter. Figure (a) is adapted from Materials Test Systems Corporation, Eden Prairie, Minnesota, and (b) from Hysitron Corporation, Minneapolis, Minnesota.

experiment. This capability also allows sensitive assessment of surface contact between the indenter and the material being probed, and it may prove convenient to study materials whose mechanical properties are time-dependent. The phase angle between the dynamic load and displacement is continuously and directly measured with the aid of a lock-in amplifier.

Figure 7.35(b) shows another design of a nanoindenter in which a three-plate capacitance configuration is utilized for the simultaneous measure of indentation load and displacement, to a resolution of approximately 100 pN and 0.1 nm, respectively. A dc voltage applied between the center and bottom plates causes the indentation load to be imposed on the specimen, whereas an ac voltage between the top and bottom plates enables a continuous record of the normal displacement of the indenter tip into the specimen; the indenter is attached rigidly to the center plate. An appealing feature of this design is that the pre-indentation profile of the surface and residual impression of the indenter on the surface following indentation can both be imaged. Similar to an atomic force microscope operating in a contact mode, the indenter tip of this apparatus is mounted on a cantilever displacement sensing system. A piezoelectric scanner moves the substrate surface in three dimensions relative to the indenter tip while maintaining a contact force, as small as 1 μN, with the aid of a feedback control loop between the cantilever beam and the piezoelectric actuator. Once the imaging process is completed, the scanning operation is

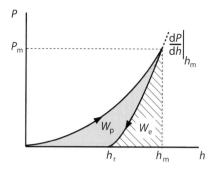

Fig. 7.36. Schematic showing the indentation load P versus penetration depth h during the loading and unloading stages.

terminated and the indentation loads are imposed on the surface. In both systems shown in Figure 7.35, the specimen is housed on a motorized stage with which it can be translated in plane by a displacement as small as 1 μm.

Figure 7.36 is a schematic of a typical indentation load P versus depth h curve for a metal.[*] As the indenter is pressed against the surface, the material deforms plastically, following an initial elastic deformation regime, and permanent deformation occurs whereby a hardness impression approximately conforming to the shape of the indenter is formed on the surface being probed. As the indenter is released from a maximum load P_m, only the elastic portion of the displacement is recovered from a maximum value of h_m. The cross-hatched region in Figure 7.36 shows the recovered elastic work W_e whereas the shaded region represents the work W_p which is dissipated through plastic deformation; the sum of these two work terms constitutes the total work which is the area under the indentation loading curve up to the maximum loading point $\{P_m, h_m\}$. The residual depth h_r is the permanent displacement left by the plastic impression upon complete removal of the indenter from the surface.

Nanoindentation is commonly used to measure two properties of a bulk or thin film material or a surface coating: the elastic modulus and the hardness. The slope of the initial portion of the elastic unloading curve is related to the elastic modulus of the indented material as

$$\left.\frac{dP}{dh}\right|_{h=h_m} = \frac{2\alpha^* E^* A_c^{1/2}}{\sqrt{\pi}}, \qquad E^* = \left\{\frac{1-v_s^2}{E_s} + \frac{1-v_{in}^2}{E_{in}}\right\}^{-1}, \qquad (7.90)$$

where E^* is the reduced elastic modulus of the indenter–material system, E and v represent, as before, the elastic modulus and Poisson ratio, respectively, with the subscripts 's' and 'in' denoting the specimen and the indenter, respectively, A_c is the projected contact area of the impression produced on the indented surface at the

---

[*] Although the symbol h usually represents film thickness, it is adopted in this subsection as the de facto standard symbol for penetration depth.

maximum load P_m, and the nondimensional constant α^*, which is on the order of unity, represents a correction factor which accounts for the shape of the indenter tip (King 1987, Doerner and Nix 1988, Oliver and Pharr 1992, Dao et al. 2001). Thus the elastic modulus of the indented material is obtained from the experimentally determined P–h curve in conjunction with (7.90) where the projected area is either estimated from post-indentation observation of the indenter impression or from detailed computational simulations of the elastoplastic deformation induced during indentation (e.g., Dao et al. (2001)), elastic properties of the commonly used diamond indenter (typically, $E_{in} = 1140$ GPa and $\nu_{in} = 0.07$) and the Poisson ratio of the indented material (typically, $\nu = 0.3$ for most metallic materials).

The indentation hardness signifies the mean pressure that can be sustained under the indentation load and is defined as

$$H = \frac{P_m}{A_c}. \tag{7.91}$$

The indentation hardness is commonly used as a convenient material parameter with which different materials can be ranked in terms of their strength values.

Depth-sensing instrumented indentation offers many possibilities for the mechanical probing of thin films and surfaces at sub-microscopic length scales, from which a number of issues related to thin film mechanical response can be pursued quantitatively. For example, nanoindentation experiments were used in Section 6.8.2 to investigate the critical conditions governing the onset of homogeneous nucleation of dislocations in an initially defect-free metallic crystal with an accuracy and flexibility for measurement during the application of a mechanical load that are difficult to achieve using other experimental techniques. It was also shown in Figure 6.40 of that section that the rising portions of the P versus h plot were indicative of the elastic response of a single crystal thin film of pure aluminum where the volume of the material sampled was as small as a few hundred nm^3. Nanoindentation experiments also provide a means to investigate quantitatively many fundamental aspects of asperity-level contact at surfaces of films and coatings, such as those used to impart protection against tribological damage. Furthermore, the response captured by the P–h curve in Figure 7.36 is known to be strongly influenced by such factors as the inelastic properties of the material (Oliver and Pharr 1992, Dao et al. 2001), equi-biaxial residual stress in the film (Tsui et al. 1996, Suresh and Giannakopoulos 1998), and substrate constraint (Laursen and Simo 1992, Tsui et al. 1999). Consequently, the indentation method provides a convenient experimental tool with which such issues as the plastic properties, internal stress fields, and the size-dependence of plastic response can be systematically investigated in bulk and thin film materials over multiple length scales. Computational simulations based on continuum elastoplastic analysis suggest that depth-sensing nanoindentation results, such as those schematically shown in Figure 7.36, for thin films can be extracted without sensitivity to the substrate when the maximum depth of penetration h_m of the indenter into the film surface is typically smaller than about 10% of the film thickness h_f (Laursen and Simo 1992, Bhattacharya and Nix 1988).

The use of depth-sensing instrumented indentation to study the inelastic deformation response of thin metal films on substrates remains a work in progress. A key underlying challenge is to estimate the connection between the projected contact area A_c and the indenter depth of penetration h for particular geometrical configurations, constraint conditions and material parameters by properly accounting for the manner in which the material flows around the indenter. References to relevant literature on this topic can be found in Section 6.8.2 as well as in reports by Oliver and Pharr (1992), Nix (1997) and Dao et al. (2001).

7.10 Exercises

1. It was noted in Section 7.1.2 that the common approach to calculation of equilibrium spacing of parallel misfit dislocations in a strained layer beyond critical thickness tacitly assumes that the dislocation array was formed as a consequence of threading dislocations gliding in perfect unison. Adopt this configuration with all threading dislocations moving at the same speed as a starting point, and assume that the speed of any individual threading dislocation always varies directly with the driving force acting on it.

    (a) Develop a qualitative argument to show that, if a single threading dislocation in the array breaks ranks with the other threading dislocations and moves a significant distance ahead of the others, its speed will be such that it will move ever further ahead.
    (b) Likewise, develop a qualitative argument to show that, if a single threading dislocation in the array falls a significant distance behind the others, its speed will be such that it will fall ever further behind.

2. The temperature change $\bar{T}_Y$ from a stress-free reference state at which plastic yielding begins during thermal excursion in a film–susbtrate bilayer of arbitrary layer thickness is given in (7.41). Assuming that the conditions which underlie the derivation of this equation hold, show that the plastic zone spreads monotonically from the film–substrate interface to the free surface of the metal film if the conditions

$$\eta^2 m \leq 1, \tag{7.92}$$

and

$$3\eta m \left[3 - 5\eta^2\right] + 3\eta^2 m^2 \left[-9 + 9\eta + 60\eta^2 + 45\eta^3 + \eta^4\right]$$
$$+ \eta^4 m^3 \left(4\eta + 3\right) \left[-27 - 27\eta + 45\eta^2 + 51\eta^3 + 4\eta^4\right] \geq 2 \tag{7.93}$$

are satisfied, where $\eta = h_f/h_s$ and $m = M_f/M_s$.

3. A bilayer is made by diffusion-bonding a metallic layer of thickness h_f to a ceramic layer of thickness h_s at a stress-free reference temperature T_0. The properties of the metallic layer are: $E_f = 210\,\text{GPa}$, $\nu_f = 0.3$, $\alpha_f = 15 \times 10^{-6}/^\circ\text{C}$ and $M_f/\sigma_{Y_f} = 3800$. The corresponding properties of the ceramic layer are: $E_s = 380\,GPa$, $\nu_s = 0.25$, and $\alpha_s = 5 \times 10^{-6}/^\circ\text{C}$. The metallic layer can be assumed to be an isotropic elastic–ideally plastic material. Assume that the properties of both layers are independent of temperature, and that $h_f/h_s = 1$.

   (a) Determine the decrease in temperature from the reference state that is necessary to initiate plastic yielding in the metallic layer at the metal–ceramic interface.
   (b) Determine the decrease in temperature from the reference state that is necessary to induce plastic yielding throughout the metallic layer.
   (c) If $h_s = 1$ mm, determine the magnitude and sign of substrate curvature when the condition in (b) is reached.

4. A trilayer is formed by first depositing an Al film on a Si(100) substrate and then passivating the film with a silicon oxide capping layer. The evolution of substrate curvature upon cooling from a stress-free reference temperature can be determined by extending the analysis for the bilayer given in Section 7.4.1 to a three layer system. The sign of the curvature can change during a monotonic reduction in temperature from the stress-free reference state for specific thickness ratios of the three layers for this configuration (Shen and Suresh 1995a), and such a reversal of curvature has been predicted for the thicknesses (in arbitrary units) of the substrate, film and passivation layer of $h_s = 5$, $h_f = 1$ and $h_p = 1$, respectively. Provide a rationale for the possible occurrence of curvature reversal in such a system during cooling. The elastic and thermal properties of the three layers can be found in the property tables in Chapters 2 and 3. The plastic yield strength of the aluminum film, which can be assumed to be elastic–ideally plastic for the purpose of this exercise, is 100 MPa.

5. The variation of indentation load as a function of penetration depth for a sphere of radius R indenting the flat surface of an elastic solid is given by the Hertzian theory of elastic contact as

$$P = \frac{4}{3} E^* R^{\frac{1}{2}} h^{\frac{3}{2}}, \qquad (7.94)$$

where P, E^* and h are defined in Section 7.9.5 (Johnson 1985). From the geometry of elastic contact, it is found that the projected contact area $A_c \approx \pi R h$. For elastic spherical indentation, show that the initial unloading slope dP/dh is of the form given by (7.90) with $\alpha^* = 1$.

6. Hardness tests are commonly used in industry to assess, on a relative basis, the strength of engineering materials and coatings. Conduct a literature search to find (a) at least four different definitions of hardness that are widely used in practice, (b) relationships, if any, among these different hardness measures, and (c) the advantages and limitations of each hardness test. Useful background information can be found in the references: D. Tabor, *The hardness of metals*, Clarendon Press, Oxford, UK, 1951, and *Metals Handbook*, Second Edition, edited by J. R. Davis, American Society for Materials International, Metals Park, Ohio, 1998.

7. Suppose that the microbeam deflection experiment described in Section 7.9.2 is applied for the case of a gold film of thickness 0.5 μm on a silicon oxide substrate of thickness 5 μm. Assume that both the film and substrate materials are isotropic and that they deform under plane strain conditions. The distance between the cantilever support and the load point is 140 μm and the width of the configuration is 20 μm.

    (a) Determine the relationship between the applied indentation force and the corresponding deflection for the case when both the film and the substrate deform elastically.

    (b) Suppose that the gold is idealized as an elastic-ideally plastic Mises material with a flow stress of 195 MPa. Determine the influence of plastic deformation in the film on the load-deflection relationship as plastic yielding, assumed to be uniform across the width, spreads from the cantilevered end of the configuration toward the load point. For purposes of the calculation, assume that the film is thin compared to the substrate so that deformation is nominally uniform across the thickness of the film.

8

Equilibrium and stability of surfaces

The role of mechanical stress in establishing the equilibrium and stability of material surfaces is the focus in this chapter. The concepts developed provide the basis for consideration of stress-driven transient evolution of surfaces, which will be taken up in the following chapter. Examples of material surfaces that are of interest include the free surface of a crystal, which may be adjacent to a vacuum or in contact with a vapor, and the interface between two crystals. In all these situations, it is assumed that each solid phase is a homogeneous elastic medium. A characterizing feature of such a solid is that there is always a well-defined *reference configuration* to which it returns when all external loads are removed from it. While this configuration is commonly viewed as being immutable, the focus in this chapter is on situations in which this reference configuration can change shape or size over time. Physical processes by which this can occur involve the addition, rearrangement and removal of mass which can proceed, for example, by means of condensation/evaporation, surface diffusion, grain boundary diffusion or grain boundary migration. These processes are typically very slow, and they are not usually taken into account in applications other than material processing, a case in which stress usually plays a minor role. However, in systems with very small size scale and high temperature, stress of large magnitude can have a significant influence on the evolution of surfaces and interfaces. All three features – small size, high temperature and large stress – are characteristic of thin film systems, where the phenomenon of stress-driven surface evolution is of considerable scientific and practical interest. The actual ranges of stress, temperature and size for which this is the case will emerge in the course of discussion.

This chapter begins with a brief summary of the basic approach and terminology used in describing the equilibrium of surfaces. The concept of chemical potential is then introduced as a means to generalize the effects of stress and mass transport in conjunction with several physical phenomena such as free surface evolution in a stressed solid, grain boundary diffusion or migration, and the advance or healing of a crack. Attention is then shifted to the issue of stability of an initially flat surface of a stressed solid under small amplitude perturbations. The mechanics of contact between impinging islands during deposition of film material is examined next. How do internal stresses develop in the evolving thin film as a consequence of such interactions between contacting islands? This is followed by consideration of a variety of phenomena dealing with the connections between free energy and surface evolution. How does the presence of misfit dislocations influence waviness and pattern formation on a surface? What is the chemical potential of the free surface of an elastic solid whose surface energy is anisotropic? How is island formation on a strained surface influenced by equilibrium thermodynamics and mechanical factors? The energetics of arrays of nanocrystalline islands of predictable size and shape are of critical importance for quantum electronics applications. Some of these topics are further pursued in Chapter 9 in the context of surface evolution that is influenced by mass transport.

8.1 A thermodynamic framework

The material systems to be considered in this chapter are deformed due to stress, in general, and are assumed to be always in mechanical equilibrium. Typically, any transients in configuration occur too slowly for material inertia effects to come into play. The systems are not necessarily in thermodynamic equilibrium, however. The thermodynamic state is understood to be represented by a deformation field defined over the bulk phase, the position and shape of the surface, and the temperature field defined over the bulk phase. For the time being, the system will be assumed to be in thermal equilibrium (relatively rapid thermal conduction) and at a fixed temperature (available heat reservoir). The appropriate thermodynamic state function is the Helmholtz free energy function (at fixed temperature), say $\mathcal{F}$. The free energy is the sum of the total elastic energy of deformation in the bulk phase and the total surface energy of the system. The elastic strain energy per unit volume of the bulk phase, say U, and the energy per unit area of the surface, say U_S, incorporate whatever constitutive models are to be used to describe material behavior. Usually, there are other contributions to the total free energy of a system. However, only those contributions that actually change in value as the surface evolves are relevant to the discussion of underlying physical processes.

Most of the discussion of this chapter is concerned with infinitesimally small deformation of the bulk phase of the material from its reference configuration, although extension to finite deformation is straightforward in most cases if expressions are assumed to be enforced in the reference configuration of the material. A second point concerns the terms *surface energy* and *surface tension*. These terms are often used interchangeably to describe a certain physical characteristic of a fluid.

Surface energy refers to the energy of cohesion of an undeformed material surface, as discussed in Chapter 1. The physical effect that does work as a material surface is deformed, over and above the work of stress within the bulk phase, will be called *surface stress*; it is the thermodynamic force that is work-conjugate to surface strain with respect to free energy. As was noted in Chapter 1, surface stress arises naturally when surface energy depends on surface strain. The term *surface tension* will not be used to describe any particular physical characteristic of the system in the course of this discussion.

Changes in magnitude of the free energy corresponding to physically realizable perturbations in deformation and/or surface shape from the current state represent the tendency for a system to depart from its current state. If the variations in free energy for all possible perturbations in shape are zero to first order in the perturbations then the current state is a thermodynamic equilibrium state. If the free energy is a local minimum for perturbations from that state, then the equilibrium configuration is *stable*; otherwise, the equilibrium configuration is not stable, and it may be *unstable*. This statement of equilibrium is a corollary to Gibbs' original equilibrium condition, which is expressed in terms of internal energy as the fundamental state function and which requires that entropy must be stationary under variations in state for equilibrium. Likewise, the definition of stability adopted here is a corollary to Gibbs' requirement that entropy must be maximum for an equilibrium state to be stable, as discussed in the introduction to Chapter 9.

If the variation in free energy for some possible perturbation in state is negative to first order, the system is not in thermodynamic equilibrium. Furthermore, it can lower its free energy by actually executing a change in state without an external stimulus; such a process is said to be *spontaneous*. A description of the way in which the system actually executes a change in state requires the introduction of additional constitutive hypotheses which relate the rates of change of deformation and/or surface configuration to the associated free energy variations; these constitutive relations are the *kinetic relations* of irreversible thermodynamics which relate the rates of change of state (here, surface shape and deformation) to thermodynamic driving forces (chemical potential and stress) in nonequilibrium configurations. A final point is that the restriction to isothermal conditions reduces the theoretical framework from that of irreversible thermodynamics to that of mechanics of dissipative systems; the framework must be consistent with thermodynamics, of course, but it does not include its full range of behavior.

In the present context, the restriction of the material of the bulk phase to be an inherently stable elastic material and the requirement that the elastic stress field is an equilibrium field imply that variations of free energy due to perturbation in the deformation field at fixed shape will vanish. This is equivalent to the above requirement that the system is always in mechanical equilibrium. The novel feature of the system, from the perspective of mechanical behavior, is that the free energy also changes due to changes in surface shape. The surface field that represents the tendency for the system to change the shape of its reference configuration is the *chemical potential field* (Gibbs 1878, Herring 1953). This surface field is defined quantitatively in the next section, where several generalizations of the concept are also considered for surfaces that define interfaces between materials. The general concept of chemical potential is also applied in Chapter 9 in situations where bulk material transport occurs.

8.2 Chemical potential of a material surface

The concept of chemical potential was introduced by Gibbs (1878) in order to describe the tendency for materials to combine with each other chemically. Suppose that there is a homogeneous mass of material, say AB, which is a compound of materials A and B. This mass has the possibility for chemical interaction with a second homogeneous mass composed of material B. The removal of a unit of mass from B and its addition to AB is accompanied by a change in the free energy of the system. Gibbs termed this change in free energy per unit mass the *potential* for the mass transfer process. The process tends to occur spontaneously if this potential is negative, whereas it does not tend to occur if the associated change in potential is positive. As stated by Gibbs, a necessary and sufficient condition for the system to be in chemical equilibrium is that the total free energy cannot be reduced further by *any possible* mass transfer process.

The concept of chemical potential as an indicator of the tendency for mass transfer to occur in a system has been generalized to a range of physical phenomena. The issue is not so much the transfer itself, but is rather the energetics of forming or breaking chemical bonds and the local state of the material as bonds are established or severed. In this section, several such generalizations are described in which stress effects are incorporated. These include evolution of a free surface of a stressed solid, mass diffusion along a grain boundary driven by a stress gradient, grain boundary migration and crack growth or healing.

8.2.1 An evolving free surface

Consider an elastic solid that occupies a region R of space in its reference configuration. The boundary S of the solid consists of two parts, a part S' that can change over the course of time and a part S'' that is fixed with respect to material points instantaneously on it. The portion S' of the surface is free of externally applied traction. The material is stressed by imposing some conditions on S'', but this is done in such a way that there is no exchange of energy between the material in R

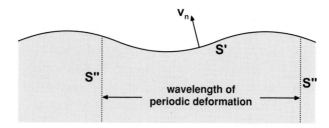

Fig. 8.1. A schematic diagram of a periodic material structure undergoing surface evolution. Such configurations are conveniently studied by focusing on a sub-region bounded by a segment S' of the evolving surface and by material surfaces S'' across which traction acts but through which net energy flux is zero. A particularly convenient choice for S'' for subject configuration are symmetry surfaces.

and its surroundings as S' changes in shape; this is not an essential requirement but is, rather, a feature assumed to make this development as transparent as possible. An example of a situation where this restriction is relevant is a system for which both the geometrical configuration and the deformation field are periodic and in which S'' encloses only a single period of this configuration, as illustrated in Figure 8.1. In this case, the net energy flux through S'' is exactly zero as a result of periodicity, even though the local flux is not necessarily zero pointwise over S''; the simplest choice would be to form S'' from among symmetry surfaces of the periodic system. A second example arises when surface shape fluctuations are localized in space and S'' is sufficiently remote so that the effect of the fluctuations in shape of S' on values of field quantities at points on S'' is negligibly small.

The elastic strain energy density $U(\epsilon_{ij})$ is determined locally by the elastic strain field ϵ_{ij}, and the energy density function incorporates constitutive assumptions about mechanical behavior. The strain energy density U does not depend explicitly on position in the material, implying that the material is homogeneous; again, this feature is assumed for simplicity but is not essential to the development. The surface energy density U_S is assumed to be constant over the entire surface S for the time being. The neglect of dependence of U_S on deformation implies that surface stress effects are ignored, and the neglect of dependence of U_S on local surface orientation implies that surface anisotropy effects are ignored. Implications of including both of these effects are examined in Section 8.8.

With these constitutive assumptions, the free energy of the system is the sum of elastic energy and surface energy,

$$\mathcal{F}(t) = \int_R U(\epsilon_{ij})\,dR + \int_{S'} U_S\,dS. \tag{8.1}$$

As noted in Section 8.1, these are the contributions to free energy that change as shape changes. The surface integral extends only over S' because this is the only part of the surface S that can change its shape with respect to the material as a function of time; it is only free energy *changes* that are significant. The time dependence of $\mathcal{F}$ is indicated explicitly on the left side of (8.1). The time dependence on the right side arises from several effects: the region R changes in time, the strain field ϵ_{ij} changes with time at each material point, and the location of the free surface S' with respect to the material changes in time.

The change of shape of S' can be expressed in terms of the normal velocity of the surface v_n, viewed as a function of position over S' for the time range of interest. This is the normal speed of the surface with respect to the material particles instantaneously on the surface, as determined in the reference configuration of the material. Then

$$\dot{\mathcal{F}}(t) = \int_R \frac{\partial U}{\partial \epsilon_{ij}} \dot{\epsilon}_{ij} \, dR + \int_{S'} v_n U \, dS - \int_{S'} v_n \kappa U_S \, dS, \qquad (8.2)$$

where κ is twice the mean curvature of the surface S', or the sum of the curvatures in any two orthogonal directions, at any point on the surface; a superposed dot denotes time derivative. The sign convention is such that the curvature is positive if the center of curvature at a point on the surface lies in the direction from the surface of the outward unit normal vector n_i. The interpretation of the three terms in (8.2) is straightforward. As the shape of S' changes, the strain at each interior material point changes; the resulting net change in free energy due to this effect is given by the first term on the right side of (8.2). As material is added ($v_n > 0$) or removed ($v_n < 0$) at any point on S', it must be strained to be compatible with the surface to which it is being added or from which it is being removed; the net strain energy change is given by the second term on the right side of (8.2). Lastly, the amount of surface area increases or decreases locally, depending on the sign of the mean curvature and the direction of surface propagation; the net change in free energy due to change in surface area is given by the third term on the right side of (8.2).

The expression in (8.2) can be obtained by direct application of the fundamental definition of a derivative. However, an intuitive basis for the second and third terms on the right side of (8.2) can be found in an examination of the sketch in Figure 8.2 for a two-dimensional situation. The sketch shows a small portion of the volume R and its free surface S' at time t. It also shows the position of S' at a slightly later time $t + \Delta t$ after movement by its normal speed v_n. Suppose that the system depicted has unit extent in the direction normal to the plane shown, and focus attention on an element of surface area dS. At time t, the curvature of that cylindrical element in the plane is κ, or the radius of curvature is $1/\kappa$. Once the surface has moved to the dashed position at time $t + \Delta t$ due to its local normal speed, its radius of curvature has been reduced to $1/\kappa - v_n \Delta t$. Thus, the area of the surface element

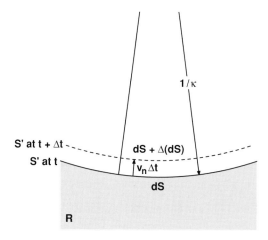

Fig. 8.2. A two-dimensional sketch of a curved surface advancing through a material that provides a simple basis for interpreting the second and third terms on the right side of (8.2).

dS at t becomes

$$dS + \Delta(dS) = \kappa\, dS(\kappa^{-1} - v_n\Delta t) = dS - v_n\kappa\, dS\,\Delta t \qquad (8.3)$$

at time $t + \Delta t$. The local change in area per unit time is $\Delta(dS)/\Delta t = -v_n\kappa\, dS$ which, upon integration over S', provides the third term on the right side of (8.2). The origin of the second term on the right side of (8.2) is seen to be a consequence of the fact that the surface element dS in Figure 8.2 sweeps out a volume $v_n\Delta t\,dS$ during elapsed time Δt; the material added in this volume as the surface advances must have the local strain energy density U. Thus, the local strain energy addition per unit time is $Uv_n\,dS$ which, upon integration over S', provides the second term on the right side of (8.2).

The quantity $\partial U/\partial\epsilon_{ij}$ is the stress σ_{ij} which is necessary to enforce the elastic strain ϵ_{ij}. The stress field is an equilibrium field which, in the absence of body forces, requires that $\sigma_{ij,j} = 0$. The identity

$$\sigma_{ij}\dot\epsilon_{ij} = \sigma_{ij}\dot u_{i,j} = \left(\sigma_{ij}\dot u_i\right)_{,j} \qquad (8.4)$$

follows immediately, where u_i is the material particle displacement field. Then, the result (8.4) is substituted into the first term on the right side of (8.2), and the divergence theorem is applied to the resulting integral. Recall that the surface S' is free of applied traction, so that

$$\sigma_{ij}n_j = 0, \qquad (8.5)$$

pointwise on S'. Furthermore, if there is to be no exchange of work between the material and its surroundings, as assumed, then

$$\int_{S''} \sigma_{ij} n_j \dot{u}_i \, dS = 0. \tag{8.6}$$

In light of (8.5) and (8.6), the rate of change of free energy (8.2) becomes

$$\dot{\mathcal{F}}(t) = \int_{S'} [U - \kappa U_S] \, v_n \, dS. \tag{8.7}$$

This expression leads naturally to a definition of chemical potential as a field over the evolving surface S'. Note that $v_n dS$ represents a local rate of material volume addition ($v_n > 0$) or removal ($v_n < 0$). If it is understood that all quantities are represented in the reference configuration of the material, then there is an unambiguous equivalence between volume of material added and mass of material added. The free energy per unit volume of material being added locally is then

$$\boxed{\chi = U - \kappa U_S,} \tag{8.8}$$

so χ is identified as the *surface chemical potential field* over S'. Nozieres (1992) has pointed out that the chemical potential as defined for a solid–vacuum surface, as in (8.8), does not have the same physical meaning as does the concept originally introduced by Gibbs. Its value at a point on the surface does not have an intrinsic physical interpretation. Instead, its significance arises through the *comparison* of values at neighboring points as represented by a gradient along the surface. Following Herring (1953), Mullins (1957) and others, the designation as a chemical potential is retained in the present discussion due to the result (8.7), which makes clear the role of χ as a configurational force that tends to drive changes in surface position.

The chemical potential for surface evolution is often expressed as the free energy per atom or molecule added to the surface. With that definition, the chemical potential is $\chi \Omega$, where Ω is the atomic or molecular volume of the material being deposited. The chemical potential is also sometimes expressed as the free energy change per mole of material added. Again, the difference introduced in this way is only a constant multiple factor of the expression in (8.8), provided that χ is defined with respect to the reference configuration of the material.

The local form of chemical potential in (8.8) is due to Herring (1953), and the inclusion of elastic stress effects in this local form is due to Asaro and Tiller (1972) and Rice and Chuang (1981). Its principal role in this form is to describe the changing shape of the reference configuration of an elastic solid subject to a spatially nonuniform stress distribution. If χ varies from point to point on the surface of the solid, then the system free energy can be lowered by moving material from portions of S' with relatively high values of χ to portions with relatively low values. This

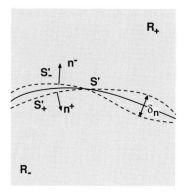

Fig. 8.3. A schematic diagram showing the surface S' (solid line) separating a volume of material into two parts R_+ and R_- at a certain instant. Subsequently, mass flow along the interface causes the material surfaces initially joined at S' to separate where material is added or to interpenetrate where it is removed. The quantity δ_n represents the volume of material added per unit area on the original surface S'.

diffusive mass transfer results in motion of S' with respect to the material particles instantaneously on it. Several examples of this kind are discussed in detail in Sections 8.3 and 8.4. The system can lower its free energy by mass rearrangement if χ varies with position over S'; it follows that $\chi = $ constant over S' is a necessary condition for a configuration to be in thermodynamic equilibrium.

In those cases in which the surface S' is an interface between two materials, rather than being the free boundary of a single material, the surface can advance by means of the *transformation* of material ahead of the interface into material behind it, rather than by surface diffusion. This process results in normal motion of the surface without mass transport. The concept of chemical potential is considered from this point of view in Section 8.2.3.

8.2.2 Mass transport along a bimaterial interface

A mass transport process that can operate to change the system free energy is mass flow along a surface that is interior to the material. A grain boundary in a polycrystalline material and a phase boundary in a multiphase material are representative examples. The basic idea is that traction associated with a state of internal stress can be transmitted across such a boundary, but that the material surfaces which together constitute the interface can change over time as mass rearrangement occurs along the interface. For the time being, it is assumed that no net mass flow occurs into or out of the bulk materials on either side of the interface.

Consider the schematic diagram in Figure 8.3 which shows a volume of material occupying a region R of space in its reference configuration. The outer boundary of R is again a material surface denoted by S''. The material volume is divided into two parts R_+ and R_- by an internal surface S'. At a certain instant of time, the material boundaries of R_+ and R_- which coincide with S' are identified as reference material surfaces S'_+ and S'_-; the displacements of material points on these surfaces S'_+ and S'_- coincide at that instant of time so that $u_i^+ = u_i^-$. At a later time, the material surfaces S'_+ and S'_- no longer coincide with S', in general, because mass flowing along S' may cause them to separate (mass addition) or interpenetrate (mass depletion). Compatibility requires that the net volume of material added per unit area to the interface, say δ_n at a point, is necessarily related to the separation by

$$\delta_n = - \left(u_i^+ n_i^+ + u_i^- n_i^- \right) . \tag{8.9}$$

This separation δ_n is a time-dependent field over S', and its history represents shape evolution associated with mass transport along the surface. It could be anticipated at this point that the corresponding work-conjugate force is the normal stress acting on the surface. This is borne out by the following calculation which establishes a chemical potential for this mass transport process.

To derive an expression for the chemical potential, suppose that the loading conditions imposed on S'' are such that there is no exchange of energy between the material in R and its surroundings outside S''. Furthermore, to avoid dissipative shear deformations across S' which are not involved in mass transport, it is assumed either that the tangential displacement is continuous across S', which requires that $u_i^+ - u_j^+ n_j^+ n_i^+ = u_i^- - u_j^- n_j^- n_i^-$, or that the shear traction vanishes on S' from either side, which requires that $\sigma_{ij}^\pm n_j^\pm - \sigma_{kj}^\pm n_j^\pm n_k^\pm n_i^\pm = 0$.

The free energy is simply

$$\mathcal{F}(t) = \int_{R_+ + R_-} U(\epsilon_{ij}) \, dR \tag{8.10}$$

in this case. No surface or interface energy term is included because the areas of surfaces and interfaces do not change under the conditions assumed. It is tacitly assumed that the chemical structure of the interface is unaltered by the transport process. Thus, even if a term representing structure would be included in $\mathcal{F}(t)$, its rate of change would be zero under these conditions. Because strain may be discontinuous across S', the rates of change of the free energies in the regions R_+ and R_- are calculated separately. Following the development in Section 8.2.1 for calculation of $\dot{\mathcal{F}}(t)$, it is found that

$$\dot{\mathcal{F}}(t) = \int_{S'} \left(\sigma_{ij}^+ n_j^+ \dot{u}_i^+ + \sigma_{ij}^- n_j^- \dot{u}_i^- - \tfrac{1}{2}(U^+ + U^-)(\dot{u}_n^+ + \dot{u}_n^-) \right) dS, \tag{8.11}$$

where it has been assumed that material is added to or removed from the bulk materials joined at the interface at equal rates.

If the tangential displacement across S' is continuous, then there are scalar functions u_n^+ and u_n^- such that $u_i^\pm = u_n^\pm n_i^\pm$ with $\delta_n = -(u_n^+ + u_n^-)$. Then, continuity of normal traction implies that $n_i^+ \sigma_{ij}^+ n_j^+ = \sigma_n = n_i^- \sigma_{ij}^- n_j^-$. Continuity of traction and of tangential displacement also imply that $U^+ = U^- = U$ on the interface. The expression for free energy change in (8.11) then reduces to

$$\dot{\mathcal{F}}(t) = -\int_{S'} (\sigma_n + U)\dot{\delta}_n \, dS. \qquad (8.12)$$

In most cases of practical interest, and particularly for mass diffusion along a grain boundary, the magnitude of U is roughly equal to the magnitude of elastic strain times the magnitude of σ_n, and the second term in the integrand of (8.12) is therefore negligible compared to the first term. Consequently,

$$\dot{\mathcal{F}}(t) = -\int_{S'} \sigma_n \dot{\delta}_n \, dS \qquad (8.13)$$

for most practical purposes.

Alternatively, if the shear tractions are zero on S', then there is a surface field σ_n such that $\sigma_{ij}^\pm n_j^\pm = \sigma_n n_i^\pm$. In light of (8.9), the expression for the rate of change of free energy (8.11) again reduces to (8.12). Following the reasoning outlined above, $\dot{\delta}_n \, dS$ is the rate at which material volume is being added locally on the surface S'; equivalently, $\dot{\delta}_n$ is the rate of addition of material volume per unit area on S'. The chemical potential field over S' for this mass transport process is then

$$\boxed{\chi = -\sigma_n} \qquad (8.14)$$

as the generalized force that is work-conjugate to material volume added locally. The form (8.14) of chemical potential is applied in Section 9.5.1.

Although a number of simplifying assumptions were introduced to facilitate the derivation of (8.14), it is important to note that the result for chemical potential (8.14) is independent of these assumptions. In other words, (8.14) provides a chemical potential for interface mass transport for any such system for which behavior is described by the same constitutive assumptions as were invoked above. This is so in cases where the system does exchange energy with its environment or in which there is dissipation due to interfacial shearing, for example.

8.2.3 Migration of a material interface

Another surface evolution process that arises during microstructure evolution is the transverse migration of an interface which is a boundary shared between two materials. The two materials might be grains of the same material which differ only in their crystallographic orientation, different phases of the same material, or a solid

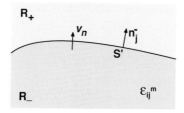

Fig. 8.4. A schematic diagram showing two materials occupying the regions R_+ and R_- with a shared interface S'. The interface propagates with normal speed v_n with respect to the material on either side as one material is transformed into the other.

solution and a solid precipitate of that solution. The phenomenon of migration of the interface between two materials does not necessarily involve mass transport; instead, motion of the boundary is a consequence of the continuous transformation of one material to the other, a process that, in effect, moves the interface with respect to the material instantaneously on it, as was noted in Section 8.2.1.

An example of such interface migration arises in the fabrication of epitaxial SiGe thin films on Si substrates by the process of solid phase epitaxy (Paine et al. 1991). In this process, Ge ions are implanted at high energy and relatively low temperature into a Si crystal near a planar surface. The incoming Ge ions distort the Si lattice in a layer near the surface to the point where the SiGe material is essentially amorphous or disordered to a depth determined by the energy of the implant, usually a few hundred nm or so, resulting in an interface between the undisturbed Si crystal and the amorphous overlayer. Then, upon subsequent heating, the layer crystallizes by propagation of the interface between the crystalline Si substrate and the amorphous SiGe layer toward the free surface. The result is a SiGe epitaxial strained layer on a Si substrate, with the Ge concentration profile established by the implantation process.

A schematic diagram of the system is shown in Figure 8.4, where the notation established earlier in this section is retained. An interior surface S' separates the two materials which occupy the regions R_+ and R_- in the reference configuration. The interface migrates with normal speed v_n into the material in R_+. As it does so, it leaves behind a mismatch strain field ϵ_{ij}^m in the material R_-. This mismatch strain field may be inhomogeneous and incompatible; only the total strain – elastic minus mismatch – must be compatible. It could also be assumed that there is a mismatch strain field in R_+ but this added complexity is not included here. In either case, there is a discontinuity in mismatch strain across S' as it propagates. Again, for convenience, it is assumed that the outer non-migrating boundary S'' is workless during interface migration.

The total free energy of the system in this case is

$$\mathcal{F}(t) = \int_{R_+} U(\epsilon_{ij}) \, dR + \int_{R_-} U(\epsilon_{ij} + \epsilon_{ij}^{\mathrm{m}}) \, dR + \int_{S'} U_{\mathrm{S}} \, dS, \qquad (8.15)$$

where U_{S} is now an energy per unit area associated with the interface and ϵ_{ij} is the *total* strain in the material, that is, elastic strain minus mismatch strain. The interface energy U_{S} is a positive scalar quantity, typically less in magnitude than the sum of the free surface energies of the two materials separately. For the time being, surface stress effects are ignored and U_{S} is assumed to be independent of orientation with respect to either material. Following steps similar to those used to obtain (8.2) and (8.11), the time rate of change of free energy is

$$\dot{\mathcal{F}}(t) = \int_{S'} \left[\sigma_{ij}^+ n_j^+ \dot{u}_i^+ + \sigma_{ij}^- n_j^- \dot{u}_i^- - U_+ v_n + U_- v_n - \kappa v_n U_{\mathrm{S}} \right] dS, \qquad (8.16)$$

where the equilibrium condition $\sigma_{ij,j} = 0$ on stress has been enforced in both R_+ and R_-. The displacement field appearing in (8.16) represents the total displacement of a material point due to both elastic deformation and stress-free misfit deformation. Equilibrium at the interface requires that the traction must be continuous, that is,

$$\sigma_{ij}^+ n_j^+ + \sigma_{ij}^- n_j^- = 0 \quad \text{on} \quad S'. \qquad (8.17)$$

Thus, if the traction exerted by the material in R_+ on the material in R_- is denoted by $T_i = \sigma_{ij}^- n_j^-$, then

$$\dot{\mathcal{F}}(t) = \int_{S'} \left[T_i [\![\dot{u}_i]\!] + v_n [\![U]\!] - \kappa v_n U_{\mathrm{S}} \right] dS, \qquad (8.18)$$

where the double square brackets are used to indicate a discontinuity in limiting values of a field quantity across the surface. For example,

$$[\![\dot{u}_i]\!] = \dot{u}_i^- - \dot{u}_i^+, \qquad (8.19)$$

where $\dot{u}_i^{\pm}$ is the limiting value of particle velocity component at a certain instant of time as the observation point approaches the surface S' in a normal direction from within $R_{\pm}$. The kinematic jump condition for discontinuous fields at moving surfaces requires that

$$[\![\dot{u}_i]\!] = -v_n [\![u_{i,j}]\!] n_j^-, \qquad (8.20)$$

which is a direct consequence of the continuity of u_i across S', or $[\![u_i]\!] = 0$. Again, it is noted that $u_{i,j}$ is the displacement gradient due to the elastic deformation and misfit deformation together. In light of (8.20), the rate of change of free energy becomes

$$\dot{\mathcal{F}}(t) = \int_{S'} \left[[\![U]\!] - T_i [\![u_{i,j}]\!] n_j^- - \kappa U_{\mathrm{S}} \right] v_n \, dS. \qquad (8.21)$$

It follows immediately that the chemical potential for migration of the interface is

$$\chi = [\![U]\!] - T_i [\![u_{i,j}]\!] n_j^- - \kappa U_S, \tag{8.22}$$

a result established by Eshelby (1970). Suppose that one of the materials, say that in R_+, is absent but that T_i is an applied traction on the surface S' of the remaining material. Then $U_+ = 0$, $U_- = U$ on S', U_S is the surface energy density of the free surface, and the result in (8.22) is a generalization of (8.2) to the case of motion of a boundary subjected to an external load. If $T_i = 0$, then the result reduces to (8.8).

If there is no internal or chemical energy change associated with the motion of the interface, it then tends to advance locally in the n_j^- direction if $\chi < 0$, thereby reducing system free energy; the interface, however, tends to move in the opposite direction if $\chi > 0$. It follows that $\chi = 0$ is a local equilibrium condition for the surface. Stable equilibrium is expected if χ increases from zero with a perturbation in position in the n_j^- direction or if χ decreases from zero with a perturbation in position in the $-n_j^-$ direction. The equilibrium condition relates the local mean curvature κ of the surface to energy densities according to

$$\kappa = \frac{[\![U]\!] - T_i [\![u_{i,j}]\!] n_j^-}{U_S}. \tag{8.23}$$

Many conditions of this general type, relating surface curvature to energy densities, are commonly identified as a *Gibbs–Thomson relationship*.

As an illustration of expression (8.22) for the chemical potential of a migrating interface, consider the spherically symmetric configuration in which a material undergoes a stress-free isotropic volumetric strain $3\epsilon_m > 0$ at a material point as the interface S' passes that point. As a result, the mismatch strain behind the interface is $\epsilon_{ij}^m = -\epsilon_m \delta_{ij}$ relative to the material in front of the interface. The system is depicted in cross-sectional view in Figure 8.5. The elastic material within $0 < r < R$ is subject to the mismatch strain ϵ_{ij}^m, but the elastic strain is partially relaxed from this level due to the compliance of the surrounding material. The material within $R < r < \infty$ is under stress due to the tendency of the transformed material to expand. Suppose that a remote isotropic tensile stress $\sigma_\infty > 0$ is also imposed on the material. Finally, it is assumed for convenience that the elastic constants of both the transformed and untransformed phases of the material are the same and are given by modulus E and Poisson ratio ν.

The spherically symmetric elasticity problem is readily analyzed to determine the elastic fields throughout the materials involved (Timoshenko and Goodier 1987);

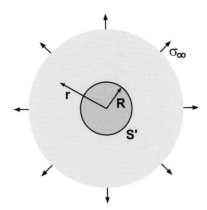

Fig. 8.5. A cross-sectional view of spherically symmetric elastic solid undergoing a stress-free isotropic expansion behind an advancing transformation front S' at $r = R$. The material is subject to a remote tension σ_∞.

the details are left as an exercise. The value of κ on the spherical interface is $-2/R$. Substitution of the limiting values of the elastic fields on the interface into (8.22) yields

$$\chi = \frac{E}{1 - \nu}\epsilon_{\mathrm{m}}^2 - 3\epsilon_{\mathrm{m}}\sigma_\infty + \frac{2}{R}U_{\mathrm{S}}. \qquad (8.24)$$

When R is very small, the third term on the right side of (8.24) is dominant and χ is positive. Consequently, spontaneous formation of transformed spherical particles is ruled out unless the interface energy U_{S} is very small.

As R becomes larger, the other contributions on the right side of (8.24) assume increasing importance. If $\sigma_\infty > E\epsilon_{\mathrm{m}}^2/3(1 - \nu)$, then the equilibrium condition $\chi = 0$ is achieved for some value of R, say R_{eq}. This equilibrium configuration is unstable, and spontaneous growth of the transformed region can proceed once particle size increases beyond R_{eq} to larger values of R. The free energy increase of the system as R increases from zero to R_{eq} is normally called the activation energy for formation of a transformed particle; it is readily calculated in terms of system parameters to be

$$\mathcal{F}_{\mathrm{eq}} = \int_0^{R_{\mathrm{eq}}} 4\pi R^2 \chi \, dR = \frac{4(1 - \nu)^2 U_{\mathrm{S}}^3}{3\epsilon_{\mathrm{m}}^2[\epsilon_{\mathrm{m}}E - 3(1 - \nu)\sigma_\infty]^2}. \qquad (8.25)$$

8.2.4 Growth or healing of crack surfaces

The process of crack growth in a material can be considered in the present context, provided that the mechanism of material separation is that of overcoming material cohesion without dissipative deformation processes operating in the bulk of the

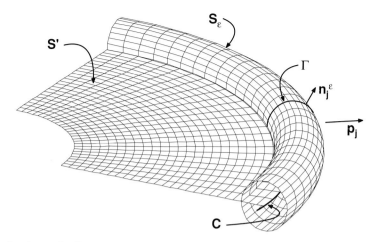

Fig. 8.6. A schematic diagram of the region near the edge of a crack in an elastic solid. The surface S' represents both faces of the crack in the reference configuration of the material. The bounding curve C of the surface S' is the crack edge. The unit vector p_j is normal to the crack edge and locally lies in the tangent plane of S'. The circular tube centered on the crack edge allows calculation of energy variations of the material outside of the tube for arbitrarily small values of the tube radius.

material. Suppose that a material occupies the region of space R in its reference configuration. The outer boundary of R is S''. The fracture surface S' is any smooth surface in R; the space curve C that bounds S' may be interior to R or it may lie in part on S''. The part of C interior to R is denoted by C' and the part on S'', if any, is denoted by C''. In either case, $C = C' + C''$. Crack growth is equivalent to expansion (or possibly contraction) of S' through the motion of C' in its normal direction. The vector that defines the direction of propagation of C' is denoted by p_i; at each point on C', it is a unit vector normal to C' that lies in the tangent plane to S' at that point and that is directed toward the exterior of S'. The normal speed of the crack edge in the direction of p_i is v_p. The region near the edge of S' is shown in Figure 8.6.

For convenience, the applied loading on S'' is assumed to be workless during growth of the crack. The process of crack growth results in a change in the system elastic energy and the system surface energy. The free energy in this case is

$$\mathcal{F}(t) = \int_R U(\epsilon_{ij}) \, dR + 2 \int_{S'} U_S \, dS. \tag{8.26}$$

The factor 2 in front of the second integral accounts for the fact that two surfaces, each with surface energy U_S, are created from the surface S' in the reference configuration. Continuum stress and deformation fields near a geometrical discontinuity such as a crack edge are potentially singular. In such a case, the differentiability requirements on these fields for interpretation of the continuum equilibrium equation or the divergence theorem may not be satisfied; consequently, a special interpretation is required (Freund 1990a).

This interpretation amounts to isolation of the singular line C' from the region R by putting a tubular surface of small cross-sectional radius around it and properly accounting for energy flux through the surface as the crack edge advances. In the end, the limiting behavior as the tubular surface is shrunk onto C' is considered. In the present situation, the tubular surface S_ϵ is formed by putting one and the same circular contour Γ of radius ϵ around the crack edge in each normal plane to C'. The radius ϵ is assumed to be small compared to all physical lengths in the problem. Points on S_ϵ are then located by giving the corresponding arclength along C' measured from some arbitrary reference point and the arclength along Γ measured from some arbitrary reference point on that contour. The unit normal to S_ϵ, say n_i^ϵ, is oriented as shown in Figure 8.6. The region R is then understood in (8.26) to include all material enclosed within S'' but outside S_ϵ. The outward normal vector to the boundary of R is n_j, so that $n_j^\epsilon = -n_j$ on S_ϵ. The time rate of change of free energy is then

$$\dot{\mathcal{F}}(t) = \int_{S_\epsilon} \left[-v_p p_k n_k^\epsilon U - \sigma_{ij} n_j^\epsilon \dot{u}_i \right] + \int_{C'} 2U_S v_p \, dC . \qquad (8.27)$$

The local field near the crack edge has an autonomous dependence on local coordinates (Irwin 1957), a feature that has many consequences. An important consequence for present purposes is that the particle velocity $\dot{u}_i$ which appears in (8.27) can be replaced by a spatial gradient in a direction normal to the crack edge C' according to

$$\dot{u}_i = -v_p u_{i,k} p_k . \qquad (8.28)$$

Then (8.27) becomes

$$\dot{\mathcal{F}}(t) = \int_{C'} \left[p_k \int_\Gamma \left[-U n_k^\epsilon + \sigma_{ij} n_j^\epsilon u_{i,k} \right] d\Gamma + 2U_S \right] v_p \, dC . \qquad (8.29)$$

The path-integral along Γ is a line integral in the plane perpendicular to the curve C' at each of its points. This integral is the negative of the vector-valued two-dimensional J-integral of fracture mechanics (Eshelby 1951, Rice 1968)

$$J_k = \int_\Gamma \left[U n_k^\epsilon - \sigma_{ij} n_j^\epsilon u_{i,k} \right] d\Gamma \qquad (8.30)$$

so that the free energy rate becomes

$$\dot{\mathcal{F}}(t) = \int_{C'} [2U_S - J_k p_k] v_p \, dC . \qquad (8.31)$$

Thus, a configurational force per unit length along the crack edge that is work-conjugate to crack edge motion with respect to free energy is

$$\chi = 2U_S - J_k p_k \, . \tag{8.32}$$

This quantity is perhaps not appropriately called a chemical potential, but it has the character of chemical potential as being developed here, where the corresponding 'material volume' is the amount of crack surface area. The equilibrium condition $\chi = 0$ corresponds to the Griffith crack growth criterion introduced in Section 4.2.1. These ideas are applied in a discussion of cohesive contact in Section 8.6.

8.3 Elliptic hole in a biaxially stressed material

In the preceding section, the chemical potential was introduced as the surface field that is work-conjugate to a geometrical or kinematic surface property (for example, shape, thickness, position, etc.) of the surface of an elastic material. In this and subsequent sections, some specific configurations of elastic materials subjected to stress are considered. Expressions for chemical potential in terms of surface shape and loading parameters are obtained for these configurations. In some cases, the rates of change of free energy with respect to variations in shape are computed as illustrations. As noted in the Introduction, an examination of the rate of change of free energy for some class of kinematic variations of surface features leads to an understanding of the configurational stability of the system under that class of kinematic variations. Constitutive assumptions governing shape evolution are discussed in Chapter 9.

Consider an isotropic elastic solid that contains a long cylindrical hole with an elliptical cross-section. A right-handed rectangular coordinate system is introduced with the z-axis coincident with the axis of the cylinder and with the x- and y-axes coincident with the principal axes of the elliptic cross-section. The extent of the hole in the x-direction is $2a$ and in the y-direction is $2b$. The material is subjected to an equi-biaxial tensile stress of magnitude σ_∞ applied far from the hole. In terms of stress components referred to rectangular coordinates, the remote state of stress is $\sigma_{xx} \to \sigma_\infty, \sigma_{xy} \to 0, \sigma_{yy} \to \sigma_\infty$ at distances far from the hole compared to the larger of a and b. The surface of the hole is free of traction, and the state of deformation is assumed to be plane strain. The two-dimensional system is depicted in Figure 8.7. The main features of shape evolution for the system have been described by Suo and Wang (1994).

With reference to the discussion in Section 8.2, the surface S' in the present case is the cylindrical surface of the hole. The equation of this surface in the plane of

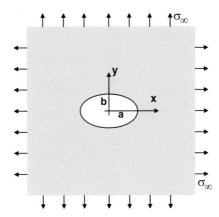

Fig. 8.7. A cylindrical hole with an elliptical cross-section in an elastic solid subjected to equi-biaxial tension under plane strain conditions.

deformation is

$$f(x, y, t) \equiv x^2/a^2 + y^2/b^2 - 1 = 0, \tag{8.33}$$

that is, the coordinates x_S, y_S of any material point on the surface of the hole in the reference configuration must satisfy this equation. Points on the surface of the hole are conveniently located in terms of a single elliptic coordinate η according to $x_S = a \cos \eta$, $y_S = b \sin \eta$ where $0 \leq \eta < 2\pi$. The normalized in-plane curvature of the boundary of the hole is

$$a\kappa \equiv \hat{\kappa}(\eta) = \frac{a^2 b}{(a^2 \sin^2 \eta + b^2 \cos^2 \eta)^{3/2}} \tag{8.34}$$

as a function of position η. Note that this curvature is everywhere positive according to the convention adopted in Section 8.2. The curvature of the surface in the out-of-plane direction is everywhere zero, so the parameter κ given in (8.34) is twice the mean curvature of the surface.

8.3.1 Chemical potential

Stress analysis of this configuration reveals that the in-plane tensile stress at a point on the boundary of the hole, acting in the direction tangent to the surface of the hole, is (Timoshenko and Goodier 1987)

$$\frac{\sigma_{\eta\eta}}{\sigma_\infty} \equiv \hat{\sigma}_{\eta\eta}(\eta) = \frac{2ab}{a^2 \sin^2 \eta + b^2 \cos^2 \eta}. \tag{8.35}$$

The in-plane tensile stress in the direction normal to the hole and the in-plane shear stress on the surface of the hole are zero because of the condition of zero applied

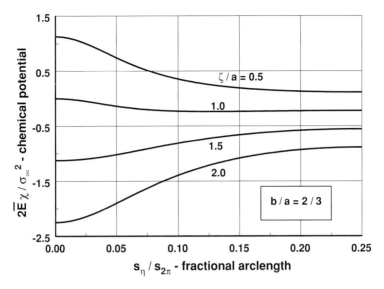

Fig. 8.8. Dependence of normalized chemical potential on position around the boundary of the elliptic hole depicted in Figure 8.7 for the aspect ratio $b/a = 2/3$. The natural length parameter ζ is defined in (8.37).

traction. The strain energy density as a function of position along the surface of the hole is $U(\eta) = \sigma_{\eta\eta}^2/2\bar{E}$ where $\bar{E} = E/(1 - \nu^2)$ is the plane strain elastic modulus. The chemical potential field over the surface of the hole is then

$$\chi(\eta) = U_\infty \left[\hat{\sigma}_{\eta\eta}(\eta)^2 - \frac{\zeta}{a}\hat{\kappa}(\eta) \right] \qquad (8.36)$$

where

$$\zeta = \frac{U_S}{U_\infty} \qquad \text{with} \qquad U_\infty = \frac{\sigma_\infty^2}{2\bar{E}}, \qquad (8.37)$$

is a characteristic length for the physical system. It is essentially the ratio of the surface energy density to a measure of the bulk elastic strain energy density. The actual physical size of the system, represented in this example by the length a, is compared to this characteristic length to assess the importance of surface effects relative to bulk deformation effects. Note that both $\hat{\sigma}_{\eta\eta}$ and $\hat{\kappa}$ depend on a and b only through the ratio b/a.

Representative graphs of normalized chemical potential $2\bar{E}\chi(\eta)/\sigma_\infty^2$ versus fractional arclength $s_\eta/s_{2\pi}$ along the surface of the hole for $0 \le \eta \le \pi/2$ with $b/a = 2/3$ are shown in Figure 8.8. Arclength s_η is measured clockwise from the point where $\eta = 0$ on the cross-section of the hole, and Figure 8.8 is established as a parametric plot with parameter η. Both the strain energy density U and the curvature κ are largest near the ends of the major principal axis of the ellipse ($x = \pm a$, $y = 0$ in this case) and are smallest near the ends of the minor principal axis ($x = 0$, $y = \pm b$ in this case), but these curvature and strain effects enter the chemical potential with opposite algebraic sign so as to compete with each other. Thus, if the system is large in size compared to its characteristic length ζ, say for $\zeta/a = 0.5$, strain effects are dominant and chemical potential is largest at $\eta = 0$, π and is smallest at $\eta = \pi/2$, $3\pi/2$. On the other hand, if the system is small compared to its characteristic length, say for $\zeta/a = 2$, then the opposite is true. Understanding this competition between strain effects and curvature effects is central to understanding certain fundamental issues in the evolution of small material structures.

8.3.2 Shape stability

It is evident from (8.7) and (8.8) that it is necessary to consider a motion of the free surface as represented by v_n in order to calculate a rate of change of free energy $\dot{\mathcal{F}}(t)$. In the case of the elliptic hole in a biaxially stressed solid, the rate of change of free energy is given by

$$\dot{\mathcal{F}}(t) = \int_0^{2\pi} \chi(\eta) v_n(\eta) \frac{ds_\eta}{d\eta} \, d\eta \tag{8.38}$$

in terms of the elliptic coordinate η. As will be discussed in Chapter 9, the evolution of the surface depends on the chemical potential field in a way prescribed by an appropriate kinetic relationship. For the time being, however, free energy rates are considered only for some hypothetical velocity distributions v_n without regard for their physical origin or connection to χ through kinetic relations describing mass transport.

For purposes of illustration, suppose that the only velocity fields v_n considered are those corresponding to the cross-sectional shape of the cavity remaining elliptical but with changing aspect ratio. This can be accomplished by adopting the equation $f(x, y, t) = 0$ as the equation of the ellipse, as in (8.33), with the additional feature that a and b are now viewed as functions of time t. The equation in (8.33) is then satisfied identically in time. For any time-dependent plane curve described by such a function f, the normal velocity is given in terms of partial derivatives of that function with respect to its arguments as

$$v_n = -\frac{f_{,t}}{\sqrt{f_{,x}^2 + f_{,y}^2}}. \tag{8.39}$$

If the resulting expression is written in terms of the elliptic coordinate η, then

$$v_n(\eta) = -\frac{\dot{a}(b/a)\cos^2\eta + \dot{b}\sin^2\eta}{[\sin^2\eta + (b/a)^2\cos^2\eta]^{1/2}}. \tag{8.40}$$

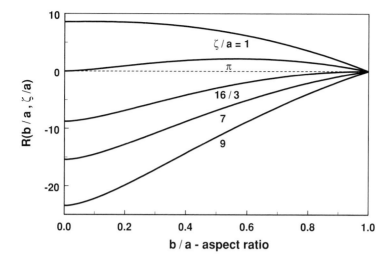

Fig. 8.9. Plots of the quantity $R(b/a, \zeta/a)$ introduced in (8.42) versus aspect ratio of the elliptical hole for several values of the ratio ζ/a. The results show that the circular shape, which is always an equilibrium shape, is stable for values of $\zeta/a > 16/3$ but is unstable for $\zeta/a < 16/3$.

Similarly, the increment of arclength s_η per unit increment in η is given by

$$\frac{ds_\eta}{d\eta} = a[\sin^2 \eta + (b/a)^2 \cos^2 \eta]^{1/2}. \tag{8.41}$$

The chemical potential is already available as a function of η in (8.36).

To simplify the calculation further, suppose that v_n is also restricted by the condition that the volume of the cavity remains unchanged. This is assured if the area $\pi a b$ of the elliptic hole is held fixed or, equivalently, if $\dot{a}b + a\dot{b} = 0$ is satisfied identically in time.

With this additional constraint, the integral in (8.38) takes the form

$$\dot{\mathcal{F}} = -\frac{\sigma_\infty^2}{\bar{E}} \dot{a} a R(b/a, \zeta/a), \tag{8.42}$$

where R is a dimensionless quantity that depends on the two ratios b/a and ζ/a. The integral form of R can be represented in terms of complete elliptic integrals; while this result is useful for establishing certain asymptotic properties, it is not sufficiently transparent to reveal any general aspects of system behavior. Rather, the behavior of the function R is illustrated graphically in Figure 8.9, where each curve shows the variation of R as a function of b/a for a fixed value of ζ/a. Only the aspect ratio range $0 \leq b/a \leq 1$ is considered; the results for $b/a > 1$ follow from symmetry of the system.

The plots in Figure 8.9 reveal significant information about the behavior of this system. As is evident from (8.42), the algebraic sign of $\dot{\mathcal{F}}$ is determined by the sign of the product $\dot{a}R$. For example, consider the stability of the configuration with a circular hole $a = b$ for various values of ζ/a. For the shape to evolve into an ellipse, it is necessary that $\dot{a} > 0$ so that b/a tends to decrease from its initial value of $b/a = 1$. From Figure 8.9, it is seen that the value of R increases from zero if $\zeta/a > 16/3$ but that it decreases from zero if $\zeta/a < 16/3$. Thus, the configuration with a circular hole is a stable equilibrium configuration for $\zeta/a > 16/3$, a range in which surface effects dominate, but it is an unstable equilibrium configuration for $\zeta/a < 16/3$, a range in which bulk deformation effects dominate.

Other aspects of system behavior can be deduced from Figure 8.9. For example, suppose the system is in the configuration with $b/a = 0.5$ and that $\zeta/a = 7$. Then the system can lower its free energy by evolving so that $\dot{a} < 0$, and it tends toward the stable circular configuration $a = b$. However, the ordinate of the graph cannot be interpreted as a configurational force because of the use of the ratio ζ/a as an indentifying parameter; this ratio changes as the aspect ratio b/a changes. An alternate approach, whereby the quantity R in (8.42) is viewed as a function of b/a and $\zeta/\sqrt{ab}$, would result in the ordinate representing the configurational force but the connection to fracture mechanics to be described next would then be obscured.

Yet another interesting observation based on Figure 8.9 concerns the case when $\zeta/a = \pi$. Any configuration with $0 < b/a \leq 1$ is unstable for this value of ζ/a. However, $\dot{\mathcal{F}} = 0$ for small variations in configuration as $b/a \to 0$ because the function $R(b/a, \pi)$ has zero slope at $b/a = 0$. In other words, the state satisfies neither the condition for stability nor the condition for instability. Furthermore, the condition $\zeta/a = \pi$ implies that

$$\pi a \frac{\sigma_\infty^2}{\bar{E}} = 2U_{\mathrm{S}}. \tag{8.43}$$

This is precisely the condition for crack growth established by Griffith (1920) on the basis of an energy balance argument for a crack of length $2a$ in a uniformly stressed solid under two-dimensional plane strain conditions; see Section 4.2.1. Such a crack is a degenerate ellipse. In his original work, Griffith considered only a uniaxial remote stress of magnitude σ_∞ acting in a direction normal to the line of the crack, as opposed to an equi-biaxial stress as in the present case. However, the stress component acting in the direction parallel to the crack line is known to have no influence on the energetics of crack advance within the small deformation framework. Thus, the result represented by (8.43) is completely equivalent to the Griffith fracture condition; it is a particular case of (8.32) with $\chi = 0$.

The formation of macroscopic cracks in a stressed single crystal at elevated temperature has been discussed by Sun et al. (1994). They observed that small voids produced in the crystal during fabrication can change shape and volume as atoms migrate under various circumstances. As shown above, the smallest voids tend to remain rounded in shape due to the dominant influence of surface energy over elastic energy. However, larger voids can become elongated and eventually develop into fractures. This is more likely to occur under uniaxial stress than under the conditions of equi-biaxial stress that were assumed for purposes of illustration here.

8.4 Periodic perturbation of a flat surface

Suppose that, in its reference configuration, an elastic material has a nominally flat surface that is free of applied traction. If the solid is then stressed, is the flat surface shape stable under perturbations in shape? These perturbations are changes in the reference configuration of the material that are achieved by rearrangement of mass, and are not a consequence of deformation of the material. The question is addressed in this section for the case of periodic perturbation of surface shape. Perturbations of very small amplitude are considered first, and this discussion is followed by consideration of higher order effects and substrate stiffness. Cases of nonperiodic surface perturbations are considered in the following section.

8.4.1 Small amplitude sinusoidal fluctuation

An isotropic elastic solid with a nominally flat, traction-free surface is subjected to an initial equilibrium stress field. Suppose that the shape of the free surface S' in the undeformed reference configuration of the material is not actually a plane but that it is slightly wavy. The nominally flat surface coincides with the plane $y = 0$ and the position of the actual surface varies with respect to $y = 0$ in the x-direction. At time t, the position of the surface at coordinate x is given by $y = h(x, t)$. For the discussion in this section, it is assumed that the slope of the surface is small everywhere, that is, $|h_{,x}| \ll 1$ at all points on the surface. The boundary condition that must be enforced on the wavy surface is that the traction is zero.

If σ_{ij} is the stress field evaluated at a point on the surface and n_j is the outward unit normal vector there then $\sigma_{ij} n_j = 0$ at that point. This condition must be enforced pointwise on the wavy surface, and this will be done for the case of small amplitude surface slope. A sketch of a small portion of the surface is shown in Figure 8.10. To lowest order in surface slope, the vector n_i has components $n_x \approx -h_{,x}, n_y \approx 1$. The tangential direction on the surface is then given by the unit vector s_i which has components $s_x = n_y$ and $s_y = -n_x$. The normal traction on the surface with unit normal vector n_i corresponding to boundary values of stress defined by its tensor components σ_{ij} in rectangular coordinates is $\sigma_n = n_i \sigma_{ij} n_j$. Similarly, the shear

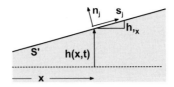

Fig. 8.10. The shape of a material surface in its reference configuration is described by means of its normal distance $h(x, t)$ from a reference plane at place x and time t. The unit vectors s_j and n_j specify directions locally tangent and normal to the surface.

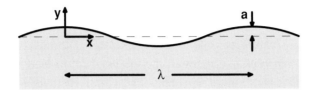

Fig. 8.11. Schematic diagram of the nominally flat material surface which deviates sinusoidally from the mean surface position. Surface deviation is described by (8.47), and the amplitude-to-wavelength ratio a/λ is assumed to be small.

traction is $\sigma_s = s_i \sigma_{ij} n_j$. Both σ_n and σ_s must vanish on $y = h(x, t)$ if the surface is to be free of applied traction. When expanded to first order in the magnitude of the surface slope, these conditions become

$$\sigma_{yy}(x, h) - 2h_{,x}\,\sigma_{xy}(x, h) = 0\,,$$
$$h_{,x}\,\sigma_{xx}(x, h) - h_{,x}\,\sigma_{yy}(x, h) - \sigma_{xy}(x, h) = 0\,. \qquad (8.44)$$

Prior to perturbation of surface shape, an equilibrium state of stress $\sigma_{ij}^{(0)}(x, y)$ exists in the material with the surface $y = 0$ being traction-free. The total stress field for the perturbed shape of the surface is of the form

$$\sigma_{ij}(x, y) = \sigma_{ij}^{(0)}(x, y) + \sigma_{ij}^{(h)}(x, y)\,. \qquad (8.45)$$

Then (8.44) indicates that the boundary conditions on the *additional* stress field $\sigma_{ij}^{(h)}(x, y)$ due to perturbation of the shape of the free surface to $y = h(x, t)$ will be

$$\sigma_{yy}^{(h)}(x, h) - 2h_{,x}\,\sigma_{xy}^{(0)}(x, 0) = 0\,, \quad h_{,x}\,\sigma_{xx}^{(0)}(x, 0) - \sigma_{xy}^{(h)}(x, h) = 0\,. \qquad (8.46)$$

Nonetheless, the analysis is pursued on the basis of the boundary conditions in the form (8.44).

To make the discussion more concrete, suppose that the surface shape is sinusoidal in the x-direction with wavelength $\lambda > 0$ and amplitude $a > 0$, so that

$$h(x, t) = a \cos \frac{2\pi x}{\lambda} \qquad (8.47)$$

as indicated in Figure 8.11. The shape is uniform in the z-direction. The restriction to surface shapes with small slope implies that $a/\lambda \ll 1$. The goal is to find the perturbation of the initial equilibrium stress field $\sigma_{ij}^{(0)}(x, y)$, denoted by $\sigma_{ij}^{(h)}(x, y)$ above, that is required to enforce the boundary conditions (8.44). For the time being, suppose that the initial stress state is spatially uniform and that it results from a remotely applied equi-biaxial tensile stress of magnitude σ_m acting in the direction parallel to the surface. In this case,

$$\sigma_{xx}^{(0)}(x, y) = \sigma_m \quad \text{and} \quad \sigma_{xy}^{(0)}(x, y) = \sigma_{yy}^{(0)}(x, y) = 0 \tag{8.48}$$

throughout the material.

For a sinusoidal perturbation of shape as given by (8.47) and for a uniform initial stress field, the additional elastic stress is also expected to be sinusoidal in x for fixed y. The stress field has the appropriate symmetry if it is derived from an Airy stress function of the form

$$A(x, y) = f(y) \cos \frac{2\pi x}{\lambda}, \tag{8.49}$$

where $f(y)$ is a function of y which is to be determined. The stress function A must satisfy the biharmonic equation, which ensures both that the stress field is an equilibrium field and that the associated strain field is compatible (Timoshenko and Goodier 1987). Furthermore, all stress components must vanish as $y \to -\infty$, which implies that

$$f(y) = \left(c_0 + c_1 \frac{y}{\lambda} \right) e^{2\pi y/\lambda} \tag{8.50}$$

where c_0 and c_1 are constants to be determined by enforcing of the boundary conditions. If the components of stress $\sigma_{ij}^{(h)}(x, y)$ are derived from (8.49) according to (6.3), and if the boundary conditions (8.44) are enforced for $y = h = a \cos(2\pi x/\lambda)$, then it is found that

$$c_0 = 0, \quad c_1 = -a \lambda \sigma_m \tag{8.51}$$

to first order in a/λ. The corresponding stress components along the surface are

$$\sigma_{xx}^{(h)} = -\frac{4\pi a}{\lambda} \sigma_m \cos \frac{2\pi x}{\lambda}, \quad \sigma_{yy}^{(h)} = 0, \quad \sigma_{xy}^{(h)} = 0. \tag{8.52}$$

It can be seen immediately from (8.52) that, as a increases from zero, the stress at the peaks ($x = 0, \pm\lambda, \ldots$) decreases in magnitude from the initial value σ_m and that the stress at the valleys in the surface profile ($x = \pm\lambda/2, \pm3\lambda/2, \ldots$) increases in magnitude from the initial value σ_m.

The strain energy density along the surface, accurate to first order in the small parameter a/λ, is then

$$U(x) = U_m \left[1 - 4\pi(1+v)\frac{a}{\lambda}\cos\frac{2\pi x}{\lambda} \right], \tag{8.53}$$

where

$$U_m = \frac{\sigma_m^2}{M} \tag{8.54}$$

is the uniform strain energy density of the remote field; $M = E/(1-v)$ is the biaxial elastic modulus expressed in terms of the elastic modulus E and Poisson ratio v.

The local increase in area of the surface due to the perturbation in shape is approximately

$$\tfrac{1}{2}h_{,x}(x,t)^2 = 2\pi^2\frac{a^2}{\lambda^2}\sin^2\frac{2\pi x}{\lambda}, \tag{8.55}$$

which is of second order in a/λ, and the local curvature is approximately

$$\kappa = h_{,xx}(x,t) = -4\pi^2\frac{a}{\lambda^2}\cos\frac{2\pi x}{\lambda}. \tag{8.56}$$

The chemical potential (8.8) then varies along the surface according to

$$\chi = U_m + 4\pi\frac{a}{\lambda}\left[\frac{\pi}{\lambda}U_S - (1+v)U_m\right]\cos\frac{2\pi x}{\lambda}. \tag{8.57}$$

Thus, the surface chemical potential varies in phase (out of phase) with the perturbation in shape when $\pi U_S/\lambda - (1+v)U_m > 0$ (< 0). This is the condition that discriminates between stable and unstable surface perturbations. To see this, let a be time-dependent and note that

$$v_n = \dot{a}\cos\frac{2\pi x}{\lambda} \tag{8.58}$$

for small a/λ. Then the rate of change of free energy per period is approximately

$$\dot{\mathcal{F}}(t) = \int_0^\lambda \chi v_n \, dx = 2\pi a\dot{a}\left[\frac{\pi}{\lambda}U_S - (1+v)U_m\right], \tag{8.59}$$

where the natural length parameter

$$\zeta = \frac{U_S}{U_m} \tag{8.60}$$

again emerges. Therefore, for any small amount of waviness represented by $a > 0$, the system can lower its free energy by moving toward a configuration with a flat surface ($\dot{a} < 0$) if $\pi U_S/\lambda > (1+v)U_m$. On the other hand, if $\pi U_S/\lambda < (1+v)U_m$, it can lower its free energy by increasing the amplitude of its waviness ($\dot{a} > 0$). A nominally flat surface is said to be stable under perturbations in surface shape in the former case and to be unstable in the latter case.

For a given set of system parameters U_S, E, ν and σ_m, the discriminating wavelength or *critical wavelength* is

$$\lambda_{cr} = \frac{\pi U_S}{(1 + \nu)U_m} = \frac{\pi U_S \bar{E}}{\sigma_m^2} . \tag{8.61}$$

A surface with sinusoidal perturbation of its nearly flat surface with wavelength that is less (greater) than λ_{cr} is stable (unstable) against spontaneous growth of the perturbation amplitude. For a surface energy density of $U_S = 1\,\text{J}\,\text{m}^{-2}$, a plane strain modulus of $\bar{E} = 10^{11}\,\text{N}\,\text{m}^{-2}$ and an applied stress of $\sigma_m = 10^9\,\text{N}\,\text{m}^{-2}$, this critical wavelength λ_{cr} is approximately 300 nm.

Note that the critical wavelength in (8.61) depends on stress σ_m only through σ_m^2. As a result, the system behavior depends on the magnitude of stress but not on its algebraic sign, that is, the behavior under tension and compression are indistinguishable according to (8.61). This is at variance with observed behavior, where materials with $\sigma_m < 0$ behave in the way described here but systems with $\sigma_m < 0$ do not consistently do so (Xie et al. 1994). This question is re-examined in Section 8.8.3 on the basis of a surface energy density that depends on surface orientation in a way dictated by the fundamental nature of crystals.

The stability of a nominally flat bounding surface of a stressed solid was investigated independently by Asaro and Tiller (1972), Grinfeld (1986) and Srolovitz (1989). The instability was first observed and described qualitatively in solid ^{4}He by Bodensohn et al. (1986). Subsequently, Torii and Balibar (1992) developed an experiment that made it possible to induce the instability in solid ^{4}He under controlled conditions and to interpret the observations quantitatively. They determined equilibrium shapes, showing surface waviness, of the interface between a pure helium crystal and a helium vapor at very low temperature. This was done by imposing a variable applied stress on the crystal by means of piezoelectric end walls of its centimeter scale container. The surface shape was monitored by directing light onto the crystal and correlating patterns obtained due to interference of light reflected from both the top surface of the crystal and the mirrored bottom surface of its container with surface shape. In the growth of epitaxial semiconductor films that have a lattice mismatch with respect to their substrates, stresses sufficiently large to induce surface instability are readily achieved through the constraint of epitaxy. A particular case is described in the next example.

8.4.2 Example: Stability of a strained epitaxial film

A $Si_{0.81}Ge_{0.19}$ alloy film is grown epitaxially on a (001) surface of the Si substrate. Estimate the critical wavelength λ_{cr} as defined in (8.61) for the nominally flat growth surface of the film. Base the calculation on the values of elastic constants given in Table 3.1 and the surface energy value of $U_S = 1.2\,\text{J}\,\text{m}^{-2}$ for the alloy.

Solution:

The mismatch strain between the film and substrate is approximated as described in (1.15) for a Ge concentration of $x = 0.19$, with the result that

$$\epsilon_m = -0.04 \times 0.19 = -0.0076\,. \tag{8.62}$$

The elastic constants to use with the isotropic model are estimated in terms of the constants for single crystal Si and Ge given in Table 3.1 by assuming that the waviness is aligned with a $\langle 100 \rangle$ direction. In this case, the elastic modulus and Poisson ratio values of both Si and Ge are determined according to

$$E = 1/s_{11} \quad \text{and} \quad \nu = -s_{12}/s_{11}\,. \tag{8.63}$$

Then, applying the linear rule of mixtures for the alloy, it follows that $E_{SiGe} = 125\,\text{GPa}$ and $\nu_{SiGe} = 0.276$. The mismatch stress is then

$$\sigma_m = \frac{E_{SiGe}}{1 - \nu_{SiGe}}\epsilon_m = -1.31\,\text{GPa}\,. \tag{8.64}$$

The value of critical wavelength is estimated from (8.61) to be

$$\lambda_{cr} = \frac{\pi \times 1.2 \times 1.25 \times 10^9}{(1 - 0.276^2) \times 1.31^2 \times 10^{18}} = 296\,\text{nm}\,. \tag{8.65}$$

A cross-sectional transmission electron microscopy image of a $Si_{0.81}Ge_{0.19}$ alloy film that was grown on the (001) surface of the Si substrate is shown in Figure 8.12. The growth temperature was $750\,°C$ and mean film thickness is approximately $50\,\text{nm}$. The ripples on the surface are aligned with a $\langle 100 \rangle$ direction of the material. The wavelength of the more or less periodic surface profile is approximately $315\,\text{nm}$. Evidently, the amplitude-to-wavelength ratio of the surface profile is not small compared to unity, but the wavelength agrees reasonably well with the value estimated in this example. The surface profile is also not a simple sinusoid but, instead, it is sharpened in the valleys, flattened at the peaks and nearly linear between the valleys and peaks. The influence of higher order effects on perturbed shape will be considered in Section 8.4.4, and aspects of the profile arising from other effects will be discussed later in this chapter and in the next chapter.

8.4.3 Influence of substrate stiffness on surface stability

In the foregoing discussion of stability of the flat surface of a stressed solid, it was assumed that the elastic material is homogeneous. It was also assumed that, prior to formation of fluctuations in surface shape, the material was homogeneously stressed, that is, the equi-biaxial stress σ_m acted throughout the material. This assumption on

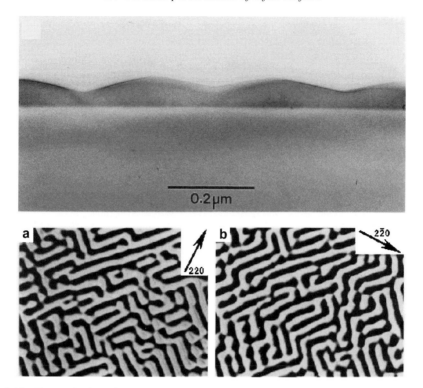

Fig. 8.12. Transmission electron microscopy cross-sectional image of a $Si_{0.81}Ge_{0.19}$ alloy film grown epitaxially on a Si substrate (top). The ridges are aligned with a $\langle 100 \rangle$ crystallographic direction. While the upper image appears to represent a fully two-dimensional configuration, the plan-view images of the film surface in the lower portion of the figure show that the regular ordering has a relatively short range. The normal distance between parallel lines in the lower images is the peak-to-peak distance in the upper image, or about 300 nm. Reproduced with permission from Cullis et al. (1992).

stress is not essential. It will be shown in Section 8.9 that the results are unaffected if the initial mismatch stress acts only to some finite depth, provided only that this depth extends beyond the roots of the valleys in surface fluctuation. Otherwise, a discontinuity in the initial stress across some plane $y =$ constant is immaterial. Discontinuities in elastic properties, however, do have an influence on stability.

Suppose that a film of thickness h is epitaxially bonded to a substrate with relatively large thickness. Both materials are assumed to be homogeneous and isotropic elastic materials, but with different elastic moduli, so that $E_s \neq E_f$. For simplicity, the two materials are assumed to have the same Poisson ratio, that is, $\nu_f = \nu_s = \nu$. The film is subjected to a homogeneous equi-biaxial stress σ_m when the free surface is flat and the substrate is initially stress-free. The central question again concerns the stability of the flat surface under sinusoidal perturbations of its shape resulting from mass rearrangement.

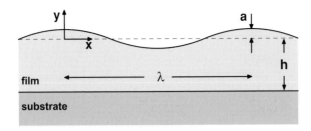

Fig. 8.13. Schematic diagram of the boundary value problem analyzed to determine the influence of substrate stiffness on the stability of a flat surface of a stressed thin film.

Qualitatively, the role of substrate stiffness is evident. The film material at the initial stress level is characterized by the natural length parameter ζ defined in (8.60). If h is large compared to this parameter then the surface material is unaware of the substrate and the substrate stiffness is expected to have only a minor influence on the critical wavelength λ_{cr} in (8.61). On the other hand, if h is very small compared to ζ, then the film material is inconsequential and the critical wavelength should be given by (8.61) but with $\bar{E}_{\mathrm{s}}$ replacing $\bar{E}_{\mathrm{f}}$, where the latter modulus is represented by $\bar{E}$ in that equation. It follows that a relatively stiff substrate enhances stability of the surface if the film is thin compared to the natural length parameter, whereas a relatively compliant substrate tends to destabilize the surface. The point of the analysis to follow is to provide the transition between these extremes in behavior.

The boundary value problem to be formulated is depicted in Figure 8.13. The formulation follows that in the preceding subsection, except that two stress functions must be determined, one in the region occupied by film material and the other in the region occupied by substrate material. The surface shape is again assumed to be given by (8.47) with amplitude $a > 0$, and the boundary conditions to be enforced on the wavy surface are given in (8.45) in terms of the stress field perturbations $\sigma_{ij}^{(\mathrm{h})}$. In addition to these boundary conditions, the particle displacement and traction associated with perturbations in surface shape must be continuous across the interface at $y = -h$, and stress components in the substrate must vanish in the limit as $y \to -\infty$.

The result of solving this boundary value problem is that the perturbation stress along the wavy surface is again given by (8.52), except that the expression for the component $\sigma_{xx}^{(\mathrm{h})}$ includes an additional factor in the form

$$\sigma_{xx}^{(\mathrm{h})} = -\frac{4\pi a}{\lambda}\sigma_{\mathrm{m}}\Lambda(m, hk, \nu)\cos\frac{2\pi x}{\lambda} \tag{8.66}$$

where $m = E_{\mathrm{s}}/E_{\mathrm{f}}$, $k = 2\pi/\lambda$ and

$$\Lambda(m, hk, \nu) = \frac{d_1 + d_2 hk\, e^{-2hk} + d_3\, e^{-4hk}}{d_1 + (d_2 + d_4 h^2 k^2)\, e^{-2hk} + d_3\, e^{-4hk}}. \tag{8.67}$$

The coefficients in this expression are

$$d_1 = 3 + 2v(m-1)^2 - v^2(m-1)^2 + 10m + 3m^2$$
$$d_2 = 4(1+v)(m-1)(3+m+v(m-1))$$
$$d_3 = (m-1)^2(-3-2v+v^2)$$
$$d_4 = 2(m-1)(3-2v(m-1)+v^2(m-1)+5m). \tag{8.68}$$

Following the reasoning that led to (8.61), it is found that the critical wavelength λ for finite h, say $\lambda_{cr}^{(h)}$, must satisfy

$$\frac{\lambda_{cr}^{(h)}}{\lambda_{cr}} \Lambda(m, 2\pi h/\lambda_{cr}^{(h)}, v) = 1, \tag{8.69}$$

where λ_{cr} is given in (8.61).

The qualitative behavior anticipated above can readily be confirmed. For any fixed value of $\lambda_{cr}^{(h)}/\lambda_{cr}$, it is evident that

$$\lim_{h/\lambda_{cr}\to\infty} \Lambda(m, 2\pi h/\lambda_{cr}^{(h)}, v) = 1. \tag{8.70}$$

Consequently, the stability condition reduces to $\lambda_{cr}^{(h)} = \lambda_{cr}$, that is, it is unaffected by the presence of the substrate. On the other hand, for any fixed value of $\lambda_{cr}^{(h)}/\lambda_{cr}$,

$$\lim_{h/\lambda_{cr}\to 0} \Lambda(m, 2\pi h/\lambda_{cr}^{(h)}, v) = m^{-1} \tag{8.71}$$

so that the stability condition reduces to

$$\lambda_{cr}^{(h)} = \frac{\pi U_s \bar{E}_s}{\sigma_m^2}. \tag{8.72}$$

Thus, the condition is the same as that for homogeneous material with modulus $\bar{E}$ except that the elastic part of the response is controlled by modulus $\bar{E}_s$ instead, as was anticipated.

The transition in behavior between these extremes is illustrated in Figure 8.14. This figure shows level curves representing the wavelength $\lambda_{cr}^{(h)}$ that discriminates between those for which surface fluctuations are unstable (larger) from those for which surface fluctuations are stable (smaller); these curves are shown in the plane of system parameter $(1-m)/(1+m)$, representing the modulus ratio between the film and the substrate, and h/λ_{cr}, the thickness of the film relative to the natural length scale for the system. In this case, the natural length scale is represented by λ_{cr} which, as is seen in (8.61), differs from ζ only by a numerical factor of order unity. The case of doubly periodic sinusoidal surface perturbations was first analyzed by Spencer et al. (1991), and the stability condition for this system was also obtained by Freund and Jonsdottir (1993) by means of finite element simulation. It will be

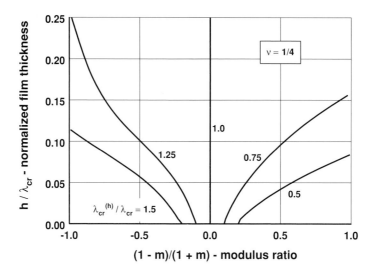

Fig. 8.14. Level curves of the wavelength $\lambda_{cr}^{(h)}/\lambda_{cr}$ that discriminates between stable and unstable surface perturbations in the plane of stiffness ratio of substrate and film represented by $(1-m)/(1+m)$ and normalized film thickness $h/\lambda_{cr}^{(h)}$.

shown in Section 8.5 that the stability condition for a doubly periodic perturbation is the same as for the case of a surface perturbation that generates only two-dimensional deformation fields, provided that the wave number of the perturbation in the latter case is replaced by the square root of the sum of the squares of the two wave numbers in the former case.

8.4.4 Second order surface perturbation

The energy change resulting from a sinusoidal perturbation of the initially flat surface of a stressed solid was considered in Section 8.4.1. The analysis was restricted to the case when the amplitude-to-wavelength ratio a/λ of the perturbation was small enough so that the stress change due to the perturbation was linear in a/λ. The material response is linear no matter how large the perturbation in surface shape might be, and the potential for nonlinear behavior arises only through the boundary conditions. In writing the approximate boundary conditions in (8.46), only those terms that are linear in the small surface slope $h_{,x}(x,t)$ were retained. In this section, the boundary conditions are extended to higher order in the surface slope in order to examine how the system departs from linearity in the course of stress relaxation. In particular, the energetics of the perturbation of the evolving sinusoidal shape can be studied for behavior into the nonlinear range. If the nominally flat surface of a stressed solid is unstable then, according to the linear theory, the amplitude of any perturbation in surface shape will grow to indefinitely large magnitude as time goes on. However, if the process is pursued into the nonlinear range, then such an unstable surface may have a nearby stable configuration, as will be demonstrated in this section.

For a surface perturbation represented by the height function $h(x, t)$, as in Section 8.4.1, the components of the unit vector normal to the surface are now

$$n_x = -h,_x \,, \quad n_y = 1 - \tfrac{1}{2}h,_x^2 \,, \tag{8.73}$$

accurate to second order in the local slope. The components of the unit vector tangent to the surface are again given by $s_x = n_y$ and $s_y = -n_x$. The total stress field for the perturbed surface is taken to be of the form (8.45) with an initial stress state specified by $\sigma_{xx}^{(0)} = \sigma_{zz}^{(0)} = \sigma_m$. The perturbed stress field $\sigma_{ij}^{(h)}$ is of order $h,_x$ so the boundary conditions of vanishing traction on $y = h(x, t)$, accurate to second order in the small slope, are

$$\sigma_{yy}^{(h)}(x, h) - 2\sigma_{xy}^{(h)}(x, h)h,_x +\sigma_m h,_x^2 = 0 \,,$$
$$\sigma_{xy}^{(h)}(x, h) - \sigma_m h,_x - \left(\sigma_{xx}^{(h)}(x, h) - \sigma_{yy}^{(h)}(x, h)\right)h,_x = 0 \,. \tag{8.74}$$

These boundary conditions are identical to the conditions given in (8.46) to first order in $h,_x$.

When a linear perturbation in surface shape is taken into account, it is sufficient to consider only a single mode in surface shape as specified by (8.47). Once the second order terms are included in the boundary conditions, however, the first harmonic mode of the fundamental mode is brought into play. Thus, it becomes necessary to consider a perturbation of the form

$$h(x, t) = a_1(t)\cos\frac{2\pi x}{\lambda} + a_2(t)\cos\frac{4\pi x}{\lambda} \tag{8.75}$$

so that there are contributions in the boundary condition to balance the higher harmonic term. The stress field $\sigma_{ij}^{(h)}$ is then expected to be composed of terms that are linear in a_1/λ plus terms that are proportional to a_1^2/λ^2 and a_2/λ. The equilibrium stress field $\sigma_{ij}^{(h)}$ in the solid is then derived from an Airy stress function of the form

$$A(x, y) = \left(c_0 + c_1\frac{y}{\lambda}\right)e^{2\pi y/\lambda}\cos\frac{2\pi x}{\lambda} + \left(c_2 + c_3\frac{y}{\lambda}\right)e^{4\pi y/\lambda}\cos\frac{4\pi x}{\lambda}\,. \tag{8.76}$$

The boundary conditions (8.74) imply that

$$c_0 = 0, \quad c_1 = -\sigma_m a_1\lambda, \quad c_2 = \tfrac{1}{4}\sigma_m a_1^2, \quad c_3 = \sigma_m(-a_2\lambda + \pi a_1^2)\,. \tag{8.77}$$

With the stress field completely determined to the desired order, the change in elastic energy which results from the perturbation in surface shape can be calculated.

The result is

$$\mathcal{E}(a_1, a_2) = \int_0^\lambda \int_{-\infty}^h [U(x, y, t) - U_m] \, dy \, dx$$

$$= \pi \frac{\sigma_m^2}{\bar{E}} \lambda^2 \left[-\frac{a_1^2}{\lambda^2} + 3\pi^2 \frac{a_1^4}{\lambda^4} + 4\pi \frac{a_1^2 a_2}{\lambda^3} \right], \tag{8.78}$$

where $U(x, y, t)$ is the strain energy density throughout the material and U_m is defined in (8.54). The quantity $\mathcal{E}$ is the elastic energy change per period of the material per unit distance in the direction perpendicular to the plane of deformation. The corresponding change in surface energy is

$$\mathcal{S}(a_1, a_2) = U_S \int_0^\lambda \sqrt{1 + h_{,x}^2} \, dx = \pi^2 U_S \lambda \left[\frac{a_1^2}{\lambda^2} - \frac{3\pi^2}{4} \frac{a_1^4}{\lambda^4} + 4\frac{a_2^2}{\lambda^2} \right]. \tag{8.79}$$

The quantity $\mathcal{S}$ is the surface energy change per period of the material per unit distance perpendicular to the plane of deformation.

For this system, which has two degrees of freedom, the rate of change of free energy is

$$\dot{\mathcal{F}} = \dot{\mathcal{E}} + \dot{\mathcal{S}} = -\mathcal{Q}_k \dot{a}_k \tag{8.80}$$

where $\mathcal{Q}_k = -\partial(\mathcal{E} + \mathcal{S})/\partial a_k$ is the generalized force that is work-conjugate to a_k. Explicit expressions for the generalized forces are extracted from (8.78) and (8.79) as

$$\mathcal{Q}_1(a_1, a_2) = \pi^2 U_S \left[2\frac{a_1}{\lambda} \left(\frac{\lambda}{\lambda_{cr}} - 1 \right) + 3\pi^2 \frac{a_1^3}{\lambda^3} \left(1 - 4\frac{\lambda}{\lambda_{cr}} \right) - 8\pi \frac{a_1 a_2}{\lambda^2} \frac{\lambda}{\lambda_{cr}} \right],$$

$$\mathcal{Q}_2(a_1, a_2) = -\pi^2 U_S \left[4\pi \frac{a_1^2}{\lambda^2} \frac{\lambda}{\lambda_{cr}} + 8\frac{a_2}{\lambda} \right], \tag{8.81}$$

where λ_{cr} is defined in (8.61). The actual path followed over the course of time in the plane of a_1 and a_2 is determined by the kinetics of the relaxation process, and it will be studied in the next chapter. Some features can be anticipated, however, by examining the nature of the generalized forces in (8.81). As a basis for doing so, assume that $\lambda > \lambda_{cr}$ so that the flat surface is indeed unstable for a sinusoidal perturbation of wavelength λ according to the criterion considered in Section 8.4. It can also be assumed that $a_1/\lambda > 0$, without loss of generality, which implies that $\mathcal{Q}_1 > 0$ if free energy is to be reduced by increasing a_1. The stability condition established in Section 8.4 is also evident from the present development by observing that $\mathcal{Q}_1(a_1, 0) > 0$ for small a_1/λ only if $\lambda > \lambda_{cr}$.

For the particular case when $a_2/\lambda = 0$, note that $\mathcal{Q}_1$ is positive for very small values of a_1/λ. As the amplitude of a_1/λ increases, the value of $\mathcal{Q}_1$ decreases, and

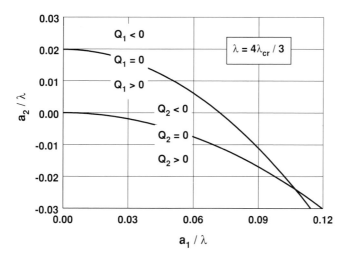

Fig. 8.15. The two curves show the locus of points in the plane of amplitudes a_1 and a_2 for which the generalized forces $\mathcal{Q}_1$ and $\mathcal{Q}_2$ vanish for the case when $\lambda = \frac{4}{3}\lambda_{cr}$.

it becomes zero when

$$\left(\frac{a_1}{\lambda}\right)^2 = \frac{2(\lambda/\lambda_{cr} - 1)}{3\pi^2(4\lambda/\lambda_{cr} - 1)}. \tag{8.82}$$

This observation implies that a sinusoidal perturbation in surface shape is *stabilized* as its amplitude increases, a fact that could not be anticipated by the linear stability analysis but that is consistent with numerical simulations (Chiu and Gao 1994). For $\lambda = \frac{4}{3}\lambda_{cr}$, the amplitude implied by (8.82) is about $a_1/\lambda = 0.07$.

The locus of points for which $\mathcal{Q}_1 = 0$ in the plane of a_1 and a_2 is shown in Figure 8.15 for $\lambda = \frac{4}{3}\lambda_{cr}$. The only portion of this plane that corresponds initially to energy reduction is the region for which both $\mathcal{Q}_1 > 0$ and $a_1 > 0$. The locus of points for which $\mathcal{Q}_2 = 0$ is also shown in Figure 8.15. There is no region in the plane for which both $\mathcal{Q}_2$ and a_2 are positive, which suggests that a_2 must decrease from its initial value of zero. Thus, the part of the plane of interest is that where both $\mathcal{Q}_2 < 0$ and $a_2 < 0$. This expectation is reinforced by the observation that $\mathcal{Q}_1$ is increased if a_2 takes on negative values. There is indeed a thin crescent-shaped region of the plane of a_1 and a_2 within which all the expected circumstances arise, bounded by the line $a_2 = 0$, the curve $\mathcal{Q}_1 = 0$ and the curve $\mathcal{Q}_2 = 0$, and within which a_2/λ is indeed small in magnitude compared to a_1/λ. Therefore, it is expected that the actual path of evolution of shape lies within that region.

Finally, the anticipated behavior corresponding to a_1/λ increasing from a small positive value and a_2/λ decreasing from zero as time goes on implies a particular trend in surface shape as represented by (8.75). With increasing time, these observations suggest that the valleys in surface shape become sharper, with curvature

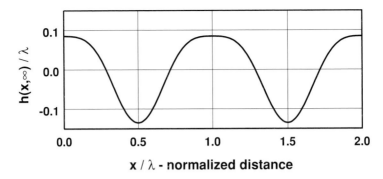

Fig. 8.16. Equilibrium shape of a surface according to (8.75) for the equilibrium configuration represented by the intersection of the curves $\mathcal{Q}_1 = 0$ and $\mathcal{Q}_2 = 0$ in Figure 8.15.

of increasing magnitude, and the hills become flatter, with curvature of decreasing magnitude. The equilibrium shape defined by the intersection of the curves $\mathcal{Q}_1 = 0$ and $\mathcal{Q}_2 = 0$ in Figure 8.15 is shown in Figure 8.16. Note that the valleys have become sharpened by the appearance of the second mode, while the peaks in the profile have become flattened, as has been observed in numerical simulation studies (Chiu and Gao 1994, Chiu and Gao 1995). Similar observations have been reported by Nozieres (1993) on the basis of a higher order analysis.

For the case of a strained thin film of mean thickness h on a relatively thick unstrained substrate, the depth of the valleys in this equilibrium configuration determine whether the material separates into distinct islands or it reaches equilibrium before doing so. The valley depth in this two degree of freedom idealization is a distance $|a_1| + |a_2|$ below the mean film thickness. Therefore, the material will separate into islands if $h < |a_1| + |a_2|$ and it will not do so otherwise. For the case of $\lambda = \frac{4}{3}\lambda_{\mathrm{cr}}$, this depth is approximately $0.135\lambda = 0.18\lambda_{\mathrm{cr}}$. This observation implies that a strained film will divide up into islands only if the mean thickness is less than about 18% of the system specific critical wavelength. The particular model is too idealized to expect that the numerical estimates will be accurate, but the underlying idea is general.

8.4.5 *Example: Validity of the small slope approximation*

Examination of the result (8.59) indicates that, for a sinusoidal perturbation in surface shape represented by amplitude a and wavelength λ, the free energy will continue to decrease indefinitely as a increases as long as $\lambda > \lambda_{\mathrm{cr}}$. This inference is based on the assumption of small surface slope, however, so its validity beyond the range for which $a/\lambda \ll 1$ is uncertain. Consider variations of elastic energy and surface energy with amplitude a for fixed wavelength λ, assuming the two-dimensional sinusoidal surface profile given by (8.47) but without a restriction on the magnitude of a/λ; otherwise, the assumptions invoked in

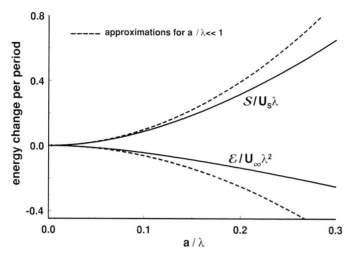

Fig. 8.17. The dependence of change in surface energy and elastic energy of a solid due to sinusoidal perturbation of surface shape versus the amplitude-to-wavelength ratio of that perturbation. The results show that the small slope approximation is reliable for values of a/λ up to 0.1 and is a fair approximation for significantly larger values.

the case of a slightly wavy surface are retained. Estimate the range of a/λ for which the small slope approximation leads to a valid estimate of energy change.

Solution:

The surface energy per period (per unit depth perpendicular to the plane of the figure represented in Figure 8.11) is λU_S when $a/\lambda = 0$. Thus, a *change* in surface energy in the interval $0 \le x < \lambda$ for $a > 0$ is

$$S(a) = \lambda U_S \int_0^1 \sqrt{1 + \left(\frac{2\pi a}{\lambda}\right)^2 \sin^2 \frac{2\pi \xi}{\lambda}} \, d\xi - \lambda U_S. \qquad (8.83)$$

The integral in (8.83) can be written in terms of elliptic functions or it can be evaluated numerically for any range of values of a/λ to obtain a result shown graphically in Figure 8.17 in the form of $S/\lambda U_S$ versus a/λ. Thus, the surface energy increases with increasing a, but at a rate smaller than that implied by the small amplitude approximation.

The elastic energy in the material decreases as a/λ increases from zero value. There is no analytical procedure for solution of the elasticity boundary value problem when a/λ is not restricted to be very small compared to unity. However, the accurate estimation of this energy reduction by means of the numerical finite element method is routine. The elastic energy reduction within the interval $0 \le x < \lambda$ from the configuration with $a/\lambda = 0$ is denoted by $\mathcal{E}$, and this quantity is normalized by $\lambda^2 U_\infty$. A plot of $\mathcal{E}/\lambda^2 U_\infty$ versus a/λ is also shown in Figure 8.17. The results indicate that the small slope approximation is reliable for values of a/λ up to 0.1, and it is a fair approximation for significantly larger values.

8.5 General perturbation of a flat surface

The boundary conditions (8.45) expressed in terms of perturbation fields suggest a direct approach to determining the additional stress field $\sigma_{ij}^{(h)}$ that results from perturbation of the flat surface of a stressed solid. This procedure is outlined here, first for the case of plane strain deformation in order to generalize the results of the preceding subsection and to demonstrate consistency with those results. Once this is accomplished, the case of a general small amplitude perturbation which varies in both directions on the surface is considered; this perturbation results in a three-dimensional state of deformation in the material. The stress field for the unperturbed surface $y = 0$ is again given by $\sigma_{xx}^{(0)}(x, y, z) = \sigma_{zz}^{(0)}(x, y, z) = \sigma_{\rm m}$, with all other stress components being zero prior to perturbation of the surface.

8.5.1 Two-dimensional configurations

For perturbations of surface shape that vary only in the x-direction, assume that the y-coordinate of the free surface becomes $y = h(x, t)$ at place x and time t. The slope is everywhere small, so that $|h_{,x}| \ll 1$. The resulting *change* in deformation associated with the change in surface shape is essentially plane strain.

If a surface traction T_i with components

$$T_x = -\sigma_{\rm m} h_{,x} , \quad T_y = 0 \tag{8.84}$$

is applied to all points of the perturbed surface, then the stress in the material would be the initial uniform stress everywhere. It follows that the stress field $\sigma_{ij}^{(h)}$ is the result of *negating* the surface traction (8.84). This is precisely the implication of (8.45). The stress field to be determined is the result of a distributed shear traction $h_{,x} \sigma_{\rm m}$ applied to the surface of a half space; the normal surface traction is zero to first order in the small slope. The stress component $\sigma_{xx}^{(h)}(x, 0)$ due to a concentrated force of magnitude P acting at $x = \xi$ is known to be $-2P/\pi(x - \xi)$ (Timoshenko and Goodier 1987). If P is identified with the force $\sigma_{\rm m} h_{,\xi} \, d\xi$ negated due to traction acting between $x = \xi$ and $x = \xi + d\xi$, and if the results are superimposed by integration over the full range of ξ, then it follows that

$$\sigma_{xx}^{(h)}(x, h) = \frac{2}{\pi} \sigma_{\rm m} \int_{-\infty}^{\infty} \frac{h_{,\xi}(\xi, t)}{\xi - x} \, d\xi . \tag{8.85}$$

The integrand is strongly singular and the integral does not converge in the usual sense. However, the integral has an unambiguous interpretation in the sense of the Cauchy principal value of such a singular integral. If $h(x, t) = a(t) \cos(2\pi x/\lambda)$, the expression (8.85) reduces to (8.52). The expression can be used to determine the surface chemical potential for small amplitude perturbations of the surface of *arbitrary* shape.

On the basis of the boundary conditions (8.45), the same procedure can be followed even when the stress distribution in the material with the unperturbed flat surface is spatially nonuniform. For example, if the nonzero stress component at the surface prior to perturbation in shape is $\sigma_{xx}^{(0)}(x, 0)$ then the additional stress along the surface due to a general perturbation is

$$\sigma_{xx}^{(h)}(x, h) = \frac{2}{\pi} \int_{-\infty}^{\infty} \frac{\sigma_{xx}^{(0)}(\xi, 0)\, h_{,\xi}\,(\xi, t)}{\xi - x}\, d\xi \,, \tag{8.86}$$

which reduces to (8.85) when the stress is uniform. The more general form was applied by Jonsdottir and Freund (1995), for example, in the study of surface waviness induced by subsurface dislocations in the material. The stress field of the dislocations provides $\sigma_{xx}^{(0)}(x, 0)$ in this case; see Section 8.7.

8.5.2 Three-dimensional configurations

The approach introduced in this section to study the change of an elastic field in a homogeneously stressed solid due to surface shape perturbation under plane strain conditions can be generalized to the case of perturbation of a nominally flat surface under three-dimensional conditions (Gao 1991, Freund 1995a, Freund 1995b). Suppose that the surface takes on a slightly altered shape, with no change in remote loading, so that the y-coordinate of the surface becomes

$$y = h(x, z)\,, \quad \sqrt{h_{,x}^2 + h_{,z}^2} \ll 1\,. \tag{8.87}$$

Although the perturbation may be time-dependent, the variable t is not included explicitly as an argument in the various fields that arise here, simply as a matter of convenience.

Guided by the development for the two-dimensional case above, it is observed that if a surface traction T_i with components

$$T_x = -\sigma_{\mathrm{m}} h_{,x}\,, \quad T_y = 0\,, \quad T_z = -\sigma_{\mathrm{m}} h_{,z} \tag{8.88}$$

is applied to all points of the perturbed surface, then the stress in the material would be the initial uniform stress everywhere. It again follows that the stress field $\sigma_{ij}^{(h)}(x, y, z)$ is the result of *negating* the surface traction (8.88). The stress field to be determined is the result of an applied distributed shear traction opposite to that in (8.88) acting on the surface of a half space; the normal surface traction is zero to first order in the small slope.

A solution can be constructed by using the Cerruti analysis of the elastic field induced in an isotropic half space by a concentrated force acting at a point on the surface and in a direction tangential to the surface (Timoshenko and Goodier 1987). Only the surface field is needed to determine the surface chemical potential. For a tangential force acting at the point $x = 0$, $z = 0$ along the surface $y = 0$ of an elastic half space $y < 0$, the components of particle displacement in the surface are

$$u_\alpha^{(h)}(x, z) = \frac{1 + \nu}{2\pi E} G_{\alpha\beta}(x, z) P_\beta, \quad \alpha, \beta = x \text{ or } z, \tag{8.89}$$

where P_β represents the concentrated force with components P_x and P_z acting in the associated surface coordinate directions, and $G_{\alpha\beta}(x, z)$ is a symmetric 2×2 matrix field with components

$$G_{xx} = \frac{2(1 - \nu)}{(x^2 + z^2)^{1/2}} + \frac{2\nu x^2}{(x^2 + z^2)^{3/2}},$$

$$G_{xz} = \frac{2\nu x y}{(x^2 + z^2)^{3/2}}, \tag{8.90}$$

$$G_{zz} = \frac{2(1 - \nu)}{(x^2 + z^2)^{1/2}} + \frac{2\nu z^2}{(x^2 + z^2)^{3/2}}.$$

By superposition, the full surface displacement field due to the perturbation is

$$u_\alpha^{(h)}(x, z) = \frac{(1 + \nu)\sigma_{\mathrm{m}}}{2\pi E} \int_{-\infty}^{\infty} \int_{-\infty}^{\infty} \Big[G_{\alpha x}(x - \xi, z - \eta) h_{,\xi}(\xi, \eta)$$

$$+ G_{\alpha z}(x - \xi, z - \eta) h_{,\eta}(\xi, \eta) \Big] d\xi \, d\eta \tag{8.91}$$

for $\alpha = x$ or z. The out-of-plane stress components at the surface are known from the boundary conditions and the in-plane components can be calculated from the displacement field in (8.91). With these stress components in hand, it is possible to determine the surface strain energy density distribution $U(x, z) = U_{\mathrm{m}} + U^{(h)}(x, z)$ to first order in gradients of surface shape. The change in surface strain energy density as a function of position on the surface is found to be

$$U^{(h)} = U_{\mathrm{m}} \frac{1 + \nu}{\pi} \int_{-\infty}^{\infty} \int_{-\infty}^{\infty} \frac{(\xi - x) h_{,\xi}(\xi, \eta) + (\eta - z) h_{,\eta}(\xi, \eta)}{\left[(\xi - x)^2 + (\eta - z)^2 \right]^{3/2}} d\xi \, d\eta. \tag{8.92}$$

This is the elastic field that is necessary to specify the chemical potential on the perturbed surface (Freund 1995b).

8.5.3 Example: Doubly periodic surface perturbation

The deformation fields arising from any small amplitude surface perturbation can be constructed from fields corresponding to doubly periodic perturbations by Fourier methods. Examine the particular case of a doubly periodic perturbation described by

$$h(x, z) = a \, \cos \frac{2\pi x}{\lambda_x} \cos \frac{2\pi z}{\lambda_z} \tag{8.93}$$

on the basis of the development in Section 8.5.2. As usual, the positive parameters a, λ_x and λ_z are limited to ranges for which

$$a\sqrt{\frac{1}{\lambda_x^2} + \frac{1}{\lambda_z^2}} \ll 1 . \tag{8.94}$$

Determine the necessary condition on the wavelengths for the surface to be stable for the perturbation (8.93).

Solution:

Substitution of (8.93) into (8.91) and evaluation of the integrals yields

$$u_x = -\frac{2\sigma_{\mathrm{m}}a\lambda_z}{\bar{E}\sqrt{\lambda_x^2 + \lambda_z^2}} \sin\frac{2\pi x}{\lambda_x} \cos\frac{2\pi z}{\lambda_z} ,$$

$$u_z = -\frac{2\sigma_{\mathrm{m}}a\lambda_x}{\bar{E}\sqrt{\lambda_x^2 + \lambda_z^2}} \cos\frac{2\pi x}{\lambda_x} \sin\frac{2\pi z}{\lambda_z} . \tag{8.95}$$

The change in elastic energy of the system per period due to formation of the perturbed shape is given by the surface integral representing the work done in creating these displacements,

$$\mathcal{E}^{(\mathrm{h})} = \int_0^{\lambda_z} \int_0^{\lambda_x} \tfrac{1}{2}\left(u_x^{(\mathrm{h})}\sigma_{xy}^{(\mathrm{h})} + u_z^{(\mathrm{h})}\sigma_{yz}^{(\mathrm{h})}\right) dx\,dz = -\frac{\pi\sigma_{\mathrm{m}}^2 a^2}{2\bar{E}}\sqrt{\lambda_x^2 + \lambda_z^2} . \tag{8.96}$$

Similarly, the change in surface energy due to formation of the perturbed shape is

$$\mathcal{S}^{(\mathrm{h})} = U_{\mathrm{S}} \int_0^{\lambda_z} \int_0^{\lambda_x} \tfrac{1}{2}\left(h_{,x}^2 + h_{,z}^2\right) dx\,dz = \frac{\pi^2 U_{\mathrm{S}} a^2}{2}\frac{\lambda_x^2 + \lambda_z^2}{\lambda_x \lambda_z} . \tag{8.97}$$

Several observations can be made on the basis of these simple three-dimensional results.

First, the stability of the flat surface shape in the presence of biaxial stress under doubly periodic perturbations of surface shape can be examined. The sum $\mathcal{E}^{(\mathrm{h})} + \mathcal{S}^{(\mathrm{h})}$ is the change in free energy corresponding to the change in shape. Thus, if this change is positive (negative) then the flat surface is stable (unstable) against such changes in shape. This line of reasoning leads to the conclusion that the condition discriminating between stability and instability is when the sum is zero, that is, when

$$\sqrt{\frac{1}{\lambda_x^2} + \frac{1}{\lambda_z^2}} = \frac{\sigma_{\mathrm{m}}^2}{\pi\bar{E}U_{\mathrm{S}}} = \frac{1}{\lambda_{\mathrm{cr}}} . \tag{8.98}$$

The stability limit obviously agrees with the result (8.61) obtained for a sinusoidal perturbation when $\lambda_z/\lambda_x \to \infty$. Furthermore, the result (8.98) shows that the stability condition depends on the wavelengths in coordinate directions in (8.93) only through the *magnitude* of the spatial wave number of the perturbation.

A second observation follows from consideration of the relative magnitudes of $\mathcal{E}^{(\mathrm{h})}$ and $\mathcal{S}^{(\mathrm{h})}$. Of particular interest is the way in which the ratio of these two energy changes depends on the wavelengths. From (8.96) and (8.97), it follows that

$$\left|\frac{\mathcal{E}^{(\mathrm{h})}}{\mathcal{S}^{(\mathrm{h})}}\right| = \frac{\lambda_{\mathrm{cr}}}{\lambda_x}\sqrt{1 + \frac{\lambda_x^2}{\lambda_z^2}} . \tag{8.99}$$

The smallest value of λ_x for which the ratio has the value unity is $\lambda_x = \lambda_{\mathrm{cr}}$ with $\lambda_z/\lambda_{\mathrm{cr}} \to \infty$.

This may explain the experimental observation that when thin films with flat but unstable surfaces are annealed, the evolving surface shape appears at first to assume a predominantly one-dimensional structure (Gao and Nix 1999). It is only after surface features become well defined and some time elapses that the surface fluctuations become two-dimensional, with the associated three-dimensional stress fields.

8.6 Contact of material surfaces with cohesion

If two solids are brought into contact with each other, the materials interact through the traction that is exerted by each solid on the other. The understanding of the contact of elastic bodies with smooth surfaces is quite complete. In the classical elastic theory, the materials come into contact over certain parts of their surfaces, and these parts conform perfectly to each other. The equal but opposite tractions acting on the material surfaces are compressive (or possibly zero) in the direction normal to the contact surface; in the classical theory, tensile traction cannot be supported across the contact area. There may or may not be shear traction transmitted across the contact surface, depending on configuration and friction characteristics of the material surfaces involved.

8.6.1 Force–deflection relationship for spherical surfaces

A theory of normal contact of isotropic elastic spheres without frictional interaction was developed by Hertz; a thorough description of this theory and its implications is provided by Johnson (1985). The main assumption of the Hertz contact theory is that the size of the contact area, usually understood to be the longest great circle arclength included within the zone of contact in the undeformed configuration, is much smaller than the initial or undeformed radius of either sphere. This assumption is exploited in the Hertz theory in the following way. As elastic spheres are pressed together, the radius of curvature of the surface of each sphere changes from the radius of the undeformed sphere to some other radius (which is typically larger in magnitude). Hertz reasoned that the mechanics of this process could be modeled by starting with a large elastic solid with a *flat* surface in its undeformed configuration, that is, the relatively simple configuration of an elastic half space with a traction-free surface. Then, by applying pressure over a part of the surface of this half space, a curvature could be induced that would be the same as the *change* in curvature in the original contact problem. Hertz assumed that the pressure distribution and contact area size predicted in this way would provide good approximations to the actual pressure and contact area size, an expectation that has been borne out by subsequent experience.

For contact of two spheres of the same isotropic elastic material with plane strain modulus $\bar{E}$ and the same undeformed radius R, the deformed contact area is flat

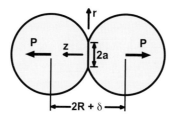

Fig. 8.18. Identical elastic spheres of radius R are brought into contact over a circular contact zone of radius a. The parameter δ characterizes the separation distance between the centers of the spheres with $\delta = 0$ when the separation distance is $2R$. The parameter P is the resultant force due to contact stress within the contact zone. The z-axis lies along the line of centers of the spheres, and $z = 0$ is at the contact plane.

and its boundary is circular due to the symmetry of the configuration. According to the Hertz point of view, the normal stress distribution predicted for the contact area $z = 0, 0 \leq r \leq a$ is

$$\sigma_{zz}^{\mathrm{H}}(r) = -\frac{2a\bar{E}}{\pi R}\left(1 - \frac{r^2}{a^2}\right)^{1/2} \tag{8.100}$$

and the radius of the contact area is

$$a = \left(\frac{3R|P|}{4\bar{E}}\right)^{1/3} \tag{8.101}$$

in terms of the magnitude of the force P acting on the spheres together; see Figure 8.18. The relative displacement of the centers of the spheres for $P \leq 0$ is approximately

$$\delta_{\mathrm{H}} = -\frac{2a^2}{R}. \tag{8.102}$$

This displacement estimate must also be obtained from the elasticity solution for the half space problem; it is exactly twice the normal displacement of the surface at the center of the contact area relative to the normal displacement at a remote point on the half space surface in the limit as $r/a \to \infty$.

Elastic contact theory can be extended to cases where the contacting surfaces of spheres interact not only through the compressive traction which each exerts on the other, but also through a tensile attraction due to influences at a sub-continuum level. Such an effect might arise, for example, from van der Waals interaction or, for clean surfaces of aligned crystals, from epitaxial bonding. In either case, the effect can be represented in a model by assuming that, prior to contact, the surface of each sphere has an energy per unit area U_{S} that is 'recovered' over the contact area when the spheres are brought into contact (Johnson et al. 1971); Hertz contact theory modified in this way is known as the Johnson–Kendall–Roberts or JKR theory. As spheres are pushed together or pulled apart under such circumstances, the perimeter of the contact area is essentially a healing or growing crack, respectively, and the results of Section 8.2.4 can be applied to study the energy variations.

It is well established in the theory of elasticity that a pressure distribution $(1 - r^2/a^2)^{-1/2}$ applied to the surface of the elastic half space $z \geq 0$ over the circular area $0 \leq r \leq a$ results in a *uniform* normal displacement of magnitude $\pi a/\bar{E}$ over $0 \leq r \leq a$. Thus, to represent the effect of cohesion without altering the symmetry of the contact of identical spheres, a total normal stress in the contact area $0 \leq r \leq a$ is assumed as

$$\sigma_{zz}(r) = \sigma_{zz}^{\mathrm{H}}(r) + \frac{B}{\sqrt{1 - r^2/a^2}}, \tag{8.103}$$

where σ_{zz}^{H} is given in (8.100) and B is a non-negative parameter with physical dimensions of stress; the value of this parameter is determined on the basis of physical features of the system. The total force acting on the contact area is then

$$P = 2\pi \int_0^a r\sigma_{zz}(r)\, dr = -\frac{4a^3 \bar{E}}{3R} + 2\pi a^2 B. \tag{8.104}$$

The corresponding relative displacement of the 'centers of the spheres', based on the solutions of the relevant elastic half space problem, is

$$\delta = \delta_{\mathrm{H}} + 2\pi a B/\bar{E} = -\frac{2a^2}{R} + \frac{2\pi a B}{\bar{E}}. \tag{8.105}$$

Finally, the role of the parameter B in determining energy variations must be considered.

The stress within the contact area as given by (8.103) is square root singular as $r \to a^-$, so that the edge of the contact zone has the character of an elastic crack (Lawn 1993); see Section 4.2.3. The elastic stress intensity factor K associated with the singular stress field along this crack edge is

$$K = \lim_{r \to a^-} \sigma_{zz}(r)\sqrt{2\pi(a-r)} = B\sqrt{\pi a}. \tag{8.106}$$

The value of the component of the J-integral (8.30) in the radial direction is then

$$J_r = \frac{K^2}{\bar{E}} = \frac{\pi a B^2}{\bar{E}}. \tag{8.107}$$

Therefore, from (8.32), the value of χ for this process is

$$\chi = -\frac{\pi a B^2}{\bar{E}} + 2U_{\mathrm{S}} \tag{8.108}$$

for any value of B. Note that χ is positive, zero or negative according to whether B is less than, equal to or greater than $\sqrt{2U_{\mathrm{S}}\bar{E}/\pi a}$, respectively.

Consider next the rate of change of free energy as a changes. Again, the case of $\delta = \text{constant}$ is considered as a matter of convenience, so that there is no exchange of energy between the system of two spheres and its surroundings. Because the configuration is axially symmetric, it follows that

$$\dot{\mathcal{F}}(t) = -2\pi a \dot{a} \chi \, . \tag{8.109}$$

The minus sign arises because the speed v_p defined in conjunction with Figure 8.6 is $-\dot{a}$ in this case. It is evident from (8.109) that, for the system to be in equilibrium, it is necessary for the driving force χ itself to be zero. This requires that a and B must be related according to

$$B = \sqrt{2U_S\bar{E}/\pi a} \, , \tag{8.110}$$

which establishes the value of the parameter introduced in (8.103) on physical grounds.

To represent the equilibrium behavior of the system, it is convenient to define a natural length scale of the material as

$$\zeta = \frac{2U_S}{\pi \bar{E}} \, . \tag{8.111}$$

The definition leads naturally to

$$\hat{P} = \frac{P}{\pi R^2 \bar{E}} \frac{R}{\zeta} \, , \quad \hat{\delta} = \frac{\delta}{R}\left(\frac{R}{\zeta}\right)^{2/3} , \quad \hat{a} = \frac{a}{R}\left(\frac{R}{\zeta}\right)^{1/3} \tag{8.112}$$

as convenient forms for normalized force, normalized deflection and normalized contact area radius, respectively. In terms of normalized variables, the relationship between the radius of the contact area and the contact force is

$$\hat{P}(\hat{a}) = -\frac{4}{3\pi}\hat{a}^3 + 2\hat{a}^{3/2} \, , \tag{8.113}$$

and the relationship between the radius of the contact area and the relative displacement of the centers of the spheres is

$$\hat{\delta}(\hat{a}) = -2\hat{a}^2 + 2\pi \hat{a}^{1/2} \, . \tag{8.114}$$

The results of Hertz contact theory are recovered if the second term on the right sides of both (8.113) and (8.114) are ignored. It is evident in both of these expressions that the classical results are recovered whenever $\hat{a} \gg 1$ but that the effect of cohesion is significant otherwise.

The relationships (8.113) and (8.114) involving contact force, contact radius and deflection are shown graphically in Figure 8.19 in the form of $\hat{a}$ versus $\hat{\delta}$ by means of a dashed curve and $\hat{P}$ versus $\hat{\delta}$ by means of a solid curve (Freund and Chason 2001). Some general features of the behavior evident from Figure 8.19 are noted.

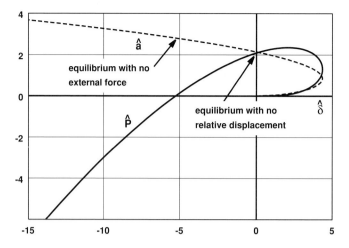

Fig. 8.19. Plots of normalized contact zone size $\hat{a}$ and normalized contact force $\hat{P}$ versus normalized relative separation distance $\hat{\delta}$ between centers of the spheres as given in (8.114) and (8.113), respectively. The points corresponding to equilibrium under zero applied force or under zero relative displacement are identified. The point on the $\hat{P}$ versus $\hat{\delta}$ curve with horizontal tangent locates a state of instability during separation of the spheres under applied force. Likewise, the point on the same curve with vertical tangent locates a state of instability during separation under imposed relative displacement.

For example, suppose two identical elastic spheres are slowly brought into close proximity with essentially no external force applied. The inference to be drawn from Figure 8.19 is that, immediately upon touching of the surfaces of the spheres as $\hat{\delta} \rightarrow 0^+$, the centers of the spheres will be drawn toward each other until the state with $\hat{P} = 0$ is reached. This is a stable equilibrium state with no external forces acting and with deflection $\hat{\delta} = -\pi(3\pi/2)^{1/3}$; the size of the contact area in this state is $\hat{a} = (3\pi/2)^{2/3}$. Suppose that two spheres are initially in this state and are then gradually separated by increasing $\hat{P}$ from zero. The relative displacement $\hat{\delta}$ first increases and $\hat{a}$ decreases as $\hat{P}$ increases up to a value of $\hat{P} = 3\pi/4$, the point on the solid curve at which the tangent line is parallel to the deflection axis, whereupon the spheres separate abruptly under the condition of imposed loading.

Alternatively, suppose that two spheres are gradually brought into close proximity with the deflection $\hat{\delta}$ controlled by means of external forces. The inference to be drawn from Figure 8.19 in this case is that, immediately upon touching at one point when $\hat{\delta} \rightarrow 0^+$, the surfaces of the spheres are drawn together and the spheres deform elastically to make this possible; the centers of the spheres are constrained to remain a distance $2R$ apart, however. This is an equilibrium configuration with zero relative displacement, as indicated in the figure. The attraction induces a force between the spheres of $\hat{P} = 2\pi/3$ and the size of the resulting contact area is $\hat{a} = \pi^{2/3}$. This is also a stable equilibrium position under small variations in $\hat{a}$ or $\hat{\delta}$. If spheres that are initially in this state are gradually separated by increasing $\hat{\delta}$ from zero, then $\hat{a}$ gradually decreases. The force $\hat{P}$ increases to a value of $3\pi/4$, and it then decreases as $\hat{\delta}$ is further increased. When $\hat{\delta}$ reaches the value $6(\pi/4)^{4/3}$, corresponding to the point on the solid curve in the figure at which the tangent line is parallel to the force axis, the separation is completed abruptly.

This model is based on the idea that the energy gain or loss from the macroscopic system per unit area of interface as a result of a singular stress field moving along this interface is the same as the work of cohesion of that interface per unit area, as long as the active nonlinear zone in which separation is actually occurring is small in size compared to the size of the contact area. The singular crack edge field provides the vehicle by which energy exchange between the elastic fields and the sub-continuum surface energy can occur. The transition behavior from cases dominated by a cohesion rule to those that are adequately represented by singular fields has been discussed by Maugis (1992) and thoroughly studied by Kim et al. (1998).

8.6.2 *Example: Stress generated when islands impinge*

During deposition of a film material onto the surface of a substrate of a different material, growth often tends to nucleate at distributed sites on the growth surface. As deposition continues, film growth proceeds by expansion of clusters of film material, or islands, from the nucleation sites. This is the Volmer–Weber growth mechanism discussed in Section 1.3. A phenomenon which has been observed when the islands become large enough to begin impinging on each other at their perimeters is that tensile stress is generated in the film (Nix and Clemens 1999, Seel et al. 2000, Floro et al. 2001). Most commonly, this stress is detected through observation of the change in curvature of the substrate which it induces. Estimate the magnitude of this stress, interpreted as a volume average stress, on the basis of the foregoing cohesive sphere model for a regular array of islands. The *volume average stress* is defined as the *spatially uniform stress* that would produce the same observed substrate curvature in a film of *uniform thickness* that has the same areal density of film material as the island array.

Solution:

Consider an array of islands on a substrate in which all the islands are identical hemispheres of radius R. The centers of the islands are fixed at the intersection points of the square grid of lines with uniform spacing $2R$ on the growth surface. It is assumed that the islands otherwise behave like free spheres with surfaces having free surface energy density U_S.

The islands are formed in this pattern with radii smaller than R, and they grow until they achieve radii of R. The constraint of the substrate is assumed to be negligible for purposes of this discussion, so the islands behave exactly as did the spheres in the preceding section, but only one-half of the volume of each is taken into account. Once the radius of each island reaches R, contact areas with cohesion form spontaneously between each island and its four

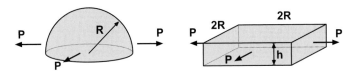

Fig. 8.20. The hemispherical island of radius R is subjected to a tensile force P on four sides due to impingement of four nearest neighbors. If the material were spread uniformly to depth $h = \frac{1}{6}\pi R$ over the substrate surface area $2R \times 2R$ then the uniform stress in the film of uniform thickness h which is the volume average of the stress in the islands is $\sigma_{\mathrm{ave}} = P/2Rh$.

immediate neighbors. Again, as a result of this cohesive contact, a tensile force is generated across each island which could be sensed by means of substrate curvature measurement.

According to the result depicted in Figure 8.19, the normalized force $\hat{P}$ that is generated upon island impingement is one-half of $2\pi/3$ or $\hat{P} = \pi/3$. In terms of the parameters that characterize the system, the force is (Freud and Chason 2001)

$$P = \frac{\pi^2}{3}R\bar{E}\zeta = \frac{2\pi}{3}R\gamma\,, \tag{8.115}$$

where ζ is the material length scale defined in (8.111). A surprising feature of the estimate in (8.115) is that it does not depend on the elastic properties of the material. This feature was noted by Johnson et al. (1971) in their original study of contact with cohesion, and experimental confirmation of features of the JKR theory is described by Israelachvili (1992).

The volume of material included in a hemispherical island upon coalescence is $\frac{2}{3}\pi R^3$. If the material in each island were to be spread uniformly over a $2R \times 2R$ square area on the substrate, as indicated in Figure 8.20, the uniform depth h would be $h = \frac{1}{6}\pi R$. The area of a lateral face of this equivalent block of material is $2R \times h$, so that the uniform tensile traction acting on this surface which produces the same resultant force as in (8.115) is $P/2Rh$ or

$$\sigma_{\mathrm{ave}} = 2\frac{\gamma}{R}\,. \tag{8.116}$$

This is the volume average tensile stress. As a representative numerical value of σ_{ave}, suppose that $\gamma = 1\,\mathrm{J\,m^{-2}}$ and that islands impinge when $R = 80$ nm. This implies that $\sigma_{\mathrm{ave}} = 25$ MPa. The physical effect described here is believed to account for the increase in tensile stress observed for films during the early stages of deposition, as described in Section 1.8.3, although the complexity of the actual island impingement process cannot be adequately represented by this simple model. For example, neither the interaction of an island with its substrate, the variation of island coalescence events across a growth surface, nor the transient role of surface diffusion of growth flux have been taken into account.

8.7 Consequences of misfit dislocation strain fields

Consider a uniformly strained film of thickness h that is epitaxially bonded to a relatively thick substrate which has essentially the same elastic properties as the film. Before any strain-relieving dislocations are formed in the film, the surface is flat and uniformly strained. It follows that the chemical potential is spatially uniform over the surface of the film; the flat surface is therefore an equilibrium shape. This equilibrium shape may be unstable, in a sense described in Section 8.4, but flat surfaces are commonly observed in strained films when the magnitude of the mismatch strain is not large. Such surfaces may be stabilized by orientation dependence of the surface energy density if the strain is not too large in magnitude, as discussed in Section 8.8.3.

The condition of uniform chemical potential is altered once misfit dislocations are formed at the film–substrate interface. The magnitude of the average elastic strain along the surface is reduced by the formation of dislocations, and the strain distribution becomes spatially nonuniform. The magnitude of elastic strain at the film surface due to dislocations is typically on the order of b/h where b is the length of the Burgers vector; this magnitude may be comparable locally to the magnitude of the mismatch strain itself. This nonuniform strain field can have any of several consequences. For example, its presence implies that the surface chemical potential becomes nonuniform; consequently, there is a configurational force acting on the nominally flat surface tending to change its shape. Alternatively, if a material with a large misfit strain is deposited epitaxially onto the surface of a dislocated film, the deposit will be most likely to attach in regions where the surface elastic strain is least, indicating a means of inducing geometrical patterns into subsequent deposits as described in Section 8.7.2. These consequences of misfit dislocation strain fields are considered in this section.

8.7.1 Surface waviness due to misfit dislocations

In this subsection, the magnitude of the fluctuations in surface shape that are required for a dislocated film to achieve an equilibrium morphology is estimated. Fluctuations in surface shape of partially relaxed strained films which correlate in position with misfit dislocations on the interface have been observed, for example, by Pinnington et al. (1997), Giannakopoulos and Goodhew (1998) and Springholz (1999). Furthermore, surface morphology has been identified as an important factor in impeding the progress of threading dislocations as they advance through the film (Samavedam and Fitzgerald 1997). Modeling of mass transport processes that might account for the formation of the fluctuations has not yet been pursued in a systematic way, but an estimate of achievable surface fluctuations can be obtained on the basis of the approximations introduced by Jonsdottir and Freund (1995).

Consider a periodic array of edge dislocations at the interface shared by the strained film and its substrate, as illustrated in Figure 8.21. For simplicity, the Burgers vector of length b of each misfit dislocation is assumed to lie in the interface in a direction corresponding to strain relief, and the dislocations are equally spaced

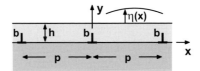

Fig. 8.21. Schematic diagram of a periodic array of edge dislocations at the interface between a strained layer of thickness h and a relatively thick substrate. The Burgers vector lies in the interface and in a direction consistent with relaxation of the mismatch strain. The dislocation array is periodic with spacing p. An array of dislocation clusters can be considered by viewing b as an appropriate integer multiple of the Burgers vector.

along the interface at interval p. The deformation is two-dimensional in this configuration. The distribution of stress along the initially flat free surface due to these dislocations is given by

$$\frac{\sigma_{xx}^{(d)}(x)}{\bar{E}} = \frac{b\left[-4h\pi + 4h\pi \cos\frac{2\pi x}{p}\cosh\frac{2h\pi}{p} + 2p\cos\frac{2\pi x}{p}\sinh\frac{2h\pi}{p} - p\sinh\frac{4h\pi}{p}\right]}{2p^2\left[\cos\frac{2\pi x}{p} - \cosh\frac{2h\pi}{p}\right]^2},$$

(8.117)

where it has been assumed that the dislocations relieve a tensile mismatch stress. If σ_m is the initial equi-biaxial mismatch stress, the strain energy density along the flat free surface is

$$U(x) = \frac{\sigma_m^2}{M} + \frac{1}{\bar{E}}\sigma_{xx}^{(d)}(x)\left[\sigma_m + \tfrac{1}{2}\sigma_{xx}^{(d)}(x)\right].$$

(8.118)

For a constant surface energy density $U_S = \gamma$, the surface chemical potential $\chi = U - \gamma\kappa$ is nonuniform in this configuration. The objective is to identify nearby shapes for which the chemical potential is again uniform.

Let $\eta(x)$ denote the local perturbation in surface shape, so that the local perturbed film thickness is $h + \eta(x)$, and assume that the perturbation is also periodic in x with period p. For fluctuations in shape that are small in amplitude compared to p, the fluctuation $\eta(x)$ is determined by the condition of uniform chemical potential along the surface, which is assured by the differential equation

$$U(x) + \Delta U(x) - \gamma\eta''(x) = \text{constant},$$

(8.119)

where $\Delta U(x)$ is the necessary modification in strain energy density along the surface so as to account for the small fluctuation $\eta(x)$. For present purposes, the modification $\Delta U(x)$ is assumed to be small compared to $U(x)$ itself, so that (8.119) is replaced by

$$U(x) - \gamma\eta''(x) = \frac{1}{p}\int_0^p U(s)\,ds.$$

(8.120)

The value of the constant incorporated into (8.120) is dictated by the condition of

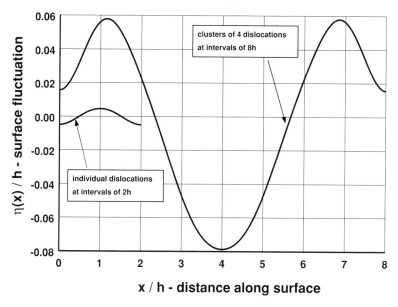

Fig. 8.22. Graphs of surface fluctuations from uniform film thickness that define equilibrium morphologies on which the chemical potential is spatially uniform, according to (8.120), for periodic arrays of dislocations at the film–substrate interface. Results are shown for two cases corresponding to the same average elastic strain in the film. In one case, individual dislocations exist at intervals of $2h$ along the interface; in the second case, clusters of four dislocations each exist at intervals of $8h$ along the interface.

periodicity. The resulting ordinary differential equation is readily solved by numerical methods, with the constants of integration being determined by the constraint of periodicity and conservation of mass.

If all lengths are normalized by the initial film thickness h, then (8.120) can be written in terms of the three nondimensional parameters

$$\hat{p} = \frac{p}{h}, \quad \hat{\gamma} = \frac{\gamma \bar{E}}{h \sigma_m^2}, \quad \hat{b} = \frac{b \bar{E}}{h \sigma_m}. \tag{8.121}$$

Results are included here for only two sets of values of these parameters, namely, $\hat{p} = 2$, $\hat{\gamma} = 1$, $\hat{b} = 1$ and $\hat{p} = 8$, $\hat{\gamma} = 1$, $\hat{b} = 4$. Note that the spatial average of elastic strain is the same in the two cases. However, this is accomplished by placing single misfit dislocations at intervals of $2h$ on the interface in the first case and by placing clusters of four misfit dislocations at intervals of $8h$ on the interface in the second case.

Plots of $\eta(x)/h$ versus distance x/h along the interface for these two cases are shown in Figure 8.22. For the case when $\hat{p} = 2$ and $\hat{b} = 1$, the total peak-to-valley fluctuation in film thickness anticipated on the basis of the approach taken here is approximately one percent of the initial film thickness, a change that is too small to be of consequence. However, for $\hat{p} = 8$ and $\hat{b} = 4$, the predicted fluctuation is approximately 15 percent of the initial film thickness. This is a significant alteration in surface profile which could indeed influence subsequent strain relaxation; it should also be noted that this result is probably beyond the range of small fluctuations adopted at the outset. Dislocations are often observed to exist in clusters, perhaps because all members in the cluster were produced from a common source. Connections between surface morphology and strain relaxation by dislocation motion have been studied by Samavedam and Fitzgerald (1997), Fitzgerald et al. (1997) and Fitzgerald et al. (1999).

8.7.2 Growth patterning due to misfit dislocations

In this subsection, it is assumed that a thin film has been deposited epitaxially onto the surface of a relatively thick substrate with lattice mismatch. The film thickness is assumed to be beyond its critical thickness, and consequently misfit dislocations have been formed at the film–substrate interface. Thereafter, additional material with a lattice mismatch with respect to the unrelaxed film material is to be deposited on the surface of the film. Subsequent deposit would be expected to nucleate most readily at places on the growth surface at which the mismatch is smallest. For example, suppose that a SiGe film with a mismatch strain ϵ_{m} is deposited onto Si(001) to a thickness beyond h_{cr} as defined in Chapter 5 and that the strain is then partially relaxed through formation of misfit dislocations. As a result, the elastic strain at the surface becomes nonuniform. Then, Ge is deposited on the surface of the dislocated film. Because the lattice parameter of Ge is larger than the effective lattice parameter of SiGe, the conditions are most favorable for attachment of the Ge deposit at points on the surface at which the remaining surface strain is the least compressive or the most relaxed.

To arrive at an estimate of the degree of relaxation that can be expected to occur, consider the surface strain distribution in the vicinity of a pair of interface misfit dislocations oriented at a right angle to each other. Suppose that the film–substrate interface coincides with the xz-plane and that the film occupies the region $0 \leq y \leq h$. One dislocation lies along the x-axis and the other along the z-axis. The Burgers vector of the former is $b_y = b_z = -b/\sqrt{2}$ and of the latter is $b_x = b_y = -b/\sqrt{2}$, where the sign of b is assumed to be the same as the sign of ϵ_{m} and its magnitude is the amount of slip across the dislocation glide plane.

The surface stress field is readily determined by application of (6.2) and (6.3), and the surface strain field follows readily by means of Hooke's law. The sum of the elastic extensional strain components in the surface due to the dislocations is

$$\epsilon_{xx}^{(\mathrm{d})} + \epsilon_{zz}^{(\mathrm{d})} = -\frac{\sqrt{2}(3 - 5\nu)}{\pi(1 - \nu)} \frac{b}{h} \left[\frac{xh^2(h + x)}{(h^2 + x^2)^2} + \frac{zh^2(h + z)}{(h^2 + z^2)^2} \right]. \tag{8.122}$$

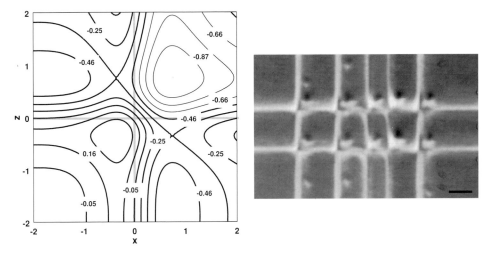

Fig. 8.23. The contour plot on the left shows level curves of the negative of the quantity in square brackets in (8.122), indicating the reduction in surface strain as a result of formation of a pair of orthogonal interface misfit dislocations lying along the coordinate axes at the film–substrate interface a distance h below the surface. The image on the right is a plan-view transmission electron micrograph of the results of depositing Ge on to a partially relaxed SiGe film on a Si(001) substrate; the scale bar represents 100 nm. The image shows that Ge islands nucleated in a particular quadrant of dislocation intersections, and an analysis of the configuration confirms that this is the quadrant of largest strain relief through interface misfit dislocation formation. Reproduced with permission from Ross (2001).

A contour plot of the factor in square brackets on the right side of (8.122) is shown in Figure 8.23. It is clear that the elastic strain of the film surface, represented by $\epsilon_m + \epsilon_{xx}^{(d)} + \epsilon_{zz}^{(d)}$, is very nonuniform. For this particular choice of Burgers vector, the relaxation is greatest at roughly the point on the surface where the glide plane surface traces of the two dislocations intersect. This particular result depends on the choice of Burgers vector, but other possible Burgers vectors can be investigated in essentially the same way.

The growth sequence outlined above as motivation for considering surface strain due to misfit dislocations was probed experimentally by Ross (2001). First, a $Si_{0.8}Ge_{0.2}$ film was deposited onto Si(001) to a thickness beyond the critical thickness of the film, and partial relaxation of the mismatch strain occurred by dislocation formation. Pure Ge was then deposited onto the surface of the partially relaxed film. A representative result is shown on the right side in Figure 8.23 as a plan-view image of the final structure. The white lines are interface misfit dislocations, and the black–white dots are Ge islands that have been nucleated on the surface of the film. The most striking feature of the image is that islands formed in the same quadrant of all dislocation intersections in the field of view. A detailed study of the results confirmed that the island centers coincided more or less with the surface traces of the dislocation glide planes, and that the quadrant in which the islands formed was indeed the quadrant of maximum strain relaxation due to dislocation formation.

8.8 Surface energy anisotropy in strained materials

The concept of a chemical potential field as it applies to surface evolution in materials was introduced in the early part of this chapter, and several explicit forms relevant to certain physical processes were discussed in Section 8.2. In each of these examples, a mathematical expression was deduced for the driving force tending to alter a feature of the system. This force represented the combined influence of surface energy and elastic strain energy. In the derivations of these force expressions, the analysis was simplified considerably by restricting consideration to systems for which the surface energy density in the reference configuration of the material is isotropic, the elastic strain is everywhere infinitesimal, and the surface energy density is independent of deformation of the surface. The main consequences of these assumptions are that the surface energy density at any point on the surface is independent of local orientation of the surface with respect to the material and that no surface stress acts locally in the tangent plane of the surface.

Perhaps the most extreme of these assumptions is the restriction to surface energy isotropy. In bulk response, material isotropy implies that the stress required to enforce any deformation specified with respect to a particular observer does not depend on the orientation of the material with respect to that observer. Likewise, surface energy isotropy implies that the value of surface energy density of a free surface with orientation specified by a particular observer does not depend on the orientation of the material with respect to that observer when the surface is created. Both ideas can be applied globally for spatially homogeneous states and locally for inhomogeneous states. Observations of cleavage behavior in single crystals of most materials, as well as of the stepped structure of macroscopically plane surfaces with orientations near the high symmetry orientations of crystals, suggest a significant degree of surface energy anisotropy in real materials (Blakely 1973). Surface stress effects have been observed in very small material structures through their influence on deformation within the bulk phase. Furthermore, the dependence of surface energy density on both surface strain and surface orientation opens the possibility of coupling between these effects. In the remainder of this section, an expression for the chemical potential of the free surface of an elastic solid is derived for a general two-dimensional configuration, taking into account the effects of surface energy anisotropy and surface stress. This result will incorporate virtually all physical effects except for finite strain that can be represented by local continuum mechanics, and the macroscopic continuum concept of surface stress arises naturally in the development.

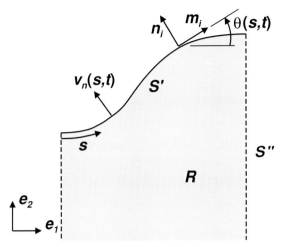

Fig. 8.24. Schematic diagram of a deformable body in its natural reference configuration, including a portion S' of the surface which evolves by mass transfer.

8.8.1 Implications of mechanical equilibrium

A portion of an elastic solid is shown in Figure 8.24 in its natural reference configuration. For plane deformation, the material occupies the area R. The portion of the bounding surface with the potential for shape change is S'. The remainder of the boundary S'' (shown dashed) represents workless constraints. Material points are located in R with respect to a fixed frame of reference by means of a set of rectangular base vectors $\mathbf{e}_k$ and coordinates x_k. Arclength s along S' is measured from some arbitrary origin. Local surface orientation is represented by the tangent line, with orientation described by a unit vector m_i in the direction of increasing s, and by a unit vector n_i; the two unit vectors satisfy $m_i n_i = 0$. It is also useful to represent local surface orientation at any time t by means of the angle $\theta(s, t) = \sin^{-1}[\mathbf{e}_3 \cdot (\mathbf{e}_1 \times \mathbf{m})]$. In terms of this angle, the surface curvature in the reference configuration is

$$\kappa = \frac{\partial \theta}{\partial s} \quad \Rightarrow \quad \frac{\partial m_i}{\partial s} = \kappa n_i \,, \tag{8.123}$$

at any position s on the surface and any time t.

The strain at a point of the surface S' is understood to be consistent with the limiting value of bulk strain ϵ_{ij} as the observation point approaches the surface from within R. The *surface strain* ϵ_{ij}^S expressed as a three-dimensional tensor field over the surface with outward unit normal n_i is

$$\epsilon_{ij}^S = \epsilon_{ij} - \epsilon_{ik} n_k n_j - n_i n_k \epsilon_{kj} + n_i n_k \epsilon_{kl} n_l n_j. \tag{8.124}$$

For example, if $n_1 = n_2 = 0$, $n_3 = 1$ then

$$\epsilon_{ij}^S = \begin{bmatrix} \epsilon_{11} & \epsilon_{12} & 0 \\ \epsilon_{12} & \epsilon_{22} & 0 \\ 0 & 0 & 0 \end{bmatrix} . \tag{8.125}$$

The expression (8.124) faithfully reproduces strain components for all material lines or directions within the surface, but includes no contribution from lines or directions normal to the surface. Thus, it is convenient for expressing energy changes of the surface associated with deformation. In particular, for the two-dimensional state being considered, the extensional strain of the surface is defined locally as

$$\epsilon(s, t) = m_i \epsilon_{ij}^S m_j = m_i \epsilon_{ij} m_j . \tag{8.126}$$

The superposed dot notation is reserved for the material time derivative of time-dependent fields. The shape evolution of the free surface of the material in its reference configuration is again described by the local normal speed v_n. Consider the neighborhood of a particular point on the surface at a certain instant of time. If v_n has a spatial gradient along the surface at the point of interest, then the infinitesimal surface element has not only a normal speed v_n but also an angular rate of rotation. The rate of change of orientation $\theta(s, t)$ is related to the gradient of normal speed along the surface by

$$\frac{\partial \theta}{\partial t} = \frac{\partial v_n}{\partial s} \quad \Rightarrow \quad \frac{\partial m_i}{\partial t} = \frac{\partial \theta}{\partial t} n_i . \tag{8.127}$$

The total time rate of change of extensional strain of the surface includes contributions due to the time dependence of strain at a material point, the time dependence of the orientation of the surface, and the motion of the surface with respect to the material particles instantaneously on it. If these effects are taken into account, then

$$\frac{d\epsilon}{dt} = \gamma_{mn} \frac{\partial v_n}{\partial s} + m_i \dot{\epsilon}_{ij} m_j + v_n \frac{\partial \epsilon}{\partial x_k} n_k , \tag{8.128}$$

where $\gamma_{mn} = 2 n_i \epsilon_{ij} m_j$ is the shear strain between material directions aligned with m_i and n_i in the reference configuration and the last term is the gradient of extensional strain parallel to the surface in a direction locally normal to the surface.

It is assumed that the system free energy is again given by (8.1) where $U_S = U_S(\theta, \epsilon)$ depends on the elastic strain field and on the local orientation of the material surface in a characteristic way for a given material. The extensional strain of the surface is related to the bulk strain field through (8.124). The process of elastic equilibration is expected to be very rapid compared to shape equilibration, which requires diffusive transport of material. Consequently, it is assumed from the outset that the elastic field is *always* in mechanical equilibrium for any surface shape. This expectation is enforced by requiring the variation in free energy due to a kinematically admissible small perturbation δu_i in the displacement field from its equilibrium distribution to be stationary, that is, to vanish to first order in the perturbation.

For any fixed surface shape, the variation in free energy due to δu_i is

$$\delta \mathcal{F} = \int_R \frac{\partial U}{\partial \epsilon_{ij}} \delta u_{i,j} \, dR + \int_{S'} \frac{\partial U_S}{\partial \epsilon} m_i \delta u_{i,j} \, m_j \, ds \qquad (8.129)$$

to first order in the perturbation. The quantity $\sigma_{ij} = \partial U / \partial \epsilon_{ij}$ is the mechanical stress field. Likewise, the quantity $\tau = \partial U_S / \partial \epsilon$, which is work-conjugate to the surface strain ϵ, is the *surface stress* in the reference configuration. It represents the sensitivity of surface energy density to surface strain, and it arises naturally as a configurational force in the description of the system. Integration by parts, followed by application of the divergence theorem to the first term on the right side of (8.129), leads to

$$\delta \mathcal{F} = -\int_R \sigma_{ij,j} \, \delta u_i \, dR + \int_{S'} \left[\sigma_{ij} n_j - \frac{\partial}{\partial s} (\tau m_i) \right] \delta u_i \, ds . \qquad (8.130)$$

The integrated terms which arise drop out because of the condition of workless boundaries. Conditions necessary to ensure that $\delta \mathcal{F} = 0$ for arbitrary δu_i are

$$\sigma_{ij,j} = 0 \quad \text{in} \quad R \qquad (8.131)$$

and

$$\sigma_{ij} n_j = m_i \frac{\partial \tau}{\partial s} + \tau \kappa n_i \quad \text{on} \quad S' . \qquad (8.132)$$

These equations are sufficient to determine the elastic strain distribution for any surface shape. The condition (8.131) is the familiar local mechanical equilibrium equation for stress throughout R and (8.132) is the traction boundary condition on S' in the presence of surface stress. The second term on the right side of (8.132) is the normal traction as usually specified through the Laplace–Young relation (Gibbs 1878); the pressure within the material balancing the quantity $-\tau \kappa$ is usually called the Laplace pressure. The first term on the right side is the shear traction induced in the material by a gradient in surface stress along the surface. The main result is that, given surface shape and surface characteristics, the equilibrium strain field is known. It has been tacitly assumed that suitable boundary conditions on the surface S'' are known from external loading conditions, symmetry or other imposed constraints. The validity of the results obtained does not hinge on S'' being a workless boundary.

In light of the equilibrium constraints summarized in (8.131) and (8.132), it would appear that the free energy $\mathcal{F}$ defined in (8.1) could be written in terms of surface fields alone by applying the divergence theorem to the term representing bulk elastic energy. This is often the case, but the process of executing the steps may involve some subtle issues. For example, if the elastically strained region R is unbounded in the model being studied, then the elastic energy is also unbounded. However, this difficulty can be circumvented by viewing U as the change in elastic energy density from an initial value throughout the region. Another issue that may require special attention arises when the mismatch strain defined throughout the region R is not compatible, a feature that would preclude transformation of the volume integral in (8.1) to a surface integral by application of the divergence theorem. However, this difficulty can also be circumvented by viewing the total strain as the sum of the mismatch strain plus any additional strain due to change in shape of the boundary of the solid. The latter contribution to the strain is necessarily compatible and, again, the portion of free energy depending on the instantaneous shape of the surface S' can be converted to a surface integral in most cases. A specific example illustrating these points is presented in Section 9.4.

This discussion is concluded with a few remarks concerning the results obtained. The consideration of states of deformation that are thermodynamic equilibrium states for fixed surface shape is an assumption based on the underlying physics and not an essential restriction. For processes of film growth from a vapor, and even for solidification from a liquid state, the process is typically slow enough so that mechanical equilibrium is maintained continuously in time. There are instances of phase boundary motion at rates large enough to warrant abandonment of this assumption as, for example, in laser ablation of solid surfaces (Davis 2001), but these are not germane in this discussion. The appearance of the surface stress τ in the course of examining energy variations associated with deformation provides a natural way to incorporate this concept in continuum modeling of surface evolution. As such, it is a macroscopic concept for describing material behavior. There is a corresponding microscopic implementation of the same physical effect, based on chemical bonding at a free solid surface, that will arise in Section 8.8.3. The boundary condition (8.132) adds a new feature to boundary value problems in continuum mechanics. In effect, this result implies that the limiting values of normal and shear traction as the bounding surface is approached from within the material are not zero, in general, even though there is no *applied* traction on the surface. In this sense, the surface takes on the character of a physical entity separate from the bulk material. Finally, the development has been carried out within the framework of small deformation theory. However, the results apply for large deformation as well, provided that all fields and orientations are interpreted in terms of the reference configuration of the material (Wu 1996, Freund 1998, Norris 1998).

8.8.2 *Surface chemical potential*

Once the assumption of mechanical equilibrium has been incorporated, the free energy expression can be reinterpreted. Instead of viewing the free energy as a functional of both strain and surface shape, it can instead be viewed as a functional of surface shape alone (Shenoy and Freund 2002). In principle, the strain is known in terms of surface shape through the solution of the boundary value problem represented by (8.131) and (8.132). The goal in this section is to write the time rate of change of free energy in the form

$$\dot{\mathcal{F}}(t) = \int_{S'} \chi(s, t) v_n(s, t) \, ds \,, \tag{8.133}$$

in which case the quantity χ can again be identified as a surface chemical potential.

Following the reasoning outlined in Section 8.2.1, the time rate of change of free energy $\mathcal{F}(t)$ is

$$\dot{\mathcal{F}} = \int_{S'} \left[\sigma_{ij} n_j \dot{u}_i + U v_n - \kappa U_S v_n \right] ds + \int_{S'} \left[\frac{\partial U_S}{\partial \theta} \frac{d\theta}{dt} + \frac{\partial U_S}{\partial \epsilon} \frac{d\epsilon}{dt} \right] ds \,. \tag{8.134}$$

The first integral expression on the right side is the same as that found in Section 8.2.1, where the boundary condition of the evolving surface was $\sigma_{ij} n_j = 0$ as stated in (8.5). Here, the surface traction on the evolving surface is not zero but, instead, is specified by (8.132). The second integral expression on the right side of (8.134) introduces, in addition, the orientation and deformation dependence of the surface free energy density. The configurational force $\tau = \partial U_S / \partial \epsilon$ has already been identified in the foregoing discussion as the surface stress at a given surface orientation and level of deformation. A second configurational force arises here, namely,

$$q = \frac{\partial U_S}{\partial \theta} \,. \tag{8.135}$$

The quantity q can be interpreted as a distributed surface torque or moment that tends to rotate the surface toward an orientation with lower surface energy.

If the relations (8.127) and (8.128) are incorporated into (8.134), and if integration by parts is invoked to isolate a factor v_n in each term, it is found that

$$\dot{\mathcal{F}} = \int_{S'} v_n \left[U - \kappa U_S - \frac{\partial}{\partial s}(q + \tau \gamma_{mn}) + \tau \frac{\partial \epsilon}{\partial n} \right] ds \,. \tag{8.136}$$

The terms obtained in the course of integration by parts have zero net value because of the assumed workless nature of the boundary constraints. The quantity γ_{mn} appearing in (8.136) is the shear strain introduced in (8.128) and $\partial \epsilon / \partial n = n_k \partial \epsilon / \partial x_k$ is again the normal derivative of the extensional strain in the direction parallel to the surface, which was also introduced in the same expression.

In light of the result (8.136), it follows from (8.133) that

$$
\chi = U - \kappa U_{\mathrm{S}} - \frac{\partial}{\partial s}(q + \tau \gamma_{mn}) + \tau \frac{\partial \epsilon}{\partial n}
\tag{8.137}
$$

is the surface chemical potential incorporating surface anisotropy and deformation dependence. Similar results have been reported by Wu (1996), Freund (1998) and Fried and Gurtin (2003). The quantities involving strain – U, γ_{mn} and ϵ – are presumed known from the solution of the mechanical boundary value problem identified in Section 8.8.1. The remaining quantities are determined by the constitutive properties incorporated in $U_{\mathrm{S}}(\theta, \epsilon)$. In writing (8.137), it has been tacitly assumed that U_{S} is twice differentiable with respect to each of its arguments, but this is not so for all common descriptions of surface behavior. The dependence of surface energy on strain and orientation is very difficult to characterize experimentally. Atomistic simulations can also provide insight into the nature of this dependence. The dependence of U_{S} on ϵ and θ for a vicinal surface is considered next.

8.8.3 *Energy of a strained vicinal surface*

For a continuum model of a single crystal, the value of energy density of a given crystallographic surface orientation is determined by the fine scale structure of that surface. For a high symmetry orientation, such as a $\{100\}$ surface of a cubic crystal, the surface is atomically flat. For orientations close to this surface, the structure usually consists of flat terraces with well defined local surface energies separated by atomic scale ledges or steps, as illustrated schematically in Figure 8.25. Such a surface is commonly called a *vicinal surface*. The steps alter the macroscopic surface energy from that of the high symmetry orientation by an amount corresponding to their energies of formation and interaction. Furthermore, the energies of individual steps and the interactions of these steps are influenced by the presence of background strain in the crystal.

Macroscopically, the surface is assumed to be smooth (as indicated by the dashed line in the figure) and to have a local surface energy density U_{S} at any point on the surface, determined by the orientation of the tangent plane and the level of elastic strain in the crystal at that point. A quantitative interpretation of the free energy of a strained crystal surface in terms of the physics of crystallographic steps has been developed by Shenoy and Freund (2002). The surface chemical potential of a stepped surface follows from (8.137), given the material specific surface energy $U_{\mathrm{S}}(\theta, \epsilon)$. The actual form of this dependence can be determined only through atomistic simulation or experimental observation (Shenoy et al. 2002).

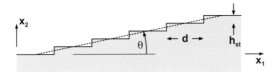

Fig. 8.25. Geometry of a vicinal surface that makes an angle of θ with a high symmetry direction. The spacing d between the steps is related to the step height h_{st} through the relation $d = h_{st}/\tan\theta$. The dashed line shows the macroscopic surface orientation.

The central idea of the continuum description is based upon an understanding of stepped crystal surfaces that is drawn from surface science (Blakely 1973, Jeong and Williams 1999). Below the characteristic roughening temperature of the material, which can be in excess of $1000\,^{\circ}\text{C}$ for semiconductors, a nominally flat surface of a crystal that is misoriented by a small angle from a high symmetry direction consists of a train of straight parallel steps, as illustrated schematically in Figure 8.25. In the absence of strain, the surface energy density of such a vicinal surface is given by

$$U_S(\theta) = \gamma_0 \cos\theta + \beta_1 |\sin\theta| + \beta_3 \frac{|\sin\theta|^3}{\cos^2\theta}, \qquad (8.138)$$

where θ is the misorientation angle, γ_0 the energy density of the atomically flat surface with $\theta = 0$ per unit area in the reference configuration, and the parameters β_1 and β_3 are related to step creation energy and step interaction energy, respectively. Specifically, β_1 is the energy needed to create a unit length of an isolated step per unit height of that step. In writing (8.138), it is tacitly assumed that the structure is symmetric under reflection in the plane $x_1 = 0$. Like any other crystallographic defects, say dislocations or vacancies, steps give rise to long-range elastic stress fields in the material; the steps interact with each other, as well as with other defects in the material, through these elastic stress fields. The amplitude of the stress field produced by a step decays with the inverse square of the distance from the step (Marchenko and Parshin 1980). The last term on the right in (8.138) arises from the interactions between steps through their stress fields. For small values of θ, it is evident that the second term on the right side of (8.138) is linear in θ, being essentially proportional to the number of steps per unit length along the surface. The last term on the right is cubic in θ, with a factor θ arising once again from step density and the additional factor θ^2 arising from the proportionality of interaction energy to the inverse square of the separation distance between steps. Finally, U_S is understood to be the energy per unit area in the reference configuration of the macroscopic mean surface.

It is shown next that an elastic strain in the crystal, arising from mismatch or some other origin, can lead to significant alteration of the surface energy of the vicinal surfaces compared to that of the nearby high symmetry orientation or compared to the unstrained vicinal surface. The ideas incorporated in (8.138) are readily generalized to the case of strained crystals by including the influence of surface strain ϵ on the coefficients for a specific material structure. The expression (8.138) for surface energy becomes

$$U_S(\theta, \epsilon) = (\gamma_0 + \tau_0 \epsilon) \cos \theta + (\beta_1 + \hat{\beta}_1 \epsilon) |\sin \theta| + \beta_3 \frac{|\sin \theta|^3}{\cos^2 \theta} \qquad (8.139)$$

where only the first order influence of strain is incorporated for small strain magnitudes. In this expression, τ_0 is the surface stress of the flat surface at the current level of strain and $\hat{\beta}_1$ is a measure of the sensitivity of the formation energy of a step to surface strain. The last term has the same interpretation as in (8.138) and its inclusion is essential for a proper description of behavior. In principle, the coefficient β_3 should also be expanded to account for the influence of surface strain. However essential behavior is captured by ignoring this possibility and, consequently, only the leading order term represented by β_3 is retained.

Suppose that the reference surface $\theta = 0$ is the {100} surface of a cubic crystal, and that the background strain ϵ_{ij}^{m} in this state is a spatially uniform equilibrium elastic field consistent with a traction-free surface. As θ increases from zero, the surface strain ϵ is given by (8.126) where $m_1 = \cos \theta$, $m_2 = \sin \theta$. The local state of strain would be altered as a result of long-range elastic effects if θ would be spatially nonuniform; in the present instance, such spatial variation is not taken into account and the surface strain is determined by θ and the uniform strain field ϵ_{ij}^{m}. For the case in which $\epsilon_{12}^{m} = 0$, substitution of the expression for ϵ into (8.139), followed by expansion of the result in powers of θ up to degree three, yields

$$U_S = \left(\gamma_0 + \tau_0 \epsilon_{11}^{m} \right) + \left(\beta_1 + \hat{\beta}_1 \epsilon_{11}^{m} \right) |\theta| + \left[\tau_0 (\epsilon_{22}^{m} - \tfrac{3}{2} \epsilon_{11}^{m}) - \tfrac{1}{2} \gamma_0 \right] \theta^2$$
$$+ \left[\beta_3 - \tfrac{1}{6} \beta_1 + \hat{\beta}_1 (\epsilon_{22}^{m} - \tfrac{7}{6} \epsilon_{11}^{m}) \right] |\theta|^3 . \qquad (8.140)$$

The first term in parentheses on the right side is the energy density of the strained surface for $\theta = 0$. The first order term in θ is usually identified as the origin of a sharp corner or *cusp* in the dependence of surface energy on surface orientation, arising through the dependence of U_S on the absolute value of θ. Such a feature is illustrated by the dotted line in Figure 8.26 for the case of zero background strain.

When strain is involved, the coefficient of $|\theta|$ in (8.140), usually presumed to be positive for unstrained crystals, has some remarkable properties. This coefficient depends *asymmetrically* on strain, assuming that $\hat{\beta}_1 \neq 0$. Furthermore, if $\hat{\beta}_1 \neq 0$,

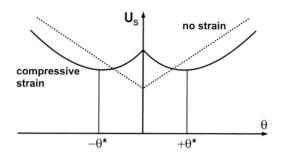

Fig. 8.26. Schematic diagram illustrating the influence of background elastic strain on the dependence of surface energy density on surface orientation for a stepped surface, as implied by (8.140). The figure is drawn under the assumption that $\hat{\beta}_1 > 0$. In this case, the dependence of surface energy on orientation suggests that the high symmetry orientation $\theta = 0$ becomes unstable for sufficiently large compressive strain. Furthermore, as a result of the strong effect of step repulsion at larger values of θ, a local minimum in surface energy emerges for some nonzero value of θ, say θ^*, suggesting the appearance of preferred surface orientations due to strain that do not arise in unstrained crystals.

then it must be either positive or negative for a fixed surface structure. It follows that $\beta_1 + \hat{\beta}_1\epsilon_{11}^{m}$ may be negative in value for compressive or tensile strain ϵ_{11}^{m}, respectively, of sufficiently large magnitude. This implies that the energy of the stepped surface may become *lower* than the energy of the surface without steps for strain of one sign or the other. A plot of the surface energy of a vicinal surface for which $\epsilon_{12}^{m} = 0$ is shown by the solid line in Figure 8.26 to illustrate this point. In this case, when strains are sufficiently compressive with $\hat{\beta}_1 > 0$, the change of surface energy with surface orientation θ is essentially

$$U_S - \left(\gamma_0 + \tau_0\epsilon_{11}^{m}\right) \approx (\beta_1 + \hat{\beta}_1\epsilon_{11}^{m})|\theta| + \beta_3|\theta|^3 \tag{8.141}$$

and it develops *minima* away from $\theta = 0$. At small angles, the step formation energy dominates the repulsive interaction energy between the steps, while at larger angles, the step interactions are dominant. The competition between these two opposing effects may result in optimum misorientation angles, denoted by $\theta = \pm\theta^*$ in the figure, for sufficiently large compressive strain. For most practical purposes, the value of this angle is essentially

$$\theta^* = \sqrt{-\frac{\beta_1 + \hat{\beta}_1\epsilon_{11}^{m}}{\beta_3}} \tag{8.142}$$

when the minima in U_S with respect to θ exist. It follows that $\hat{\beta}_1$ is a key quantity in determining the surface morphology of strained crystals. The physical origin of this effect can be found in the details of surface reconstruction for the vicinal surface arrangement as proposed originally by Khor and Das Sarma (1997). Tension is

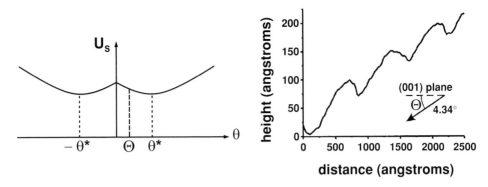

Fig. 8.27. The graph on the right shows an atomic force microscopy scan in the miscut direction of the surface of a $Si_{0.5}Ge_{0.5}$ deposit, nominally 2.5 nm in thickness, on a vicinal Si(001) surface miscut at 4.34°; the deposition temperature was 650 °C. Reproduced with permission from Schelling et al. (2001). The growth surface retains the average slope of the miscut surface, represented by Θ in the sketch on the left, but it is divided into segments with slopes of either $+\theta^*$ or $-\theta^*$, representing the minima in surface energy for the strained stepped surface.

induced in all of the most severely strained bonds in the vicinity of the surface steps in the course of reconstruction (Shenoy et al. 2002). This results in a relatively large tensile surface stress on the vicinal surface. Furthermore, because the stretching is predominantly tensile, a superimposed background compressive strain can result in a dramatic reduction in macroscopic surface energy as illustrated schematically in Figure 8.26.

Continuum theory provides no basis for estimating the sub-continuum parameters β_1, $\hat{\beta}_1$ and β_3 for any particular material, even though these parameters are well-defined characteristics of behavior. They derive strictly from the discreteness of the material, so values can be estimated only when the model of the material includes that discreteness. Such estimates require the adoption of a particular interatomic potential to simulate the material and, in the case of the surface, an atomic structure for that surface.

The features of the dependence of surface energy on vicinal surface orientation illustrated in Figure 8.26 are evident in experimental results reported by Schelling et al. (2001). They deposited SiGe onto Si(001) miscut by 4.34°. For relatively high Ge content in the deposited material, a surface morphology developed with average slope of 4.34° but with the surface divided into more or less regular segments of slope $+11°$ and $-11°$. A representative scan of the surface in the miscut direction resulting from deposition of a 2.5 nm thick layer of $Si_{0.5}Ge_{0.5}$ at 650 °C is shown in Figure 8.27. The proportion of the surface at angle $+11°$ to that at angle $-11°$ is determined by the proportion of the angle difference $\Theta + \theta^*$ to the difference $\theta^* - \Theta$.

8.8.4 Example: Stepped surface near (001) for strained Si

Interpret the surface energy expression (8.140) for the case of a strained vicinal surface on a Si crystal near the high symmetry (001) orientation.

Solution:

Commonly used potentials for Si and Ge are the Stillinger–Weber potential (Stillinger and Weber 1985, Ding and Andersen 1986) and the Tersoff potential (Tersoff 1988, Tersoff 1989). The most commonly adopted structure for the (001) surface of Si or Ge is based on a 2×1 reconstruction whereby long rows of rebonded dimer pairs aligned with the [110] direction form in the [1$\bar{1}$0] direction, or rows of dimers each aligned with the [1$\bar{1}$0] direction form in the [110] direction, with these two reconstructions alternating with atomic plane through the crystal.

The introduction of steps on the surface complicates the picture significantly. Terraces can alternate between the two dimer reconstructions indicated for step heights of a single atomic plane, or one-fourth of the unit cell dimension in a diamond cubic material. However, the search continues for minimum energy structures that interrupt the dimer arrays and reconstruct to eliminate as many dangling bonds as possible. A promising structure for steps aligned with the [100] direction has been identified and analyzed by Shenoy et al. (2002) and Fujikawa et al. (2002) for Si and Ge. In effect, steps are first formed in this orientation in the simplest way. Then, alternate atoms along the step edge are removed. Finally, reconstructions are introduced that significantly reduce the step energy and that lead to a behavior consistent with observations. This configuration has been verified experimentally by atomic force microscopy observations of the growth surface of Ge deposited epitaxially on Si(001) (Fujikawa et al. 2002).

Calculations based on the Tersoff potential for strained Si crystals with mean surface orientations $(10n)$, where $n = 1, 3, 5, 7, \ldots$, and imposed biaxial strain varying between -0.05 and 0.05 have been reported by Shenoy et al. (2002). For the case of reconstructed steps of height equal to a single atomic plane or spacing, the values of the parameters were found to be $\beta_1 = 0.093 \, \mathrm{J \, m^{-2}}$, $\hat{\beta}_1 = 16.8 \, \mathrm{J \, m^{-2}}$ and $\beta_3 = 3.48 \, \mathrm{J \, m^{-2}}$. These values are consistent with the dependence of surface energy on orientation shown schematically in Figure 8.26. Plots of surface energy versus strain for (105) stepped surfaces are illustrated in Figure 8.28.

The angle θ^* implied by these parameter values at strain $\epsilon_{11}^{\mathrm{m}}$ is about 8.5°. The actual crystal can assume only those angles characteristic of $(10n)$ orientations, where $n = 5, 9, 13, \ldots$, for this reconstruction, and 8.5° is about midway between the (105) and (109) orientations. Long-range elastic effects, which have not been taken into account in the discussion in this section, would tend to drive the angle toward the larger angle orientation.

8.9 Strained epitaxial islands

In Section 8.4.1 it was shown that the flat surface of a stressed solid with isotropic surface energy is unstable under perturbations in shape of the surface of the solid in its reference configuration. The shape perturbations may be a consequence of mass rearrangement at fixed total mass, for example. For sufficiently high stress

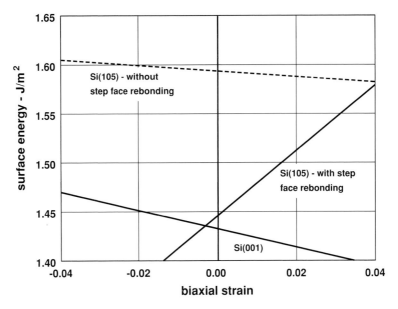

Fig. 8.28. Plots of surface energy density versus biaxial mismatch strain for (105) vicinal surfaces with steps aligned with the [010] direction on (001)Si. The stepped edges are 'serrated' in both cases by removing every other step edge atom, which leaves two possibilities for dimer reconstruction on the terraces. In the configuration proposed by Khor and Das Sarma (1997), the dimer nearest the step edge rebonds directly into the step face, whereas the configuration proposed by Mo et al. (1990) presumes the alternate dimer reconstruction. The plot is based on computations by Shenoy et al. (2002) using the Tersoff potential.

and temperature, the configurational driving force tending to increase the amplitude of fluctuations in surface shape is large enough to cause the fluctuations to grow. For the case of a thin film deposited epitaxially on a substrate in the presence of lattice mismatch, the flat surface of the stressed film is also unstable. Indeed, if the elastic properties of the film and the substrate are the same, then the stability condition is the same as that obtained in Section 8.4.1. On the other hand, if the substrate is elastically stiffer (more compliant) than the film material, the flat surface configuration is destabilized at a larger (smaller) internal stress than for an elastically homogeneous material. The influence of substrate stiffness was considered quantitatively in Section 8.4.3 and, for the case of an isolated island, it is revisited at the end of Section 8.9.1.

In cases where the flat film surface is unstable, perturbations in surface shape may grow in time by means of some mechanism of mass rearrangement. In cases of growth, the valleys in surface topography may reach the film–substrate interface, resulting in clusters of film material being isolated from one another by regions of exposed substrate surface or perhaps by regions of the substrate covered only by a very thin wetting layer of film material. Conditions under which this can occur

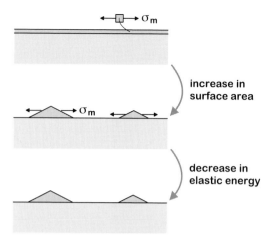

Fig. 8.29. In order to visualize the competing energy effects involved in converting a uniformly strained film into partially relaxed islands on a substrate, the process is imagined to occur in two steps. In the first step, uniformly strained film material is gathered into clusters, thereby increasing surface area without relaxing elastic energy. In the second step, the surface traction that was artificially introduced to maintain elastic strain is relaxed, thereby reducing elastic energy without changing the surface area.

were discussed briefly in Section 8.4.4. These isolated clusters are commonly called *islands*.

To see that there is a natural tendency for the film material to agglomerate into islands in the presence of mismatch strain, imagine the sequence of events shown schematically in Figure 8.29. Initially, a film of uniform thickness is formed on a substrate of relatively large thickness. The film is assumed to be subject to an extensional mismatch strain ϵ_m, isotropic in the plane of the interface. The strain energy per unit volume for an elastically isotropic film material is $M_f \epsilon_m^2$ where M_f is the biaxial modulus of the film. Imagine that the film material is then gathered into clusters on the surface of the substrate, while maintaining the epitaxial bond with the substrate. Furthermore, suppose that this is done in such a way that tractions are applied to the lateral faces of the clusters which maintain the mismatch strain in the film material at its initial value of ϵ_m. At the end of this step, the total surface area of the material structure has been increased from its value at the beginning of this step but the total elastic energy has been left unchanged. If the surface energy densities of the film material and the substrate material are comparable or, alternatively, if a very thin wetting layer of film material remains in the gaps between the islands, then the total surface energy of the system has been increased. The increase in surface energy scales with a surface energy density, say γ_f.

Next, suppose that the artificial tractions, which were introduced in order to maintain the strain level everywhere in the film material, are relaxed. For any stable elastic material, the work done on the system by these tractions as they are relaxed is *negative*. Consequently, once the tractions have been completely removed, the total elastic energy has been decreased from its value at the start of the step. This decrease in elastic energy scales with the initial elastic energy density $M_f \epsilon_m^2$.

Thus, the total change in free energy in forming any particular island arrangement from a strained thin film may be positive or negative, depending on whether the surface energy increase or the elastic energy reduction dominates the process. If the net change in free energy is negative then the change may occur spontaneously, but otherwise it may not. The ratio of the scaling factors for the two steps in the imaginary process, that is, the ratio

$$\zeta = \frac{\gamma_f}{M_f \epsilon_m^2}, \qquad (8.143)$$

is a system parameter with the physical dimensions of length, analogous to that introduced in Section 8.3 and elsewhere in this chapter. Whether or not a particular island configuration will tend to arise spontaneously will depend on the values of characteristic lengths of that configuration relative to ζ. This basic idea is developed more fully and in quantitative form in the sections that follow.

8.9.1 An isolated island

Consider a single island of film material bonded to a relatively large substrate. Even for this relatively simple material system, there are many system parameters that can influence variations in total free energy. These include size and shape of the island, elastic properties of the materials involved, surface energies of the materials, surface energy anisotropy and proximity to other islands, to name a few possibilities. To facilitate understanding of the influence of size and shape on the free energy of the island, each of these particular basic features are given one-parameter characterizations; consideration of the remaining influences identified is postponed to later discussion. Thus, a configuration that is completely characterized by two geometrical parameters is considered. A simple shape that is suitable for this purpose and that includes three-dimensional effects is a right circular cone, as illustrated in Figure 8.30. The size and shape of the conical island is completely specified by the radius r of its base and the altitude h. An alternate parameterization, and one that is more useful for present purposes, is to specify the volume V of the cone and its aspect ratio a; in terms of the dimensions illustrated in Figure 8.30, these are

$$V = \tfrac{1}{3}\pi r^2 h, \quad a = h/r. \qquad (8.144)$$

For the time being, the film and substrate materials are assumed to be elastically isotropic and to have identical elastic properties, say the biaxial modulus M and

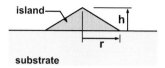

substrate

Fig. 8.30. Schematic diagram of an elastically strained island on a relatively large substrate. The island is assumed to be in the shape of a right circular cone for purposes of illustration, and it is restrained from relaxing by the substrate.

Poisson ratio ν. The total elastic strain energy of the system is

$$W = \int_R \tfrac{1}{2} c_{ijkl} \epsilon_{ij} \epsilon_{kl} \, dR \,, \tag{8.145}$$

where R is the region of space occupied by the island and its substrate, c_{ijkl} is the array of elastic constants of the materials as introduced in Chapter 3, and ϵ_{ij} is the nonuniform elastic strain field throughout R. The strain arises solely from the mismatch in lattice parameters between the two materials. Suppose that all lengths in the system are rescaled by the dimension $V^{1/3}$, so that $x_i = V^{1/3} \hat{x}_i$, for example, where $\hat{x}_i$ is a nondimensional spatial coordinate. If the particle displacement is also normalized in the same way, then the strain itself depends on position in the island only through the normalized coordinate $\hat{x}_i$. Furthermore, strain ϵ_{ij} must depend linearly on ϵ_m and it must vanish everywhere when $\epsilon_m = 0$. Therefore, there exists a strain field $\hat{\epsilon}_{ij}(\hat{x}_1, \hat{x}_2, \hat{x}_3)$ that is independent of both ϵ_m and V and that is defined by

$$\hat{\epsilon}_{ij}(\hat{x}_1, \hat{x}_2, \hat{x}_3) = \epsilon_{ij}(x_1, x_2, x_3)/\epsilon_m \,. \tag{8.146}$$

Note that $\hat{\epsilon}_{ij}(\hat{x}_1, \hat{x}_2, \hat{x}_3)$ is the strain field for an island of unit volume with a unit mismatch strain with respect to the substrate. In addition, the array of elastic moduli c_{ijkl} can be rewritten as

$$c_{ijkl} = M \hat{c}_{ijkl} \,, \tag{8.147}$$

where the nondimensional components of $\hat{c}_{ijkl}$ may depend on ν but are independent of M. Once the scaling rules are introduced into (8.145), that expression reduces to

$$W = M \epsilon_m^2 V \int_{\hat{R}} \tfrac{1}{2} \hat{c}_{ijkl} \hat{\epsilon}_{ij} \hat{\epsilon}_{kl} \, d\hat{R} = M \epsilon_m^2 V \hat{W}(a, \nu) \,, \tag{8.148}$$

where $\hat{W}(a, \nu)$ is a dimensionless function of the island aspect ratio a and Poisson ratio ν. Its value must be determined numerically for specific values of these parameters, in general.

To consider the corresponding surface energy variation with aspect ratio a for fixed volume V, assume that the surface energy density of the film material γ_f is independent of surface orientation. The energy densities of the substrate material and the film–substrate interface are denoted by γ_s and γ_{fs}, respectively. A suitable expression for the net surface energy is

$$\Gamma = \gamma_f \pi r^2 \sqrt{1 + h^2/r^2} + (\gamma_{fs} - \gamma_s)\pi r^2 \,. \tag{8.149}$$

This expression provides the description of change in surface energy with respect to change in aspect ratio. The value of total surface energy is not useful because the amount of surface area involved in this configuration is essentially unbounded.

The energy γ_{fs} per unit area of the epitaxial interface between the film and the substrate materials is expected to be relatively small compared to γ_f for a fully coherent interface, and this energy contribution is neglected for convenience throughout the present discussion. If h and r are expressed in terms of V and a according to (8.144), then

$$\Gamma = \pi \left(\frac{3V}{\pi a}\right)^{2/3} \left[\gamma_f \sqrt{1 + a^2} - \gamma_s\right] = \gamma_f V^{2/3} \hat{\Gamma}(a)\,, \tag{8.150}$$

where $\hat{\Gamma}(a)$ is a dimensionless function of aspect ratio. Note that the behavior of $\hat{\Gamma}(a)$ near $a = 0$ is consistent with the notion of surface wetting. For example, if $\gamma_f < \gamma_s$ then there is benefit in having the substrate covered with film material and, accordingly, there is a deep energy well in $\hat{\Gamma}(a)$ as $a \to 0^+$. Likewise, if $\gamma_f > \gamma_s$ then there is benefit in having the substrate surface exposed, consistent with a minimum value in $\hat{\Gamma}(a)$ at some positive value of a.

To pursue the discussion in further detail, with particular emphasis on the influence of elastic energy, assume that $\gamma_f = \gamma_s = \gamma$. The total free energy change due to formation of the island from the uniformly strained film configuration is $W - M\epsilon_m^2 V + \Gamma$, or from (8.148)

$$\mathcal{F} = M\epsilon_m^2 V \left[\hat{W}(a, v) - 1\right] + \gamma V^{2/3} \hat{\Gamma}(a)\,. \tag{8.151}$$

Consequently, the total free energy change normalized by the initial elastic energy of a flat film is

$$\frac{\mathcal{F}}{M\epsilon_m^2 V} = \hat{W}(a, v) - 1 + Z\hat{\Gamma}(a)\,, \tag{8.152}$$

where the nondimensional ratio Z is defined to be

$$Z = \frac{\gamma}{M\epsilon_m^2 V^{1/3}}\,. \tag{8.153}$$

An alternative, and more useful, interpretation of (8.152) is as the free energy change of the system per unit volume of island material, normalized by the initial elastic energy per unit volume. All dimensional parameters appear in this expression through the single quantity Z which relates the size of the island, namely

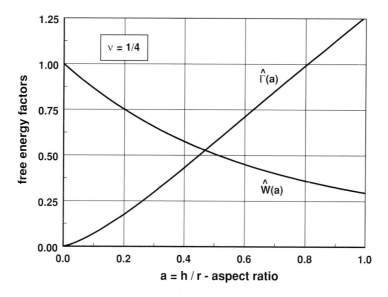

Fig. 8.31. Graphs of the elastic energy shape factor $\hat{W}$ defined in (8.148) and the surface energy shape factor $\hat{\Gamma}$ defined in (8.150) versus island aspect ratio $a = h/r$ for a Poisson ratio of $1/4$.

$V^{1/3}$, to the characteristic length ζ of the system as defined in (8.143). The normalized free energy change is specified for any ν by the aspect ratio a and the size parameter Z.

The dependence of the function $\hat{W}(a, \nu)$ on a for $\nu = 1/4$ is readily established by appeal to numerical simulations using the finite element method. The island and substrate were analyzed as a rotationally symmetric elastic structure. The axis of the cone is the axis of rotational symmetry of the structure, and the substrate is assumed to be a right circular cylinder with both radius and height equal to $10r$. The effect of the incompatibility strain due to lattice mismatch is simulated as a thermoelastic deformation. A nonzero coefficient of thermal expansion is assigned to the island material, and a temperature change is imposed so that the stress-free isotropic thermal expansion strain of the island would be $-\epsilon_m$ if it were not constrained. The fact that the expansion of the island is resisted by the substrate gives rise to the nonuniform elastic strain denoted by ϵ_{ij} in (8.145). There is no net force acting on the island. Consequently, the stress field decays in amplitude in the substrate with increasing distance from the island. The energy calculation is insensitive to the details of the boundary conditions on the lateral surface or the base of the cylinder representing the substrate. The result is shown in Figure 8.31. The dependence of the function $\hat{\Gamma}(a)$, defined in (8.150), on a is also shown in the same figure. The two functions $\hat{W}$ and $\hat{\Gamma}$ represent the influence of elastic energy

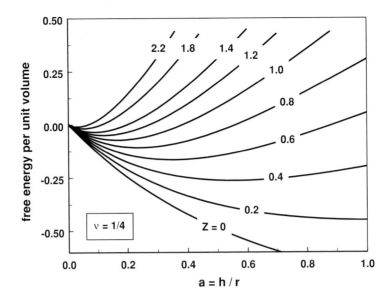

Fig. 8.32. Free energy change per unit volume of island of material due to reorganization of uniformly strained film material to a nonuniformly strained conical island. The parameter Z, defined in (8.153), represents the size of the island relative to the natural system length ζ.

and surface energy on total free energy reduction, respectively, and the parameter Z represents the influence of surface energy relative to elastic energy.

The normalized free energy change per unit volume versus a is shown in Figure 8.32 for a range of values of Z. To provide an indication of the range of physical parameters involved, note that if $\gamma = 1\,\mathrm{J\,m^{-2}}$, $M = 10^{11}\,\mathrm{N\,m^{-2}}$ and $\epsilon_m = 0.01$ then $Z = 1$ implies $V^{1/3} = 100\,\mathrm{nm}$. Each curve in Figure 8.32 represents the variation of the free energy change per unit volume of the island material with aspect ratio for a *fixed* volume of island material. The island volume V varies inversely with Z according to (8.153); therefore, each curve represents the free energy per unit volume of an island of a particular volume.

Several significant observations can be made concerning the results in Figure 8.32. All curves leave the origin with negative slope, implying that the initial flat film is unstable. It is also evident that each curve has a local minimum for some nonzero aspect ratio, and that there is a configurational force tending to drive any particular material volume to the shape with this particular aspect ratio. The configurational force tending to change the aspect ratio of an island of a given volume is the negative slope of the curve corresponding to that volume at the current aspect ratio. Furthermore, the value of the aspect ratio at the free energy minimum increases with increasing volume V or with decreasing Z. Thus, larger islands tend to have

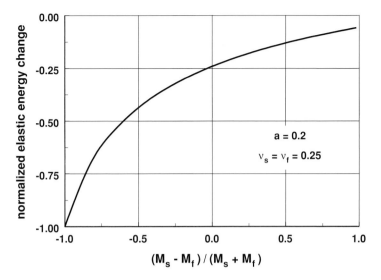

Fig. 8.33. Dependence of elastic energy change in the strained system due to island formation on the ratio of the elastic stiffness of the substrate to that of the film. Results are shown
for island aspect ratio of 0.2 and a common Poisson ratio of 0.25.

higher aspect ratios at equilibrium than do smaller islands, within the framework
of the present model. This is so because more elastic energy can be released by
increasing the aspect ratio while the penalty associated with increasing surface
energy is rendered less significant by the small value of Z for large islands.

The calculation of free energy change in the transition of a uniformly strained
film into an isolated epitaxial island which led to Figure 8.32 was based on the
assumption that the elastic properties of the film material and substrate material are
identical. If this is not the case, the modulus difference between the materials can
have a significant influence on the change in elastic energy associated with island
formation. To illustrate this point, consider the formation of an isolated conical
island from a uniform film with biaxial elastic modulus M_f and Poisson ratio v_f.
The corresponding elasticity parameters of the substrate are M_s and v_s. Suppose that
attention is limited to a particular aspect ratio, say $a = 0.2$, and that $v_s = v_f = 0.25$.
The normalized elastic energy reduction per unit volume of the island, which was
denoted by $\hat{W}(a, v) - 1$ in (8.152), can then be calculated for fixed M_f and ϵ_m, and
for a range of values of modulus ratio M_s/M_f. The result for $a = 0.2$ is shown in
Figure 8.33, which illustrates the influence of modulus difference on energy change.
For example, when $M_s = M_f$ the elastic energy reduction in island formation is
about 24%, as shown in Figure 8.32. For purposes of comparison, when $M_s = 2M_f$
it is only about 12% and when $M_s = \frac{1}{2}M_f$ it is about 40%. The influence of modulus
difference for other aspect ratios is similar.

8.9.2 Influence of an intervening strained layer

In the foregoing study of elastic energy change during island formation, it was demonstrated that the elastic stiffness of the substrate has a relatively strong influence on the proportion of elastic misfit energy that can be released to drive the process. In the analysis, it was tacitly assumed that the substrate itself was not subject to a misfit strain. In other words, virtually *all* material subjected to misfit with respect to the unstrained substrate is ultimately gathered into the island. This need not be so. It is conceivable that an island will be formed on the surface of a layer which itself is subject to the same misfit strain. This strained layer, in turn, may be bonded to the surface of a relatively thick substrate. This is the situation considered here. The main point to be made is that the misfit strain in such an intervening layer, lying between a thick substrate and a surface island, has no effect whatsoever on the energy reduction associated with island formation.

The conclusion follows directly from superposition of linear elastic fields without the need for detailed calculation. The system under discussion is depicted in part (a) of Figure 8.34. The figure shows a relatively thick elastic substrate with a film of some uniform thickness bonded to its surface. The film supports a spatially uniform mismatch strain, presumably due to the constraint of epitaxy. The elastic properties of the film material may be different from those of the substrate material. Some of the film material has gathered into an isolated epitaxial island on the surface of the strained layer. The island material also supports the same mismatch strain. The lateral faces of the island are free of applied traction. As a result, the elastic strain field is spatially nonuniform and the elastic strain energy is partially relaxed from the uniform density of the flat film. For purposes of this development, the island may have any shape, either two-dimensional or three-dimensional. For any particular island size and shape, how does the presence of misfit strain in the layer between the island and the substrate influence elastic energy reduction upon island formation?

The answer to the question becomes evident if the elastic state depicted in part (a) of Figure 8.34 is reduced, by means of linear superposition, to the sum of states shown in parts (b) and (c) of the same figure. In part (b), the substrate supports no mismatch strain, while the intervening uniform layer and the island are both subject to the spatially uniform mismatch strain $\epsilon_{ij}^{\mathrm{m}}$. In this case, however, the lateral faces of the island are subjected to precisely the right traction to prevent elastic relaxation in the island, that is, it is subjected to the traction $\sigma_{ij}^{\mathrm{m}} n_j$ where n_i is the outward unit normal vector to the island surface and $\sigma_{ij}^{\mathrm{m}} = c_{ijkl}\epsilon_{kl}^{\mathrm{m}}$. Thus, the uniform elastic strain in both the island and the intervening layer is still $\epsilon_{ij}^{\mathrm{m}}$, and the elastic strain in the substrate is everywhere zero. The elastic properties of the substrate may be different from those of the strained layer and the island which may, in turn, be

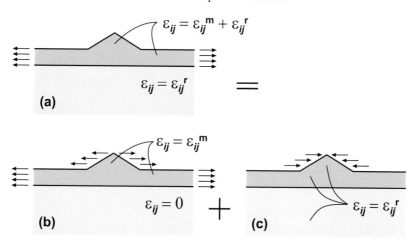

Fig. 8.34. Linear superposition of elastic deformation fields associated with island formation in a strained layer system. The configuration with a fully formed, partially relaxed
island in part (a) is decomposed into island formation without relaxation in part (b), which
involves introduction of artificial surface tractions on the island surface, plus negation of
the artificial tractions in part (c) to result in an island with free lateral surfaces.

different from each other. Furthermore, the mismatch strain in the island and in the
intervening strained layer may also be different from each other, as will be clear
from the arguments to follow.

The elastic state in part (c) of Figure 8.34 must be added to that in part (b)
to recover the state of interest in part (a). The elastic properties of the materials
and the geometrical configuration are identical in parts (a), (b) and (c). In part (c),
however, none of the material components carries mismatch strain; all the mismatch
strain is already taken into account in part (b). To render the lateral surface of the
island traction-free, it is necessary to impose a surface traction $-\sigma_{ij}^m n_j$ on the island
surface in part (c). The elastic strain in all material components in part (c) is solely
the strain that arises from relaxation of the lateral faces, say ϵ_{ij}^r. It is a compatible,
spatially nonuniform strain field.

The superposition argument is based on recognition that the total elastic strain
is the sum of the mismatch strain and a relaxation strain $\epsilon_{ij} = \epsilon_{ij}^m + \epsilon_{ij}^r$, and it
separates the influence of these two strain contributions to energy change. There
is no elastic energy change associated with island formation in part (b). Thus, the
elastic energy *change* associated with island formation is represented by the fields
of part (c) alone, but this state does not involve the elastic mismatch strain in the
intervening layer in any way. It follows that the energy change is *independent* of this
mismatch strain. Furthermore, the conclusion does not depend on the thickness of
the intervening layer, on the shape of the island or on the relative elastic properties

of the materials. The same conclusion can be reached by direct appeal to elastic field theory; the details are left as an exercise.

8.9.3 Influence of surface energy anisotropy

In the description of formation of a strained epitaxial island in Section 8.9.1, it was assumed that the surface energy density of the film material was independent of strain and of surface orientation. It was recognized in Section 8.8.3 that, if the lateral faces of the island are vicinal surfaces of the crystalline material, then the surface energy density may depend on both orientation and strain. In this section, the question of free energy change in the system during island formation is re-visited for the case when the surface energy density varies linearly with surface slope for slopes of small amplitude, as indicated by the general expression (8.138) with $a = \tan|\theta|$. How does this surface energy anisotropy influence island formation? The corresponding issue for large values of slope is pursued in subsequent sections.

The principal change from the development of Section 8.9.1 is that the factor γ in the first term on the right side of (8.149) must be replaced by $(\gamma_0 + \beta_1 a)/\sqrt{1 + a^2}$, neglecting the step interaction contribution for small values of a. The change in surface energy of the system can again be written as $\Gamma = \gamma_0 V^{2/3} \hat{\Gamma}(a)$, but in this case

$$\hat{\Gamma}(a) = \pi \left(\frac{3}{\pi a}\right)^{2/3} \left[\left(1 + \frac{\beta_1}{\gamma_0} a\right) - 1\right] = \pi \left(\frac{3}{\pi}\right)^{2/3} \frac{\beta_1}{\gamma_0} a^{1/3}. \tag{8.154}$$

This expression differs from the corresponding result for the case of isotropic surface energy density in one particularly important aspect.

The expression (8.154) implies that, as aspect ratio a increases from zero, the increase in surface energy always dominates the corresponding decrease in elastic energy, which is essentially linear in a for any material volume V. In other words, the appearance of the factor $a^{1/3}$ in (8.154), instead of $a^{4/3}$ as in (8.150), implies that the flat surface of the strained film is stable against formation of island-like perturbations of its surface for any material volume.

Although the foregoing result was obtained in the course of studying a particular configuration, that result is quite general in scope. Surface energy anisotropy of the kind represented by the plot labeled 'no strain' in Figure 8.26 has a strong stabilizing influence on the flat surface of a strained crystal in a high symmetry orientation. This behavior is consistent with observations for situations in which the strain magnitude is very small. However, formation of epitaxial islands on high symmetry surfaces is common in strained layer systems, and the apparent reason for this behavior is that strain itself can have pronounced influence on surface energy density for

surface orientations that are close to high asymmetry directions, as described in
Section 8.8.3.

If the influence of strain is also taken into account in describing the surface
energy of the lateral faces of an isolated island for small aspect ratio a, then (8.154)
is replaced by

$$\hat{\Gamma}(a) = \pi \left(\frac{3}{\pi}\right)^{2/3} \frac{\beta_1 + \hat{\beta}_1 \epsilon_m}{\gamma_0} a^{1/3} \qquad (8.155)$$

for small values of a, where ϵ_m is the equi-biaxial mismatch strain of the film
material with respect to the substrate material. For values of the parameters β_1 and
$\hat{\beta}_1$ appropriate for the Si/Ge material systems, and for compressive mismatch strain
ϵ_m greater in magnitude than about one-half of one percent, the coefficient of $a^{1/3}$
in (8.155) becomes *negative*. This implies that the (001) strained surface is unstable
against island formation of any volume.

The dependence of surface energy density of a vicinal surface on strain given in
(8.140), as well as its implications for island formation reflected in (8.155), point
to an asymmetry of behavior with respect to the sign of strain that drives island
formation. If a nominally flat surface becomes unstable under small perturbations
in its shape for a particular strain, either tensile or compressive, it will not exhibit an
instability for an imposed strain of the same magnitude of opposite sign. This is in
contrast to the classical stability result described in Section 8.4.1 where mismatch
strain ϵ_m enters the description only through the quantity ϵ_m^2; as such, the stability
condition for tensile strain is indistinguishable from the corresponding condition
for compressive strain.

The asymmetry in behavior with respect to the sign of strain was anticipated
in the experimental results reported by Xie et al. (1994). They deposited 5 nm
films of $Si_{0.5}Ge_{0.5}$ on relaxed substrates of $Si_{1-x}Ge_x$ at 650 °C, with x varying
within the range $0 < x < 1$. This implies that the film growth surface was always
the same material. As a result, the mismatch strain of the film varied within the
range $-0.02 < \epsilon_m < 0.02$. The root-mean-square surface roughness of the films
after deposition was determined by means of atomic force microscopy. For tensile
mismatch strains, the roughness was found to be no greater than the roughness of
the surface of the substrate prior to growth. For compressive mismatch strain, on the
other hand, the measured surface roughness was significantly larger in general, and
as much as 20 times larger for compressive mismatch strains greater in magnitude
than about one percent. The effect was attributed to the energies of reconstructed
dimer pairs on the terraces near the step edges.

Experimental observations reported by Sutter and Lagally (2000) and Tromp et al.
(2000) on the growth of $Si_{1-x}Ge_x$ films on Si(001) substrates, with Ge concentration
x in the range $0.1 \leq x \leq 0.4$, have clearly revealed that shallow mounds emerge

during early stages of growth. These mounds are the nuclei for an intrinsic morphological instability of the strained film material. Because these mounds assumed crystallographic features from the outset, it was assumed that their lateral faces were vicinal surfaces with widely spaced steps. As more material was deposited, the spacing between the steps gradually decreased until the lateral surfaces reached a certain crystallographic orientation. This island formation behavior arises as a natural surface instability, as implied by (8.155). Approaches to surface instability that account for neither orientation dependence nor strain dependence of surface energy have been unable to explain the observed growth mode. It is evident that any assumed surface behavior of the form (8.155) will lead to a similar description of island growth; the development of this form within the context of stepped surfaces provides a fundamentally consistent picture. A similar growth mode was also observed in Ge films grown on Si(001) substrates by Vailionis et al. (2000) for which widely spaced steps were the three-dimensional features that first appear during growth; these features were precursors to the faceted islands that formed eventually. The emergence of islands with crystallographically stepped lateral surfaces, arising as a natural surface instability, can be explained by incorporating the physics of steps, in particular their interactions and the dependence of their formation energies on the mismatch strain, as was done in Section 8.8.3.

8.9.4 Nucleation barrier for islands on stable surfaces

From the foregoing discussion, it is evident that there is invariably an elastic energetic driving force for formation of an island on the surface of a strained layer. Elastic strain energy is always reduced in the course of island formation from a strained film with an initially flat surface. In arriving at the description of free energy variations with island shape illustrated in Figure 8.32, it was assumed that the surface energy density of the island faces was independent of surface orientation. If the surface energy density of the material does depend on surface orientation, then the process of island formation can be altered significantly. If the surface energy density is smaller for surface orientations that are inclined slightly to the growth surface than for the growth surface itself, as suggested for some vicinal surfaces in connection with (8.155), then the dependence of free energy per unit volume of island material on island aspect ratio illustrated in Figure 8.32 departs from $a = 0$ with an even steeper negative slope. In such a case, the flat growth surface is unstable and islands can form spontaneously from the outset. On the other hand, if the growth surface orientation is stable in this sense, the question arises as to whether or not islands can still form. The goal in this section is to illustrate that island formation is still possible, but that a nucleation barrier must be overcome in order to achieve formation.

To pursue this issue, consider the dependence of the free energy change associated with island formation on both island volume V and aspect ratio a, retaining the notation and configuration introduced in Section 8.9.3. In the present instance, however, the surface energy density depends on a. The surface energy density for $a = 0$ is denoted by γ_0 for both the substrate or growth surface and for the flat surface of the film material. For surfaces inclined to the growth surface, the surface energy density depends on the angle of inclination in some way; examples are given in Section 8.9.3. For example, for the surface energy density of a vicinal surface as given in (8.138), the nondimensional surface energy factor $\hat{\Gamma}(a)$ introduced in Section 8.9.1 takes the form (8.154), which implies that the flat growth surface is *stabilized* against small perturbations from $a = 0$. For any such surface energy factor, the normalized free energy of the island–substrate system is

$$\hat{\mathcal{F}}(v, a) = \frac{\mathcal{F}}{M\epsilon_m^2 \zeta_0^3} = v^3 \left[\hat{W}(a, v) - 1 + v^{-1}\hat{\Gamma}(a) \right] \qquad (8.156)$$

in nondimensional form, where

$$\zeta_0 = \gamma_0 / M\epsilon_m^2 \quad \text{and} \quad v = V^{1/3}/\zeta^0. \qquad (8.157)$$

The function $\hat{\mathcal{F}}(v, a)$ defines a surface over the plane of $v > 0$ versus $a > 0$ that represents the energy 'landscape' in terms of configuration.

For any growth surface orientation that is stable under small fluctuations in shape, $\hat{\mathcal{F}} = 0$ at $v = a = 0$ and $\hat{\mathcal{F}} > 0$ for all configurations in a small neighborhood of that point. It is known that the elastic energy effects can overwhelm surface energy influences for sufficiently large islands. In many cases, the energy surface defined by (8.156) has the form of a 'saddle' with large values of $\mathcal{F}$ for $v \to \infty$, $a \to 0^+$ and for $v \to 0^+$, $a \to \infty$, but with more moderate values between these extremes. In such cases, the stationary point at the root of the 'saddle' defines an energy barrier that must be overcome to create an island. The energy barrier height represents an activation energy for island formation on material surfaces that are stable under small fluctuations due to surface energy anisotropy. A related issue was described in detail in Section 1.3.5.

The values $v = v_*$, $a = a_*$ of volume and aspect ratio that locate the stationary point are found among the roots of the two equations

$$\frac{\partial \mathcal{F}}{\partial v} = 0, \quad \frac{\partial \mathcal{F}}{\partial a} = 0, \qquad (8.158)$$

which are nonlinear, in general. The corresponding energy level $\hat{\mathcal{F}}_* = \hat{\mathcal{F}}(v_*, a_*)$ at that point is found by substituting the pertinent roots into (8.156). It is interesting to note that the equations in (8.158) can be uncoupled to obtain a single equation $-3\hat{\Gamma}'(a)\left[\hat{W}(a, v) - 1\right] + 2\hat{\Gamma}(a)\hat{W}'(a, v) = 0$ that can first be solved for a_*; the prime denotes differentiation with respect to aspect ratio. With that result in hand, v_* can be determined from $v = -\hat{\Gamma}'(a)/\hat{W}'(a, v)$. For the surface energy factor $\hat{\Gamma}(a)$ given in (8.154) with $\beta_1/\gamma_0 = 0.1$, for example, it is found that $v_* = 0.30$, $a_* = 0.73$ and $\mathcal{F}_* = 0.0082$. The details of the calculation are left as an exercise.

A similar approach to that discussed in this section was followed by Tersoff and LeGoues (1994) in the study of island nucleation. In that work, it was assumed that only discrete values of aspect ratio a were admissible. The nature of behavior under this constraint can be seen by adding a vertical line to Figure 8.32 marking the smallest nonzero admissible value of a, a pre-assigned value. It is immediately obvious that islands of small volume would have to increase their energies to form, whereas islands of relatively large volume could form with a net decrease in free energy. The discriminating value of island volume between these two types of behavior implies the existence of a critical island volume and a corresponding energy barrier to island formation in the sense discussed in this section.

8.9.5 Shape transition for preferred side wall orientations

Suppose that only certain discrete values of slope of the lateral surface of the island are possible. In this subsection, implications of this assumption are examined for circumstances in which more than one such preferred lateral face orientation exists. The surface energy density of the substrate surface or flat growth surface is γ_0. Once it is nucleated at a certain aspect ratio, say a_1, the volume V of an island continues to increase, due perhaps to deposition or coarsening, but with fixed aspect ratio a_1. The surface energy density of the orientation is γ_1. Within the context of the model of cone-shaped islands, the free energy per unit volume of material in the island is

$$\frac{\mathcal{F}(a_1, Z, \gamma_1/\gamma_0)}{M\epsilon_m^2 V} = \hat{W}(a_1, v) - 1 + Z\hat{\Gamma}(a_1, \gamma_1/\gamma_0), \qquad (8.159)$$

where Z and $\hat{\Gamma}$ are defined by

$$Z = \frac{\gamma_0}{M\epsilon_m^2 V^{1/3}}, \quad \hat{\Gamma}(a_1, \gamma_1/\gamma_0) = \left(\frac{9}{\pi a_1^2}\right)^{1/3} \left[\frac{\gamma_1}{\gamma_0}\sqrt{1 + a_1^2} - 1\right]. \qquad (8.160)$$

As the volume of the island grows larger, the free energy per unit volume diminishes. If there is another admissible surface orientation, say $a = a_2 > a_1$ with surface energy density γ_2, then the free energy per unit volume for that aspect ratio is also given by $\mathcal{F}(a_2, Z, \gamma_2/\gamma_0)/M\epsilon_m^2 V$ as defined in (8.159). Eventually, V becomes large enough so that circumstances will favor an island with a larger aspect ratio. The transition first becomes possible when $\mathcal{F}(a_1, Z, \gamma_1/\gamma_0) = \mathcal{F}(a_2, Z, \gamma_2/\gamma_0)$ for some Z. This transition is illustrated in Figure 8.35 for the case when $a_1 = 0.2$, $a_2 = 0.4$, $\gamma_1/\gamma_0 = 1$ and $\gamma_2/\gamma_0 = 1, 1.1$ or 1.2. The figure shows graphs of free energy reduction as a result of island formation versus island size, represented by

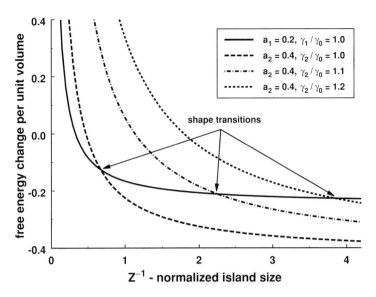

Fig. 8.35. Representative plots of free energy reduction per unit island volume versus island size for several combinations of relative surface energy density and island aspect ratio. The crossover points of the graphs identify conditions under which an island shape with larger aspect ratio becomes more favorable than the smaller aspect configuration.

the normalized length Z^{-1}. For example, for the case when $\gamma_2/\gamma_0 = 1.1$, the shaped transition from $a = a_1$ to $a = a_2$ first becomes possible when Z^{-1} has increased to the value of approximately 2.2. It is evident from Figure 8.35 that the transition size is strongly dependent on the relative surface energy densities of the lateral face orientations involved in the shaped transition.

The foregoing discussion is based on the assumption that the only possible island shapes are truncated solid cones with any of a discrete set of aspect ratios. The transitions from one to another aspect ratio implied by this simple picture are abrupt. A more realistic situation involving gradual transition might emerge if it is assumed that the islands take on shapes with surface orientation constrained by the slopes a_1 and a_2, but that these orientations can coexist. For example, an intermediate shape might consist of a frustum of cone with lateral surface slope a_2 plus a cone with lateral surface slope $a_1 < a_2$. A sequence of such shapes is illustrated in Figure 8.36 with the transition accomplished by increasing the portion of surface area at slope a_2 and decreasing the portion at slope a_1. Presumably, the proportion of surface of each orientation would be determined by means of a requirement that any configuration must be a minimum energy configuration within the family of admissible shapes. Behavior maps for describing such shape changes, analogous to phase diagrams, have been introduced by Daruka et al. (1999). Observations of shape transitions in the Si/Ge material systems are described in the next section.

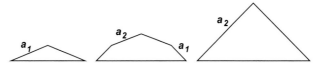

Fig. 8.36. Island shape transitions are represented in Figure 8.35 under the assumption that only abrupt transitions from one conical shape with relatively small aspect ratio a_1 to another with relatively large aspect ratio a_2 are possible. This diagram suggests a more gradual transition from one conical shape to another, effected by having the steeper orientation a_2 gradually expand on the lateral face of the island until the transition is completed.

Fig. 8.37. Surface profiles of $Si_{0.8}Ge_{0.2}$ grown epitaxially on the (001) surface of a Si substrate at $755\,°C$ obtained by means of atomic force microscopy. The plan view area of each image is $2\,\mu m \times 2\,\mu m$, and the vertical scales are identical. The mean deposition depths are $25\,\text{Å}$ in the upper image, where widely spaced incipient islands are evident, and $100\,\text{Å}$ in the lower image, where an array of islands with $\{105\}$ lateral faces has been formed. Reproduced with permission from Floro et al. (1997, 1998).

8.9.6 Observations of island formation

Images of islands that formed spontaneously during deposition of the alloy $Si_{0.8}Ge_{0.2}$ on Si(001) at $755\,°C$, obtained by means of atomic force microscopy after the deposition was interrupted, are shown in Figure 8.37. The upper image shows very small

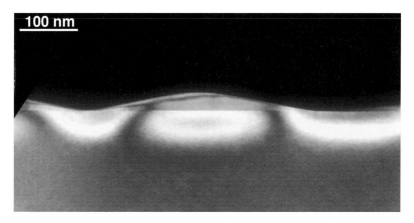

Fig. 8.38. Cross-sectional image of one of the islands in the lower portion of Figure 8.37, obtained by means of transmission electron microscopy. The lateral faces of the island are {105} surfaces, typical for this material system. Reproduced with permission from Floro et al. (1997).

islands forming on the growth surface during the early stage of deposition. In the lower image, islands have grown nearly to the point of mutual impingement. The edges of the square island bases are aligned with $\langle 100 \rangle$ directions on the growth surface. Note that the lateral faces of the islands all have approximately the same slope. An enlarged view of a cross-section of one of the islands in the lower portion of Figure 8.37, obtained by transmission electron microscopy, is shown in Figure 8.38. This image reveals that the lateral faces of the island appear to be {105} surfaces.

The evolution of island shapes and the transition from a pyramidal shape to a dome shape has been studied in detail for the SiGe/Si(001) material system by Ross et al. (1999). The observations were carried out in an apparatus that integrated low energy electron microscopy (LEEM) with *in situ* film growth capabilities. Geometrical features of evolving island configurations are distinguishable in LEEM images due to their relatively strong contrast. The shape transition for $Si_{0.75}Ge_{0.25}$ deposited at $730\,^{\circ}C$ is summarized in Figure 8.39. The flat growth surface in this case is unstable as described in Section 8.9.3, and the deposit agglomerates into islands with vicinal lateral faces approaching the {105} orientation, the so-called pyramidal (P) shape. As the island becomes larger in volume, it transforms to a truncated pyramid (TP) shape as {113} faces appear at the corners of the square base and grow in area; a schematic of the TP shape is illustrated on the right in Figure 8.39. Once the {113} faces become fully established, additional faces emerge which appear to flatten edges at face boundaries to a certain extent. The configuration proceeds through successive dome-like shapes (D1, D2, D3) until the final shape D4 is reached, consisting of {113} and {15 3 23} lateral surfaces plus remnants of the {105} faces at the top. The transitions all appear to be temperature dependent.

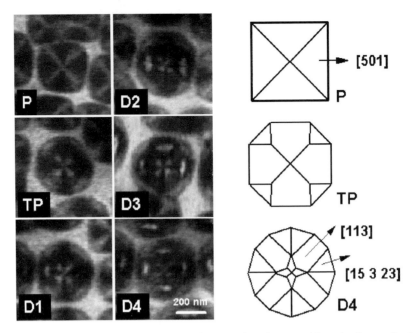

Fig. 8.39. Low energy electron microscopy images showing transitions in shape of epitaxial SiGe islands deposited on a Si(001) substrate. Reproduced with permission from Ross et al. (1999).

Although the plots in Figure 8.32 are based on the assumption of a single island which is isolated by distance from any other islands on the same surface of the substrate, they imply that there is a driving force for a *coarsening* of any island distribution on the surface. Any two islands with individual volumes V_1 and V_2 can lower the free energy per unit volume of the material involved by combining into a single larger island with volume $V_1 + V_2$. This tendency prevails at any size scale and hence there is a driving force for coarsening which persists at all size scales. This simple picture overlooks any elastic interaction of islands in close proximity to each other, as well as the transient mechanisms which must come into play to actually transfer material from one island to another or to cause islands to coalesce. Processes by which material microstructures coarsen or 'ripen' in response to such an energetic driving force are often identified collectively as Ostwald ripening processes (Ostwald 1887, Zinke-Allmang et al. 1992).

The influence of elastic interactions can be anticipated, to a certain extent. As noted in the discussion above, the elastic energy change due to formation of an isolated island of some shape and size depends on the elastic compliance of the substrate. If this same island is formed in close proximity to other islands bonded to the surface of the substrate, then the substrate invariably appears to be less compliant than it would if the island were isolated from all others. This implies that there will

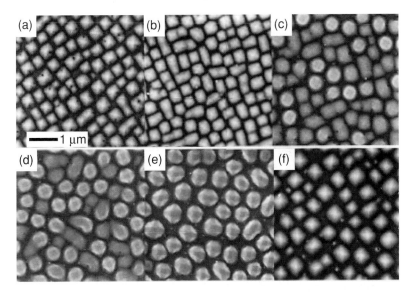

Fig. 8.40. Scanning electron micrographs, shown at the same magnification, of islands formed during the MBE growth of $Si_{0.8}Ge_{0.2}$ onto a $Si(100)$ substrate during six different depositions. The images were taken after film deposition to a mass equivalent thickness of (a) 10.4, (b) 13, (c) 17, (d) 23 and (e) 27.5 nm. The image shown in Figure (f) is a scanning electron micrograph of a film grown to a thickness of 10.4 nm, following which it was annealed at 755 °C for 45 min. Reproduced with permission from Floro et al. (1998).

be less relaxation of elastic energy due to formation of closely spaced islands than is the case for formation of widely spaced islands. This, in turn, implies the existence of a repulsive configurational force tending to increase separation distance between closely spaced islands, assuming both have mismatch strain of the same sign. This force opposes the tendency for structural coarsening of island arrays, and can influence the stress driven evolution of islands (Floro et al. 2000).

The role of elastic interactions in influencing island shape transitions in $Si_{0.8}Ge_{0.2}$ films was studied by Floro et al. (1998) using two different methods. They monitored film stress by recourse to *in situ* substrate curvature measurements that involved the multi-beam optical stress sensor, which was described in Section 2.3.2, during the deposition of the films to different thicknesses by molecular beam epitaxy onto $Si(100)$ surfaces at a temperature of 755 ± 10 °C and growth rate of 0.1 Å s^{-1}. An optical scattering technique, relying on the peak in the optical scattering spectrum produced by the islands which act as a diffraction grating during the broadband illumination of the substrate, was used to estimate the mean spacing of the island arrays through spectroscopic detection of the backscattered light.

Figures 8.40(a)–(f) are scanning electron microscope images showing the evolution of island morphology after six different depositions. The first five images

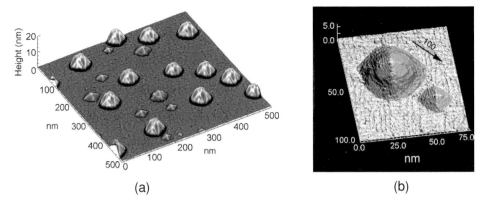

Fig. 8.41. (a) Scanning tunneling microscope image showing the topography of Ge islands formed on the surface of Si(100) during physical vapor deposition in ultra-high vacuum at a substrate temperature of 600 °C. This image was obtained after deposition of eight equivalent monolayers of Ge where one equivalent monolayer corresponds to 6.3×10^{18} atoms m^{-2}. (b) A higher magnification image showing the coexistence of a hut or pyramid and a dome. Reproduced with permission from Medeiros-Ribeiro et al. (1998).

in this figure correspond to $Si_{0.8}Ge_{0.2}$ mass equivalent film thicknesses of 10.4, 13, 17, 23 and 27.5 nm, respectively. Figure (f) is an image of a film grown to a thickness of 10.4 nm following which it was annealed at 755 °C for 45 minutes to facilitate static coarsening of the islands. It was found that such static coarsening reduced the areal coverage of the island array by nearly one half while the mean island volume was approximately doubled compared to continuous growth from 10.4 to 13 nm, that is, for continuous growth that occurred between Figures (a) and (b). These observations suggest that when the growth surface is heavily covered by SiGe islands, the equilibrium transition volume as well as the activation barrier for the transition from the pyramid-shaped islands bound by {501} surfaces, that is, the so-called hut morphology, to the more isotropic island shape bound by {311} facets, the so-called dome morphology, can be strongly influenced by the elastic interactions between the islands.

Studies of Ge island formation on Si(100) substrates during physical vapor deposition at approximately 600 °C in ultra-high vacuum have been performed by Medeiros-Ribeiro et al. (1998), who observed the nascent islands using an *in situ* scanning tunneling microscope (STM) with a lateral resolution of 1.25 nm. Figure 8.41(a) is an STM image showing the island height as a function of position where the smaller nanocrystals were identified to be square-based pyramidal huts and the larger ones multifaceted domes. The size distribution of islands observed in these experiments was similar to that seen by Kamins et al. (1997) and Williams et al. (2000) during the chemical vapor deposition of Ge onto Si(100) at approximately the same substrate temperature and growth rate, although the background pressures in these two deposition conditions differed by as much as 11 orders of magnitude. In both cases, a bimodal distribution of island distribution was found with the smaller, pyramid-shaped island huts, up to 6 nm in height, forming at lower volumes of deposited material, typically less than 5,000 nm^3; for deposition volume in excess of 10,000 nm^3, larger nanocrystals with multifaceted domes, which were up to 15 nm in height, were observed. An example of coexistent shapes of Ge islands on a Si(001) substrate is illustrated in Figure 8.41(b). The smaller island is a hut or pyramid shape with {105} lateral faces and a square base aligned with ⟨100⟩ directions on the growth surface. The larger island is a so-called dome shape with lateral surface consisting of a number of different orientations, as suggested by the intermediate stage in the diagram of Figure 8.36.

The description of the energetics of island formation as developed in this chapter does not lead to a prediction of a preferred island size for a given material system. In order to identify a preferred size for pyramidal epitaxial islands, Shchukin et al. (1995) included a contribution in the elastic energy change associated with island formation that accounts for the influence of the unbalanced surface stress at edges where flat lateral faces are joined on the distribution of elastic strain. This energy scales with the linear dimension of an island, whereas the surface energy scales with area and the bulk elastic energy scales with the volume. It has been suggested that the additional natural length scale can account for the existence of a preferred island volume and for the narrow distributions of island size that have been observed for some material systems. Alternatively, it could also be argued that the growth of an array of islands is restricted by the fact that the surface energy density γ_f of the lateral faces is less than the surface energy density γ_s of the growth surface. In the process of coarsening, islands increase the amount of the area with surface energy γ_s and decrease the amount with surface energy γ_f, which represents a net increase in surface energy of the system. Yet another limitation on island size is that mass transfer from one island to another is very restricted, unless the islands are very close to each other.

The foregoing experimental observations reveal that the actual formation of islands in a particular material system is a complicated transient process, with islands evolving through mass transport in the presence of a deposition flux. Consequently, temperature, deposition rate and other factors not included in the present discussion can have a strong influence. For example, Copel et al. (1989) have shown that use of As as a surfactant suppresses the formation of Ge islands on Si(001) whereas the work of Kamins et al. (2001) indicates that the addition of P as a dopant drastically alters the shape of the islands formed in the same system. Experiments have also shown that the crystal structure of the deposited material can have a strong effect on the shape and orientation of the nanocrystalline islands formed on the Si substrate. The addition of rare earth metals with hexagonal crystal structure, such as

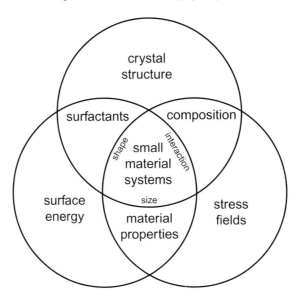

Fig. 8.42. Schematic illustration of the interactions among crystal structure, surface energy and stress fields and their collective influence on the evolution of surfaces in small-volume structures. Adapted from a diagram by R. S. Williams, Hewlett-Packard Corp.

Er, results in the formation of stable silicides at elevated temperatures and these silicide nanocrystals grow along certain preferential crystallographic directions on the substrate surface. Figure 8.42 schematically illustrates the interplay among crystal structure, surface energy and stress fields which can influence the evolution of nanocrystalline islands on surfaces; the multitude of factors shown here also points to the flexibility with which the size, shape and spatial distribution of such islands can be controlled by recourse to materials chemistry.

The analyses presented in this chapter lead to the firm conclusion that there is a significant driving force for formation of epitaxial islands in strained material systems. The manner in which the surface energy of the material influences the response to this driving force determines the nature of island growth. Determining how this surface energy is modified by a variety of geometrical, chemical and material parameters remains a work in progress. The rich variety of physical phenomena that influence the thermodynamics and kinetics of surface evolution also provides potential opportunity for the controlled synthesis of patterned surfaces for many functional applications.

8.10 Exercises

1. In the study of surface stability in Section 8.4.1, it was shown that the flat surface of a strained elastic solid is unstable under small amplitude sinusoidal perturbations in shape for isotropic surface energy density U_S. Repeat the calculation for an anisotropic surface energy that depends on the the magnitude of the local surface

Fig. 8.43. A ridge formed on the surface of a strained elastic material under plane strain conditions is shown on the left. The diagram on the right shows a pair of identical ridges with their centers separated by a distance d.

slope according to $U_S = \gamma + \beta|h_{,x}|$, where γ and β are material constants with physical dimensions of energy per unit area. Show that the surface is stable under small amplitude sinusoidal perturbations of shape at all wavelengths if $\beta > 0$; the parameter γ is always positive.

2. The expression (8.85) provides a means for calculating the reduction in elastic energy of a biaxially strained material with an initially flat surface when some material is formed into a ridge as depicted in Figure 8.43, provided that $h/a \ll 1$.

 (a) For an isolated ridge, determine the fractional reduction in elastic energy, measured per unit length of the ridge, as a function of h/a.
 (b) Consider the case of parallel ridges on the surface, as depicted on the right in Figure 8.43. Determine the fractional reduction in elastic energy for $h/a = 0.15$ as a function of separation distance d in the range $0 \le d \le 3a$.

3. The time rate of change of free energy of an elastic solid with an evolving surface given in (8.7) was obtained on the basis of heuristic reasoning. Apply the basic definition of a derivative with respect to time to the free energy expression (8.1) to formally confirm the result.

4. The free energy change due to formation of an isolated epitaxial island on a strained surface is given approximately by (8.156) for a particular surface energy anisotropy as represented by (8.154). This particular anisotropy stabilizes the flat surface configuration and gives rise to an energetic barrier to island formation.

 (a) Determine the values of island volume v and aspect ratio a that locate a saddle point in the free energy $\mathcal{F}(v, a)$ viewed as a surface over the plane represented by $v > 0, a > 0$.
 (b) Determine an expression for the value of free energy at the saddle point in terms of system parameters. This value of free energy is the activation energy for island formation.

5. As was noted in Section 8.3 , the ordinate of the plots in Figure 8.9 cannot be interpreted as a configurational force because of the use of the ratio ζ/a as a plotting parameter. To get a picture of the configurational force tending to change the aspect ratio b/a of the ellipse at fixed area πab, express the integral in (8.42) representing $\dot{\mathcal{F}}$ in the alternative form

$$\dot{\mathcal{F}} = -\frac{\sigma_\infty^2}{\bar{E}}\dot{a}a\tilde{R}(b/a, \zeta/\sqrt{ab}) \qquad (8.161)$$

where $\tilde{R}$ versus b/a for fixed $\zeta/\sqrt{ab}$ does represent the configurational force that is work-conjugate to aspect ratio. Evaluate the integral defining $\tilde{R}$ numerically and plot the result versus b/a for a range of values of $\zeta/\sqrt{ab}$. Does this point of view alter the stability condition for a circular hole established in Section 8.3?

6. Consider the advance of a material interface as depicted in Figure 8.5.
   (a) Carry out a stress analysis for the spherically symmetric configuration in order to show that the chemical potential defined in (8.22) reduces to (8.24) for this configuration under the conditions described in Section 8.2.3 .
   (b) Determine the value of R_{eq} when $\sigma_\infty = E\epsilon_m$, $\epsilon_m = 0.01$, $E = 10^2$ GPa, $\nu = 0.25$ and $U_S = 2$ J/m^2 .

7. Continuous measurements of substrate curvature were made during the deposition of a thin film onto a relatively thick (100) Si substrate; the substrate thickness is 600 μm. It was determined that film growth follows the Volmer–Weber mechanism, and that the process of stress evolution could be approximated by the island impingement model presented in Section 8.6.2. The surface energy per unit area of the thin film is approximately 1 J/m^2. The magnitude of substrate curvature at the instant when the islands impinge on the surface of the film was measured to be 5.8×10^{-5} m^{-1}.
   (a) Estimate the average radius of the islands when they impinge.
   (b) What is the sign of the curvature of the substrate?
   (c) Calculate the volume-averaged stress in the film.

8. The values of the parameters in (8.139) that characterize the dependence of energy U_S of a strained surface on step density have been estimated to be $\beta_1 = 0.162$ J/m^2, $\beta_1 = 12.5$ J/m^2 and $\beta_3 = 2.29$ J/m^2 for Ge on the basis of the Tersoff interatomic potentials (Shenoy et al. 2002). Determine the angle θ^* defined in (8.142) for epitaxial growth of Ge on Si(001).

9. Consider the chemical potential expression (8.137) for the special case when the surface energy density U_S depends on orientation θ but not on surface strain ϵ. Verify that the chemical potential expression reduces to $\chi = U - \kappa(U_S + U_S'')$, where the prime denotes differentiation with respect to θ. The quantity in parentheses is sometimes called *surface stiffness*; its existence presumes the function $U_S(\theta)$ is indeed twice differentiable.

10. The change in elastic energy due to island formation is the elastic energy of the state in part (a) of Figure 8.34 minus that of the state in part (b), or

$$\Delta W = \int_R \tfrac{1}{2} c_{ijkl} \epsilon_{ij} \epsilon_{kl} \, dR - \int_R \tfrac{1}{2} c_{ijkl} \epsilon_{ij}^m \epsilon_{kl}^m \, dR , \qquad (8.162)$$

where R is the volume of space occupied by the material system and c_{ijkl} is the array of elastic constants which may have different values within the distinct material components. Use the idea of superposition as depicted in Figure 8.34 to show that the expression for ΔW can be expressed in terms of only the relaxation field in part (c), thereby corroborating the conclusions of Section 8.9.2.

9

The role of stress in mass transport

In Chapter 8, connections between the energetic state of a small material volume subjected to stress and the configuration of the bounding surface of that volume were considered. The dependence of the free energy of the material system on its configuration was established, and the nature of the variation of system free energy in the course of alterations in configuration was examined. Attention was limited to processes that are slow enough so that the system is always in mechanical equilibrium. This point of view leads naturally to the notion of a thermodynamic equilibrium state of a small material volume as a state in which *any* admissible variation in configuration results in an increase in the free energy of the system. If this is true only for small excursions in configuration then the equilibrium configuration is metastable or locally stable; if it is true for admissible fluctuations of any magnitude then the equilibrium configuration is globally stable.

The fundamental criteria of thermodynamic equilibrium and stability are usually stated in somewhat different terms, but the equivalence to the fundamental criteria is readily established for the physical systems of interest here. The first law of thermodynamics states that the rate of change of internal energy U of a material system equals the rate at which work W is being done on it plus the rate at which heat Q is being added to it, or

$$\dot{U} = \dot{W} + \dot{Q}, \tag{9.1}$$

where the dot denotes time rate of change.

In general, work and heat can be added through boundaries or through volumetric influences – body forces, radiation absorption and so on. In addition, work can be dissipated internally as heat through deformation for some materials. The entropy $\mathcal{H}$ of the system is the portion of the internal energy that cannot be recovered as work and that exists as heat within the system according to

$$\dot{\mathcal{H}} = \frac{\dot{Q}}{T},\qquad(9.2)$$

where T is the absolute temperature.

The first law of thermodynamics (9.1) states that the sum of energies added to the material system in the various forms in which it can exist equals the increase in internal energy. The second law of thermodynamics has the form of a constraint on the nature of changes that can occur in a material system that is *not* in equilibrium. While it can be stated in various ways, all expressions are equivalent to the statement that any changes in the state of a closed system lead to an increase in entropy, or possibly leave entropy unchanged. In other words, if a nonequilibrium system exchanges neither work nor heat with its surroundings in the course of changing its configuration, so that $\dot{\mathcal{U}} = 0$, then it can do so only with

$$\dot{\mathcal{H}} \geq 0.\qquad(9.3)$$

An immediate consequence is that the closed system will be in stable equilibrium if and only if the entropy of that system is maximum under all admissible variations of its configuration. The stability can be local or global, depending on whether the variations of state used to probe stability are infinitesimal or arbitrary in magnitude. This is the form of the original statement on equilibrium and stability of material systems (Gibbs 1876).

The discussion of energy variations in preceding chapters has been expressed entirely in terms of free energy, or Helmholtz free energy at constant temperature. In approximate terms, free energy is the portion of internal energy that can be recovered from a material system to perform work, or the excess of internal energy over entropic energy, that is,

$$\mathcal{F} = \mathcal{U} - T\mathcal{H}.\qquad(9.4)$$

To demonstrate the validity of the statement of equilibrium and stability in terms of free energy adopted in Chapter 8 for the system studied, consider a closed material system in which the total volume of material is very large compared to the volume of material undergoing a change in configuration; the large volume serves as a heat reservoir to maintain the overall temperature constant through conduction, even though entropy is produced over a small portion of the material. For a closed system ($\dot{\mathcal{U}} = 0$) at constant temperature ($\dot{T} = 0$), the rate of change of free energy is

$$\dot{\mathcal{F}} = -T\dot{\mathcal{H}}\qquad(9.5)$$

where absolute temperature T is non-negative.

If the criterion for a stable equilibrium configuration is that entropy assumes a maximum value, it follows immediately from (9.5) that an equivalent statement for the class of systems described is that a stable equilibrium configuration is one in which free energy assumes a minimum value. Furthermore, it follows from the second law (9.3) that, if a nonequilibrium system of this class alters its configuration spontaneously, it must do so with

$$\dot{\mathcal{F}} \leq 0 . \tag{9.6}$$

In light of these observations on equivalence of various statements of stability and equilibrium, evolution of configurations of material systems is discussed in terms of free energy variations in the sections to follow.

The particular evolution phenomena in material systems considered in this chapter include the transition from a nominally flat surface to a wavy surface in a stressed solid, the spontaneous growth of epitaxial islands due to deposition of a material on a substrate with lattice mismatch, stress relaxation by grain boundary diffusion, the role of stress in altering compositional variations in solid solutions, and stress-assisted diffusion in the presence of an electric field or electromigration.

9.1 Mechanisms of surface evolution

The surface chemical potential field of a solid material, as developed in Chapter 8, represents a local free energy change accompanying addition of material locally on the surface; the change is measured per unit volume of added material. In describing the energetics of surface evolution, it becomes necessary to say something about the energy state of that material *before* it is added. This can be done in a fairly straight-forward way for two common mechanisms of surface evolution, namely, surface diffusion and condensation–evaporation. In the case of surface diffusion, material moves from one location on the material surface to another, and the resultant change in free energy is represented in terms of the gradient of chemical potential along the surface. In the case of condensation, on the other hand, the material added to the solid surface is drawn from an adjacent vapor. Thus, it becomes necessary to relate the thermodynamic state of the vapor to that of the solid in order to draw conclusions about the free energy change associated with condensation. The basic notions underlying these two mechanisms are introduced in this section. These two physical processes could contribute to shape change simultaneously, but they will be discussed separately for the most part in this chapter. The basic features of surface diffusion will be considered first.

9.1.1 Surface diffusion

Consider a stressed solid with a free surface S' in its reference configuration, un-der circumstances in which the body does not exchange mass or energy with its surroundings, as depicted in Figure 9.1. The chemical potential field over the free

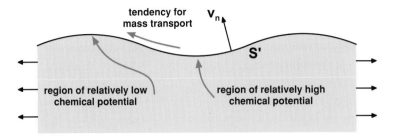

Fig. 9.1. Schematic diagram showing a stressed solid with free surface S'. Variation in chemical potential along the surface provides a driving force for mass transport by means of surface diffusion from regions of high chemical potential to regions of low chemical potential.

surface S' is given in terms of the elastic constants, the state of deformation, the surface shape and the surface energy density by (8.8). Variation in this surface field along the surface represents the tendency for the material system to alter its surface shape spontaneously in order to reduce the free energy of the system.

At elevated temperature, atoms at the surface of a crystal tend to move about on the surface in a random manner in the absence of stress and/or curvature, and possibly to exchange positions with atoms in the nearby subsurface region. The role of a chemical potential gradient is to bias this random motion, resulting in a net directional drift of material along the surface; this net drift is a mechanism of mass transport usually termed *surface diffusion* of the material. From the definition of chemical potential as the energy change per unit volume of material deposited locally, it follows that the physical dimensions of the components of the chemical potential gradient are force/length^3.

The flux of material along the surface S' is represented by a surface vector field $\mathbf{j}$ that is everywhere tangent to the surface S'. If τ^S is any unit vector that is tangent to the surface at a point, then the inner product $\mathbf{j} \cdot \tau^S$ is the volume flow rate of material in the direction τ^S per unit distance along the surface in the direction perpendicular to τ^S. From the definition of material flux as volume of material passing a line per unit length of that line per unit time, it follows that the components of the flux vector have physical dimensions of length^2/time. The flux vector is assumed to be related to the chemical potential gradient through a *kinetic relationship*, a postulate on material behavior that must be introduced as a constitutive assumption. Unless stated to the contrary, it will be assumed that this relationship is linear and that surface transport is isotropic, so that the kinetic relationship has the form

$$\mathbf{j} = -m_S \nabla_S \chi \,, \tag{9.7}$$

where $m_S > 0$ is a surface mobility coefficient, a parameter characteristic of the system, and ∇_S is the interior gradient operator on the surface. The surface mobility parameter, which incorporates effects of surface diffusivity, concentration of diffusing species and temperature, is discussed further below. If surface diffusion would be anisotropic, the mobility would be represented by a second rank tensor and mass flux would be expressed as the negative of a quantity obtained by the operation of this tensor on the vector-valued surface gradient of chemical potential.

Conservation of mass at each point on the surface requires that the net rate of material accumulation at that point due to surface flux is the negative of the divergence of the surface flux at that point. The net rate of material accumulation determines the surface speed v_n, which is the normal speed of the free surface relative to the material points currently on that surface, all referred to the reference configuration. Thus, in the present context,

$$v_n = -\nabla_S \cdot \mathbf{j} . \qquad (9.8)$$

When coupled with (9.7), the surface normal speed is related to the surface chemical potential through

$$v_n = \nabla_S \cdot (m_S \nabla_S \chi) . \qquad (9.9)$$

This result provides the normal velocity of the surface in terms of local elastic strain energy density, the local mean curvature and the surface energy density.

The transport equation (9.7) is known by several names, with Nernst–Einstein equation and Fick's law being among them. The former identifier commonly refers to charge transport through an electrode and the latter refers to concentration driven transport. The first application in the study of mass transport along a surface was that reported by Mullins (1957) in a theory of thermal grooving on material surfaces. Consequently, the relationship (9.7) will be termed the Mullins equation in the discussion to follow.

The surface mobility coefficient m_S is a phenomenological parameter that represents the net mass flow per unit gradient in surface chemical potential. The physical mechanisms of mass transport underlying this parameter are not well understood quantitatively, but some dependencies can be surmised. Surely, mobility must depend on temperature, binding energy and concentration of the diffusing species. A qualitative argument that leads to an estimate of the mobility coefficient proceeds as follows.

Mass transport by surface diffusion is a dissipative process, with the dissipation rate being

$$\mathcal{D} = \frac{1}{m_S} \int_S \mathbf{j} \cdot \mathbf{j} \, dS . \qquad (9.10)$$

Note that the dissipation varies inversely with the surface mobility. On a vicinal surface, as described in Section 8.8.3, energy is dissipated in transport either over the terraces or across the steps at the terrace boundaries, and the mobility parameter should reflect these two influences in some way. The dissipation occurring in the course of diffusion on terraces is understood better, and this mechanism is discussed first. This is followed by similar observations on dissipation due to transport across crystallographic steps. The separate mobility coefficients for these two mechanisms are denoted as m_{S_1} and m_{S_2}, respectively.

Consider the thermally activated motion of adatoms on the surface of a crystal. For purposes of this discussion, the surface is imagined to be a regular array of energy wells, each separated from its neighbors by an energy barrier of height $\mathcal{E}_d$. An adatom spends most of the time oscillating in one or another of the energy wells on the surface. From statistical mechanics, the probability that the energy of an adatom equals or exceeds the barrier height is $\exp[-\mathcal{E}_d/kT]$, where k is the Boltzmann constant and T is the absolute temperature. The rate of successful attempts by the adatom to hop out of a well is $\omega_d \exp[-\mathcal{E}_d/kT]$ where ω_d is the frequency of vibration. Viewing the vibration as that of an isolated harmonic oscillator, the stiffness is roughly Ea, where E is the elastic modulus and a is the well-to-well spacing; the mass of the adatom is roughly $\rho\Omega$, where ρ is the mass density of the material and Ω is its atomic volume. If the direction in which the adatom hops from one well to the next is random, then the frequency of hops in a particular direction is about $\frac{1}{4}\omega_d \exp[-\mathcal{E}_d/kT]$. Motion is as likely to occur in any one direction as in any other, and there is no net mass transport.

Suppose now that the energy wells become slightly shallower by some amount $\Delta\mathcal{E}$ from one well to the next in a certain direction, so that the barrier height in that direction is $\mathcal{E}_d + \Delta\mathcal{E}$; the barrier height in the opposite direction is $\mathcal{E}_d - \Delta\mathcal{E}$. Then, the net rate of well-to-well hops in that direction for a given well is

$$\frac{\omega_d}{4}e^{-(\mathcal{E}_d+\Delta\mathcal{E})/kT} - \frac{\omega_d}{4}e^{-(\mathcal{E}_d-\Delta\mathcal{E})/kT} \approx -\frac{\omega_d}{2}\frac{\Delta\mathcal{E}}{kT}e^{-\mathcal{E}_d/kT} \qquad (9.11)$$

times the fraction of wells that are occupied, where the result has been linearized for small values of $\Delta\mathcal{E}/kT$. Then, if $\Delta\mathcal{E}$ is identified with $\Omega a \nabla_S \chi$, hopping rate with $a\mathbf{j}/\Omega$, and fraction of wells occupied with $a^2 c_d$ where c_d is the number of diffusing particles per unit area of the surface, it follows that

$$m_{S_1} \sim \frac{a^2 \omega_d c_d \Omega^2}{2kT}e^{-\mathcal{E}_d/kT}. \qquad (9.12)$$

The activation barrier $\mathcal{E}_d$ is usually considered to be a fraction of the chemical bonding energy $\mathcal{E}_f$ (Zangwill 1988).

Suppose now that diffusion on the terraces is very rapid, implying that all atoms on a particular terrace have the same energy state, and that transport is controlled by movement across steps. If it is assumed that hopping up and down a step are equally likely, then (9.11) again applies but with a value of $\mathcal{E}_d$ representing the energy barrier imposed by the step. The energy distribution of atoms on any given terrace is now uniform, and $\Delta\mathcal{E}$ represents the difference in this energy level from one terrace to the next in a particular direction. To make the connection to the macroscopic representation of transport (9.7) in this case, denote the terrace width by w_d. Then, if $\Delta\mathcal{E}$ is identified with $\Omega w_d \nabla_S \chi$, hopping rate with $a\mathbf{j}/\Omega$, fraction of sites occupied along the step with $a^2 c_d$, and a/w_d with the magnitude of the local slope of the surface $|\nabla_S S|$, it follows that

$$ m_{S_2} \sim \frac{a^2 \omega_d c_d \Omega^2}{2kT |\nabla_S S|} e^{-\mathcal{E}_d/kT} . \tag{9.13} $$

The two effects represented by (9.12) and (9.13) can be combined into a single mobility parameter by assuming that

$$ m_S = \frac{m_{S_1} m_{S_2}}{m_{S_1} + m_{S_2}}, \tag{9.14} $$

which assures that transport is controlled by the *more* dissipative mechanism of the two possibilities discussed here. For example, if $m_{S_1} \ll m_{S_2}$ then $m_S \approx m_{S_1}$. A number of interesting observations on this dissipation process associated with surface diffusion are discussed by Nozieres (1992). Among these is the so-called Schwoebel effect whereby diffusing species encounter significantly greater resistance in crossing steps in the downward direction than in the upward direction (Schwoebel 1969).

9.1.2 Condensation–evaporation

The process of condensation or evaporation at a solid surface involves a phase transformation of the material. As a result, the mass of the solid is not conserved in the course of deposition, in general, unlike the situation for the process of shape change by means of surface diffusion. In light of this difference, it becomes necessary to indicate the thermodynamic state of the material being added or removed from the surface of the solid before it is added or after it is removed, respectively. Throughout the discussion of condensation, it is assumed that the solid surface of interest, customarily denoted by S', is in contact with a spatially uniform vapor of the same material. Furthermore, the state of the vapor as represented by pressure, temperature and density remains the same throughout the condensation process. The thermodynamic state of the vapor is represented in terms of the state of an auxiliary solid body of material, identical to that in contact with the vapor but which is homogeneously deformed and in contact with the vapor over a planar surface. The pressure in the vapor is taken to be the equilibrium vapor pressure, as defined in Section 1.3.1, for the case in which the auxiliary solid is subjected to a spatially uniform equi-biaxial stress σ_v acting in the plane parallel to the solid–vapor interface. The free energy per atom of the vapor and auxiliary solid are identical in this situation, which is the basis on which equilibrium vapor pressure is defined. If ρ_s and ρ_v are the mass densities of the solid and vapor, respectively, then the free energy per unit volume of the vapor is $\chi_v = \rho_v \sigma_v^2 / M\rho_s$, where M is the biaxial modulus of the elastic solid.

Recall that the chemical potential χ of a solid surface, as defined throughout Chapter 8, can be interpreted as the local free energy of the solid per unit reference volume of the material added to the surface S'. The free energy per unit volume of this same material prior to its addition to the solid was different from χ, in general. The basic idea is that the surface S' tends to move spontaneously with respect to the material on it in a way to reduce system free energy. Thus, in the present instance, it tends to incorporate *more* material onto the solid through condensation at points at which $\chi < \chi_v$; likewise, the surface tends to recede at places on S' at which $\chi > \chi_v$. A simple kinetic relationship representing surface response to chemical potential variations on S' is (Mullins 1959, Nozieres 1992)

$$v_n = -c_S(\chi - \chi_v),$$

(9.15)

where c_S is a material constant controlling the rate of transformation with physical dimensions of length/(time $\times$ force). It could be argued that it is only the difference $\chi - \chi_v$ that should be called a chemical potential in this context. The relationship (9.15) is commonly adopted in situations that are close to equilibrium situations, that is, where χ differs only slightly from χ_v. The difference in chemical potentials appearing on the right side of (9.15) is consistent with the interface chemical potential obtained in (8.22) under circumstances for which the equilibrium vapor pressure is small enough so that the traction T_i on the interface is negligibly small. In the special case when $\sigma_v = \sigma_m$ and elastic energy effects are ignored, (9.9) and (9.15) represent the two extreme cases of diffusion-limited surface motion and attachment limited surface motion, which have been discussed by Cahn and Taylor (1994).

9.2 Evolution of small surface perturbations

In this section, situations are considered for which the surface of a stressed solid is initially flat, or nearly so, and for which the slope of the evolving surface is everywhere small in magnitude throughout the evolution process. Chemical potential for a one-dimensional sinusoidal surface shape was developed in Section 8.4.1, for a two-dimensional sinusoidal shape in Section 8.5.3, and for a general small amplitude surface profile in Section 8.5.2. These results are used to examine surface evolution by either the mechanism of surface diffusion or condensation, as described in Section 9.1. In all cases considered in this section, surface energy is assumed to have the constant value γ_0, independent of surface orientation and surface strain. Implications of surface energy anisotropy and strain dependence are examined subsequently.

9.2.1 One-dimensional sinusoidal surface

Consider a biaxially stressed solid with a free surface described in a rectangular coordinate system by its height

$$y = h(x, t) = \bar{h}(t) + a(t) \cos \frac{2\pi x}{\lambda} \tag{9.16}$$

with respect to the plane $y = 0$, where λ is a constant wavelength, $\bar{h}$ is the mean surface height and a is a time-dependent amplitude of the fluctuation with respect to mean height. The material occupies the region $y \le h(x, t)$ for all x and z. The remote equi-biaxial stress of magnitude σ_m acts in the x, z-plane. The chemical potential χ for this case is given in (8.57).

Assuming that the sinusoidal surface evolves into a sinusoidal surface of the same wavelength under generalized plane strain conditions, the normal speed of the evolving surface with respect to the material instantaneously on it is

$$v_n = \frac{\dot{\bar{h}} + \dot{a}\cos(2\pi x/\lambda)}{\sqrt{1 + (2\pi a/\lambda)^2 \sin^2(2\pi x/\lambda)}} \approx \dot{\bar{h}} + \dot{a}\cos\frac{2\pi x}{\lambda}. \tag{9.17}$$

The speed is positive locally if material is being added there. Substitution of the normal surface speed v_n and chemical potential χ for this particular case into the general relationship (9.9) governing surface diffusion yields the ordinary differential equations

$$\dot{\bar{h}}(t) = 0, \quad \dot{a}(t) + \frac{16\pi^3 m_S}{\lambda^3} \left[\frac{\pi}{\lambda} \gamma_0 - (1+\nu)U_m \right] a(t) = 0 \tag{9.18}$$

for the amplitude and mean height. Conservation of mass implies immediately that $\bar{h}(t) = 0$ in this case. If the second equation is augmented by an initial condition, say $a(0) = a_0 > 0$, then it is evident that $a(t)$ grows (decays) exponentially in time if the quantity in square brackets in (9.18) is negative (positive). If both amplitude and wavelength are rescaled by the natural length parameter $\zeta = \gamma_0/U_m$ defined in (8.60), then this differential equation reduces to

$$\frac{d\hat{a}}{d\hat{t}}(\hat{t}) + \frac{16\pi^3}{\hat{\lambda}^4} \left[\pi - (1+\nu)\hat{\lambda} \right] \hat{a}(\hat{t}) = 0 \tag{9.19}$$

in nondimensional form, where

$$\hat{a} = a/\zeta, \quad \hat{\lambda} = \lambda/\zeta, \quad \hat{t} = m_S\gamma_0 t/\zeta^4. \tag{9.20}$$

The quantity

$$\tau = \zeta^4 / m_S \gamma_0 \qquad (9.21)$$

emerges as a natural time parameter for the evolution process.

The sinusoidal surface is stable or unstable, depending on whether $\hat{a}$ grows indefinitely with time or decays continuously with time. The value of the wavelength λ discriminating between these two possibilities is the value that reduces the quantity in square brackets in (9.19) to zero; this is precisely the critical wavelength λ_{cr} defined in (8.61) on the basis of equilibrium considerations.

It is evident from the evolution equation (9.19) that the rate of growth is zero when $\lambda = \lambda_{cr}$, and it is also zero in the limit as $\lambda/\lambda_{cr} \to \infty$. Because the growth rate is everywhere positive between these extremes, it must have an absolute maximum in value for some finite wavelength larger than λ_{cr}. Examination of the coefficient of $\hat{a}$ in (9.19) leads immediately to the conclusion that the growth rate is maximum when $\hat{\lambda} = 4\pi/3(1 + \nu)$. If the corresponding value of wavelength λ is denoted by λ_{gr}, then this fastest-growing wavelength is (Srolovitz 1989)

$$\lambda_{gr} = \tfrac{4}{3} \lambda_{cr} . \qquad (9.22)$$

Study of the evolution of this same material system under conditions of condensation–evaporation for the case when $\chi_v = U_m$ leads to the same evolution equations for amplitude and mean height, except that the characteristic time for amplitude evolution is determined by different material parameters. Derivation of these evolution equations is left as an exercise. If $\chi_v > U_m$, then the sinusoidal fluctuation in surface shape is superimposed on the mean surface speed $\dot{\bar{h}} = -c_S(U_m - \chi_v)$.

9.2.2 Example: The characteristic time

The characteristic time defined in (9.21) establishes a timescale for surface evolution of the kind discussed in the preceding section. Its definition depends on a number of parameter values that are not measurable and, therefore, are not known with any certainty. To get some idea of its magnitude, estimate the value of the characteristic time for the particular case of a Si surface with a mismatch strain of $\epsilon_m = 0.008$ at a temperature of $T = 600\,°C$. Base the estimate on the unit cell dimension of $a = 0.5431\,nm$ for the diamond cubic crystal structure, and on the following values of macroscopic material parameters: an elastic modulus of $E = 130\,GPa$, a Poisson ratio of $\nu = 0.25$, a mass density of $\rho = 2328\,kg$ m^{-2}, and a surface energy of $\gamma_0 = 2\,J\ m^{-2}$. Assume that 10% of the surface atoms are involved in the mass transport process at any instant so that $c_d a^2 = 0.1$.

Solution:

By combining (8.60) and (9.7) with (9.21), the expression for the characteristic time becomes

$$\tau = \left(\frac{\gamma_0}{M \epsilon_m^2} \right)^4 \frac{2kT}{\gamma_0 a^2 c_d \omega_d \Omega^2} e^{\mathcal{E}_d / kT} . \qquad (9.23)$$

The biaxial modulus is given by $M = E/(1 - \nu) = 173\,\text{GPa}$ and γ_0 is given, so the characteristic length in this case is $\zeta = 180\,\text{nm}$. The atomic volume Ω of the diamond cubic crystal structure is related to the unit cell dimension a by $\Omega = a^3/8$. As in Section 9.1.1, vibrational frequency ω_d can be estimated by assuming that an atom vibrates as a simple harmonic oscillator with the elastic stiffness estimated to be $E\Omega^{1/3}$ and the oscillator mass estimated to be $\rho\Omega$, which implies that

$$\omega_d = \frac{1}{2\pi\,\Omega^{1/3}}\sqrt{\frac{E}{\rho}} = 3.75 \times 10^{12}\ \text{cycles s}^{-1}. \tag{9.24}$$

The height of the diffusion barrier $\mathcal{E}_d$ is assumed to be 30% of the binding energy $\mathcal{E}_f$, which is given roughly by $\gamma_0\Omega^{2/3}$. Finally, the thermal energy kT is $1.21 \times 10^{-20}\,\text{J}$ at $600\,°\text{C}$. Combining these estimates of the parameters involved, it follows that the characteristic time is estimated to be $\tau = 5.03\,\text{s}$. The result implies that an initial shape imperfection on an unstable surface can grow to a size e times larger than the initial size in a few seconds for this system.

9.2.3 General surface perturbations

Building on the results obtained in Sections 8.5.2 and 8.5.3, it would be possible at this point to treat the evolution of the doubly periodic surface perturbation directly, in much the same way as was done in Section 9.2.1. Alternatively, this case can be viewed as a special case of general small amplitude surface perturbations, and this is the point of view adopted here. In this section, a general solution for evolving surface shape under conditions of surface diffusion is derived. This solution is obviously restricted by the condition that the slope is small in magnitude everywhere. It is also limited to surface shapes having certain reflective symmetries but this is not an essential restriction. Rather, it is adopted because a general solution for evolving surface shape can be obtained in a far more transparent way than if it were not invoked. The restriction excludes no general features of behavior, and the way to relax it becomes evident in the course of the analysis. The approach has been generalized by Kim et al. (1999) in the study of chemical etching of surfaces.

Once again, suppose that the surface S' is defined in a rectangular coordinate system with reference to a flat surface with the interior coordinates x, z. The y-direction is normal to this flat reference surface. The nearly flat material surface can be specified by giving its y-coordinate as a function of position x, z at any time t, say

$$y = h(x, z, t). \tag{9.25}$$

The restriction to small amplitude fluctuations requires that

$$\sqrt{h_{,x}^2 + h_{,z}^2} \ll 1 \tag{9.26}$$

for all x, z and t. If this is the case, then all surface fields can be regarded as functions of x, z and t, the surface normal speed v_n is approximately $h_{,t}$, the surface gradient operator ∇_S is the two-dimensional gradient operator ∇ in the x, z-plane, and the curvature field κ is approximately $\nabla^2 h$. With these approximations, the surface evolution equation (9.9) becomes

$$h_{,t}(x, z, t) = m_S \nabla^2 \left[U(x, z, t) - \gamma_0 \nabla^2 h(x, z, t) \right] \tag{9.27}$$

with $U_S = \gamma_0$ as noted above.

The block of material is subjected to an equi-biaxial remote traction so that, when the surface is perfectly flat with $h(x, z, t) \equiv 0$, the state of stress everywhere is $\sigma_{xx} = \sigma_{zz} = \sigma_m$ and other stress components are zero. The equi-biaxial strain corresponding to the stress is $\epsilon_m = \sigma_m / M$. This simple state of stress and deformation is then perturbed by formation of surface waviness. The perturbed stress field determines the surface strain energy density $U(x, z, t)$ in (9.27).

A procedure by which the perturbed stress field can be determined has already been described for both plane strain deformation and general three-dimensional deformation in Chapter 8, all subject to the limitation of small amplitude surface fluctuations. To lowest order in the surface fluctuation, the strain energy density distribution over the surface is

$$U(x, z, t) = U_m + U^{(h)}(x, z, t) \tag{9.28}$$

where the second term on the right side of (9.28) is given in (8.88). Therefore, the dependence of the field $U(x, z, t)$ in (9.27) on the surface shape $h(x, z, t)$ is already in hand. It is linear in $h(x, z, t)$ and, at each point x, z on the surface, it depends on the behavior of h over the entire surface. In spite of this apparent complexity, solutions can be found for some cases of interest.

Based on experience with the simpler case of sinusoidal perturbations, it is again convenient to reduce the governing equation (9.27) to a nondimensional form by introducing the natural length scale ζ and the natural timescale τ, similar to (9.20). Nondimensional variables are again denoted by a superposed hat. The normalizations that are incorporated are

$$\hat{U} = \frac{U}{U_m}, \quad \hat{x} = \frac{x}{\zeta}, \quad \hat{h} = \frac{h}{\zeta}, \quad \hat{t} = \frac{m_S \gamma_0 t}{\zeta^4} \tag{9.29}$$

and similarly for other length parameters. In terms of these parameters, the surface evolution equation (9.27) reduces to

$$\frac{\partial \hat{h}}{\partial \hat{t}}(\hat{x}, \hat{z}, \hat{t}) = \hat{\nabla}^2 \left[\hat{U}(\hat{x}, \hat{z}, \hat{t}) - \hat{\nabla}^2 \hat{h}(\hat{x}, \hat{z}, \hat{t}) \right] . \tag{9.30}$$

The goal of the analysis is to determine solutions of this equation for various choices of initial surface shape

$$\hat{h}(\hat{x}, \hat{z}, 0) = \hat{h}_I(\hat{x}, \hat{z}) \tag{9.31}$$

and possibly for specified asymptotic behavior of $\hat{h}(\hat{x}, \hat{z}, \hat{t})$ as $\sqrt{\hat{x}^2 + \hat{z}^2} \to \infty$; such behavior must be dictated by physical features in any specific case.

Because of the form of $U^{(h)}$ as a convolution integral, the governing equation is readily solved by means of Fourier transforms. If attention is restricted to surface shapes with the reflective symmetry prescribed by $h(x, -z, t) = h(-x, z, t) = h(x, z, t)$, then the real Fourier cosine transform can be used; this feature accounts for the relative transparency of the steps that follow. Suppose that $H(\alpha, \beta, \hat{t})$ is the time-dependent double Fourier cosine transform of $\hat{h}(\hat{x}, \hat{z}, \hat{t})$ in both x and z, that is,

$$H(\alpha, \beta, \hat{t}) = \frac{4}{\pi^2} \int_0^\infty \int_0^\infty \hat{h}(\hat{x}, \hat{z}, \hat{t}) \cos \alpha \hat{x} \cos \beta \hat{z} \, d\hat{x} d\hat{z} \qquad (9.32)$$

with transform parameters α and β corresponding to the physical coordinates x and z, respectively. The result of applying this transform to each term of the governing equation is the ordinary differential equation

$$\frac{\partial H}{\partial \hat{t}}(\alpha, \beta, \hat{t}) = -(\alpha^2 + \beta^2)^{3/2} \left[(\alpha^2 + \beta^2)^{1/2} - 2(1 + v) \right] H(\alpha, \beta, \hat{t}). \qquad (9.33)$$

The solution of this differential equation, which depends parametrically on the transform variables α and β, is

$$H(\alpha, \beta, \hat{t}) = H_1(\alpha, \beta) e^{-\omega(\alpha, \beta)\hat{t}} \qquad (9.34)$$

where

$$\omega(\alpha, \beta) = (\alpha^2 + \beta^2)^{3/2} \left[(\alpha^2 + \beta^2)^{1/2} - 2(1 + v) \right] \qquad (9.35)$$

and $H_1(\alpha, \beta)$ is the double cosine transform of the initial shape $\hat{h}_1(\hat{x}, \hat{z})$. Thus, the solution of the governing equation (9.27) subject to the initial condition (9.31) is obtained as (Freund 1995a)

$$\hat{h}(\hat{x}, \hat{z}, \hat{t}) = \int_0^\infty \int_0^\infty H_1(\alpha, \beta) e^{-\omega(\alpha, \beta)\hat{t}} \cos \alpha \hat{x} \cos \beta \hat{z} \, d\alpha d\beta. \qquad (9.36)$$

As is usual in the application of integral transform methods to solve boundary value problems, it is tacitly assumed that all the integrals which have been written actually converge. This must be confirmed in any particular case before the solution implied by (9.36) can be accepted as an actual solution.

A feature that is central to the behavior of the solution $\hat{h}$ in (9.36) is the form of $\omega(\alpha, \beta)$ in the exponent of the integrand. If $\omega(\alpha, \beta) < 0$ for any part of the range of α and β in transform space, then a solution may grow without bound as $\hat{t} \to \infty$. Exploiting the fact that this function depends on α and β only through the combination $\alpha^2 + \beta^2$, it is plotted in Figure 9.2 in the form of $\omega(\alpha, \beta)/16(1 + v)^4$ versus $\sqrt{\alpha^2 + \beta^2}/2(1 + v)$ where it is evident that $\omega < 0$ for $0 < \sqrt{\alpha^2 + \beta^2} < 2(1 + v)$ and that the local minimum in ω occurs for $\sqrt{\alpha^2 + \beta^2} = 3(1 + v)/2$. The

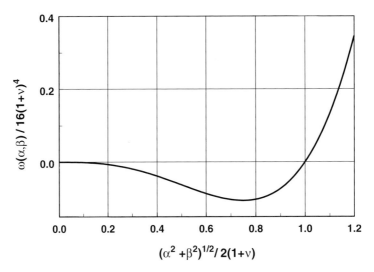

Fig. 9.2. Graph of the function $\omega(\alpha, \beta)$ defined in (9.35). This function appears in the exponent of the general solution for surface evolution given in (9.36), expressed as a Fourier integral. Adapted from Freund (1995a).

implication of these observations is that, if any portion of the spectrum $H_I(\alpha, \beta)$ of the initial shape falls within the region where $\omega < 0$, then that portion will grow indefinitely large as time goes on. The way in which this occurs depends on the details of any particular case, and some illustrative cases are considered in the sections that follow.

For example, study of the doubly periodic function

$$h_I(x, z) = h_0 \cos \frac{2\pi x}{\lambda_x} \cos \frac{2\pi z}{\lambda_z} , \qquad (9.37)$$

where $\lambda_x > 0$ and $\lambda_z > 0$ are wavelengths in surface topography in the x and z directions and $h_0 > 0$ is an amplitude that is small compared to either wavelength, leads once again to the stability condition given in (8.98) and provides a basis for determining the wavelength for which the amplitude grows most rapidly. The case of an isolated surface mound or shallow island is considered in the next subsection.

9.2.4 An isolated surface mound

Suppose that the initial surface shape is a shallow mound, circular in plan view, that is described by the rotationally symmetric function

$$h_I(x, z) = h_0 e^{-(x^2+z^2)/a^2} , \qquad (9.38)$$

where $a > 0$ is a length parameter and $h_0 > 0$ is an amplitude which is small compared to a. A cross-section of the shape represented by (9.38) is included in

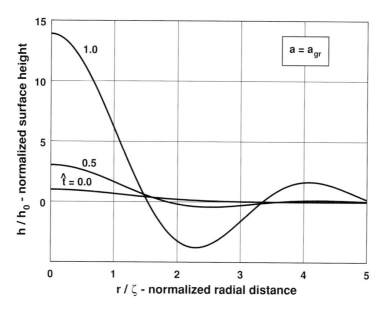

Fig. 9.3. The initial cross-sectional profile $\hat{t} = 0.0$ of an isolated surface mound, as given by (9.38), and the shapes of the mound resulting from mass transport by surface diffusion at two later times are illustrated for the case when $a = a_{gr}$, which exhibits the most rapid growth. The profiles for times $\hat{t} = 0.5$ and 1.0 were determined by numerical evaluation of (9.43).

Figure 9.3. The surface is essentially flat everywhere outside of a circle of radius about $2a$ in the x, z-plane that is centered at the origin. Within this circular region, there is a smooth, rotationally symmetric mound of material with center height of h_0. Given this initial shape, the goal is to determine $h(x, z, t)$ for $t > 0$.

The Fourier transform of the initial shape in normalized variables is

$$H_1(\alpha, \beta) = \frac{\hat{a}^2 \hat{h}_0}{\pi} e^{-\frac{1}{4}\hat{a}^2(\alpha^2 + \beta^2)} \tag{9.39}$$

where $\hat{a} = a/\zeta$. The transient solution for surface shape is then given by (9.36) and it is restricted to be rotationally symmetric. The fact that α and β appear in the transformed solution only in the combination $\alpha^2 + \beta^2$ suggests a transformation from rectangular coordinates α and β to polar coordinates, say ρ and φ, in the transformed plane according to

$$\alpha = \rho \cos\varphi, \quad \beta = \rho \sin\varphi. \tag{9.40}$$

A comparable coordinate transformation to polar coordinates, say $\hat{r}$ and θ, in the physical plane is

$$\hat{x} = \hat{r} \cos\theta, \quad \hat{z} = \hat{r} \sin\theta. \tag{9.41}$$

The cosine factors in the integrand are then rewritten in terms of polar coordinates by means of the identity

$$\cos \alpha \hat{x} \cos \beta \hat{z} = \tfrac{1}{2} \cos \left[\hat{r}\rho \cos(\theta + \varphi)\right] + \tfrac{1}{2} \cos \left[\hat{r}\rho \cos(\theta - \varphi)\right]$$

$$= J_0(\hat{r}\rho) + 2 \sum_{k=1}^{\infty} (-1)^k J_{2k}(\hat{r}\rho) \cos 2k\theta \, \cos 2k\varphi , \qquad (9.42)$$

where J_m is the ordinary Bessel function of the first kind of integer order m. The area element transforms as $d\alpha \, d\beta \to \rho \, d\rho \, d\varphi$ with the integration range becoming $0 < \rho < \infty, 0 < \varphi < \pi/2$.

The only place in the integrand where φ appears is through the factor $\cos 2k\varphi$ in each term of the infinite series in (9.42). The integral of each such term over φ is zero. Consequently, the solution is independent of θ, as was expected, and it takes the form

$$\hat{h}(\hat{r}, \hat{t}) = \hat{h}_0 \frac{\hat{a}^2}{2} \int_0^{\infty} e^{-\frac{1}{4}\rho^2 \hat{a}^2 - \omega(\rho)\hat{t}} J_0(\hat{r}\rho) \, \rho \, d\rho . \qquad (9.43)$$

Under the change of variables (9.40), $\omega(\alpha, \beta)$ is a function of only a single variable ρ. Nevertheless, the same symbol $\omega(\rho)$ is again used to represent this function of ρ, with the intended independent variables being evident from the context.

In general, the integral (9.43) must be evaluated numerically for given values of $\hat{r}$ and $\hat{t}$, which is a straightforward task. However, some useful information can be obtained in closed form by considering the initial rate of change of island height at its center, that is, $\partial \hat{h}/\partial \hat{t}$ for $\hat{r} = 0$ and $\hat{t} \to 0^+$. The result of evaluating the integral in (9.43) for this special case is

$$\frac{\partial \hat{h}}{\partial \hat{t}}(0, 0) = \hat{h}_0 \frac{12\sqrt{\pi}\hat{a}(1 + \nu) - 32}{\hat{a}^4} . \qquad (9.44)$$

From this result, a critical value of the size parameter a, which is proportional to the characteristic dimension ζ of the system, can be identified as

$$a_0 = \frac{8\zeta}{3\sqrt{\pi}(1 + \nu)} = \frac{8\gamma_0}{3\sqrt{\pi}(1 + \nu)U_m} . \qquad (9.45)$$

For $a < a_0$, the height of the island initially diminishes from h_0 whereas, for $a > a_0$, the height initially increases. All other things being equal, the value of a for which the height has the greatest rate of growth, is

$$a_{gr} = \frac{32\zeta}{9\sqrt{\pi}(1 + \nu)} = \frac{32\gamma_0}{9\sqrt{\pi}(1 + \nu)U_m} . \qquad (9.46)$$

Azimuthal sections of the surface shapes determined from (9.43) at normalized times $\hat{t} = 0$, 0.5 and 1.0 and for five values of the size parameter of $\hat{a} = a_{gr}/2$, $a = a_{gr}$ and $a = 2a_{gr}$ are shown in Figure 9.3; Poisson's ratio was taken to be $\nu = 1/4$ in the calculations.

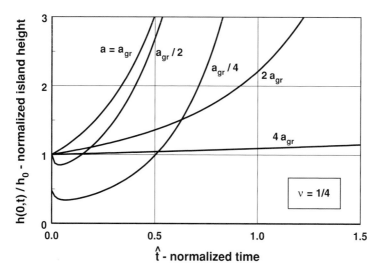

Fig. 9.4. The height of the surface of the center of the mound above the reference plane, normalized by the initial height, versus normalized time for several values of the size parameter a. As was anticipated in (9.46), the most rapid growth of the mound occurs for the case when $a = a_{gr}$. These results were obtained by evaluation of (9.43) for $r = 0$.

Because the Fourier transform of the initial shape includes all possible wavelengths to some degree, including contributions for which $(\alpha^2 + \beta^2)^{1/2} < 2(1 + \nu)$, the height of the island eventually grows larger for *any* value of a. This is illustrated in Figure 9.4 which shows $\hat{h}(0, \hat{t})/\hat{h}_0$ versus $\hat{t}$ for values of a equal to $\frac{1}{4}, \frac{1}{2}, 1, 2, 4$ times a_{gr}. It is evident from this figure that islands within a fairly narrow size range near a_{gr} grow much more rapidly than islands which are significantly larger or smaller. All of these observations are based on behavior of small amplitude fluctuations in surface shape, and there is no reason to expect this behavior to persist far into the regime of large amplitude fluctuations. Finally, other simple surface shapes, such as a pair of mounds in close proximity, can be analyzed in the same way as long as surface slopes are small. If this is not the case, then a computational approach must be developed as a means of studying surface evolution (Kukta and Freund 1997).

9.3 A variational approach to surface evolution

A partial differential equation governing the evolution of the surface of a strained elastic solid due to surface diffusion appears in (9.9). Given the instantaneous elastic state of the deformed solid and the instantaneous shape of the evolving surface, this partial differential equation specifies the rate of change of surface shape. The principal assumption underlying the equation is that the system free energy is the sum of elastic strain energy and surface energy. The equation applies for general shape changes and for general anisotropic elastic properties and/or surface energy density. When the shape change is only a small perturbation from a simple smooth shape and the surface energy density is constant, the governing equation reduces to a linear partial differential equation along the unperturbed boundary shape; an illustration of such a linear equation is given by (9.27) for perturbations of an initially flat free surface.

If the evolving shape itself must be determined over the course of time as part of the solution, the underlying boundary value problem is no longer linear. Analysis of surface evolution in such situations must rely on approximate methods, in general. The two most common approaches are: (i) to determine approximate solutions to the governing partial differential equations by numerical methods and (ii) to express the evolving shape in terms of a small number of modes, each with its own time-dependent amplitude. In the former approach, the solution proceeds incrementally. An elasticity problem is solved for the configuration of fixed shape at time t, yielding the distribution of strain energy density U over the evolving surface. The surface curvature κ is known from the shape and hence the chemical potential field χ is known over the surface. Then, during a small time increment Δt, the surface in the reference configuration moves locally in the normal direction a distance $v_n \Delta t$ where v_n is given by (9.9), thereby defining a new configuration to be analyzed in the same way starting at time $t + \Delta t$, and so on. On the other hand, in the modal approach, the objective is to determine ordinary differential equations governing the time evolution of the mode amplitudes. In fact, this was the approach followed in Section 8.4 for a cosine mode shape with amplitude $a(t)$.

For either numerical solution of the field equations by means of the finite element method or determination of a system of ordinary differential equations for modal amplitudes, the existence of a variational statement or weak form of the field equations is essential. For the complementary aspect of the problem concerned with the elastic field for a fixed boundary configuration, the powerful minimum potential energy theorem is available (Fung 1965). The purpose here is to introduce a variational principle as a basis for describing the rate of shape evolution for a fixed shape and a fixed elastic field.

9.3.1 A variational principle for surface flux

The starting point for this development is the expression

$$\dot{F} = \int_S \chi v_n \, dS \tag{9.47}$$

for the rate of change of free energy in terms of the surface chemical potential field χ and the instantaneous normal velocity v_n of the evolving free surface; (9.47) follows from the definition of chemical potential in (8.7) and (8.8). Recall that, in writing (9.47), it is assumed that the surface S represents the evolving part of the boundary of the solid and that there is no exchange of energy between the deforming solid and its surroundings other than a possible addition or reduction of material at S. The form of (9.47) motivated the introduction of the constitutive equation (9.7) relating the local rate of material transport $\mathbf{j}$ over the surface S to the local surface gradient in chemical potential $\nabla_S \chi$. The local surface divergence of mass flux then prescribes the rate of normal advance v_n of the surface S as a result of conservation of mass, which is equivalent to conservation of material volume in the reference configuration. These results can be recast into variational form, leading to a result that provides the weak form of the governing equations that is the basis for the variational methods outlined.

Suppose that the mass conservation equation is used to replace v_n in (9.47) by its equivalent form in terms of $\mathbf{j}$. Then, after application of a vector identity in the integrand,

$$\dot{F} = - \int_S \nabla_S \cdot (\chi \mathbf{j}) \, dS + \int_S \mathbf{j} \cdot \nabla_S \chi \, dS \,. \tag{9.48}$$

The first term on the right side of (9.48) is the integral of a divergence expression, so it can be rewritten as an integral along the bounding curve of S. The role of this contribution from the bounding curve of S is discussed further later in this section. However, the value of the contribution is exactly zero in a number of cases of interest. For example, if S is a closed surface then it has no bounding curve and the value is zero. If the fields and surface shape are spatially periodic and if S includes exactly one period of the system, then the value of the integral is also zero. This follows from the observation that χ would have the same value on opposite sides of a segment of S in this case due to periodicity, and the mass flux out of S on one side would exactly cancel the mass flux out on the opposite side for the same reason. Similarly, if the surface S extends to indefinitely remote regions compared to the size of the field of view, and if the evolving portion of S is small enough so that the magnitude of $\chi \mathbf{j}$ decays appropriately with distance into remote regions, the contribution to $\dot{F}$ from the bounding curve of S will vanish even though the length of that bounding curve may become indefinitely large. Thus, this integral over S of a divergence expression is neglected here and elsewhere in this development. If the constitutive equation (9.7) is used to replace the chemical potential gradient $\nabla_S \chi$ in favor of $\mathbf{j}$ in (9.48),

$$\dot{F} = -\frac{1}{m_S} \int_S \mathbf{j} \cdot \mathbf{j} \, dS < 0 \,. \tag{9.49}$$

This result implies that any choice of surface mass flux $\mathbf{j}$ on S derived from a chemical potential results in a *reduction* in free energy, provided only that $m_S > 0$.

This is an essential feature of spontaneous surface evolution, and the result also provides an important ingredient for a variational statement.

Next, suppose that $\mathbf{j}$ and χ in (9.7) are the actual mass flux field and chemical potential field on S. Furthermore, suppose that $\mathbf{q}$ is any possible mass flux field on S, that is, it shares any properties of periodicity, smoothness and remote decay characteristics with $\mathbf{j}$, but is otherwise arbitrary. Form the inner product of each side of (9.7) with $\mathbf{q}$ and then integrate each side over S. The result can be written in the form

$$\int_S \chi \, \nabla_S \cdot \mathbf{q} \, dS - \frac{1}{m_S} \int_S \mathbf{j} \cdot \mathbf{q} \, dS = 0 \qquad (9.50)$$

where an integral of a divergence expression has again been set equal to zero.

The two results in (9.49) and (9.50) provide the basis for an extremum principle that can be expressed as either a minimum principle or a maximum principle; the two forms are equivalent and the latter option is adopted here. For a given surface shape and elastic field, define a functional over the range of all trial surface flux fields $\mathbf{j}^*$ as

$$\Phi[\mathbf{j}^*] = \int_S \left[\chi \, \nabla_S \cdot \mathbf{j}^* - \frac{1}{2m_S} \mathbf{j}^* \cdot \mathbf{j}^* \right] dS, \qquad (9.51)$$

where χ is the actual chemical potential of the state being considered at the present instant. Then (9.50) implies that this functional is stationary under arbitrary variations of surface flux $\mathbf{j}^*$ when the flux field is the actual field, that is, when $\mathbf{j}^* = \mathbf{j}$. This follows immediately from (9.50) if $\mathbf{q}$ is interpreted as the perturbation in flux $\delta\mathbf{j}^*$ in the usual variational notation. The arbitrariness of $\delta\mathbf{j}^*$ and the fundamental theorem of the calculus of variations require that $\mathbf{j}$ must satisfy (9.7) pointwise on S. The fact that $\Phi[\mathbf{j}^*]$ is not only stationary when $\mathbf{j}^* = \mathbf{j}$ but is also an absolute maximum under variations in $\mathbf{j}^*$ follows from consideration of the quantity $\Phi[\mathbf{j}^* + \delta\mathbf{j}^*] - \Phi[\mathbf{j}^*]$ in light of the requirement that $m_S > 0$.

Next, a situation in which the first term on the right side of (9.48) comes into play is considered briefly. The situation is depicted in Figure 9.5. The surface S consists of two smooth surface segments, denoted by S_1 and S_2, which intersect along the space curve C. The chemical potential fields $\chi^{(1)}$ and $\chi^{(2)}$ exist over the surfaces S_1 and S_2, respectively, with the corresponding surface mass flux vectors $\mathbf{j}^{(1)}$ and $\mathbf{j}^{(2)}$ being related to the corresponding chemical potentials through the kinetic relations

$$\mathbf{j}^{(k)} = -m_{S_k} \nabla_S \chi^{(k)} \qquad (9.52)$$

over the surfaces. On either surface, the mass flux is a vector field which, at any surface point, lies in the tangent plane to the surface at that point.

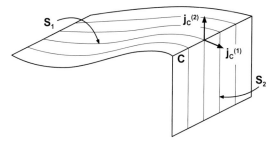

Fig. 9.5. Schematic diagram of a surface consisting of two smooth segments S_1 and S_2 joined along a space curve C. The component of surface mass flux along C labeled $j_C^{(k)}$ represents the rate of mass loss from surface S_k due to transport through C for either $k = 1$ or 2.

The local mass flux out of S_1 at a point on C is a component of the limiting value of $\mathbf{j}^{(1)}$ at that point on C from within S_1. This component is determined by the properties that it must be normal to C, must be directed away from S_1 and must lie in the tangent plane to S_1 at that point. This mass flux is labeled $j_C^{(1)}$ in Figure 9.5. Similarly, $j_C^{(2)}$ is the mass flux out of S_2 through C. If no material is gained or lost along C, conservation of mass requires that

$$j_C^{(1)} + j_C^{(2)} = 0 . \tag{9.53}$$

The limiting values of chemical potential on C from within S_1 and S_2 are different, in general, and this difference can be assumed to drive mass flow past C according to a constitutive equation that might have the form

$$j_C^{(1)} = -m_C \left(\chi_C^{(2)} - \chi_C^{(1)} \right) , \tag{9.54}$$

where m_C is a mobility parameter. In this case, the functional

$$\Phi[\mathbf{j}^*] = \int_S \left[\chi \nabla_S \cdot \mathbf{j}^* - \frac{1}{2m_S} \mathbf{j}^* \cdot \mathbf{j}^* \right] dS - \int_C \frac{1}{2m_C} \left(j_C^{(1)} \right)^2 dC \tag{9.55}$$

replaces the functional defined in (9.51) in the statement of the variational principle. In this compact form of the functional, it is presumed that the surface S is divided into the parts S_1 and S_2, and that the appropriate form of chemical potential and surface mobility is selected for each part. If m_C is large compared to either m_{S_1} or m_{S_2} divided by a typical linear dimension of the body, then the influence of the added term in (9.55) can be ignored and chemical potential is continuous across C.

A variational principle equivalent to (9.51) expressed in terms of a functional of the scalar normal velocity field on S, rather than the vector surface transport field, can be constructed in the same way. The functional in this case is

$$\Psi[v_n^*] = \int_S \left[m_S v_n^* \nabla_S^2 \chi - \tfrac{1}{2} v_n^{*2} \right] dS . \tag{9.56}$$

For a specified shape of the surface S and a prescribed chemical potential field χ over S, this functional is maximum when the trial velocity field v_n^* is in fact that actual velocity field v_n defined by (9.9). The variational approach in terms of surface flux was introduced in a numerical study of grain boundary transport and cavitation by Needleman and Rice (1980), for which the grain boundary chemical potential is given by (8.14). It was introduced in the way as described here by Suo (1997) and Zhang et al. (1999) as a basis for study of surface evolution.

9.3.2 Application to second order surface perturbation

In Section 8.4.4, the higher order perturbation of the shape of the surface of the biaxially stressed elastic solid was considered under two-dimensional conditions of generalized plane strain. The material response is linear in this case, but the boundary conditions of vanishing traction on the perturbed surface were expanded to second order in the small slope of the perturbed surface shape given by (8.75). These boundary conditions are given in (8.74). The appearance of the second order contributions in the boundary conditions dictates the introduction of a second term in the assumed surface shape which is the first harmonic of the dominant sinusoidal term, as seen in (8.75). The generalized forces $\mathcal{Q}_1$ and $\mathcal{Q}_2$ that are work-conjugate to the mode amplitudes a_1 and a_2, respectively, were determined; they appear in (8.81). The variational principle introduced in Section 9.3.1 provides a means for determining the system of nonlinear ordinary differential equations that govern $a_1(t)$ and $a_2(t)$. These equations are determined in this section, and a numerical solution is presented for a particular case.

The surface flux vector $\mathbf{j}^*$ appearing in the functional defined in (9.51) is completely determined by the amplitudes $a_k(t)$ appearing in (8.75). The functional has the form

$$\Phi[\dot{a}_1, \dot{a}_2] = Q_k(a_1, a_2)\dot{a}_k - \frac{1}{2m_S} \int_S \mathbf{j}^* \cdot \mathbf{j}^* \, ds . \tag{9.57}$$

The task remaining is to write the dissipation term on the right side of this expression in terms of the amplitudes a_k and the rates $\dot{a}_k$.

The normal speed of the surface can be expressed in terms of the time-dependent surface shape $h(x, t)$ given in (8.75) as

$$v_n(x, t) = h_{,t} \left(1 - \tfrac{1}{2} h_{,x}^2 \right) , \tag{9.58}$$

accurate to second order in the small slope. If $j(x, t)$ is the magnitude of the surface mass flux along the wavy surface, then $v_n = -\partial j/\partial s$. If this relationship is expressed to second order in surface slope, it reduces to the simple form

$$h_{,t} = -j_{,x} \ . \tag{9.59}$$

The flux itself is then readily determined by elementary integration to yield

$$j(x,t) = -\dot{a}_1(t)\frac{\lambda}{2\pi}\sin\frac{2\pi x}{\lambda} - \dot{a}_2\frac{\lambda}{4\pi}\sin\frac{4\pi x}{\lambda} \ . \tag{9.60}$$

This expression is then substituted into the last term in (9.57) according to the identification $\mathbf{j}^* \cdot \mathbf{j}^* = j(x,t)^2$, and the integral is evaluated for the interval $0 \le x \le \lambda$, retaining terms of second order in the small slope in changing the variable of integration from arclength s to rectangular coordinate x.

Differential equations for $a_k(t)$ are then obtained by imposing the requirement that the optimal choice of the amplitudes is that which renders the functional a maximum, that is, the optimal choice is defined by

$$\frac{\partial \Phi}{\partial \dot{a}_k}[\dot{a}_1, \dot{a}_2] = 0 \,, \quad k = 1, 2 \,. \tag{9.61}$$

The resulting equations can be written compactly in terms of normalized time $\hat{t}$ and normalized amplitudes $\hat{a}_k$ defined by

$$\hat{t} = m_S U_S t/\lambda_{\mathrm{cr}}^4 \,, \quad \hat{a}_k = a_k/\lambda \tag{9.62}$$

for constant surface energy density U_S and surface mobility m_S. The natural length parameter λ_{cr} is defined in (8.61). In terms of these quantities, the differential equations are

$$\frac{1}{16}\left[2\pi^{-2} + 3\hat{a}_1^2\right]\hat{a}_1' + \frac{1}{4}\hat{a}_1\hat{a}_2\hat{a}_2' = \frac{\lambda_{\mathrm{cr}}^4(1+\nu)^4}{\lambda^4\pi^4 U_S}\mathcal{Q}_1(\hat{a}_1, \hat{a}_2)$$

$$\frac{1}{4}\hat{a}_1\hat{a}_2\hat{a}_1' + \frac{1}{32}\left[\pi^{-2} + \hat{a}_1^2\right]\hat{a}_2' = \frac{\lambda_{\mathrm{cr}}^4(1+\nu)^4}{\lambda^4\pi^4 U_S}\mathcal{Q}_2(\hat{a}_1, \hat{a}_2) \tag{9.63}$$

where the prime denotes differentiation with respect to $\hat{t}$. If the amplitude a_2 is set equal to zero for the moment and the differential equation (9.63) is linearized for small a_1, it reduces to

$$\hat{a}_1' = \frac{16(1+\nu)^4}{\pi^2}\frac{\lambda_{\mathrm{cr}}^4}{\lambda^4}\left(\frac{\lambda}{\lambda_{\mathrm{cr}}} - 1\right)\hat{a}_1 \ . \tag{9.64}$$

The amplitude $\hat{a}_1$ decays in time according to this equation from its initial value if $\lambda < \lambda_{\mathrm{cr}}$, whereas it grows without bound if $\lambda > \lambda_{\mathrm{cr}}$, as expected. Furthermore, the greatest rate of amplitude growth occurs for $\lambda = \frac{4}{3}\lambda_{\mathrm{cr}}$, as anticipated in the results of the linearized approach; see (9.22). In retrospect, this result on rate of growth

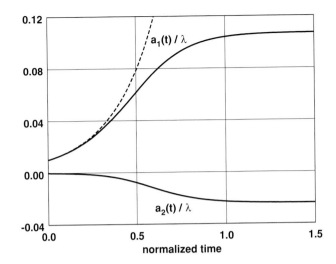

Fig. 9.6. Evolution of normalized amplitudes $a_1(t)/\lambda$ and $a_2(t)/\lambda$ according to the differential equation (9.63). The dashed curve is the solution of the linearized differential equation $(9.63)_1$ for $a_1(t)/\lambda$. The normalized timescale is $(1 + \nu)^4 \hat{t}/\pi^4$ where $\hat{t}$ is defined in (9.62).

was the basis for choosing the value of $\frac{4}{3}$ for λ/λ_{cr} in doing numerical calculations in Section 8.4.4.

The nonlinear coupled ordinary differential equations (9.63) are readily solved numerically. For the case when $\lambda = \frac{4}{3}\lambda_{cr}$, $\hat{a}_1(0) = 0.01$ and $\hat{a}_2 = 0$, numerical results are shown in Figure 9.6. The solid lines represent $a_1(t)/\lambda$ and $a_2(t)/\lambda$ as indicated; the dashed line represents the solution of the *linear* equation (9.64) for $a_1(t)\lambda$ for the same values of system parameters. For very small normalized time on the scale of the plot, the system response is consistent with the linear theory. The amplitude $a_1(t)\lambda$ begins to increase exponentially, following the solution of the linear problem which is shown by the dashed curve in the figure; the first harmonic amplitude remains essentially zero during this time. After the amplitude $a_1(t)/\lambda$ becomes large enough to bring second order effects into play, the response diverges dramatically from linear behavior, mainly due to the geometry dependence of the driving forces. After a relatively long time, the solution approaches an equilibrium state that corresponds to the intersection of the two curves $\mathcal{Q}_1 = 0$ and $\mathcal{Q}_2 = 0$ in Figure 8.15. The trajectory of the configuration in the plane of a_2 and a_1 passes through the crescent-shaped region of Figure 8.15 as anticipated in Section 8.4.4, starting from $a_1(0)/\lambda = 0.01$, $a_2(0)/\lambda = 0$ and terminating at $a_1(\infty)/\lambda = 0.11$, $a_2(\infty)/\lambda = -0.025$.

Although the development in this section is nonlinear, it is sufficiently transparent to reveal several significant features of behavior. For one thing, it provides quantitative information on the limit to the range of *linear* behavior, as illustrated in Figure 9.6. Secondly, the departure in surface shape from a simple sinusoid to a morphology with sharper valleys and flatter peaks emerges. Finally, the results illustrate a situation in which a system that is globally unstable from the point of view of linear perturbation evolves to a nearby stable equilibrium configuration according to the nonlinear formulation. The general case of evolution of the periodic surface shape under plane strain conditions, as governed by (9.9) but without restriction on the magnitude of accessible surface slopes, must be analyzed numerically. This has been done by means of different methods but with consistent results by Chiu and Gao (1994), Yang and Srolovitz (1994) and Spencer and Meiron (1994). The numerical solutions showed that the valleys in the periodic surface become narrower or sharper as time goes on, eventually becoming cusp-like as the surface tangent line becomes nearly perpendicular to the original nominally flat surface orientation.

9.4 Growth of islands with stepped surfaces

A description of surface energy anisotropy for stepped or vicinal crystallographic surfaces was outlined in Section 8.8.3. Key factors in determining the difference in surface energy between a vicinal surface and that of the nearby high symmetry surface are the step formation energy and the interaction of steps through their long-range elastic fields. These factors are represented by the parameters β_1 and β_3 in the expression (8.138) for dependence of surface energy density on angular orientation with respect to the high symmetry orientation. Values of these parameters for particular materials to be used in a continuum model must be inferred from observations or estimated through atomistic simulations. Once these values are known, no additional parameters are needed to develop a continuum picture of relaxation. For unstrained crystals, the surface energy of a vicinal surface is invariably higher than that of the nearby high symmetry surface.

Because values of step formation energy β_1 depend on changes in lengths of chemical bonds, it is likely that the values will be sensitive to imposed background strain in the crystal. This point was pursued in Section 8.8.3 where it was demonstrated that there is a potential for the step formation energy to become negative for strain of one sign or the other, if that strain is sufficiently large in magnitude. This difference has important ramifications for island formation and evolution. It was shown in Section 8.9.3 that a positive step formation energy has a strong stabilizing effect on surfaces in atomically flat, high symmetry orientations, thereby overriding the type of surface instability in strained materials discussed in Section 8.4.1 for isotropic surface energies. An incipient island, if it is to form at all in this case, must overcome a nucleation barrier. However, under conditions for which the step formation energy is negative due to strain, the surface energy has a destabilizing

influence and islands can form spontaneously from strained deposits on a surface
of high symmetry orientation.

The purpose in this section is to illustrate island formation and evolution under
circumstances in which strain destabilizes the growth surface, following the work of
Shenoy and Freund (2002). This is approached by adopting the description of surface
energy developed throughout Section 8.8.3, incorporating it into the representation
of free energy in Section 8.8.1, and applying the variational principle of Section 9.3
to solve for the evolving surface profiles as seemingly faceted islands emerge.

9.4.1 Free energy change

The analysis is based on a physical system consisting of an isotropic elastic half-
space $y \leq 0$ onto which a mismatched isotropic elastic material is deposited at
certain flux rate f starting at time $t = 0$. This flux, which is assumed to be constant,
has physical dimensions of volume per unit area per unit time. As time increases,
the mean deposited film thickness is ft. The analysis is limited to two-dimensional
configurations with all fields being uniform in the z-direction, but this is not an
essential restriction. The profile of the material surface at any time is $y = h(x, t)$.
No features of behavior are precluded by limiting consideration to surfaces inclined
by small angles from the growth surface, so it is assumed throughout that $|h,_x| \ll 1$
for all x and t.

The deposited material is assumed to have an equi-biaxial elastic mismatch ϵ_m
in the plane of the interface with respect to the substrate, and the elastic constants
of the film are taken to be the same as those of the substrate. The total stress (strain)
at any material point is the sum of the mismatch stress (strain), which is zero in the
substrate, and a contribution due to change in shape of the film surface, that is,

$$\sigma_{xx} = \sigma_{xx}^{(h)} + \sigma_m, \quad \sigma_{yy} = \sigma_{yy}^{(h)}, \quad \sigma_{xy} = \sigma_{xy}^{(h)}$$

$$\epsilon_{xx} = \epsilon_{xx}^{(h)} + \epsilon_m, \quad \epsilon_{yy} = \epsilon_{yy}^{(h)} - \frac{2v}{1-v}\epsilon_m, \quad \epsilon_{xy} = \epsilon_{xy}^{(h)}, \tag{9.65}$$

where it is understood that σ_m and ϵ_m vanish for $y \leq 0$. To first order in surface
slope, the components of the surface normal vector n_i are $n_x = -h,_x, n_y = 1$ and
of tangent vector m_i are $m_x = 1, m_y = h,_x$.

The surface energy density for small slope is adapted from (8.141) as

$$U_S = (\gamma_0 + \tau_0\epsilon_m) + (\beta_1 + \hat{\beta}_1\epsilon_m)|h,_x| + \beta_3|h,_x|^3, \tag{9.66}$$

where γ_0 is the surface energy density of the high symmetry surface in the absence
of strain, τ_0 is the surface stress of the high symmetry surface, β_1 is the step formation energy on the vicinal surface in the absence of strain, $\hat{\beta}_1$ is the sensitivity of

step formation energy to strain and β_3 represents the step interaction energy. The contribution proportional to surface slope raised to the third power is the dominant factor limiting step density; even though this is a higher order contribution to the surface energy, it is *essential* that this term be retained even when considering the small slope approximation because it is the dominant effect resisting slope increase; linearization would eliminate this important physical effect. The term independent of slope does not enter into the description of surface evolution. The surface stress for small slope is

$$\tau = \tau_0 + \hat{\beta}_1 |h_{,x}| . \tag{9.67}$$

With these small slope approximations made explicit, the traction boundary conditions (8.132) on the evolving surface can be written in terms of surface shape and system parameters as

$$\sigma_{xy}^{(h)} = \sigma_m h_{,x} + \hat{\beta}_1 \mathrm{Sgn}[h_{,x}] h_{,xx} \equiv p_x , \quad \sigma_{yy}^{(h)} = \tau_0 h_{,xx} \equiv p_y , \tag{9.68}$$

where $\mathrm{Sgn}[\,\cdot\,]$ denotes the algebraic sign of its argument. The physical interpretation of the contributions to these boundary conditions is relatively straightforward. The first term in (9.68) is necessary to negate the traction induced on the growth surface by the elastic mismatch field. As is to be expected for a stepped surface, the remaining terms in the boundary traction are proportional to the surface curvature $h_{,xx}$. The surface slope arises from a distribution of surface steps, and the influence of these steps is captured in the continuum model by representation as a distribution of dipoles. A spatially uniform distribution of surface dipoles corresponding to $h_{,xx} = 0$ has no resultant force locally, whereas a nonuniform dipole distribution does give rise to a local force distribution that is proportional to the local curvature.

The difference between the value of total free energy $\mathcal{F}$ of the system with a wavy surface and the corresponding value for a flat surface at the same time t, say $\mathcal{F}_0$, is then given as a functional of the surface shape by

$$\mathcal{F}(t) = \mathcal{F}_0(t) + \int_{S'} \Big[(\beta_1 + \hat{\beta}_1 \epsilon_m) |h_{,x}(x, t)| + \beta_3 |h_{,x}(x, t)|^3$$
$$+ \tfrac{1}{2} \big(p_x(x, t) u_x^{(h)}(x, t) + p_y(x, t) u_y^{(h)}(x, t) \big) \Big] dx , \tag{9.69}$$

where $u_i^{(h)}(x, t)$ is the equilibrium surface displacement due to application of the surface traction $p_i(x, t)$ for any surface shape. The displacement is related to the traction through the surface Green's function as established by the Flamant solution of the theory of elasticity. Consequently, the free energy can be calculated for any surface shape $h(x, t)$.

9.4.2 *Formation and interaction of islands*

For surface transport dominated by resistance to diffusion on the terraces of the stepped surfaces, the surface mass flux $j(x, t)$ is related to surface chemical potential according to (9.7). In the present instance, the dependence of the chemical potential on surface orientation is discontinuous at the high symmetry orientation. As a result, the gradient of the chemical potential along the surface is strongly singular and direct numerical solution is complicated; Spohn (1993) and Hager and Spohn (1995) have shown that the evolution can be approached as a free boundary problem. The variational principle developed in Section 9.3 provides an attractive alternative as the basis for numerical analysis of the evolution. It can be used without the need for an ad hoc regularization of the singularity in chemical potential gradient, and it can lead to a numerical solution of high accuracy in the form of a series representation of the surface shape, provided that convergence in time can be demonstrated.

It is assumed that the surface profile for $t > 0$ is described by the finite series

$$h(x, t) = \sum_{k=0}^{N_h} a_k(t) \cos \frac{2\pi k x}{\lambda} \qquad (9.70)$$

within an interval $0 \le x \le \lambda$, where the parameter N_h is a specified integer. Obviously, the solution is assumed to be spatially periodic with period λ. The idea is to select λ to be large enough (or, equivalently, to select N_h to be large enough) to include many incipient islands within the range. The objective is to understand how these incipient islands grow and interact with each other over the course of time. The Fourier cosine series is a complete representation of all moderately smooth symmetric functions in the interval $0 \le x \le \lambda$. The concern in application of the variational approach is the behavior of $a_k(t)$ as $t \to \infty$ for large values of k.

For any surface shape (9.70), the mass flux distribution is related to shape change through local mass conservation. The surface mass conservation requirement is given in (9.8) in the absence of a deposition flux. In the present instance, the mass conservation relation is modified through the inclusion of deposition flux, and it takes the form

$$h_{,t} + j_{,x} = f , \qquad (9.71)$$

where the small slope approximation $v_n \approx h_{,t}$ has also been incorporated. In light of (9.70), the surface mass flux can be determined by means of (9.71) to be

$$j(x, t) = \sum_{k=1}^{N_h} \dot{a}_k(t) \frac{\lambda}{2\pi k} \sin \frac{2\pi k x}{\lambda} . \qquad (9.72)$$

The mean surface height is $a_0(t) = ft$.

A key observation in applying the variational principle is in recognizing that the first term on the right side of (9.51) which defines the functional Φ is the time rate of change of free energy $\mathcal{F}$. Furthermore, in light of the explicit spatial dependence of surface mass flux in the representation (9.72), the second term in the definition of Φ can be readily integrated. Recognizing that the free energy is completely determined by the values of the coefficients $a_k(t)$, the functional of surface shape Φ is reduced to an ordinary function of the time rates of change $\dot{a}_k$ of the amplitudes representing surface shape in (9.70), that is,

$$\Phi(\dot{a}_1, \dot{a}_2, \ldots) = \mathcal{F}(a_0, a_1, a_2, \ldots) + \frac{1}{2m_S} \sum_{k=1}^{N_h} \frac{\pi}{k} \left(\frac{\lambda}{2\pi} \right)^3 \dot{a}_k^2. \qquad (9.73)$$

The optimum choice of rates $\dot{a}_k$ is the set that renders Φ a maximum with respect to variations in those rates, which leads to a system of nonlinear ordinary differential equations for $a_k(t)$ of the form

$$\frac{\partial \Phi}{\partial \dot{a}_k} = 0 \quad \Rightarrow \quad \dot{a}_k(t) = -\frac{m_S k}{\pi} \left(\frac{2\pi}{\lambda} \right)^3 \frac{\partial \mathcal{F}}{\partial a_k} (a_0(t), a_1(t), \ldots) \qquad (9.74)$$

for $k = 1, \ldots, N_h$. These differential equations, augmented by suitable initial conditions at $t = 0$, must be integrated numerically. The integration is fairly straightforward if the mismatch strain ϵ_m is allowed to extend indefinitely into the substrate. On the other hand, it is significantly more difficult to execute if the mismatch strain is restricted to the deposited material only, that is, if it is restricted to $y \geq 0$. In the latter situation, the value of σ_m must be set to zero for any surface points extending into $y < 0$. Only results of calculations in which the substrate is *unstrained* are included here (Shenoy and Freund 2002).

The numerical integration of (9.74) was carried out using the fourth-order Runge–Kutta procedure with adaptive step-size control as described in detail by Press et al. (1992). The parameters in the continuum description of the energy were chosen to be $\beta_1 = 0.03\,\mathrm{J\,m^{-2}}$, $\hat{\beta}_1 = 15\,\mathrm{J\,m^{-2}}$, $\tau_0 = 1\,\mathrm{J\,m^{-2}}$, $\epsilon_m = -0.01$ and $\beta_3 = 2.86\,\mathrm{J\,m^{-2}}$. It is assumed that the elastic modulus and the Poisson ratio are $10^{11}\,\mathrm{N\,m^{-2}}$ and 0.3, respectively, for both the film and substrate materials. For the compressively strained film, the surface energy of the film (sketched schematically in Figure 8.26) attains a minimum when $\theta^* = 0.12$, which implies that the side walls of the stepped mounds would evolve naturally toward this angle.

The initial profile of the film was chosen to be a sinusoid with a wavelength of $\lambda = 400\,\mathrm{nm}$ and with an amplitude of $0.4\,\mathrm{nm}$; the only non-vanishing Fourier component amplitudes at $t = 0$ were a_0 and a_1. In these calculations, sixteen Fourier coefficients were used to represent the surface shape, so $N_h = 16$. The evolution of the deposited film is shown in Figure 9.7. As the deposition flux is turned on at $t = 0$, the material on the surface quickly gathers into five stepped mounds with

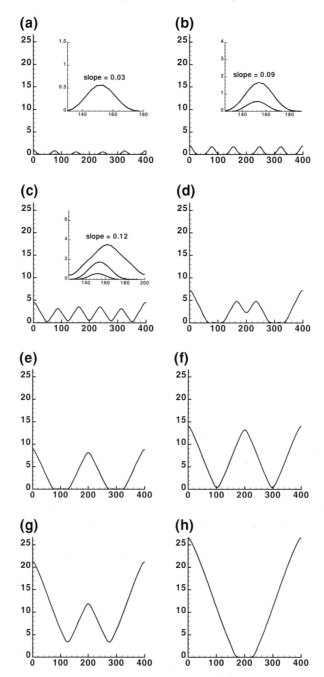

Fig. 9.7. The figure shows a time sequence of surface profiles of $h(x, t)$ versus x for a strained film with a constant deposition flux onto a lattice-mismatched substrate. All the dimensions are in nanometers. The insets in (a)–(c) show the evolution of the third island from the right. To aid in the comparison of shapes at different times, the island shape from (a) has been included in (b). Similarly, the island shapes from (a) and (b) in (c). The slope of the largest island in each of the smaller insets is indicated.

slopes much smaller than the optimum value of 0.12 as shown in Figure 9.7. The insets in Figures 9.7(a)–(c) track the evolution of one of these mounds. As more material is deposited, the side walls of the mound being tracked become steeper until they reach the optimum angles $\theta = \pm 0.12$ in Figure 9.7(c). It can also be observed that the center of mass of the mound shifts gradually to the right in going from Figure 9.7(a) to Figure 9.7(c). This shift can be understood by looking at the interactions between the islands. Since elastic relaxation is achieved for widely spaced islands, the islands tend to repel each other (Floro et al. 2000); examples of such interactions were also illustrated in Figure 8.40. The island in Figure 9.7(a) is located closer to the island on its left and therefore would tend to shift towards the right via diffusion of atoms from the side wall on the left to the one on the right. Once the side walls of the islands reach the optimum orientation, they grow in a more or less self-similar fashion until they come in contact with their neighbors as shown in Figure 9.7(c). At this point, self-similar coarsening is initiated, which leads to a decrease in the fraction of the surface of the substrate covered by the film, as is evident in Figure 9.7(e). Here there are two islands, with side walls at $\theta = \pm 0.12$, separated by 50 nm. As more material is deposited, these further grow in size and, eventually, one of the islands grows at the expense of the other island as shown in Figure 9.7(f).

The evolution of the film in the early stages of island growth is in agreement with the observations of Sutter and Lagally (2000) and Tromp et al. (2000) on Ge/Si(001). The key result of these experiments is that the islands evolve as a *natural instability* without any nucleation barrier. As was shown in Section 8.9.3, there is no barrier for nucleating stepped islands in this material system if the strains are compressive. Islands with orientations below the optimum face orientation lower the energy of the film and provide a kinetic pathway for obtaining faceted islands that is free of any nucleation barrier. This is indeed what is seen in experiments and during early stages of growth in Figures 9.7(a–c). Furthermore, observations indicate that evolving islands show crystallographic features from the outset, which implies that the island surfaces evolve as vicinal surfaces.

It can be seen from Section 8.9 that the energy of the islands is dominated by the surface energy at small island volumes and by the elastic energy at large volumes. Using the parameters adopted in these calculations, it can be seen that the crossover between these two regimes takes place when the base width of the island is about 200–300 nm. When the island sizes are smaller than this value, the side walls are oriented at angles close to the optimum angle that minimizes the surface energy. With increasing base width, it is known that the islands undergo shape transitions (Medeiros-Ribeiro et al. 1998, Ross et al. 1999), where the side wall angles change to a steeper orientation, which is usually a combination of low energy crystallographic orientations as discussed in Section 8.8. The slope of

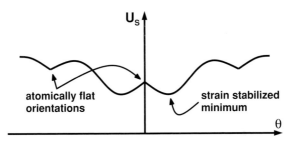

Fig. 9.8. A schematic of surface energy density versus surface orientation in a single plane, suggesting the possibility of multiple local minima in the dependence. In this particular diagram, the minima closest to the reference singular or atomically flat surface orientation at $\theta = 0$, which is destabilized by strain, are smooth minima representing stable stepped surfaces. Other singular orientations, represented as stable in this diagram, occur at larger angular deflections from the reference surface. The faces corresponding to combinations of local minima, perhaps some strain-stabilized vicinal orientations and other stable singular orientations, could coexist to form the island dome structures discussed in Section 8.9.5.

the side walls of the large islands in Figure 9.7(e) is about 0.15, which is about 25% larger than the optimum slope. This indicates that elastic energy variations of these larger islands are becoming comparable to the surface energy variations. There is no fundamental impediment to including additional low energy surface orientations in order to study steeper side wall facets. For example, these might appear as additional relative minima in the variation of $U_S(\theta)$ with orientation θ at angles greater in magnitude than θ^*, as shown in the sketch in Figure 9.8. Pioneering measurements of the dependence of energies of surfaces of strained crystals on orientation are discussed by Blakely (1973); such data are central to understanding the evolution of small crystal structures and this is an area requiring further study.

Island growth behavior of the type described here is typical of the SiGe/Ge material systems in the presence of compressive mismatch strain in the deposited film. Images illustrating the dominance of the surface orientation of about 11 degrees with respect to the growth surface as a result of the strain-induced surface energy minimum have been shown in Figures 8.12 and 8.38. A particularly distinct hut structure image with nearly perfect (105) faces is shown in Figure 9.9 (Medeiros-Ribeiro et al. 1998), the material system in this case is Ge/Si(001). The left front edge of the pyramid, where adjacent {105} faces intersect, appears to have features twice as large as those on the {105} faces. A possible interpretation is that the faces are {001} vicinal surfaces with single height steps aligned with ⟨100⟩ directions, whereas the edge region is a {001} vicinal surface with double-height steps aligned with a ⟨110⟩ direction (Shenoy et al. 2002). A number of intriguing images from experiments in the SiGe/Si(001) material system were also reported by Rastelli et al. (2001). In a series of experiments, they deposited Ge onto Si (001) and, as expected, the deposited Ge evolved into the islands with quite steep and complicated

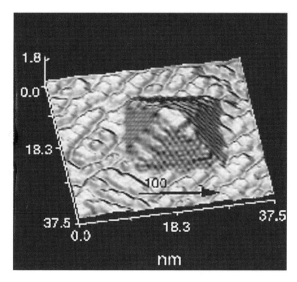

Fig. 9.9. An island configuration observed by scanning tunneling microscopy in the course of depositing Ge islands on a Si(001) surface. The base of the island covers an area of 18×18 square nm, the height of the peak above the base is 1.8 nm, the lateral surfaces are $\{105\}$ surfaces and the island incorporates approximately 6,000 atoms. Reproduced with permission from Medeiros-Ribeiro et al. (1998).

faces almost immediately. Then, Si was deposited onto the islanded surface. In the course of deposition of Si, the islands changed through a sequence of shapes. This presumably occurred because the Si was absorbed into the islands, thereby reducing the mismatch strain magnitude which, in turn, increased the role of surface energy relative to elastic energy. It is also possible that the varying composition influenced the surface energy in ways that are not yet understood.

9.5 Diffusion along interfaces

The role of grain boundary diffusion in stress relaxation in a thin polycrystalline film on a substrate is considered here as a vehicle for illustrating the use of the variational principle for mass flux. This process, an example of which is illustrated in Figure 7.28, is rich in physical phenomena and includes the competition of various nonequilibrium processes to achieve stress relaxation; only the most rudimentary aspects will be incorporated in the course of developing an illustration. Those aspects of the relaxation process that are being overlooked will be noted along the way. More realistic models have been discussed by Thouless (1993) and Gao et al. (1999). The discussion of grain boundary diffusion is followed by an example of a significantly enhanced diffusion along a deformation band during the application of stress whereby nanocrystalline particles form locally.

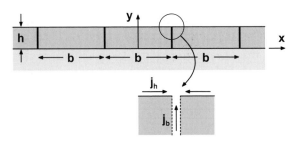

Fig. 9.10. Schematic diagram of a polycrystalline thin film of thickness h on a substrate. Grains are of width b and the system is periodic in the x-direction. Initially, the film supports a mismatch tensile stress in the direction of the interface, which induces a configurational force tending to drive mass transport from the free surface into the grain boundaries. The mass fluxes j_h along the free surface and j_b along the interface are estimated through the use of the variational principle based on the functional in (9.51).

9.5.1 Stress relaxation by grain boundary diffusion

Consider a polycrystalline film of initial thickness h_0 on a substrate. The state of deformation is assumed to be plane strain. All grains have the same initial width b_0, as indicated in the sketch in Figure 9.10, and have relatively large extent in the direction normal to the plane of the sketch. The grain boundaries are perpendicular to the plane of the film, and the configuration is periodic in the direction along the film. The top surfaces of the grains are free; the bottom faces are confined to slide along the substrate surface but are otherwise unconstrained. Any deformation of the substrate is neglected. Initially, the film supports a uniaxial extensional elastic strain ϵ_0. This strain will be assumed to be positive to facilitate the discussion, but the case $\epsilon_0 < 0$ is identical except for the algebraic signs of all first-order fields representing physical quantities. The mechanical response of each grain is assumed to be linearly elastic for the time being, with plane strain modulus $\bar{E}$.

According to (8.14), the initial tensile stress $\bar{E}\epsilon_0$ acting on the grain boundary implies a layer of large negative chemical potential. On the free surface of the film, on the other hand, the strain energy density is everywhere positive and the curvature of the surface is zero. Consequently, the chemical potential is positive and there is a thermodynamic driving force tending to move material from the free surface and into the grain boundary.

Suppose that a mass transport process actually takes place in response to this configurational force. Mass flow along the free surface in the positive x-direction is represented by the flux $j_h(x, t)$ and mass flow along the grain boundary in the positive y-direction is represented by $j_b(y, t)$. No other transport processes are active. The objective of this example is to arrive at estimates of the fluxes by appeal to the functional (9.51) underlying the variational principle stated earlier in this section.

In constructing an admissible mass flux field, it is assumed that the free surface of the grain remains essentially flat, corresponding to a relatively high surface energy compared to the interfacial energy of the grain boundary. The unstrained film thickness at time t is $h(t)$, where $h(0) = h_0$. It is also assumed that the surface flux is symmetric with respect to the center of the grain, with the flux being toward the nearer grain boundary at all points on the surface. A surface flux consistent with these characteristics is

$$j_h(x, t) = -\dot{h}(t)x . \tag{9.75}$$

Within each grain boundary, it is assumed that the rate of material addition to each grain is spatially uniform, which requires the base of the grain to slide freely along the substrate surface. This is an extreme assumption that can be overcome in a number of ways by allowing for spatially nonuniform deformation. However, the complexity added by taking into account nonuniform deformation obscures the main points being illustrated, so the assumption of an unconstrained base remains in place. At the junction of the grain boundary and the film–substrate interface, the grain boundary flux must vanish. It is also assumed that the grain boundary neither distorts out of its plane nor migrates along the film. The unstrained grain width at time t is $b(t)$, with $b(0) = b_0$. A grain boundary flux consistent with these characteristics is

$$j_b(y, t) = -\dot{b}(t)y . \tag{9.76}$$

At the junction of the grain boundary and the free surface, mass conservation requires that the fluxes are related according to

$$2 j_h(b/2, t) + j_b(h, t) = 0 , \tag{9.77}$$

which implies that the unstrained cross-sectional area of each grain $b(t)h(t) = b_0 h_0$ remains unchanged.

Under the assumptions adopted, the chemical potentials are spatially uniform over both the free surface and the grain boundary. The free surface chemical potential is denoted by $\chi_h(x, t)$ and is determined according to (8.8), while the grain boundary chemical potential is denoted by $\chi_b(y, t)$ and is determined according to (8.14). In terms of instantaneous grain dimensions, these are

$$\chi_h = \frac{1}{2}\bar{E}\left(\epsilon_0 - \frac{b(t) - b_0}{b_0}\right)^2 , \quad \chi_b = -\bar{E}\left(\epsilon_0 - \frac{b(t) - b_0}{b_0}\right) , \tag{9.78}$$

where the term $(b(t) - b_0)/b_0$ is the stress-free strain along the film due to mass rearrangement.

Finally, the mass fluxes j_h and j_b are assumed to be related to gradients in the corresponding chemical potentials according to a transport equation of the form (9.7) with mobility parameters m_h and m_b, respectively. In order to consider the implications of the variational principle, values of the free surface mobility parameter m_h and the grain boundary mobility parameter m_b are required. Only the relative magnitude is important for this illustration. For copper at 327 °C, these were estimated by Thouless (1993) to have the values

$$m_h = 2.35 \times 10^{-32} \, m^6 (J \, s)^{-1}, \quad m_b = 5.27 \times 10^{-33} \, m^6 (J \, s)^{-1}. \tag{9.79}$$

For present purposes, it is only important to note that the surface mobility is usually found to be about 4 to 10 times larger than the grain boundary mobility.

The functional in (9.51) is evaluated over the region $0 < x < b, 0 < y < h$ which includes a volume of material that exchanges no energy with its surroundings, due to symmetry in this case; the condition of no energy change with external sources satisfies one of the important constraints on the variational principle as developed here. The two rate variables $\dot{h}$ and $\dot{b}$ are related through the mass conservation relation (9.77), and $\dot{b}$ is chosen arbitrarily as the independent variable in the argument of the functional. The result is

$$\Phi[\dot{b}(t)] = -\chi_h \dot{h} b - \chi_b \dot{b} h - \frac{\dot{h}^2 b^3}{24 m_h} - \frac{\dot{b}^2 h^3}{6 m_b}, \tag{9.80}$$

where $h = b_0 h_0 / b$ and $\dot{h} = -\dot{b} b_0 h_0 / b^2$. The property that the functional is stationary for the optimum transport field, within the range of admissible fields, means that $b(t)$ should be chosen to render $\partial \Phi / \partial \dot{b} = 0$, which implies that it should satisfy the ordinary differential equation

$$\frac{1}{2} \bar{E} \left(\epsilon_0 - \frac{b - b_0}{b_0} \right)^2 + \bar{E} \left(\epsilon_0 - \frac{b - b_0}{b_0} \right) = \frac{\dot{b} h_0 b_0}{12 m_h} + \frac{\dot{b} b_0^2 h_0^2}{3 m_b b^2}. \tag{9.81}$$

Upon comparing the two terms on the left side of this equation, it is noted that the first term is essentially an elastic strain times the second term. Because elastic strain is assumed to be very small compared to one, the first term is negligible compared to the second and its contribution is ignored. If the differential equation is expressed in terms of the stress-free strain along the film $\eta(t) = (b(t) - b_0) / b_0$ and it is then linearized for $\eta(t) \ll 1$, it is found that

$$\eta(t) = \epsilon_0 \left(1 - e^{-t/t_0} \right), \tag{9.82}$$

where

$$t_0 = \frac{h_0 b_0}{48 \bar{E}} \left(\frac{b_0}{m_h} + \frac{16 h_0}{m_b} \right) \tag{9.83}$$

is the characteristic time for the process. It is clear from the characteristic time that, if the grain size is comparable to film thickness, the relaxation process is dominated by grain boundary mobility.

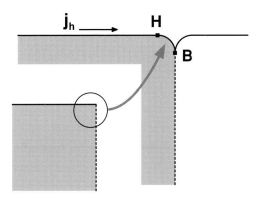

Fig. 9.11. Schematic diagram showing a possible physical structure for the transition from the free surface of a grain to the grain boundary for the case of a polycrystalline film illustrated in Figure 9.10. The rapid transition in chemical potential through this small region provides little resistance to mass transport in this case.

The influence of a contribution to the functional that arises from the 'corner' where the free surface of the grain meets the grain boundary, represented by the integral over C in (9.55), has been ignored in this example. This may be justified by considering the structure of this corner on a smaller size scale. For example, the corner may have the appearance suggested by Figure 9.11 where the radius of curvature of the arc from point H to the point B is everywhere very small compared to grain size and film thickness. The chemical potential must vary continuously from χ_h at point H to χ_b at point B, and the mass flux is essentially uniform throughout the arc with value j_h. Integration of (9.52) over this arc yields

$$j_h \Delta s = -m_h (\chi_b - \chi_h) , \qquad (9.84)$$

where Δs is the arclength from point H to B. On a larger scale, for which the details of the corner structure are indiscernible, the corner transport is described by (9.54) which, in the notation of the example being developed in this section, is

$$j_h = -m_C (\chi_b - \chi_h) . \qquad (9.85)$$

But consistency of these two results requires that

$$m_C = \frac{m_h}{\Delta s} = \frac{m_h}{b} \frac{b}{\Delta s} . \qquad (9.86)$$

Because $b/\Delta s \gg 1$, m_C is very large compared to m_h/b and the corner effect is indeed negligible in this case. Other lines of reasoning must be followed if the physical structure of the corner differs from that imagined in this illustration.

A model of the process that includes the resistance of an elastically deforming substrate and also includes nonuniform mass flux within the grain boundary has been analyzed by Gao et al. (1999). These effects both retard the process compared to the simple picture developed here. Thouless (1993) showed that surface diffusion acts in series with other diffusive mechanisms of mass transport. Nonequilibrium surface curvatures that develop during grain creep were found to induce back stresses that inhibit grain boundary diffusion, thus retarding the relaxation process. This rate limiting role of surface diffusion would probably be enhanced if three-dimensional effects were taken into account, which would increase the amount of grain boundary per unit area of film–substrate interface; the additional grain boundaries would compete for mass diffusing along the free surfaces of grains and the relaxation process would be further retarded.

In the foregoing discussion, it was assumed that deposition of film material was not occurring during relaxation by grain boundary diffusion and that the free surface of the film was an equilibrium surface at the prevailing temperature. It was observed in Chapter 1 that a growth flux provides a supersaturated distribution of adatoms on the growth surface. These atoms have free energy in excess of that of equilibrium surface atoms. This excess energy may be sufficient to drive material into the grain boundaries, even in the absence of elastic strain to be relaxed. The incorporation of excess atoms into the grain boundaries tends to produce a compressive stress in the film. In the absence of flux, (9.82) implies that elastic strain eventually decays to zero amplitude. In the presence of the growth flux, however, it is conceivable that film stress would become compressive, even though this implies a positive chemical potential within the grain boundaries with respect to an equilibrium film surface. Such stress evolution phenomena in thin polycrystalline films are not yet fully understood.

9.5.2 Diffusion along shear bands during deformation

The discussion up to this point has focused on the role of free surfaces and internal interfaces, such as grain boundaries, in mass diffusion. Surfaces produced internally in the material as a consequence of permanent deformation and damage induced by stress can also serve, in some cases, as paths along which enhanced atomic diffusion may occur. In amorphous solids undergoing active plastic flow, such increased atomic mobility along shear bands can result in the formation of nanocrystalline particles locally at the bands. An example of such crystallization processes is illustrated in this section for the case of a bulk amorphous metallic alloy subjected to quasi-static nanoindentation at room temperature.

Glassy metal alloys, in thin film or bulk form, are not in a state of stable equilibrium. Consequently, they undergo crystallization when heated. Subjecting thin ribbons of amorphous glasses to severe bending or high energy ball milling or mechanical alloying is also known to result in the formation of shear bands within

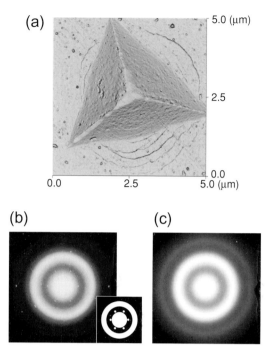

Fig. 9.12. (a) Atomic force microscope image of the impression created on a Zr–17.9Cu–14.6Ni–10Al–5Ti (atomic percent) bulk metallic glass alloy which was subjected to nanoindentation at a maximum load of 60 mN. Discontinuous shear bands encompass the indent. (b) SAD patterns showing diffraction spots which were produced by the formation of nanocrystalline particles at the indents and in the shear bands. The inset schematically shows six diffraction spots which were associated with the (111) plane of tetragonal Zr_2Ni particles. (c) A small distance away from the indent only halo ring patterns characteristic of a fully amorphous structure are seen. Reproduced with permission from Kim et al. (2002).

which nanocrystalline particles nucleate (Chen et al. 1994). Substantial local heating due to adiabatic shear is possible in thin ribbons of amorphous solids during severe plastic deformation, dynamic loading and fracture, and consequently, the occurrence of crystallization during plastic flow in these loading situations has often been associated with local temperature rise. Experiments, however, have been performed by Kim et al. (2002) in bulk amorphous alloys stressed to form crack-like shear bands that are connected to ideal heat sinks around quasi-statically produced nanoindents where the possibility of any appreciable local heating is essentially fully suppressed. Figure 9.12(a) shows an atomic force microscope image of the impression produced on the surface of a Zr-based bulk amorphous alloy which was subjected to a slow indentation to a maximum load of 60 mN; application of this indentation load to the alloy caused the pyramid-shaped diamond indenter tip to penetrate the surface to a maximum depth of approximately 720 nm (see

Section 7.9.5). Discontinuous shear bands formed in the vicinity of the indentation are clearly evident in this figure. Figure 9.12(b) is a selected area diffraction (SAD) pattern obtained in an electron microscope in the center region of a similarly produced indenter impression where the diffraction spots associated with the presence of nanocrystalline particles are clearly visible; the inset in this figure schematically shows the location of the diffraction spots. Electron diffraction results in both a diffused halo ring pattern, which is indicative of the amorphous material, and distinct spots, which are indicative of nanocrystalline particles, because the aperture size for SAD is about the same as the indent size. The six spots visible in Figure 9.12(b) were close to the exact Bragg condition, and analysis by Kim et al. revealed that they were produced by diffraction from the (111) plane of tetragonal Zr_2Ni nanocrystalline particles.

Preferential crystallographic alignment of these particles appears to have been induced by nanoindentation. Transmission electron microscope (TEM) observations showed the presence of these nanoparticles, about 10 to 40 nm in diameter with a median size of approximately 20 nm, along the faces of the indent and in the regime directly beneath the center of the impression. These crystalline particles were also found to be the same as those nucleated in the same alloy at an annealing temperature of 783 K in the absence of any mechanical deformation (Wang et al. 2000). In regions a small distance away from the indent, no crystalline particles form, and the SAD pattern shown in Figure 9.12(c) reveals only halo rings representative of a fully amorphous structure. An identical glassy structure with diffuse ring patterns and no diffraction spots was also seen in the as-received, unindented metallic glass on which TEM observations were made using thin foil preparation techniques that were identical to the ones employed to produce the information in Figure 9.12.

Plastic deformation in a disordered, amorphous solid at temperatures well below the glass transition temperature T_g is expected to engender flow dilatation which is associated with intense shear localization along well defined shear bands such as those shown in Figure 9.12(a) (Spaepen 1977). Direct diffusion measurements on amorphous polymers at temperatures well below T_g have further shown that the atomic level topological features and diffusional mobilities inside actively deforming shear bands are nearly the same as those at T_g (Zhou et al. 2001). In particular, it was shown that the rate of penetration of a low molecular weight diluent into a high molecular weight glassy polymer subjected to active plastic deformation at 386 K was essentially the same as the penetration rate of the former into the latter at the glass transition temperature of 486 K in the absence of any plastic flow. This deformation-induced enhancement in diffusion constant was estimated by Zhou et al. to be as high as four orders of magnitude. On the other hand, an increase in local dilatation or free volume due to an increase in temperature would be expected to result in a marked decrease in interatomic interactions. Although comparable direct measurements of diffusion rates during deformation have not been performed in amorphous metals, the nanocrystallization phenomenon produced during deformation in the indentation experiments shown in Figure 9.12 also appears to arise from the pronounced increase in atomic diffusion within shear bands of the metallic glass as a result of flow dilatation.

9.6 Compositional variations in solid solutions

Throughout the foregoing chapters, frequent reference has been made to the behavior of alloy materials, either compounds or solid solutions, in thin films and layered materials. In particular, alloying is the origin of the elastic mismatch strain ϵ_m in epitaxial films in some cases. The mismatch strain provides the driving force for dislocation formation, surface morphology change and other physical processes. The goal of this section is to introduce phenomena involving alloys with spatially nonuniform composition and to examine the role of mechanical stress in determining the stability of compositions or in influencing the time evolution of compositional variations.

Attention is focused mainly on the behavior of a *solid solution*, which is a continuous sequence of substances with compositions varying between those of two chemically distinct homogeneous materials at the extremes of the range of composition. Perhaps the most thoroughly studied semiconductor solid solution is $Si_{1-\xi}Ge_{\xi}$ where ξ is the atomic fraction or volume per unit volume of Ge in the solution. The symbol for atomic fraction is changed from the customary x to ξ because of the need to consider spatially varying compositions for which atomic fraction ξ may be a function of spatial coordinate x. The properties of this material are assumed to vary continuously from those of elemental Si when $\xi = 0$ to those of elemental Ge when $\xi = 1$. A more complex semiconductor solid solution is $In_{\xi}Ga_{1-\xi}As$ where ξ is the atomic fraction of group III sites in the lattice of the zinc blende structure occupied by In atoms. The properties in this case are assumed to vary continuously between those of the limiting homogeneous zinc blende phases for GaAs and InAs. In both of these examples, as well as in all examples to be discussed in this section, the two chemically distinct materials defining the extremes are crystals of the same class. Furthermore, the solutions considered are of the substitutional type, in which any given atom can occupy only particular sites. In this case, compositional changes can occur only by exchanges of atoms between sites, assuming that there is no creation or annihilation of vacancies. The likelihood of such exchanges is enhanced near lattice vacancies, and it is possible that the role of vacancies is essential in executing exchanges.

It is also assumed that the atoms are distributed among the possible sites throughout the lattice in a more or less random way. In other words, the probability that any particular site in the diamond cubic lattice of homogeneous $Si_{1-\xi}Ge_\xi$ alloy is occupied by the Ge atom is ξ. Solid solutions with this characteristic are normally said to be disordered. The notions of lattice unit cell and lattice translational invariance are not strictly valid in a disordered solid solution. Nonetheless, an effective or average unit cell dimension can be adopted for a solid solution at any composition between the limiting chemically homogeneous materials at the end points of the compositional range. For example, the average unit cell dimension of $Si_{1-\xi}Ge_\xi$ in the absence of applied stress is given by the linear interpolation between values for Si and Ge as in (1.14) (Christian 1975).

9.6.1 *Free energy of a homogeneous solution*

Consider a statistically homogeneous solid solution, with homogeneous and chemically distinct materials A and B as its end-point compositions, at some fixed temperature T. Any intermediate composition is characterized by the parameter ξ which is the fraction of lattice sites occupied by the atoms of material B; as such, this parameter is in the range $0 \leq \xi \leq 1$. In the SiGe system, the number of lattice sites involved in this comparison is eight per unit cell, which represents all possible sites. On the other hand, in the InGaAs system, only four atom sites per unit cell are occupied by either In or Ga atoms. The free energy per unit volume of the homogeneous stress-free solution at temperature T is commonly expressed in terms of the free energy densities of the pure substances under the same conditions of stress and temperature, say φ_A and φ_B, and the volume fraction ξ as

$$\varphi(\xi) = (1 - \xi)\varphi_A + \xi\varphi_B + \varphi_{mix}(\xi), \qquad (9.87)$$

where φ_{mix} represents the *free energy of mixing* of A and B. If the two component materials do not interact chemically with each other in solution, then the internal energy of mixing is zero and the free energy of mixing is only the configurational entropy of the atomic arrangement. For a random arrangement of constituents, the entropic free energy per atom is $kT[(1-\xi)\ln(1-\xi) + \xi \ln \xi]$; the coefficient is replaced by kT/Ω to represent this quantity as a free energy per unit volume (Christian 1975). A convenient empirical form to adopt for additional contributions to the free energy of a less ideal solution is an algebraic function of ξ that vanishes at the end points. The simplest among these is $\xi(1 - \xi)$, which would commonly have a (possibly temperature-dependent) positive coefficient. Tsao (1993) adopts this form for the SiGe solutions where the coefficient is estimated to be 0.045 eV/atom.

Possible dependence of free energy per unit volume $\varphi(\xi)$ on composition ξ for a binary solution is illustrated by the two examples in Figure 9.13. The main feature distinguishing one case from the other is that the curvature has the same algebraic

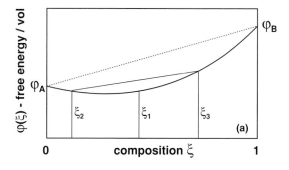

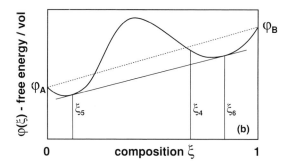

Fig. 9.13. Sample diagrams of free energy density $\varphi(\xi)$ versus fractional composition ξ of constituent B for a disordered substitutional solid solution $A_{1-\xi}B_\xi$. The free energy variation in the upper diagram has the property that $\varphi''(\xi) > 0$ over the full range which implies unconditional stability of the solution, whereas the variation illustrated in the lower diagram has points of inflection which implies more complex characteristics.

sign for all ξ in (a) whereas the curve has points of inflection in (b). The dashed line in each case represents the dependence of free energy on composition for a *perfect solution*, that is, one for which the free energy of mixing is zero.

The rate of change of free energy of a homogeneous unit volume of material, which is understood to include a fixed number of lattice sites, is

$$\dot{\varphi}(\xi) = \varphi'(\xi)\dot{\xi} \tag{9.88}$$

as composition ξ changes. The quantity $\dot{\xi}$ is the rate of increase of the volume of B within the unit volume, and it follows that the local slope $\varphi'(\xi)$ is the generalized force that is work-conjugate to the volume of B with respect to free energy, that is, it is the chemical potential of material B.

For the solution characterized by Figure 9.13(a), consider a case in which the average composition is ξ_1, but in which the volume is divided into clusters each having either composition ξ_2 or ξ_3. The free energy of the clustered volume lies somewhere on the straight line connecting the points (ξ_2, φ_2) and (ξ_3, φ_3). The chemical potential of clusters with composition ξ_2 is less than that for clusters with composition ξ_3. Consequently, there is a driving force tending to remove volume of material B from the latter clusters and to add it to the former clusters, and vice versa for material A. This situation prevails until both types of clusters approach composition ξ_1, representing a spatially uniform solution. This behavior will be established as a general result in the next subsection, that is, it will be shown that any spatially uniform solution will tend to remain spatially uniform for the case (a) in Figure 9.13.

Behavior is more complex for case (b) in Figure 9.13, which is distinguished by reversals in the sign of curvature of the curve defining $\varphi(\xi)$ over $0 < \xi < 1$. Compositions ξ_5 and ξ_6 locate the points of tangency of the only line having multiple points of tangency with the particular free energy curve illustrated. For compositions in the ranges $0 \leq \xi < \xi_5$ and $\xi_6 < \xi \leq 1$, the behavior is as described for case (a); in particular, the uniform composition in an isolated system will tend to remain uniform. On the other hand, for a uniform composition in the range $\xi_5 < \xi < \xi_6$, say ξ_4, there is a tendency toward a nonuniform redistribution of material B in the solution. If attention is limited to a combination of clusters with either of two compositions, it is evident that segregation into clusters will proceed by exchange of materials A and B in equal amounts until each cluster has a composition of either ξ_5 or ξ_6. In this configuration, the chemical potentials of the two types of clusters are equal to each other and there is no longer a driving force tending to reorganize the solution. For a solution characterized by the free energy distribution in Figure 9.13(b), the compositions ξ_5 and ξ_6 are commonly called the *limits of solubility*, and homogeneous solutions in the range $\xi_5 < \xi < \xi_6$ are unstable.

The general issue of stability of composition of a solid solution is pursued further in the next subsection. Two potentially important physical effects are not taken into account in the discussion of energy variations with composition above. One of these effects arises from the possibility of atomic misfit of one species in the solution with respect to the other. The average unit cell dimension of a solid solution may depend on the composition, so that there is a stress-free volume change (or a more complex stress-free strain, perhaps) with change in concentration. For a spatially nonuniform composition, the associated stress-free strain field will be incompatible, in general, giving rise to a residual stress distribution.

The second effect to be incorporated arises from the observation that steep spatial gradients in composition imply local energies higher than those of homogeneous samples of the same average composition. This effect is most evident at interfaces between homogeneous samples, which have an excess free energy over and above that of the homogeneous constituent solutions even with ideal epitaxial bonding. Cahn and Hilliard (1958) introduced a way to account for the role of spatial gradients in composition in the free energy of a solution, and their result is incorporated here.

Thus, the total free energy of a spatially nonuniform solid solution is the sum of three energy densities: an energy of solution determined by the local composition as discussed above, an additional energy of solution determined by the local gradient of composition, and an elastic energy density arising from misfit.

9.6.2 Stability of a uniform composition

Following the pioneering work of Cahn (1961), the Helmholtz free energy at fixed temperature of a solid solution occupying region R is given as a functional of composition $\xi(x_1, x_2, x_3, t)$ by

$$\mathcal{F}(t) = \int_R \left[\varphi(\xi) + \kappa_{ij}^{\mathrm{ch}} \xi_{,i}\, \xi_{,j} + \tfrac{1}{2} c_{ijkl} \epsilon_{ij} \epsilon_{kl} \right] dR. \tag{9.89}$$

The first term in the integrand is the free energy density based on a homogeneous solution with composition equal to the local density, as described in the preceding subsection. The second term in the integrand is an adjustment to the local free energy arising from the local gradient in composition $\xi_{,k}$ as introduced by Cahn and Hilliard (1958), where the phenomenological tensor-valued Cahn–Hilliard parameter $\kappa_{ij}^{\mathrm{ch}}$ has physical dimensions of energy/length. Only isotropic systems are considered here, in which case $\kappa_{ij}^{\mathrm{ch}} = \kappa_{\mathrm{ch}} \delta_{ij}$; the parameter κ_{ch} is a scaler quantity of order $kT_{\mathrm{ch}}/\Omega^{1/3}$ where k is the Boltzmann constant, T_{ch} is the absolute temperature beyond which a planar interface in the material system cannot be maintained, and Ω is the atomic volume (Cahn 1961). The last term in the integrand of (9.89) is the elastic energy density of the material. The strain field ϵ_{ij} represents the elastic strain only, and it does not include any stress-free strain associated with composition change. The inclusion of the elastic energy corrects the oversight noted in the preceding subsection, where it was tacitly assumed that composition change occurred without volume change. In the present discussion, only isotropic elastic response which is independent of composition is considered. The elastic strain is conveniently separated into two contributions, one of which is the (possibly incompatible) contribution that elastically restores the material to the mean lattice spacing at $\xi = 0$; the second contribution is the additional (compatible) elastic strain $\epsilon_{ij}^{\mathrm{r}}$ representing the extent of relaxation accommodated in achieving equilibrium and satisfying the boundary conditions. If the effect of composition change is to induce a pure volume change characterized by ξ and an extensional mismatch strain ϵ_1, then the stress-free volume change per unit volume is $-3\xi\epsilon_1$. The subscript 1 is intended to reflect the fact that ϵ_1 is the mismatch of $A_{1-\xi}B_\xi$ with respect to material A when the fractional content of material B in the alloy is unity, that is, when $\xi = 1$. For example, for the alloy $Si_{1-\xi}Ge_\xi$, the value of ϵ_1 is approximately -0.04. In these cases, the total elastic strain is expressed as

$$\epsilon_{ij} = \epsilon_{ij}^{\mathrm{r}} + \xi \epsilon_1 \delta_{ij} . \tag{9.90}$$

In discussing epitaxial systems in preceding chapters, the quantity represented by $\xi \epsilon_1$ was customarily denoted by ϵ_{m}.

Next, the time rate of change of free energy of the material in R that arises from time dependence of composition ξ at any point in R is examined. Differentiation of (9.89) with respect to time for a fixed material volume R with bounding surface S, followed by application of the divergence theorem to the result, yields

$$\dot{\mathcal{F}}(t) = \int_R \left[\varphi' - 2(\kappa_{\mathrm{ch}} \xi_{,k})_{,k} + \epsilon_1 \sigma_{jj} \right] \dot{\xi} \, dR + \int_S \left[\kappa_{\mathrm{ch}} \dot{\xi} \xi_{,k} n_k + \dot{u}_i^{\mathrm{r}} \sigma_{ij} n_j \right] dS , \tag{9.91}$$

where the superposed dot denotes the material time derivative, n_i is the outward unit normal vector to S, and σ_{ij} is the elastic stress. The first term in the integrand of the surface integral implies that either the gradient $\xi_{,k}$ or the rate $\dot{\xi}$ must be specified on the boundary S in order to analyze a boundary value problem. The second term in the surface integral is the rate of external mechanical work on the material in R. For the time being, it is assumed that R can be chosen so that both surface contributions vanish, perhaps due to symmetry, periodicity or remoteness of S.

The factor $\dot{\xi} \, dR$ is the rate of change of volume of material B in dR. It follows that the quantity in square brackets in the integral over R in (9.91) is the generalized force that is work-conjugate to this local volume change. This field thus represents the local chemical potential of material B in the solution,

$$\chi = \varphi' - 2(\kappa_{\mathrm{ch}} \xi_{,k})_{,k} + \epsilon_1 \sigma_{jj} . \tag{9.92}$$

This generalizes (9.88) to include composition gradient effects and elastic energy effects. Unless the material is intrinsically nonhomogeneous, each term in this expression depends on spatial position and time only through the dependence of composition on spatial position and time. Thus, the first term on the right side of the equal sign in (9.91) provides a basis for directly assessing the stability of a particular composition field ξ with a prescribed rate of change of that field. The case of stability of a particular spatially uniform composition under small amplitude perturbations is considered next.

Consider an elastic material subjected to an equi-biaxial state of stress, characteristic of a uniform epitaxial thin film coherently bonded to a substrate in the presence of lattice mismatch. This is the state of stress that develops in the film when the alloy $A_{1-\xi_*} B_{\xi_*}$ with spatially uniform composition ξ_* is deposited onto the planar substrate surface of material A. The mismatch of B with respect to A in the plane of the interface is ϵ_1. With reference to an xyz rectangular coordinate system with the y-direction normal to the interface, the stress in the film material is $\sigma_{xx} = \sigma_{zz} = M\epsilon_1\xi_*$, $\sigma_{yy} = 0$ where M is the biaxial elastic modulus of the film; all shear stress components are zero. If this state of stress prevails throughout an elastic solid solution, is the composition stable under small amplitude perturbations in the spatial distribution of material B?

To answer this question, consider the change in system free energy corresponding to the spatially nonuniform perturbed composition

$$\xi = \xi_* + \xi_\lambda \cos \frac{2\pi x}{\lambda} \cos \frac{2\pi z}{\lambda}, \tag{9.93}$$

where λ is a fixed wavelength and the perturbation is small, that is, $|\xi_\lambda|/\xi_* \ll 1$. Any two-dimensional perturbation of small amplitude can be represented as a linear superposition of sinusoidal perturbations of the kind (9.93) by means of Fourier methods. For compositions near to ξ_*, the quantity $\varphi'(\xi)$ in (9.91) can be approximated by the first two terms of a Taylor series expansion about the value $\xi = \xi_*$, which yields the locally linear function

$$\varphi'(\xi) \approx \varphi'(\xi_*) + \varphi''(\xi_*)(\xi - \xi_*). \tag{9.94}$$

The perturbed stress field that arises as a result of the nonuniform composition can be found by straightforward application of the stress equilibrium equations and Hooke's law of linear response for an isotropic elastic material, subject to the constraints that the perturbation alters neither the mean extensional strain in any direction in the xz-plane nor the zero net force per wavelength in the y-direction. The mean normal stress implied by these constraints is

$$\sigma_{kk} = 2M\epsilon_1\xi_* + 2\bar{E}\epsilon_1 \xi_\lambda \cos \frac{2\pi x}{\lambda} \cos \frac{2\pi z}{\lambda}, \tag{9.95}$$

where $\bar{E}$ is the plane strain modulus of the material.

The rate of change of free energy in a block of material occupying $0 \le x \le \lambda$, $0 \le y \le 1$, $0 \le z \le \lambda$ is, according to (9.91),

$$\dot{\mathcal{F}}(t) = \tfrac{1}{4}\lambda^2 \left[\varphi''(\xi_*) + 2\kappa_{\text{ch}} \left(\frac{2\pi}{\lambda} \right)^2 + 2\bar{E}\epsilon_1^2 \right] \xi_\lambda \dot{\xi}_\lambda. \tag{9.96}$$

If $\xi_\lambda > 0$, then the system is unstable if it is possible for $\dot{\xi}_\lambda > 0$ to proceed spontaneously, that is, with $\dot{\mathcal{F}} < 0$. This is possible only if the quantity enclosed within the square brackets in (9.96) is negative. Alternatively, stability requires that this quantity must be positive for all wavelengths λ, which will be so if

$$\varphi''(\xi_*) + 2\bar{E}\epsilon_1^2 > 0 \tag{9.97}$$

for the composition ξ_*, assuming that all wavelengths in the range $0 < \lambda < \infty$ are possible. If the solid is of limited extent, then there is a corresponding upper bound on the range of accessible wavelengths and a corresponding stabilization; this effect is minor in all but the very smallest material systems. The result (9.97) is essentially the condition established by Cahn (1961).

Several general observations on the stability of a spatially uniform composition follow from (9.97):

- Without lattice mismatch, the quantity ϵ_1 vanishes and stability of composition is determined solely by the local curvature of $\varphi(\xi)$ versus ξ. If $\varphi''(\xi_*) > 0$ then the composition is stable at fixed temperature under small perturbations, whereas it is unstable if $\varphi''(\xi_*) < 0$. The locus of points in the phase diagram of the alloy swept out by the condition $\varphi''(\xi_*) = 0$ as temperature is varied is called the *spinodal* or *spinodal curve* of the alloy; the historical origin of the terminology is described by Cahn (1968).
- If material B is perfectly soluble in material A but it alters the mean lattice parameter, the energy of mixing is zero but ϵ_1 is not zero. Under these circumstances, a uniform composition is always stable. This outcome is attributable to the positive definite and super-linear (that is, quadratic) dependence of strain energy on elastic strain.
- In the absence of a kinematic constraint, the first term on the right side of the equal sign in (9.95) is absent but the contribution in the second term arises nonetheless. It is the latter contribution that contributes to the free energy change. In this sense, the kinematic constraint is irrelevant in calculating free energy variations. Elasticity effects arise through incompatible deformations associated with nonuniform composition.
- If the dependence of free energy on composition illustrated in Figure 9.13 is concave downward at $\xi = \xi_*$, that is, if $\varphi''(\xi_*) < 0$, and if elastic mismatch is present, the effect of elasticity is to *stabilize* the composition. This stabilization is effective to the degree that local concavity of $\varphi(\xi)$ versus ξ can be accommodated without loss of stability, which is the case as long as

$$\varphi''(\xi_*) > -2\bar{E}\epsilon_1^2 . \tag{9.98}$$

The question of stability of an alloy composition is examined for other particular cases in the exercises, and the issue of elastic stabilization is considered further in Section 9.6.3.

The discussion of elastic stabilization of a uniform composition that led to the criterion (9.97) was based on the assumption of an unbounded material, an idealization which overlooks any influence of boundaries. The corresponding issue of a strained thin film deposited onto a relatively thick substrate was considered by Glas (1987). The film was assumed to be a solution and the substrate to be a pure substance representing one of the extremes in composition of that solution. The change in free energy of the system that results from a sinusoidal perturbation from the initially uniform composition of the film was considered. As might be expected in such a situation, the stabilizing influence of elasticity was found to be weaker than that for an unbounded solid. The reason for the difference is that the incompatibility in deformation induced by the nonuniform composition in the film can be accommodated by elastic deformation of both the film and the substrate material, thus mitigating the elastic incompatibility that underlies the stabilizing influence. Quantitative estimates of the magnitude of the effect were obtained for a number of III–V semiconductor film–substrate material systems.

9.6.3 Example: Elastic stabilization of a composition

Consider an $In_{\xi}Ga_{1-\xi}As$ solution with spatially uniform composition ξ. The elastic modulus and Poisson ratio of the cubic alloy are $E = 85.3(1 - \xi) + 51.4\xi$ GPa and $\nu = 0.31(1 - \xi) + 0.35\xi$, respectively; these isotropic estimates are based on a linear rule of mixtures and elastic constants for stressing along a cube edge. The stress-free mean lattice parameters of GaAs and InAs at room temperature are 0.56532 nm and 0.60584 nm, respectively; these values are adopted at all temperatures, thereby neglecting any influence of thermal expansion. The value of ϵ_1 implied by the difference in lattice parameters is -0.072 in this case. The free energy of mixing per unit volume of a homogeneous alloy at composition ξ is assumed to be

$$\varphi_{mix}(\xi) = (kT/\Omega)\left[(1 - \xi)\ln(1 - \xi) + \xi \ln \xi\right] + C_{mix}\xi(1 - \xi) \qquad (9.99)$$

where k is the Boltzmann constant, T is absolute temperature, Ω is the atomic volume of the material, and C_{mix} is an empirically determined parameter with an estimated value of 450×10^6 J m^{-3}. The atomic volume of the solution is approximated by that of GaAs throughout the calculation.

(a) At very high temperature T, the free energy of mixing has the property that $\varphi''_{mix}(\xi) > 0$ throughout the range $0 < \xi < 1$. Determine the critical temperature T_{cr} in degrees K that defines the lower limit of this range.

(b) Estimate the *additional* range of T below T_{cr} for which a uniform composition is stabilized by elastic effects according to the criterion (9.97).

Solution:

(a) The bound on the range of temperatures for which the uniform composition is stable, neglecting elastic effects, is determined as the lowest temperature for which $\varphi''_{mix}(\xi) = 0$ for any value of ξ. From (9.99), it is seen that

$$\varphi''_{mix}(\xi) = \frac{kT}{\Omega}\frac{1}{\xi - \xi^2} - 2C_{mix}. \qquad (9.100)$$

This quantity can first become zero at $\xi = 1/2$, and the temperature at which this occurs is the critical temperature

$$T_{cr} = C_{mix}\Omega/2k \approx 368 \text{ K}. \qquad (9.101)$$

(b) If elastic energy effects are included, the stability criterion (9.97) takes the form

$$\frac{kT}{\Omega} \frac{1}{\xi - \xi^2} - 2C_{\text{mix}} + 2\bar{E}\epsilon_1^2 > 0. \tag{9.102}$$

The lower bound on the temperature range over which this condition is satisfied is again reached at $\xi = 1/2$. The corresponding value of temperature is given by

$$T = \left(C_{\text{mix}} - \bar{E}\epsilon_1^2\right) \Omega/2k \approx 43\,\text{K}. \tag{9.103}$$

The result implies that the influence of elastic strain is to stabilize the uniform composition over the full temperature range. This strong effect could be significantly mitigated by the presence of geometrical boundaries of the solid, as well as by diffusion in cases in which the uniform composition prevails only over a part of the solid.

9.6.4 Evolution of compositional variations

The study of stability of a spatially uniform composition in a solid solution under periodic small-amplitude fluctuations pursued in Section 9.6.2 was based on equilibrium considerations. A composition was considered to be stable or unstable, depending on whether the system free energy tended to increase or decrease as a result of small fluctuations. The approach leads to no conclusion on the rate of increase in amplitude of an unstable fluctuation or on the rate of decay of a stable fluctuation. The question of time evolution of spatial fluctuations in composition is pursued here, following the ideas introduced by Hillert (1961) and Cahn (1961).

The local chemical potential field corresponding to a spatial distribution ξ of component B in an $A_{1-\xi}B_\xi$ substitutional solid solution was defined in (9.92). Spatial variation of the chemical potential field implies the existence of a driving force tending to reduce system free energy by redistributing component B. If it is assumed that the flux $\mathbf{j}$ of component B is proportional to the gradient in chemical potential throughout the isotropic material volume R, then the transport equation relating flux to chemical potential is

$$\mathbf{j} = -m_R \nabla \chi \tag{9.104}$$

in R, where m_R is a positive mobility parameter. In general, m_R depends on the local composition. The inner product of the flux vector $\mathbf{j}$ with a unit vector at a point in the material is the volume of B material passing through a plane normal to this unit vector at the material point per unit area per unit time. The composition ξ is the fraction of lattice sites that can possibly be occupied by B atoms that are in fact occupied by B, so local mass conservation requires that

$$\dot{\xi} = -\nabla \cdot \mathbf{j} \tag{9.105}$$

throughout the region of interest. It should be kept in mind that the number of possible lattice sites per unit volume of the solution is determined by the unstrained lattice of component A, although the consequences of this distinction are usually minor. Equations (9.104) and (9.105) combine to yield an evolution equation for the composition field in the form

$$\dot{\xi} = \nabla \cdot \left(m_R(\xi) \nabla \left[\varphi'(\xi) - 2 \nabla \cdot (\kappa_{ch}(\xi) \nabla \xi) + \epsilon_1 \sigma_{kk}(\xi) \right] \right). \tag{9.106}$$

This is a nonlinear partial differential equation for ξ as a function of time and of position throughout the region R. Analytical solutions of the equation are not available, in general.

For the particular situation in which ξ represents only a small departure from the spatially uniform composition ξ_*, the equation (9.106) can be linearized and solutions are readily obtained. In this situation, $m_R(\xi) \approx m_R(\xi_*) = m_R^*$, $\kappa_{ch}(\xi) \approx \kappa_{ch}(\xi_*) = \kappa_{ch}^*$ and $\varphi'(\xi)$ is approximated as in (9.94). The partial differential equation then reduces to

$$\dot{\xi} = m_R^* \nabla^2 \left[\varphi_*''(\xi - \xi_*) + \epsilon_1 \sigma_{kk}(\xi) - 2\kappa_{ch}^* \nabla^2 \xi \right]. \tag{9.107}$$

Furthermore, if it is assumed that the composition varies periodically about the mean value ξ_* with spatial period λ as in (9.93), and that the corresponding mean normal stress is given by (9.94), the partial differential equation is reduced to the ordinary differential equation

$$\dot{\xi}_\lambda = -m_R^* \left(\frac{2\pi}{\lambda} \right)^2 \left[\varphi_*'' + 2\bar{E}\epsilon_1^2 + 2\kappa_{ch}^* \left(\frac{2\pi}{\lambda} \right)^2 \right] \xi_\lambda. \tag{9.108}$$

Some general aspects of behavior of the solution of this equation are evident. For example, if $\varphi_*'' + 2\bar{E}\epsilon_1^2 > 0$ then the amplitude ξ_λ decays in magnitude from any nonzero initial value for all values of wavelength λ. This observation is equivalent to the criterion of unconditional stability in (9.97). If $\varphi_*'' + 2\bar{E}\epsilon_1^2 < 0$, then the amplitude ξ_λ increases indefinitely in time for all wavelengths λ greater than the value

$$\lambda_{cr} = 2\pi \sqrt{\frac{-2\kappa_{ch}^*}{\varphi_*'' + 2\bar{E}\epsilon_1^2}} \tag{9.109}$$

and decays to ever smaller amplitude for any smaller wavelength. Finally, the value of wavelength for which the rate of growth of amplitude ξ_λ is largest is

$$\lambda_{gr} = \sqrt{2}\lambda_{cr}. \tag{9.110}$$

These implications are based on the assumption of a one-dimensional perturbation of composition from its mean value ξ_*. Similar conclusions can be drawn through consideration of perturbations in two or three dimensions; details are left as an exercise.

9.6.5 *Coupled deformation–composition evolution*

There are a number of evolutionary processes involving very small material structures in the course of which changes in composition induce stress fields and/or changes in shape. The latter effects, in turn, give rise to driving forces for alteration of the composition distribution. A situation of practical interest is the stress-driven self-assembly of alloy nanostructures. For example, consider the stress-driven formation of SiGe quantum dots during epitaxial deposition of the alloy onto a Si (001) substrate. As was seen in Section 8.8.3, the flat growth surface is unstable for fractional Ge concentration greater than about 0.2. The deposit evolves into epitaxial islands by means of surface diffusion; initially, the lateral faces of small islands are vicinal surfaces with slopes up to about 11–12 degrees with respect to the growth surface. As the island volume increases, the lateral faces can become significantly steeper, as described in Section 8.9.5. It is tacitly assumed in describing the process in this way that the Ge is uniformly distributed throughout an island and that the composition is the same as the composition of the growth flux. However, the state of stress in the island is such that it is energetically preferable for Ge to be near the apex of the island where mean normal compressive stress is smaller than it is near the island–substrate interface, the island is relatively unrelaxed and the mean normal compressive stress is relatively large. It is also generally preferable for the Ge to be in the substrate than in the island. These observations imply the existence of a thermodynamic driving force tending to redistribute the Ge toward the apex of the island and perhaps to drive it into the substrate. If such a redistribution occurs by compositional evolution as described in the preceding subsection, then the elastic energy due to mismatch is effectively reduced. This implies that the driving force which caused the island faces to become steeper in the first place has been diminished, and the height-to-width aspect ratio of the island would also tend to undergo a reversal and would tend to diminish.

This sequence of shape changes during SiGe island evolution has been observed directly by Henstrom et al. (2000), who also reported evidence of average mismatch strain reduction in the later stages of the process. Direct observation of compositional variations is extraordinarily difficult, but such variations are the most likely cause of the shape reversals detected. Strong support for this interpretation was provided subsequently in the experiments of Rastelli et al. (2001). They deposited Ge onto Si (001), a situation in which the flat growth surface is unstable. Islands formed in the early stages of growth, as expected. The lateral faces of the islands steepened to the {105} orientation and, as the island volumes increased further, the lateral faces steepened and the islands assumed the so-called dome shape. At this point in the experiments, the Ge growth flux was stopped and the Ge islands were then exposed to an incident Si growth flux. The Si was apparently absorbed into the islands and the effective mismatch strain was reduced accordingly. As a result, the sequence of island shapes observed during growth was reversed and the aspect ratios of the islands decreased.

These examples illustrate the interaction of composition distribution and stress field in a deformed solid solution. The mathematical structure exists for analysis of such phenomena, but the governing equations are inherently nonlinear; analysis is very difficult if shape changes are taken into account and systematic study has not yet been undertaken. The purpose here is to formulate and analyze a physical situation involving coupled deformation and composition evolution that serves as a reasonably transparent vehicle for presenting the underlying ideas, but that avoids the complexity of evolution of shape.

Consider an elastic layer occupying the region $-\infty < x, y < \infty$, $-\frac{1}{2}h \leq z \leq \frac{1}{2}h$. Attention is limited to fields that vary only through the thickness of the layer. The slab material is an $A_{1-\xi}B_\xi$ solid solution as described in Section 9.6.1. The composition is assumed to vary through the thickness according to

$$\xi(z, t) = \xi_* + \beta(t) \sin \frac{\pi z}{h}, \qquad (9.111)$$

where ξ_* is the mean composition within the layer and $\beta(t)$ is the time-dependent amplitude of a sine-shaped perturbation; the perturbation magnitude is not assumed to be small in this case. The flux of material B in the z-direction corresponding to (9.111) is

$$j(z, t) = \frac{h}{\pi} \dot\beta(t) \cos \frac{\pi z}{h}, \qquad (9.112)$$

which necessarily vanishes at $z = \pm \frac{1}{2}h$.

The layer is flat for any uniform composition and it remains so if there is no lattice mismatch between constituents A and B. Assume that constituent B has an isotropic extensional mismatch strain ϵ_1 with respect to constituent A. As a result, the midplane of the layer is curved, in general, for a nonuniform distribution ξ. From (2.58), it is known that the state of stress inducing this curvature is

$$\sigma_{xx}(z, t) = \sigma_{yy}(z, t) = M \left[\epsilon_1(\xi(z, t) - \xi_*) - \kappa(z - z_{np}) \right] \qquad (9.113)$$

where M is the elastic bulk modulus, κ is the spherical curvature of the midplane of the layer, and z_{np} is the location of the neutral plane of the layer. As indicated in (2.54), the in-plane stress components, which are identical, must satisfy

$$\int_{-h/2}^{h/2} \sigma_{xx}(z, t)\, dz = 0, \qquad \int_{-h/2}^{h/2} z\sigma_{xx}(z, t)\, dz = 0 \qquad (9.114)$$

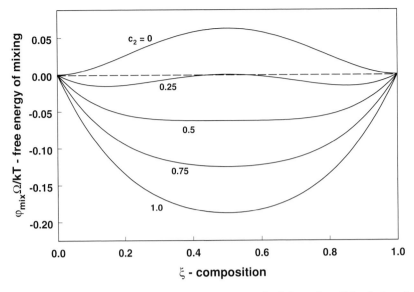

Fig. 9.14. Illustration of the dependence of free energy of mixing of a solid solution defined in (9.115) for several values of the parameter c_2.

to ensure equilibrium; in writing (9.114), it has been presumed that the remote edges of the layer are free of constraints. Enforcement of the equilibrium conditions in this way yields expressions for κ and z_{np} in terms of ξ_* and β. In the system under discussion, it is known *a priori* that $z_{np} = 0$ as a result of symmetry.

The chemical potential representing the tendency for redistribution of composition is defined in (9.92). The mean normal stress σ_{jj} appearing in this expression is $\sigma_{xx} + \sigma_{yy}$. The stress component σ_{zz} and all shear stress components vanish throughout the layer in this case as a result of symmetry, translational invariance and vanishing of traction on the faces $z = \pm\frac{1}{2}h$ of the layer. The parameter κ_{ch} in the chemical potential comes into play only when spatial gradients in ξ are very steep, and its effect is neglected for the distribution given in (9.111).

The first term in the expression (9.92) for chemical potential is essentially the derivative of the free energy of mixing of the solution with respect to composition. In the present example, this concentration dependent measure of miscibility is assumed to be

$$\varphi_{mix}(\xi) = \frac{kT}{\Omega} \left[\xi^2(1-\xi)^2 - c_2\xi(1-\xi) \right], \tag{9.115}$$

where c_2 is a numerical constant. Plots of $\varphi_{mix}\Omega/kT$ versus ξ for several values of c_2 are shown in Figure 9.14. If the results of enforcing equilibrium are used to eliminate curvature κ and neutral plane location z_{np} from the equations, the consequence of enforcing conservation of mass (9.105) is the ordinary differential equation

$$\dot{\beta}(t) = -\frac{kT m_R}{\Omega h^2}\pi^2 \left[2\left(1 - \frac{96}{\pi^4}\right)\epsilon_1^2 \frac{M\Omega}{kT} \right.$$

$$\left. +2\left(1 + c_2 - 6\xi_* + 6\xi_*^2\right) + 3\beta(t)^2 \right]\beta(t) \qquad (9.116)$$

governing $\beta(t)$, where the transport mobility m_R has been assumed to be independent of concentration.

Several general observations follow directly from the form of the differential equation (9.116). In the absence of an elasticity effect (first term in square brackets) and a nonlinear effect (third term in square brackets), the stability of the configuration is determined solely by the second term in square brackets on the right side. This term is negative for all values of ξ_* if $c_2 > \frac{1}{2}$ and is non-negative for some values of ξ_* otherwise. Thus, a uniform composition ξ_* of any magnitude is stable under these conditions if $c_2 > \frac{1}{2}$, consistent with the general observations of Section 9.6.1. For any value of $c_2 < \frac{1}{2}$, the limits of solubility of the composition are established by the condition

$$1 + c_2 - 6\xi_*(1 - \xi_*) = 0, \qquad (9.117)$$

which is also consistent with the observations in Section 9.6.1.

A novel aspect of the behavior of this simple system is that the stabilizing influence of elasticity on composition is almost completely negated by the coupling to deformation. In other words, if curvature of the layer is suppressed (by constraining the layer between smooth rigid platens, say), then the coefficient of the strain energy term in the square brackets on the right side of (9.116) would be 2. As a result of the coupling, however, the coefficient is reduced to the value 0.029, which is essentially negligible by comparison. The origin of this coupling effect is readily identified in the present situation. To see this, suppose for the moment that $\epsilon_1 > 0$ and that the composition begins to redistribute with $\xi > \xi_*$ above the midplane and $\xi < \xi_*$ below. Without bending, this change induces a tensile mean normal stress above the midplane and a compressive stress below; this state of stress would retard further migration of B and is the origin of the stabilizing influence of elasticity. If unconstrained, the layer takes on a positive spherical curvature in response to this stress, with a resulting decrease in mean normal stress above the midplane and an increase below. Consequently, the influence of deformation is to directly counteract the stabilizing influence of stress, a behavior that becomes evident by solving the governing equation (9.116) for this simple example.

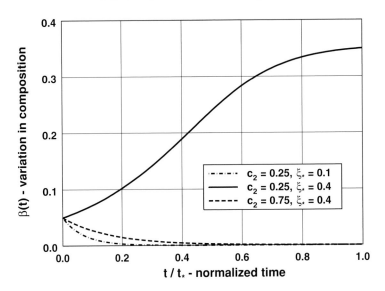

Fig. 9.15. Representative plots of the solution (9.118) for $\beta(t)$ introduced in (9.111) versus
t for several combinations of c_2 and ξ_*. The other system parameters are fixed at the values
$\beta_0 = 0.05$ and $\epsilon_1^2 M\Omega/kT = 0.1$ in all three plots.

For the initial condition $\beta(0) = \beta_0$, the solution of the ordinary differential
equation (9.116) for $t > 0$ is

$$\beta(t) = \beta_0 \left[\frac{C_*}{3\pi^2 \beta_0^2 + (C_* - 3\pi^2 \beta_0^2) e^{-2C_* t/t_*}} \right]^{1/2} , \qquad (9.118)$$

where

$$C_* \equiv -2\pi^2 \left[\left(1 - \frac{96}{\pi^4}\right) \epsilon_1^2 \frac{M\Omega}{kT} + (1 + c_2 - 6\xi_* + 6\xi_*^2) \right] \qquad (9.119)$$

and the characteristic time for the relaxation process is

$$t_* = \Omega h^2/kT m_R . \qquad (9.120)$$

It is readily verified that the limiting behavior of the solution as $t \to \infty$ is

$$\beta(t) \to \begin{cases} \sqrt{C_*/3\pi^2} , & \text{if } C_* > 0 \\[2mm] 0, & \text{if } C_* < 0. \end{cases} \qquad (9.121)$$

On physical grounds, the value of $\beta(t)$ is restricted to the range $|\beta(t)| \leq \xi_*$ for all
t. The behavior of the solution is illustrated for several combinations of c_2 and ξ_*
with $\beta_0 = 0.05$ and $\epsilon_1^2 M\Omega/kT = 0.1$ in Figure 9.15.

The class of possible composition fields has been limited to the simple form (9.111) for purposes of illustration in the present discussion. There is no fundamental limitation of this kind, and more general forms, such as

$$\xi(z, t) = \xi_* + \sum_{i=1}^{N_\beta} \beta_i(t) \sin \frac{(2i - 1)\pi z}{h},\tag{9.122}$$

could be adopted with N_β large enough to provide accurate representations of the evolving composition profile. The ordinary differential equations for the coefficients $\beta_i(t)$ in (9.122) are most readily established through a straightforward generalization of the variational principle introduced in Section 9.3. The details are left as an exercise.

9.7 Stress-assisted diffusion: electromigration

Metallic interconnect lines, such as those made of Al or Cu in an integrated circuit, are conduits for high electric current densities that are more than four orders of magnitude greater than the current density of 100 A cm^{-2} carried by typical residential or industrial wiring. Bulk metal wires typically reach their melting points as a consequence of Joule heating when subjected to current densities of approximately 10^4 A cm^{-2}. The ability of the thin metal lines to carry much higher current densities without melting is mainly a consequence of the role of the surrounding dielectric materials as an effective heat sink for removal of heat from the interconnect lines. Figure 9.16 shows an example of the metal interconnect structure in a CMOS circuit of a computer chip microprocessor manufactured by the International Business Machines Corporation.

In the course of transporting high current densities through the metal interconnects during operation of an integrated circuit, significant momentum transfer from moving electrons to nominally stationary metal atoms takes place. This momentum transfer has the potential for inducing the diffusion of metal atoms in the direction of electron transport, which is counter to the direction of the conventional applied electric field. *Electromigration* in the interconnect lines is the current-driven, biased self-diffusion of metal atoms. Loss of current carrying capability and the possible occurrence of electrical short circuits due to electromigration are major concerns of practical interest which can limit the reliability of integrated circuits (D'Heurle 1971, Blech 1976, Suo 2003). All metallic components of integrated circuits that carry electric current are potentially susceptible to damage and failure by electromigration. These components include not only interconnect lines, as already noted, but also powered electrical contacts and vias as well. Damage induced at contacts by atomic diffusion can be exacerbated by electromigration in essentially any device. The interfaces between vias and interconnects are also potential sites of atomic flux divergence and void growth.

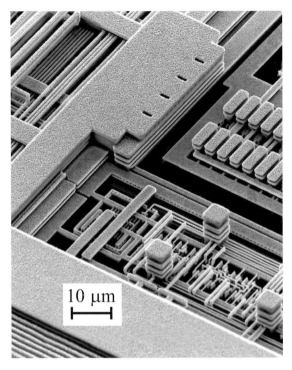

Fig. 9.16. A CMOS circuit with six levels of metallization where the dielectric surrounding the metal interconnects is removed in order to image the embedded line features. Reproduced with permission from IBM Corporation, New York.

9.7.1 Atom transport during electromigration

During transport of electrons through a metal conducting line in response to an applied electric field, metal ions are subjected to several forces having the potential for dislodging them from their structural equilibrium positions. Among these is the electrostatic force due to the applied field which tends to move the positively charged ions in a direction opposite to the direction of flow of the electrons. In addition, trajectories of moving electrons are deflected by the ions due to electrostatic interaction forces between the ions and electrons. Momentum is conserved in such encounters, and as a result, some momentum is invariably transferred from the electrons to the ions. If the density of flowing electrons is high and they have sufficiently high speed, then there is a potential for forcing the metal ions to move in the same direction as electron flow. The force acting on each metal ion is similar in nature to the pressure exerted by a gas on the walls of its container, in the sense that it is a consequence of momentum transfer and that it represents a time average over many individual encounters. In any case, the latter effect is usually dominant and the result is a net drift of metal ions along the conductor in the direction of the electron flow. It follows that, because ions reside in shallower energy minima at grain boundaries, defects, free surfaces and interfaces, ion transport is more likely to occur along such paths than through well ordered regions. However, the difference is only one of degree and ion transport is possible in single crystals as well. Electromigration behavior is described here in a macroscopic or phenomenological way, in large part to indicate the role of stress in mass transport, and the discussion is restricted to one-dimensional behavior.

The electromigration force Q acting on each metal ion is usually expressed with respect to some fixed direction in the material as

$$Q = z^* e \mathcal{E}, \qquad (9.123)$$

where $\mathcal{E}$ is the electric field strength in that direction, e is the electron charge and z^* is the effective valence of the ions. The electric potential can be expressed in terms of the electric current i and the electrical resistance ρ as $\mathcal{E} = \rho i$. The parameter z^* is the nominal valence of the metal ions in the absence of current flow, reduced by the influence of dynamic coupling between the ions and moving electrons which induces a force in the direction *opposite* to the direction of the electric field. Experimentally, it is found that z^* is typically negative and large in magnitude compared to the nominal valence, implying that the indirect dependence on $\mathcal{E}$ through the dynamic coupling dominates the direct effect of force on this ion due to its own charge. The quantity Q has physical dimensions of force per ion.

The flux of metal ions in response to the force Q induced on each is customarily written as

$$j_a = \frac{m_s Q}{\Omega} \qquad (9.124)$$

where j_a is volume flow rate in the direction of the electric field per unit area of a planar material surface transverse to the flow, Ω is an atomic volume, and m_s is a transport mobility parameter for self-diffusion under the prevailing conditions. The physical dimensions of j_a are volume/(area×time) = length/time and the mobility parameter m_s has physical dimensions of length^4/(time×force). The mobility is commonly expressed as a diffusivity divided by kT where k is the Boltzmann constant and T is absolute temperature.

Consider an initially homogeneous metal conducting line that is confined to occupy a cylindrical volume of fixed length and fixed cross-sectional area. If the line is uniform along its length, any material flux due to current will result in an accumulation of material at one end of the line and a depletion of material toward the other end. In a confined line, a compressive stress is induced in regions of material accumulation and a tensile stress in regions of depletion. The compressive stress has the potential for driving cracks into the material surrounding the conductor and/or for inducing plastic deformation, while the tensile stress has the potential for nucleating voids within the conducting line or at its interface with the surrounding material. Both crack formation and void formation are nucleation-controlled processes, and the stress levels required to activate these processes can be estimated. In addition to these potential material failure modes, the induced gradient in stress also tends to drive diffusive mass transport in the opposite direction, that is, from the region of compressive stress toward the region of tensile stress. If a sufficiently large stress gradient can be sustained, mass transport due to electromigration can be completely negated by its effect. This possibility is discussed more quantitatively later in this section. Note that, if electric current is driven through an unconfined cylindrical conductor, the conducting line will tend to move steadily in the direction of the electron flow by drawing material from one end and depositing it on the surface at the other end.

Microscopically, electromigration is assumed to be the hopping of metal atoms from one equilibrium position to another in a certain direction determined by the flow of electrons through the material. The change in system energy associated with the microscopic redistribution of mass was discussed in some detail in Section 9.6.2. In particular, it was observed through the expression (9.91) that the rate of increase of free energy of an element of material held at fixed volume per unit volume of excess material added per unit time is $-\frac{1}{3}\sigma_{kk}$, the negative mean normal stress in the material. Recall that $\epsilon_1 > 0$ in that discussion implied *tensile* misfit or, in the present context, material *removal*, an observation that accounts for the negative sign here. In that case, the mismatch was uniquely defined by the difference in stress-free volume of the two homogeneous materials at the extremes of the range of a substitutional solid solution. In the case of electromigration, which involves self-diffusion by interstitial migration within grains or by excess atom flow along grain boundaries, the measure of mismatch is not nearly so well defined. Customarily, the measure of excess material $-\epsilon_1$ is assumed to have the value unity, even though the distribution is tacitly assumed to be dilute, so the chemical potential for stress-driven transport is

$$\chi_\sigma = -\sigma_{kk} \qquad (9.125)$$

or three times the negative of the mean stress. In one-dimensional configurations involving only a single component of stress, say $\sigma(x, t)$, and a single spatial coordinate x, the chemical potential is $\chi_\sigma = -\sigma$. The associated flux j_σ of excess material due to a gradient in the stress is assumed to be

$$j_\sigma = -m_s \frac{\partial \chi_\sigma}{\partial x} = m_s \frac{\partial \sigma}{\partial x}, \qquad (9.126)$$

where m_s is again the mobility parameter for self-diffusion. Note that the flux given in (9.126) represents mass flow from regions of material accumulation toward regions of material depletion, or mass flow in a direction opposite to the flux j_a introduced in (9.124). The total flux of metal atoms accounting for electromigration and stress-driven diffusion together is then $j = j_a + j_\sigma$.

As was noted in the study of evolution of compositional variations in Section 9.6.4, the local rate of mass accumulation due to a mass flux vector is the negative of the flux divergence. The local addition of material is represented in continuum modeling as a stress-free volumetric strain. At fixed overall strain, the rate of material addition induces a proportional rate of increase in mean normal compressive stress. In the present one-dimensional configuration, this connection is expressed as

$$-\frac{\partial j}{\partial x} = -\frac{1}{B}\frac{\partial \sigma}{\partial t} \tag{9.127}$$

at each section along the line, where B is a relevant elastic constant. For uniaxial stress in an isotropic material, the value of B is essentially the elastic modulus E. If the expressions for mass flux due to electromigration (9.124) and due to stress gradient (9.126) are incorporated, it follows that

$$\frac{\partial \sigma}{\partial t} = B\frac{\partial}{\partial x}\left[m_s\left(\frac{\partial \sigma}{\partial x} + \frac{z^* e \rho i}{\Omega}\right)\right]. \tag{9.128}$$

If the self-diffusion mobility is assumed to be constant, then the equation reduces to the standard form of the diffusion equation in one dimension for stress, that is,

$$\frac{\partial \sigma}{\partial t} = B m_s \frac{\partial^2 \sigma}{\partial x^2}. \tag{9.129}$$

In spite of the fact that they do not appear in this partial differential equation, the electrical parameters are still central to the description of the process because the boundary conditions commonly are expressed in terms of mass flux.

There are two particular configurations in which (9.128) or (9.129) can be applied that have provided simple analytical results which have proven useful in correlating experimental observations. One of these configurations is a very long one-dimensional conducting line occupying the region $0 < x < \infty$. The electric field vector of magnitude $\mathcal{E}$ acts in the direction of decreasing x and the electromigration mass flux is in the direction of increasing x, or vice versa. At its end $x = 0$, the line is confined in such a way that electric current flows freely but that there is no net mass flux, that is,

$$\frac{\partial \sigma}{\partial x}(0, t) + \frac{z^* e \rho i}{\Omega} = 0 \tag{9.130}$$

for all $t > 0$. In the remote region,

$$\frac{\partial \sigma}{\partial x}(x, t) \to 0 \quad \text{as} \quad x \to \infty. \tag{9.131}$$

The stress is initially constant over the full length of the line, and this constant value is taken to be zero for present purposes; therefore, the initial condition is

$$\sigma(x, 0) = 0, \quad 0 < x < \infty. \tag{9.132}$$

The boundary value problem consisting of the partial differential equation (9.129), along with its initial and boundary conditions, has a well-known solution that includes the time-dependence of stress at the confined end of the line given by

$$\sigma(0, t) = \frac{2z^* \rho i}{\Omega} \sqrt{\frac{m_s B t}{\pi}}. \tag{9.133}$$

Note that z^* is negative in this case and that i has the same sign as $\mathcal{E}$. Consequently, the stress at the end of the line is tensile or compressive, depending on whether nominal current i flows in the direction of increasing x or decreasing x, respectively. Suppose it is postulated that failure in the line occurs when $|\sigma(0, t)|$ has increased to the value σ_{cr}, the smallest stress magnitude that is required to activate any of the possible line failure mechanisms. The time required for the stress to reach this level is determined from (9.133) to be

$$t_{fail} = \frac{\pi}{4m_s B} \left(\frac{\Omega \sigma_{cr}}{z^* e \rho} \right)^2 \frac{1}{i^2}. \tag{9.134}$$

Analysis of this one-dimensional model leads to the conclusion that the time to failure under the circumstances assumed is inversely proportional to the electric current density squared. Experiments have shown time to failure to be proportional to i^{-n}, as a rule, where the exponent has been found to fall within the range $1.3 \leq n \leq 3.0$ (Thompson and Lloyd 1993). The self-diffusion process of electromigration is believed to be thermally activated, and consequently the mobility has an exponential dependence on an activation energy and on absolute temperature as expressed in the familiar Arrhenius form. This feature accounts for the principal temperature-dependence of the time to failure in interconnect lines.

A second one-dimensional configuration that leads to results which have been useful in interpretation of observations is that of a conducting line of moderately small length, occupying the interval $0 \leq x \leq L$ say. It is again presumed that a steady electric current i flows along the line and both ends of the line are impermeable to mass flux, so that $j(0, t) = 0$ and $j(L, t) = 0$. When the current first begins to flow, there is some period of transient response during which mass is redistributed along the line. After some time has passed, however, the governing equations admit a steady-state solution with $\partial \sigma(x, t)/\partial t = 0$, implying a fixed or time-independent stress profile. From (9.128), it follows that $j(x, t) = 0$ throughout the length of the line, once steady state is achieved, and that the steady-state stress profile is a

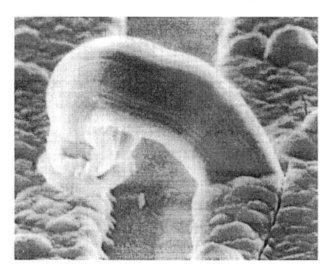

Fig. 9.17. An extruded hillock in an Al interconnect line undergoing electromigration damage causing a short. Reproduced with permission from Devaney (1989).

uniform gradient with value

$$\frac{\partial \sigma}{\partial x}(x, t) = -\frac{z^* e \rho i}{\Omega} . \qquad (9.135)$$

In effect, (9.135) implies that the stress gradient is in equilibrium with a distribution of body force along the line generated in the course of momentum transfer from the flowing electrons to the metal ions.

If it is assumed that this stress distribution is antisymmetric with respect to the midpoint of the line, which is the distribution implied by a linear theory in the absence of an initial stress, then the largest gradient that can be sustained without electromigration damage occurring in the line is $2\sigma_{cr}/L$. Because the value of the steady-state stress gradient is fixed by other parameters in this case, it follows that there is a threshold length given by

$$L_{th} = \left| \frac{2\Omega \sigma_{cr}}{z^* e \rho i} \right| \qquad (9.136)$$

below which electromigration damage cannot arise. The existence of this threshold length for onset of electromigration damage was first recognized by Blech (1976), and evidence for its existence is noted in the discussion of the drift test in Section 9.7.2.

An illustration of damage due to electromigration is shown in Figure 9.17. This is a scanning electron microscope (SEM) image of a hillock formed in an Al metal interconnect which has forced an opening through a silicon oxide passivation layer; interconnect material has been extruded through the opening. Cracks emanating

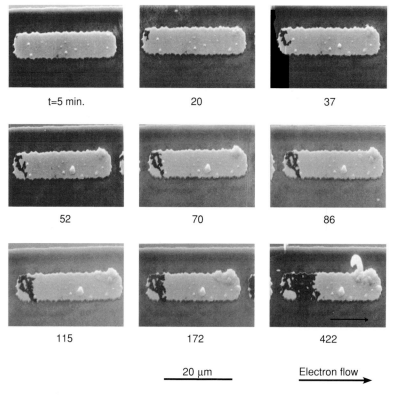

t=5 min.	20	37
52	70	86
115	172	422

20 μm   Electron flow

Fig. 9.18. The evolution of electromigration-induced voids in the wake of electron transport and the formation of hillocks at the opposite end in pure Al strips subjected to a current density, $i = 6.5 \times 10^5$ A cm^{-2} at 340 °C. The numbers beneath each micrograph indicate the time, in minutes, after the commencement of the electromigration experiment. (Courtesy of I. A. Blech. Reproduced with permission.)

from the opening can be observed in the image. The extruded metal has extended far enough to make contact with an adjacent aluminum line, thereby resulting in an electric short circuit.

9.7.2 The drift test

A particularly significant development in the study of electromigration in thin films was the introduction by Blech (1976) of an experimental configuration commonly known as the 'drift test'. In this experiment, a stripe of a metal with relatively low electrical resistance, Al for example, whose electromigration response is to be investigated, is patterned over a length L on a metal film such as TiN which has a relatively high melting point and which is not susceptible to electromigration under the experimental conditions, as shown schematically in Figure 9.18. The film, in turn, is deposited on a relatively thick nonconducting substrate. As an electric bias

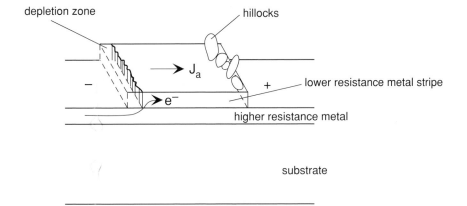

Fig. 9.19. Schematic of a typical 'drift test'. A low-resistance metal stripe is patterned, over a length L, on top of a higher-resistance stripe through which an electric current is applied. As the current shunts through the lower-electrical-resistance metal, current-induced atomic drift leads to the formation of a depleted zone in the wake of electron transport at one end of the stripe and hillocks and extrusions at the other end.

voltage is imposed across the length of the film, the resulting electric current shunts through the lower-resistance stripe over its entire length. When the current density is large enough to induce electromigration in the metal stripe, atoms of the metal drift from the cathode to the anode. As a consequence, the entire stripe is displaced over time in the direction of electron motion, as seen in Figure 9.19. The leading edge of the stripe is populated with 'hillocks' (see Figure 7.25, for example) as a result of the spatially nonuniform accumulation of metal atoms. The average drift velocity v_{dr} is equal to the mass flux j across any cross-section of the stripe (Huntington and Grone 1961). A representative sequence of micrographs shown in Figure 9.19 illustrates the evolution of voids and hillocks over the course of time in an Al stripe subjected to electric current in the drift test configuration.

Blech (1976) measured the average drift velocity v_{dr} as a function of current density for electromigration in a 115-μm long and 460-nm thick Al stripe on a 260-nm thick TiN layer which was sputtered on an oxidized Si substrate. Over the current range 0.5×10^5 A cm$^{-2} \le i \le 4.0 \times 10^5$ A cm^{-2} at 350 °C, he observed a linear relationship between v_{dr} and i. However, there was a threshold current density of approximately 1.1×10^5 A cm^{-2} below which no drift was detected. The threshold current density was also found to be inversely proportional to the stripe length over the range $30 \, \mu$m $\le L \le 150 \, \mu$m. The threshold value of i was observed to increase with decreasing temperature. Enclosing the Al line in a silicon nitride passivation layer was also found to increase the value of threshold current for onset of electromigration damage. Figure 9.20 shows the results of an electromigration experiment conducted on thin Al stripes of different lengths deposited on a TiN

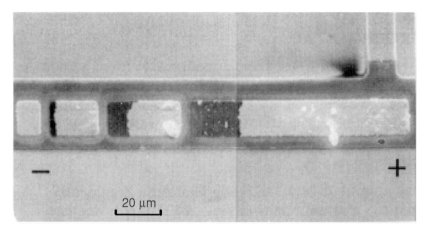

Fig. 9.20. Electromigration experiment conducted at room temperature on four aluminum stripes of different lengths in series with $j = 3.7 \times 10^5$ A cm^{-2}. The aluminum stripes were heat treated at 350 °C for 20 h prior to the experiment. Note that the shortest stripe on the left does not show any drift, whereas the increase in length of the other three stripes shows an increase in drift. Reproduced with permission from Blech (1976).

layer, where the increase in the stripe length is seen to promote an increase in atom drift. The shortest stripe in this micrograph has a subcritical length and hence does not undergo any electromigration damage.

9.7.3 Effects of microstructure on electromigration damage

In most polycrystalline thin films, columnar grains typically span the thickness of the film. When a metal stripe or line is patterned from a thin film, there is a continuous network of grain boundaries along the direction of current-induced atomic flux, if the width of the lines is larger than the grain size. These boundaries provide a relatively easy path for atomic diffusion because the mobility for self-diffusion is larger within a grain boundary than in the interior of grains. However, when the line thickness becomes comparable to or smaller than the average grain size, there are no longer continuous grain boundary paths from one end of the line to the other. Consequently, alternative diffusion paths along the metal–oxide interface through the bulk of the grains become more prominent in electromigration processes.

With the miniaturization of integrated circuits and the attendant reduction in feature dimensions, along with cleaner and higher temperature deposition processes which promote larger grains, there is a greater propensity for the predominant evolution of so-called bamboo structures. Configurations in which nearly all grain boundaries extend across the full width of the line, are more or less perpendicular to the line and are spaced at more or less regular intervals along the line are called

$\longrightarrow$ [001]

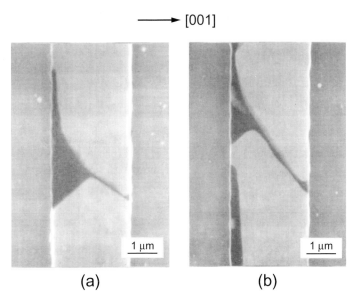

Fig. 9.21. Formation of slit-like voids whose geometry is determined by crystallographic texture during the electromigration of unpassivated, single crystalline Al lines of (110) texture. (a) Voids bound by {111} planes in ⟨$\bar{1}$12⟩ directions, and (b) void in the ⟨$\bar{1}$11⟩ direction, bound by {011} planes. Reproduced with permission from Joo and Thompson (1997).

bamboo grain structures because, in plan view, images of such lines have the general appearance of stalks of bamboo. The bamboo structure significantly impedes atomic flux which leads, in turn, to a marked increase in electromigration time to failure in interconnect lines. Post-pattern annealing of Al interconnect lines, which promotes the evolution of bamboo grains, has also been shown to provide a substantial rise in the time to electromigration failure. The bamboo grain structures, however, can be prone to transgranular failure processes which develop during electromigration. These cracking patterns and the associated stress voiding are known to be strongly influenced by the crystallographic texture of the interconnect lines (Sanchez et al. 1990, Joo and Thompson 1997).

The effect of crystallographic texture on the evolution of slit-like cracks during electromigration of unpassivated (110)-textured Al single crystal lines is shown in the micrographs of Figure 9.21. These observations were made from wafer-level accelerated testing for electromigration, where the tests were carried out at constant voltage while the current was being measured. The experiments were carried out at a temperature within the range 350–400 °C with current densities in the range $1.5–2.7 \times 10^5$ A cm^{-2}. The micrograph in Figure 9.21(a) corresponds to the most commonly observed voids, which are bounded by {111} planes in ⟨$\bar{1}$12⟩ directions;

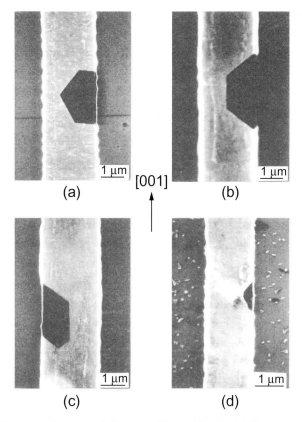

Fig. 9.22. SEM images of stress voiding and slit cracking in Al single crystals of different crystallographic orientations. (a)-(c) Voids in slowly cooled, {110}-textured Al lines. The voids are bounded by {111} planes. (d) Voids in air-cooled {110}-textured lines. Reproduced with permission from Joo and Thompson (1997).

Figure 9.21(b) shows a slit-like void in the $\langle\bar{1}11\rangle$ direction, bounded by {011} planes.

Stresses induced by electromigration, in conjunction with those produced by thermal fluctuations, collectively lead to diffusion of vacancies and inelastic deformation processes which cause stress voiding and slit cracking in metal interconnects (Sanchez et al. 1992, Joo and Thompson 1997). Such failure processes are also strongly influenced by the crystallographic texture of the current carrying interconnect lines. Figure 9.22 shows examples of stress-induced voids and cracks that form in single crystal aluminum lines subjected to thermal loading. Note the pronounced effect of crystallographic texture on the geometry of these voids.

Several approaches for enhancing the resistance to electromigration failure through microstructural control and design have been proposed.

- It is known that the addition of Cu to Al significantly increases the resistance to electromigration and the interconnect life by raising the activation energy for Al atomic flux. As a result, Al–Cu alloys have been commonly used as interconnect metals in integrated circuits.
- The use of high temperature materials, which have lower mobilities for self-diffusion at service temperatures, is also expected to enhance resistance to electromigration. For example, the use of Cu instead of Al provides this beneficial effect. Other materials with high melting points, such as W, Ti and Ta, have lower electrical resistivity.
- Narrow lines which promote bamboo-like grain structures are likely to provide enhanced electromigration life. In addition, post-patterning anneal treatments which have been shown to increase lifetimes of Al interconnect lines could also offer the possibility of extending the lifetime of Cu and other metals subjected to electromigration (Thompson and Lloyd 1993).
- The use of refractory metal shunt layers of materials such as TiN between an interconnect line and its surrounding dielectric material provides a mechanism for the current in an interconnect interrupted by a void to bypass the void once the electrical resistance reaches unacceptably large levels. Likewise, liners introduced between Cu lines and the surrounding dielectric materials could be used to protect the functionality of circuits with Cu interconnect lines.

9.7.4 Assessment of interconnect reliability

On the basis of experimental observations of electromigration failure in Al interconnect lines, Black (1969) postulated a simple relationship for failure time t_{fail} in terms of current density, i, and temperature, T, in the form

$$t_{\text{fail}} = A_{\text{em}} \cdot i^{-n_{\text{em}}} \cdot \exp\left(\frac{Q_{\text{em}}}{kT}\right), \qquad (9.137)$$

where A_{em}, Q_{em} and n_{em} are empirically determined constants. For wide aluminum lines (for which the line width is greater than the grain size), Black found that $n_{\text{em}} = 2$ and $Q_{\text{em}} = 0.5$ eV. The constants in the Black equation are affected by a number of factors including the properties of the interconnect metal, line geometry, grain structure and crystallographic texture. For example, the value of Q_{em} is typically about 0.7 eV for wide Al–Cu lines, 0.9 eV for narrow or bamboo-like Al–Cu lines, and 1.0 eV for wide or bamboo-like pure Cu lines (Srikar and Thompson 1999). Likewise, the geometry of the lines can influence the value of n_{em}: it varies from about 2 for wide pad-to-pad lines, to a value in the range 1–2 for narrow via-to-via lines, and to a value widely ranging above 1 for short, narrow via-to-via lines.

Nevertheless, (9.137) has remained a basis for most electromigration reliability analyses during the past several decades.

A key challenge in the assessment of the mechanical and electrical reliability of interconnect lines prone to electromigration damage is to derive the long-term failure times for complex circuits on the basis of accelerated, simple laboratory or wafer-level tests. In addition, the small population of samples tested in the simulated environment should provide a realistic representation of a substantially larger population encountered in service. There are several different types of electromigration tests that are commonly used to assess the lifetime of interconnects (Ohring 1998). One type of test entails raising the temperature in steps in a packaged metal stripe while an electric current is simultaneously imposed. The resulting temperature-dependence of resistance increase due to electromigration is empirically determined for analysis. In another type of test, the electric current is raised in steps to very high values, the electrical resistance is monitored at each current step level and the temperature is calculated. In many standard test structures that have been used, a test line containing four contact pads is employed, with the two outer pads imposing an electric current and the two inner pads monitoring the voltage of the lines. Failure is often defined as an open circuit failure due to a through-thickness void, a short-circuit failure due to an extrusion, or a percent change in electrical resistance.

Once a test structure and type of test are chosen, an acceleration factor for the simulated test is selected. For example, let t_{design} be the minimum acceptable failure time for a particular design characterized by a current i and temperature T in service. If the duration of the accelerated simulation test is chosen as t_{fail}, the current i_{test} corresponding to a temperature T_{test} is chosen from the acceleration factor AF which is derived from Black's equation to be

$$\text{AF} = \frac{t_{design}}{t_{fail}} = \left\{ \frac{i_{test}}{i} \right\}^2 \cdot \exp\left(-\frac{Q_{em} n_{em}}{kT} \left[\frac{1}{T_{test}} - \frac{1}{T} \right] \right). \tag{9.138}$$

If a wafer-level test leads to the result that $t_{fail} > t_{design}/\text{AF}$, then the wafer could be accepted as having a sufficiently long life expectancy for electromigration.

Such simulated tests are carried out to record electromigration lifetimes for a large population of lines. The statistical characteristics of these sample tests are then analyzed, and appropriate scaling laws, such as (9.134), (9.137) and (9.138), are applied to develop models for circuit-level reliability. Although electromigration tests on straight line segments have traditionally been the standard for fundamental research, there is growing interest in the study of electromigration reliability of interconnect tree test structures comprising, for example, L-shaped and T-shaped line segments (Srikar and Thompson 1999, Hau-Riege and Thompson 2000, Suo 2003). Typically, the median time to failure, usually denoted by t_{50} to represent the time below which 50% of the lines fail, is used in reliability analyses. The standard unit of measure to characterize interconnect reliability is the FIT, where one FIT equals one failure in 10^9 device hours.

9.8 Exercises

1. The growth or decay in amplitude of a sinusoidal surface profile on a uniformly stressed solid as a result of surface diffusion is considered in Section 9.2.1 for the case of isotropic surface energy. Consider the case in which the change in surface profile occurs by evaporation-condensation, rather than by surface diffusion. Express the characteristic time for the surface evolution process in terms of system parameters, and show that the wavelength with the highest rate of growth is again given by (9.22).

2. Suppose that a small volume of material in the shape of an inverted cone (as depicted in Figure 8.30, for example) is removed from near the surface of a strained crystal; the base of the cone is originally part of the crystal surface. The material element is then flipped over and its base is bonded epitaxially to the surface of the crystal at a small distance from the conical pit created by its removal. Assume that all surfaces have the same constant surface energy density γ and that the elastic strain relaxed in this process is roughly $M\epsilon_m^2$ times the volume of material removed, where M is the biaxial elastic modulus and ϵ_m is the initially uniform equi-biaxial elastic strain. Estimate the range of values of $\gamma/M\epsilon_m^2$ for which the island and pit are subject to a mutually repulsive configurational force; express the result in terms of the volume and aspect ratio of the conical form. Estimate the magnitude of the effect for the island shown in Figure 9.9.

3. Derive the expression for the perturbed mean normal stress given in (9.95); the constraints on the material sample which provide boundary conditions are described in Section 9.6.2.

4. The compositional stability of an $In_\xi Ga_{1-\xi}As$ solid solution was considered in Section 9.6.3 for the form of free energy of mixing given by (9.99). In particular, the temperatures at which a composition would become unstable for any value of ξ was determined with elastic energy effects neglected, and then the change in this estimate found by incorporating the elastic energy effect was determined. For the case with composition $\xi = 0.2$, determine the lower bound on the range of temperatures for which the composition is stable if the effect of elastic energy is ignored, and then estimate how this bound changes if the effect of elastic energy is included.

5. Suppose that a conical island as depicted in Figure 8.30 undergoes a gradual change in its aspect ratio a at fixed material volume V. Let the variable q denote the distance of a point on the island surface from the apex, measured along a surface generator.

   (a) Determine the normal speed $v_n(q)$ of the surface as a function of distance from the apex for some instantaneous value of $\dot{a}$. Express the result in terms of the volume V and the current value of a.

   (b) Determine the mass flux vector $j(q)$ along a generator of the cone corresponding to the normal speed determined in part (a). Note that conservation of mass requires that $j(0) = 0$. Evaluate $j(q)$ at the base of the conical island. Is the result consistent with evolution of shape at constant volume?

   (c) Assuming the existence of a surface chemical potential distribution, evaluate the functional defined in (9.51). Maximize the result with respect to $\dot{a}$ in order to obtain an evolution equation for island shape. Confirm that the result is a differential equation for a as a function of time of the form

$$\frac{\partial \mathcal{F}}{\partial a}(a) + \mathcal{L}(a)\dot{a} = 0$$

where $\mathcal{F}$ is given by (8.151) and $\mathcal{L}$ is a shape-dependent quantity.

6. The coupling between deformation and driving force for change in compositional distribution was considered for the case of a layer with through-the-thickness variation in composition in Section 9.6.5. For the case of composition evolution represented by the solution in (9.118), determine the corresponding dependence of curvature of the layer on time.

7. Examine the influence of the elastic energy on the contact angle or wetting angle given by Young's equation (1.4). Adopt the configuration with a rigid substrate as depicted in Figure 1.10, but include elastic energy W in the total free energy of the island by assuming that $W(V, \theta) = M\epsilon_{\mathrm{m}}^2(1 - \theta/\pi)V$ where M is the biaxial elastic modulus and ϵ_{m} is the mismatch strain.

   (a) Determine the range of system parameters for which the elastic energy has a significant influence on the wetting angle for fixed island volume V.
   (b) Suppose that the island evolves at a fixed rate $\dot{V}$. Determine the corresponding rate of change $\dot{\theta}$ of the wetting angle if alteration in shape is due to surface diffusion with surface mobility m_{S}.

References

Abermann, R. (1990), Measurements of the intrinsic stress in thin metal films, *Vacuum* **41**, 1279–1282.

Abermann, R. and Koch, R. (1986), In situ study of thin film growth by internal stress measurement under ultrahigh vacuum conditions: Silver and copper under the influence of oxygen, *Thin Solid Films* **142**, 65–76.

Abermann, R., Kramer, R. and Mäser, J. (1978), Structure and internal stress in ultra-thin silver films deposited on MgF$_2$ and SiO substrates, *Thin Solid Films* **52**, 215–229.

Alcala, J., Gaudette, F., Suresh, S. and Sampath, S. (2001), Instrumented spherical micro-indentation of plasma-sprayed coatings, *Materials Science and Engineering* **A316**, 1–10.

Alexander, H. (1986), Dislocations in covalent crystals, *Dislocations in Solids*, edited by F. R. N. Nabarro, Elsevier, 113–234.

Allen, H. G. (1969), *Analysis and Design of Structural Sandwich Panels*, Pergamon Press, Oxford.

Anderson, G. P., Devries, K. L. and Williams, M. L. (1974), Mixed mode stress field effect in adhesive fracture, *International Journal of Fracture* **10**, 565–583.

Andreussi, F. and Gurtin, M. E. (1977), On the wrinkling of a free surface, *Journal of Applied Physics* **48**, 3798–3799.

Andrews, E. H. and Kinloch, A. J. (1973), Mechanics of adhesive failure. I, *Proceedings of the Royal Society (London)* **A332**, 385–399.

Arakawa, T., Tsukamoto, S., Nagamune, Y., Nishioka, M., Lee, J.-H. and Arakawa, Y. (1993), Fabrication of InGaAs strained quantum wire structures using selective area metal-organic chemical vapor deposition growth, *Japanese Journal of Applied Physics* **32**, L1377–1379.

Argon, A. S., Gupta, V., Landis, H. S. and Cornie, J. A. (1989), Intrinsic toughness of interfaces between SiC coatings and substrates of Si or C fibre, *Journal of Materials Science* **24**, 1207–1218.

Arzt, E. (1998), Size effects in materials due to microstructural and dimensional constraints: A comparative review, *Acta Materialia* **46**, 5611–5626.

Asaro, R. J. and Tiller, W. A. (1972), Interface morphology development during stress corrosion cracking: Part I. Via surface diffusion, *Metallurgical Transactions* **3**, 1789–1796.

Ashby, M. F. (1970), The deformation of plastically nonhomogeneous materials, *Philosophical Magazine* **21**, 399–424.

713

Atkinson, C., Smelser, R. and Sanchez, J. (1982), Combined mode fracture via the cracked Brazilian disk test, *International Journal of Fracture* **18**, 279–291.

Audoly, B. (1999), Stability of straight delamination blisters, *Physical Review Letters* **83**, 4124–4127.

Auld, B. A. (1973), *Acoustic Fields and Waves in Solids*, Volume I, Wiley, New York.

Bagchi, A., Lucas, G. E., Suo, Z. and Evans, A. G. (1994), A new procedure for measuring the decohesion energy for thin ductile films on substrates, *Journal of Materials Research* **9**, 1734–1741.

Baker, S. P. and Nix, W. D. (1994), Mechanical properties of compositionally modulated Au–Ni thin films – Nanoindentation and microcantilever deflection experiments, *Journal of Materials Research* **9**, 3131–3145.

Balk, T. J., Dehm, G. and Arzt, E. (2001), Dislocations and deformation mechanisms in thin films and small structures, Materials Research Society Symposium, Vol. 673, 2.7.1–2.7.6.

Baral, D., Ketterson, J. B. and Hilliard, J. E. (1985), Mechanical properties of composition modulated Cu–Ni foils, *Journal of Applied Physics* **57**, 1076–1083.

Barenblatt, G. I. (1959), Concerning equilibrium cracks forming during brittle fracture: The stability of isolated cracks, relationship with energetic theories, *Applied Mathematics and Mechanics (PMM)* **24**, 1273–1282.

Barnett, D. M. and Lothe, J. (1974), An image force theorem for dislocations in anisotropic bicrystals, *Journal of Physics. F: Metal Physics* **4**, 1618–1635.

Bauer, E. (1958), Phenomenologische theorie der kristallabscheidung an oberflachen I, *Zeitschrift fur Kristallographie* **110**, 372–394.

Bauschinger, J. (1886), Ueber die veränderungen der elastizitäwtsgrenze und der festigkeit des eisens und stahls durch strecken, quetschen, erwärmen abkuehlen und durch oftmals widerholte Belastung, *Mitt: Mech-Tech Laboratory, Munich* **XIII**.

Bean, J. C., Feldman, L. C., Fiory, A. T., Nakahara, S. and Robinson, I. K. (1984), Ge_xSi_{1-x}/Si strained layer superlattice grown by molecular beam epitaxy, *Journal of Vacuum Science and Technology* **A2**, 436–440.

Beardsley, M. B. (1997), Thick thermal barrier coatings for diesel engines, *Journal of Thermal Spray Technology* **6**, 181–186.

Beltz, G. E. and Freund, L. B. (1993), Nucleation of a dislocation loop at the free surface of a crystal, *Physica Status Solidi* **108b**, 303–313.

Beltz, G. E. and Freund, L. B. (1994), Analysis of the strained layer critical thickness concept based on a Peierls–Nabarro model of a threading dislocation, *Philosophical Magazine* **69**, 183–202.

Beuth, J. L. (1992), Cracking of thin bonded films in residual tension, *International Journal of Solids and Structures* **29**, 1657–1675.

Bhattacharya, A. K. and Nix, W. D. (1988), Analysis of elastic and plastic deformation associated with indentation testing of thin films on substrates, *International Journal of Solids and Structures* **24**, 1287–1298.

Biot, M. A. (1965), *Mechanics of Incremental Deformation*, Wiley Publishing, New York.

Black, J. R. (1969), Electromigration – A brief survey and some recent results, *IEEE Transactions on Electron Devices* **ED16**, 338–347.

Blakely, J. M. (1973), *Crystal Surfaces*, Pergamon Press, Oxford.

Blech, I. A. (1976), Electromigration in thin aluminum films on titanium nitride, *Journal of Applied Physics* **47**, 1203–1208.

Blech, I. A. and Cohen, U. (1982), Effects of humidity on stress in thin silicon dioxide films, *Journal of Applied Physics* **53**, 4202–4207.

Blech, I. A. and Meieran, E. S. (1967), Enhanced x-ray diffraction from substrate crystals coating discontinuous surface films, *Journal of Applied Physics* **38**, 2913–2919.

Blech, I. A., Petroff, P. M., Tai, K. L. and Kumar, V. (1975), Whisker growth in Al thin films, *Journal of Crystal Growth* **32**, 161–169.

Bodensohn, J., Nicolai, K. and Leiderer, P. (1986), The growth of atomically rough ^{4}He crystals, *Zeitschrift Physik* **B64**, 55–64.

Bogy, D. B. (1971), Two edge bonded elastic wedges of different materials and wedge angles under surface traction, *Journal of Applied Mechanics* **38**, 377–386.

Bowden, N., Brittain, S., Evans, A. G., Hutchinson, J. W. and Whitesides, G. M. (1998), Spontaneous formation of ordered structures in thin films of metals supported on elastomeric polymer, *Nature* **393**, 146–149.

Bragg, W. and Nye, J. F. (1947), A dynamic model of a crystal structure, *Proceedings of the Royal Society (London)* **A190**, 474–481.

Budiansky, B., Hutchinson, J. W. and Evans, A. G. (1993), On neutral holes in tailored, layered sheets, *Journal of Applied Mechanics* **60**, 1056–1058.

Cahn, J. W. (1961), On spinodal decomposition, *Acta Metallurgica* **9**, 795–801.

Cahn, J. W. (1968), Spinodal decomposition, *Transactions of the Metallurgical Society of AIME* **242**, 166–180.

Cahn, J. W. (1980), Surface stress and the chemical equilibrium of small crystals – I. The case of the isotropic surface, *Acta Metallurgica* **28**, 1333–1338.

Cahn, J. W. and Hilliard, J. E. (1958), Free energy of a nonuniform system. I. Interfacial free energy, *Journal of Applied Physics* **28**, 258–267.

Cahn, J. W. and Taylor, J. E. (1994), Surface motion by surface diffusion, *Acta Metallurgica et Materialia* **42**, 1045–1063.

Cammarata, R. C. (1994), Surface and interface stress effects in thin films, *Progress in Surface Science* **46**, 1–38.

Cammarata, R. C. and Sieradzki, K. (1989), Surface stress effects on the critical thickness for epitaxy, *Applied Physics Letters* **55**, 1197–1198.

Cammarata, R. C., Trimble, T. M. and Srolovitz, D. J. (2000), Surface stress model for intrinsic stresses in thin films, *Journal of Materials Research* **15**, 2468–2474.

Cao, H. C. and Evans, A. G. (1989), An experimental study of the fracture resistance of bimaterial interfaces, *Mechanics of Materials* **7**, 295–304.

Chai, H. (1990), Three-dimensional fracture analysis of thin film debonding, *International Journal of Fracture* **46**, 237–256.

Charalambides, P. G., Lund, J., Evans, A. G. and McMeeking, R. M. (1989), A test specimen for determining the fracture resistance of bimaterial interfaces, *Journal of Applied Mechanics* **56**, 77–82.

Chason, E., Sheldon, B. W., Freund, L. B., Floro, J. A. and Hearne, S. J. (2002), Origin of compressive residual stress in polycrystalline thin films, *Physical Review Letters* **88**, 156103-(4).

Chaudhari, P. (1971), Grain growth and stress relief in thin films, *Journal of Vacuum Science and Technology* **9**, 520–522.

Chaudhari, P. (1979), Plastic properties of polycrystalline thin films on a substrate, *Philosophical Magazine* **39**, 507–516.

Chen, H., He, Y., Shiflet, G. J. and Poon, S. J. (1994), Deformation-induced nanocrystal formation in shear bands of amorphous alloys, *Nature* **367**, 541–544.

Chiu, C.-H. and Gao, H. (1994), Numerical simulation of diffusion controlled surface evolution, *Mechanisms of Thin Film Evolution*. MRS vol. 317, edited by S. M. Yalisove et al., Materials Research Society, Pittsburgh, 369–374.

Chiu, C.-H. and Gao, H. (1995), A numerical study of stress controlled surface diffusion during epitaxial film growth, *Thin films: Stresses and Mechanical Properties V*. MRS vol. 356 edited by S. P. Baker et al. Materials Research Society, Pittsburgh, 33–44.

Choi, H. C. and Kim, K. S. (1992), Analysis of the spontaneous interfacial decohesion of a thin surface film, *Journal of the Mechanics and Physics of Solids* **40**, 75–103.

Choi, Y. and Suresh, S. (2002), Size effects on the mechanical properties of thin polycrystalline films on substrates, *Acta Materialia* **50**, 1881–1883.

Choquette, K. D. (2002), Vertical cavity surface emitting lasers: Light for the information age, *MRS Bulletin* **27**, 507–511.

Chou, Y. T. (1966), Screw dislocations in and near lamellar inclusions, *Physica Status Solidi* **17**, 509–516.

Christensen, R. J., Toplygo, V. K. and Clarke, D. R. (1997), The influence of reactive element yttrium on the stress in alumina scales formed by oxidation, *Acta Materialia* **45**, 1761–1766.

Christian, J. W. (1975), *Transformations in Metals and Alloys*, Part I, Pergamon Press, Oxford, 2nd Edition.

Clemens, B. M. and Bain, J. A. (1992), Stress determination in textured thin films using x-ray diffraction, *MRS Bulletin* **17**, 46–51.

Copel, M., Reuter, M. C., Kaxiras, E. and Tromp, R. M. (1989), Surfactants in epitaxial growth, *Physical Review Letters* **63**, 632–635.

Corcoran, S. G., Colton, R. J., Lilleodden, E. T. and Gerberich, W. W. (1997), Anomalous plastic deformation at surfaces: nanoindentation of gold single crystals, *Physical Review B* **55**, 16057–16060.

Cornella, G., Lee, S.-H., Nix, W. D. and Bravman, J. C. (1997), An analysis technique for extraction of thin film stresses from x-ray data, *Applied Physics Letters* **71**, 2949–2951.

Cottrell, A. H. (1964), *The Mechanical Properties of Matter*, John Wiley & Sons, New York.

Craighead, H. G. (2000), Nanoelectromechanical systems, *Science* **290**, 1532–1535.

Cullis, A. G., Robbins, D. J., Pidduck, A. J. and Smith, P. W. (1992), The characteristics of strain-modulated surface undulations formed upon epitaxial $Si_{1-x}Ge_x$ alloy layers on Si, *Journal of Crystal Growth* **123**, 333–343.

Cullity, B. D. (1978), *Elements of X-ray Diffraction*, Addison-Wesley, Reading, MA.

Currey, J. D. (1976), Further studies on the mechanical properties of mollusk shell material, *Journal of the Zoological Society of London* **180**, 445–453.

Dao, M., Chollacoop, N., Van Vliet, K. J., Venkatesh, T. A. and Suresh, S. (2001), Computational modeling of the forward and reverse problems in instrumented sharp indentation, *Acta Materialia* **49**, 3899–3918.

Daruka, I., Tersoff, J. and Barabasi, A.-L. (1999), Shape transition in growth of strained islands, *Physical Review Letters* **82**, 2753–2756.

Dauskardt, R. H., Lane, M., Ma, Q. and Krishna, N. (1998), Adhesion and debonding of multi-layer thin film structures, *Engineering Fracture Mechanics* **61**, 141–162.

Davis, S. H. (2001), *Theory of Solidification*, Cambridge University Press, Cambridge.

Dehm, G., Balk, T. J., von Blanckenhage, B., Gumbsch, P. and Arzt, E. (2002b), Dislocation dynamics in sub-micron confinement: recent progress in Cu thin film plasticity, *Zeitschrift für Metallkunde* **93**, 383–391.

Dehm, G., Wagner, T., Balk, T.J. and Arzt, E. (2002a), Plasticity and interfacial dislocation mechanisms in epitaxial and polycrystalline Al films constrained by substrates, *Journal of Materials Science and Technology* **18**, 113–117.

Delannay, F. and Warren, P. (1991), On crack interaction and crack density in strain-induced cracking of brittle films on ductile substrates, *Acta Metallurgica et Materialia* **39**, 1061–1072.

Devaney, J. R. (1989), Failure mechanisms in active devices, *Electronic Materials Handbook, Volume I: Packaging*, ASM International, Materials Park, Ohio, 1006–1017.

DeWolf, I., Ignat, M., Pozza, G., Maniguet, L. and Maes, H. E. (1999), Analysis of local mechanical stresses in and near tungsten lines on silicon substrate, *Journal of Applied Physics* **85**, 6477–6485.

DeWolf, I., Vanhellemont, J., Romano-Rodriguez, A., Norstrom, H. and Maes, H. E. (1992), Micro-Raman study of stress distribution in local isolation structures and correlation with transmission electron microscopy, *Journal of Applied Physics* **71**, 898–906.

D'Heurle, F. M. (1971), Electromigration and failure in electronics: An introduction, *Proceedings of IEEE* **59**, 1409–1418.

Ding, K. and Andersen, H. C. (1986), Molecular-dynamics simulation of amorphous germanium, *Physical Review B* **34**, 6987–6991.

Dodson, B. W. and Tsao, J. Y. (1987), Relaxation of strained-layer semiconductor structures via plastic flow, *Applied Physics Letters* **51**, 1325–1327.

Doerner, M. F. and Brennan, S. (1988), Strain distribution in thin aluminum films using x-ray depth profiling, *Journal of Applied Physics* **63**, 126–131.

Doerner, M. F., Gardner, D. S. and Nix, W. D. (1986), Plastic properties of thin films on substrates as measured by submicron indentation hardness and substrate curvature techniques, *Journal of Materials Research* **1**, 845–851.

Doerner, M. F. and Nix, W. D. (1988), Stresses and deformation processes in thin films on substrates, *CRC Critical Reviews in Solid State and Materials Sciences* **14**, 225–268.

Doljack, F. A. and Hoffman, R. W. (1972), The origins of stress in thin nickel films, *Thin Solid Films* **12**, 71–74.

Dundurs, J. (1969), Edge-bonded dissimilar orthogonal elastic wedges, *Journal of Applied Mechanics* **36**, 650–652.

Eaglesham, D. J., Kvam, E. P., Maher, D. M., Humphreys, C. J. and Bean, J. C. (1989), Dislocation nucleation near the critical thickness in GeSi/Si strained layers, *Philosophical Magazine* **A59**, 1059–1073.

Elsasser, W. M. (1969), Convection and stress propagation in the upper mantle, *The application of modern physics to the earth and planetary interiors*, edited by S. K. Runcorn, Wiley-Interscience, London, 223–246.

Erdogan, F. (1995), Fracture mechanics of functionally graded materials, *Composite Engineering* **5**, 753–770.

Erdogan, F. and Gupta, G. D. (1972), On the numerical solution of singular integral equations, *Quarterly Journal of Applied Mathematics* **30**, 525–534.

Eshelby, J. D. (1951), The force on an elastic singularity, *Philosophical Transactions of the Royal Society (London)* **A244**, 87–112.

Eshelby, J. D. (1957), The determination of the elastic field of an ellipsoidal inclusion, and related problems, *Proceedings of the Royal Society (London)* **A241**, 376–396.

Eshelby, J. D. (1970), Energy relations and the energy-momentum tensor in continuum mechanics, *Inelastic Behavior of Solids*, edited by M. F. Kanninen et al., McGraw-Hill, New York, 77–115.

Eshelby, J. D. (1979), Boundary problems, *Dislocations in Solids*, edited by F. R. N. Nabarro, North-Holland Publishing, Amsterdam, 167–221.

Eskin, S. G. and Fritze, J. R. (1940), Thermostatic bimetals, *Transactions of the American Society of Mechanical Engineers* **62**, 433–440.

Evans, A. G., Drory, M. D. and Hu, M. S. (1988), The cracking and decohesion of thin films, *Journal of Materials Research* **3**, 1043–1049.

Evans, A. G., He, M. Y. and Hutchinson, J. W. (1998), Effect of interface undulations on the thermal fatigue of thin films and scales on metal surfaces, *Acta Materialia* **45**, 3543–3554.

Evans, A. G. and Hutchinson, J. W. (1995), The thermomechanical integrity of thin films and multilayers, *Acta Metallurgica et Materialia* **43**, 2507–2530.

Evans, A. G., Hutchinson, J. W. and Wei, Y. (1999), Interface adhesion: effects of plasticity and segregation, *Acta Materialia* **47**, 4093–4113.

Evans, A. G., Mumm, D. R., Hutchinson, J. W., Meier, G. H. and Pettit, F. S. (2001), Mechanisms controlling the durability of thermal barrier coatings, *Progress in Materials Science* **46**, 505–553.

Farris, R. J. and Bauer, C. L. (1988), A self-delamination method of measuring the surface energy of adhesion of coatings, *Journal of Adhesion* **26**, 293–300.

Finot, M., Blech, I. A., Suresh, S. and Fujimoto, H. (1997), Large deformation and geometric instability of substrates with thin-film deposits, *Journal of Applied Physics* **81**, 3457–3464.

Finot, M. and Suresh, S. (1996), Small and large deformation of thick and thin film multi-layers: effects of layer geometry, plasticity and compositional gradients, *Journal of the Mechanics and Physics of Solids* **44**, 683–721.

Fiory, A. T., Bean, J. C., Hull, R. and Nakahara, S. (1984), Thermal relaxation of metastable strained layer Ge_xSi_{1-x}/Si epitaxy, *Physical Review B* **31**, 4063–4065.

Fitzgerald, E. A. (1995), GeSi/Si nanostructures, *Annual Reviews of Materials Science* **25**, 417–454.

Fitzgerald, E. A., Currie, M. T., Samavedam, S. B., Langdo, T. A., Taraschi, G., Yang, V., Leitz, C. W. and Bulsara, M. T. (1999a), Dislocations in relaxed SiGe/Si heterostructures, *Physica Status Solidi* **171**, 227–238.

Fitzgerald, E. A., Kim, A. Y., Currie, M. T., Langdo, T. A., Taraschi, G. and Bulsara, M. T. (1999b), Dislocation dynamics in relaxed graded composition semiconductors, *Materials Science and Engineering* **B67**, 53–61.

Fitzgerald, E. A., Samavedam, S. B., Xie, Y. H. and Giovane, L. M. (1997), Influence of strain on semiconductor thin film epitaxy, *Journal of Vacuum Science and Technology* **A15**, 1048–1056.

Fitzgerald, E. A., Watson, G. P., Proano, R. E., Ast, D. B., Kirchner, P. D., Pettit, G. D. and Woodall, J. (1989), Nucleation mechanisms and the elimination of misfit dislocations at mismatched interfaces by reduction in growth area, *Journal of Applied Physics* **65**, 2220–2237.

Fitzgerald, E. A., Xie, Y. H., Green, M. L., Brasen, D., Kortan, A. R. and Michel, J. (1991), Totally relaxed Ge_xSi_{1-x} layers with low threading dislocation densities grown on Si substrates, *Applied Physics Letters* **59**, 811–813.

Fleck, N. A. and Hutchinson, J. W. (1997), Strain gradient plasticity, *Advances in Applied Mechanics* **33**, 295–361.

Fleck, N. A., Muller, G. M., Ashby, M. F. and Hutchinson, J. W. (1994), Strain gradient plasticity: Theory and experiment, *Acta Metallurgica et Materialia* **42**, 475–487.

Flinn, P. A. and Chiang, C. (1990), X-ray diffraction determination of the effect of various passivations on stress in metal films and patterned lines, *Journal of Applied Physics* **67**, 2927–2931.

Flinn, P. A., Gardner, D. S. and Nix, W. D. (1987), Measurement and interpretation of stress in aluminum-based metallization as a function of thermal history, *IEEE Transactions on Electron Devices* **ED34**, 689–699.

Floro, J. A. and Chason, E. (1996), Measuring Ge segregation by real-time stress monitoring during $Si_{1-x}Ge_x$ molecular beam epitaxy, *Applied Physics Letters* **69**, 3830–3832.

Floro, J. A., Chason, E., Lee, S. R., Twesten, R. D., Hwang, R. Q. and Freund, L. B. (1997), Real-time stress evolution during $Si_{1-x}Ge_x$ heteroepitaxy: Dislocations, islanding and segregation, *Journal of Electronic Materials* **26**, 969–979.

Floro, J. A., Hearne, S. J., Hunter, J. A., Kotula, P., Chason, E., Seel, S. C. and Thompson, C. V. (2001), The dynamic competition between stress generation and relaxation mechanisms during coalescence of Volmer–Weber thin films, *Journal of Applied Physics* **89**, 4886–4897.

Floro, J. A., Lucadamo, G. A., Chason, E., Freund, L. B., Sinclair, M., Twesten, R. D. and Hwang, R. Q. (1998), SiGe island shape transitions induced by elastic repulsion, *Physical Review Letters* **80**, 4717–4720.

Floro, J. A., Sinclair, M., Chason, E., Freund, L. B., Twesten, R. D., Hwang, R. Q. and Lucadamo, G. A. (2000), Novel SiGe island coarsening kinetics: Ostwald ripening, and elastic interactions, *Physical Review Letters* **84**, 701–705.

Frank, F. C. and Van der Merwe, J. H. (1949a), One-dimensional dislocations: I. Static theory, *Proceedings of the Royal Society (London)* **A198**, 205–216.

Frank, F. C. and Van der Merwe, J. H. (1949b), One-dimensional dislocations: II. Misfitting monolayers and oriented growth, *Proceedings of the Royal Society (London)* **A198**, 216–225.

Freund, L. B. (1978), Stress intensity factor calculations based on a conservation integral, *International Journal of Solids and Structures* **14**, 241–250.

Freund, L. B. (1987), The stability of a dislocation threading a strained layer on a substrate, *Journal of Applied Mechanics* **54**, 553–557.

Freund, L. B. (1990a), *Dynamic Fracture Mechanics*, Cambridge University Press, New York.

Freund, L. B. (1990b), A criterion for arrest of a threading dislocation in a strained epitaxial layer due to an interface misfit dislocation in its path, *Journal of Applied Physics* **68**, 2073–2080.

Freund, L. B. (1990c), The driving force for glide of a threading dislocation in a strained epitaxial layer on a substrate, *Journal of the Mechanics and Physics of Solids* **38**, 657–679.

Freund, L. B. (1993), The stress distribution and curvature of a general compositionally graded semiconductor layer, *Journal of Crystal Growth* **132**, 341–344.

Freund, L. B. (1994), The mechanics of dislocations in strained-layer semiconductor materials, *Advances in Applied Mechanics* **30**, 1–66.

Freund, L. B. (1995a), Evolution of waviness on the surface of a strained elastic solid due to stress-driven diffusion, *International Journal of Solids and Structures* **32**, 911–923.

Freund, L. B. (1995b), On the stability of a biaxially stressed elastic material with a free surface under variations in surface shape, *Zeitschrift für angewandte Mathematik und Physik* **46**, S185–S200.

Freund, L. B. (1996), Some elementary connections between curvature and mismatch strain in compositionally graded thin films, *Journal of the Mechanics and Physics of Solids* **44**, 723–736.

Freund, L. B. (1998), A surface chemical potential for elastic solids, *Journal of the Mechanics and Physics of Solids* **46**, 1835–1844.

Freund, L. B. (2000), Substrate curvature due to thin film mismatch strain in the nonlinear deformation range, *Journal of the Mechanics and Physics of Solids* **48**, 1159–1174.

Freund, L. B., Bower, A. and Ramirez, J. C. (1989), Mechanics of elastic dislocations in strained layer structures, MRS Symposium Proc. Vol. 130, 139–152.

Freund, L. B. and Chason, E. (2001), Model for stress generated upon contact of neighboring islands on the surface of a substrate, *Journal of Applied Physics* **89**, 4866–4873.

Freund, L. B., Floro, J. A. and Chason, E. (1999), Extensions of the Stoney formula for substrate curvature to configurations with thin substrates and large deformations, *Applied Physics Letters* **74**, 1987–1989.

Freund, L. B. and Gosling, T. J. (1995), A critical thickness condition for growth of strained quantum wires in substrate V-grooves, *Applied Physics Letters* **66**, 2822–2824.

Freund, L. B. and Hull, R. (1992), On the Dodson–Tsao excess stress for glide of a threading dislocation in a strained epitaxial layer, *Journal of Applied Physics* **71**, 2054–2056.

Freund, L. B. and Jonsdottir, F. (1993), Instability of a biaxially stressed thin film on a substrate due to material diffusion over its free surface, *Journal of the Mechanics and Physics of Solids* **41**, 1245–1264.

Freund, L. B. and Nix, W. D. (1996), A critical thickness condition for a strained compliant substrate/epitaxial film system, *Applied Physics Letters* **69**, 173–175.

Fried, E. and Gurtin, M. E. (2003), The role of the configuration of force balance in the nonequilibrium epitaxy of films, *Journal of the Mechanics and Physics of Solids* **51**, 487–517.

Friesen, C. and Thompson, C. V. (2002), Reversible stress relaxation during precoalescence interruptions of Volmer–Weber thin film growth, *Physical Review Letters* **89**, 126103–4.

Fritz, I. J. (1987), Role of experimental resolution in measurements of critical layer thickness for strained-layer epitaxy, *Applied Physics Letters* **51**, 1080–1082.

Frost, H. J. and Ashby, M. F. (1982), *Deformation-Mechanisms Maps*, Pergamon Press, New York.

Frost, H. J., Thompson, C. V. and Walton, D. T. (1990), Simulation of thin film grain structures – I. Grain growth stagnation, *Acta Metallurgica et Materialia* **38**, 1455–1462.

Fujikawa, Y., Akiyama, K., Nagao, T., Sakurai, T., Lagally, M. G., Hashimoto, T., Morikawa, Y. and Terakura, K. (2002), Origin of the stability of Ge (105) on Si: A new structure model and surface strain relaxation, *Physical Review Letters* **88**, 176101.

Fukuda, Y., Kohama, Y., Seki, M. and Ohmachi, Y. (1988), Misfit dislocation structures at MBE-grown $Si_{1-x}Ge_x$/Si interfaces, *Japan Journal of Applied Physics* **27**, 1593–1598.

Fung, Y. C. (1965), *Foundations of Solid Mechanics*, Prentice-Hall, Inc.

Gao, H. (1991), Stress concentration at slightly undulating surfaces, *Journal of the Mechanics and Physics of Solids* **39**, 443–458.

Gao, H. and Nix, W. D. (1999), Surface roughening of heteroepitaxial thin films, *Annual Review of Material Science* **29**, 173–209.

Gao, H., Zhang, L., Nix, W. D., Thompson, C. V. and Arzt, E. (1999), Crack-like grain boundary diffusion wedges in thin metal films, *Acta Materialia* **47**, 2865–2878.

Gaudette, F. G., Suresh, S., Evans, A. G., Dehm, G. and Rühle, M. (1997), The influence of chromium addition on the toughness of γ-Ni/α-Al$_2$O$_3$ interfaces, *Acta Materialia* **45**, 3503–3513.

Gaudette, F. G., Suresh, S. and Evans, A. G. (2000), Effects of sulfur on the fatigue and fracture resistance of interfaces between γ-Ni (Cr) and α-Al$_2$O$_3$ interfaces, *Metallurgical and Materials Transactions* **A31**, 1977–1983.

Gent, A. N. and Lewandowski, L. H. (1987), Blow-off pressures for adhering layers, *Journal of Applied Polymer Science* **33**, 1567–1577.

Giannakopoulos, A. E., Blech, I. A. and Suresh, S. (2001), Large deformation of layered thin films and flat panels: Effects of gravity, *Acta Materialia* **49**, 3671–3688.

Giannakopoulos, A. E. and Suresh, S. (1997), Indentation of solids with gradients in elastic properties: Part I. Point force, Part II. Axisymmetric indenters, *International Journal of Solids and Structures* **34**, 2357–2428.

Giannakopoulos, A. E., Suresh, S., Finot, M. and Olsson, M. (1995), Elastoplastic analysis of thermal cycling: Layered materials with compositional gradients, *Acta Metallurgica et Materialia* **43**, 1335–1354.

Giannakopoulos, K. P. and Goodhew, P. J. (1998), Striation development in CBE-grown vicinal plane InGaAs layers, *Journal of Crystal Growth* **188**, 26–31.

Gibbs, J. W. (1876), On the equilibrium of heterogeneous substances, *Transactions of the Connecticut Academy of Sciences* **3**, 108–248; reprinted by Oxbow Press, Woodbridge, CT (1992).

Gibbs, J. W. (1878), On the equilibrium of heterogeneous substances, *Transactions of the Connecticut Academy of Sciences* **3**, 343–524; reprinted by Oxbow Press, Woodbridge, CT (1992).

Gillard, V. T., Nix, W. D. and Freund, L. B. (1994), Role of dislocation blocking in limiting strain relaxation in heteroepitaxial films, *Journal of Applied Physics* **76**, 7280–7287.

Gille, G. and Rau, B. (1984), Buckling instability and adhesion of carbon layers, *Thin Solid Films* **120**, 109–121.

Gioia, G. and Ortiz, M. (1997), Delamination of compressed thin films, *Advances in Applied Mechanics* **33**, 119–192.

Glas, F. (1987), Elastic state and thermodynamical properties of inhomogeneous epitaxial layers: Application to immiscible III–V alloys, *Journal of Applied Physics* **62**, 3201–3208.

Gore, G. (1858), On the properties of electro-deposited antimony, *Philosophical Transactions of the Royal Society (London)* **148**, 185–197.

Gore, G. (1862), On the properties of electro-deposited antimony (concluded), *Philosophical Transactions of the Royal Society (London)* **152**, 323–331.

Gosling, T. J. (1996), The stresses due to arrays of inclusions and dislocations of infinite length in an anisotropic half space: Application to strained semiconductor structures, *Philosophical Magazine* **A73**, 11–45.

Gosling, T. J., Jain, S. C. and Harker, A. H. (1994), The kinetics of strain relaxation in lattice-mismatched semiconductor layers, *Physica Status Solidi* **A146**, 713–734.

Gosling, T. J. and Willis, J. R. (1995), Mechanical stability and electronic properties of buried strained quantum wire arrays, *Journal of Applied Physics* **77**, 5601–5610.

Gouldstone, A., Koh, H.-J., Zeng, K. Y., Giannakopoulos, A. E. and Suresh, S. (2000), Discrete and continuous deformation during indentation of thin films, *Acta Materialia* **48**, 2277–2295.

Gouldstone, A., Van Vliet, K. J. and Suresh, S. (2001), Simulation of defect nucleation in a crystal, *Nature* **411**, 656–656.

Gourley, P. L., Fritz, I. J. and Dawson, L. R. (1988), Controversy of critical layer thickness for InGaAs/GaAs strained layer epitaxy, *Applied Physics Letters* **52**, 377–379.

Griffith, A. A. (1920), The phenomenon of rupture and flow in solids, *Philosophical Transactions of the Royal Society (London)* **A221**, 163–198.

Grinfeld, M. (1986), Instability of the separation boundary between a nonhydrostatically stressed elastic body and a melt, *Soviet Physics Doklady* **31**, 831–834.

Guillaume, C. E. (1897), Recherches sur les aciers au nickel. Dilatations aux temperatures elevees; resistance electrique *Reports of the French Academy of Science* **125**, 35–128.

Hager, J. and Spohn, H. (1995), Self-similar morphology and dynamics of periodic surface profiles below the roughening transition, *Surface Science* **324**, 365–372.

Hau-Riege, S. P. and Thompson, C. V. (2000), The effects of mechanical properties of the confinement material on electromigration in metallic interconnects, *Journal of Materials Research* **15**, 1797–1802.

He, M.-Y., Evans, A. G. and Hutchinson, J. W. (1994), Crack deflection at an interface between dissimilar elastic materials: Role of residual stresses, *International Journal of Solids and Structures* **31**, 3443–3455.

He, M. Y., Evans, A. G. and Hutchinson, J. W. (1997), Convergent debonding of films and fibers, *Acta Materialia* **45**, 3481–3489.

He, M. Y., Evans, A. G. and Hutchinson, J. W. (1998), Effects of morphology on the key cohesion of compressed films, *Materials Science and Engineering* **A245**, 168–181.

He, M.-Y. and Hutchinson, J. W. (1989a), Crack deflection at an interface between dissimilar elastic materials, *International Journal of Solids and Structures* **25**, 1053–1067.

He, M.-Y. and Hutchinson, J. W. (1989b), Kinking of a crack out of an interface, *Journal of Applied Mechanics* **56**, 270–278.

He, M. Y., Turner, M. R. and Evans, A. G. (1995), Analysis of the double cleavage drilled compression specimen for interface fracture energy measurements over a range of mode mixities, *Acta Metallurgica et Materialia* **43**, 3453–3458.

Head, A. K. (1953), The interaction of dislocations and boundaries, *Philosophical Magazine* **44**, 92–94.

Hellemans, A. (2000), Cantilever tales, *Science* **290**, 1529.

Henstrom, W. L., Liu, C. P., Gibson, J. M., Kamins, T. I. and Williams, R. S. (2000), Dome-to-pyramid shape transition in Ge/Si islands due to strain relaxation by interdiffusion, *Applied Physics Letters* **77**, 1623–1625.

Herman, H., Sampath, S. and McCune, R. (2000), Thermal spray: current status and future trends, *MRS Bulletin* **25**, 17–25.

Herring, C. (1951a), Surface tension as a motivation for sintering, *The Physics of Powder Metallurgy*, edited by W. E. Kingston, McGraw-Hill, 143–179.

Herring, C. (1951b), Some theorems on the free energies of crystal surfaces, *Physical Review* **82**, 87–93.

Herring, C. (1953), The use of classical macroscopic concepts in surface-energy problems, *Structure and Properties of Solid Surfaces*, edited by R. Gomer and C. S. Smith, University of Chicago Press, 5–72.

Herrman, H. (1920), Thermostatic metal, *General Electric Review* **23**, 57–59.

Hill, R. (1962), Acceleration waves in solids, *Journal of the Mechanics and Physics of Solids* **10**, 1–16.

Hillert, M. (1961), A solid-solution model for inhomogeneous systems, *Acta Metallurgica* **9**, 525–535.

Hinkley, J. A. (1983), A blister test for adhesion of polymers to SiO_2, *Journal of Adhesion* **16**, 115–125.

Hirth, J. P. and Lothe, J. (1982), *Theory of Dislocations*, John Wiley & Sons, New York.

Hobart, K. D., Kub, F. J., Fatemi, M., Twigg, M. E., Thompson, P. E., Kuan, C. K. and Inoki, T. S. (2000), Compliant substrates: A comparative study of the relaxation mechanisms of strained films bonded to high and low viscosity oxides, *Journal of Electronic Materials* **29**, 897–900.

Hoffman, R. W. (1966), The mechanical properties of thin condensed films, *Physics of Thin Films* **3**, 211–273.

Hoffman, R. W. (1976), Stresses in thin films: The relevance of grain boundaries and impurities, *Thin Solid Films* **34**, 185–190.

Hoffman, D. W. and Thornton, J. A. (1977), The compressive stress transition in aluminum, vanadium, zirconium, niobium and tungsten methyl films sputtered at low working pressure, *Thin Solid Films* **45**, 387–396.

Hoffman, D. W. and Thornton, J. A. (1980), Compressive stress and inert gas in molybdenum films sputtered from a cylindrical post magnetron with neon, argon, krypton and xeon, *Journal of Vacuum Science and Technology* **17**, 380–383.

Hommel, M. and Kraft, O. (2001), Deformation behavior of thin copper films on deformable substrates, *Acta Materialia* **49**, 3935–3947.

Hommel, M., Kraft, O. and Arzt, E. (1999), Deformation behavior of thin copper films on deformable substrates, *Journal of Materials Research* **14**, 2373–2376.

Hosford, W. F. (1993), *The Mechanics of Crystals and Textured Polycrystals*, Oxford University Press, Oxford.

Houghton, D. C. (1991), Strain relaxation kinetics in $Si_{1-x}Ge_x$/Si heterostructures, *Journal of Applied Physics* **70**, 2136–2151.

Huang, Y., Liu, C. and Stout, M. G. (1996), A Brazilian disk for measuring fracture toughness of orthotropic materials, *Acta Materialia* **44**, 1223–1232.

Huang, H. and Spaepen, F. (2000), Tensile testing of free-standing Cu, Ag and Al thin films and Ag/Cu multilayers, *Acta Materialia* **48**, 3261–3269.

Hull, R. and Bean, J. C. (1989), Thermal stability of Si/Ge_xSi_{1-x}/Si heterostructures, *Applied Physics Letters* **55**, 1900–1902.

Hull, R., Bean, J. C., Bahnck, D., Peticolas, L. J., Short, K. T. and Unterwald, F. C. (1991), Interpretation of dislocation propagation velocities in strained Ge_xSi_{1-x}/Si(100) heterostructures by the diffusive kink pair model, *Journal of Applied Physics* **70**, 2052–2065.

Hull, R., Bean, J. C., Eaglesham, D. J., Bonar, J. M. and Buescher, C. (1989), Strain relaxation phenomena in Ge_xSi_{1-x}/Si strained structures, *Thin Solid Films* **183**, 117–132.

Huntington, H. B. and Grone, A. R. (1961), Current-induced marker motion in gold wires, *Physics and Chemistry of Solids* **20**, 76–87.

Hurtado, J. and Freund, L. B. (1999), The force on a dislocation near a weakly bonded interface, *Journal of Elasticity* **52**, 167–180.

Hutchinson, J. W. (1996), *Mechanics of Thin Films and Multilayers*, Technical University of Denmark, Lyngby.

Hutchinson, J. W. and Evans, A. G. (2000), Mechanics of materials: Top-down approaches to fracture, *Acta Materialia* **48**, 125–135.

Hutchinson, J. W., He, M. Y. and Evans, A. G. (2000), The influence of imperfections on the nucleation and propagation of buckling driven delaminations, *Journal of the Mechanics and Physics of Solids* **48**, 709–734.

Hutchinson, J. W. and Suo, Z. (1992), Mixed mode cracking in layered materials, *Advances in Applied Mechanics* **29**, 63–191.

Hutchinson, J. W., Thouless, M. D. and Liniger, E. G. (1992), Growth and configurational stability of circular, buckling-driven film delaminations, *Acta Metallurgica et Materialia* **40**, 295–308.

Hyer, M. W. (1981a), Some observations on the curved shape of thin unsymmetric laminates, *Journal of Composite Materials* **15**, 175–194.

Hyer, M. W. (1981b), Calculations of the room-temperature shapes of unsymmetric laminates, *Journal of Composite Materials* **15**, 296–310.

Ilic, B., Czaplewski, D., Craighead, H. G., Neuzil, P., Campagnolo, C. and Batt, C. (2000), Mechanical resonant immunospecific biological detector, *Applied Physics Letters* **77**, 450–452.

Irwin, G. R. (1948), Fracture dynamics, *Fracturing of Metals*, American Society of Metals, Cleveland, 147–166.

Irwin, G. R. (1957), Analysis of stresses and strains near the end of a crack traversing a plate, *Journal of Applied Mechanics* **24**, 361–364.

Irwin, G. R. (1960), Fracture mechanics, *Structural Mechanics*, edited by J. N. Goodier and N. J. Hoff, Pergamon Press, Oxford, 557–594.

Israelachvili, J. N. (1992), *Intermolecular and Surface Forces*, Academic Press, San Diego.

Jansson, C. (1974), Specimen for fracture mechanics studies on glass, *Proceedings of the 10th International Conference on Glass*, Ceramic Society of Japan, Tokyo, 10.23–10.30.

Jawarani, D., Kawasaki, H., Yeo, I.-S., Rabenberg, L., Starck, J. P. and Ho, P. S. (1997), *In situ* transmission electron microscopy study of plastic deformation in passivated Al–Cu thin films, *Journal of Applied Physics* **82**, 171–181.

Jensen, H. M. (1991), The blister test for interface toughness measurement, *Engineering Fracture Mechanics* **40**, 475–486.

Jensen, H. M. (1993), Energy release rates and stability of straight-sided, thin-film delaminations, *Acta Metallurgica et Materialia* **41**, 601–607.

Jensen, H. M. (1998), Analysis of mode mixity in blister tests, *International Journal of Fracture* **94**, 79–88.

Jeong, H. C. and Williams, E. D. (1999), Steps on surfaces: experiment and theory, *Surface Science Reports* **34**, 171–294.

Jesson, D. E., Pennycook, S. J. and Baribeau, J. M. (1991), Direct imaging of interfacial ordering in ultra-thin $(Si_m Ge_n)_p$ superlattices, *Physical Review Letters* **66**, 750–753.

Jesson, D. E., Pennycook, S. J., Baribeau, J.-M. and Houghton, D. C. (1993), Direct imaging of surface cusp evolution during strained layer epitaxy and implications for strain relaxation, *Physical Review Letters* **71**, 1744–1747.

Ji, H., Was, G. S. and Thouless, M. D. (1998), Measurement of the niobium/sapphire interface toughness via delamination, *Engineering Fracture Mechanics* **61**, 163–171.

Jitcharoen, J., Padture, N. P., Giannakopoulos, A. E. and Suresh, S. (1998), Hertzian crack suppression in ceramics with elastic-modulus-graded surfaces, *Journal of the American Ceramic Society* **81**, 2301–2308.

Johnson, H. T. and Freund, L. B. (1997), The mechanics of coherent and dislocated island morphologies in strained epitaxial material systems, *Journal of Applied Physics* **81**, 6081–6090.

Johnson, H. T., Freund, L. B., Akyuz, C. D. and Zaslavsky, A. (1998), Finite element analysis of strain effects on electronic and transport properties in quantum dots and wires, *Journal of Applied Physics* **84**, 3714–3725.

Johnson, K. L. (1985), *Contact Mechanics*, Cambridge University Press, Cambridge.

Johnson, K. L., Kendall, K. and Roberts, A. D. (1971), Surface energy and the contact of elastic solids, *Proceedings of the Royal Society (London)* **A324**, 301–313.

Jonsdottir, F. and Freund, L. B. (1995), Equilibrium surface roughness of a strained epitaxial film due to surface diffusion induced by interface misfit dislocation, *Mechanics of Materials* **20**, 337–349.

Joo, Y.-C. and Thompson, C. V. (1997), Electromigration induced transgranular failure mechanisms in single crystal aluminum interconnects, *Journal of Applied Physics* **81**, 6062–6072.

Kamat, S., Su, X., Ballarini, R. and Heuer, A. H. (2000), Structural basis for the fracture toughness of the shell of the conch *Strombus gigas*, *Nature* **405**, 1036–1040.

Kamins, T. I., Carr, E. C., Williams, R. S. and Rosner, S. J. (1997), Deposition of the three-dimensional Ge islands on Si (001) by chemical vapor deposition at atmospheric and reduced pressures, *Journal of Applied Physics* **81**, 211–219.

Kamins, T. I., Ohlberg, D. A. A. and Williams, R. S. (2001), Effect of phosphorus on Ge/Si (001) island formation, *Applied Physics Letters* **78**, 2220–2222.

Kaysser, W. A. (1999), *Functionally Graded Materials 1998*, Trans Tech Publications Ltd., Hampshire, UK.

Keller, R.-M., Baker, S. P. and Arzt, E. (1999), Quantitative analysis of strengthening mechanisms in thin Cu films: effects of film thickness, grain size and passivation, *Journal of Materials Research* **13**, 1307–1317.

Keller-Flaig, R. M., Legros, M., Sigle, W., Gouldstone, A., Hemker, K. J., Suresh, S. and Arzt, E. (1999), In situ transmission electron microscopy investigation of threading dislocation motion in passivated aluminum lines, *Journal of Materials Research* **14**, 4673–4676.

Kelly, A. and Groves, G. W. (1970), *Crystallography and Crystal Defects*, The Longman Group, New York.

Kesler, O., Finot, M., Suresh, S. and Sampath, S. (1997), Determination of processing induced stresses and properties of layered and grated coatings: Experimental method and results for plasma sprayed NiAl$_2$O$_3$, *Acta Materialia* **45**, 3123–3134.

Khor, K. E. and Das Sarma, S. (1997), Surface morphology and quantum dot self-assembly in growth of strained layer semiconducting films, *Journal of Vacuum Science and Technology* **B15**, 1051–1055.

Kiely, J. D. and Houston, J. E. (1998), Nanomechanical properties of Au (111), (001) and (110) surfaces, *Physical Review B* **57**, 12588–12594.

Kim, D.-K., Heiland, B., Nix, W. D., Deal, M. D. and Plummer, J. D. (2000), Microstructure of thermal hillocks on blanket Al thin films, *Thin Solid Films* **371**, 278–282.

Kim, J.-J., Choi, Y., Suresh, S. and Argon, A. S. (2002), Nanocrystallization during nanoindentation of a bulk amorphous metal alloy at room temperature, *Science* **295**, 654–657.

Kim, J. O., Achenbach, J. D., Mirkarimi, P. B. and Barnett, S. A. (1993), Acoustic microscopy measurements of the elastic properties of TiN/(V$_x$Nb$_{1-x}$)N superlattice films, *Physical Review B* **48**, 1726–1737.

Kim, K. S. and Aravas, N. (1988), Elastoplastic analysis of the peel test, *International Journal of Solids and Structures* **24**, 417–435.

Kim, K. S., Hurtado, J. A. and Tan, H. (1999), Evolution of the surface-roughness spectrum caused by stress in nanometer scale chemical etching, *Physical Review Letters* **83**, 3872–3875.

Kim, K. S., McMeeking, R. M. and Johnson, K. L. (1998), Adhesion, slip, cohesive zones and energy fluxes for elastic spheres in contact, *Journal of the Mechanics and Physics of Solids* **46**, 243–266.

King, R. B. (1987), Elastic analysis of some punch problems for a layered medium, *International Journal of Solids and Structures* **23**, 1657–1664.

Kittel, C. (1971), *Introduction to Solid State Physics*, Wiley, New York.

Klokholm, E. and Berry, B. S. (1968), Intrinsic stress in evaporated metal films, *Journal of the Electrochemical Society* **115**, 823–826.

Kobayashi, K., Inoue, Y., Nishimura, T., Hirayama, M., Akasaka, Y., Kato, T. and Ibuki, S. (1990), Local-oxidation-induced stress measured by Raman microprobe spectroscopy, *Journal of the Electrochemical Society* **137**, 1987–1998.

Kobrinsky, M. J. and Thompson, C. V. (2000), Activation volume for any elastic deformation in polycrystalline Ag thin films, *Acta Materialia* **48**, 625–633.

Koch, R. (1994), The intrinsic stress of polycrystalline and epitaxial thin metal films, *Journal of Physics: Condensed Matter* **6**, 9519–9550.

Koch, R., Leonhard, H., Thurner, G. and Abermann, R. (1990), A UHV-compatible thin film stress measuring apparatus based on the cantilever beam principal, *Review of Scientific Instruments* **61**, 3859–3862.

Koiter, W. J. (1955), On the diffusion of load from a stiffener into a sheet, *Quarterly Journal of Mechanics and Applied Mathematics* **8**, 164–178.

Kraft, O., Freund, L. B., Phillips, R. and Arzt, E. (2002), Dislocation plasticity in thin films, *MRS Bulletin* **27**, 30–37.

Kraft, O. and Nix, W. D. (1998), Measurement of the lattice thermal expansion coefficients of thin metal films on substrates, *Journal of Applied Physics* **83**, 3035–3038.

Kuan, T. S. and Murakami, M. (1982), Low temperature strain behavior of Pb thin films on a substrate, *Metallurgical Transactions* **A13**, 383–391.

Kukta, R. V. and Freund, L. B. (1997), Minimum energy configurations of epitaxial material clusters on a lattice-mismatched substrate, **45**, 1835–1860.

Kurdjumov, G. and Sachs, G. (1930), The mechanism of hardening steel, *Zeitschrift für Physik* **64**, 325–343.

Kvam, E. P. and Hull, R. (1993), Surface orientation and stacking fault generation in strained epitaxial growth, *Journal of Applied Physics* **73**, 7407–7411.

Laugier, M. (1981), Intrinsic stress in thin films of vacuum evaporated LiF and ZnS using an improved cantilevered plate technique, *Vacuum* **31**, 155–157.

Laursen, T. A. and Simo, J. C. (1992), A study of the mechanics of microindentation using finite elements, *Journal of Materials Research* **7**, 618–626.

Lawn, B. (1993), *Fracture of Brittle Solids*, 2nd Edition, Cambridge University Press, Cambridge.

Lee, H., Rosakis, A. J. and Freund, L. B. (2001), Full field optical measurement of curvatures in ultrathin film-substrate systems in the range of geometrically nonlinear deformations, *Journal of Applied Physics* **89**, 6116–6129.

Legros, M., Dehm, G., Keller-Flaig, R.-M., Arzt, E., Hemker, K. J. and Suresh, S. (2001), Dynamic observations of Al thin films plastically strained in a TEM, *Materials Science and Engineering* **A309**, 463–467.

Legros, M., Hemker, K. J., Gouldstone, A., Suresh, S., Keller-Flaig, R.-M. and Arzt, E. (2002), Microstructural evolution in passivated Al thin films on Si substrates during thermal cycling, *Acta Materialia* **50**, 3435–3452.

Li, J., Van Vliet, K. J., Zhu, T., Yip, S. and Suresh, S. (2002), Atomistic mechanisms governing elastic limits and incipient plasticity in crystals, *Nature* **418**, 307–310.

Liang, J., Huang, R., Yin, H., Sturm, J. C., Hobart, K. D. and Suo, Z. (2002), Relaxation of compressed elastic islands on a viscous layer, *Acta Materialia*, **50**, 2933–2944.

Liechti, K. M. and Chai, Y. S. (1992), Asymmetric shielding in interfacial fracture under in-plane loading, *Journal of Applied Mechanics* **59**, 295–304.

Luttinger, J. M. and Kohn, W. (1955), Motion of electrons and holes in perturbed periodic fields, *Physical Review* **97**, 869–883.

Ma, Q., Clarke, D. R. and Suo, Z. (1995), High-resolution determination of the stress in individual interconnect lines and the variation due to electromigration, *Journal of Applied Physics* **78**, 1614–1622.

Machlin, E. S. (1995), *The Relationships Between Thin Film Processing and Structure*, Giro Press, Croton-on-Hudson, New York.

Macionczyk, M. and Brückner, W. (1999), Tensile testing of AlCu thin films on polyimide foils, *Journal of Applied Physics* **86**, 4922–4929.

Mahajan, S. (2000), Defects in semiconductors and their effects on devices, *Acta Materialia* **48**, 137–149.

Marchenko, V. I. and Parshin, A. Ya. (1980), Elastic properties of crystal surfaces, *Soviet Physics JETP* **52**, 129–131.

Martin, J. B. (1975), *Plasticity*, The MIT Press, Cambridge, MA.

Martinez, R. E., Augustyniak, W. M. and Golovchenko, J. A. (1990), Direct measurement of crystal surface stress, *Physical Review Letters* **64**, 1035–1038.

Masters, C. B. and Salamon, N. J. (1993), Geometrically nonlinear stress-deflection relations for thin film/substrate systems, *International Journal of Engineering Science* **31**, 915–925.

Matthews, J. W. (1975), Defects associated with the accommodation of misfit between crystals, *Journal of Vacuum Science and Technology* **12**, 126–133.

Matthews, J. W. and Blakeslee, A. E. (1974), Defects in epitaxial multilayers I. Misfit dislocations, *Journal of Crystal Growth* **27**, 118–125.

Matthews, J. W. and Blakeslee, A. E. (1975), Defects in epitaxial multilayers II. Dislocation pile-ups, threading dislocations, slip lines and cracks, *Journal of Crystal Growth* **29**, 273–280.

Matthews, J. W., Mader, S. and Light, T. B. (1970), Accommodation of misfit across the interface between crystals of semiconducting elements or compounds, *Journal of Applied Physics* **41**, 3800–3804.

Maugis, D. (1992), Adhesion of spheres: JKR-DMT transition using a Dugdale model, *Journal of Colloid and Interface Science* **150**, 243–269.

Mays, C. W., Vermaak, J. S. and KuhlmannWilsdorf, D. (1968), On surface stress and surface tension: II. Determination of the surface stress of gold, *Surface Science* **12**, 134–140.

McHenry, M. E. and Laughlin, D. E. (2000), Nano-scale materials development for future magnetic applications, *Acta Materialia* **48**, 223–238.

Medeiros-Ribeiro, G., Bratkovski, A. M., Kamins, T. I., Ohlberg, D. A. A. and Williams, R. S. (1998), Shape transition of Ge nanocrystals on Si (001) surface from pyramids to domes, *Science* **279**, 353–355.

Mehregany, M., Allen, M. G. and Senturia, S. D. (1997), The use of micromachined structures for the measurement of mechanical properties and adhesion of thin films, *Micromechanics and MEMS*, edited by W. S. Trimmer, IEEE Press, New York, 664–666.

Mehregany, M., Howe, R. T. and Senturia, S. D. (1987), Novel microstructures for the in situ measurement of mechanical properties of thin films, *Journal of Applied Physics* **62**, 3579–3584.

Michalske, T. A., Smith, W. L. and Chen, E. P. (1993), Stress intensity calibration for the double cleavage drilled compression specimen, *Engineering Fracture Mechanics* **45**, 637–642.

Minor, A. M., Morris, J. W. and Stach, E. A. (2001), Quantitative in situ nanoindentation in an electron microscope, *Applied Physics Letters* **79**, 1625–1627.

Mizawa, N., Ohba, T. and Yagi, H. (1994), Planarized copper multilevel interconnections for ULSI applications, *MRS Bulletin* **19**, 63–67.

Mo, Y. W., Savage, D. E., Swartzentruber, B. S. and Lagally, M. G. (1990), Kinetic pathways in Stranski–Krastanov growth of Ge on Si(001), *Physical Review Letters* **65**, 1020–1023.

Moon, M.-W., Chung, J.-W., Lee, K.-R., Oh, K. H., Wang, R. and Evans, A. G. (2002), An experimental study of the influence of imperfections on the buckling of compressed thin films, *Acta Materialia* **49**.

Movchan, B. A. and Demchishin, A. M. (1969), Structure and properties of thick vacuum-condensates of nickel, titanium, tungsten, aluminum oxide and zirconium oxide, *Fizika Metallov I Metallovedenie* **28**, 653–660.

Mullins, W. W. (1957), Theory of thermal grooving, *Journal of Applied Physics* **28**, 333–339.

Mullins, W. W. (1958), The effect of thermal grooving on grain boundary motion, *Acta Metallurgica* **6**, 414–427.

Mullins, W. W. (1959), Flattening of a nearly plane solid surface due to capillarity, *Journal of Applied Physics* **30**, 77–83.

Müllner, P. and Arzt, E. (1998), *Thin Films: Stresses and Mechanical Properties VII*, edited by R. C. Cammarata, M. Nastasi, E. P. Busso and W. C. Oliver, Materials Research Society, Pittsburg, 149–155.

Mulville, D. R., Hunston, D. L. and Mast, P. W. (1978), Developing failure criteria for adhesive joints under complex loading, *Journal of Engineering Materials and Technology* **100**, 25–31.

Murakami, M. (1978), Residual strains of Pb thin films deposited onto Si substrates, *Acta Metallurgica* **26**, 175–183.

Murakami, M. (1982), Thermal stability of Pb-alloy Josephson junction electrode materials. VI: Effects of film edges on strain distribution of Pb–Bi counterelectrodes, *Journal of Applied Physics* **53**, 3560–3565.

Murakami, M., Kuan, T.-S. and Blech, I. A. (1982), Mechanical properties of thin films on substrates, *Treatise on Materials Science and Technology* **24**, 164–210.

Muskhelishvili, N. I. (1953), *Singular Integral Equations*, P. Noordhoff Ltd., Groningen.

Nakamura, T. and Kamath, S. M. (1992), Three dimensional effects in thin film fracture mechanics, *Mechanics of Materials* **13**, 67–77.

Needleman, A. (1990), An analysis of tensile decohesion along an interface, *Journal of the Mechanics and Physics of Solids* **38**, 289–324.

Needleman, A. and Rice, J. R. (1980), Plastic creep flow effects in diffusive cavitation of grain boundaries, *Acta Metallurgica* **28**, 1315–1332.

Nicola, L., Van der Giessen, E. and Needleman, A. (2003), Discrete dislocation analysis of size effects in thin films, *Journal of Applied Physics* **93**, 5920–5928.

Nir, D. (1984), Stress relief forms of diamond-like carbon thin films under internal compressive stress, *Thin Solid Films* **112**, 41–49.

Nishiyama, G. (1934), X-ray investigation of the mechanism of transformation from face-centered cubic lattice to body-centered cubic, *Scientific Report of Tohoku University* **23**, 638.

Nix, W. D. (1989), Mechanical properties of thin films, *Metallurgical Transactions* **20A**, 2217–2245.

Nix, W. D. (1997), Elastic and plastic properties of thin films on substrates: nanoindentation techniques, *Materials Science and Engineering* **A234**, 37–44.

Nix, W. D. (1998), Yielding and strain hardening of thin metal films on substrates, *Scripta Materialia* **39**, 545–554.

Nix, W. D. and Clemens, B. M. (1999), Crystallite coalescence: A mechanism for intrinsic tensile stresses in films, *Journal of Materials Research*, **14**, 3467–3473.

Nix, W. D., Noble, D. B. and Turlo, J. F. (1990), Mechanisms and kinetics of misfit dislocation formation in heteroepitaxial thin films, *Materials Research Society Symposium Proceedings* **188**, 315–330.

Noble, D. B., Hoyt, J. L., Gibbons, J. F., Scott, M. P. and Laderman, S. S. (1989), Thermal stability of Si/Si$_{1-x}$Ge$_x$/Si heterojunction bipolar transistor structures grown by limited reaction processing, *Applied Physics Letters* **55**, 1978–1980.

Norris, A. N. (1998), The energy of a growing elastic surface, *International Journal of Solids and Structures* **35**, 5237–5252.

Nozieres, P. (1992), Shape and growth of crystals, *Solids Far From Equilibrium 1*, Cambridge University Press, 1–154.

Nozieres, P. (1993), Amplitude expansion for the Grinfeld instability due to the axial stress at a solid surface, *Journal de Physique I* **3**, 681–686.

Nurmikko, A. and Han, J. (2002), Blue and near-ultraviolet vertical-cavity surface-emitting lasers, *MRS Bulletin* **27**, 502–506.

O'Dowd, N. P., Shih, C. F. and Stout, M. G. (1992a), Test geometry for measuring interfacial fracture toughness, *International Journal of Solids and Structures* **29**, 571–589.

O'Dowd, N. P., Stout, M. G. and Shih, C. F. (1992b), Fracture toughness of alumina–niobium interfaces: experiments and analyses, *Philosophical Magazine* **66**, 1037–1064.

Ogawa, K., Okhoshi, T., Takeuchi, T., Mizoguchi, T. and Masumoto, T. (1986), Nucleation and growth of stress relief patterns in sputtered molybdenum films, *Japanese Journal of Applied Physics* **25**, 695–700.

Oh, T. S., Rödel, J., Cannon, R. M. and Ritchie, R. O. (1988), Ceramic/metal interfacial crack growth: Toughening by controlled microcracks and interfacial geometries, *Acta Materialia* **36**, 2083–2093.

Ohring, M. (1992), *The Materials Science of Thin Films*, Academic Press, San Diego.

Ohring, M. (1998), *Reliability and Failure of Electronic Materials and Devices*, Academic Press, San Diego, CA.

Oliver, W. C. and Pharr, G. M. (1992), An improved technique for determining hardness and elastic modulus using load and displacement sensing indentation experiments, *Journal of Materials Research* **7**, 1564–1583.

O'Reilly, E. P. (1989), Valence band engineering in strained-layer structures, *Semiconductor Science and Technology* **4**, 121–137.

Orowan, E. (1940), Problems in plastic gliding, *Philosophical Transactions of the Royal Society (London)* **52**, 8–22.

Orowan, E. (1955), Energy criteria of fracture, *Welding Journal Research Supplement* **34**, 157s–160s.

Osterberg, P. M. and Senturia, S. D. (1997), M-TEST: A test chip for MEMS material property measurement using electrostatically actuated test structures, *Journal of Microelectromechanical Systems* **6**, 107–118.

Ostwald, W. (1887), *Lehrbuch der allgemeinen Chemie*, vol. 2, W. Engelmann, Leipzig.

Padture, N. P., Gell, M. and Jordan, E. H. (2002), Thermal barrier coatings for gas turbine engine applications, *Science* **296**, 280–284.

Paine, D. C., Howard, D. J. and Stoffel, N. G. (1991), Strain relief in compositionally graded Si$_{1-x}$Ge$_x$ formed by high dose ion implantation, *Journal of Electronic Materials* **20**, 735–746.

Pan, J. T. and Blech, I. A. (1984), In situ measurement of refractory silicides during sintering, *Journal of Applied Physics* **55**, 2874–2880.

Park, T. S. and Suresh, S. (2000), Effects of line and passivation geometry on curvature evolution during processing and thermal cycling in copper interconnect lines, *Acta Materialia* **48**, 3169–3175.

Perovic, D. D., Weatherly, G. C., Baribeau, J. M. and Houghton, D. C. (1989), Heterogeneous nucleation sources in molecular beam epitaxy grown Ge_xSi_{1-x}/Si strained layer superlattices, *Thin Solid Films* **183**, 141–156.

Peterson, K. E. (1982), Silicon as a mechanical material, *Proceedings of the IEEE* **70**, 420–457.

Pimpinelli, A. and Villain, J. (1998), *Physics of Crystal Growth*, Cambridge University Press, Cambridge.

Pinnington, T., Lavoie, C. and Tiedje, T. (1997), Effect of growth conditions on surface ripening of relaxed InGaAs on GaAs, *Journal of Vacuum Science and Technology* **B15**, 1265–1269.

Press, W. H., Teukolsky, S. A., Vetterling, W. T. and Flannery, B. P. (1992), *Numerical Recipes: The Art of Scientific Computing*, Cambridge University Press, Cambridge.

Ramaswamy, V., Phillips, M. A., Nix, W. D. and Clemens, B. M. (2001), Observation of the strengthening of Pt layers in Ni/Pt and Pd/Pt multilayers by in situ substrate curvature measurement, *Materials Science and Engineering* **A319**, 887–892.

Rastelli, A., Kummer, M. and von Kanel, H. (2001), Reversible shape evolution of Ge islands on Si (001), *Physical Review Letters* **87**, 6101–6104.

Rayleigh, Lord (1890), On the theory of surface forces, *Philosophical Magazine* **30**, 285–298.

Read, D. T. (1998), Young's modulus of thin films by speckle interferometry, *Measurement Science and Technology* **9**, 676–685.

Retajczyk, T. F. and Sinha, A. K. (1980), Elastic stiffness and thermal expansion coefficient of BN films, *Applied Physics Letters* **36**, 161–163.

Rice, J. R. (1968), Mathematical analysis in the mechanics of fracture, *Fracture*, Vol. II, edited by H. Liebowitz, Academic Press, New York, 191–311.

Rice, J. R. (1970), On the structure of stress–strain relations for time dependent plastic deformation in metals, *Journal of Applied Mechanics* **37**, 728–737.

Rice, J. R. (1976), The localization of plastic deformation, *Theoretical and Applied Mechanics* edited by W. T. Koiter, North Holland, Amsterdam, 207–220.

Rice, J. R. and Beltz, G. E. (1994), The activation energy for dislocation nucleation at a crack, *Journal of the Mechanics and Physics of Solids* **42**, 333–360.

Rice, J. R. and Chuang, T.-J. (1981), Energy variations in diffusive cavity growth, *Journal of the American Ceramic Society* **64**, 46–53.

Rice, J. R. and Thomson, R. (1973), Ductile versus brittle behavior of crystals, *Philosophical Magazine*, 73–97.

Rich, T. A. (1934), Thermo-mechanics of bimetal, *General Electric Review* **37**, 102–105.

Rosakis, A. J., Singh, R. B., Tsuji, Y., Kolawa, E. and Moore, N. R. (1998), Full field measurements of curvature using coherent gradient sensing – application to thin-film characterization, *Thin Solid Films* **325**, 42–54.

Ross, F. M. (2001), Dynamic studies of semiconductor growth processes using in situ electron microscopy, *MRS Bulletin* **26**, 94–101.

Ross, F. M., Tromp, R. M. and Reuter, M. C. (1999), Transition states between pyramids and domes during Ge/Si island growth, *Science* **286**, 1931–1934.

Ruud, J. A., Josell, D., Spaepen, F. and Greer, A. L. (1993), A new method for tensile testing of thin films, *Journal of Materials Research* **8**, 112–117.

Samavedam, S. B. and Fitzgerald, E. A. (1997), Novel dislocation structure and surface morphology effects in relaxed Ge/Si–Ge(graded)/Si structures, *Journal of Applied Physics* **81**, 3108–3116.

Sanchez, J. E., Kraft, O. and Arzt, E. (1992), Electromigration induced transgranular slit failures in near bamboo Al and Al–2% Cu thin film interconnects, *Applied Physics Letters* **61**, 3121–3123.

Sanchez, J. E., McKnelly, L. T. and Morris, Jr., J. W. (1990), Morphology of electromigration-induced damage and failure in Al alloy thin film conductors, *Journal of Electronic Materials* **19**, 1213–1220.

Sarikaya, M., Liu, J. and Aksay, I. A. (1995), Nacre: Properties, crystallography, morphology and formation, *Biomimetics: Design and Processing of Materials*, edited by M. Sarikaya and I. A. Aksay, American Institute of Physics, 35–90.

Schadler, L. S. and Noyan, I. C. (1995), Quantitative measurement of the stress transfer function in nickel/polyimide thin film/copper thin film structures, *Applied Physics Letters* **66**, 22–24.

Scheeper, P. R., van der Donk, A. G. H., Olthuis, W. and Bergveld, P. (1994), A review of silicon microphones, *Sensors and Actuators* **A44**, 1–11.

Schelling, C., Mühlberger, M., Springholz, G. and Schäffler, F. (2001), $Si_{1-x}Ge_x$ growth instabilities on vicinal Si(001) substrates: Kinetic versus strain-induced effects, *Physical Review B* **64**, 041301–4.

Schell-Sorokin, A. J. and Tromp, R. M. (1990), Mechanical stresses in (sub) monolayer epitaxial films, *Physical Review Letters* **64**, 1039–1042.

Schneeweiss, H. J. and Abermann, R. (1992), Ultra-high vacuum measurements of the internal stress of PVD titanium films as a function of thickness and its dependence on substrate temperature, *Vacuum* **43**, 463–465.

Schwaiger, R. and Kraft, O. (2003), Size effects in the fatigue behavior of thin Ag films, *Acta Materialia* **51**, 195–206.

Schwartzman, A. and Sinclair, R. (1991), Metastable and equilibrium defect structure of II–VI/GaAs interfaces, *Journal of Electronic Materials* **20**, 805–814.

Schwarz, K. W. and Tersoff, J. (1996), Interaction of threading and misfit dislocations in a strained epitaxial layer, *Applied Physics Letters* **69**, 1220–1222.

Schweinfest, R., Ernst, F., Wagner, T. and Rühle, M. (1999), High-precision assessment of interface lattice offset by quantitative HRTEM, *Journal of Microscopy* **194**, 142–151.

Schwoebel, R. L. (1969), Step motion on crystal surfaces II, *Journal of Applied Physics* **40**, 614–618.

Seel, S. C., Thompson, C. V., Hearne, S. J. and Floro, J. A. (2000), Tensile stress evolution during deposition of Volmer–Weber thin films, *Journal of Applied Physics* **88**, 7079–7088.

Sharpe, W. N. (1982), Applications of the interferometric strain/displacement gauge, *Optical Engineering* **21**, 483–488.

Shchukin, V. A., Borovkov, A. I., Ledentsov, N. N. and Bimberg, D. (1995a), Tuning and breakdown of faceting under externally applied stress, *Physical Review B* **51**, 10104–10118.

Shchukin, V. A., Ledentsov, N. N., Kopev, P. S. and Bimberg, D. (1995b), Spontaneous ordering of arrays of coherent strained islands, *Physical Review Letters* **75**, 2968–2971.

Shen, Y.-L. and Suresh, S. (1995a), Elastoplastic deformation of multilayered materials during formal cycling, *Journal of Materials Research* **10**, 1200–1215.

Shen, Y.-L. and Suresh, S. (1995b), Thermal cycling and stress relaxation response of Si–Al and Si–Al–SiO_2 layered thin films, *Acta Metallurgica et Materialia* **43**, 3915–3926.

Shen, Y.-L. and Suresh, S. (1996), Steady-state creep of metal–ceramic multilayered materials, *Acta Materialia* **44**, 1337–1348.

Shen, Y. L., Suresh, S. and Blech, I. A. (1996), Stresses, curvatures and shape changes arising from patterned lines on silicon wafers, *Journal of Applied Physics* **80**, 1388–1398.

Shen, Y.-L., Suresh, S., He, M. Y., Bagchi, A., Kienzle, O., Rühle, M. and Evans, A. G. (1998), Stress evolution in passivated thin films of Cu on silica substrates, *Journal of Materials Research* **13**, 1928–1937.

Shenoy, V. B., Ciobanu, C. and Freund, L. B. (2002), Strain induced stabilization of stepped Si and Ge surfaces near (001), *Applied Physics Letters* **81**, 364–366.

Shenoy, V. B. and Freund, L. B. (2002), A continuum description of the energetics and evolution of stepped surfaces in strained nanostructures, *Journal of the Mechanics and Physics of Solids* **50**, 1817–1841.

Shetty, D. K., Rosenfeld, A. and Duckworth, W. (1987), Mixed mode fracture in biaxial stress state: Application of the diametral-compression (Brazilian disk) test, *Engineering Fracture Mechanics* **26**, 825–840.

Shield, T. W. and Kim, K. S. (1992), Beam theory models for thin film segments cohesively bonded to an elastic half space, *International Journal of Solids and Structures* **29**, 1085–1103.

Shintani, K. and Fujita, K. (1994), Effect of anisotropy on the excess stress and critical thickness of $Si_{1-x}Ge_x$ strained layers, *Journal of Applied Physics* **75**, 7842–7846.

Shull, A. L. and Spaepen, F. (1996), Measurements of stress during vapor deposition of copper and silver thin films and multilayers, *Journal of Applied Physics* **80**, 6243–6256.

Shuttleworth, R. (1950), The surface tension of solids, *Proceedings of the Physical Society* **A63**, 444–457.

Singh, J. (1993), *Physics of Semiconductors and their Heterostructures*, McGraw-Hill, New York.

Singh, R. K. and Bajaj, R. (2002), Advances in chemical–mechanical planarization, *MRS Bulletin* **27**, 743–783.

Small, M. K. and Nix, W. D. (1992), Analysis of the accuracy of the bulge test in determining the mechanical properties of thin films, *Journal of Materials Research* **7**, 1553–1563.

Smith, B. L., Schaffer, T. E., Viani, M., Thompson, J. B., Frederick, N. A., Kindt, J. Belcher, A., Stucky, G. D., Morse, D. E. and Hansma, P. K. (1999), Molecular mechanistic origin of the toughness of natural adhesives, fibers and composites, *Nature* **399**, 761–763.

Spaepen, F. (1977), Microscopic mechanism for steady-state inhomogeneous flow in metallic glasses, *Acta Metallurgica* **25**, 407–415.

Spaepen, F. (1996), Substrate curvature resulting from the capillary forces of a liquid drop, *Journal of the Mechanics and Physics of Solids* **44**, 675–681.

Spaepen, F. (2000), Interfaces and stresses in thin films, *Acta Materialia* **48**, 31–42.

Spearing, S. M. (2000), Materials issues in microelectromechanical systems (MEMS), *Acta Materialia* **48**, 179–196.

Spencer, B. J. and Meiron, D. I. (1994), Nonlinear evolution of the stress-driven morphological instability in a two-dimensional semi-infinite solid, *Acta Metallurgica et Materialia* **42**, 3629–3641.

Spencer, B. J., Voorhees, P. W. and Davis, S. H. (1991), Morphological instability in epitaxially strained dislocation free solid films, *Physical Review Letters* **67**, 3696–3699.

Spohn, H. (1993), Surface dynamics below the roughening transition, *Journal de Physique* **3**, 69–81.

Springholz, G. (1999), Observation of large scale surface undulations due to inhomogeneous dislocation strain fields in lattice-mismatched epitaxial layers, *Applied Physics Letters* **75**, 3099–3101.

Sridhar, N., Srolovitz, D. J. and Suo, Z. (2001), Kinetics of buckling of a compressed film on a viscous substrate, *Applied Physics Letters* **78**, 2482–2484.

Srikar, V. T. and Thompson, C. V. (1999), Diffusion and electromigration of copper in SiO$_2$ passivated single crystal aluminum interconnects, *Applied Physics Letters* **74**, 37–39.

Srolovitz, D. J. (1989), On the stability of surfaces of stressed solids, *Acta Metallurgica* **37**, 621–625.

Stach, E. A. and Hull, R. (2001), Enhancement of dislocation velocities by stress-assisted kink nucleation at the native oxide/SiGe interface, *Applied Physics Letters* **79**, 335–337.

Stach, E. A., Hull, R., Tromp, R. M., Reuter, M. C. and Copel, M. (1998), Effect of the surface upon misfit dislocation velocities during growth and annealing of SiGe/Si (001) heterostructures, *Journal of Applied Physics* **83**, 1931–1937.

Stach, E. A., Hull, R., Tromp, R. M., Ross, F. M., Reuter, M. C. and Bean, J. C. (2000), In situ transmission electron microscopy studies of the interaction between dislocations in a strained SiGe/Si(001) heterostructure, *Philosophical Magazine* **A80**, 2159–2200.

Stillinger, F. H. and Weber, T. A. (1985), Computer simulation of local order in condensed phases of silicon, *Physical Review B* **31**, 5262–5271.

Stölken, J. S. and Evans, A. G. (1998), A microbend test method for measuring the plasticity length scale, *Acta Materialia* **46**, 5109–5115.

Stoney, G. G. (1909), The tension of metallic films deposited by electrolysis, *Proceedings of the Royal Society (London)* **A82**, 172–175.

Storåkers, B. (1988), Nonlinear aspects of delamination in structural members, *Proc. 17th International Congress on Theoretical & Applied Mechanics*, edited by P. Germain et al., Elsevier, Amsterdam, 315–336.

Stringfellow, R. G. and Freund, L. B. (1993), The effect of interfacial friction on the buckle-driven spontaneous delamination of a compressed thin film, *International Journal of Solids and Structures* **30**, 1379–1396.

Sun, B., Suo, Z. and Evans, A. G. (1994), Emergence of cracks by mass transport in elastic crystals stressed at high temperatures, *Journal of the Mechanics and Physics of Solids* **42**, 1653–1677.

Suo, Z. (1997), Motions of microscopic surfaces in materials, *Advances in Applied Mechanics* **33**, 193–293.

Suo, Z. (2003), Reliability of interconnect structures, *Interfacial and Nanoscale Failure*, edited by W. W. Gerberich and W. Yang, Elsevier, Amsterdam.

Suo, Z. and Hutchinson, J. W. (1989), Steady-state cracking in brittle substrates beneath adherent films, *International Journal of Solids and Structures* **25**, 1337–1353.

Suo, Z. and Hutchinson, J. W. (1990), Interface crack between two elastic layers, *International Journal of Fracture* **43**, 1–18.

Suo, Z. and Wang, W. (1994), Diffusive void bifurcation in stressed solid, *Journal of Applied Physics* **76**, 3410–3421.

Suresh, S. (1998), *Fatigue of Materials*, 2nd Edition, Cambridge University Press, Cambridge.

Suresh, S. (2001), Graded materials for resistance to contact deformation and damage, *Science* **292**, 2447–2451.

Suresh, S. and Giannakopoulos, A. E. (1998), A new method for estimating residual stresses by instrumented sharp indentation, *Acta Materialia* **46**, 5755–5767.

Suresh, S., Giannakopoulos, A. E. and Olsson, M. (1994), Elastoplastic analysis of thermal cycling: Layered materials with sharp interfaces, *Journal of the Mechanics and Physics of Solids* **42**, 979–1018.

Suresh, S. and Mortensen, A. (1998), *Fundamentals of Functionally Graded Materials*, Institute of Materials, London.

Suresh, S., Nieh, T.-G. and Choi, B. W. (1999a), Nanoindentation of copper thin films on silicon substrates, *Scripta Materialia* **41**, 951–957.

Suresh, S., Olsson, M., Giannakopoulos, A. E., Padture, N. P. and Jitcharoen, J. (1999b), Engineering the resistance to sliding-contact damage through controlled gradients in elastic properties at contact surfaces, *Acta Materialia* **47**, 3915–3926.

Suresh, S., Shih, C. F., Morrone, A. and O'Dowd, N. P. (1990), Mixed mode fracture toughness of ceramic materials, *Journal of the American Ceramic Society* **73**, 1257–1267.

Suresh, S., Sugimura, Y. and Ogawa, T. (1993), Fatigue cracking in materials with brittle surface coatings, *Scripta Metallurgica et Materialia* **29**, 273–242.

Sutter, P. and Lagally, M. G. (2000), Nucleationless three-dimensional island formation in low misfit heteroepitaxy, *Physical Review Letters* **84**, 4637–4640.

Tabor, D. (1951), *The Hardness of Metals*, Clarendon Press, Oxford.

Tada, H., Paris, P. and Irwin, G. (1985), *The Stress Analysis of Cracks Handbook*, Del Research Corp., St. Louis.

Tersoff, J. (1988), Empirical interatomic potential for silicon with improved elastic constants, *Physical Review B* **38**, 9902–9905.

Tersoff, J. (1989), Modeling solid-state chemistry: Interatomic potentials for multicomponent systems, *Physical Review B* **39**, 5566–5568.

Tersoff, J. and LeGoues, F. K. (1994), Competing relaxation mechanisms in strained layers, *Physical Review Letters* **72**, 3570–3573.

Thompson, C. V. (1993), The yield stress of polycrystalline thin films, *Journal of Materials Research* **8**, 237–238.

Thompson, C. V. (2000), Structural evolution during processing of polycrystalline films, *Annual Reviews in Materials Science* **30**, 159–190.

Thompson, C. V. and Carel, R. (1995), Texture development in polycrystalline thin films, *Materials Science and Engineering* **B32**, 211–219.

Thompson, J. M. T. and Hunt, G. W. (1973), *A General Theory of Elastic Stability*, John Wiley & Sons Ltd., London.

Thompson, C. V. and Lloyd, J. R. (1993), Electromigration and IC interconnects, *MRS Bulletin* **18**, 19–25.

Thornton, J. A. (1977), High rate thick film growth, *Annual Review of Materials Science* **7**, 239–260.

Thornton, J. A., Tabock, J. and Hoffman, D. W. (1979), Internal stresses in metallic films deposited by cylindrical magnetron sputtering, *Thin Solid Films* **64**, 111–119.

Thouless, M. D. (1990a), Fracture of a model interface under mixed-mode loading, *Acta Metallurgica* **38**, 1135–1140.

Thouless, M. D. (1990b), Crack spacing in brittle films on elastic substrates, *Journal of the American Ceramic Society* **73**, 2144–2146.

Thouless, M. D. (1993), Effect of surface diffusion on the creep of thin films and sintered arrays of particles, *Acta Metallurgica et Materialia* **41**, 1057–1064.

Thouless, M. D., Evans, A. G., Ashby, M. F. and Hutchinson, J. W. (1987), The edge cracking and spalling of brittle plates, *Acta Metallurgica* **35**, 1333–1341.

Thouless, M. D., Hutchinson, J. W. and Liniger, E. G. (1992a), Plane strain, buckling-driven delamination of thin films: Model experiments and mode II fracture, *Acta Metallurgica et Materialia* **40**, 2639–2649.

Thouless, M. D. and Jensen, H. M. (1992b), Elastic fracture mechanics of the peel-test geometry, *Journal of Adhesion* **38**, 185–197.

Thouless, M. D., Olsson, E. and Gupta, A. (1992), Cracking of brittle films on elastic substrates, *Acta Metallurgica et Materialia* **40**, 1287–1292.

Thurner, G. and Abermann, R. (1990), Internal stress and structure of ultrahigh vacuum evaporated chromium and iron films and their dependence on substrate temperature and oxygen partial pressure, *Thin Solid Films* **192**, 277–285.

Tien, N. C. (1998), Silicon micromachined thermal sensors and actuators, *Microscale Energy Transport*, edited by C. L. Tien, A. Majumdar and F. M. Gerner, Taylor & Francis, 369–386.

Timoshenko, S. (1925), Analysis of bi-metal thermostats, *Journal of the Optical Society of America* **11**, 233–255.

Timoshenko, S. P. and Goodier, J. N. (1987), *Theory of Elasticity*, 3rd Edition, McGraw-Hill.

Torii, R. H. and Balibar, S. (1992), Helium crystals under stress: The Grinfeld instability, *Journal of Low Temperature Physics* **89**, 391–400.

Trantina, G. C. (1972), Combined mode crack extension in adhesive joints, *Journal of Composite Materials* **6**, 271–385.

Tromp, R. M., Ross, F. M. and Reuter, M. C. (2000), Instability-driven SiGe island growth, *Physical Review Letters* **84**, 4641–4644.

Tsao, J. Y. (1993), *Materials Fundamentals of Molecular Beam Epitaxy*, Academic Press, San Diego.

Tsao, J. Y. and Dodson, B. W. (1988), Excess stress and the stability of strained heterostructures, *Applied Physics Letters* **53**, 848–850.

Tsao, J. Y., Dodson, B. W., Picraux, S. T. and Cornelison, D. M. (1987), Critical stresses for Si_xGe_{1-x} strained layer plasticity, *Physical Review Letters* **59**, 2455–2458.

Tsui, T. Y., Oliver, W. C. and Pharr, G. M. (1996), Influences of stress on the measurement of mechanical properties using nanoindentation. Part I. Experimental studies in an aluminum alloy, *Journal of Materials Research* **11**, 752–759.

Tsui, T. Y., Vlassak, J. and Nix, W. D. (1999), Indentation plastic displacement field: Part I. The case of soft films on hard substrates; Part II. The case of hard films on soft substrates, *Journal of Materials Research* **14**, 2196–2209.

Tuppen, C. G. and Gibbings, C. J. (1989), Misfit dislocations in annealed $Si_{1-x}Ge_x$/Si heterostructures, *Thin Solid Films* **183**, 133–139.

Tuppen, C. G. and Gibbings, C. J. (1990), A quantitative analysis of strain relaxation by misfit dislocation glide in $Si_{1-x}Ge_x$/Si heterostructures, *Journal of Applied Physics* **68**, 1526–1534.

Tuppen, C. G., Gibbings, C. J. and Hockly, M. (1989), The effects of misfit dislocation nucleation and propagation on $Si/Si_{1-x}Ge_x$ critical thickness values, *Journal of Crystal Growth* **94**, 392–404.

Tvergaard, V. and Hutchinson, J. W. (1993), The influence of plasticity on mixed mode interface toughness, *Journal of the Mechanics and Physics of Solids* **41**, 1119–1135.

Vailionis, A., Cho, B., Glass, G., Desjardins, P., Cahill, D. G. and Greene, J. E. (2000), Pathway to a strain driven two-dimensional to three-dimensional transition during growth of Ge on Si (001), *Physical Review Letters* **85**, 3672–3675.

Van der Merwe, J. H. and van der Berg, N. G. (1972), Misfit dislocation energy in epitaxial overgrowths of finite thickness, *Surface Science* **32**, 1–15.

Van Mellaert, L. J. and Schwuttke, G. H. (1972), Feedback control scheme for x-ray topography, *Journal of Applied Physics* **43**, 687–692.

Venables, J. A. (2000), *Introduction to Surface and Thin Film Processes*, Cambridge University Press, Cambridge, UK.

Venkatraman, R. and Bravman, J. C. (1992), Separation of film thickness and grain boundary strengthening effects in Al thin films on Si, *Journal of Materials Research* **7**, 2040–2048.

Venkatraman, R., Bravman, J. C., Nix, W. D., Davies, P. W., Flinn, P. A. and Fraser, D. B. (1990), Mechanical properties and microstructural characterization of Al–0.5%Cu thin films, *Journal of Electronic Materials* **19**, 1231–1237.

Vinci, R. P., Zielinski, E. M. and Bravman, J. C. (1995), Thermal strain and stress in thin copper films, *Thin Solid Films* **262**, 142–153.

Vlassak, J. and Nix, W. D. (1992), A new voltage test technique for the determination of Young's modulus and Poisson ratio of thin films, *Journal of Materials Research* **7**, 3242–3249.

Volkert, C. A. (1991), Stress and plastic flow in silicon during amorphization by ion-bombardment, *Journal of Applied Physics* **70**, 3521–3527.

Von Mises, R. (1913), Mechanik der festen körper in plastisch deformablen zustand, *Göttinger Nachr Math Phys Kl*, 582–592.

Wang, P. C., Cargill III, G. S., Noyan, I. C. and Hu, C. K. (1998), Electromigration-induced stress in aluminum conductor lines measured by x-ray diffraction, *Applied Physics Letters* **72**, 1296–1298.

Wang, J. G., Choi, B. W., Nieh, T. G. and Liu, C. T. (2000), Crystallization and nanoindentation behavior of a bulk ZrAlTiCuNi amorphous alloy, *Journal of Materials Research* **15**, 798–807.

Wang, J.-S. and Suo, Z. (1990), Experimental determination of interfacial fracture toughness using Brazil-nut sandwiches, *Acta Metallurgica et Materialia* **38**, 1279–1290.

Wang, R. Z., Suo, Z., Evans, A. G., Yao, N. and Aksay, I. A. (2001), Deformation mechanisms in nacre, *Journal of Materials Research* **16**, 2485–2493.

Wasserman, G. (1933), Influence of the α–γ transformation of an irreversible nickel steel on crystal orientation and tensile strength, *Archive Eisenhuettenwesen* **6**, 347–351.

Weeks, R., Dundurs, J. and Stippes, M. (1968), Exact analysis of an edge dislocation near a surface layer, *International Journal of Engineering Science* **6**, 365–372.

Weihs, T. P., Hong, S., Bravman, J. C. and Nix, W. D. (1988), Mechanical deflection of cantilever microbeams: a new technique for testing the mechanical properties of thin films, *Journal of Materials Research* **3**, 931–942.

Weiss, D., Gao, H. and Arzt, E. (2001), Constrained effusion will creep in UHV-produced copper thin films, *Acta Materialia* **49**, 2395–2403.

Whitcomb, J. D. (1986), Parametric analytical study of instability-related delamination growth, *Composite Science and Technology* **25**, 19–48.

Wikström, A., Gudmundson, P. and Suresh, S. (1999a), Thermoelastic analysis of periodic thin lines deposited on a substrate, *Journal of the Mechanics and Physics of Solids* **47**, 1113–1130.

Wikström, A., Gudmundson, P. and Suresh, S. (1999b), Analysis of average thermal stresses in passivated metal interconnects, *Journal of Applied Physics* **86**, 6088–6095.

Wilcock, J. D. and Campbell, D. S. (1969), A sensitive bending beam apparatus for measuring the stress of evaporated thin films, *Thin Solid Films* **3**, 3–12.

Wilcock, J. D., Campbell, D. S. and Anderson, J. C. (1969), Internal stress in evaporated silver and gold films, *Thin Solid Films* **3**, 13–34.

Williams, R. S., Medeiros-Ribeiro, G., Kamins, T. I. and Ohlberg, D. A. A. (2000), Thermodynamics of the size and shape of nanocrystals: Epitaxial Ge on Si (001), *Annual Review of Physical Chemistry* **51**, 527–551.

Wilson, C. A. (1858), *Apparatus for Warming by Steam*, U.S. Patent No. 24896, U.S. Patent and Trademark Office, Washington.

Wu, C. H. (1996), The chemical potential for stress-driven surface diffusion, *Journal of the Mechanics and Physics of Solids* **44**, 2059–2077.

Wu, X. and Weatherly, G. C. (1999), Cracking phenomena in $In_{0.25}Ga_{0.75}As$ films on InP substrates, *Acta Materialia* **47**, 3383–3394.

Wu, X. and Weatherly, G. C. (2001), The first stage of stress relaxation in tensile strained $In_{1-x}Ga_xAs_{1-y}P_y$ films, *Philosophical Magazine* **A81**, 1489–1506.

Wulff, G. (1901), Zur frage der geschwindeigkeit des Wachsthung und der auflösung der krystallslächen, *Zeitschrift fur Kristallographie* **34**, 449–530.

Xia, Z. C. and Hutchinson, J. W. (2000), Crack patterns in thin films, *Journal of the Mechanics and Physics of Solids* **48**, 1107–1131.

Xie, Y. H., Gilmer, G. H., Roland, C., Buratto, S. K., Chang, J. Y., Fitzgerald, E. A., Kortan, A. R., Schuppler, S., Marcus, M. A. and Citrin, P. H. (1994), Semiconductor surface roughness: Dependence on sign and magnitude of bulk strain, *Physical Review Letters* **73**, 3006–3009.

Yang, W. and Srolovitz, D. J. (1994), Surface morphology evolution in stressed solids: surface diffusion controlled crack initiation, *Journal of the Mechanics and Physics of Solids* **42**, 1551–1574.

Yang, M. Y., Sturm, J. C. and Prevost, J. (1997), Calculation of band alignments and quantum confinement effects in zero- and one-dimensional pseudomorphic structures, *Physical Review B* **56**, 1973–1980.

Yin, W. L. (1985), Axisymmetric buckling and growth of a circular delamination in a compressed laminate, *International Journal of Solids and Structures* **21**, 503–514.

Young, T. (1805), An essay on the cohesion of fluids, *Philosophical Transactions of the Royal Society (London)* **95**, 65–87.

Yu, E. T., McCaldin, J. O. and McGill, T. C. (1992), Band offsets in semiconductor heterostructures, *Solid State Physics* **46**, 1–146.

Yu, H. H., He, M. Y. and Hutchinson, J. W. (2001), Edge effects in thin film delamination, *Acta Materialia* **49**, 93–107.

Yu, H. H. and Hutchinson, J. W. (2002), Influence of substrate compliance on buckling delamination of thin films, *International Journal of Fracture* **113**, 39–55.

Zak, A. R. and Williams, M. L. (1963), Crack point singularities at a bi-material interface, *Journal of Applied Mechanics* **30**, 142–143.

Zangwill, A. (1988), *Physics at Surfaces*, Cambridge University Press, Cambridge.

Zhang, Y. W., Bower, A. F., Xia, L. and Shih, C. F. (1999), Three dimensional finite element analysis of the evolution of voids and thin films by strain and electromigration induced surface diffusion, *Journal of the Mechanics and Physics of Solids* **47**, 173–199.

Zhang, H. and Yang, W. (1993), Three dimensional dislocation loops generated from a weak inclusion in a strained material heterostructure, *Journal of the Mechanics and Physics of Solids* **42**, 913–930.

Zhou, Q. Y., Argon, A. S. and Cohen, R. E. (2001), Enhanced case II diffusion of diluents into glassy polymers undergoing plastic flow, *Polymer* **42**, 613–621.

Zhuk, A. V., Evans, A. G., Hutchinson, J. W. and Whitesides, G. M. (1998), The adhesion energy between polymer thin films and self-assembled monolayers, *Journal of Materials Research* **13**, 3555–3564.

Zinke-Allmang, M., Feldman, L. C. and Grabow, M. H. (1992), Clustering on surfaces, *Surface Science Reports* **16**, 377–463.

Author index

Subject index

745